Dedicated to

My Mother

for initiating me into the world of knowledge

on her 75th birth anniversary

Preface

Mobile computing has revolutionised the way in which we work, connect, and communicate to the world. Applications of mobile computing are uncountable. It offers mobility with computing power and facilitates a large number of applications on a single device. The earliest mobile communication devices in the true sense were the portable radio transmitters introduced during the Second World War for military deployment. Popularly known as walkie-talkies, these backpacked (and handheld) transmitters launched by Motorola used vacuum tubes and high-voltage dry cell batteries.

The latter half of the twentieth century saw exponential development in the field of wireless communication and computing technology so much so that from the multitude of electronic gadgets that came and went, arose the star player—the do-it-all mobile phone. Known as Smartphones in technical jargon, the present-day mobile phones offer amalgamated functionalities of PocketPCs, palmtops, camera, media players, radios, and even TVs. There is a plethora of mobile devices available, besides mobile phones, which perform numerous complex functions. But the most revolutionary and spectacular prototype of mobile devices remains the ever versatile cellular phone.

Advances in cellular communication networks and mobile device technology have unearthed the twin concepts of pervasive computing and ubiquitous computing. The future is looking towards smart devices, smart appliances, and increasing capabilities of artificial intelligence. The day is not far when computing systems will pervade our everyday lives and simplify complex tasks.

About the Book

In the present context, mobile computing and its offshoots are the most important players in the computing arena. *Mobile Computing* attempts to comprehensively introduce the reader to the various aspects of computing in mobile environments. Primarily designed to serve as a textbook for undergraduate students enrolled in the disciplines of computer science, electronics, and communications engineering, this book will also prove to be a useful reference for postgraduate students and research scholars.

Undergraduate students of engineering need a text which covers 2G and 3G communication systems, databases in mobile systems, methods of data dissemination and synchronization, and programming languages and operating systems used in mobile devices. Written in a lucid, student-friendly manner, the book includes detailed discussions on all the relevant topics. Simplicity of style and language, and many examples spread through the text allow the reader to grasp the concepts easily. A special attempt has been made to include topics which are a part of curricula of courses offered by a large cross-section of educational establishments.

Each chapter is divided into independent sections. Suitable figures are provided in all chapters to enable easy understanding of the important concepts. Important concepts are explained with the help of tables and flowcharts. A large number of examples and illustrations are interspersed through the text. Sample program segments have been provided to assist the reader in understanding the coding procedures of XML and related languages. Key terms introduced in each chapter are listed at the end of the chapter along with their definitions. End-chapter exercises include review questions and objective type questions, which comprehensively test student's understanding of the concepts discussed. The appendices on mobile satellite communication networks and Java programs presented towards the end of the book are useful supplements to the text. A Select Bibliography is provided at the end of the book for those interested in further reading.

Content and Structure

This book contains 14 chapters which provide comprehensive discussions on the various aspects of mobile computing.

Chapter 1 provides an introduction to mobile communication and mobile computing. The concepts presented in this chapter are discussed in detail in the ensuing chapters.

Chapter 2 describes the various kinds of mobile devices available in the market as also their features and limitations. It includes comprehensive sections on smart systems, automotive systems, and the operating systems used in mobile devices.

Chapter 3 elucidates GSM systems, including the interfaces, protocols, and processes, which form the basis of GSM communication. A description of related network architectures, such as GPRS, HSCSD, and DECT, is also included.

Chapter 4 details methods of wireless medium access control and provides an in-depth introduction to CDMA-based communication systems. Topics discussed here include OFDMA, cdmaOne, IMT-2000, and i-mode.

Chapter 5 discusses the protocols and processes of the mobile IP network layer. Special emphasis is laid on handover and location management, registration and route optimization, and the dynamic host configuration protocol (DHCP).

Chapter 6 serves to describe the functions of transport layer and the protocols used in mobile environments. It discusses the pros and cons of the conventional TCP and the need for new versions of TCP. Indirect TCP, snooping TCP, TCP over 2.5G/3G mobile networks, and mobile TCP are some of the topics reviewed here.

Chapter 7 introduces databases in mobile computing systems. Database hoarding techniques and caching mechanisms are comprehensively discussed. The chapter develops a focussed understanding of client–server computing, transaction models, query processing, data recovery, and quality of service.

Chapter 8 illustrates data dissemination mechanisms and broadcast models. It also describes the digital audio and video broadcast systems with a special introduction to mobile TV and DVB-H technology.

Chapter 9 focusses on data synchronization in mobile computing systems. It includes description of the various synchronization software, such as HotSync, ActiveSync, and Intellisync. Synchronization protocols and the synchronization mark-up language (SyncML) are also described in relevant detail.

Chapter 10 serves to illustrate the concepts of mobile agents, servers, service discovery, device management, and security. Application servers such as the Sun Java System Web Server 6, the IBM WebSphere MQe, and the Oracle Application Server; gateways and portals; service discovery; and the CODA mobile file system are some of the topics discussed in this chapter.

Chapter 11 offers details on mobile ad-hoc and sensor networks. It provides a comprehensive introduction to the various aspects of MANET as well as wireless sensor networks and their applications.

Chapter 12 covers wireless LAN, mobile Internet connectivity, and personal area networks. Beginning with WLAN architecture and protocols, the chapter goes on to encompass a variety of topics such as WAP, XHTML (extensible hypertext markup language), Bluetooth, IrDA, and ZigBee.

Chapter 13 is dedicated to programming languages for mobile applications and includes discussions on XML, Java, J2ME, and JavaCard.

Chapter 14 discusses operating systems for mobile devices. The architecture, functioning, and application development of PalmOS, Windows CE, and Symbian OS are explained in detail. A short introduction to Linux as a viable operating system for mobile devices is also provided.

Appendix 1 describes the use of satellite networks in mobile communication. Appendix 2 gives several sample Java programs for developing mobile computing applications. Appendix 3 provides solutions to the objective type questions given in each chapter of the book.

Acknowledgements

I am immensely grateful to my teachers at the Indian Institute of Technology Delhi (1966–72) and the University of Upsala, Sweden (1978–79, 1984) for teaching me the importance of self-learning and the essence of keeping up with emerging technology. I would like to thank Prof. M.S. Sodha, F.N.A., for his support and blessings throughout my academic life. I acknowledge my colleagues—Dr P.C. Sharma, Dr P.K. Chande, Dr Sanjeev Tokekar, Mrs Vrinda Tokekar, Dr A.K. Ramani, Dr M.K. Sahu, Dr Maya Ingle, Mr Sanjay Tanwani, Dr Preeti Saxena, Mrs Shraddha Masih, Ms Aparna Dev, and Ms Vasanti G. Parulkar—for their constant encouragement and appreciation. I am thankful to the editorial team at Oxford University Press India for their reviews and suggestions. Finally, I would like to acknowledge my wife Sushil Mittal and my family members—Needhi Mittal, Dr Atul Kondaskar, Dr Shilpi Kondaskar, and Arushi Kondaskar—for their love, understanding, and support during the writing of this book.

Any suggestions, comments, and feedback for further improvement of the text are welcome. A link for the purpose can be found on the author's website at www.rajkamal.org.

Raj Kamal

Contents

Abbreviations and Acronyms

2G second generation
2.5G+ enhancement of 2G
3G third generation
3GPP 3G partnership project
3GPP2 3G partnership project 2
4-QPSK four-phase quadrature phase shift keying
6lowpan IPv6 over low power personal area network
8-PSK eight-phase shift keying
9iAS Oracle 9i application server
16-QAM sixteen quadrature amplitude modulation
π/4-QPSK π/4 quadrature phase shift keying
AAA authentication, authorization, and auditing
AAC advanced audio coding, and also Apple audio communication data format
ACID atomicity, consistency, isolation, and durability
ACL asynchronous connectionless link
ADO .net active-X data objects in dot net
AES advanced encryption standard
AIDA adaptive information dispersal algorithm
AMA active member address
AODV ad-hoc on-demand distance vector routing protocol
APDU application protocol data unit
API application program interface (software)
ARCNET attached resource computer network
ARP address resolution protocol
ARQ automatic repeat request
ASK amplitude shift keying
ASP active server page
ATM automatic thread migration in agent; asynchronous transfer mode protocol
AuC authentication center
BER bit error rate
BES BlackBerry enterprise server
BFSK binary frequency shift keying
BMC broadcast and multicast control protocol
BPSK binary phase shift keying
BSAC bit slice arithmetic coding
BSC base station controller
BSS basic service set
BTS base transceiver station
BWA broadband wireless access
c-HTML compact HTML
CCXML call control extensible markup language
CCD charge coupled device
CCK complementary code keying
CDC connected device configuration
CLDC connection limit device configuration
CDMA code division multiple access
CDMA 1xEV-DO cdmaone 1.2288 Mchip/s channel evolution for high speed integrated data optimised
CDMA 1xEV-DV CDMA 1.2288 Mchip/s channel evolution for high speed integrated data and voice
CDMA 3x CDMA three 1.2288 Mchip/s channels
CIF common interleaved frame
CDPD cellular digital packet data for dipital radio
CGI common gateway interface
CGSR cluster-head gateway switch routing
CHAP challenge handshake authentication protocol
COA care-of-address
CODEC coder-decoder, compression-decompression
COFDM code orthogonal frequency division multiple access
CORMOS communication-oriented runtime system for sensor networks
CM call management
CN correspondent node
CRC cyclic redundancy check
CSI channel stream identification
CSMA carrier sense multiple access
CSMA/CA carrier sense multiple access collision avoidance
CSMA/CD carrier sense multiple access collision detect
CVM coherent virtual machine
DAB digital audio broadcasting
DACK duplicate acknowledgement
DB2 IBM DB2 database server
DB2e IBM DB2Everyplace
DBMS database management system
DECT digital enhanced cordless telecommunication
DES data encryption standard
DHCP dynamic host configuration protocol
DLS dynamic label segment
DMSP designated multicast service provider

DOM document object model
DPCH dedicated physical channel
DPDCH dedicated physical data channel
DSMA digital sense multiple access
DSDV destination sequenced distance vector algorithm
DSSS direct sequence spread spectrum
DS-CDMA direct sequence CDMA
DSP digital signal processor
DSR dynamic source routing
DTD data type definition
DTMF dual tone multiple frequency
DVB digital video broadcasting
DVB-H DVB for handheld devices
DVB-H+ DVB-H with hybrid satellite/terrestrial architecture
DVB-VBI DVB with vertical blanking interval

EBSN explicit bad state notification
ECN explicit congestion notification
EDGE enhance data rate for GSM evolution
EIR equipment identity register
ELC embedded linux consortium
ERI extensible remote invocation
ESS extended service set
ETSI European Telecommunication Standard Institute

FA foreign agent
FEC forward error correction
FDMA frequency division multiple access
FDD FDMA duplex (separate carrier for uplink and downlink)
FHSS frequency hopping spread spectrum
FIC fast information channel
FIFO first in first out
FM frequency modulation
FOMA freedom of mobile multimedia access
FRA fixed radio access
FRR fast retransmission/fast recovery
FRT fixed phone radio interface
FSK frequency shift keying

GEO geosynchronous earth orbit
GGSN gateway GPRS support node
GMM GPRS mobility management
GMSK Gaussian minimum shift keying
GMSC gateway mobile service center
GPRS general packet radio service
GRE generic routing encapsulation
GSM global system for mobile communication (from European Groupe Speciale Mobile)
GTK graphic tool kit language
GTP gateway tunneling protocol
GUI graphic user interface

HD high-definition radio
HDTV high-definition TV
HEC header check field
HEO highly elliptical earth orbit
HiperLAN high performance LAN
HLR home location register
HA home agent
HMAC hash message authentication code
HSCSD high speed circuit switched data
HTTP hyper-text transfer protocol
HTML hyper-text markup language
HTTPS HTTP with SSL

I-component in-phase component
i-Mode Internet in mobile mode
I-Q in-phase and quadrature phase components (modulation or demodulation)
I-pilot in-phase pilot
IBSS independent basic service set
ICMP Internet control message protocol
IDA information dispersal algorithm
IEEE Institute of Electrical and Electronics Engineers
IETF Internet Engineering Task Force
IIOP Internet inter-ORB protocol
IMEI international mobile equipment identity
IMPS instant messaging and presence service
IMSI international mobile subscriber identity
IMT International Mobile Telecommunications
IPP interleaved push-and-pull
IrDA infrared data association
IrLAN infrared LAN access
ISDN integrated services digital network
ISR interrupt service routine
IST interrupt service thread
ITU International Telecommunication Union
IWF inter-working function

JCF Jataayu client framework
JCRE Java Card runtime environment
JDBC Java Database connectivity
JINI Java Intelligence Network Infrastructure[1]
JNDI Java naming and directory interface
JNDS Java naming and directory service
JNI Java native interface
JSP Java server page
JSR168 Java specification recommendation 168
JVM Java virtual machine
J2ME Java 2 Micro Edition
J2SE Java 2 Standard Edition
J2EE Java 2 Enterprise Edition

[1] JINI is not an acronym according to SUN developer team

KVM kilo virtual machine[2]

L2CAP logical link control and adaptation protocol
LAPD link access protocol D-channel
LDAP light weight directory access protocol
LAI location area identification
LAN local area network (of computing device)
LCD liquid crystal display
LEO low earth orbit
LFSR linear feedback shift register
LM link manager
LOS line of sight
LW long wave

M-sequence maximum length sequence
M-TCP mobile TCP
MAC message authentication code (for security)[3]
MANET mobile ad-hoc network
MEO medium earth orbit
MobiTex mobile text protocol
MCC mobile country code
MD 5 message digest algorithm 5
MESI modified, exclusive, shared, and invalid states
MIDP mobile information device profile
MIMO multiple-input multiple-output communication
MM mobility management
MMC multimedia card
MMS multimedia messaging service
MMU memory management unit
MNC mobile network code
MoM mobile multicast
MOT multimedia object transfer protocol
MQe IBM WebSphere Messaging and Queuing Everyplace
MSC mobile services switching center
MSIN mobile subscriber identity number
MSISDN mobile subscriber ISDN number
MSP mobile service provider
MSPADS MSP application distribution system
MTP message transfer part
MTU maximum transferable unit
MW medium wave

NSS network and switching subsystem

OBEX object exchange protocol at transport layer
OFDM orthogonal frequency division multiple access
OLSR optimized link state routing protocol
OMA open mobile alliance
OMC operation and maintenance center
OQPSK offset quadrature phase shift keying
ORB object request broker
OS operating system (software)
OSI open system interconnection
OSS operating subsystem
OTA over the air
OVSF orthogonal variable spread factor

PACK partial acknowledgement
PAS personal area synchronization
PC personal desktop computer
PDA personal digital assistant
PDCP packet data convergence protocol
PCM pulse code modulation
PCS point coordination support
PIM personal information manager
PIN personal identification number
PLCP physical layer convergence protocol
PLMN public land mobile network
PMA parked member address
PMD physical media dependent layer
PN pseudo noise
PN_I PN in-phase component
PN_Q PN quadrature component
PPE parallel page engine
PPM pulse position modulation
PPP point to point protocol
PsiF PLCP signaling field
PSF PLCP service field
PSK phase shift keying
PSTN public switching telephone network
PSPDN public switching packet data network
PTS priority token scheduling
PUK personal identification number unblocking key
PUMA protocol for unified multicasting through announcements
PWT public wireless telephone

Q-component quadrature-component
Q-pilot quadrature phase-shifted pilot
QAM quadrature amplitude modulation
QoS quality-of-service
QPSK quadrature phase-shift keying

RADIUS remote authentication dial in user service
RARP reverse address resolution protocol
RFID radio frequency identification devise
RLC radio link control
RLP radio link layer protocol
RMI remote method invocation
RPC remote procedure call
RRC radio resource control

[2]Virtual machine of kB size in place of MB in Java 2.
[3]It is also used for medium access control during communication.

RRM radio resource management
RSA Rivest, Shamir and Adleman
RSS radio subsystem
RTT round trip time

SAP service access point
SACK selective acknowledgement
SAX simple API for XML
SBR spectral band replication
SCCP signalling connection control protocol
SCI synchronization capsule indicator
SCO synchronous connection oriented
SDIO secure digital input/output memory card
SDK software development kit
SDMA space division multiple access
SDP service discovery protocol
SEQN sequence number field
SESBS satellite earth station and broadcast service
SFN single frequency network
SGML standard generalized markup language
SGSN serving GPRS support node
SHA 1 secure hash algorithm 1
SHF super high frequency
SIM subscriber identity module
SLP service location protocol
SM session management
SMIL synchronized multimedia integration language
SMS short message service
SNDCOP sub-network dependent convergence protocol
SNR signal-to-noise ratio
SOAP simple object access protocol
SOM start of message
SOP sum of products
SP service provider
SPA service provider address
SPS speech processing system
SRGS speech recognition grammar specification
SRS speech recognition system
SS supplementary services
SS7 signalling system 7
SSDP simple service discovery protocol
SSL secure socket layer
SSML speech synthesis markup language
STT speech-to-text
SW short wave
SynC synchronization channel
SyncML synchronization markup language
Sync4J synchronization language for Java (also called Funambol)

T-TCP transaction oriented TCP protocol
TCP transport control protocol
TCP/IP transport control protocol/Internet protocol suite
TDD time division duplex
TDMA time division multiple access
TIA Telecommunications Industry Association
TLS transport layer security
TMSI temporary mobile subscriber identity
TORA temporary ordered routing algorithm
TPDU transport protocol data units
TTL time-to-live
TTS text-to-speech

UDP user datagram protocol
UHF ultra high frequency
UMTS Universal Mobile Telecommunication System
UnPnP universal plug and play
URI universal resource identifier
URL universal resource locator
UTRA UMTS (or universal) terrestrial radio access
UTRAN UTRA network

VANET vehicular ad-hoc network
VHF very high frequency
VOFDM vector OFDM
VUI voice user interface
VXML voice XML

WAE wireless application environment
WAP wireless application protocol
WBXML WAP binary XML
WCMP wireless control message protocol
WDP wireless datagram protocol
WEA IBM WebSphere Everyplace access
WEP wired equivalent privacy in 802.1b
WiFi wireless fidelity
WiMax worldwide interoperability for microwave access
WLAN wireless local area network
WLL wireless in local loop
WML wireless markup language
WOFDM wideband OFDM
WPAN wireless personal area network
WSP wireless session protocol
WSRP web services for remote protlets
WTA wireless telephony application
WTCP wireless TCP
WTLS wirless transport layer security
WTP wireless transport protocol

XForms XML form
XHTMLMP extensible hypertext markup language mobile profile
XML extensible markup language
XOR exclusive OR logic operation
XUI XML user interface

Chapter 1

Mobile Communications: An Overview

This chapter serves as an introduction to mobile communications and mobile computing. The chapter outlines the following concepts which are discussed in greater detail in the subsequent chapters.

- Recapitulation of the basic concepts of wireless transmission—ranges of frequencies which are used for transmitting signals, propagation of signals by antennae, modulating and shift keying of amplitude, frequency, and phase before transmission of the signals
- Multiplexing of signals from multiple sources as a function of space, time, frequency, and code
- Differences between circuit switching and packet switching techniques
- Common technologies used for mobile communication
- Mobile computing and system architecture used in pervasive computing
- Operating systems, languages, and protocols used in mobile computing
- Mobile devices, systems, and networks

1.1 Mobile Communication

Mobile communication entails transmission of data to and from handheld devices. Out of the two or more communicating devices, at least one is handheld or mobile. The location of the device can vary either locally or globally and communication takes place through a wireless, distributed, or diversified network (Sections 1.5.1 and 1.1.4).

Communication is a two-way transmission and reception of data streams. Voice, data, or multimedia streams are transmitted as signals which are received by a

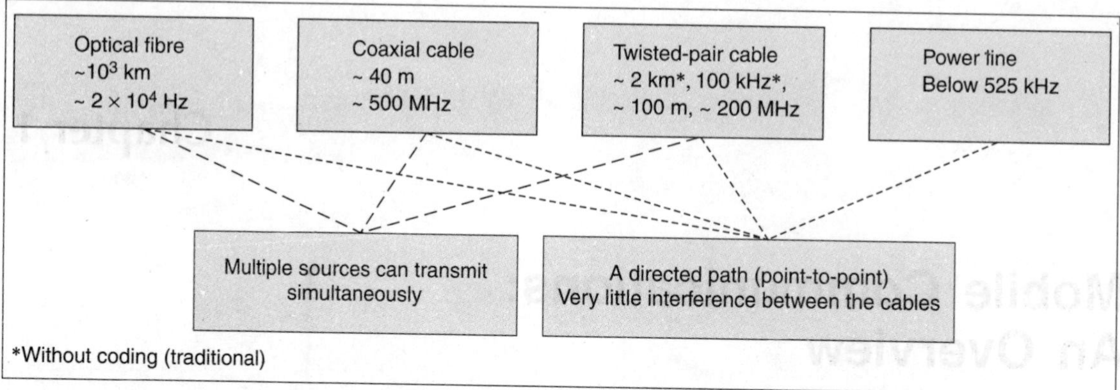

Fig. 1.1 Fibre- and wire-based transmissions and their ranges (without using repeater), frequencies, and properties

receiver. Signals from a system can be transmitted through a fibre, wire, or wireless medium. During the transmission process, the transmitter sends the signals according to defined regulations, recommended standards, and protocols.

1.1.1 Guided Transmission

Metal wires and optical fibres are used in guided or wired transmission of data. Figure 1.1 shows the fibre- and wire-based communication frequencies and the main properties of this mode of transmission. Guided transmission of electrical signals takes place using four types of cables—(i) optical fibre for pulses of wavelength 1.35–1.5 μm, (ii) coaxial cable for electrical signals of frequencies up to 500 MHz and up to a range of about 40 m, (iii) twisted wire pairs for conventional (without coding) electrical signals of up to 100 kHz and up to a range of 2 km, or for coded signals of frequencies up to 200 MHz and a range of about 100 m, and (iv) power lines, a relatively recent advent in communication technology, are used for long-range transmission of frequencies between 10 kHz and 525 kHz.

The advantages of cable-based transmission are—(a) transmission is along a directed path from one point to another, (b) there is practically no interference in transmission from any external source or path, and (c) using multiplexing and coding, a large number of signal sources can be simultaneously transmitted along an optical fibre, a coaxial cable, or a twisted-pair cable.

Significant disadvantages of transmission through cables are:

(a) Signal transmitter and receiver are fixed (immobile). Hence there is no mobility of transmission and reception points.

(b) The number of transmitter and receiver systems limits the total number of interconnections possible.

1.1.2 Unguided Transmission

Wireless or unguided transmission is carried out through radiated electromagnetic energy.

1.1.2.1 Signal Propagation Frequencies

Electrical signals are transmitted by converting them into electromagnetic radiation. These radiations are transmitted via antennae that radiate electromagnetic signals. There are various frequency bands within the electromagnetic spectrum and all have different transmission requirements. Figure 1.2 shows the VHF and UHF frequencies for wireless transmission and their transmission properties. Frequency, f, in MHz and wavelength, λ, in meters of electromagnetic radiation are related by the classical formula

$$f = c/\lambda = (300/\lambda) \tag{1.1}$$

Here the velocity of signal propagation, c, is 300×10^6 m/s (this is the velocity of electromagnetic waves in vacuum or air; for other media, such as water, this velocity

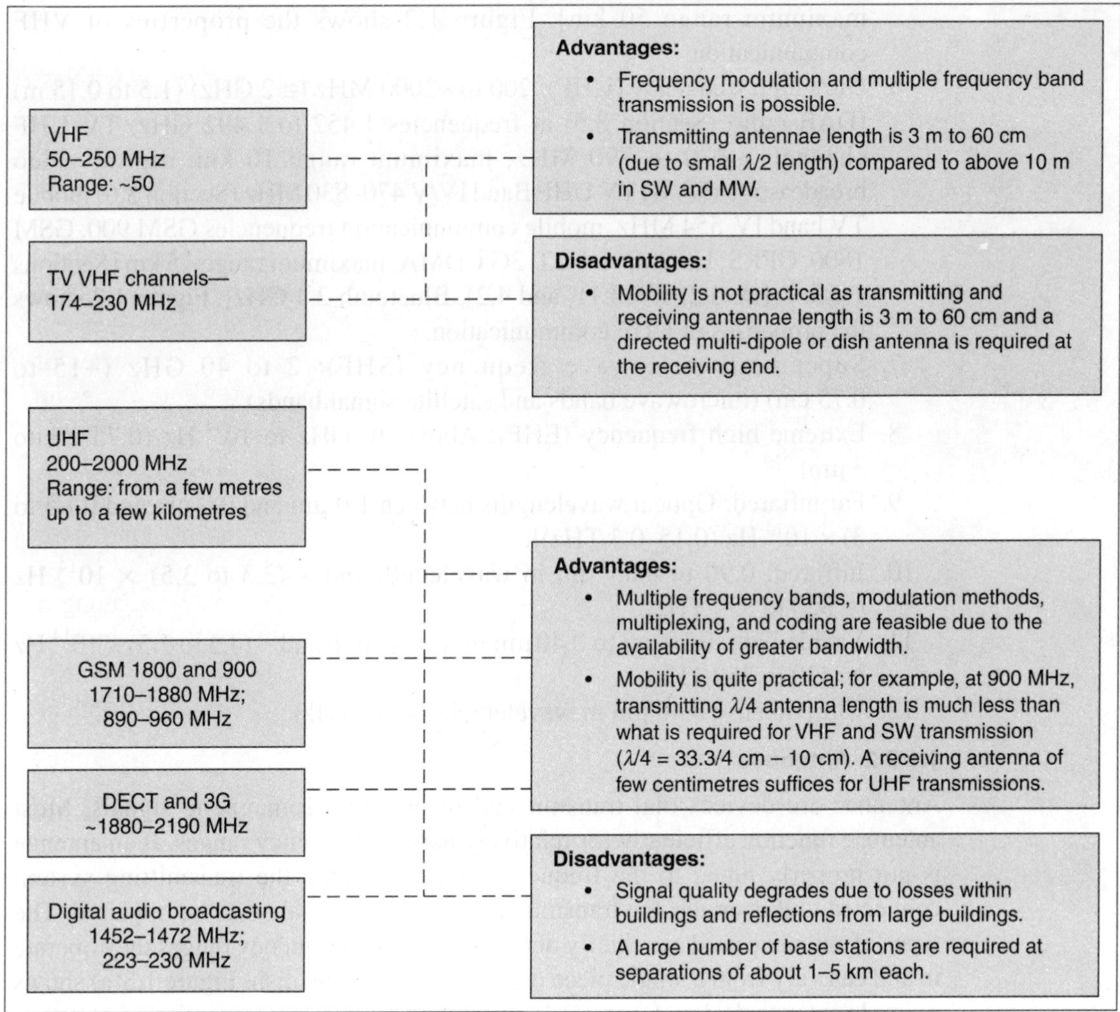

Fig. 1.2 Wireless transmission in VHF and UHF ranges: frequencies and properties

will be different). The frequencies, and thus wavelengths, of transmitters for various ranges are as follows:

1. Long-wavelength radio, very low frequency (LW): 30 kHz to 1 MHz (10,000 to 300 m).
2. Medium-wavelength radio, medium frequency (MW): 0.5 to 2 MHz (600 to 150 m).
3. Short-wavelength radio, high frequency (SW): 6 to 30 MHz (50 to 10 m).
4. FM radio band frequency (FM): 87.5 to 108 MHz (3.4 to 2.8 m), maximum range 50 km.
5. Very high frequency (VHF): 50 to 250 MHz (6 to 1.2 m) [digital audio broadcasting (DAB) band III VHF 174 to 240 MHz (Section 8.5), 226 ± 4 MHz, maximum range 50 km, TV VHF channels—174 to 230 MHz, maximum range 50 km]. Figure 1.2 shows the properties of VHF communication.
6. Ultra high frequency (UHF): 200 to ~2000 MHz (≅ 2 GHz) (1.5 to 0.15 m) [DAB radio (Section 8.5) at frequencies 1.452 to 1.492 GHz, TV UHF channels—470 to 790 MHz, maximum range 10 km, digital video broadcasting (DVB) TV UHF Band IV/V 470–830 MHz (Section 8.6) mobile TV band IV, 554 MHz, mobile communication frequencies GSM 900, GSM 1800, GPRS, HSCSD, DECT, 3G CDMA, maximum range ~5 km (Sections 1.1.3, 1.1.4, 3.2, 3.9–3.11, and 4.2), Bluetooth 2.4 GHz]. Figure 1.2 shows the properties of UHF communication.
7. Super high microwave frequency (SHF): 2 to 40 GHz (~15 to 0.75 cm) (microwave bands and satellite signal bands).
8. Extreme high frequency (EHF): Above 40 GHz to 10^{14} Hz (0.75 cm to 3 μm).
9. Far infrared: Optical wavelengths between 1.0 μm and 2.0 μm and [(1.5 to 3) × 10^{14} Hz (0.15–0.3 THz)].
10. Infrared: 0.90 to 0.85 μm in wavelength and ~ (3.3 to 3.5) × 10^{14} Hz (≅ 350 to 330 THz).
11. Visible light: 0.70 μm to 0.40 μm in wavelength and ~ (4.3 to 7.5) × 10^{14} Hz (≅ 430 to 750 THz).
12. Ultraviolet: < 0.40 μm in wavelength (> 750 THz).

1.1.2.2 Antennae

Antennae are devices that transmit and receive electromagnetic signals. Most antennae function efficiently for relatively narrow frequency ranges. If an antenna is not properly tuned to the frequency band in which the transmitting system connected to it operates, the transmitted or received signals may be impaired. The types of antennae used are chiefly determined by the frequency ranges they operate in and can vary from a single piece of wire to a parabolic dish. Figure 1.3(a) shows a simple antenna design. It is a $\lambda/2$-long antenna for wireless transmission of waves

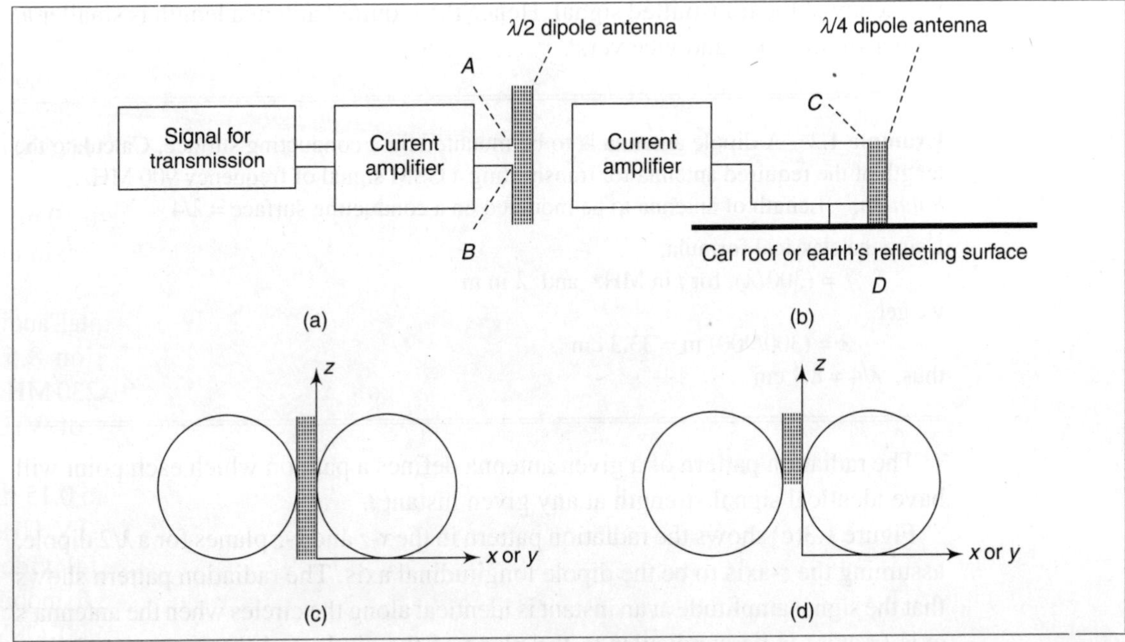

Fig. 1.3 (a) $\lambda/2$ dipole antenna (b) $\lambda/4$ dipole antenna (c) Radiation pattern in *z-y* and *x-z* planes for $\lambda/2$ dipole (d) Radiation pattern in *y-z* and *x-z* planes for $\lambda/4$ dipole

of wavelength λ. It is also called a dipole antenna because, at any given instant, both ends *A* and *B* are 180° out of phase.

Example 1.1 A 200 MHz to 2000 MHz UHF signal is to be transmitted wirelessly. Calculate the length of the dipole antenna required for transmission.

Solution: Length of the dipole antenna = $\lambda/2$

$\lambda = 300/f$ (Here λ is in m and f is in MHz)

Therefore, λ = [300/(200 to 2000)] m

= 1.5 to 0.15 m = 150 to 15 cm

$\lambda/2$ = 75 to 7.5 cm

Therefore, the length of the required antenna will be 75 cm down to 7.5 cm.

Figure 1.3(b) shows an antenna equivalent to the $\lambda/2$ antenna—an antenna of length $\lambda/4$ mounted on a long conducting surface, for example, the roof of a car or a moist ground surface. At any given instant, the end *C* and surface *D* are 180° out of phase. The original and reflected waves thus superimpose and create the same electrical effects as in the $\lambda/2$ antenna. The length of a $\lambda/4$ antenna for the 200–2000 MHz UHF transmission of Example 1.1 is just 37.5 cm down to 3.75 cm. In general, the length of an antenna is directly proportional to the wavelength and, therefore [using the formula λ = (300/f MHz) m], inversely proportional to the frequency of the transmitted signal. Hence the required antenna length is smaller at higher frequencies and vice versa.

Example 1.2 A dipole antenna is to be mounted on a conducting surface. Calculate the length of the required antenna for transmitting a GSM signal of frequency 900 MHz.
Solution: Length of antenna to be mounted on a conducting surface = $\lambda/4$

Using the classical formula,

$$f = (300/\lambda), \text{ for } f \text{ in MHz and } \lambda \text{ in m}$$

we get

$$\lambda = (300/900) \text{ m} = 33.3 \text{ cm}$$

thus, $\lambda/4 = 8.1$ cm

The radiation pattern of a given antenna defines a path on which each point will have identical signal strength at any given instant *t*.

Figure 1.3(c) shows the radiation pattern in the *x-z* and *y-z* planes for a $\lambda/2$ dipole, assuming the *z*-axis to be the dipole longitudinal axis. The radiation pattern shows that the signal amplitude at an instant is identical along the circles when the antenna's axis (*z*-axis) is perpendicular to the plane of the circle and the antenna axis is a tangent to both the circles (circles directed along the *x* and –*x* axes or the *y* and –*y* axes). The radiation pattern in the *x-y* plane (perpendicular to both *x-z* and *y-z* planes) will be a single circle with the *z*-axis passing through its centre and perpendicular to the plane it lies in. Figure 1.3(d) shows that a $\lambda/4$ dipole mounted on a conducting surface will also have an identical radiation pattern. The radiation patterns in the *x-y*, *y-z*, and *x-z* planes are identical to the one shown in Fig. 1.3(c).

The radiation pattern is an important feature of an antenna. A circular pattern means that radiated energy, and thus signal strength, is equally distributed in all directions in the plane. A pattern in which the signal strength is directed along a specific direction is shown in Fig. 1.4. A directed radiation pattern is required between a mobile user and the base station.

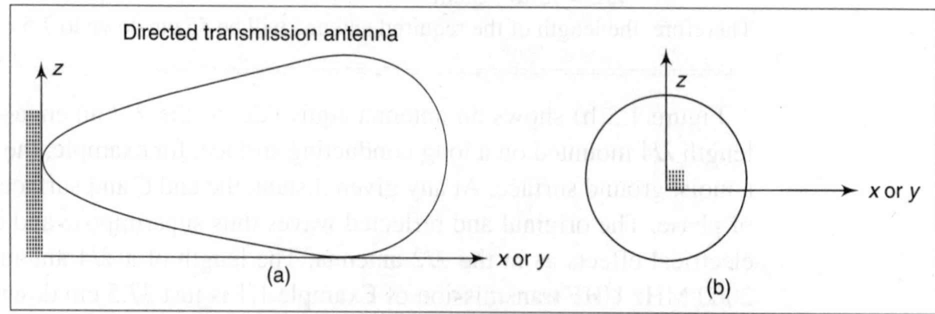

Fig. 1.4 (a) Radiation pattern along the *z-y* and *z-x* planes in a directed transmission case antenna (e.g., between a base station and mobile phone) (b) Radiation pattern of the same antenna along the *x-y* plane

1.1.2.3 Propagation of Signals

Wireless propagation of signals faces many complications. Mobile communication renders reliable wireless transmission much more difficult than communication between fixed antennae. The antenna height and size at mobile terminals are generally quite small. Therefore, obstacles in the vicinity of the antenna have a significant influence on the propagated signal. To minimize such impairing effects on the signal, propagation routes have to be specially designed and calculated taking into account the various types of propagation losses. Also, the propagation properties vary with place and, for a mobile terminal, with time. Nowadays, statistical propagation models are used whereby no specific data paths are considered, rather the channel parameters are modelled as stochastic variables (probability-based random variables).

(a) *Line-of-sight propagation* is the transmission of signals without refraction, diffraction, or scattering in between the transmitter and the receiver. But even in such an ideal scenario, transmission losses do occur. Let us look back at the radiation pattern represented in Fig. 1.3. Signal strength in free space decreases as the square of the distance from the transmitter. This is because at larger distances, the radiated power is distributed over a larger spherical surface area.

Example 1.3 A transmitter sends a signal which has a strength of 9 $\mu W/cm^2$ at a distance of 500 m. Assuming free space propagation in line of sight, calculate the signal strength at 1500 m.

Solution: The strength of the transmitted signal is inversely proportional to the square of the distance from the transmitter.

Therefore, the strength of the signal at 1500 m = $(500/1500)^2 \times 9\ W/cm^2 = 1\ \mu W/cm^2$

(b) Signal strength also decreases due to *attenuation* when obstacles in the path of the signal are greater in size than the wavelength of the signal. A few examples of attenuation of signals are as follows—(i) If an FM radio transmitter sends out a 90 MHz ($\lambda = 3.3$ m) FM band signal, then the signal will be attenuated by objects of size 10 m and above (size much greater than λ), (ii) If a transmitter sends a GSM 900 MHz ($\lambda = 33$ cm) signal, then it will face attenuation in objects of size 1 m and above (size much greater than 33 cm).

(c) A signal *scatters* when it encounters an obstacle of size equal to or less than the wavelength. For example, a GSM signal, about 33 cm in wavelength, is scattered by an object of 30 cm or less. Figure 1.5(a) shows the scattering of a transmitted signal. Only a small part of the scattered signal reaches the receiver.

(d) A signal bends as a result of *diffraction* from the edges of an obstacle of size equal to or less than the wavelength. For example, a GSM signal of wavelength 33 cm will diffract from an object of 33 cm or less. Figure 1.5(b) shows a diffracted signal. A diffracted signal may or may not reach its destination; it depends on the

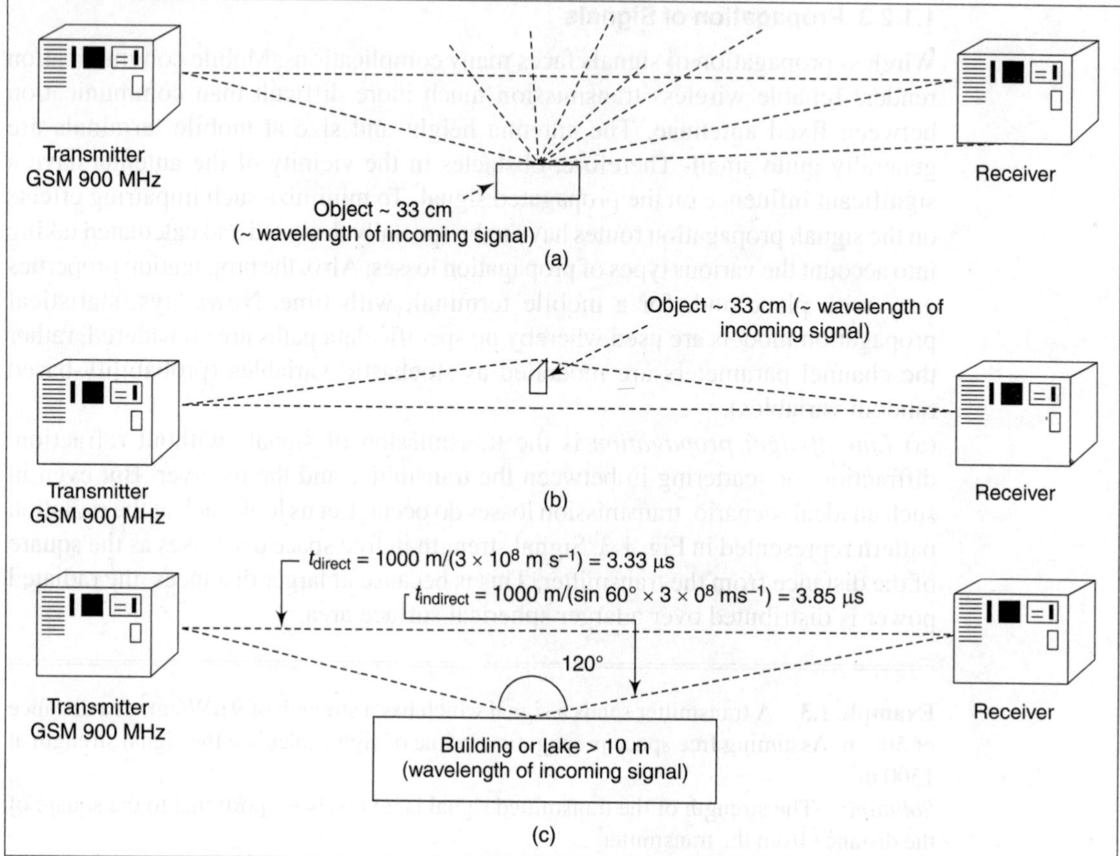

Fig. 1.5 (a) Scattering of signal (b) Diffraction of signal (c) Reflected and direct signals from a 900 MHz transmitter and calculation of delay

geometry of the obstacle and the separation between the object, the receiver, and the transmitter.

(e) The signal may also be *reflected* from the surface of an obstacle, the earth's surface, or a water body of size greater than the wavelength of the signal. For example, if a transmitter sends out a GSM 900 MHz (λ = 33 cm) signal, then the transmitter signal reflects from an object of size 10 m and above (much greater than λ) .The reflected signal suffers a delay in reaching its destination. Figure 1.5(c) shows the reflection and the delay. The delay is more pronounced in case of multi-hop paths. Delayed signals have distorted waveforms and cause misrepresentation of information encoded in the signal. There are digital signal processing techniques to eliminate the distortions due to delays from direct and multiple paths so that the original signal can be recovered. The delay in the reflected signal with respect to the original direct signal is given as

$$\text{Delay} = t_{\text{indirect}} - t_{\text{direct}} = \frac{\text{additional path travelled in meters}}{3 \times 10^8 \, \text{m s}^{-1}} \tag{1.2}$$

Example 1.4 A receiver receives two signals, one directly in line-of-sight and the other after a reflection of 120° from a transmitter at a distance of 1000 m (Fig. 1.5(c)). Calculate the delay in the reflected signal with respect to the direct signal.

Solution: Direct path time, $t_{direct} = \dfrac{1000 \text{ m}}{3 \times 10^8 \text{ ms}^{-1}} = 3.33\ \mu s$

Reflected path time, $t_{indirect} = \dfrac{1000 \text{ m}}{\sin(120°/2) \times (3 \times 10^8 \text{ ms}^{-1})} = 3.85\ \mu s$

Delay in reflected signal with respect to direct signal $= t_{indirect} - t_{direct}$
$= 3.85–3.33\ \mu s = 0.52\ \mu s$

A transmitter can simultaneously transmit a signal $s(t)$ and a reference signal $s_{ref}(t)$. From the received superimposed reference of the two path signals, a signal processor can recover $s_{ref}(t)$ and the reflected path reference signal $s'_{ref}(t)$. Using these, the processor can determine the delay parameters of the reflected paths. Now the signal processor can separate the direct path signal $s(t)$ from the received superimposed signal including $s(t)$ and reflected path signal $s'(t)$.

Figure 1.5 shows how reflection, scattering, and diffraction help a signal reach the points not reachable by the line-of-sight path but they also cause degradation of signal quality. The distortion of the signal may or may not be compensated for by the digital signal processor (DSP) circuits. The compensation depends upon the signal-to-noise ratio in the received signals. (Noise refers to the unwanted, randomly varying signals of other sources, which are received along with the original signal due to interference.)

1.1.2.4 Modulation

We have seen that the sizes of antennae required for wireless transmission are inversely proportional to the frequencies of transmitted signals. This means that low frequency signals need very large antennae for their transmission. For example, voice signals have frequencies between 0.1 kHz and 8 kHz. Music-signal frequencies lie between 0.1 kHz and 16 kHz. These ranges are unsuitable for wireless transmission due to the requirement of abnormally large-sized antennae. Moreover, due to the properties of the signal-propagating medium (air or vacuum), ultralow frequency signals cannot be transmitted across long distances without a significant loss of signal strength. Also, video-signal frequencies lie in the range between 10 kHz and 2 MHz. But this range of frequencies is reserved for MW (medium wave) and SW (short wave) radio broadcasts. Therefore, independent (without modulation) wireless transmission of voice, music, video, and data signals is not very practical. Modulation is required to make wireless transmission practical by increasing the compatibility of the transmitted signal and the medium of transmission.

Modulation is the process of varying one signal, called the carrier, according to the pattern provided by another signal (modulating signal). The carrier is usually an analog signal selected to match the characteristics of a particular transmission system. The amplitude, frequency, or phase angle of a carrier wave is varied in proportion to the variation in the amplitude variation of the modulating wave (message signal).

Let us now consider the instantaneous change in the value of the amplitude of a signal, $s(t)$. The signal $s(t)$, at an instant t, is given by the classical sinusoidal equation

$$\begin{aligned} s(t) &= s_0 \sin\,[(2\pi \times c/\lambda \times t) + \varphi_{t0}] \\ &= s_0 \sin\,[(2\pi \times f \times t) + \varphi_{t0}] \end{aligned} \tag{1.3}$$

where s_0 is the peak amplitude (amplitude varies between s_0 and $-s_0$), c is the velocity of the transmitted wave, φ_{t0} is the phase angle of the signal at $t = 0$ (a reference point with respect to which t is considered), and f is the signal frequency. At the instant t_0 ($t = 0$), the signal amplitude is $s_0 \sin(\varphi_{t0})$ and this value is repeated after every interval of $T = f^{-1}$, i.e., f times in one second. The phase angle term, $(2\pi \times f \times t) + \varphi_{t0}$, varies between 0 and 2π radian during the interval t to $t + T$. The angular frequency, ω, is $2\pi \times f = 2\pi/T$.

Consider that a carrier frequency, f_c, is used to modulate a signal for its wireless transmission. The carrier amplitude and its instantaneous amplitude, $s_c(t)$, is as per Eqn (1.3). Let the carrier parameters be as follows: peak amplitude = s_{c0}, phase angle = φ_{ct0} when $t = 0$. Now, modulation is a technique by which f_c or a set of carrier frequencies is used for wireless transmission such that peak amplitude, s_{c0}, frequency, f_c, or phase angle, φ_{ct0}, varies with t in proportion to the peak amplitude of the modulating signal $s_m(t)$, which is the voice or data signal to be communicated. The modulation is referred to as amplitude, frequency, or phase modulation, depending upon which parameter of the carrier is varied. Digital modulation is a technique by which amplitude, frequency, or phase angle parameters of carrier or subcarrier frequencies are varied according to the variation in the modulating signal bit 1 or 0, the modulating bit-pair, or set of bits. The modulation is then called amplitude, frequency, or phase-shift keying (ASK, FSK, or PSK) modulation, respectively.

In general, any periodic wave can also be defined by another classical equation called the Fourier equation

$$s(t) = \Sigma a_n \sin(2\pi \times n \times f \times t) + \Sigma b_n \cos(2\pi \times n \times f \times t) + 0.5a_0 \tag{1.4}$$

where a_0 is the constant part of the signal amplitude (dc part) and n is an integer varying from 1, 2, 3, ..., ∞. From the Fourier equation it can be concluded that any signal (sinusoidal or non-sinusoidal), which repeats itself every interval of $T = f^{-1}$, can be considered to be consisting of a sinusoidal signal of fundamental frequency $f(n = 1)$ of amplitude given by constants a_1 and b_1 and harmonics (sinusoidal signals) of frequencies which are the integral multiples of $f(n \times f;\ n = 2, 3, ..., \infty)$, the amplitude of each given by constants a_n and b_n. Equation (1.3) for a sinusoidal wave is a special case, when $a_1 \neq b_1$; $a_0, a_2, a_3, \ldots$ and $b_2, b_3, \ldots$ are 0.

Amplitude Modulation The equation for an amplitude-modulated signal $s(t)$ after modulation with a sine wave of modulating frequency f_m is as follows:

$$s(t) = s_{c0}\sin(2\pi \times f_c \times t + \varphi_{t0}) \tag{1.5}$$

where $s_{c0} = k_{am} \times s_m(t) = k_{am} \times s_{m0}\sin[(2\pi \times f_m \times t) + \varphi_{mt0}]$ (1.6)

where k_{am} is a constant, s_{m0} is modulating signal peak amplitude, f_m is modulating signal frequency, and φ_{mt0} is the phase angle at $t = 0$. The modulated signal of Eqn (1.5) can be analysed mathematically to show that the transmitted signal frequency consists of three sinusoidal components of frequencies, $f_c - f_m$, f_c, and $f_c + f_m$. The $f_c - f_m$ and $f_c + f_m$ components are known as lower and upper sidebands of f_c. Figure 1.6(a) shows an amplitude-modulated signal. Each vertical line represents a sinusoidal component.

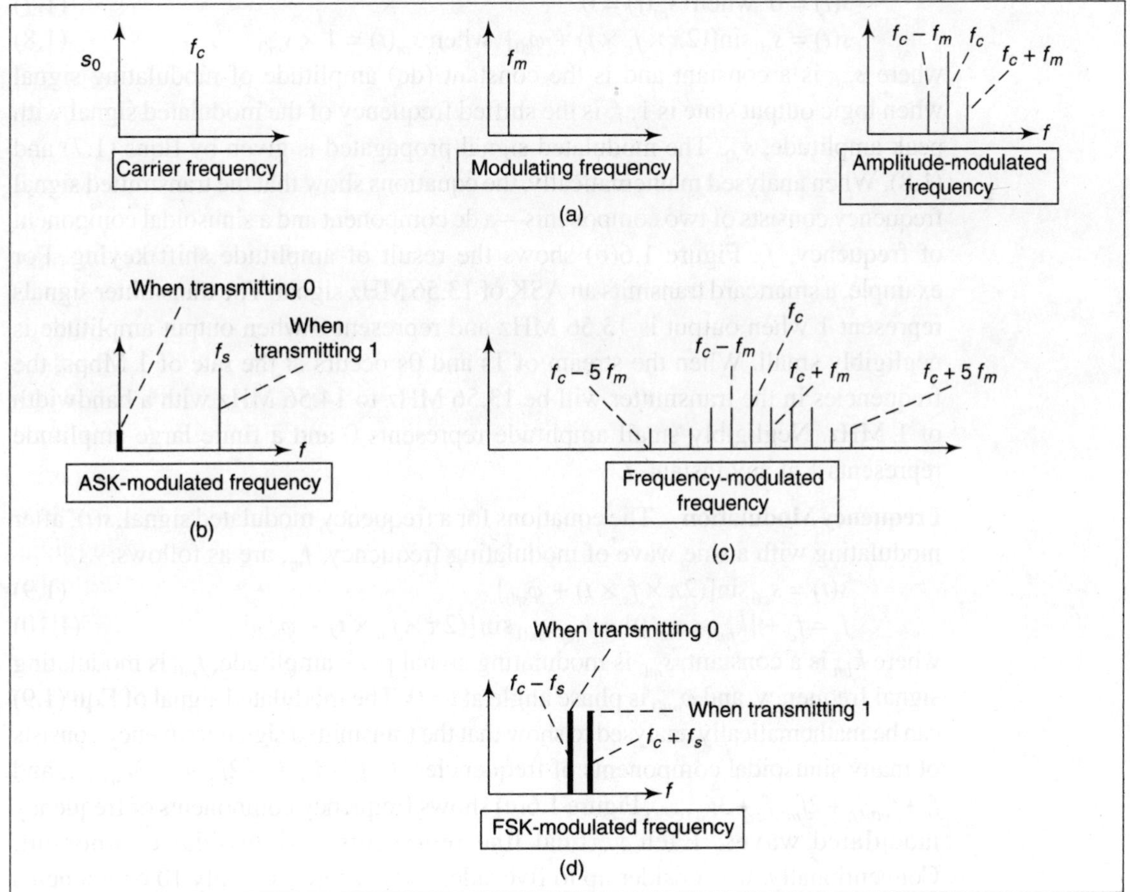

Fig. 1.6 (a) AM signal (b) ASK signal (c) FM signal (d) FSK signal (all plots are shown in frequency domain. Each vertical line represents a sine wave of Eqn (1.1). The *x*-axis depicts the frequency and *y*-axis, the peak amplitude)

Example 1.5 An SW transmitter sends out a 6.000 MHz signal after amplitude modulation of frequency 5 kHz at an instant. What are the three sinusoidal components in the transmitter at that instant? Calculate the bandwidth of the transmitted signal.

Solution: The three sinusoidal components are

$f_c - f_m = 5.995$ MHz,

$f_c = 6.000$ MHz, and

$f_c + f_m = 6.005$ MHz.

Bandwidth of the transmitted signal = 6.005 MHz – 5.995 MHz = 10 kHz

Amplitude Shift Keying The equations for amplitude shift keying of a sine wave of frequency f_c after modulating with signals of 0s and 1s are as follows:

$$s(t) = 0 \text{ when } s_m(t) = 0 \quad (1.7)$$

$$s(t) = s_{s0}\sin[(2\pi \times f_c \times t) + \varphi_{t0}] \text{ when } s_m(t) = 1 \times s_{m0} \quad (1.8)$$

where s_{m0} is a constant and is the constant (dc) amplitude of modulating signal when logic output state is 1, f_s is the shifted frequency of the modulated signal with peak amplitude, s_{s0}. The modulated signal propagated is given by Eqns (1.7) and (1.8). When analysed mathematically, the equations show that the transmitted signal frequency consists of two components—a dc component and a sinusoidal component of frequency, f_s. Figure 1.6(b) shows the result of amplitude shift keying. For example, a smartcard transmits an ASK of 13.56 MHz signal. The transmitter signals represent 1 when output is 13.56 MHz and represent 0 when output amplitude is negligibly small. When the stream of 1s and 0s occurs at the rate of 1 Mbps, the frequencies in the transmitter will be 13.56 MHz to 14.56 MHz with a bandwidth of 1 MHz. Negligibly small amplitude represents 0 and a finite large amplitude represents 1 at any instant, *t*.

Frequency Modulation The equations for a frequency modulated signal, $s(t)$, after modulating with a sine wave of modulating frequency, f_m, are as follows:

$$s(t) = s_{c0}\sin[(2\pi \times f_s \times t) + \varphi_{st0}] \quad (1.9)$$

$$f_s = f_c + [k_{fm} \times s_m(t)] = k_{fm}\, s_{m0} \sin[(2\pi \times f_m \times t) + \varphi_{mt0}] \quad (1.10)$$

where k_{fm} is a constant, s_{m0} is modulating signal peak amplitude, f_{m0} is modulating signal frequency, and φ_{mt0} is phase angle at $t = 0$. The modulated signal of Eqn (1.9) can be mathematically analysed to show that the transmitted signal frequency consists of many sinusoidal components of frequencies: $f_c, f_c - f_m, f_c - 2f_m, f_c - 3f_m, \ldots$, and $f_c + f_m, f_c + 2f_m, f_c + 3f_m, \ldots$. Figure 1.6(c) shows frequency components of frequency modulated waves. Each vertical line represents a sinusoidal component. Conventionally, we consider up to five sidebands. Therefore, only 10 components are shown.

Example 1.6 An FM transmitter sends a 90 MHz signal after frequency modulation of frequency 5 kHz at an instant. Considering one carrier, five lower, and five upper side bands, what are the 11 sinusoidal components in the transmitter? Calculate the bandwidth of the signal. What will be the bandwidth for a voice-signal frequency modulation of 8 kHz?

Solution: The 11 sinusoidal components in the transmitter at that instant are

Lower sidebands: $f_c - 5f_m$ = 89.975 MHz, $f_c - 4f_m$ = 89.980 MHz, $f_c - 3f_m$ = 89.985 MHz, $f_c - 2f_m$ = 89.990 MHz, $f_c - f_m$ = 89.995 MHz

Carrier frequency: f_c = 90.000 MHz

Upper sidebands: $f_c + f_m$ = 90.005 MHz, $f_c + 2f_m$ = 90.010 MHz, $f_c + 3f_m$ = 90.015 MHz, $f_c + 4f_m$ = 90.020 MHz, $f_c + 5f_m$ = 90.025 MHz

Bandwidth of the transmitted signal = 90.025 MHz – 89.975 MHz = 50 kHz

For an 8 kHz voice-signal frequency modulation, the bandwidth = 10 × 8 kHz = 80 kHz

---

Frequency Shift Keying The equation for frequency shift keying of a sine wave with modulating signal of 0s and 1s are

$$s(t) = s_{s0}\sin[\{2\pi \times (f_c - f_s) \times t\} + \varphi_{t0}] \text{ when } s_m(t) = 0 \tag{1.11}$$

$$s(t) = s_{s0}\sin[\{2\pi \times (f_c + f_s) \times t\} + \varphi_{t0}] \text{ when } s_m(t) = 1 \times s_{m0} \tag{1.12}$$

where s_{m0} is a constant which represents peak amplitude of the modulating signal when logic output state is 1, f_s is shifting frequency of the modulated signal with peak amplitude, s_{s0}. The modulated signal given by Eqns (1.11) and (1.12) can be analysed to show that the transmitted signal frequency consists of a component of frequency $(f_c - f_s)$ and one sinusoidal component of frequency $(f_c + f_s)$, both having identical amplitudes. Figure 1.6(d) displays the result of FSK. For example, a wireless modem transmits an FSK 902.000 MHz signal, then the transmitter signals 1 when output is 902.032 MHz and signals 0 when output is 901.968 MHz signal. When the stream of 1s and 0s occurs at a rate of 64 kbps, the frequencies in the transmitter will be 901.968 MHz to 902.032 MHz with a bandwidth of 64 kHz.

Phase Modulation The equations for a phase-modulated signal, $s(t)$, with a modulating sine wave of modulating frequency, f_m, are as follows:

$$s(t) = s_{c0}\sin[(2\pi \times f_c \times t) + \varphi_{st0}] \tag{1.13}$$

$$\varphi_{st0} = k_{pm} \times s_m(t) = k_{pm} \times s_{m0}\sin[(2\pi \times f_m \times t) + \varphi_{mt0}] \tag{1.14}$$

where k_{pm} is a constant, s_{m0} is modulating signal peak amplitude, f_{m0} is modulating signal frequency, and φ_{mt0} is the phase angle at $t = 0$. The modulated signal of Eqn (1.13) is then propagated and can be analysed mathematically to show that the transmitted signal frequency consists of many sinusoidal components of frequencies: $f_c, f_c - f_m, f_c - 2f_m, f_c - 3f_m, f_c - 4f_m, \ldots$, and $f_c + f_m, f_c + 2f_m, f_c + 3f_m, f_c + 4f_m, \ldots$

Phase Shift Keying Phase shift keying (PSK) is a digital modulation technique in which data is transmitted by varying or modulating the phase of the carrier wave. There are various types of PSK techniques, each using a finite number of phases. Each of these phases corresponds to a unique binary number or a sequence of bits. The different methods for phase shift keying are described below.

Binary phase shift keying The equations for binary phase shift keying (BPSK) of a sine wave of modulating signal of 0s and 1s at frequency f_c are as follows:

$$s(t) = s_{s0}\sin[(2\pi \times f_c \times t) + \varphi_{t0} + \pi] \text{ when } s_m(t) = 0 \quad (1.15)$$

$$s(t) = s_{s0}\sin[(2\pi \times f_c \times t) + \varphi_{t0}] \quad \text{when } s_m(t) = 1 \times s_{m0} \quad (1.16)$$

where s_{m0} is a constant and is the amplitude of the modulating signal. When the logic output state is 0, there is a phase-shift of 360° in the modulated signal and peak amplitude is s_{s0}. The modulated signal is given by Eqns (1.15) and (1.16). The transmitted signal frequency phase advances or decreases by 180° at the instant when 1 changes to 0 or 0 changes to 1, respectively. The sinusoidal components have identical amplitudes. Figure 1.7(a) shows the BPSK output phases when the modulating signal is ⟨01⟩.

Gaussian minimum shift keying During transmission by BPSK, the carrier frequency abruptly advances or retards the phase by 180° when 0 follows 1 or 1 follows 0, respectively. These abrupt changes bring about high-frequency components in $s(t)$. [Refer to the Fourier equation, Eqn (1.4).] The periodically varying abrupt changes induce the higher harmonics in the transmitted frequency. Using minimizing techniques and DSP-based Gaussian low pass filter, these components can be filtered out so that changes at the transitions 1–0 and 0–1 are smoothened out. Such a BPSK signal is called GMSK (Gaussian minimum shift keying).

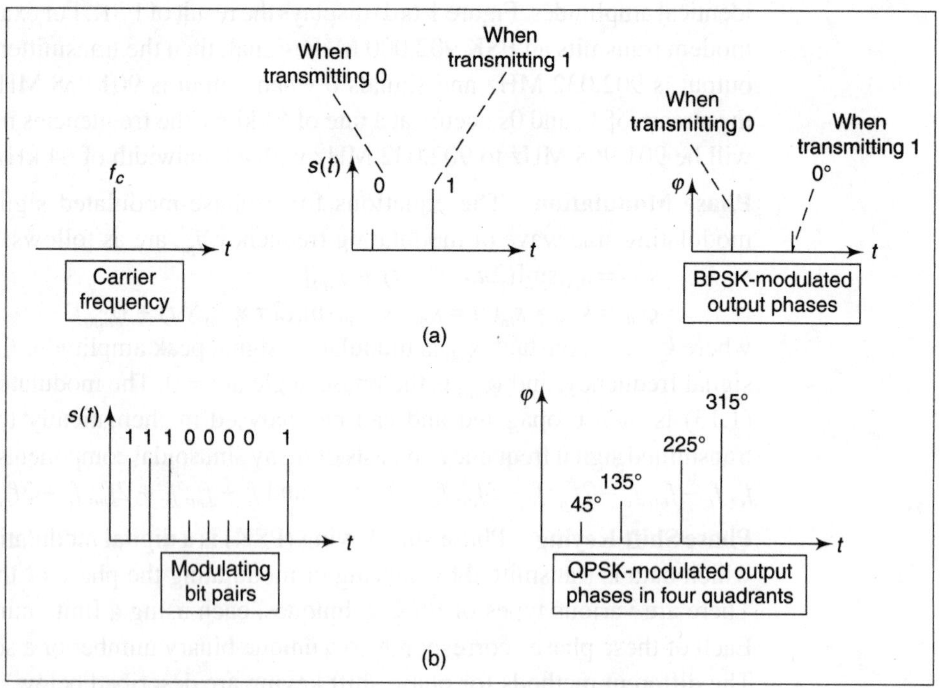

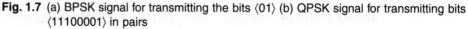

Fig. 1.7 (a) BPSK signal for transmitting the bits ⟨01⟩ (b) QPSK signal for transmitting bits ⟨11100001⟩ in pairs

Quadrature phase shift keying Quadrature phase shift keying (QPSK) is a modulation technique in which the phase angle of the carrier is shifted in one of the four quadrants between 0° and 360° and the shift depends on whether the modulating signal pair of bits are (1, 0), (0, 1), (0, 0), or (1, 1). For example, let us consider the case where pairs of bits 0 and 1 are being transmitted at rate T^{-1}. Bit pattern is <u>10</u> <u>00</u> <u>01</u> <u>11</u>. The phase angle of the transmitted signal $s(t)$ will be $3\pi/4$, $-3\pi/4$, $-\pi/4$, $+\pi/4$, after each successive time interval T. The frequency of the signal remains f_c and peak amplitude s_{s0}.

The equations for QPSK of a sine wave of modulating signal of 0s and 1s at a frequency f are as follows:

$$s(t) = s_{s0} \sin\left[(2\pi \times f_c \times t) + \varphi_{t0} + \left(\frac{3\pi}{4}\right)\right] \quad \text{when } s_m(t) = 1 \text{ and } s_m\left(t + \frac{T}{2}\right) = 0 \tag{1.17}$$

$$s(t) = s_{s0} \sin\left[(2\pi \times f_c \times t) + \varphi_{t0} - \left(\frac{\pi}{4}\right)\right] \quad \text{when } s_m(t) = 0 \text{ and } s_m\left(t + \frac{T}{2}\right) = 1 \tag{1.18}$$

$$s(t) = s_{s0} \sin\left[(2\pi \times f_c \times t) + \varphi_{t0} - \left(\frac{3\pi}{4}\right)\right] \quad \text{when } s_m(t) = 0 \text{ and } s_m\left(t + \frac{T}{2}\right) = 0 \tag{1.19}$$

$$s(t) = s_{s0} \sin\left[(2\pi \times f_c \times t) + \varphi_{t0} + \left(\frac{\pi}{4}\right)\right] \quad \text{when } s_m(t) = 1 \text{ and } s_m\left(t + \frac{T}{2}\right) = 1 \tag{1.20}$$

Figure 1.7(b) shows the QPSK signal output phases when the modulating signal is $\langle 11100001 \rangle$.

Eight phase shift keying In the eight phase shift keying (8-PSK) modulation technique, the phase angles of a carrier frequency can be one of eight different angles corresponding to one of the eight combinations of a set of three consecutive bits (000, 001, ..., 111). The triplets of bits are thus transmitted in varying phases so that triplet combinations can be simultaneously transmitted, in place of one bit at a time. For example, let us assume that a triplet of bits is being transmitted at rate T^{-1}. Bit pattern is <u>101</u> <u>000</u> <u>110</u> <u>011</u> <u>100</u> <u>111</u>. The phase angle of the transmitted signal $s(t)$ will be $-5\pi/8$, $\pi/8$, $-3\pi/8$, $7\pi/8$, $-7\pi/8$, and $-\pi/8$, after each successive time interval of T. The frequency of signal remains f_c and peak amplitude, s_{s0}.

Let us assume that the carrier is a sine wave of frequency, f_c, and it modulates a signal of 0s and 1s (binary signal). Then the eight equations for 8-PSK are as follows:

$$s(t) = s_{s0} \sin[(2\pi \times f_c \times t) + \varphi_{t0} + \varphi] \tag{1.21}$$

where $\varphi = \pi/8, 3\pi/8, 5\pi/8, 7\pi/8, -7\pi/8, -5\pi/8, -3\pi/8, -\pi/8$ corresponding to the triplets of bits 000, 001, 010, ..., 111. The triplet $\langle n_1 n_2 n_3 \rangle$ can represent these combinations where n_1, n_2, and n_3 each takes value 0 or 1. When $s_m(t) = n_1$, then $s_m(t + T/3) = n_2$, and then $s_m(t + 2T/3) = n_3$.

Quadrature amplitude modulation Quadrature amplitude modulation uses quadrature phase shift keying (4-PSK). The 16-PSK, three-stage amplitude modulation can be defined by 16 equations. It is better known as 16-QAM (16-quadrature amplitude modulation). The 16-QAM modulation technique uses QPSK (4-PSK) modulation. There are 16 distinct quadruplets each of which has a phase angle in one of the four quadrants. A quadruplet has one of three possible peak amplitudes A_1, A_2, or A_3. Four quadruplets have amplitude A_1, eight have A_2, and the remaining four have A_3. Each quadruplet is in one of four quadrants between 0° and 360° according to one of the 16 four-bit combinations from 0000 to 1111.

Let n_1, n_2 specify a quadrant, and n_3, n_4 specify a phase angle in that quadrant. The equations for the 16 quadruplets in QAM are given as

$$s(t) = s_{s0}\sin[(2\pi \times f_c \times t) + \varphi_{t0} + \varphi]$$

where $\varphi = \pi/4$, $3\pi/4$, $-3\pi/4$, and $-\pi/4$ correspond to the quadruplets 1111, 1011, 0011, and 0111 (in general $\langle n_1n_2n_3n_4\rangle$), respectively, when

$$s_m(t) = A_1 \times n_1,\ s_m(t + T/4) = A_1 \times n_2,\ s_m(t + T/2) = A_1 \times n_3,$$

and $$s_m(t + 3T/4) = A_1 \times n_4 \tag{1.22}$$

$\varphi = \pi/8, 3\pi/8, 5\pi/8, 7\pi/8, -7\pi/8, -5\pi/8\ -3\pi/8, -\pi/8$ for the quadruplets 1100, 1110, 1000, 1010, 0000, 0010, 0100, 0110 (in general $\langle n_1n_2n_3n_4\rangle$), respectively, when

$$s_m(t) = A_2 \times n_1,\ s_m(t + T/4) = A_2 \times n_2,\ s_m(t + T/2) = A_2 \times n_3 \text{ and } s_m(t + 3T/4) = A_2 \times n_4 \tag{1.23}$$

$\varphi = \pi/4$, $3\pi/4$, $-3\pi/4$, and $-\pi/4$ for the $\langle n_1n_2n_3n_4\rangle$ quadruplets of bits 1101, 1001, 0001, and 0101, respectively, when

$$s_m(t) = A_3 \times n_1,\ s_m(t + T/4) = A_3 \times n_2,\ s_m(t + T/2) = A_3 \times n_3,$$

and $$s_m(t + 3T/4) = A_3 \times n_4 \tag{1.24}$$

1.1.2.5 Multiplexing

Multiplexing means that different channels, users, or sources can share a common space, time, frequency, or code for transmitting data.

Space Division Multiplexing Space division multiple access (SDMA) means division of the available space so that multiple sources can access the medium at the same time. SDMA is a technique in which a wireless transmitter transmits the modulated signals and accesses a space slot and another transmitter accesses another space slot such that signals from both can propagate in two separate spaces in the medium without affecting each other. For example, if there are four groups, A, B, C, and D, of mobile users and four different regional space slots, R_1, R_2, R_3, and R_4, then group A uses R_1, B uses R_2, C uses R_3, and D uses R_4 for transmitting and receiving signals to and from a base station. Sections 3.2.1 and 4.1.5 describe SDMA in the context of GSM and CDMA, respectively.

Time Division Multiplexing Time division multiplexing entails different sources using different time-slices for transmission of signals. TDMA (time division multiple access) is an access method in which multiple users, data services, or sources are allotted different time-slices to access the same channel. The available time-slice is divided among multiple modulated-signal sources. These sources use the same medium, the same set of frequencies, and the same channel for transmission of data. For example, if there are eight radio-carriers (e.g., mobile phones) C_1, C_2, C_3, C_4, C_5, C_6, C_7, and C_8 transmitting signals in a GSM channel, then the GSM channel provides for eight TDMA time-slices, one for each radio carrier. Radio interfaces of carrier waves in eight phones can simultaneously transmit in the same frequency band (channel). The time-slice allotted to each is 577 μs. C_1 transmits in the first time-slice, C_2 in the second, and so on. Therefore, each transmitter uses the channel after time intervals of (577 × 8) μs or 4.615 ms. TDMA is discussed in the context of GSM and CDMA in Sections 3.2.2 and 4.1.5, respectively.

Frequency Division Multiplexing Frequency multiplexing requires a separation of frequency bands used for transmission by different channels. The available frequency range is divided into bands which are used by multiple sources or channels at the same time. Various channels are allotted distinct frequency bands for transmission. FDMA (frequency division multiple access) is an access method which entails assignment of different frequency-slices to different users for accessing the same carrier (Section 4.1.7). For example, GSM900 operates at an 890–915 MHz uplink from user to the base station and a 935–960 MHz downlink from the base to the user. Each channel uses a 200 kHz bandwidth. A 124-channel uplink needs 200 kHz × 124 = 24.8 MHz in the 890–915 MHz uplink range. Similarly, the 124-channel downlink requires 200 kHz × 124 = 24.8 MHz in the 935–960 MHz downlink range. Therefore, there are 124 radio channels operating at frequencies f_{ch1} ± 100 kHz, f_{ch2} ± 100 kHz, …, f_{ch123} ± 100 kHz, f_{ch124} ± 100 kHz. The frequency-slice allotted to each is 200 kHz. Channel 1 transmits in the 890.1 MHz ± 100 kHz range, channel 2 in the 890.3 MHz ± 100 kHz range, and so on.

Code Division Multiplexing CDMA (code division multiple access) is an access method in which multiple users are allotted different codes (sequences of symbols) to access the same channel (set of frequencies). [A *symbol* is a bit (0 or 1) which is transmitted after encoding and processing bits of data such as text, voice, pictures, or video.]

Each code is uniquely made up of n symbols and is used for transmitting a signal of frequencies f_{c0}, $f_{c0} + f_s$, $f_{c0} + 2f_s$, …, $f_{c0} + (n-2)f_s$, $f_{c0} + (n-1)f_s$ by the same channel. These frequencies are also called chipping frequencies. The rate at which a symbol XORs with the modulating signal to generate is called the chipping rate. Chipping rates are in the multiples of f_s. XORing of the modulating and chipping signals can be explained in terms of the resultant modulated carrier frequency

Table 1.1 XORing of the chipping and modulating signals

Chipping signal amplitude (A)	*Modulating signal amplitude (B)*	*Modulated carrier amplitude (A XOR B)*
0	0	0
0	1	1
1	0	1
1	1	0

amplitude, $s(t)$. The possible amplitude combinations of the modulating and chipping signals and their XORed output representing the amplitude of the modulated carrier wave are listed in Table 1.1.

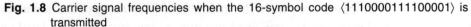

It is clear from Table 1.1 that if the modulating and chipping amplitudes are 1-1 or 0-0, the modulated amplitude is 0, but when the modulating and chipping amplitudes are dissimilar (0-1 or 1-0) then the modulated amplitude is 1. Multiple channels can transmit at the same set of chipping frequencies and for the same f_{c0} and f_s values but they must transmit via carriers with distinct codes. The transmitter performs the XOR operation between the chipping signals (carrier) and the modulating signal bit at any given instant t. Distinct codes are assigned to multiple sources, channels, or radio carriers (mobile phones). For example, the 16-bit code ⟨1110000111100001⟩ consists of 16 symbols. It is a sequence of 0s and 1s at 16 frequencies $f_{\text{ch1}}, f_{\text{ch1}} \pm f_s, f_{\text{ch2}} \pm 2f_s, \ldots, f_{\text{ch15}} \pm 14f_s, f_{\text{ch16}} \pm 15f_s$, for the 16 symbols. The amplitude in the carrier signal is 0 when the code bit or symbol is 0 and is finite and constant when the symbol is 1. The bandwidth of a channel is $16 \times f_s$ in the above example. Figure 1.8 shows the carrier signal chipping frequencies when the code ⟨1110000111100001⟩ is transmitted. Section 4.2 will discuss CDMA in greater detail.

Orthogonal Frequency Division Multiplexing Two frequency signals, $s_1(t)$ and $s_2(t)$, are said to be orthogonal if $s_1(t)$ has maximum amplitude at the instant when $s_2(t)$ has zero amplitude and vice versa. To elucidate, if at an instant $t = t'$, $s_1(t')$ has

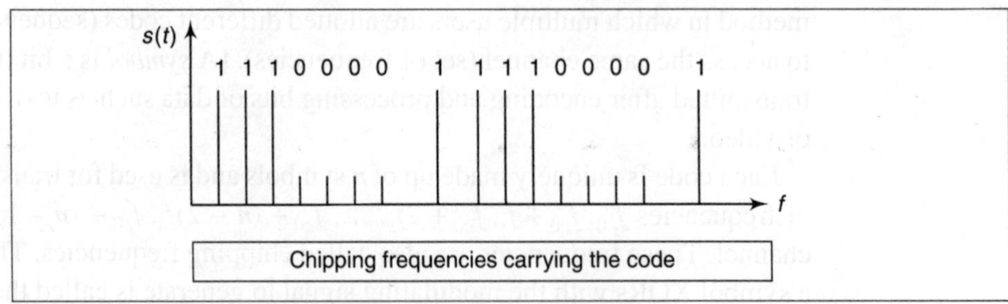

Fig. 1.8 Carrier signal frequencies when the 16-symbol code ⟨1110000111100001⟩ is transmitted

maximum amplitude and $s_2(t')$ has amplitude = 0 and at $t = t''$, $s_2(t'')$ has maximum amplitude and the amplitude of $s_1(t'')$ is 0, then $s_1(t)$ and $s_2(t)$ are orthogonal. OFDMA (orthogonal frequency division multiple access) (also called COFDM (code OFDM)) is a multi-carrier, multi-tone access method for transmitting multiple carriers for a set of symbols. Each carrier transmits a distinct set of subcarriers and each set of subcarriers is assigned a code which is orthogonal to another. For example, if a CDMA device transmits using a carrier frequency f_{c0} with N symbols in a second, where $N \cong f_{c0} + (n/2) f_s$, then instead of a single carrier frequency, f_{c0}, transmitting all the symbols, a transmitter can transmit a set of n_{sc} subcarrier frequencies $(f_{c0}/n_{sc})f_g + (f_{c0}n_{sc}{}^{-1} + n_{sc}{}^{-1}f_s), f_g + (f_{c0}n_{sc}{}^{-1} + 2n_{sc}{}^{-1}f_s), \ldots, f_g + (f_{c0}n_{sc}{}^{-1} + (n_{sc} - 1)$ $n_{sc}{}^{-1}f_s), f_g + (f_{c0}n_{sc}{}^{-1} + n_{sc}n_s{}^{-1}f_s)$. Each set of subcarriers then transmits N/n_{sc} symbols per second. The symbol f_g represents the frequency separation to guard against any overlap due to transmitter or other drifts. (A sub-carrier frequency is much lower than the carrier frequency used in a CDMA carrier.)

OFDMA is an access method in which the adjacent sets of subcarriers, $\{[f_{c0}/n_{sc}f_g + (f_{c0}n_{sc}^{-1} + n_{sc}^{-1}f_s), \ldots], [f'_{c0}/n_{sc}f_g + (f'_{c0}n_{sc}^{-1} + n_{sc}^{-1}f'_s), \ldots], [f'_{c0}/n_{sc}f_g{}''' + (f_{c0}{}'''' n_{sc}^{-1} + n_{sc}^{-1}f_s{}''), \ldots], \ldots\}$, that are carrying a subset of symbols are orthogonal. Each subset of symbols and carrier is allotted a different set of subcarriers and each adjacent set of subcarriers satisfies the orthogonality condition for the code.

1.1.2.6 Circuit and Packet Switching

Switching means establishing a communication channel so that data can transmit from the source to destination till the switch disconnects. There are two switching methods.

Circuit Switching Circuit switching is a method of data transmission in which a circuit (communication channel or path) once established, continues to be used till the transmission is complete. Let us assume, for example, that there are four routers or base stations, *A*, *B*, *C*, and *D*. A radio carrier (mobile device), T_1, wants to communicate with a receiver, R_1. Circuiting provides a path from the beginning to the termination of the transmission. In a circuit-switched transmission, a circuit is established and a path T_1–*B*–*D*–R_1 is provided. Each data frame transmitted along the path will take the same interval of time and the channel *B*–*D* is not available to any other circuit till the present circuit (interconnection) is terminated, irrespective of whether T_1 is idle during certain time intervals.

Packet Switching Packet switching is a means of establishing connection and transmitting data in which the message consists of packets containing the data frames. A *packet* is a formatted series of data, which follows a distinct path directed by a router from among a number of paths available at that instant. Each packet can be routed through different channels, carriers, or routes. Due to this the packets reach their destination with variable delays. Delays in reaching the receiver depend on the path followed through the routers. The message is recovered by assembling the packets as per the original sequence. Packet switching improves the efficiency of the available transmission capacity. Let us go back to the example cited for circuit switching. If the transmission is packet switched, then sequences of data from T_1 are framed into packets. Let us say that the path T_1–B–D–R_1 is available and at the same time the path T_1–A–B–C–D–R_1 is also available. In such a scenario T_1 simultaneously transmits one packet through T_1–A–B–C–D–R_1 and another through T_1–B–D–R_1. The packets following the first and the second paths take different amounts of time and hence have different delays. Also, the packets may not reach their destination in a sequential manner. However, a unique number called the sequence number is also transmitted along with each packet. Using these sequence numbers the receiver sequentially arranges the messages and constructs full sequences of messages transmitted through multiple packets.

1.1.3 Voice-oriented Data Communication Standards

First generation wireless devices communicated only voice signals. The symbol 1G refers to the voice-only communication. Nowadays voice transmissions also carry data. Second generation (2G) devices communicate voice as well as data signals. 2G devices came onto the market in 1988 and have data rates of up to 14.4 kbps. The 2.5G and 2.5G+ are enhancements of the second generation and sport data rates up to 100 kbps. The 3G (third generation) mobile devices communicate at even higher data rates and support voice, data, and multimedia streams. 3G facilitates data rates of 2 Mbps or higher for short distances and 384 kbps for long distance transmissions. High data rates in 3G devices enable transfer of video clips and faster multimedia communication.

Figure 1.9 shows GSM- and CDMA-based standards. It also shows a mobile communication network used for long distance communication. The following subsections (1.1.3.1 and 1.1.3.2) describe the GSM and CDMA standards used in wireless communication through mobile devices.

1.1.3.1 GSM

The global system for mobile communications (GSM) was developed by the Groupe Spéciale Mobile (GSM) which was founded in Europe in 1982. The GSM is a standard for mobile telecommunication through a cellular network at data rates of up to 14.4 kbps. Nowadays it consists of a set of standards and protocols for mobile telecommunication. Table 1.2 summarizes the standards based on GSM. Chapter 3 will discuss GSM in detail.

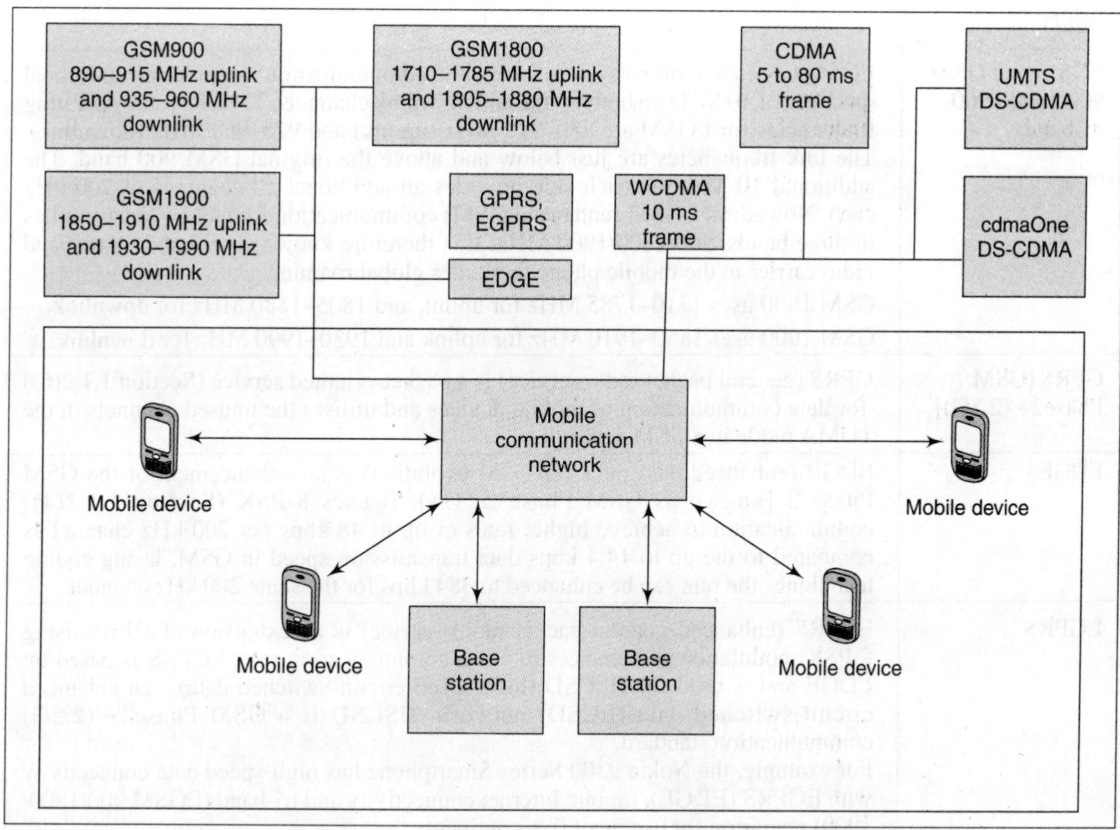

Fig. 1.9 GSM- and CDMA-based standards and a mobile communication network for long distance communication

Table 1.2 Standards based on the GSM

Extension/ Enhancement	*Description*
GSM900	The GSM (Groupe Spéciale Mobile) was founded in Europe in 1982. This led to a communication standard being founded in 1988, which is also known as GSM (global system for mobile communications). GSM uses GMSK (Section 1.1.2(4)) for transmitting 1s and 0s. GSM uses FDMA for channels and TDMA for user access in each deployed channel. Up to eight radio-carrier analog signals are transmitted using one common digital channel of bandwidth 200 kHz (Section 1.1.2(5)). GSM900 operates at 890–915 MHz for uplink and 935–960 MHz for downlink. Uplink and downlink frequency bands of 25 MHz each provide FDMA access for each channel. Each link thus provides 124 channels, each of 200 kHz. Each channel provides eight TDMA access slots, thus providing channel access to each user (radio carrier) every 4.615 µs. Users have time-slices of 577 ms each. The data rates of a GSM mobile-communication network are at maximum 14.4 kbps (kilobytes per second).

(contd)

(*contd*)

EGSM and GSM 900/1800/1900 tri-band	EGSM (extended global system for mobile communication) provides an additional spectrum of 10 MHz on both uplink and downlink channels. Therefore, the operating frequencies for EGSM are 880–915 MHz (uplink) and 925–960 MHz (downlink). The link frequencies are just below and above the original GSM 900 band. The additional 10 MHz on each side provides an additional 50 channels of 200 kHz each. Nowadays EGSM (enhanced GSM) communication frequency spectrum lies in three bands, 900/1800/1900 MHz. It is therefore known as tri-band. A tri-band radio carrier in the mobile phone facilitates global roaming. GSM 1800 uses 1710–1785 MHz for uplink and 1805–1880 MHz for downlink. GSM 1900 uses 1850–1910 MHz for uplink and 1930–1990 MHz for downlink.
GPRS [GSM Phase2+ (2.5G)]	GPRS (general packet radio service) is a packet-oriented service (Section 1.1.2(6)) for data communication of mobile devices and utilises the unused channels in the TDMA mode in a GSM network.
EDGE	EDGE (enhanced data rates for GSM evolution) is an enhancement of the GSM Phase 2 [known as GSM Phase 2.5G+]. It uses 8-PSK (Section 1.1.2(4)) communication to achieve higher rates of up to 48 kbps per 200 kHz channel as compared to the up to 14.4 kbps data transmission speed in GSM. Using coding techniques the rate can be enhanced to 384 kbps for the same 200 kHz channel.
EGPRS	EGPRS (enhanced general packet radio service) is an extension of GPRS using 8-PSK modulation. It enhances the data communication rate. EGPRS is based on EDGE and is used for HSCSD (high-speed circuit-switched data)—an enhanced circuit-switched data (ECSD) network. HSCSD is a GSM Phase2+ (2.5G) communication standard. For example, the Nokia 9300 Series Smartphone has high-speed data connectivity with EGPRS (EDGE), mobile Internet connectivity and tri-band (EGSM 900/1800/1900) operation for use in all five continents.

1.1.3.2 CDMA

Besides GSM, CDMA is the most popular mobile communication standard (Fig. 1.9). The initial evolution of CDMA was as 2.5G. It started in 1991 as cdmaOne (IS-95). Nowadays CDMA supports high data rates and is considered 3G. CDMA devices transmit voice as well as data and multimedia streams. CDMA 2000, IMT-2000, WCDMA, and UMTS, like the GSM, also support cellular networks (Section 1.5.1). Table 1.3 summarizes the standards based on CDMA. Sections 4.2–4.7 will describe CDMA standards in detail.

1.1.4 Wireless Personal Area Network

The previous section dealt with the long distance wireless mobile network communication standards GSM and CDMA. A wireless personal area network (WPAN) enables wireless communication between devices that are at short distances from each other. Figure 1.10 shows a wireless-based personal area network. It facilitates communication of mobile devices with home computers and with other devices at short distances.

Table 1.3 CDMA based standards

Extension / Enhancement	*Description*
2.5G+ standards	
cdmaOne/IS-95	CdmaOne founded in 1991, developed by QUALCOM, USA, belongs to 2G+, also known as IS-95 (interim standards 95). IS-95 operates at 824–849 MHz and 869–894 MHz. A CDMA channel can transmit analog signals from multiple sources and users.
3G standards	
3GPP (WCDMA)	The 3GPP (3G partnership project), also known as WCDMA (Wide CDMA), supports asynchronous operations and has a 10 ms frame length with 15 slices. It has a smaller end-to-end delay in the 10 ms frame as compared to 20, 40, or 80 ms frames. Each frame length is modulated by QPSK both for uplink and downlink. It uses DS (direct sequence) CDMA. It supports a 3.84 Mbps chipping rate (Section 4.3). Both short and long scrambling codes are supported, but for uplink only.
3GPP2 (IMT-2000, CDMA 2000)	3GPP2 (3G partnership project 2) started in 2001. It is compatible with CDMA 2000 and CDMA 2000 1x. The 3GPP2 chipping rates are in multiples of f_s = 1.2288 Mbps. 3G IMT 2000 carrier frequency f_{c0} = 2 GHz (Section 1.2.6). This is included in UMTS. The CDMA 2000 1x f_s = 1.2288 Mbps and is also backward compatible to 2.5G cdmaOne IS-95. 3GPP2 is used for voice communication, for circuit as well as packet-switched communication (Section 1.1.2.6), internet protocol (IP) packet transmission, and multimedia and real-time multimedia applications. It supports higher data rates, synchronous operations, and 5, 10, 20, 40, or 80 ms frame length. Each frame length is modulated by QPSK and BPSK—for uplink and down link, respectively. CDMA 2000 1x EVDO (evolution for data optimized) and CDMA 2000 1x EVDV (evolution for high-speed integrated data and voice) are enhancements, accepted as standards in 2004. CDMA 2000 3x uses three 1.2288 Mbps channels.
UMTS	UMTS (universal mobile telecommunication system) supports both 3GPP and 3GPP2. It communicates at data rates of 100 kbps to 2 Mbps. It combines CDMA for bandwidth efficiency and GSM for compatibility. It supports several technologies for transmission and gives a framework for security and management functions. It uses DS (direct sequence) CDMA and supports a 3.84 Mbps chipping rate (Section 4.2.1). For example, the BlackBerry 7130e device has CDMA 2000 1x EVDO (evolution data optimized) support and can also be used as a wireless tethered modem. (Tethered modem means when a laptop or PC is connected to the device, the device works as wireless modem for it.)

A WPAN standard is the Bluetooth IEEE 802.15.1. It operates at a frequency of 2.4 GHz radio spectrum, which is identical to that of the IEEE 802.11b WLAN standard. Bluetooth provides short distance (1 m to 100 m range as per the radio spectrum) mobile communication. The data rates between the wireless electronic devices are up to 1 Mbps. Examples of such communications are transmissions

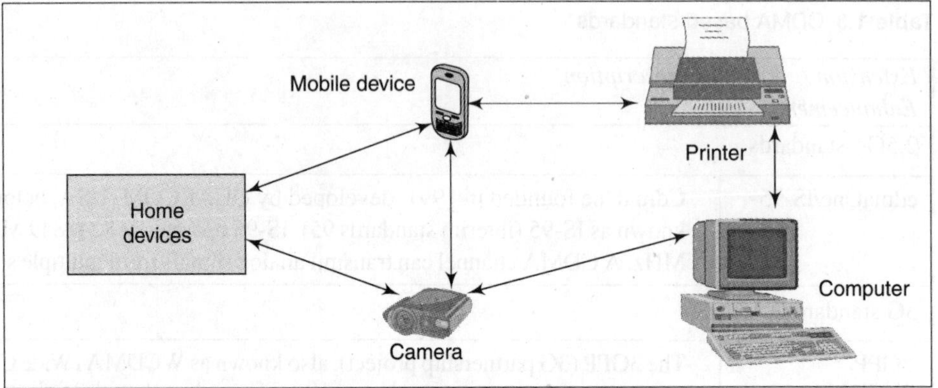

Fig. 1.10 Wireless personal area network using Bluetooth, ZigBee, or IrDA protocols

between the mobile phone handset and headset for hands-free talking, between the computer and printer, or computer and mobile phone handset. Bluetooth facilitates object exchanges. An object can be a file, address book, or presentation. Bluetooth thus enables user mobility in a short space with other Bluetooth-enabled devices or computers in the vicinity. It uses FHSS (frequency hopping spread spectrum) (Section 4.3.2). Sections 12.4 and 12.5 will delve into the details of Bluetooth. The following are a few examples of Bluetooth communication between devices.

- A Bluetooth-enabled digital camera when placed near a Bluetooth-enabled PC exchanges information with the PC and downloads pictures or video clips onto the PC.
- A Bluetooth-enabled PC downloads MP3 files from broadband Internet and when an iPod is placed near it, it transfers the media files to the iPod so that the user can listen to selected music when mobile.
- A Bluetooth-enabled mobile device and its Bluetooth-enabled headset provide hands-free wireless connectivity between the headset, ear buds, and the mobile handset. (Ear buds are wireless devices that when inserted into the ears enable hearing of music or phone calls.)

A WPAN standard that is IEEE 802.15.4-based is called ZigBee. It has lower stack size (28 kB) in the protocol and thus a lower network-joining latency when compared to Bluetooth (250 kB). Low transmitting power systems use ZigBee. It is an interoperable standard based on RF wireless communication, designed for robotic control, industrial, home, and monitoring applications. ZigBee is expected to provide large-scale automation and the remote controls up to a range of 70 m with data rates of 250 kbps, 40 kbps, and 20 kbps at the spectra of 2.4 GHz, 902 to 928 MHz, and 868 to 870 MHz, respectively. It uses DSSS (direct sequence spread spectrum) (Section 4.3). Section 12.9 will describe ZigBee in detail. Examples of ZigBee communication are as follows:

- A ZigBee-enabled electric meter communicates electricity consumption data to the mobile meter reader.
- A ZigBee-enabled home security system alerts the mobile user of any security breach at the home.

Infrared-based communication devices are also used for short distance communication when there are no obstacles, such as walls, between the devices. The devices use the IrDA (infrared data association) 1.0 protocol for data rates up to 115 kbps. IrDA 1.1 supports data rates of 1.152 to 4 Mbps. Section 12.8 will describe IrDA.

1.1.5 Wireless Local Area Network and Internet Access

Figure 1.11 shows a wireless local area network (WLAN). The network establishes communication between mobile devices and the Internet.

Sets of popular standards have been recommended for WLAN in mobile communication. These are the IEEE 802.11a, 802.11b, and 802.11g standards. Section 12.1 will discuss these in detail. WLAN is a WiFi (wireless fidelity)-based service. Table 1.4 summarizes the WLAN standards and protocols.

Three popular standards for mobile network services are as follows:

(i) WiMax (worldwide interoperability for microwave access) defines a specification for new generation innovative technology that delivers high-speed broadband, fixed, and mobile services wirelessly to large areas with much less infrastructure using the IEEE 802.16 standard.

(ii) WAP (wireless application protocol) provides web contents to small-area-display devices in mobile phones. The service providers format contents in the WAP format.

(iii) i-Mode (Internet in mobile mode) was developed by NTT DoCoMo, Japan. It is a vastly popular wireless Internet service for mobile phones.

Sections 12.1, 12.2, and 4.4 will describe the WLAN, WAP, and i-mode.

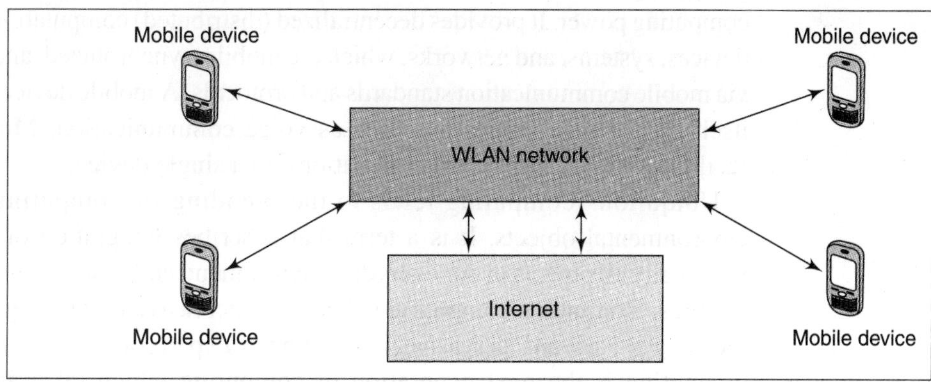

Fig. 1.11 Mobile communication using an 802.11 WLAN standard

Table 1.4 IEEE 802.11-based standards

Standard	*Extension/ Enhancement*	*Description*
802.11	802.11a	MAC layer operations are such that there can be multiple physical layers in 5 GHz (infrared, two 2.4 GHz physical layers). The layers support infrastructure-based architecture as well as mobile ad hoc network (MANET)-based architecture. (Refer to Chapter 11 for a description of the MANET.) Modulation is OFDM (Section 1.1.2.5) at data rates of 6 Mbps, 9 Mbps, ... Data rates supported are from 54 kbps to a few Mbps.
	802.11b	It operates at 54 Mbps and at 2.4 GHz. Modulation is DSSS/FHSS. It also supports short distance wireless networks using Bluetooth (IEEE 802.15.1)-based applications and the HIPERLAN2 (HIPERformance LAN 2) OFDMA physical layer. It provides protected WiFi access. The data rates are 1 Mbps (Bluetooth), 2 Mbps, 5.5 Mbps, 11 Mbps, and 54 Mbps (HIPERLAN 2).
	802.11g	It operates at 54 Mbps and at 2.4 GHz. It is also used for many new Bluetooth applications and is compatible with 802.11b. It uses DSSS (Section 4.8) in place of OFDMA.
	802.11i	It provides the AES and DES security standards (Section 12.1).

1.2 Mobile Computing

Mobile computing is the process of computation on a mobile device. In such computing, a set of distributed computing systems or service provider servers participate, connect, and synchronize through mobile communication protocols. [Wikipedia (the free encyclopaedia on the Web) defines mobile computing as a technology to wirelessly connect to and use centrally located information and/or application software through the application of small, portable, and wireless computing and communication devices.] Mobile computing offers mobility with computing power. It provides decentralized (distributed) computations on diversified devices, systems, and networks, which are mobile, synchronized, and interconnected via mobile communication standards and protocols. A mobile device does not restrict itself to just one application such as voice communication. Mobile computing facilitates a large number of applications on a single device.

Ubiquitous computing refers to the blending of computing devices with environmental objects. It is a term that describes integration of computers into practically all objects in our everyday environment, endowing them with computing abilities. Ubiquitous computing is based on pervasive computing. The dictionary meaning of the word 'pervasive' is 'existing in all parts of a place or thing'. Pervasive computing is the next generation of computing which takes into account the environment in which information and communication technology is used everywhere, by everyone, and at all times. Pervasive computing assumes information and communication technology to be an integrated part of all facets of our environment such as toys, computers, cars, homes, factories, and workareas. It takes into account the use of the integrated processors, sensors, and actuators connected through high-speed networks and combined with new devices for viewing and display.

Mobile computing is also called pervasive computing when a set of computing devices, systems, or networks have the characteristics of transparency, application-aware adaptation, and have an environment sensing ability. (Here *transparency* means without the impact of location due to mobility, the access protocol deployed, or software or hardware component failure. *Adaptability* means adaptability to the environment. *Environment* refers to the present location, surrounding devices and computing systems, present data, present network, or software and hardware components.)

Pervasive computing is a trend towards increasingly ubiquitous computing and it entails computing by devices connected to the environment by a convergence of advanced electronic and wireless technologies and the Internet. Pervasive computing devices are not PCs. These are handheld, very tiny, or even invisible devices which are either mobile or embedded in almost any type of object.

1.2.1 Novel applications

Mobile computing systems have a large number of applications. Mobile computing has very recently made mobile TV realizable. (In March 2006, at CeBIT, Hanover, Germany, big companies like Lucent and Nokia announced new deals to build the infrastructure for offering mobile TV and strive to make TV viewing on tiny cell phones easier and cheaper.) The ultra-mobile PC, unveiled in March 2006, has the size of a paperback book. Some of the recent novel applications of mobile computing are described in the following sections.

1.2.1.1 Smartphones

A Smartphone is a mobile phone with additional computing functions so as to enable multiple applications. For example, the BlackBerry 7130e (a device from Research in Motion, Inc.) has additional computational abilities which enable the following applications:

- SMS (short message service), MMS (multimedia messaging service), phone, e-mail, address book, web browsing, calender, task-to-do list, pad for memos
- Compatibility with popular Personal Information Management (PIM) software
- Integrated attachment viewing
- SureType keyboard technology with QWERTY-style layout
- Dedicated Send and End keys
- Bluetooth® capability for hands-free talking via headset, ear buds, and car kits

- EvDO* support, which enables the tethered modem capability, letting you use the device as a wireless modem for your laptop or PC
- Speaker phone
- Polyphonic ring tones for personalizing your device
- Bright, high-resolution display, supporting over 65,000 colours
- 64 MB memory

1.2.1.2 Enterprise Solutions

Enterprises or large business networks have huge database and documentation requirements. The term 'enterprise solutions' therefore refers to business solutions for corporations or enterprises. This may include specialized hardware or software programming to help an enterprise in finding solutions for various needs such as management of storage, security, revision, control, retrieval, distribution, preservation and destruction of documents and content, etc. These days mobile devices are being increasingly used to provide enterprise solutions. The Nokia 9300 (Section 2.4.3) and the BlackBerry 7130e are examples of mobile devices designed for use in enterprises.

Figure 1.12 shows enterprise-solution architecture using BlackBerry devices. It gives access to enterprise employees and connects to an enterprise server. It provides

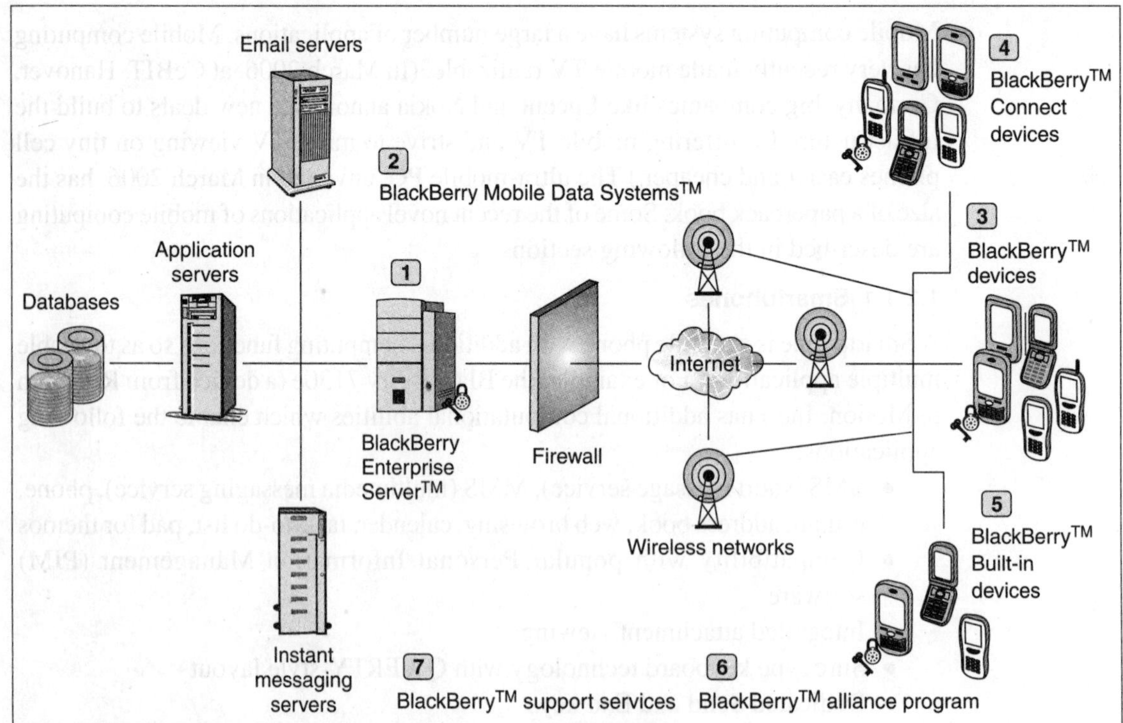

Fig. 1.12 Enterprise-solution architecture in a BlackBerry device

the company mobile users with a secure wireless access to their enterprise email and other critical business applications.

1.2.1.3 Music and Video

An instance of the novel innovative applications of mobile computing is the iPod-nano from Apple. The Apple iPods have made it possible to listen to one's favourite tunes anytime and anywhere. Besides storing music these players can also be used to view photo albums, slide shows, and video clips. Section 2.2 will describe the features and functionalities of the iPod.

A recent iPod enhancement is the 30 GB video iPod MA146LL/A0. It possesses the following features in addition to those found in the nano—it supports H.264 video, transfers data at rates up to 768 kbps, has 320 × 240 LCD screen resolution, shows 30 frames per second, baseline profile up to level 1.3 with AAC-LC up to 160 kbps, 48 kHz, plays stereo audio in .m4v, .mp3, and .mov file formats. Another version of the iPod offers MPEG-4 video, up to 2.5 mbps, 480 × 480 resolution, 30 frames per second, simple profile with AAC-LC up to 160 kbps, 48 kHz, stereo audio in .m4v, .mp3, and .mov file formats.

The Sony Network Walkman is another example. It can store and play 33 hours of music from the net without charging the battery and stores music equivalent to 11 CDs.

1.2.1.4 Mobile Cheque

An mCheque is a mobile-based payment system employed during a purchase. The service is activated through text-message exchanges between the customer, a designated retail outlet, and the mobile service provider. The service provider authenticates the customer and activates the customer account to transfer money to the retailer account. Such mobile devices are changing the way in which payments are made for purchases. Customers do not need to carry credit cards in their wallets for shopping anymore.

1.2.1.5 Mobile Commerce

An example of m-commerce is as follows. Mobile devices are used to obtain stock quotes in real time or on demand. The stock purchaser or seller first sends an SMS for the trading request, then the stock trading service responds in the same manner, requesting authentication. The client sends, through SMS, the user ID and password. The client is then sent a confirmation SMS to proceed further. The client sends an SMS for a specific stock trade request. The service provider executes the trade at the stock exchange terminal. The process is completed online within a minute or two.

Another example of m-commerce is that of a purchaser relating their intention to buy a product to the mobile purchase services provider through SMS. The service provider sends the prices of that product in ascending order of the price at different stores for the same product. The customer then requests the service provider to place the order to the cheapest and nearest supplier.

Mobile devices are also being increasingly used for e-ticketing, i.e., for booking cinema, train, flight, and bus tickets.

1.2.1.6 Mobile-based Supply Chain Management

Unless chocolates are manufactured, the distributors cannot sell them. The manufacturer cannot make the chocolates unless the distributor orders for the chocolates. This producer–consumer problem is called the supply chain management problem. Leading IT companies have developed mobile device software for supply chain management systems. The sales force and the manufacturing units use such mobile devices to maintain the supply chain.

1.2.2 Limitations

There are some limitations to mobile computing:

(a) *Resource constraints* Battery needs and recharging requirements are the biggest constraint of mobile computing. Recently, new processors have been developed, which dissipate less power for a given application. The use of these processors reduces the battery power requirement.

(b) *Interference* There may be interference in wireless signals, affecting the quality of service (QoS).

(c) *Bandwidth* There may be bandwidth constraints due to limited spectrum availability at a given instant, thereby causing connection latency. (Spectrum means the permitted range of frequencies for transmission.)

(d) *Dynamic changes in communication environment* There may be variations in signal power within a region. There may be link delays and connection losses.

(e) *Network issues* Due to ad hoc networks (networks with mobile nodes) there may be issues relating to the discovery of the connection service to destination and also those relating to connection stability.

(f) *Interoperability issues* The varying protocol standards prescribed and available between different regions may lead to interoperability glitches.

(g) *Security constraints* Protocols conserving privacy of communication may be violated. Also physical damage to or loss of a mobile device is more probable than a static computing system (Section 1.8).

1.3 Mobile Computing Architecture

This section outlines the architectural requirements for programming a mobile device. It will provide an overview of programming languages used for mobile system software (Section 1.3.1). An operating system is required to run the software components onto the hardware. Section 1.3.2 gives an overview of the operating system functions. The following subsections will summarize the middleware components deployed in mobile devices and systems and layered structure arrangement of mobile computing components. Finally, this section discusses the protocols and layers used for transmission and reception of data in a network of the mobile devices and systems.

1.3.1 Programming Languages

A variety of programming languages is used in the mobile computing architecture. One popular language used for mobile computing is Java. This is because of the most important characteristic of Java, i.e., platform independence—the program codes written in Java are independent of the CPU and OS used in a system. This is due to a standard compilation into byte codes.

Java 2 has a standard edition called J2SE. It has two limited memory sized editions—J2ME (Java2 Micro edition) and JavaCard (Java for smartcard). These two are the most used languages in mobile computing and for developing applications for a mobile device platform. The Java 2 enterprise edition (J2EE) is used for web and enterprise server-based applications of mobile services. Chapters 13 and 14 will describe the language features and development tools in detail.

C and C++ are other widely used programming languages. For these languages, program compilation depends on the CPU and OS used. One can use the in-line-assembly codes directly in a C/C++ program. The advantage of C/C++ is that it gives compact machine-specific codes. Visual C++ and Visual Basic are also used for developing applications for a pocket PC with a Windows platform.

1.3.2 Operating Systems

An operating system (OS) enables the user to run an application without considering the hardware specifications and functionalities. The OS also provides OS functions which are used for scheduling multiple tasks in a system. The OS provides management functions (such as creation, activation, deletion, suspension, and delay) for tasks and memory. It provides the functions required for the synchronization of multiple tasks in the system. A task may have multiple threads. The OS provides for synchronization and priority allocation of threads.

An OS also provides interfaces for communication between software components at the application layer, middleware layers, and hardware devices. It facilitates execution of software components on diversified hardware. An OS provides configurable libraries for the GUI (graphic user interface) in the device. User application's GUIs, VUI (voice user interface) components, and phone API (application programming interface) are a must in many user-operated devices.

The OS supplies these. The OS also provides the device drivers for the keyboard, display, USB, and other devices.

Sections 14.2–14.4 will describe some popular operating systems used in mobile computing (such as the Symbian OS, the Windows CE, and the PalmOS).

1.3.3 Middleware for Mobile Systems

Middleware are the software components that link the application components with the network-distributed components. A few examples of middleware applications are:

- To discover the nearby Bluetooth device (Section 1.1.4)
- To discover the nearby hot spot (Section 1.5.2)
- To achieve device synchronization with the server or an enterprise server (Sections 9.1–9.2)
- To retrieve data (which may be in Oracle or DB2) from a network database (Chapter 7)
- For service discovery
- For adaptation of the application to the platform and service availability

1.3.4 Mobile Computing Architectural Layers

Mobile computing architecture refers to defining various layers between the user applications, interfaces, device, and network hardware. A well-defined architecture is necessary for systematic computations and access to data and software objects in the layers. Figure 1.13 shows a computing architecture for a mobile device (client). An application deploys the software components and APIs (API refers to the application program interfaces); for example, the communication components in a mobile Smartphone, such as Internet, PIM (personal information manager), SMS, MMS, Bluetooth stack, security, and communication protocol stack, are all APIs. The client APIs can be considered at a layer in the architecture (uppermost layer in Fig. 1.13).

There are middleware components which discover the service, link the client and network service, update database, manage the device using remote-server software, perform client–server synchronization, and adapt the application to the given platform and server (Section 1.3.3).

An OS (operating system) is a layer in between the application and hardware. It facilitates the running of the program, hiding the specifications of the hardware, and provides many OS functions (e.g., device drivers) (Section 1.3.2).

An application can use an OS function directly. As an example, the function time_Delay (3000), delays the execution of further instructions by 3000 clock ticks.

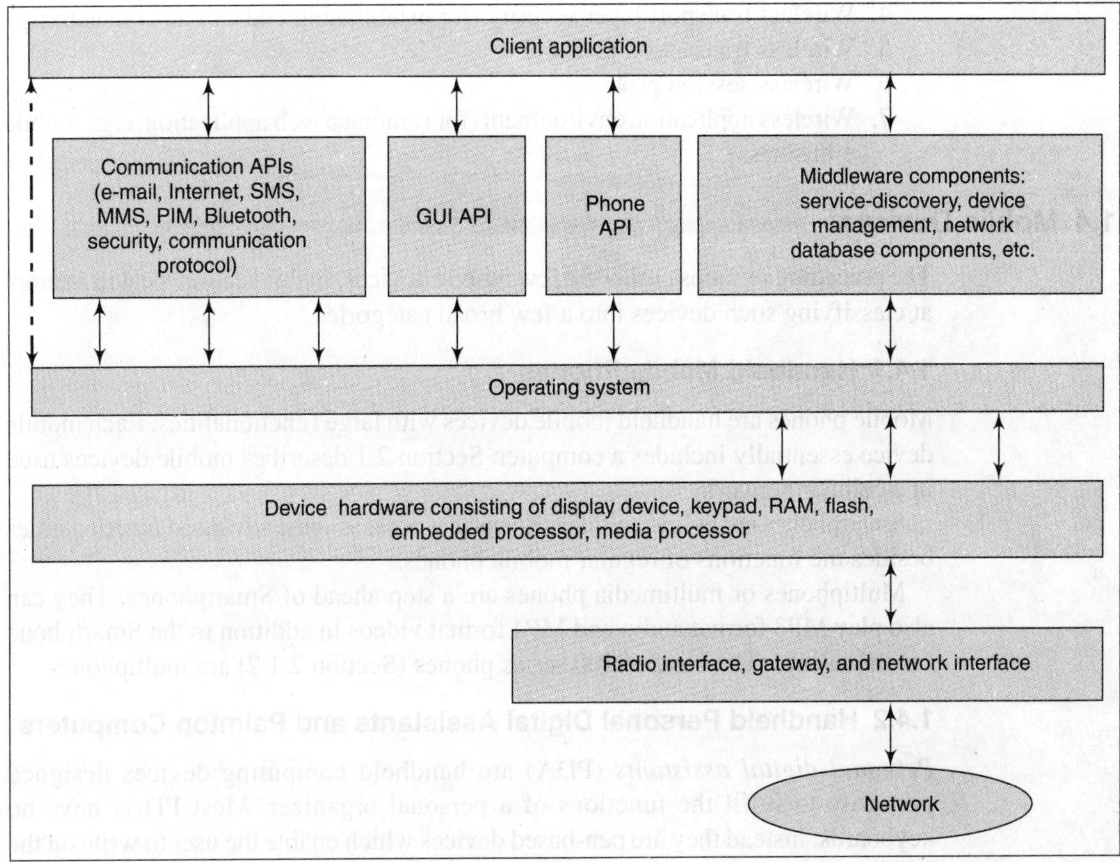

Fig. 1.13 Mobile computing architecture for a mobile device

1.3.5 Protocols

Interchanges between two diversified and distributed components need protocols and standards. Mobile computing services use a number of mobile communication protocols (Fig. 1.9), such as GSM900, GSM900/1800/1900, UMTS, and i-Mode. Chapter 12 describes WPAN protocols (e.g., Bluetooth, IrDA, and Zigbee) and WLAN protocols (e.g., 802.11a and 802.11b) and the wireless application protocol (WAP) is the communication protocol for enabling web pages on a mobile device.

1.3.6 Layers

There are different layers in network transmission and reception or in interchange of information, such as the WAP protocol layers. The OSI (open standard for interchange) seven-layer format is as follows:

1. Physical for sending and receiving signals (e.g., TDMA or CDMA coding)
2. Data-link (e.g., multiplexing)
3. Networking (for linking to the destination)

4. Wireless transport layer security (for establishing end-to-end connectivity)
5. Wireless transaction protocol
6. Wireless session protocol
7. Wireless application environment (for running a web application, e.g., mobile e-business)

1.4 Mobile Devices

The preceding sections outlined a few mobile devices. In this section we will attempt at classifying such devices into a few broad categories.

1.4.1 Handheld Mobile Phones

Mobile phones are handheld mobile devices with large functionalities. Each mobile device essentially includes a computer. Section 2.1 describes mobile devices used in a cellular network.

Smartphones are hybrid cellular phones that possess some advanced functionalities besides the functions of regular mobile phones.

Multiphones or multimedia phones are a step ahead of Smartphones. They can also play MP3 format audio and MP4 format videos in addition to the Smartphone functionalities. The Nokia 9300 series phones (Section 2.1.2) are multiphones.

1.4.2 Handheld Personal Digital Assistants and Palmtop Computers

Personal digital assistants (PDA) are handheld computing devices designed primarily to fulfil the functions of a personal organizer. Most PDAs have no keyboards; instead they are pen-based devices which enable the user to write on the LCD screen using a pen-like stick called stylus. Handwriting recognition is made possible by software specially created for the purpose. PDAs include applications such as address book, appointment scheduler, notepad, calculator, and calendar. Some PDAs also enable communication via email or fax.

A *palmtop* is a lightweight handheld programmable computer. Most palmtops include word processing and spreadsheet software in addition to the address book, calculator, and calendar functions. Other programs can be loaded and data transfer to and from desktop or laptop computers is possible. A palmtop is unlike a PDA in that it has more memory, a keyboard (some palmtops allow both touch-screen and keyboard inputs), and a larger number of available programs.

Handheld computing devices such as palmtops, PDAs, and smartphones use operating systems specially designed for them. One example of such an operating system is the PalmOS from Palm Inc. Handheld computers also use the Windows CE operating system from Microsoft. It is a real-time operating system. Section 2.4 will describe PalmOS and Windows CE-based handheld computers in detail.

1.4.3 Smartcards

A smartcard is, in fact, a mobile computer chip sandwiched within a very small space between the layers of the card. Section 2.5.1 explains the computing systems of smartcards in detail.

1.4.4 Smart Sensors

Sensors in a mobile device enable it to interact better with its surroundings. For example, a sensor for background noise can be used to control voice amplification while receiving a call. Deployed smart sensors connect and convey essential information to mobile control systems. Section 2.5.4 describes such sensors in detail.

1.5 Mobile System Networks

Mobile networks are networks of mobile devices, servers, and distributed computing systems. There are three types of mobile networks. These are described in the following sections.

1.5.1 Cellular Network

Figure 1.14 shows the cellular network architecture. A cell is the coverage area of a base station, connected to other stations via wire or fibre or wirelessly through

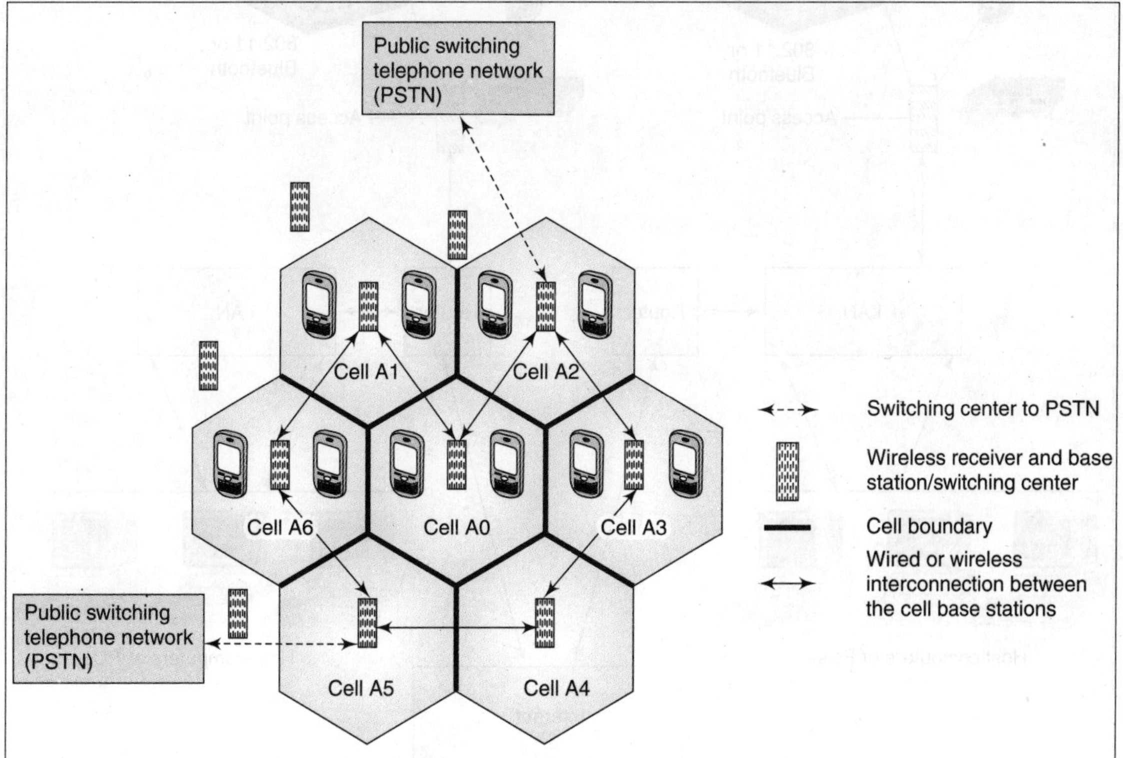

Fig. 1.14 Mobile communication using a cellular network

switching centres. The coverage area defines a cell and its boundaries. Each cell has a base station. A base station functions as an access point for the mobile service. Each mobile device connects to the base station of the cell which covers the current location of the device. All the mobile devices within the range of a given base station communicate with each other through that base station only. Section 2.1.3 will detail the functioning of a cellular network.

1.5.2 WLAN Network and Mobile IP

Figure 1.15 shows the WLAN network architecture. Sections 5.1 and 12.1 will discuss the details of mobile IP and WLAN networks, respectively. A mobile device, such as a pocket computer or a laptop, connects to an access point called a hotspot. The access point, in turn, connects to a host LAN which links up to the Internet through a router. Thus, connectivity is established between the Internet, two LANs, mobile devices, and computers.

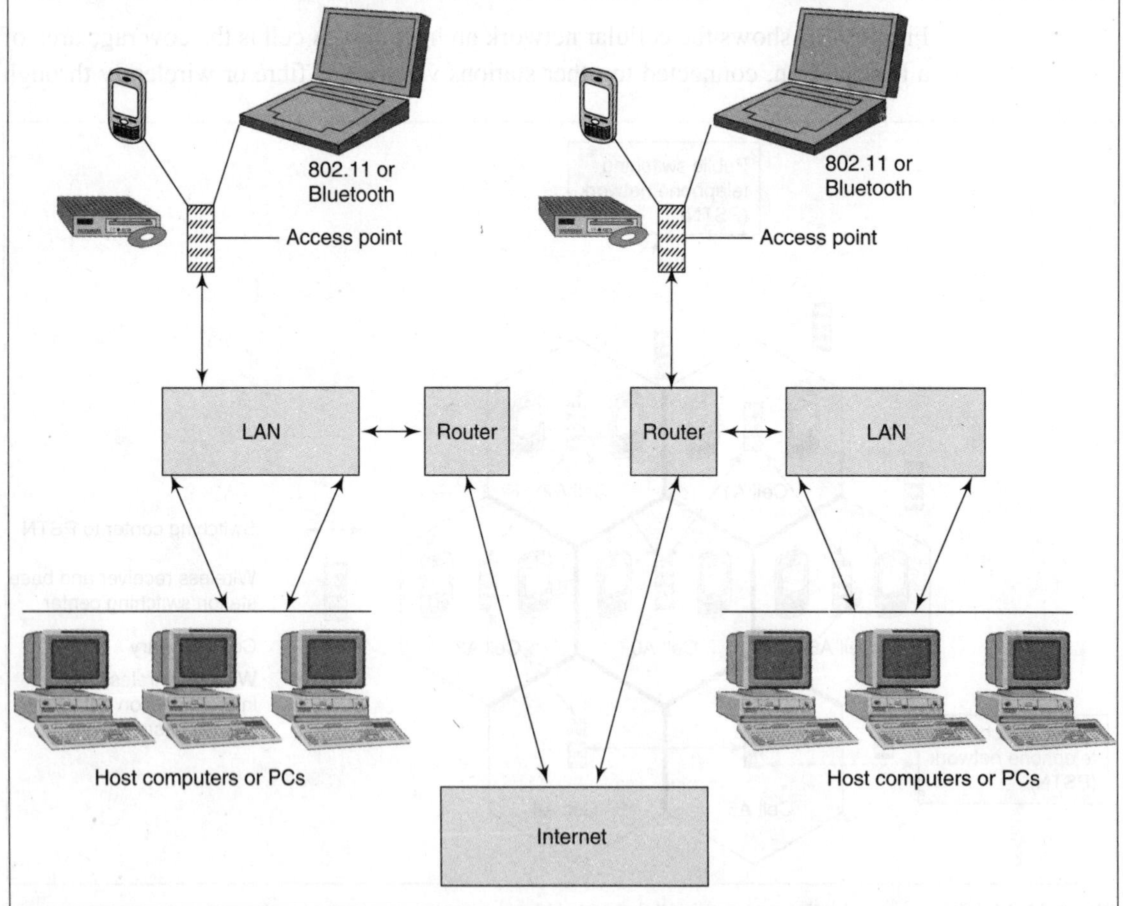

Fig. 1.15 Communication between mobile devices using a WLAN network through access points (hotspots)

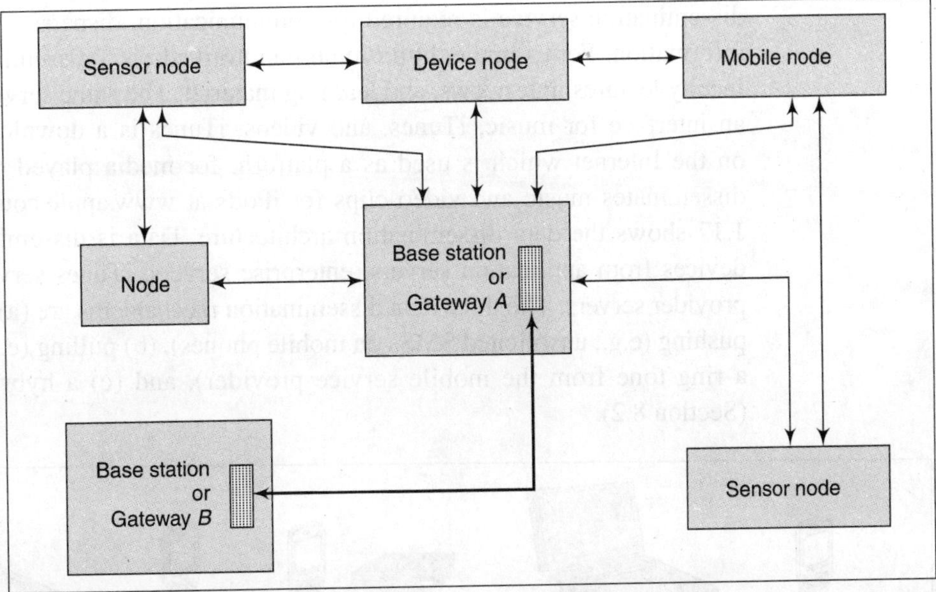

Fig. 1.16 Communication of mobile nodes and sensor nodes directly and using a base station as a gateway

Mobile IP is an open standard, defined by the Internet Engineering Task Force (IETF) RFC 2002. It is based on the IP (Internet protocol); therefore, all media supporting IP also support mobile IP. A mobile IP network provides the mobile IP service using home agents and foreign agents. Chapter 5 describes the mobile IP network.

1.5.3 Ad hoc Networks

Figure 1.16 shows ad hoc network architecture. Sections 11.2–11.4 will give the detailed features of ad hoc networks. The figure shows that the nodes, mobile nodes, and sensor nodes communicate among themselves using a base station. The base stations function as gateways. The ad hoc networks are deployed for routing, target detection, service discovery, and other needs in a mobile environment.

1.6 Data Dissemination

A mobile phone also acts as a data access device for obtaining information from the service provider's server. Smartphones used in enterprise networks work as enterprise data access devices. An enterprise server disseminates the data to the enterprise mobile device, such as the BlackBerry device.

An iPod is a data access device for accessing music or video. It links up to download files which can then be saved and played. These days students also use the iPod for replaying faculty lectures and retrieving e-learning material. A data dissemination service is required for communication, dispersion, or broadcast of information. For example, http://itune.standford.edu is a dissemination server for faculty lectures, interviews, and learning material. The same service also provides an interface for music, iTunes, and videos. iTunes is a downloadable software on the Internet which is used as a platform for media played on iPods. Apple disseminates music and video clips for iPods at www.apple.com/itunes/. Figure 1.17 shows the data dissemination architecture. Data is disseminated to mobile devices from application servers, enterprise servers, iTunes servers, and service provider servers. The three data dissemination mechanisms are (a) broadcasting or pushing (e.g., unsolicited SMSs on mobile phones), (b) pulling (e.g., downloading a ring tone from the mobile service provider), and (c) a hybrid of push–pull (Section 8.2).

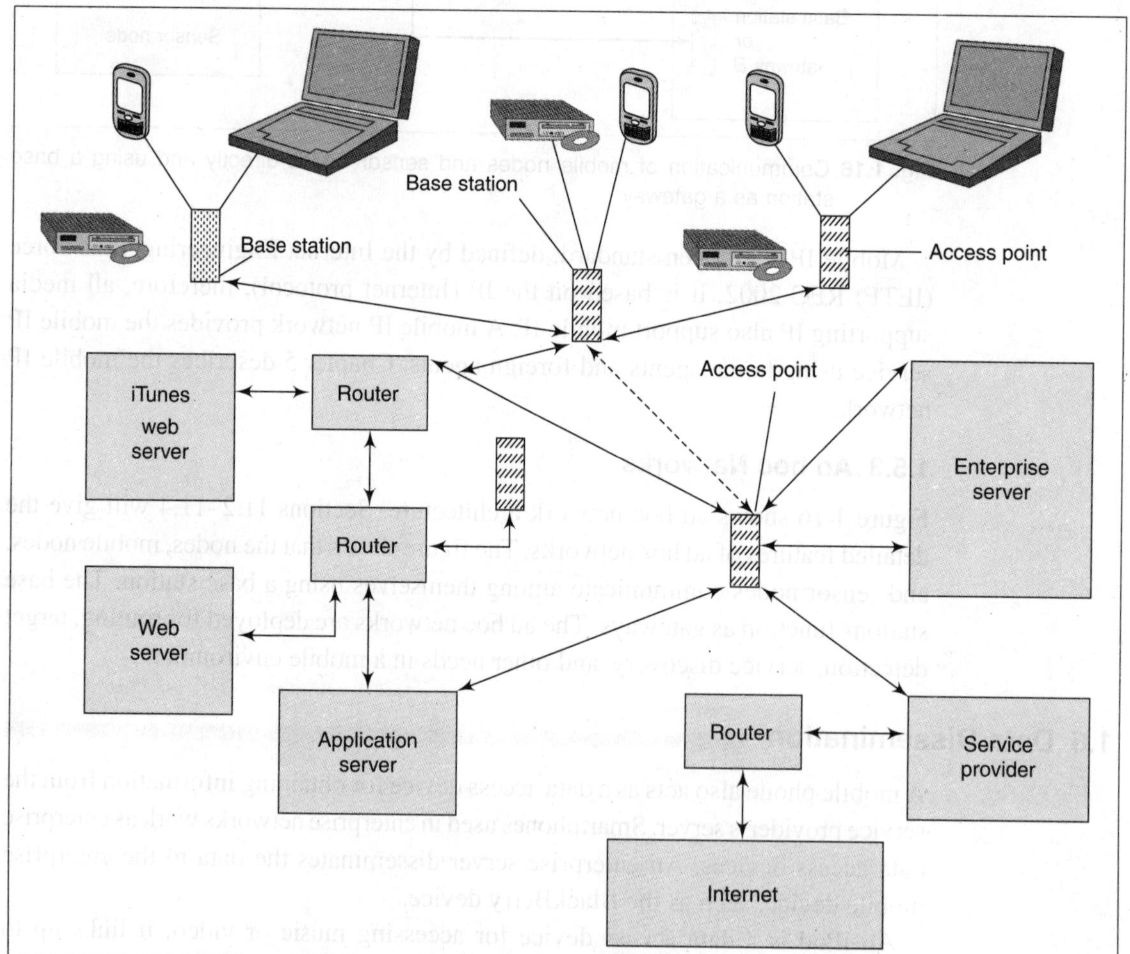

Fig. 1.17 Data dissemination by servers through base stations and access points

A new form of broadcasting is *podcasting*. It is a new innovative method by which multimedia files are distributed. For example, files for audio programs or music videos are distributed via the Internet in formats which enable their playback on mobile devices or PCs. Middleware is required for receiving the multicast or unicast data which is disseminated on a mobile network. Middleware also provides application adaptation and data management service transparency. (Transparency means that the user does not need to configure the underlying protocols.)

1.6.1 Synchronization

Data synchronization (or simply synchronization) is 'the ability for data in different databases to be kept up-to-data so that each repository contains the same information'.[1] For example, if a new popular ringtone is added to one of the servers of a mobile service provider, then data synchronization entails that all the servers of the service provider will have identical sets of ringtones. Moreover, all the devices connected to the server should be updated about the availability of any new data. This means that the ringtone databases available to all the mobile phones include a copy of the title of that tone.

1.6.1.1 One-to-one Synchronization

Data synchronization between two ends means any change in data at one end gets reflected at the other. Two ends should either have identical or consistent information. For example, if the address of a mobile device user changes and the user enters the new address in the address book of the device, then in a data-synchronized enterprise network the change is reflected at the enterprise server(s). The synchronization, in this case, is accomplished through a GSM or CDMA network. However, if an iPod placed near a PC synchronizes the iTunes downloaded on the PC, then the synchronization is by either Bluetooth or USB communication.

1.6.1.2 One-to-many Synchronization

One-to-many data synchronization means a data change or update at a node or server must get reflected at all the other (or target) nodes or servers. The copies of data should remain consistent or identical between the server (one) and the nodes (many). The communicating node or server and other nodes or servers should not have inconsistent copies of the same piece of information.

1.6.1.3 Many-to-many Synchronization

Many-to-many data synchronization between multiple nodes means that an update or change in information, at any node or server, must get reflected at all the other (target) nodes or servers. The copies of data at all nodes should remain consistent or identical. A set of nodes or servers should not have inconsistent copies of the

[1]http://www.bitpipe.com/tlist/Data-Synchronization.html

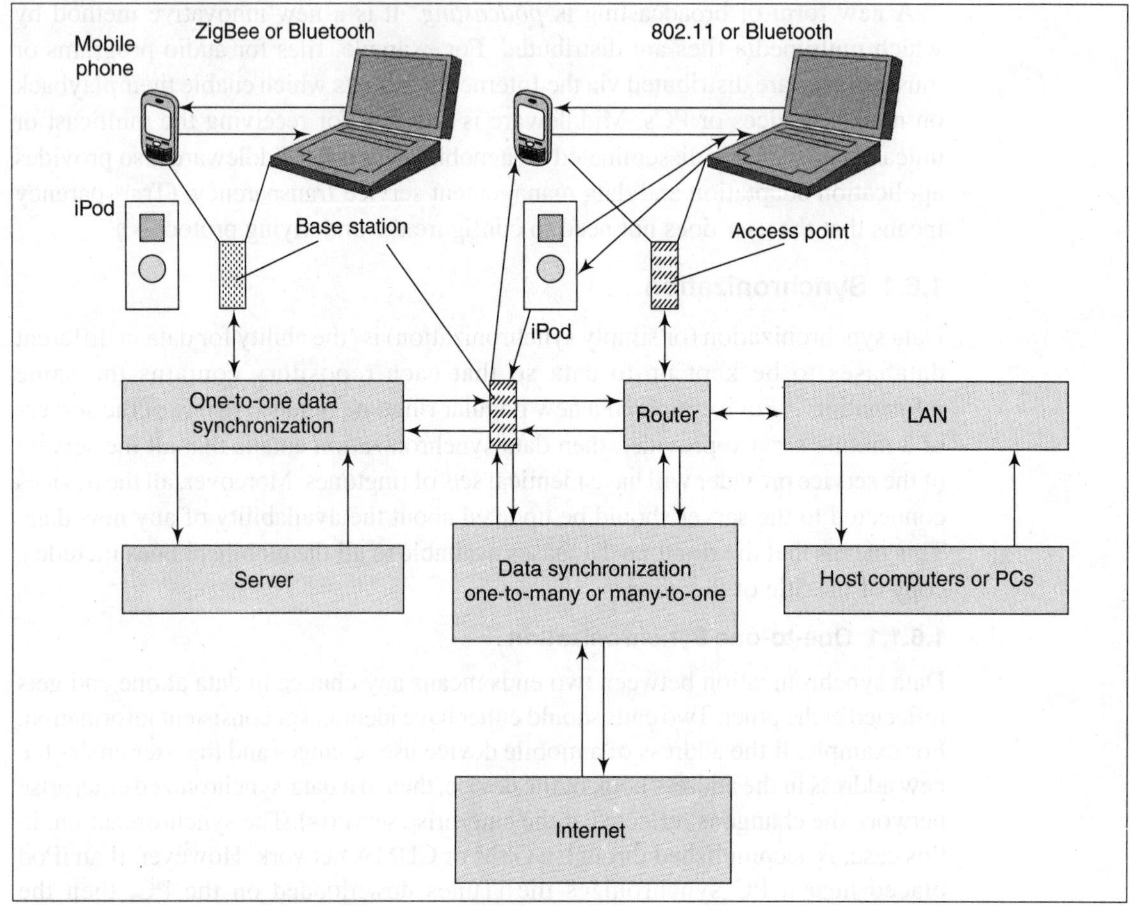

Fig. 1.18 Data synchronization paths in a mobile network

same information. Figure 1.18 shows the various paths for synchronization. Sections 9.1 to 9.4 will give a description of protocols and languages for synchronization of mobile devices and networks.

1.7 Mobility Management

Mobility management means maintaining uninterrupted (seamless) signal connectivity when a mobile device changes location from within a cell, C_i (Fig. 1.14), or network, N_i, to a cell, C_j, or network N_j (Fig. 1.15). The following are essential to ensure constant connectivity:

- *Infrastructure management* for installation and maintenance of the infrastructure that connects cell C_i to C_j or network N_i to N_j.
- *Location management and registration management by handoff* for cell transfer when a mobile device's connection with the ith cell is transferred

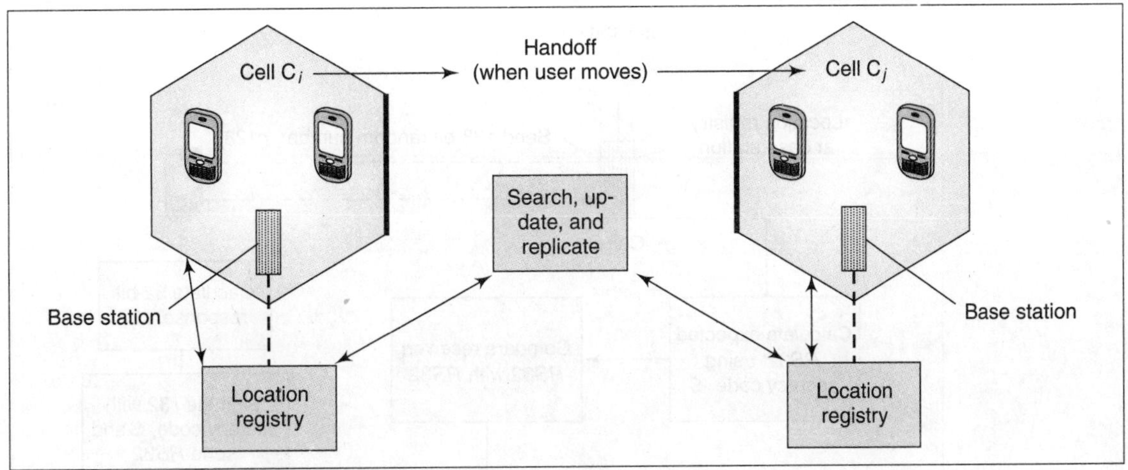

Fig. 1.19 Location registration and handoff processes

(on handoff from the ith cell) and registered at the new (j^{th}) cell. Figure 1.19 depicts localization, handoff, and registration. Sections 3.4 to 3.6 will describe management of localization, calling, hand off, and handover.

1.8 Security

Security is important for maintaining privacy and for mobile e-business transactions. Wireless security mechanisms must provide security of the data transmitted from one end point to another. It must provide for wire-equivalent privacy and non-repudiation when some data is sent to an end point. There must be no denial of service to authenticated object(s).

A serving station must be authenticated before it can provide service to mobile devices. Figure 1.20 shows the authentication method of security in case of GSM. The location registry at the home base-station sends a 128-bit random number, $n128$. The mobile device calculates the response, $r32$, using $n128$. Then the secrecy code, S, at the SIM (subscriber identity module) and the $r32$ are used to calculate $RS32$. The station receives the $RS32$ and compares it with the calculated (from S and $n128$) signed number $RS32'$. If they are equal, the mobile device present in that cell stands authenticated. Both, the mobile device and the station, use 64-bit encryption when transmitting the $RS32$ and $n128$, respectively. Only messages for authentication are encrypted in GSM. Section 3.7 will give the details.

The WLAN security standard is IEEE 802.11b. It provides WEP (Wired equivalent privacy). The 802.11i caters to enhanced security needs. There is WAP security at the transport layer (Section 12.2.3).

1.8.1 Cryptography Algorithms

The purpose of cryptography is to keep private information from getting into the hands of unauthorized agents. Encryption is the transformation of data into coded

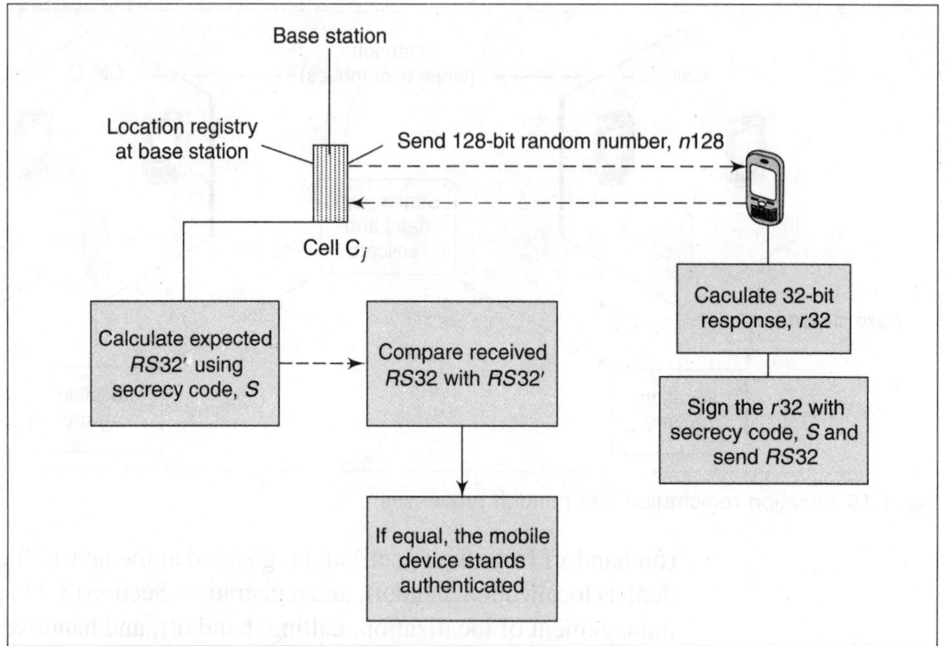

Fig. 1.20 Authentication process in GSM

formats. Encrypted data can be decrypted (transformed back to an intelligible form) at its destination.

Various cryptography algorithms are used for encryption and decryption of transmitted data. These algorithms enable the receiver and the sender to authenticate data as well as discover if data security has been compromised during transmission. Cryptography algorithms generally use a secret key to encrypt data into secret codes for transmission. For example, the RSA (Rivest, Shamir, Adleman) algorithm is a cryptography algorithm used for private key generation. Cryptography algorithms can be classified into two categories—symmetric and asymmetric (Section 10.2).

Cryptographic algorithms are used to create a *hash* of the message or an *MAC* (message authentication code). A hash function is used to create a small digital fingerprint of the data to be transmitted. This fingerprint is called the hash value, hash sum, or simply, hash. The hash of the message is a set of bits obtained after applying the hash algorithm (or function). This set of bits is altered in case the data is modified during transmission. Message authentication codes (MAC) are also used to authenticate messages during transmission. The MAC of a message is created using a cryptographic MAC function which is similar to the hash function but has different security requirements. The receiver reviews the hash or the MAC of the received message and returns it to the sender. This exchange enables the sender and the receiver to find out if the message has been tampered with and thus helps verify message integrity and authenticity. Table 1.5 provides various security standards.

Table 1.5 Cryptography algorithm

Standard	*Description*
(a) Symmetric key-based algorithms	
DES	Data encryption standard (DES) uses 56 bits for a key plus 8 bits for parity. Block length is 64 bit. (Maximum block size = 2^{64} bits.) Triple DES is an enhanced version of DES in which there can be multiple encryptions or encryption–decryption–encryption steps in the cryptic message. There is a different key at each step for cryptic message creation.
AES	Advanced encryption standard (AES) has nine possible combinations of key lengths and block lengths. The key length can be chosen as 128, 192, or 256 bits. The block lengths can also be chosen as 128, 192, or 256 bits. (Block length of 128 bits means maximum block length = 2^{128} bits.)
(b) Asymmetric key-based standards	
RSA	The RSA (Rivest, Shamir, Alderman) algorithm uses 128-, 256-, 512-, or 1024-bit prime numbers for encryption.

1.8.2 Digital Signatures and Digital Certificates

Digital signatures and digital certificates are used to enable verification of records. A DSA (digital signature algorithm) is used to sign a record before transmitting. It provides for a variable key length of maximum 512 or 1024 bits. The DSS (digital signature standard) is based on the DSA. Signatures enable identification of the sender, identify the origin of the message, and check message integrity.

A digital certificate is an electronic certificate used to establish the credentials of a data set. It is issued by a certification authority and contains the certificate holder's name, a copy of the certificate holder's public key, a serial number, and expiration dates. It also includes the digital signature of the certificate-issuing authority for verification of the authenticity of the certificate. The certification authority distributes a digital certificate, which binds a public key to a specific sender.

Keywords

8-PSK (8-phase shift keying): A modulation technique in which the phase angle of a carrier frequency shifts to one of eight different angles in one of eight octets between 0° and 360° as per the eight combinations, 000, 001 ... 111 in a set of three consecutive bits.

16-QAM (16-quadrature amplitude modulation): A modulation technique in which there is 4-PSK (QPSK) modulation with 16 distinct quadruplets. Each quadruplet has its phase angle in one of the four quadrants and has one of the three possible peak amplitudes A_1, A_2, A_3.

Bluetooth: A standard for object exchanges that provides short distance (1 m or 100 m range as per radio spectrum) mobile communication between wireless electronic devices (e.g., between a mobile phone handset and headset for hands-free talking, for connecting the computer or printer, etc.) and has data rates up to 1 Mbps.

CDMA: An access method in which multiple carriers, channels, or sources are allotted different codes (sequences of symbols) to access the same channel (set of frequencies at the same time in same space).

Cellular network: A network in which space is divided into cells such that each cell has a base station for providing service to mobile devices and when a mobile device roams into another cell, a handoff takes place from the previous station and handover takes place to the new base station.

Circuit switching: A method of establishing connection and data transmission in which a circuit (communication channel or path), once established, continues till the end of the communication.

Digital modulation: A technique in which amplitude, frequency, or phase angle parameters of carrier frequency or subcarrier frequencies are varied with time according to the variation in modulating signal bit 1 or 0, modulating signal pair, or set of bits with time.

EDGE (enhanced data rates for GSM evolution): An enhancement of the GSM Phase 2 which uses 8-PSK modulation to achieve higher transmission rates of up to 48 kbps per 200 kHz channel compared to the up to 14.4 kbps data transmission speed in GSM.

FDMA (frequency division multiple access): A technique in which a wireless transmitter channel transmits a modulated signal and accesses using a frequency-slice in a band and another modulated signal accesses via another frequency-slice in the band. The available frequency in a band is divided into slices for use by multiple sources. Therefore, signals from two sources or of two data services can use the same time interval and channel and propagate in the medium without affecting each other.

GPRS (general packet radio service): A packet-oriented service for mobile devices' data communication which utilizes the unused channels in TDMA mode in a GSM network and also sends and receives packets of data through the Internet.

GSM (global system for mobile): A standard for mobile telecommunication through a cellular network at the data rates of 14.4 kbps. It was developed by Groupe Spéciale Mobile (GSM) founded in Europe in 1982.

Mobile computing: Computation during an application done on a mobile device in which a set of distributed computing systems or service provider servers participate, connect, and synchronize through mobile communication protocols.

Modulation: The process of varying one signal, called the carrier, according to the information provided by another signal (modulating signal). The carrier is usually an analog signal selected to match the characteristics of a particular transmission system. The amplitude, s_{c0}, frequency, f_c, or phase angle, φ_{ct0}, of a carrier wave are varied in proportion to the variation in the amplitude of the modulating wave or bits (the data or information signal).

Packet: A packet is a part of data that can take a distinct path from other packets from the same source and each packet can have variable delays.

Packet switching: A method of establishing connection and packet transmission in which the message consists of packets containing the data frames. Each packet can be routed through different channels, carriers, or routes. Thus different packets reach their destination with variable delays. Delay in reaching the receiver depends on the path followed through the routers. The message is recovered by assembling the packets as per the original sequence.

Pervasive computing: A new generation of computing which takes into account the environment in which information and communication technology is used everywhere, by everyone, and at all times.

Podcasting: A method by which multimedia files (e.g., files for audio programs or music videos) are distributed over the Internet using formats which enable their playback on mobile devices or PCs.

Protocol: A standard recommended procedure for controlling and regulating data transmission and rules governing the syntax, semantics, and synchronization communication between two computing systems, establishing connection, formatting and sequencing of data, addressing the destination and sources, and termination of connection.

OFDM (orthogonal frequency division multiplexing): A multi-carrier, multi-tone access method for transmitting multiple carriers for a set of symbols such that each carrier transmits a distinct set of subcarriers and each set of subcarriers has a code which is orthogonal to another.

QPSK (quadrature phase shift keying): A modulation technique in which phase angle of the carrier shifts in one of four quadrants between 0° and 360° and the shift depends on whether the modulating signal pair of bits is (1, 0), (0, 1), (0, 0) or (1, 1).

SDMA (space division multiple access): A technique in which a wireless transmitter transmits the modulated signal and accesses a space slot and another transmitter accesses another space slot such that signals from both can propagate in two separately spaced media without affecting each other.

Symbol: A bit 0 or 1 which is to be transmitted after encoding and processing the bits of a data service, voice, picture, or video.

TDMA (time division multiple access): A technique in which a wireless transmitter channel transmits a modulated signal and accesses in a time-slice and another modulated signal accesses in another time-slice such that the signals from two sources or of two data services can propagate in the medium without affecting each other and use the same frequency band and the same channel. The available time-slice is divided for use by multiple sources.

Ubiquitous computing: A term which describes the integration of computations with environment objects having computing abilities.

WiFi (wireless fidelity): An access service that describes a WLAN network access IEEE technology or application using 802.11 standards.

WiMax (worldwide interoperability for microwave access): Specification for new generation innovative technologies for delivering high-speed, broadband, fixed, and mobile services wirelessly to large areas with much less infrastructure using 802.16 standards.

WLAN: A wireless LAN using sets of popular standards IEEE 802.11a, 802.11b, and 802.11g in communication.

WPAN (wireless personal area network): A network-based on IrDA, ZigBee, or Bluetooth IEEE 802.15.1.

Review Questions

1. IS-95 (interim standards 95) uses the 824–849 MHz and 869–894 MHz frequencies. What is the range of wavelengths used? What are the ways by which the signals are received at the base station from the transmitting devices?
2. A signal $s(t) = s_0 \sin(2\pi \times 2000 \times t)$ is to be transmitted. Find the time period, frequency, and phase angle at $t = 0$, 0.25 ms, 0.5 ms, 0.75 ms, 1 ms, 1.25 ms, and 3 ms for the signal.
3. A signal $s(t) = a_1 \sin(2\pi \times f \times t) + 0.5\, a_1 \sin(2\pi \times 3 \times f \times t) + 0.25\, a_1 \sin(2\pi \times 5 \times f \times t)$. Show the frequency spectrum of the signal. If signal is amplitude modulated with a signal $s(t) = +\,0.5\, a_1 \sin(2\pi \times 2000 \times t)$, what will be the new frequency spectrum?
4. Assume that a 24-symbol code ⟨111000011110000111100001⟩ is used with chipping frequencies of 900.000, 900.020, 900.040, …, MHz.
   (a) What will be the bandwidth used in CDMA transmission of the signal?
   (b) Assume that the symbols are transmitted after 8-PSK modulation. What will be the new carrier signal bandwidth? What will be the phase angles and amplitudes of the resulting signal?
   (c) Assume that the symbols are transmitted after QAM modulation. What will be the new carrier signal bandwidth? What will be the phase angles and amplitudes of the resulting signal?
   (d) Assume that the four subcarriers are used for transmission. What will be the frequencies in the frequency spectrum of the transmitted signal?
5. What are the advantages and disadvantages of using a wireless transmission as compared to a fibre or wire transmission?
6. Describe the signal propagation at UHF frequencies. List the UHF frequencies for the TV, DAB, DECT, and GSM applications.
7. Describe amplitude and frequency modulation.
8. Explain the features of ASK, FSK, BPSK, QPSK, and 8-PSK. Describe quadrature amplitude modulation.
9. Describe various multiplexing techniques. Explain why a given bandwidth is used most efficiently in CDMA.
10. Explain the differences between 1G, 2G, 2.5G, 2.5G+, and 3G mobile communications.
11. Describe the modulation frequencies and modulation methods in GSM 900.
12. Describe 3GPP2 communication.
13. Describe the protocols used for a wireless personal area network (WPAN). Suggest some applications at home for a WPAN network.
14. What are the differences in features provided by Bluetooth and ZigBee?
15. Describe a novel application of a mobile device that fascinates you. What are the additional software components used in this application?

16. What are the software layers and components needed in a mobile computing device?
17. Explain the need for software layers and components using the BlackBerry 7130e as an example.
18. Describe the functions of an OS in a mobile system.
19. Describe the middleware components in a tri-band EDGE device.
20. Describe the features expected in a handheld computer.
21. Describe the functions of data dissemination. How is data dissemination necessary in maintaining a mobile service?
22. Explain data synchronization and its importance with examples.
23. How is mobility managed in a mobile system?
24. Explain, using a block diagram, the process of authentication in a GSM service.
25. Describe the various cryptographic algorithms. What is the difference between a digital signature and a digital certificate? Why do you need a third party in digital certification?

Objective Type Questions

Pick the correct or most appropriate statement among the choices given:

1. UHF 900 MHz frequencies are used in mobile communication due to
   (a) line-of-sight propagation making the waves accessible directly anywhere.
   (b) small antenna length requirement, line of sight and the reflected signals ensure that the mobile device signal reaches the base station at the receiver.
   (c) non-availability of frequencies less than 900 MHz as these have been allotted to radio, FM radio, and TV VHF/UHF.
   (d) regulations and standards.
2. Modulation of a modulating signal with a very large carrier frequency in wireless transmission is necessary due to
   (a) antenna requirements, signal propagating medium properties, and need to multiplex the multiple channels and users at the transmitter.
   (b) smaller antenna size at high frequencies.
   (c) little bending of the beams at high frequencies.
   (d) mobility requirements.
3. A GSM service uses
   (a) FDMA for multiple users.
   (b) FDMA for multiple channels and TDMA for the multiple users of a channel.
   (c) different uplink and downlink modulation methods.
   (d) TDMA for multiple channels.
4. 8-PSK and QPSK are the modulation techniques in which
   (a) the phase angle of the carrier frequency can be in one of eight or four different angles as per one of the eight or four combinations of three or two bits, respectively.
   (b) the carrier frequency can use one of eight or four channels as per one of the eight or four combination of three or two bits, respectively.
   (c) the carrier frequency time-slices can use one of eight or four angles as per one of the eight or four bits, respectively.
   (d) the phase angle of the carrier frequency can be in one of eight or four different angles as per one of the eight or four bits, respectively.
5. (a) Circuit switching results in uniform delays between the end points for each messages.
   (b) Packet switching results in uniform delays between the end points for all packets of the messages.
   (c) Packet switching results in variable delays between the end points for the packets of the messages.
   (d) Circuit switching results in variable delays between the end points for all messages.
6. WLAN service uses
   (a) 802.11 protocol and long distance communication at high data rates.

(b) 802.16 protocol and long distance communication at small data rates.
(c) 802.16 protocol and short distance communication at high data rates.
(d) 802.11 protocol and short distance communication at high data rates.

7. Bluetooth provides
(a) short distance (1 to 100 m range as per radio spectrum) mobile communication. The data rates are up to 1 Mbps. It uses FHSS coding.
(b) Short as well as long distance 1 m to 1 km, data rates 1.25 Mbps and DSSS coding.
(c) Short distances 1 m to 0.1 km, data rates 2 Gbps and FHSS coding.
(d) Short distances 1 m to 0.1 km, data rates 1 Mbps and DSSS coding.

8. Mobile computing differs from other forms of distributed computing by limitations
(a) battery, memory resources, long distance bandwidth constraints, and network and interoperability issues.
(b) use of radio-frequency cellular communication.
(c) use of radio frequencies in 100–2000 of MHz.
(d) inaccessibility of web pages.

9. (a) BlackBerry 7130e device has CDMA 2000 1x EvDO (Evolution Data Optimized), application, and enterprise servers support.
(b) BlackBerry 7130e device has CDMA 2000 3x and enterprise servers support.
(c) Nokia 9300 series Smartphone has high-speed data connectivity with EGPRS (EDGE), mobile internet connectivity, and tri-band (EGSM 900/1800/1900) operation for use in five continents.
(d) Nokia 9300 series device has CDMA 2000 1x EvDO (evolution data optimized), application, and enterprise servers support.

10. Middleware is used
(a) for service discovery, application adaptability, and retrieving backend databases.
(b) for application adaptability and retrieving backend databases.
(c) for connecting to a mobile service.
(d) as a software component between the OS and hardware.

11. (a) Base station is an access point for the mobile devices in a cell and it keeps the location registry of all devices presently active in the cell.
(b) Access point is an access point for the mobile devices in a wireless LAN.
(c) Base station is an access point for the mobile devices and for the other base stations or PSTN channel switching centre in a cell and it keeps the location registry of all devices presently active in the cell.
(d) Base station and access points are the access points for the mobile devices and for the other base stations or PSTN channel switching centre, respectively, in a cell.

12. Data dissemination mechanisms are
(a) pushes of voice, data, or multimedia by the server or pulling the voice, data, or multimedia by the server on demand.
(b) pushes of database by the server.
(c) pulling the database on demand.
(d) pulling the voice, data, or multimedia by the server on demand.

13. Data synchronization means
(a) broadcasting data to all nodes simultaneously.
(b) asynchronously broadcasting the data to all nodes.
(c) maintaining the copies that are either identical or consistent pieces of voice, data, multimedia, or software components.
(d) maintaining the copies that are either identical or consistent pieces of database or software components.

14. A wireless security mechanism should provide
(a) integrity and wire-equivalent privacy, non-repudiation, and non-denial of service to authenticated object(s).
(b) integrity, confidentiality, and non-denial of service to authenticated object(s).
(c) integrity, privacy, and non-repudiation of service to mobile object(s).
(d) secure access of a mobile device to the mobile service provider.

Chapter 2

Mobile Devices and Systems

Mobile devices such as Smartphones, smartcards, sensors, and handheld and automotive computing systems are introduced in this chapter. It describes the basic features, functions, and technologies related to these. This chapter will cover the following topics:

- Features and functions of mobile phones and portable music players
- Features and applications of handheld computers including PalmOS and Windows CE-based computing systems
- Smartcards, labels, sensors, actuators, appliances, and setup boxes
- Constraints of mobile devices
- Automotive computing systems

2.1 Mobile Phones

Most mobile phones communicate with other phones using a cellular service-provider network. These days mobile phones are packed with smart functions and are available in smaller sizes. The applications of mobile phones are no longer confined to telephonic communication; rather, nowadays a mobile phone can synchronize, upload, and download data to and from PCs. It provides email and Internet connectivity. It can send faxes, and even click pictures and prepare albums. It includes a personal information manager (PIM), a handheld computer, and an entertainment device. New generation mobile phones pack in everything from a computer to an FM radio and from video recording to TV viewing. The following subsections discuss the main features of the present generation of mobile phones and the technologies incorporated in them.

2.1.1 Smartphones and Multimedia Phones

A mobile phone is a handheld computing device. It is essentially a network-connected computer. Smartphones have the functionalities of mobile phones plus several advanced functions. They offer a number of applications besides telephonic talking. An example is the Blackberry 7130e series phone described in Section 1.2.1.1. The main features of a Smartphone are as follows:

- A GSM, CDMA, or tri-band (Section 1.1.3.1) wireless radio interface to a cellular network (Fig.1.14) provided by a mobile service provider.
- A smart T9 keypad [A smart keypad is one that remembers previously keyed entries. T9 stands for 'text-on-nine-keys'. It is a text input system that offers an alternative to multi-tapping for entering textual characters on a numeric keypad (multi-tapping means tapping a single key multiple times to get the desired letter or combination of letters). For example, to type the word 'bye' one would have to enter 2-2-9-9-9-3-3 in multi-tapping mode but only 2-9-3 in the T9 mode. In case of multiple words formed using the same combination of keys, a default word is displayed first and the user can choose from a list.] Smart T9 keypads are useful for creating SMS messages and entering contact information.
- A small area LCD display.
- Functions as a phone as well as a PIM to provide applications such as phone diary, address book, task list, calculator, alarm, and calendar.
- Ability to send and receive SMS messages of up to 160 characters.
- Ability to send and receive MMS (multimedia messaging service), i.e., messages for transmission of digital images, video clips, and animations.
- WAP enabled for Web page access, download, and other Web-based applications through a WAP gateway or proxy (Section 12.2).
- Provisions for games, e-commerce, and e-ticketing.
- Bluetooth communication with PCs and neighbouring devices.
- Integration of location information.

2.1.1.1 Multimedia Phones

A multimedia phone (sometimes referred to as a multiphone) offers multimedia functionalities. Besides the functions of a Smartphone, a multiphone can also play MP3 format audio and MP4 format video files (some phones may also support other formats such as WMA, AAC, etc.). Many multimedia phones include cameras for still pictures and video recording. Some phones also offer picture-editing software which enables the user to edit, crop, and refine pictures on their cell phone handsets. Nowadays it is also possible to watch TV on a mobile phone using 3G/EDGE/ EGPRS connectivity. Many mobile service providers link up with various TV channels and enable users to enjoy mobile TV on the LCD screens of their cell phones (Sections 1.2.1 and 8.6.2). Mobile phone manufacturers are creating bigger and better display screens in phones to enhance picture clarity and quality for TV and video viewing. (The Nokia 6682, for example, has a screen that measures about 2.25 inches diagonally. It enables larger LCD screen view of pictures and videos.)

Another popular application of multimedia phones is mobile phone gaming. Increasingly sophisticated games are continuously being introduced in the market and some phones are designed with special hardware and software components to support better game-related graphics and faster interactions among the players. The Nokia N-Gage is an example of a gaming-oriented phone. It enables users to play networked multiplayer games.

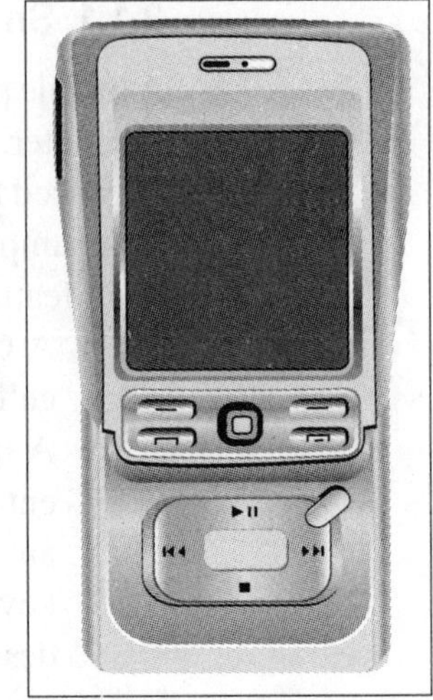

Fig. 2.1 Nokia N91

The Nokia N-series phones are designed to perform specific multimedia functions. The N91 belongs to this series and focuses on music and media playing. Some of the features of the N91 (Fig. 2.1) are as follows:

- GSM/GPRS EDGE 900/1800/1900 MHz connectivity
- Advanced voice calling functions such as an integrated handsfree speaker, voice dialling, voice recording, and conference calling
- Up to 4 GB of internal dynamic memory for multimedia functions and an additional 30 MB for storing calendar, contacts, and text messages
- A music player optimized for listening to music. It can play audio files in MP3, AAC, AAC+, eAAC+, RealAudio, WAV, WMA, M4A, True Tones, AMR-WB, and AMR-NB formats and video files in formats such as MPEG4 and RealVideo
- External speakers using a stereo audio jack
- FM radio and visual radio
- 2 Megapixel camera for video recording and still pictures
- 176 × 208 pixel display with up to 262,144 colours
- PIM for managing features such as calendar, contacts, task lists, and PIM printing
- WLAN 802.11b/g (Section 12.1) for hotspot connectivity, Bluetooth version 1.2 for wireless connectivity, and XHTML browser for Internet browsing
- Nokia PC Suite to synchronize data with the PC using a USB port or Bluetooth
- Battery with a digital talk time of up to 4 hours and standby time of up to 7.9 days

2.1.2 Cellular Networks for Mobile Phones

Figure 1.14 showed mobile communication using a cellular network. A mobile service region is divided into cells. Each cell has a base station with cell boundaries defined by the coverage area. A base station functions as an access point for the mobile service. The mobile device within a base station communicates through the base station only. Each cell has cells adjoining it in various directions. Adjacent cells have distinct frequencies. This avoids interference between the signals transmitted by different cells. Interference takes place when the frequencies are equal or integral multiples of each other. The base stations connect among themselves through either guided (wire- or fibre-based) or wireless networking. A station can also connect to a public switching telephone network (PSTN).

A multi-cell cellular network entails that when the transceivers (mobile phones) move from place to place, they will also have to switch from cell to cell. When a mobile device moves and reaches a cell boundary, there is *handoff* by a station. Switching on to next cell occurs by *handover* of the device connection to neighbouring base station. There are various mechanisms that are used for handoff and handover depending upon the type (GSM or 3G CDMA) of cellular network. The mobile device switches from the channel currently in use to a new channel and the transition is completed without disrupting the ongoing communication. Sections 3.4–3.6 and 4.2.2 will describe the device localization, calling, and handover processes.

2.1.3 Apple iPhone

The iPhone was announced in January 2007 by Apple CEO Steve Jobs. The iPhone brings together the features of an iPod, a Smartphone, a digital camera, and a handheld computer. The iPhone is a multimedia and Internet-enabled mobile phone. The main features/functionalities of the iPhone are listed below:

- It comes in two versions—4 GB and 8 GB flash memory versions.
- It has a wide, touch-sensitive, 3.5-inch display screen, which has a resolution of 320 × 480 pixels at 160 pixels-per-inch display. It has an ambient light sensor which senses the lights in proximity and automatically adjusts screen brightness to save power.
- A proximity sensor shuts down the display and touch screen when the phone is held to the ear.
- It incorporates *multi-touch sensing*[1] and a virtual keypad. The virtual keypad has automatic spell checking, predictive word capabilities, and a dynamic dictionary.

[1]Multi-touch sensing is a relatively new technology in the field of human–computer interaction. Jeff Han, a consultant research scientist at the New York University, says "while touch sensing is commonplace for single points of contact, *multi-touch* sensing enables a user to interact with a system with more than one finger at a time, as in chording and bi-manual operations. Such sensing devices are inherently also able to accommodate *multiple users* simultaneously, which is especially useful for larger interaction scenarios such as interactive walls and tabletops" (Han 2006).

- The iPhone has only a single physical button, called '*home*'. The user controls the iPhone by sliding a finger across its touch-sensitive 3.5-inch display. No stylus is needed, nor can one be used. The touch screen requires bare skin to operate.
- It makes phone call by simply pointing the finger at a name or number in a call log, address book or favorites list. Another innovative use of the contacts list is that, using a new technology, the iPhone automatically synchronizes all contacts from a PC, Mac, or Internet service.
- A special phone-call feature automatically adjusts music volume with incoming phone calls.
- An easy-to-use conference call feature lets users connect two calls with one touch of the screen. The iPhone allows conferencing, call holding, call merging, and caller ID.
- It sports the Visual Voicemail feature which allows users to skip directly to voicemails they want to hear.
- SMS text messaging on the iPhone is similar to iChat, with user dialogue encased in bubbles with familiar iChat sounds, and a touch keyboard appears below for entering text.
- It seamlessly synchronizes data for Internet HTML web browsing service and e-mail service and it works with any Mac or PC.
- It supports full iTunes integration. It provides for iPod audio and photo file formats and functions, for example, shuffling of songs, repeat one or all, sound check on or off, and 20 equalizer settings.
- It synchronizes the music and videos from iTunes, contacts, calendars, photos, notes, bookmarks, and e-mail accounts with the IMAP or POP3 e-mail service.
- It supports SMS, calendar, photo, camera (2-megapixel camera), calculator, charts (e.g., stocks), maps (integrated Google maps functionality lets users look up locations, search for local businesses, and view satellite imagery), weather, note pad, clock, settings, phone, mail (including Yahoo free push e-mail service), Web browsing, multimedia player, and iPod.
- It connects to a computer via 30-pin iPod dock connector or IEEE 802.11b/g WiFi (Section 1.1.5) or Bluetooth 2.0 (Sections 1.1.4 and 12.4) capabilities.
- It runs on Mac OS X operating system.
- Presently it supports data downloads up to 220 kbps and deploys quad-band GSM (850 MHz, 900 MHz, 1800 MHz, and 1900 MHz) + EDGE phone (Section 1.1.3.1).
- 3G features (Section 1.1.3.2) are expected to be introduced soon.
- Its physical dimensions are 115 × 61 × 11.6 mm and weight is 135 g.

2.2 Digital Music Players

Mobile computing is a term that encompasses an extremely wide range of applications for a user on the move and this includes entertainment. Digital (mobile) music players have revolutionized the way people listen to music. These players include software that play music files encoded in formats such as MP3, WMA, Realmedia, etc. on mobiles, PCs, and laptops. Realplayer, Windows Media Player, QuickTime Player, etc., are some examples of media-playing software. Digital music players also include media-playing hardware or handheld (portable) music players that play an assortment of digital media file formats such as MP3, AAC, WMA, and WAV. Such players use flash memory. The capacity of these portable music players may vary from about 128 MB to over 80 GB. Some flash-memory-stick-based players can store 15,000 songs or more. Most flash-based players can serve as storage devices. Present day media players enable video as well as audio playback. The market leader in the handheld media player market is the Apple iPod. The following subsections describe the iPod range of devices and, in particular, the features and functions of the iPod Nano.

2.2.1 Apple iPod

The iPod family of devices includes flash-based players. The iPods have simple user interfaces and are mostly designed around a central scroll wheel. The fifth generation iPod incorporates a video player. The iPods use the Apple iTunes software for transferring, storing, managing, and playing music, photos, and videos. Recent versions of iTunes offer photo and video synchronization features.

The iPod Nano is one of the latest additions in the iPod family. By the end of 2005, the iPod Nano (Fig. 2.2) had become quite a rage. The iPod Nano sports a flash-based memory with a colour screen. Detailed features of the Apple iPod MA350LL/A (popularly known as iPod Nano White) are as follows:

- It stores 240 songs in 1 GB, 500 songs in 2 GB, and 1,000 songs in the 4 GB version.
- It supports Apple audio communication by AAC files between 16 kbps and 320 kbps.
- It has a battery life of up to 14 hours for music playback and up to 4 hours for slide shows with music.
- It has a fast charging battery (battery charges in ~1.5 hours up to 80% capacity and up to full capacity in ~3 hours).
- It has a 1.5 inch (diagonal) colour LCD with blue-white LED backlight.
- It provides customized main menu to create multiple On-the-Go playlists,

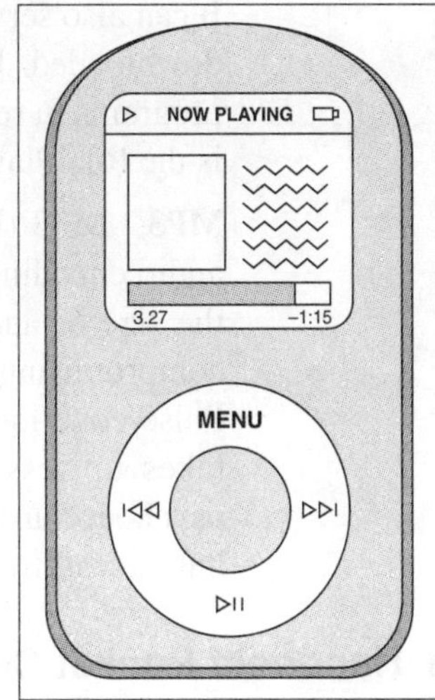

Fig. 2.2 iPod Nano mobile device

adjust audio-book playback speed, clicker playback through headsets, rate the songs, shuffle the songs or albums, repeat one or all songs, sound check (on or off) and use features such as 20 equalizer settings, sleep timer, and multilingual display.

- It supports MP3, VBR, WAV, and AIFF file formats. It supports JPEG file photo display and download. It syncs iPod-viewable photos in BMP, TIFF, PSD (Mac only), and PNG formats.
- It supports protected AAC files from the iTunes Music Store. It supports Audible (formats 2, 3, and 4).
- It supports web browsing.
- It sports a calendar and task-to-do lists.
- It supports ear-bud headphones and a speaker phone.
- It has a ports dock connector, a stereo mini jack, a USB through dock connector, and a USB cable.

2.2.2 Popular Audio File Formats

Some popular audio file formats and their characteristics are discussed below.

WAVE The WAVE (short for Waveform audio) format is an audio format standard for storing audio. It is one of the methods for storing data in 'chunks'. WAVE files usually contain uncompressed audio in the PCM (pulse-code modulation) format but they are also capable of holding compressed audio. The WAVE audio format can, therefore, be used by professionals for maximum audio quality and it can also be manipulated quite easily for storing compressed audio. It supports a variety of bit resolutions, sample rates, and channels of audio.

RealAudio The RealAudio format was developed by RealNetworks. It uses both low-bitrate formats for use over dialup modems and high-fidelity formats for music. It can also serve as a streaming audio format which can be played while it is being downloaded. RealAudio is used by many Internet radio stations to stream their programs in real time. The official playing software for the RealAudio file format is the RealPlayer.

MP3 MP3 (MPEG-1 audio layer 3) is perhaps the most popular of the digital audio encoding and compression formats available today. It is designed to reduce the size of audio files to about 10% of the original uncompressed files without compromising too much on sound quality. It became an ISO/IEC standard in 1991. Pulse-code-modulation-encoded audio is represented using the MP3 format. MP3 takes up less space than straightforward encoding methods because it uses psychoaccoustic models to remove components, which are almost inaudible to the human ear, from audio files.

2.3 Handheld Pocket Computers

Handheld computing devices come in many manifestations. One of these is the Smartphone. Handheld computers or palmtops can also be used on the move like

Smartphones. However, these pocket-sized PCs differ from Smartphones and multimedia phones in that they can be programmed for customized applications. They offer a variety of application and programming tools not included in new generation mobile phones. Unlike Smartphones, which usually use the text-on-nine-keys format, handheld computers have full text keypad or a touch screen keypad. A stylus is generally used to enter data into handheld devices such as PDAs and palmtops. Some handheld devices allow the user to write on the screen using a stylus and incorporate special software for handwriting recognition. Palmtops are programmable pocket computers. They include word processors and spreadsheet software as well as PIM (Section 1.3.4) software. Handheld computers may include QWERTY keyboards or touch screens with stylus for data inputs. Pocket computers are different from laptops, notebooks, and sub-notebooks in respect to the following features:

- Pocket PCs do not have CD drives and hard disks. Instead, they use flash memory. Most handheld computers allow the insertion of a memory stick as secondary memory. (A memory stick is a removable flash memory card.)
- Clock speeds of pocket computer processors are limited up to 200 MHz due to considerations regarding battery life.
- Unlike laptops and notebooks, which use regular microcomputer-operating systems, pocket computers have specially designed operating systems which are scaled to the requirements of the software, hardware, and peripherals used in handheld computers. A few examples of such operating systems are Windows CE and PalmOS.

2.4 Handheld Devices: Operating Systems

Till about a decade ago, Windows 95 was the most sensational of operating systems and Linux was far from popular. In the last 10 years, not only has there been a transformation of titanic proportions in computing devices, but the operating systems used in these devices have also been revolutionized. Sections 14.1 to 14.3 will describe the three most popular mobile OSs (Symbian, Windows CE, and PalmOS). These operating systems provide interfaces, perform allocation and management functions, and act as platforms for running the increasingly sophisticated software that are created for mobile computing devices. In this section we will take a look at some these operating systems and the functions they enable in a device.

2.4.1 Windows CE

Windows CE is an operating system from Microsoft. It is a real-time operating system meant for handheld computers and embedded systems. The Windows CE kernel is different from the kernel of the desktop versions of Windows. It is meant for computing devices with low storage and can be run in about 1 MB of memory.

But the Windows CE OS memory needs are larger as compared to PalmOS (Section 2.4.2). However, Windows CE can support a wider range of hardware than PalmOS. There are various versions of Windows CE to support different CPUs such as NEC MIPS, Intel StrongARM, AMD X86, etc. Windows CE is also designed to support multitasking on handheld devices. Some of the features in Windows CE devices are as follows:

- High resolution colour/display, touch screen, and stylus keypad
- Complex APIs in Windows CE. It gives the user a PC-like feel and Windows-like GUIs
- PIM, MS Office, Internet Explorer features on handheld mobile system
- The CompactFlash card slots to extend memory and extension card slots
- OS memory requirement is large but scales to the requirement of the device peripherals
- Built-in microphone for voice recording
- USB port. A cradle connects the handheld device to PC (A cradle is an attachment on which the handheld device can rest near a PC and connect to the PC via a USB or serial port, Bluetooth, or infrared)
- Infrared port to communicate with mobile phones and external modems
- Digital camera card
- Games
- Microsoft Windows Media Player and other media players
- ActiveSync for synchronizing mobile data with PC using a USB, serial port, PC infrared port, or Ethernet LAN for interfacing. (ActiveSync is a synchronization software from Microsoft. It resolves conflicts in versions of the files during data exchange and facilitates the use of multiple service providers and service managers. Sections 9.1 and 9.2 will describe data synchronization and its tools.)
- Needs high processor clock speed

Windows Mobile (formerly known as PocketPC) is a suite of basic applications for handheld devices along with a compact operating system. Windows Mobile is based on the Windows CE platform. There are many different versions of Windows Mobile. The application software suite includes the pocket (small screen display) versions of Excel, MSWord, PIM, Internet Explorer, and Outlook. It supports JavaScript and ActiveX programs. It also includes the Windows Media Player for playing files of various audio and video formats.

2.4.2 PalmOS

PalmOS is an operating system from Palm Inc. It is used in Smartphones and handheld computing devices. PalmOS is optimized to support a very specific range of hardware. CPU, controller chips, and screens of PalmOS-based devices cannot much differ from the hardware reference platform designed by Palm Computing without major changes in the operating system itself. PalmOS is advantageous in that it is compiled for a specific set of hardware and, hence, its performance is very finely tuned. However at the same time, the inability to adapt to different sorts of hardware may also be considered a limitation for this operating system. Also PalmOS does not support multitasking and is definitely not a great platform for running multimedia applications. It works efficiently when running small productivity programs but does not offer much expandability. PalmOS devices usually have wide screens and the input of data is facilitated by a touch screen. Some of the highlighting features of PalmOS devices are as follows:

- Simple APIs compared to Windows CE.
- OS memory requirement is low (16 MB memory in the system suffices).
- It needs lesser processor clock speed and, therefore, has lesser energy requirements.
- PIM, address book, data book for tasks-to-do and organizing, memo pad, SMTP (simple mail transfer protocol) e-mail download, offline creating and sending POP3 (post office protocol 3) e-mail, Internet browsing functions, Windows organizer, and PDA (personal digital assistant).
- Wireless communications including email, messaging, browsing the web, and multimedia applications such as playing music.
- A cradle connects handheld devices to PCs.
- HotSync (Sections 9.1 and 9.2) for synchronizing with PCs through a serial port or infrared port (HotSync resolves conflicts in different versions of files during data exchange.)
- Infrared port for communication with mobile phones and external modems.
- Extension card slots.
- The PalmOS is most compatible with the Dragonball processor from Motorola.
- Most PalmOS devices offer a display resolution 160 × 160 and 256 colour touch screen.
- PalmOS devices can be integrated with GSM/CDMA cellular phones.
- These devices can easily serve as platforms for third party applications such as games, travel and flight planner, calculators, graphic drawings, preparing slide shows, etc.

The PalmOne Tungsten T5 handheld uses the PalmOS and has the following features:

- Palm desktop software for both Windows and Mac and other essential software
- 256 MB internal flash memory
- Expansion slot that supports MMC (multimedia card), SD (secure digital) memory card, and SDIO (secure digital input/output) memory card
- Doubles as a flash drive that enables quick drag and drop of files from a PC to the handheld device

2.4.3 Symbian OS

The Symbian OS is the most widely used operating system for smartphones. It runs exclusively on ARM processors. The structure of Symbian OS is much like that of some desktop operating systems. It offers pre-emptive multitasking, multithreading, and memory protection. Because the Symbian OS was initially designed for handheld devices with limited resources, it strongly emphasizes on memory conservation. Also Symbian OS embodies event-based programming and when applications are not directly concerned with events, the CPU is switched off. Such techniques are very useful in conserving battery life. Some of the features of a recent version of Symbian OS are as follows:

- Support for WLAN (Section 12.1)
- Improves and speeds up Smartphone performance
- High-end security enhancement features
- Graphics support including support for 3D rendering
- Hindi and Vietnamese language support to serve a larger range of consumers
- Native support for Wi-Fi
- Support for FOTA (firmware over-the-air)
- Improved memory management
- Low boot-time
- Native support for Push-to-talk

The Nokia 9300 Smartphone (Fig. 2.3) runs on the Symbian OS. It offers organizational and multimedia applications. It has the following functionalities:

- It provides high-speed data-connectivity using EGPRS (EDGE) (Table 1.2).
- It has advanced voice features such as a hands-free speakerphone and conference calling capability.

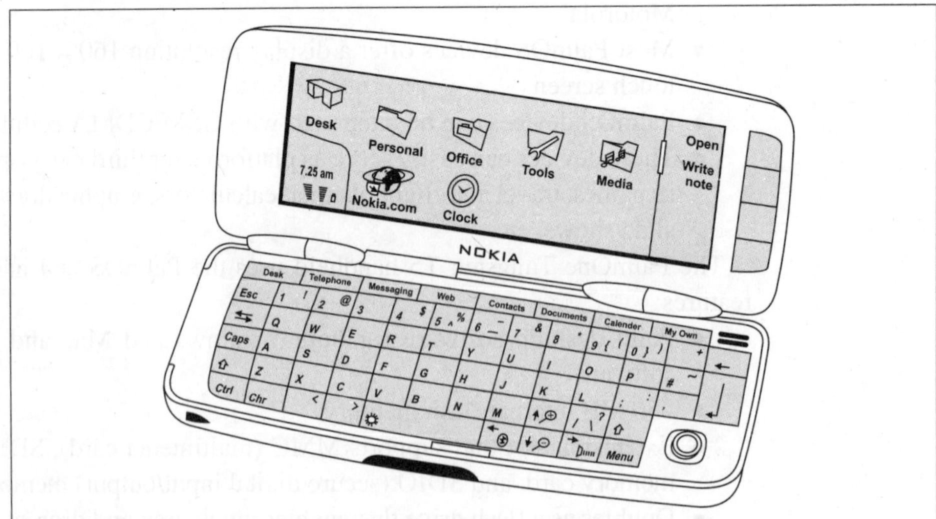

Fig. 2.3 Nokia 9300 phone based on the Symbian OS

- It has a large storage memory which includes 80 MB of built-in memory plus multimedia card (MMC) slot.
- It is compatible with most Lotus and Windows programs.
- It supports Microsoft Office formats (MS Office 97 onwards). It supports viewing of slide shows.
- It has PIM interfaces for applications such as calendar, contacts, Microsoft Outlook 98, 2000, 2002, 2003, Microsoft Outlook Express/Windows' Address Book, Intellisync Wireless e-mail, Lotus Notes (5.x and 6.x), and Lotus Organizer (5.x, and 6.x).
- It has Internet connectivity for Web browsing.
- It has the PC synchronization feature (Section 9.1). It synchronizes and chains to a PC in the vicinity.
- It integrates corporate solutions such as IBM WebSphere Everyplace Access, BlackBerry Connect, Oracle Collaboration Suite, and secure mobile connections via VPN client.
- It provides the Symantec Client Security 3.0 and the Fujitsu mProcess Business Process Mobilizer.
- It includes the Adobe Acrobat Reader for accessing PDF files.
- It sports the HP Mobile Printing software which enables Bluetooth connectivity with compatible printers for wireless printing.

2.4.4 Linux for Mobile Devices

Most operating systems used in mobile devices are designed for use on specific hardware and offer a platform for only select software applications. Linux, however, can be modified easily to suit different sorts of hardware and software applications. Being an open source OS, it enables the user to customize the device to suit specific needs. The Embedded Linux Consortium (ELC) is an association which promotes Linux and develops standards for Linux in embedded systems. They also develop standards for designing user interfaces, managing power consumption in devices, and real-time operation of embedded Linux. The ELC platform specification (ELCPS) is also a result of this consortium. Linux is also considered to be more secure than most other operating systems. Also Linux support is easily available from the many forums and associations that promote this OS. Many international mobile phone manufacturers are turning to Linux for their OS requirements. The Motorola ROKR E2 music phone is an example of a Linux-based mobile phone. Its main features are listed below:

- A 240 × 320 TFT display with 262,000 colours
- USB 2.0 PC networking for fast 'drag-and-drop' data transfer
- Built-in FM radio
- Support for Motorola's iRadio service
- Support for Bluetooth

- 1.3 megapixel camera for video capture and playback
- MMS (multimedia messaging service) enabled
- Opera web browser
- 'Airplane mode' for safely listening to music when aboard an aircraft
- PIM (personal information manager) with picture caller ID

2.5 Smart Systems

This section discusses smart systems that have embedded computational devices. Smart systems provide comfort, efficiency, and remote access to devices and appliances. Smartcards, labels, and tokens are widely used in consumer goods and service industries. Smartcards have multiple applications in our day-to-day lives in their numerous forms such as credit cards, identification cards, key cards, etc. Sensors and actuators are electronic devices that make automated systems possible. This section also covers automated robotic systems and smart appliances. We will also review these in the following subsections.

2.5.1 Smartcards

Smartcards (also known as integrated circuits cards or ICC) are small, pocket-sized cards with electronic processing circuits embedded in them. Some smartcards are simply memory cards, meant for storing data. These cards function as rewriteable memory devices for storing and updating data. Smartcards may also have embedded microprocessor circuits. Smartcards have been used since the 1970s. Earlier smartcards were used as telephone cards in European payphones. Smartcards maybe divided into contact smartcards and contact-less smartcards. Contact smartcards have small gold-coated pins on the chip that provide contact with the electrical circuits of the card reader when the card is inserted in it. The size, shape, electrical characteristics, communications protocols, command formats, and functions of smartcards are defined by the ISO/IEC 7816 and ISO/IEC 7810 standards. Examples of contact smartcards are the telephone cards used in public telephones in some European countries. Contact-less cards communicate with the card readers using the RF (radio frequency) (Section 2.5.2) induction technology. These cards have to be held close to the reader antenna. The standard for contact-less smartcard communications is ISO/IEC 14443.

The chip embedded in a smartcard is a circuit including a computer, memory, and transceiver. The chip lies in between the inner layers of the card. A transceiver is a circuit for transmitting and receiving signals. Smartcards have secured hardware. For example, certain memory sections are accessible only to the program and the OS once the card is personalized. JavaCard is used to program smartcards. Some applications of smartcards are as follows:

- A card is used for financial transactions as a credit card or ATM/debit card.
- A card can store personal ID (even photo) and personal information.

- A card can store the medical records of the holder (This may provide doctors a faster way of accessing patient records by reading this card through a PC).
- An employee in an enterprise uses a card to open the security locks at work and log in.
- A student uses a card to get books issued from the college library.

Smartcards do not have batteries. The energy is provided by the card reader. The computer in the card is activated by power (through radiation in a contact-less card and through the IC pins on the card surface in a contact card) from a nearby reader (also called host). The card can then communicate with the host after appropriate interchanges for authentication. ASK 13.56 Mbps (Section 1.1.2.4) is used for contact-less communication at data rates of ~1 Mbps. A metallic squared foil at the side of the card surface acts as the antenna for transmitting ASK signals to a host. The host is a device which reads the card and performs requested transactions. The host connects to a PC or remote server through a phone line, Internet leased line, or a fibre line. Figure 2.4(a) shows a smartcard and its host.

A card has a fabrication key, personalization key, and utilization lock embedded in it. Fabrication key identifies a card uniquely. The host machine remote server (e.g., a bank) uses the personalization key that the server inserted to activate and program the card for enabling future transactions. The utilization lock is used by the server to lock or unlock the use of the card. For example, on card expiry the server locks the card. On using a wrong password several times, the server locks the card temporarily in order to prevent use by an unauthorized user. Application protocol data unit (APDU) is an accepted standard for card–host communication. Computer communication to a host as well as the card hardware is secured. These deploy cryptographic protocols for interchange of the messages to and from the host.

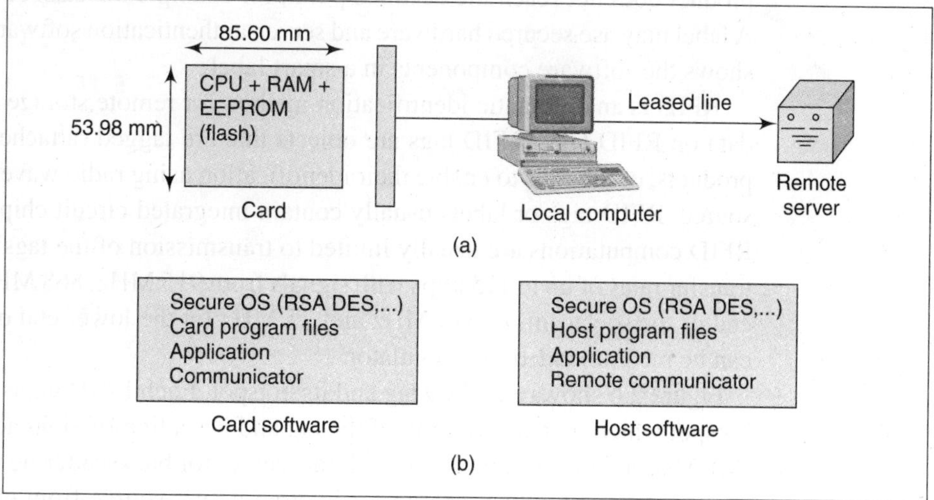

Fig. 2.4 (a) Smartcard and its host (b) Software components in a smartcard

Figure 2.4(b) shows the software components in a smartcard. There are two programs: one for in-card applications and other for transactions through the host. Some examples of in-card application are storing the balance available in the bank account after the last transaction, storing the medical history, storing reward points granted by the credit card issuing service, etc. Off-card applications include communicating the card info and previously stored history to a host for authentication and running applications on the host.

2.5.2 Smart Labels

Generally, a label serves the purpose of identifying the contents of a package. For example, a barcode label on a book packs in information about the publisher, title, author, publishing date, and reprint edition of a book. Barcode labels are also used in stores so that a reading machine can identify the product and its price. A label differs from a card in terms of thickness and visibility. A label using wireless means for product identification can be concealed inside the product. A smart label has a processor, memory, transceiver, and antenna similar to a contact- less smartcard. Smart labels are essentially an earlier version of the now popular RFID (radio frequency identification) tags. They are powered by the received signals just like smartcards. Since wireless communication is used, the label need not be visible when implanted into a product or package.

Figure 2.5(a) shows a network of labels. The smart labels are networked together using a central reading and computational device (host) or PC. Cluster of labels form a network similar to a LAN network. A collision-sense-and-avoidance protocol is used so that multiple labels are not allocated the same ID tag and the central server can uniquely identify each one. The central server can also detect the removal of a labelled product or packet from a product-shelf and raises alarm in case the product does not reach the destined point, for example the cash counter in a store. A label may use secured hardware and server-authentication software. Figure 2.5(b) shows the software components in a smart label.

RFID is an automatic identification method for remote storage and retrieval of data on RFID tags. RFID tags are objects that are tagged (attached) onto people, products, or animals to enable their identification using radio waves from a nearby source. RFID tags or labels usually contain integrated circuit chips and antennas. RFID computations are usually limited to transmission of the tags' contents. Data transfer rates of up to 115 kbps with signals from 915 MHz, 868 MHz (at the higher end of the spectrum) to 315 MHz and 27 MHz (at the lower end of the spectrum) can be recommended by a regulator.

Figure 2.6 shows an RFID tag and its hotspot. Each RFID tag is monitored by a hotspot which is in the vicinity of the tag and has a line-of-sight access to it. Each RFID has a processor, memory, and transceiver for backscattering of hotspot RF, a charge pump (to collect charge for internal power source from electrical current

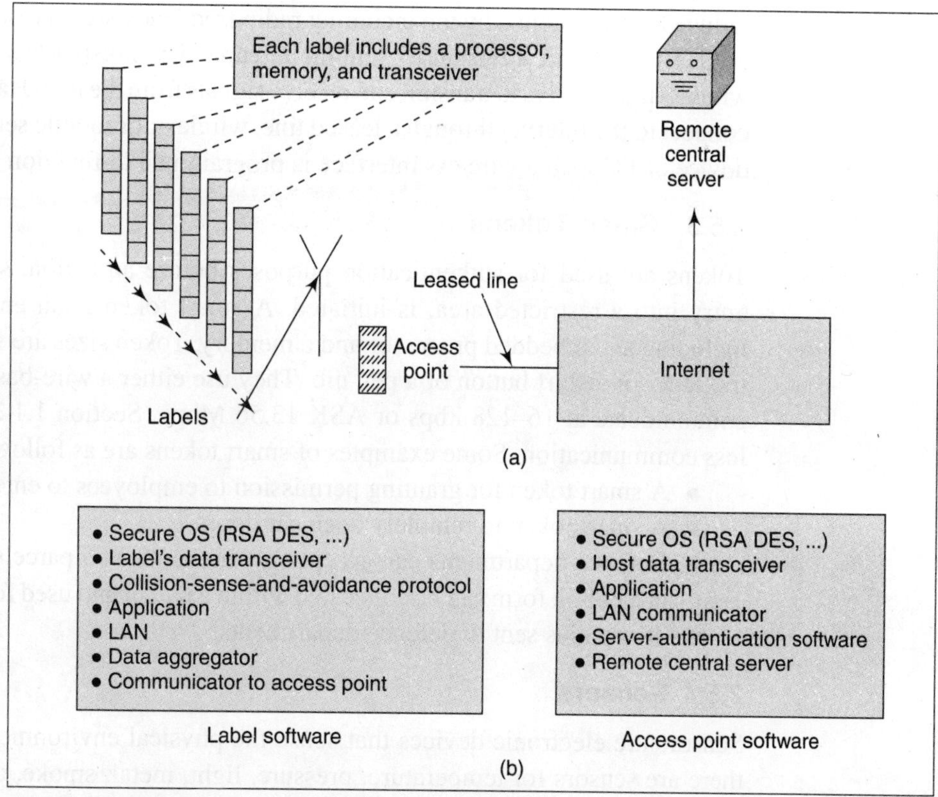

Fig. 2.5 (a) Network of labels (b) Software components in a smart label

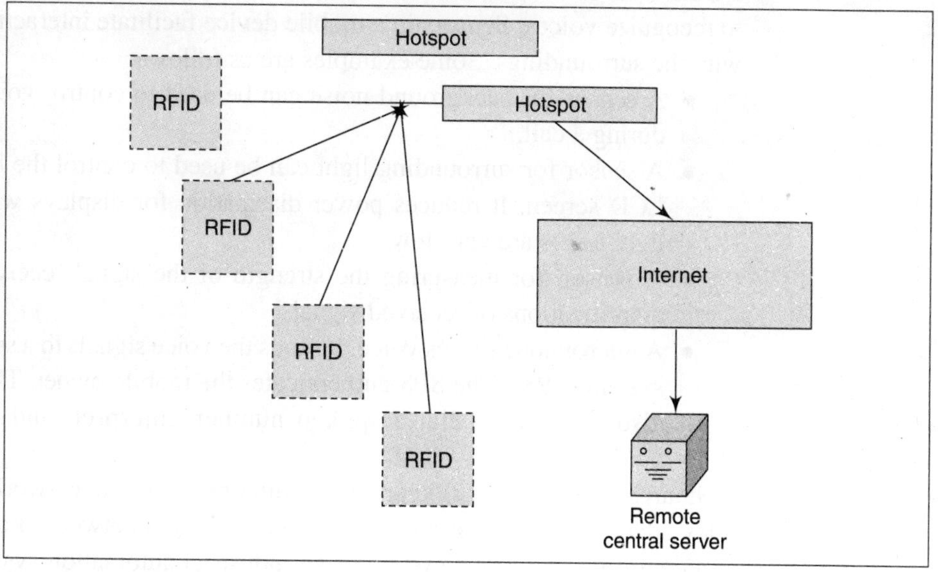

Fig. 2.6 RFID and hotspot

induced in the antenna by the incoming radio frequency signal from hotspot power up and to transmit a response), and an antenna. The hotspot has a computer and wireless transceivers to transmit and receive signals from the RFID tags. The hotspots connect to the Internet through a leased line, wireless, or mobile services. A mobile device or PC with a wireless interface is programmed to function as the hotspot.

2.5.3 Smart Tokens

Tokens are used for authentication purposes before an action, such as granting entry into a restricted area, is initiated. A smart token is an encapsulated chip including an embedded processor and a memory. Token sizes are small, usually of the order of a shirt button or a pen nib. They use either a wire-based protocol and communicate at 16–128 kbps or ASK 13.56 Mbps (Section 1.1.2.4) for contact-less communication. Some examples of smart tokens are as follows:

- A smart token for granting permission to employees to enter a work place
- A smart token to remotely open car doors
- Defence departments can accept only authenticated parcels. A smart token in a button form can be concealed within a parcel and used for authentication of supplies sent to defence departments.

2.5.4 Sensors

Sensors are electronic devices that sense the physical environment, for example, there are sensors for temperature, pressure, light, metal, smoke, and proximity to an object. The sensor sends the signals to a computer or controller. A sensor may be a CCD (charge-coupled device) camera to identify various objects or a microphone to recognize voices. Sensors in a mobile device facilitate interaction of the device with the surroundings. Some examples are as follows:

- A sensor for background noise can be used to control voice amplification during a call.
- A sensor for surrounding light can be used to control the brightness of the LCD screen. It reduces power dissipation for displays when the ambient light levels are very low.
- A sensor, for measuring the strength of the signal received, controls the amplifications of received signals.
- A microphone senses voice. It sends the voice signals to a speech processing system (SPS). The SPS authenticates the mobile owner. Then, the SPS can also be used to dial a spoken number, interpret, and execute spoken commands.

Smart sensors have computational, communication, and networking capabilities. They are deployed to communicate information to a network, a central computer, or a controller. A robotic system or an industrial automation system has multiple smart sensors embedded in it. A smart sensor consists of the sensing device, processor, memory, analog-to-digital converter (ADC), signal processing element, wireless or infrared receiver and transmitter, and performs communicational as well as computational functions. Smart sensors are generally programmed using assembly language or C.

2.5.5 Actuators

An actuator receives the signals from a controller or central computer and accordingly activates a physical device, appliance, or system. Examples of such physical devices are—a servomotor in a robot's hand, loudspeaker, power transistor supply current to an oven, solenoid-valve actuator, a transmitting device in a sensor network, etc. A smart actuator receives the commands or signals from a network, mobile device, computer, or controller and accordingly activates the physical device or system. Sensor–actuator pairs are used in control systems. For example, a temperature sensor and current actuator pair controls the oven temperature, a light sensor and bulb current actuator pair controls the light levels, and a pressure sensor and valve actuator pair controls the pressure. Industrial plants have large numbers of pairs of sensors and actuators. A set of smart sensors and actuators are networked using a control area network bus (CAN bus), for example, in an automobile or industrial plant. Smart sensors can be programmed in assembly language or C using development tools.

2.5.6 Sensors and Actuators for Robotic Systems

Robotic systems incorporate a variety of overlapping technologies from the fields of artificial intelligence and mechanical engineering. Robotic systems are essentially programmable devices consisting of mechanical actuators and sensory organs that are linked to a computer embedded in them. The mechanical structure might involve manipulators, as in industrial robotics, or might concern the movement of the robot as a vehicle, as in mobile robotics. Some examples of sensors used in robotic systems are as follows:

- Acceleration and force sensors in the right and left feet
- Infrared distance sensors at the head and hands
- CCD camera in eyes
- Angular rate sensor at the middle
- Microphones in ears
- Pinch detection at the belly
- Thermo sensors and touch sensors at shoulders, hands, and head

Some of the actuators used in a robot are:

- At the mouth, there can be a speaker to let a robot issue commands to other robots or relay sensed information via spoken messages.
- At each moving joint—feet, knee, waist, neck, shoulder, hand, and gripper palm, there are actuators and motors.

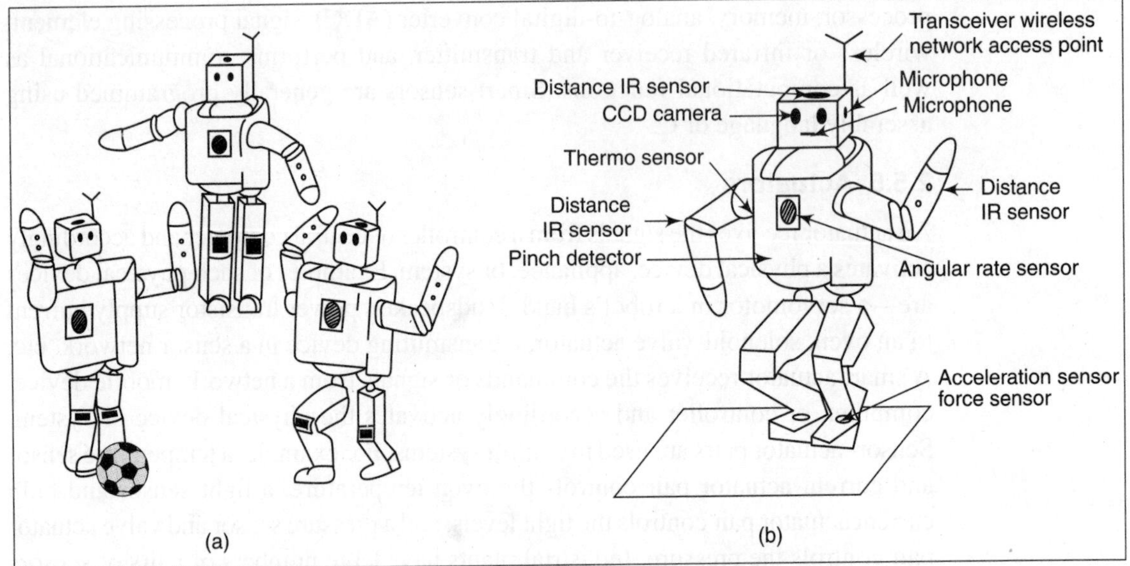

Fig. 2.7 (a) Robots playing robo-soccer (b) Sensors, actuators, and transceivers in a robot

The sensors transmit, through internal wires, the signals to the embedded processors at the central computer chip in the robot. The robot wirelessly communicates data to a central server when the actions of a group of robots need to be synchronized. Wirelessly communicating robots are mobile and are used in industrial plants for moving in areas not easily accessible by humans. Master–slave systems of robots can be used for a variety of purposes. The master robot, in such a system, sends commands to the other (slave) robots. Figure 2.7(a) shows an arrangement of robots playing robo-soccer. Figure 2.7(b) shows sensors, actuators, and transceivers in a robot. Robot sensors can be programmed using either assembly language or C.

2.5.7 Smart Appliances

With the present generation of automation technology, it is possible to control home appliances and security systems using a cell phone or computer. Home appliances can be networked using power lines. Signals of frequencies up to 525 kHz can be induced in such lines. These signals can be communicated from one appliance to another, thus forming a network. The devices can also communicate though a central server. Home appliances can also be networked using very short-range wireless protocols, such as Bluetooth or Zigbee (Sections 1.1.4, 12.5–12.7, and 12.9).

Smart home and office appliances are web-enabled devices. A smart appliance can be allotted a web address. The appliance then connects to the Internet through a residential gateway. The gateway enables the user to access devices, such as the home computer, home MP3 player, security locks, etc., from outside using the WLAN

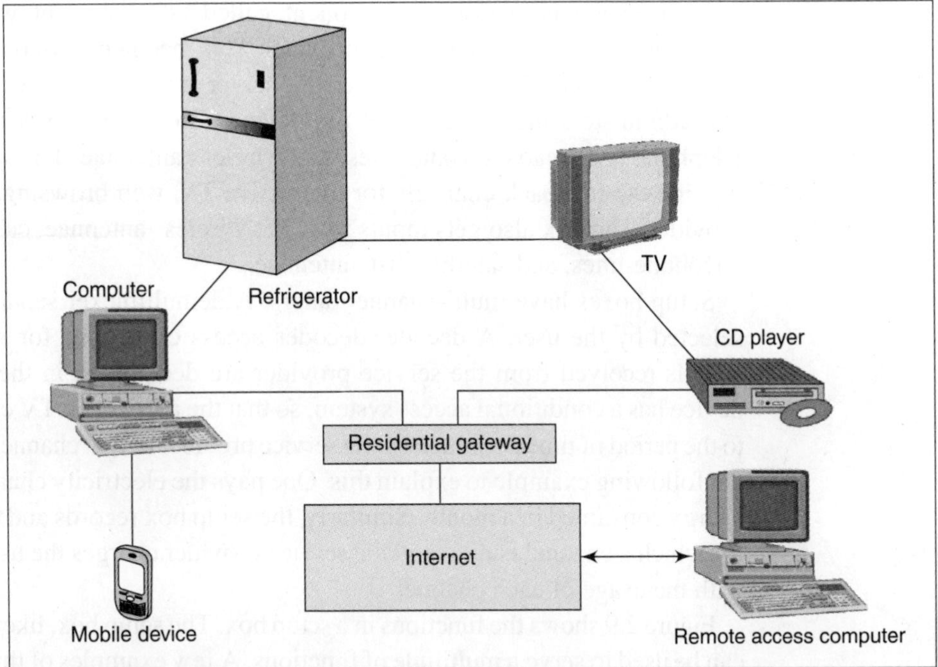

Fig. 2.8 Network of appliances

Internet or a mobile service provider access points. Figure 2.8 shows a network of appliances. A residential gateway is a system that interconnects the home appliances (e.g., media player, computer, security-locks, lights, oven, refrigerator, and air conditioner) and the Internet. An authentication process is first carried out. Once a user is authenticated, the gateway allows access from outside to the home devices or vice versa. The gateway can also use a service-provider server for the networking functions. A smart appliance can also be allotted a number by the mobile service provider. It can then be controlled from a Smartphone using the SMS service. For example, a smart AC can be switched off by an SMS in case one forgets to switch it off while leaving home. The AC can also be switched on using an SMS so that one finds the room cool on getting back home.

A new concept that is currently under development is e-Maintenance of an appliance. The maintenance service provider can access the appliance through the web and diagnose the appliance for any malfunction.

2.5.8 Setup Boxes

A setup box (also known as a set-top box) is a sophisticated computer-based device. It has data, media, and network processing capabilities. It interconnects the home TV and the broadcasting service network. Java is the most commonly used programming language in a setup box. Setup boxes run deciphering and encrypting software. There is a software component, called a device agent, which administers the device on behalf of the service provider. This mechanism of operation is similar to that of a mobile phone device, where the server of mobile service provider manages and administers the operation of the device. The setup box sends its output to telephone lines, cable coaxial lines, and wireless antennae. The setup box outputs provide the feedback channels for interactive TV, web browsing, and the service provider. The box also gets inputs from the wireless antennae, cable coaxial lines, telephone-lines, and satellite-dish antennae.

Setup boxes have multi-channel tuners. A demultiplexer separates the channel selected by the user. A decoder decodes access conditions for the channel. The signals received from the service provider are deciphered in the setup box. The device has a conditional access system, so that the access to a TV channel is limited to the period of time permitted by the service provider of that channel. Let us consider the following example to explain this. One pays the electricity charges according to energy consumed in a month. Similarly, the setup box records and tracks the period for which a channel is used and the service provider charges the user in accordance with the usage of each channel.

Figure 2.9 shows the functions in a setup box. The setup box, like the Smartphone, can be used to serve a multitude of functions. A few examples of this are as follows:

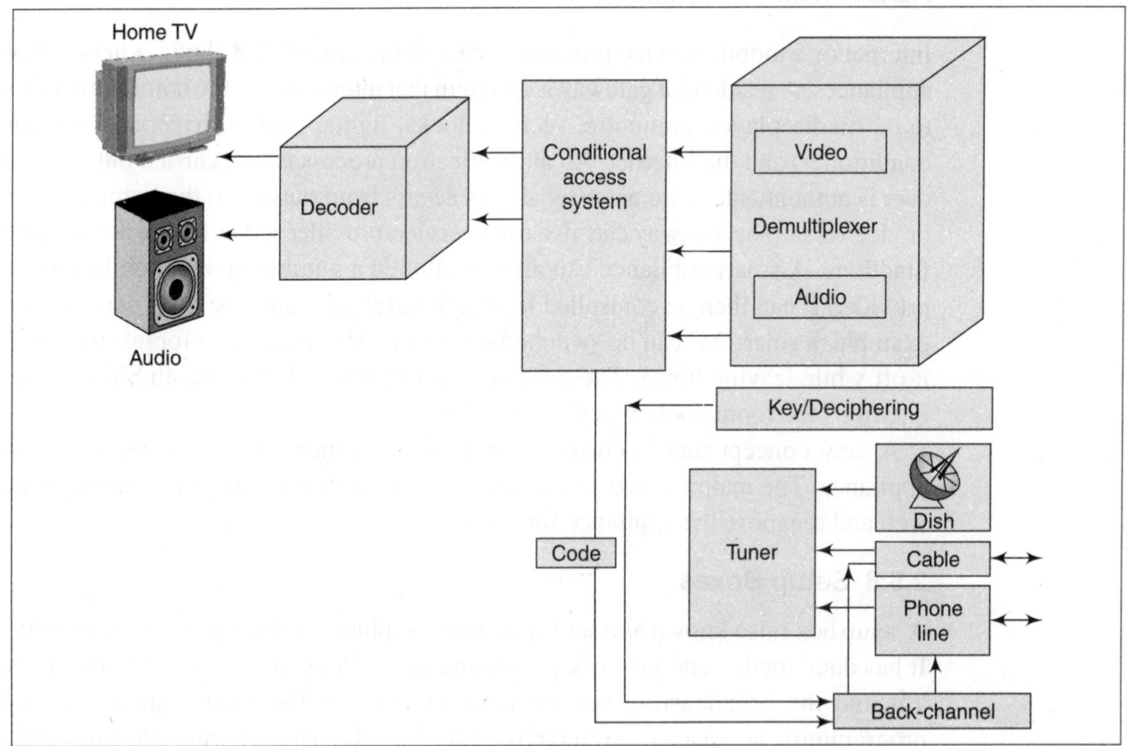

Fig. 2.9 Functions in a setup box

- The setup box provides a great platform for Java-based multimedia games.
- Some setup boxes include wireless keypads to interact with the TV for selecting, tuning, adjusting contrast, picture, and sound quality, for playing video games, and for Internet browsing.
- Setup boxes have hard disks and a CD-ROM drive.
- A setup box can be connected to a PC or printer via a USB port.

2.6 Limitations of Mobile Devices

Mobile devices have gained immense popularity in past two decades and the task of upcoming technologies is to offer more and more innovative applications and contribute to its growth. The primary goals that a mobile device is expected to fulfil are the twin objectives of efficiency and convenience. A mobile device is successful only if the endusers accept it as a helpful tool that increases their productivity and provides for a more convenient way of life. However, the use of mobile devices has its flip side too. Mobile devices are limited in various ways such as hardware limitations, quality of mobile service, security during communication and transmission, connectivity troubles, etc. This section will discuss constraints in mobile devices.

2.6.1 Quality and Security of Service

Technical restrictions and practical considerations render it difficult for service providers and device manufactures to ensure that the mobile device operations run uninterrupted. The greatest challenge facing mobile computing is maintaining quality of service along with the provisioning of seamless access to all users.

2.6.1.1 Accessibility

Each mobile device is limited by accessibility constraints. Smart labels on packages have limited access because their transmitted signals are low in power. These labels can only be read within very short ranges. An RFID access is limited to ranges within line of sight. Also, RFID transmissions require hotspots close by due to low transmitted signal strength. (A hotspot is an access point for an RFID. An access point is a wireless system which acts as an interface for mobile systems, sensing systems, and embedded systems to connect to a mobile network, wireless LAN, or the Internet.)

2.6.1.2 Range

Signal strength is inversely proportional to the square of the distance from the transmission source. In addition to this, there is degradation of signal quality due to reflection, scattering, and diffraction (Section 1.1.2.3). Therefore, the access to a mobile device is limited to the range up to which the signal strength is such that it can be separated from the noise and up to which multi-path delays can be compensated for by digital signal processing techniques to restore signal quality.

2.6.1.3 Connectivity

There may be connectivity loss or intermittent connectivity in certain situations. The atmospheric conditions and changes in environment affect signal strength. Water has a refractive index of 1.33 compared to 1 in the case of air. The velocity of the waves reduces from 3×10^8 m/s to about 2.25×10^8 m/s and attenuation of UHF and near microwave region frequencies also takes place in water. As an example, in the event of heavy rain, there may be complete loss of connectivity.

2.6.1.4 Security

A mobile device has security constraints (Section 10.8). Unsolicited advertisements and unwanted messages may be drop-delivered to a device. Virus attacks on mobile devices can cause a software crash or even corrupt the hardware. Hackers may hack into a device and render it functionless or threaten integrity and security of the data stored on the device. Noise signals transmitted by an attacker can jam a mobile device. Repeated transmission of unwanted signals by an attacker can drain the resources of the device. Energy resources are depleted when computations are forced and authentication algorithms are run repeatedly.

2.6.1.5 Mobility

Non-availability of an access point (Section 1.5.2) or base station (Fig. 1.14) and other infrastructural issues restrict the mobility of a device. In remote areas, for example, there may be no base stations or Wi-Fi hotspots to provide connectivity and access to the internet for sensors, labels, automotive systems, RFID tags, and cell phones. The use of different standards in different regions may limit the operability of a device. For example, a GSM phone may not be operable in all continents hence hampering global roaming for the user. Also, some service providers may not be able to provide connectivity in all parts of the country or in other continents, etc.

2.6.2 Energy Constraints in Devices

All mobile devices have limited energy stored in their battery. Also, battery power is limited due to considerations such as size, weight, and bulk of mobile devices. The devices, therefore, need to be recharged after short periods of time. In this way energy availability also limits device mobility. Some devices such as smartcards, smart labels, remote sensors, and actuators do not even have a battery of their own. They derive their energy from the radiation received from a wireless source in vicinity. Such devices, therefore, require these sources to operate.

2.6.2.1 Mobile Computing Strategy

A processor circuit dissipates higher energy when its clock frequency is higher. Computational speed is higher at higher clock frequency. A device is, therefore, programmed so that only computations such as graphic image processing run at full processor speed. The clock frequency is reduced for the other computations to save power. The clock is activated only when a device interrupts or starts processing instructions.

Many innovative mobile computing strategies are adopted to mitigate the effects of energy constraints on mobile computing. Some of these are as follows:

- The Zigbee protocol (Sections 1.1.4 and 12.9) has a lesser stack size as compared to Bluetooth (Section 12.4 to 12.6). Use of a communication protocol that has less protocol stack overheads also reduces the energy requirement. This is due to lesser computational requirements.
- A host or hotspot may be seeking certain data from a device frequently. A program adapts itself so that the frequently required data is calculated and stored in a buffer from where it can be sent at slow clock frequencies on demand from the host. A program can also just transmit any changes in the data with respect to previous data.
- Communication scheduling strategies are adopted and the frequently required data is transmitted as per a schedule. This saves the host energy which would otherwise be required for sending commands and also saves the devices energy that would be dissipated in processing the commands.

2.6.2.2 Processor Design

Innovative circuits of mobile device processors have been designed and are continuously improved upon so that the same program instructions process with lesser energy dissipation per unit computational speed. Examples of energy efficient processors are ARM and TigerSharc.

2.6.2.3 Transceiver Design and Programming

Innovative transceiver circuits have been designed such that (a) signals of just sufficient strengths are transmitted to the receiver. Just sufficient strength means that the signal strength is low but clearly distinguishes noise and maintains message integrity; (b) control commands from the host are sent at lower signal frequencies and once the device is ready and gets powered up, the transceiver transmits the data for operation; (c) the signal strength reduces according to the inverse square law. Multi-hop routing helps in reducing the distance up to which a signal is required to travel.

2.6.3 Hardware Limitations

Besides energy limitations due to battery size as described in the previous section, all mobile devices face various other hardware constraints too.

2.6.3.1 Memory

There are constraints on memory availability. Most mobile devices do not support hard-disk drives and CD drives due to size limitations. Innovative forms of memory have been designed and are continuously improved upon. Internal flash drives and card slots for external memory are used. Memory stick is used to enhance the memory in the device. Some examples of large memory capacity in mobile devices are the Sony Network Walkman which has a memory equivalent to that of 11 CDs for storing music and a recent enhancement of the Apple iPod that offers a 30GB video memory.

2.6.3.2 Bandwidth

Availability of bandwidth is limited by the frequency spectrum that a regulator allots to a service provider. A service provider is not permitted to air signals at any random frequency and signal strength. A regulator regulates the frequencies and signal strengths permitted. The service must use the frequency spectrum allotted to it in an efficient manner. Multiplexing and coding techniques help in achieving an efficient transmission. The technology in use also limits the spectrum efficiency. For example, CDMA has a higher spectrum efficiency as compared to GSM.

Limited bandwidth may become an obstacle to seamless connectivity and quality of signals aired when a large number of mobile devices simultaneously demand network connectivity. For example, the bandwidth available in a GSM 900 uplink channel is 125 channels. There can be 8 users per channel (Table 1.2). However, when multiple users try to use the network simultaneously, for example, on events like New Year's Eve, etc., networks are jammed or unable to offer connectivity due to bandwidth constraints.

2.7 Automotive Systems

Car engines have seen an extensive use of computing and processing units since the late 1960s for improving automobile stability, transmission and braking processes, and driving comfort and ease. In the last decade or so, a revolution has been brought about in automotive systems. From sophisticated information-oriented technology, such as GPS navigation, reverse sensing, and night vision, to communication systems such as email access, voice control, traffic congestion information, smartcard security control, and collision avoidance sensors, the present day automobiles have got it all. This section gives an overview of a few of the computing systems which are embedded in automobiles. Figure 2.10 shows the mobile computing architecture in an automobile.

2.7.1 Speech Recognition System

An application programmer can program a speech recognition system (SRS) in an automobile. The automobile can start by the driver's commands after recognizing their voice through the SRS. Application software can be programmed such that the driver can command the automobile to halt, maintain the current speed, or stay under a given speed limit. The SRS uses a digital signal processor.

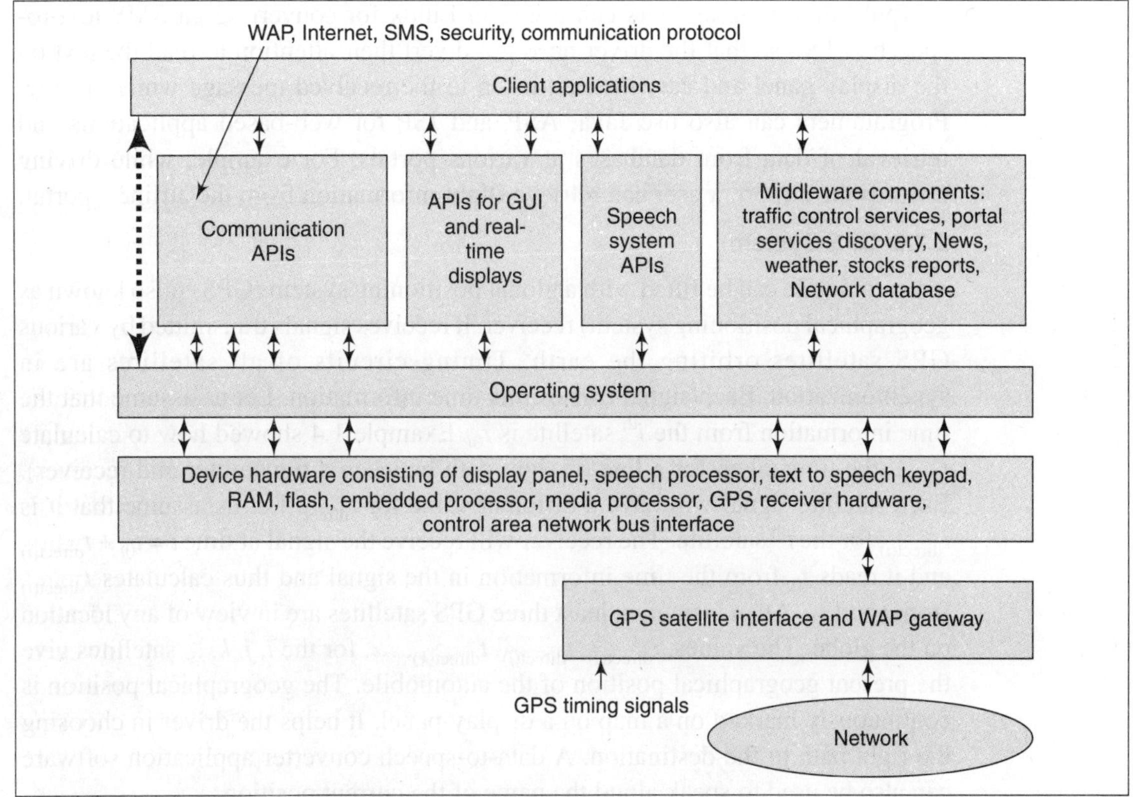

Fig. 2.10 Mobile computing architecture in an automobile

2.7.2 Messaging System

A WAP (wireless application protocol) (Section 12.2) device in an automobile enables it to connect to the Internet. A service provider can transmit, in real time, the news, weather data, and stock reports. Road maps can also be accessed using WAP and can be stored in the memory of a computer embedded into the automobile, so that they can be retrieved when the need arises.

A traffic control service sends traffic reports. The automobile owner can subscribe to a traffic control service provider which provides SMS messages about traffic slowdowns and blockages at various points in the city. The messages are then converted to speech using a text-to-speech (TTS) converter software and can be heard by the driver. It enables the driver to select roads that will provide a faster, hurdle-free passage.

An anti-collision system can warn the driver if the automobile gets too close to another. It can also sense objects which are not visible to the driver using a laser, infrared, or RADAR system. Collision avoidance systems can also take control of the vehicle to avoid colliding with other objects.

Application programmers can use C in Linux for converting an SMS text-to-speech (TTS), so that the driver need not divert their attention to read the text on the display panel and can, instead, listen to the received message while driving. Programmers can also use Java, ASP, and JSP for web-based applications and retrieval of data from databases at various portals. For example, while driving towards the airport, a user can retrieve flight information from the airline's portal.

2.7.3 GPS System

An automobile can be fitted with a global positioning system (GPS) (also known as geographical positioning system) receiver. It receives signals transmitted by various GPS satellites orbiting the earth. Timing circuits of all satellites are in synchronization. Each signal carries this time information. Let us assume that the time information from the ith satellite is t_{0i}. Example 1.4 showed how to calculate t_{direct} (the time taken for a line-of-sight path between a transmitter and receiver). Each satellite signal will have a different value for t_{direct}. Let us assume that it is $t_{\text{direct}(i)}$ for the ith satellite. The receiver will receive the signal at time $t = t_{0i} + t_{\text{direct}(i)}$ and it reads t_{0i} from the time information in the signal and thus calculates $t_{\text{direct}(i)}$ from t and t_{0i}. At an instant, at least three GPS satellites are in view of any location on the globe. The values, $t_{\text{direct}(i)}$, $t_{\text{direct}(j)}$, $t_{\text{direct}(k)}$, ..., for the i, j, k, ... satellites give the present geographical position of the automobile. The geographical position is continuously marked on a map on a display-panel. It helps the driver in choosing the right path to the destination. A data-to-speech converter application software can also be used to speak aloud the name of the current position.

Application programmers can use GTK (graphic tool kit) language or C in Linux for drawing, in real time, the road map on the display panel with the automobile's position suitably marked on the map on a real-time basis. The real-time application changes the map on the screen in case the automobile moves into another zone and also continuously shifts the marked position as the automobile moves.

2.7.4 Automobile Start and Malfunction Logins

A smartcard or smart token can be used in place of a key to start the automobile. The card inserted into the host not only starts the car but also logs the data for the malfunctions recorded during driving. At the service workshop, a card reader attached to a PC reads the card and retrieves the logged data as well as the service history details. The workshop can render a more efficient service using this information. The service provider's PC can then write the details of the service provided onto the card memory for future reference. JavaCard can be used for developing the start and malfunction logging applications.

2.7.5 Sensor and Actuator Programming

An automobile has a number of sensors and actuators. For example, pressure sensors communicate, to the display panel, warnings about tyre pressures. Sensors and actuators connect the embedded systems inside an automobile. The systems connect to the CAN bus.

2.7.6 Entertainment Systems

An automobile can be fitted with a number of entertainment systems, for example, FM radio, media players to play Wave (WAV), RealAudio (RA), and MPEG-1 audio layer 3 (MP3) files. Application programmers can develop programs for downloading music from the Internet in formats such as WAV, RA, and MP3 using a WAP gateway. A USB port can be used to download files from another system. A Bluetooth device can be used to download data from PDAs, Smartphones, and pocket PCs.

2.7.7 Real-time Applications Programming

A programmer can use the Windows CE OS (Section 14.2) for running real-time applications on the PC fitted into an automobile. Applications can be coded using any of the programming languages such as Win32 API, Visual C, and Visual Basic. The OS provides functions for the multiple threads, networking, and communication protocol APIs.

Real-time applications for the Java platform can also be developed using OSEK [*offene Systeme und deren Schnittstellen für die Elektronik in Kraftfahrzeugen* (German for 'open systems and their interfaces for the electronics in motor vehicles')]. OSEK is a small operating system used for microcontrollers in engine control units of automobiles.

Keywords

Access point: A device that provides wireless LAN or Internet connectivity or wireless connectivity through mobile service provider network to mobile systems and embedded systems. An access point may also act as an interface between wireless and wired networks.

ActiveSync: A synchronization software developed by Microsoft for synchronization of data on mobile devices used primarily in Windows CE devices. A USB or serial port, infrared port, or Ethernet LAN may be used for this purpose. It is used for resolving the conflicts in different versions of files during data exchange and using software for multiple service providers and service managers.

Actuator: A device which takes action after it receives signals from a controller or central computer and accordingly activates a physical device, appliance, or system. An actuator may or may not have a computer embedded in it.

Base station: A transceiver which connects, on one hand, to a number of mobile devices or access points wirelessly and, on the other hand, to mobiles and other networks by wire or fibre.

GPS (global positioning system or geographical positioning system): A satellite navigation system used for synchronizing multiple computing systems. GPS satellites are installed in orbits around the earth at multiple locations and devices that are fitted with GPS receivers which can catch the satellite signals and read time stamps from the transmitted signals. Using delays of the stamped time in signals received from the satellites, the receiver computes the geographical location.

HotSync: A PalmOS-based synchronization software in which a cradle connects the mobile device to a PC via a serial port or an infrared port. It is used for resolving conflicts in different versions of the files during data exchange.

iPod: A portable media playing device from Apple. The iPod can be used to playback files podcasted through the Internet. It can also connect to the Internet or a computer network for downloading multimedia files such as audio programs or music videos.

Label: A small, flat strip stuck over or concealed in a package used for identifying its contents. Smart labels can wirelessly interact with access points and are powered by wireless radiation from a local computer or access point. A label also forms a LAN system with a nearby cluster of labels.

MMS (multimedia message service): A multimedia messaging service for transmission of digital images, video and audio clips, and animations.

Pocket Computer: A handheld computer which stores distributed files, performs application development, includes programming tools, and offers mobile services.

PIM (personal information manager): A software which provides interfaces for calendar, contacts, Microsoft Outlook 98, 2000, 2002, 2003, Microsoft Outlook Express/Windows Address Book, Intellisync wireless email, Lotus Notes (5.x and 6.x), and Lotus Organizer so that the user can efficiently organize personal information.

RFID: A radio frequency-based identification system. RFID tags look like small size strips over a package or concealed in the package, used to identify the contents and interact with an access point wirelessly and powered by wireless radiation from an access point.

Sensor: A device which senses the physical environment, for example, temperature, pressure, light, metal, smoke, and proximity to an object. It connects to a computer or controller or may embed a computer for wireless communication.

Setup Box: A sophisticated computer-based device which has data, media, and network processing capabilities. It interconnects the home TV and the broadcasting service network and uses Java as a programming language.

Smartcard: A card-like mobile device which has an embedded chip with a circuit consisting of a computer, memory, and transceiver. The chip is sandwiched between the inner layers of the card.

Smart Token: A device that is meant for its authentication before an action initiates and which encapsulates a chip that embeds a processor and memory. Its size is small (of the order of shirt button or pen nib).

SMS (short message service): A service for sending text messages of up to 160 characters.

Transceiver (transmitter–receiver): A circuit for transmitting and receiving signals.

Review Questions

1. Describe the various features of a multimedia phone.
2. How is a handheld computer different from a PC? State the main points of difference between a handheld computer and a Smartphone.
3. Draw the mobile computing layer architecture of an iPod. Take Fig. 1.13 as an example.
4. Describe the functions and features of a latest PalmOS device. How do these differ from those of Windows CE devices?
5. Describe the functioning of a smartcard. Why is secured hardware and software required for a smartcard?
6. Differentiate between the functions of labels, tags, and cards. How do smartcards, smart labels, smart tokens, and RFID tags work, if they have no internal battery?
7. How are smart labels networked?
8. What are the sensors used in the pervasive computing Smartphone devices?
9. Explain the working of a sensor–actuator pair by giving an example. List the sensors and actuators used in orchestra-playing robots.
10. What are the various constraints of working with mobile devices?

11. Discuss the solutions for the energy constraint problem in mobile devices.
12. What are the security constraints in mobile devices?
13. Show the subunits in a setup box.
14. Draw the mobile computing layer architecture of a car with a GPS receiver, WAP, automatic parking lights control, smartcard-based start, and speech recognition. Take Fig. 2.10 as an example.

Objective Type Questions

Pick the correct or most appropriate statement among the choices given:

1. A mobile device is placed on a cradle. It can synchronize with a PC using
   (a) HotSync when using Windows CE.
   (b) ActiveSync when using Windows CE.
   (c) Bluetooth or ZigBee.
   (d) Serial port and infrared port.
2. An iPod device
   (a) can be used for downloading music and video clips.
   (b) can be used as a handheld computer.
   (c) can be used for downloading faculty lectures, music, photos, and video clips and for viewing slide shows with music.
   (d) cannot be used for downloading faculty lectures.
3. A cell has
   (a) one base station which interconnects to mobile devices.
   (b) one base station which interconnects to mobile devices and performs handover to the neighbouring base station when the device moves and uses a frequency band which is distinct from the neighbouring cell.
   (c) one base station and one access point which connects to mobile devices.
   (d) one base station connects to mobile devices and performs handover to the neighbouring base station when the device moves and uses a frequency band which is the same as the neighbouring cell to ensure mobility of the device in another cell.
4. Handheld computers
   (a) cannot support large memories due to energy constraints.
   (b) support large memories with memory sticks.
   (c) cannot support large memories due to use of specialized OS—Window CE or PalmOS.
   (d) support large memories including internal flash memories and extension slots for flash cards and memory sticks.
5. A smartcard uses
   (a) three keys—fabrication, personalization, and utilization.
   (b) a single key for fabrication, personalization, and utilization.
   (c) host supplied keys.
   (d) two keys—fabrication and personalization.
6. A label uses
   (a) a LAN-like network with a collision avoidance protocol and also uses a nearby label reader.
   (b) does not use a LAN-like network and connects directly to access point.
   (c) a nearby label reader.
   (d) secret codes.
7. A smart token is
   (a) for identification when on the move.

(d) used when requesting permission to access a mobile device.
(c) for identification and initiation of appropriate action.
(d) used when requesting permission to access a computer.

8. A smart actuator acts as per
(a) controller, mobile device, or central computer commands.
(b) controller, mobile device, or central computer commands issued according to the data from the sensors.
(c) its own embedded processor commands.
(d) according to the data from the sensors.

9. A setup box is
(a) a computer connected to the TV used to set the channels.
(b) a computer connected to the TV used to set the channels and play video games.
(c) a conditional access system for cable TV, games, Internet, telephone line, and dish.
(d) a conditional access system for cable TV, video games, and dish-access system.

10. Energy is conserved in mobile devices by
(a) running the processor at low frequency and transmitting signals at low frequencies.
(d) restricting the number of applications, running the processor at low clock speed, and using appropriate protocol.
(c) restricting the number of applications and running the processor at low clock speed, using an appropriate radio interface, and an appropriate communication and data transfer strategy.
(d) using energy-efficient processors, optimizing the code used for running an application, running the processor at full clock speed only when processing graphics like computational intensive software, using an appropriate radio interface, and using an appropriate communication and data transfer strategy.

11. Security issues in a mobile device are encountered due to
(a) hacking of data and virus attacks.
(b) eavesdropping, virus attacks, hacking, jamming, and forcefully exhausting the energy resources.
(c) jamming of the incoming signals.
(d) eavesdropping, hacking, and virus attacks.

12. (i) OSEK real-time operating system (ii) GPS (iii) Display panel of sub notebook size (iv) CAN bus (v) smartcard (vi) WAP (vii) Media player.
Which of the following is true in an automobile computing system?
(a) All of the above are used.
(b) (iv) and (v) are not used.
(c) (ii), (iv), (v), and (vi) are not used.
(d) (i) and (v) are not used.

Chapter 3

GSM and Similar Architectures

GSM (global system for mobile communications) is a 2G standard for cellular telecommunication. This chapter gives a description of GSM and similar systems such as GPRS, HSCSD, and DECT. This chapter will discuss the following concepts in detail.

- Services available in a GSM communication system
- Architecture of GSM systems and subsystems (such as radio, network, and switching services) used for operation and maintenance of a GSM network
- Radio interface and higher layer protocols in GSM systems
- Processes used for mobility management—localization, calling, and handover
- Data services provided by GPRS, HSCSD, and DECT systems

3.1 GSM—Services and System Architecture

The global system for mobile communications (GSM) is one of the most popular mobile communication standards. GSM communication uses cellular networks. The GSM standard operates in the frequency ranges of 900, 1800, and 1900 MHz. Tri-band (operable in GSM 900/1800/1900) phones enable easy international roaming in GSM networks. GSM is a second generation (2G) communication standard. Chapter 1 discussed multiplexing techniques used in (Section 1.1.2.5) and the various standards that fall under the GSM communication system (Table 1.2). This section describes GSM services.

3.1.1 Services

GSM provides integrated services for voice and data. Figure 3.1 shows the three types of services that integrate into a GSM system. The three kinds of services

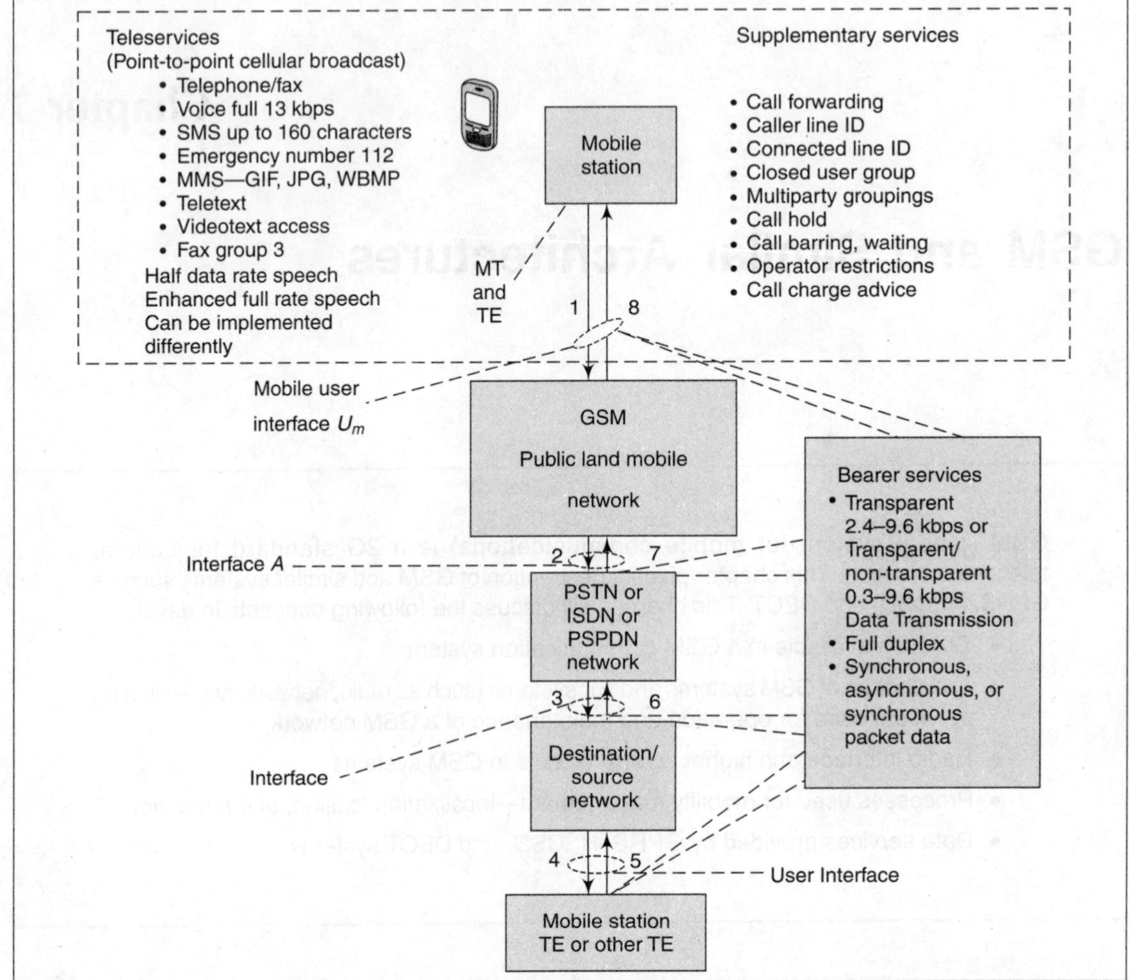

Fig. 3.1 Integration of teleservices, bearer services, and supplementary services in a GSM system

delivered by a GSM system are teleservices, supplementary services, and bearer services. The figure also shows the connection of a caller mobile station to a destination (terminal).

3.1.1.1 Connection

A mobile station, MS, has a mobile terminal, MT. A mobile terminal is a sub-unit which encodes voice and data signals from a terminal for transmission. An MT also decodes received signals back into voice or data so that the user at the terminal can comprehend them. A mobile terminal acts as an interface between a communications network (e.g., a GSM public land mobile network) and a terminal, TE. The TE is used by a caller to connect and talk (communicate). The MT is responsible for mobile communication and TE is the source or destination of the service.

A connection is established between two TEs—the source and the destination. The destination TE may or may not belong to a GSM network. It depends on the source–destination network which may be a GSM, PSTN (public-switched telephone network), ISDN (integrated services digital network), PSPDN (public-switched public data network), or any other network carrying the data to the end point TE. Figure 3.1 shows a caller TE transmitting through interface 1 to a GSM public land mobile network, through interface 2 to a PSTN network, through interface 3 to a source–destination network, and through interface 4 to a terminal or mobile station TE. (In place of the PSTN network shown in the figure, there may be an ISDN or PSPDN network.) The connected TE (shown at the bottom of Fig. 3.1), then, communicates back by transmitting through interfaces 5, 6, 7, and 8. There are four sets of interfaces (1,8), (2,7), (3,6), and (4,5). There is a transceiver in each set. The symbol U_m (user mobile interface) conventionally denotes the interface (1,8). *A* (in Fig. 3.1) denotes a mobile network interface (2,7) to a PSTN or other wired network. Each of the four transceivers can transmit as well as receive in full-duplex mode (shown by the up and down arrows in Fig. 3.1). Full-duplex mode means simultaneous two-way transmission. The MT interface can also be a half-duplex transmission (Section 3.2.2). Half-duplex means that two-way transmission is possible but not simultaneously. Some examples of establishing connection in a GSM network are as follows:

- A GSM Smartphone connects to a PSTN phone (old landline phone). In this case, a mobile station at the caller end interfaces the GSM landline mobile network through a user interface U_m. The destination for the call and thus source for the reply is the landline phone TE which interfaces through a source–destination network (PSTN in this case).
- A GSM Smartphone connects to another GSM phone. In this case, a mobile station TE at the caller end interfaces to a GSM public landline or mobile network (Fig. 3.1). The destination for the call (and source for the reply) is another mobile station TE which interfaces the source–destination network (a GSM public land mobile network in this case).
- A landline phone connects to a GSM phone. The source TE at the caller end interfaces to the PSTN network. The destination for the call (and source for the reply) is a mobile station TE which interfaces to a GSM public land network in this case.

3.1.1.2 Teleservices

Teleservices are services offered by a mobile service network to a caller (TE). Some of the teleservices provided by GSM service providers are telephonic-voice at full data rate (13.4 kbps), fax, SMS, emergency number (international SOS number 112) for emergency calls, and MMS [supporting GIF, JPG, WBMP, teletext, and videotext access (GIF, JPG, and WBMP are formats of files that store pictures)].

Teleservices are point-to-point which means that from a TE to another TE. A point-to-point service is implemented using cellular communication (Section 1.5.1). Additional teleservices (introduced in phase 2 of GSM) are half data-rate speech or enhanced full-rate speech services, and these may or may not be rendered by cellular and point-to-point access systems. For example, a GSM Smartphone, which connects to a GSM public land mobile network, offers a number of teleservices including phone, voice-data (e.g., recorded message played on auto-answer of incoming calls), SMS, and MMS to another GSM or PSTN network.

3.1.1.3 Supplementary Services

In addition to teleservices, a mobile service network also provides supplementary services to the caller and destination TEs. Some examples of supplementary services are caller line forwarding (redirection) and caller line identification at the connected terminal (TE). Additional services (added in phase 2 of GSM) are connected line identification to the caller, closed user group formation, multiparty groupings (e.g., in an enterprise), call holding, call waiting, and barring calls from specified numbers or groups. Phase 2 supplementary services also include restricted provisioning of certain services to the users (e.g., Internet and email access are granted on special requests from users) and providing information regarding call charges, remaining phone account balance, etc. Some examples of supplementary services are listed below:

- The caller's number flashes on the display panel of the GSM Smartphone, when it receives a call.
- A GSM multiphone with enterprise solutions subscribes to supplementary services such as caller ID, connection ID, call forwarding, multiparty groupings, and call barring.

3.1.1.4 Bearer Services

Bearer services are responsible for transmission of data (voice signals are also transmitted as data) between two user network interfaces [(1,8) and (4,5) in Fig. 3.1] using the intermediate interfaces [(2,7) and (3,6) in Fig. 3.1] in a mobile network. Bearer means a set of data which is transmitted from or received by a TE, i.e., the voice-data or data set that has been formatted in certain specified formats. This data is then transmitted at certain standardized rates through the interfaces. (Voice-data is data that is obtained after digitizing, coding, encoding, appending error detection and correction bits, and encrypting of a voice signal.)

Figure 3.1 shows bearer services and the interfaces used by bearer services for data transmission and reception. Each TE has a user interface. The interface (1,8) (Fig. 3.1) of a mobile station connects the MT to a GSM public land mobile network. The interface (4,5) (Fig. 3.1) of a PSTN phone connects to a PSTN network. An intermediate PSTN network acts as an interface for a GSM public land mobile network. (In place of PSTN, there may be ISDN, PSPDN, or some other network.)

A bearer service (service for transmission and reception through the interfaces) is either (a) transparent and uses data rates of 2.4 kbps, 4.8 kbps, or 9.6 kbps or (b) non-transparent and uses lower data rates (300 bps to 9.6 kbps). Bearer services are also classified as using synchronous data transfer, asynchronous data transfer, or synchronous data packet transfer.

3.1.1.5 Transparent Data Transfer

Transparent data layer is said to be transparent since the interface for the service uses only physical layer protocol. Physical layer is the layer which transmits or receives data after formatting or multiplexing using a wired (wire or fibre) or wireless (radio or microwave) medium. The physical layer protocol in a GSM bearer service also provides for FEC (forward error correction). FEC entails insertion of redundant bits along with the data to be transmitted. This redundant data allows the receiver to detect and correct errors. FEC requires higher bandwidths but is advantageous in situations where retransmission is not convenient. The FEC transmission reduces data rate, but helps in broadcasting without handshaking. (Handshaking refers to interchange of acknowledgements between two networks or systems once the connection is established between them.) It also enables broadcast to multiple destinations from a single source. If m redundant bits are appended in a data stream of n bits, then the total numbers of data bits transmitted from the sender's end is $(n + m)$ bits. At the receiving end, an algorithm is employed to detect and correct transmission errors (error means 0 received as 1 or 1 received as 0). This algorithm extracts the original n-bit streams from the received $(n + m)$-bit sequences. Therefore, for every $(n + m)$ bits sent by the sender, the receiver receives only n bits of actual data. This means that if the transmission channel offers a data rate r, then the actual data transmission rate with FEC is $r \times n/(n + m)$.

Example 3.1 Data with FEC is transmitted through a channel at the rate of 9.6 kbps. If redundant data inserted is three times more than the original data, calculate the rate at which the receiver actually receives the relevant data.

Solution: Actual data rate $= \dfrac{r \times n}{(n+m)}$

$$r = 9.6 \text{ kbps}, m = 3n$$

Therefore,

$$\text{Actual data rate} = \left[\frac{9.6 \times n}{(n+3n)}\right] \text{kbps} = \left[\frac{9.6 \times n}{(4n)}\right] \text{kbps} = 9.6/4 \text{ kbps} = 2.4 \text{ kbps}$$

3.1.1.6 Non-transparent Data Transfer

Non-transparent means that the service interface uses physical layer, data link layer, and flow control layer protocols. Data link layer is the layer which frames the data and appends additional bits to ensure very small error rate. Framing refers to combining and appending additional bits and header. The data link layer performs other functions too (Section 3.3). Flow control layer controls the flow of data by selecting or rejecting erroneous data transmitted and by re-transmitting erroneous data. The protocols for data link and flow control layers provide for (a) error detection and correction and (b) selecting, rejecting, and re-transmitting the data, respectively.

When data transmits at 9.6 kbps the data error rates are high. This is because when non-transparent data is transmitted at 9.6 kbps, there is no retransmission and erroneous data just gets rejected. The data error rate becomes negligibly small at slow data rates (300 bps). This is because when non-transparent data transmits at 300 bps, then the erroneous data is corrected or gets retransmitted at data link and flow control layers. A special error correction facility called RLP (radio link protocol), used in GSM networks, is an example of a *non-transparent* communication protocol. RLP results in a more robust transmission with very small BER (bit error rate).

3.1.1.7 Synchronous, Asynchronous, and Synchronous Packet Data Transmissions

Synchronous data transmission means that data is transmitted from a transceiver at a fixed rate, as a result constant phase differences (and thus time intervals) are maintained between data bursts or frames (Section 3.2). The receiver must synchronize the clock rate according to the incoming data bits. The receiver must also synchronize data bits coming in from multiple paths or stations and compensate for the varied delays in received signals. Handshaking is not required in synchronous transmission of data. Synchronous data transmission is fast because there is no waiting period during data transfer. Some examples of synchronous data transfer in a GSM system are as follows:

- Voice is converted into bits after coding in a GSM system and the bits are transferred at data rates of 13 kbps as synchronous data. There are no in-between acknowledgements or waiting periods in this faithful transmission of bits.
- An SMS is transmitted through a GSM channel as synchronous data. There are no in-between acknowledgements and any transmission errors are corrected using FEC.

Asynchronous data transmission means that data is transmitted by the transceiver at variable rates, as a result constant time intervals are not maintained between consecutive bursts or frames (Section 3.2). There is usually handshaking or acknowledgement of data in asynchronous data transfer. However, even if there is no acknowledgement, data flow maintained by using the FEC and buffers can still be asynchronous. Use of buffers causes variable delays in reception. Some examples of acknowledgement messages are 'receiver ready', 'receiver not ready', unnumbered acknowledgement of acceptance of data at the receiver, rejects, set asynchronous balance mode, or disconnect. As an example, program files containing middleware for mobile devices have to be transmitted by the mobile service while maintaining full data integrity. In such cases, in-between acknowledgements of faithful transmission of bits and reporting of errors during the transmission are important. Therefore, there is non-transparent data flow. An acknowledgement is sent by the receiver for each data set to the effect that the data set received is identical to the one transmitted. Some time is, therefore, spent in implementing appropriate algorithms for data set integrity checks and acknowledgements. This results in asynchronous data transmission.

Synchronous packet transmission takes place after formation of packets (Section 1.1.2.6). Different packets are transmitted through different interfaces, routes, channels, or time-slots to reach a common destination. At the destination, various packets are arranged in their original sequence. A sequence number transmitted along with each packet helps in sequential arrangement of packets at the receiver. Each packet flow is transmitted as synchronous data. There is no handshaking or acknowledgement of data during the flow of packets.

Example 3.2 *N* bits of data are to be transmitted as packet-switched data. The packets can have a maximum of *n* bits each. The data transmission rate is n/T and the data to be transmitted is formatted into four packets *A*, *B*, *C*, and *D*. If three different routes are available for transmission, calculate the time taken to complete the synchronous packet transmission. What will be the time taken if $N = 4n$? How much time will it take to transmit the same data through one single path?

Solution: A data rate of n/T implies that it takes time *T* to transmit *n* bits.

The packets *A*, *B*, and *C* will be simultaneously transmitted along the three routes available. Since each packet is of *n* bits, the time taken for the simultaneous transmission is *T*. Now, the last packet *D* will have $(N - 3n)$ bits.

$$\text{Time taken to transmit } D = \frac{(N-3n)}{(n/T)}$$

$$\text{Therefore, total time taken to transmit } N \text{ bits of given data} = T + \left[\frac{(N-3n)}{(n/T)}\right]$$

If $N = 4n$, then

$$\text{Time taken to transmit } D = \frac{(4n-3n)}{(n/T)} = \frac{n}{(n/T)} = T$$

The total time taken to transmit *A*, *B*, *C*, and $D = T + T = 2T$

If this data was transmitted using a single path, then time taken in transmission is

$$\frac{N}{(n/T)} = \frac{4n}{(n/T)} = 4T$$

3.1.2 GSM System Architecture

Figures 1.9, 1.14, and 3.1 show mobile communication using base stations in cellular networks and how a mobile station, MS, communicates with a GSM public land mobile network (PLMN) which, in turn, may connect to a PSTN network. The PSTN connects to a source–destination network which acts as an interface for the destination terminal, TE. This basic outline of a GSM network, however, does not provide a detailed picture of the GSM network architecture. Figure 3.2 shows the GSM network architecture.

The network is divided into three subsystems namely, radio subsystem (RSS), network subsystem (NSS), and operation subsystem (OSS). The RSS consists of a number of base station controllers (BSC) and each BSC connects to a number of base transceiver stations (BTS) which, in turn, provide radio interfaces for mobile devices. The NSS consists of a number of mobile services switching centres (MSC). Each MSC of the NSS interfaces to a number of BSCs in the RSS. There are also home location registers (HLR) and visitor location registers (VLR). The OSS consists

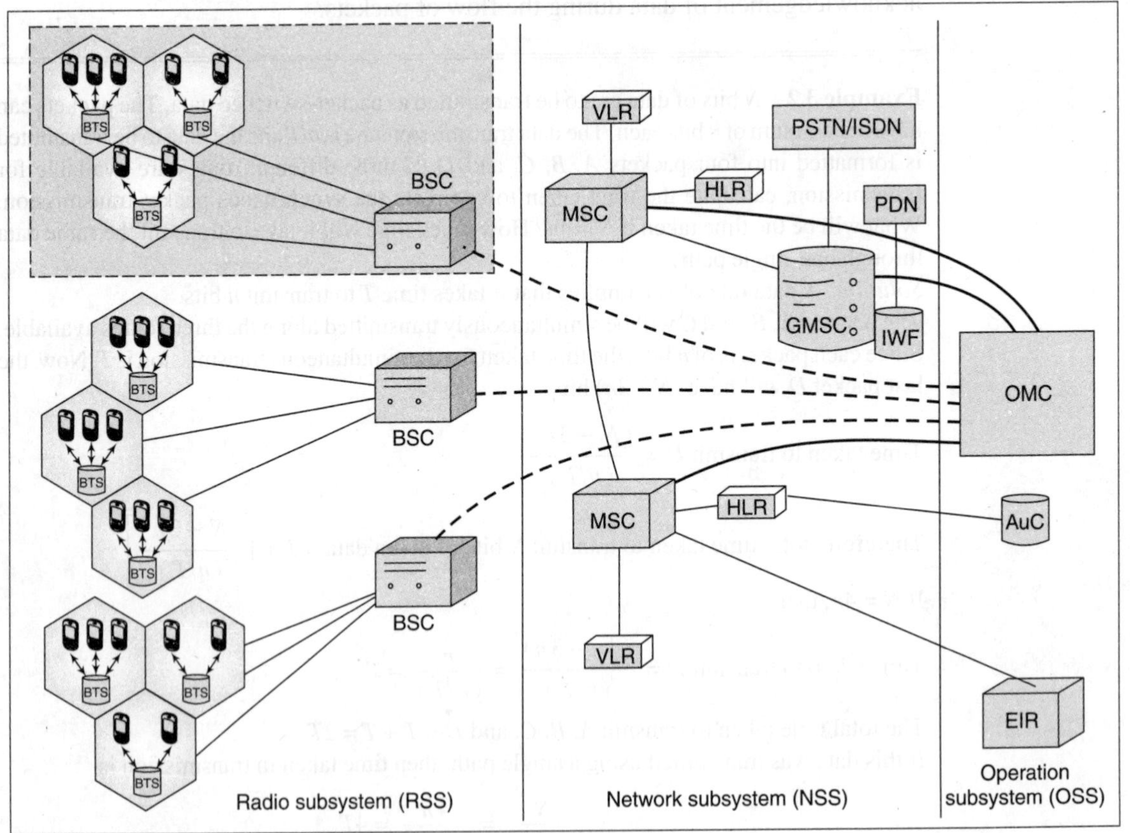

Fig. 3.2 GSM network architecture

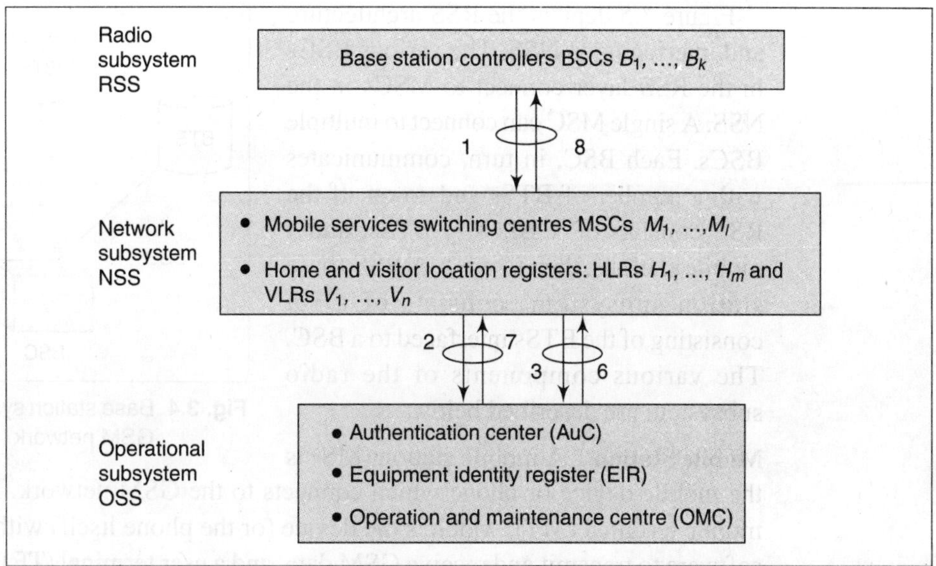

Fig. 3.3 Interfacing between different subsystems of the GSM architecture

of the authentication centre (AuC), equipment identity register (EIR), and operation and maintenance centre (OMC). Figure 3.3 shows the interfacing between the three subsystems in a GSM network. If the number of BSCs in the RSS is k and there are l MSCs in the NSS, then $k > l$, and each MSC connects to a number of BSCs. The detailed architecture of GSM network subsystems and the functions of various components in a GSM network are discussed in the following subsections. An example of the operation of various subsystems and their intercommunication is detailed below.

When a mobile station MS_x communicates to another mobile station MS_y, a switching centre MSC_i establishes (switches) a connection (channel) between (i) MS_x interfaced to the BTS_p, then to the BSC_q, then to MSC_r, and (ii) MS_y interfaced to the BTS_u, BSC_v, and MSC_w. The RSS and NSS provide a radio subsystem for the communication. MSCs must have location registries to enable the NSS to discover a path (route or channel) between MS_x and MS_y. The OSS facilitates the operations of MSCs.

3.1.2.1 Radio Subsystem

The radio subsystem (RSS) comprises all radio-specific entities in the GSM network. The RSS performs the basic functions of connecting the mobile station (mobile device or phone) to the network. The RSS consists of a number of base station controllers (BSC). Each BSC connects to a number of base transceiver systems (BTS) and each BTS connects to a number of mobile stations (MS). A BSC along with the BTSs connected to it together form a base station system (BSS) (Fig. 3.4).

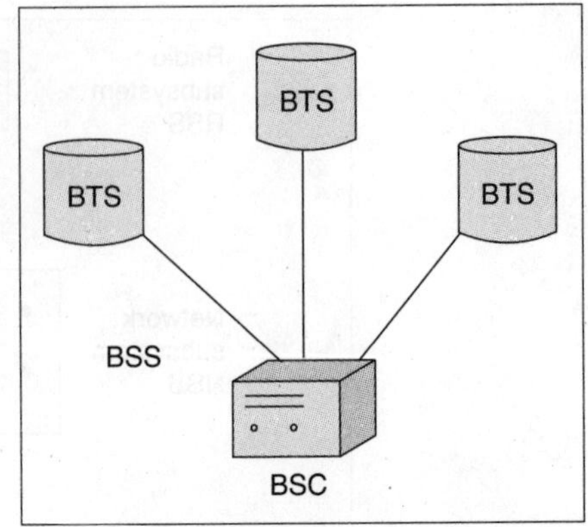

Fig. 3.4 Base station system in a cellular GSM network

Figure 3.5 depicts the RSS architecture and interfacing to NSS. The various BSCs in the RSS layer connect to MSCs in the NSS. A single MSC can connect to multiple BSCs. Each BSC, in turn, communicates with a number of BTSs and so on. If the RSS consists of k BSCs, j BTSs, and i mobile stations, then $i > j > k$. A BSS (base station subsystem) consists of a set consisting of the BTSs interfaced to a BSC. The various components of the radio subsystem are described below.

Mobile Station A mobile station (MS) is the mobile device or phone which connects to the GSM network. It consists of a mobile terminal (MT), which is the device (or the phone itself) with hardware and software to transmit and receive GSM data, and a user terminal (TE) through which the user receives and sends the data (this is the radio transmission system used in

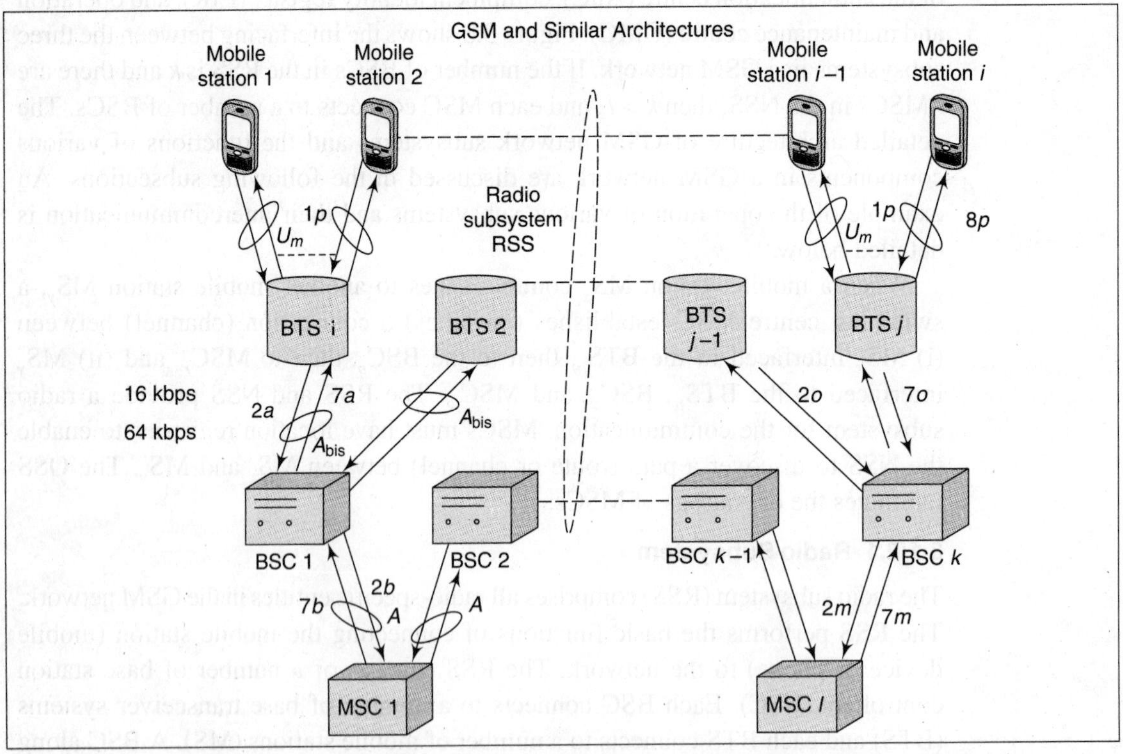

Fig. 3.5 Connection interfaces in the RSS subsystem between (a) BTS and a number of MSs, (b) BSC and a number of BTSs, and (c) MSC (in the NSS layer) to a number of BSCs ($i > j > k > l$)

mobile phones). The MT transmits through the interface U_m (Fig. 3.6) at a power of 1–2 W. Each MS has a subscriber identity module (SIM). SIM is a card inserted into the MS. It is provided by the GSM service provider. The SIM uniquely identifies the user to the service and enables the MS to connect to the GSM network. Some important functions of the SIM are as follows:

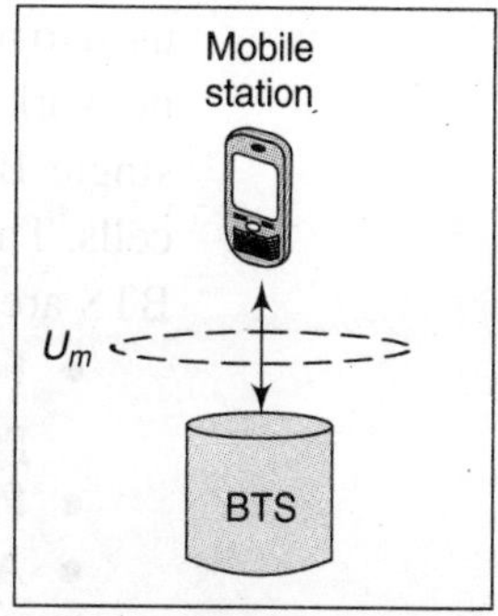

Fig. 3.6 Mobile station to BTS interface in a GSM cell

- When the MS connects to the GSM subsystems, the SIM stores a temporary mobile (dynamic) cipher key for encryption, temporary mobile subscriber identity (TMSI), and location area identification (LAI).
- SIM contains the following information which does not change when the MS moves into another location—(i) international mobile subscriber identity (IMSI), (ii) card serial number and type.
- The SIM contains a PIN (personal identification number). Using the PIN, the MS is unlocked when it seeks connection to another MS. The user can use the PIN to lock or unlock the MS.
- It stores the PUK (PIN unblocking key) which enables the subscriber to unlock the SIM if it is accidentally locked due to some reason.
- It stores a 128-bit authentication key provided by the service provider. The MS is authenticated by a switching centre through an algorithm using this key and a 128-bit random number dynamically sent by authentication centre. If the MS is not authenticated, the service to that number is blocked.
- The SIM also stores the international mobile subscriber identity (IMSI). The IMSI is a unique 15 digit number allocated to each mobile user. The IMSI has three parts—a three digit mobile country code (MCC), a mobile network code (MNC) consisting of two digits, and the mobile subscriber identity number (MSIN) with up to 10 digits. Mobile service providers all over the globe use an identical coding scheme for the IMSI. IMSI helps service providers in identifying and locating an MS. It helps the MS in obtaining the cipher key, TMSI, and LAI from the mobile service provider during connection setup. A TMSI is used to identify an MS during a connection for protecting the user ID from hackers or eavesdroppers (Section 3.7).

Base Transceiver Station A base transceiver station (BTS) connects to a number of mobile stations (MSs) (Fig. 3.5). The connection between the BTS and each MS is established through the user interface U_m [(1,8) in Figs 3.1 and 3.5]. U_m is the ISDN U interface for mobile. The BTS to MS connection through U_m in a GSM cellular network is shown in Fig. 3.6. A BTS is also connected to a BSC through the A_{bis} interface (Fig. 3.7). An A_{bis} transceiver transmits and receives data with four multiplexed channels of 16 kbps or with a 64 kbps channel. Usually a BTS is used to manage one cell in the GSM cellular network, but using a sectorized antenna, a single BTS can be used to manage many cells. The main functions performed by the BTS are as follows:

- Formation of cells using appropriately directed antennae
- Processing of signals
- Amplification of signals to acceptable strength so that they can be transmitted without loss of data
- Channel coding and decoding (e.g., coding voice into bits so that it can be transmitted at 13 kbps and decoding received coded signals back to voice)
- Frequency hopping so that multiple channels for various mobile stations can operate simultaneously using different channel band frequencies (Table 1.2)
- Encryption and decryption of data
- Adapting to the rate of data [e.g., in synchronous data transmission (Section 3.1.1.5), the receiver clock of the transceiver at one end of an interface adapts itself according to transmitter clock of the transceiver at the other end]
- Paging (Section 3.2.5)

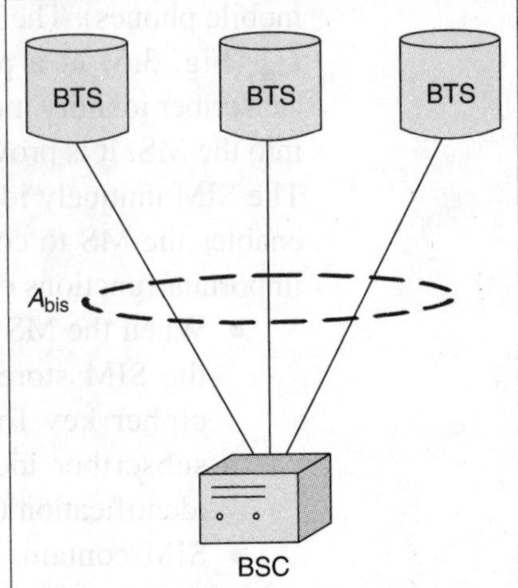

Fig. 3.7 BTS-to-BSC interface in a GSM network

Base Station Controller A base station controller (BSC) manages a number of BTSs. It uses the A_{bis} interface to connect to BTSs (Fig. 3.7). BSCs reserve radio frequencies for communication and manage handovers between BTSs. A BSC, along with the BTSs connected to it and the mobile stations managed through it, forms a base station system (BSS) shown in Fig. 3.7. The BSC is also connected to an MSC in the networking and switching layer using an interface *A* (Fig. 3.8). Some important functions performed by the BSC are as follows:

- Processing of signals
- Controlling signals to the connected BTSs and control of handover (Section 3.6) of signals from one BTS to another within a BSS (Fig. 3.7)
- Control and handover (Section 3.6) of the signals from BSC to MSC (Fig. 3.8)
- Mapping the signals of a channel (for example, a BSC at a given instant receives signals from a BTS at 16 kbps through A_{bis} and interfaces them to an MSC at 16 kbps. Alternatively, it may have to interface to a PSTN switching centre at 64 kbps through a fixed line network. Mapping, in this case, entails assigning a 16 kbps channel for 64 kbps signals and vice versa.)

- Reserving radio frequencies.
- Frequency hopping (e.g., multiple BTSs operate simultaneously by using different frequencies at a given instant.)
- Traffic control by continuous measurement of the frequency channel spectrum being used at a given instant
- Authentication, encryption, and decryption of data
- Updating location registry for the MSs
- Paging (Section 3.2)

3.1.2.2 Network Subsystem

The network subsystem (NSS) acts as an interface between wireless and fixed networks. It mainly consists of switches and databases and manages functions such as handovers between BSSs, worldwide user localization, maintenance of user accounts and call charges, and management of roaming. Figure 3.8 shows the architecture of the NSS and the interface between the NSS components and the AuC as well as the OMC in the OSS. The NSS consists of *l* mobile services switching centres (MSC), *m* and *n* home and visitor location registers, gateway MSCs (GMSC),

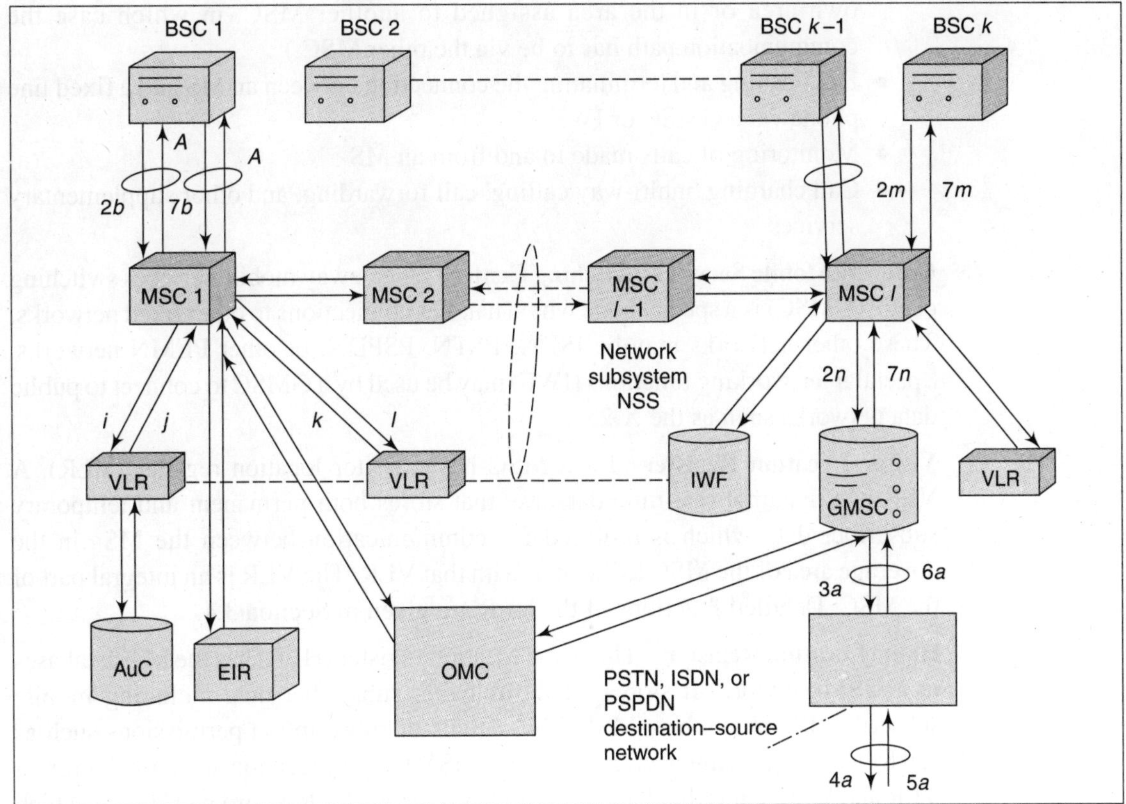

Fig. 3.8 Connection interfaces in the NSS between MSC and a number of BSCs and MSC, VLR, and OMC

and inter-working functions (IWF) with the mobile switching centres. GMSCs and IWFs connect to the other networks (e.g., PSTN, ISDN, or PSPDN). The basic connections and components in the NSS can be summed up as follows:

- Each MSC in the NSS can manage several base station systems.
- Every MSC has a home location register (HLR) and a visitor location register (VLR). An MSC can connect to another MSC, GMSC, and IWF.
- An HLR connects to an AUC in the OSS.
- A GMSC can connect to an OMC in the OSS.
- GMSCs are also used to connect to a PSTN, ISDN, or PSPDN network.

Mobile Services Switching Centre A mobile services switching centre (MSC) consists mainly of high-performance digital ISDN switches. It connects to a number of BSCs over the *A* interface. MSCs also connect to other MSCs and to fixed-line networks through GMSCs. An MSC is used to manage BSCs in a geographical area. The various functions performed by an MSC are as follows:

- Processing of signals
- Establishing and terminating the connection between various mobile stations via BSCs. (The mobile stations to be connected may fall in a given MSCs own area or in the area assigned to another MSC, in which case the communication path has to be via the other MSC.)
- Establishing and terminating the connection between an MS and a fixed line phone via a GMSC or IWF
- Monitoring of calls made to and from an MS
- Call charging, multi-way calling, call forwarding, and other supplementary services

Gateway Mobile Services Switching Centre A gateway mobile services switching centre (GMSC) is a special node which handles connections to other fixed networks. These other networks may be ISDN, PSTN, PSPDN, or other PLMN networks. Special inter-working functions (IWF) may be used by a GMSC to connect to public data networks such as the X.25.

Visitor Location Register Each MSC has a visitor location register (VLR). A VLR is a dynamic real-time database that stores both permanent and temporary subscriber data which is required for communication between the MSs in the coverage area of the MSC associated with that VLR. The VLR is an integral part of the MSC. Detailed functions of the VLR are given in Section 3.4.

Home Location Register The home location register (HLR) has the MT databases in a GSM network. It stores all the relevant subscriber data including mobile subscriber ISDN number (MSISDN), details of subscription permissions such as call forwarding, roaming, etc., subscriber's ISMI, user's location area, user's current VLR and MSC status. Each mobile user has only one HLR record worldwide, which is updated constantly on a real-time basis. Each MS must register at a specific HLR of a specific MSC. The HLR contacts AuC in the OSS for authentication. Each HLR is associated to an MSC so that when an MS registered at a certain HLR moves to another location area (LA), serviced by another MSC, the user's home MSC update the user's current VLR. Section 3.4 describes how HLRs function and help in location updating of a mobile station.

3.1.2.3 Operation Subsystem

The operation subsystem (OSS) administers the operation and maintenance of the entire network. Figure 3.9 shows the OSS architecture and the interfaces to the NSS and the RSS. The main components of the OSS are the authentication centre (AuC), the operation and maintenance centre (OMC), and the equipment identity register (EIR). Each AuC is associated with an HLR in the NSS and each EIR connects to an MSC. An OMC at OSS can connect to an MSC or a GMSC in the NSS and to a BSC at RSS.

Operation and Maintenance Centre The operation and maintenance centre (OMC) monitors and controls all other network entities through the *O* interface. The OMC's typical tasks include management of status reports, traffic monitoring, subscriber security management, and accounting and billing.

Authentication Centre The authentication centre (AuC) is used by the HLR to authenticate a user. The AuC may also be a secured partitioned part of the HLR

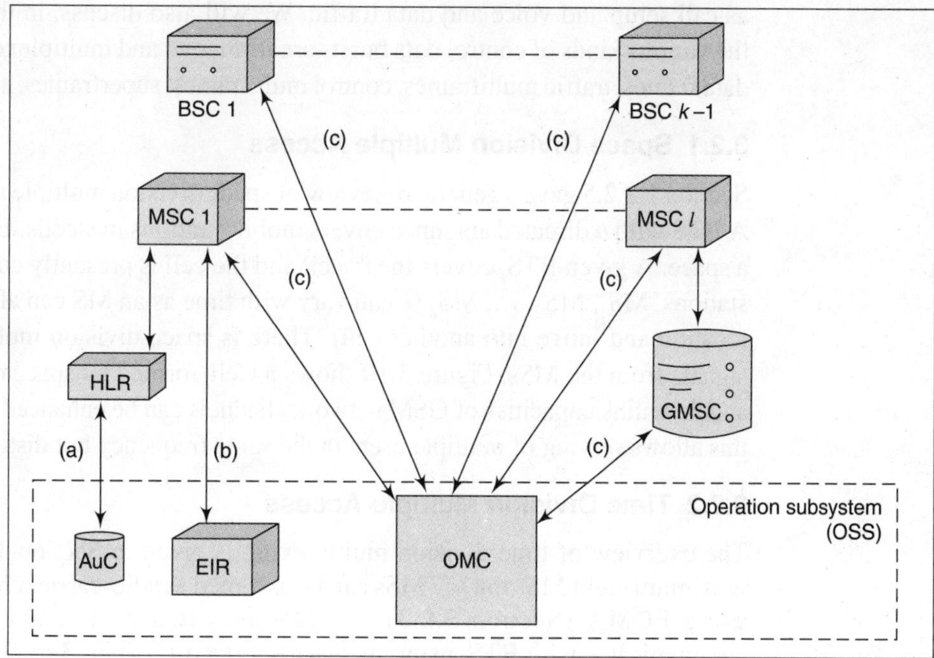

Fig. 3.9 OSS system architecture and connections between (a) AuC and HLR, (b) EIR and MSC, and (c) OMC and BSC, MSC, and GMSC

itself. Since mobile networks are quite vulnerable to attacks, the GSM standard specifies that the algorithms for key generation should be separated out as an OSS network entity. This entity is the AuC. The AuC database stores subscriber authentication keys. Other tasks carried out by the AuC include calculation of authentication parameters and then conveying these to the HLR.

Equipment Identity Register The equipment identity register (EIR) stores the international mobile equipment identity (IMEI) numbers for the entire network. The IMEI enables the MSC in identifying the type of terminal, mobile equipment manufacturer, and model and helps the network in locating the device in case it is stolen or misplaced. The EIR contains three different types of lists:

- A *black list* that includes mobile stations which have been reported stolen or are currently locked due to some reason
- A *white list* which records all MSs that are valid and operating
- A *grey list* including all those MSs that may not be functioning properly

3.2 Radio Interfaces

The U_m interface (Fig. 3.6) discussed in the previous section is a radio interface between the MS and the base transceiver BTS. This section will describe the space, frequency, and time division multiple access techniques. The base transceiver and the mobile stations communicate across the U_m interface for managing tasks such as call setup and voice and data traffic. We will also discuss, in this section, how the various kinds of control data bursts are formatted and multiplexed in the TDMA data frames, traffic multiframes, control multiframes, superframes, and hyperframes.

3.2.1 Space Division Multiple Access

Section 1.1.2.5 gave a general overview of space division multiple access (SDMA). A BTS with *n* directed antennae covers mobile stations in *n* cells. Each cell defines a space. A given BTS_j covers the ith cell and the cell is presently covering *k* mobile stations, MS_1, MS_2, …, MS_k (*k* can vary with time as an MS can always change its location and move into another cell). There is space division multiplexing of the signals from the MSs. Figure 3.10 shows a $Cell_i$ formed by space division. Uplink and downlink capacities of GSM network channels can be enhanced using SDMA as this allows serving of multiple users in the same frequency but distinct time-slots.

3.2.2 Time Division Multiple Access

The overview of time division multiplexing is given in Section 1.1.2.5. A set of maximum eight MSs out of *l* MSs can be assigned a radio-carrier channel by a BTS_j using FDMA (Section 3.2.3). An MS uses that radio-carrier channel and communicates with BTS_j using the user interface U_m (Fig. 3.6). The U_m access is such that each MS in the set (of presently communicating *k* mobile stations within

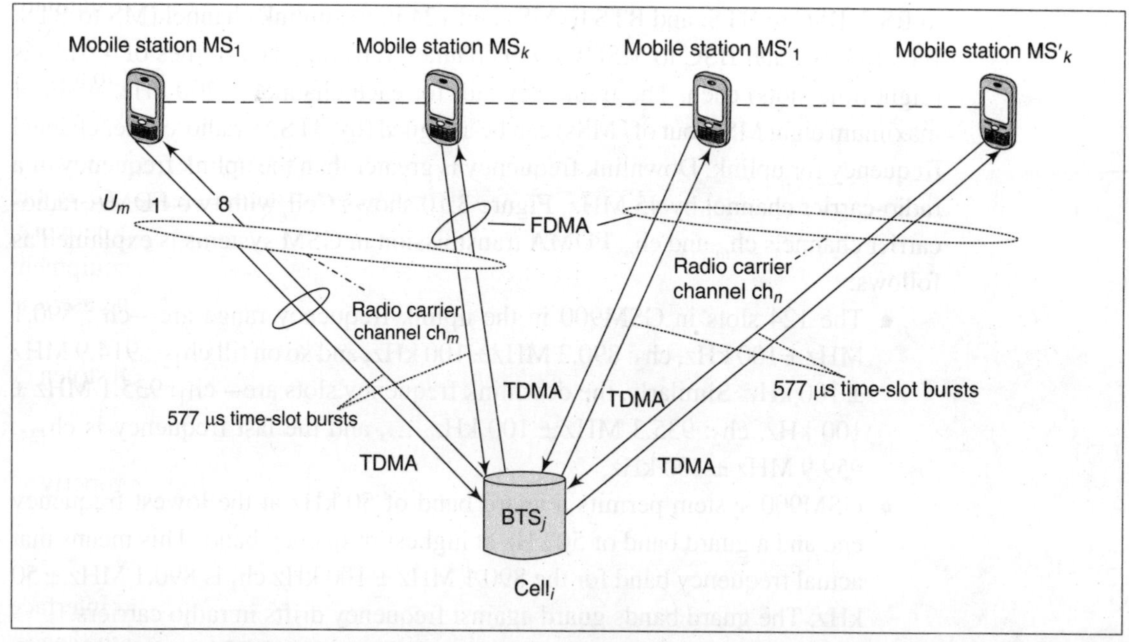

Fig. 3.10 Cell$_i$ with two radio-carrier channels ch$_m$ and ch$_n$ using FDMA and each MS in each channel transmitting bursts in 577 μs time-slots using TDMA

the cell) uses a distinct time-slot. An MS can use one of the eight distinct time-slots, SL_0, SL_1, ..., SL_7, each of 577 μs. A set of data bits in an SL is known as a *burst*. A set of eight data bursts defines a data *frame*. As an example, three mobile stations, MS_1, MS_2, and MS_3, are using the same radio-carrier channel ch$_m$. Let us assume that B_1 is assigned SL_0, B_2 is assigned SL_1, SL_4, and SL_7, and B_3 is assigned SL_2 and SL_6. (Here B_1, B_2, and B_3 are the data bursts of MS_1, MS_2, and MS_3, respectively). At an instant, a data frame can have bursts $B_1, B_2, B_3, X, B_2, B_3, X, B_2$ transmitted in eight time slots SL_0–SL_7, respectively. X represents unassigned slots for access by either BTS$_j$ or other MSs that are using the same radio-carrier channel. Since a time slot, SL = 577 μs, each data frame transmits in 8 × 577 μs = 4.615 ms (Section 1.1.2.5 and Table 1.2). Figure 3.10 shows transmission of data bursts through TDMA slots of 577 μs each. The format of a 577 μs TDMA burst is described in Section 3.2.4.

Half-duplex Transmission The transceiver of a mobile device can function in half-duplex mode when the uplink time slot, t_u and downlink time slot, t_d, are assigned separately by a BTS and $t_u - t_d$ is constant at 577 μs.

3.2.3 Frequency Division Multiple Access

As discussed in Section 1.1.2.5 frequency division multiple access (FDMA) means dividing the allotted or available bandwidth into different frequency channels for communication by multiple sources (sets of MTs). A set of maximum 124 radio-carrier channels each of 200 kHz can be used in GSM 900 downlink channel (MSC to BSC, BSC to BTS, and BTS to MS) and 124 in the uplink channel (MS to BTS, BTS to BSC, and BSC to MSC). Each channel transmits data frames of 4.615 ms (eight time-slots) each. The frequency-slot for each channel is 200 kHz. A set of maximum eight MSs (out of l MSs) can be assigned (by BTS_j) a radio-carrier channel frequency for uplink. Downlink frequency is greater than the uplink frequency of a radio-carrier channel by 45 MHz. Figure 3.10 shows $Cell_i$ with two FDMA radio-carrier channels ch_m and ch_n. FDMA transmission in GSM systems is explained as follows:

- The 124 slots in GSM900 in the uplink frequency range are—ch_1: 890.1 MHz ± 100 kHz, ch_2: 890.3 MHz ± 100 kHz, and so on till ch_{124}: 914.9 MHz ± 100 kHz. Similarly, the downlink frequency slots are—ch_1: 935.1 MHz ± 100 kHz, ch_2: 935.3 MHz ± 100 kHz, …, and the last frequency is ch_{124}: 959.9 MHz ± 100 kHz.
- GSM900 system permits a guard band of 50 kHz at the lowest frequency end and a guard band of 50 kHz at highest frequency band. This means that actual frequency band for the 890.1 MHz ± 100 kHz ch_1 is 890.1 MHz ± 50 kHz. The guard bands guard against frequency drifts in radio carriers.
- Total number of channels assigned to a BTS is 11. A BTS is permitted to use up to 10 subscriber channels and one channel is reserved for organizational data. A GSM system station is permitted use ch_2 to ch_{123} only. Therefore, 122 channels are available in GSM 900.
- All the BTSs taken together can communicate over 90 channels (ch_0, …, ch_{89}) available in GSM band. Maximum number of channels which can be allotted at a given instant to a BTS is 10). The mobile service provider can reserve one channel per BTS for transmission to MS or BSC. Total number of reserve channels can be 32 for the data transmission of mobile service provider.

Frequency Hopping in Data Frames There may be some specific frequency values that result in signal fading and do not provide expected signal strengths during transmission. A data frame frequency channel assigned to an MS by the BTS can be changed (hop) to select frequencies at a certain rate according to a predetermined sequence. This helps in ensuring better signal quality for most of the period. GSM hopping rates are 207.6 hop/s.

3.2.4 Format of a Data Burst

Section 1.1.2.3 described wireless signal propagation and the various delays encountered by a signal (Fig. 1.5(c)). There can be variable delays during transmission as the reflected signals take different amounts of time. Original signals can be reconstructed using a digital signal processor (DSP), but the DSP spends computational time in processing the signals. Therefore, at the beginning and end of every data burst of 577 μs, guard spaces of 15.25 μs (equal to 4.125-bit transmission time interval) each are reserved to account for delays in the reflected signal and computational time. The effective transmission time for the data bits is, therefore, [577 – (2 × 15.25)] = 546.5 μs. 148 bits are transmitted in 546.5 μs. The data transmission rate is (8 × 48) bits/4.615 ms = 256.555 kbps. A GMSK signal (Section 1.1.2.4) modulates and transmits at 256.555 kbps (3.898 μs/bit).

- Six bits, three at the head (*H*) and three at the tail (*T*) [conventionally, both are called tail bits (TB). However, to distinguish these while explaining the format, the terms *H* and *T* have been used here] of the 148-bit burst are 000. Now 142 bits are left in the middle.
- 26 bits in the middle of the burst are transmitted as training (*TR*) bits. The *TR* bits enable the receiver to (a) synchronize using *H*, *TR*, and *T* bits, and (b) select the strong components of the signals. (Direct path or wide reflection angle signals (Fig. 1.5(c)) are the strongest ones as they travel the least distance between the transmitter and the receiver.) Now (142 – 26)/2 = 58 bits each are left after *H* and before *T*.
- Data in the burst can be of two kinds—MS data or mobile-service NSS control data. On either side of the *TR* bits, an S bit can be placed to specify whether the source is the MS or NSS control data. Now (58 – 1) = 57 bits each remain after the *H* bits and before the *T* bits for transmitting user (subscriber) data. The useful data bits are 57 after *H* and 57 before *T*.

There can be total 114 bits (57 + 57) for the user data in a data burst (time-slot). Total number of bits per second will be 114/4.615 bit/ms = 24.7 kbps assuming that only one time-slot is used in a data frame of eight slots when transmitting voice and assuming that the only data bursts are voice-data bursts. The frequency correction and synchronization bursts require additional slots. There are 12 slots which are used for user data. User data is followed by one slot for control signals data. Therefore, due to need for frequency correction and synchronization bursts, the voice-data (user data) rates are 12/13 × 24.7 kbps = 22.8 kbps The control data slot is replaced by an empty slot *X* in every alternate set of 13 frames. Therefore, there are a total 26 data frames in one *traffic multiframe* in which there are one control data, one empty, and 24 user data frames. Traffic multiframes transmit TCH, FACCH, and SACCH data (Section 3.2.5). Control channel capacity within a traffic multiframe is 1/26 × 24.7 kbps = 950 bps. A traffic multiframe transmits in 26 × 4.615 ms = 120 ms intervals.

3.2.4.1 Interleaving in a Traffic Multiframe

Interleaving means inserting in-between. The packets, each consisting of 456 bits in a 20 ms time-slot, are interleaved in a traffic multiframe for voice traffic. To clarify this, let us consider the following example—assume that there are two MSs,

MS_i and MS_j, multiplexed in TDMA slots. There are 57 bits after *H* and 57 bits before *T* in the data bursts. TCH/F (Section 3.2.5) transmission rate is 22.8 kbps. Therefore, there are 456 (= 8 × 57) bits per 20 ms in voice traffic from two MSs. When 20 ms packets of MS_i and MS_j interleave, then all the 57 bit time-slots after *H* in each data burst are used by MS_i and all the 57 bits before *T* in each data burst are used by MS_j. Interleaving distributes the effects of channel characteristics variation with time on multiple MSs.

3.2.4.2 Four Types of Control Data Bursts

Different types of control data bursts are discussed in this section.

(a) The call setup takes place while setting the initial connection using a burst called *access burst*. The channel in which this burst is sent is called AGCH (access grant channel). It is part of the common control channel (CCCH) (Section 3.2.5).

(b) The *TR* bits help the receiver in correcting path changes. All the MSs that are communicating with the BTS must be synchronized. The total time durations of forward and return paths vary, as some MSs are closer than the others. A *timing advance* is required for synchronization when a BTS receives a signal from a far off MS compared to a short distance MS. The advance is of maximum 0.24 ms (63 × 3.692 μs period for 63 bits, because each bit is transmitting in a 3.692 μs interval). A *synchronization burst* of 64 *TR* bits helps in synchronizing the transmitter and receiver time-slots and in timing advance. The data bits after header and before tail in the burst are [(142 – 64)/2] – 1 = 38 bits in place of 57 bits. The channel in which this burst is sent is called SCH (synchronization channel). It is part of the broadcast control channel (BCCH) (Section 3.2.5).

(c) There may occur a deviation in the frequency of a radio carrier. There may also be interference with the neighbouring channel frequency. During synchronous data transmission (Section 3.1.1.7), the receiver must synchronize the clock rate according to the incoming data bits. A *frequency correction burst* from the BTS helps in correcting the carrier frequency. In place of the *TR*, *S*, and user data bits, a 142-bit sequence between *H* and *T* is deployed. The channel in which this burst is sent is called FCCH (frequency correction channel). It is also a part of BCCH (Section 3.2.5).

(d) When no useful burst is being transmitted from an MS or a BTS after a connection setup, then a *dummy burst* is transmitted.

3.2.5 Traffic and Control Data Channels

One characteristic of GSM voice and control data channels (called logical channels) is that if the transmitted data is not correctable at the receiver, it is blocked for further processing. The following subsections describe the traffic and control data channels.

3.2.5.1 Traffic Data Channels

Voice is coded using a codec. Codec is short for coder–decoder which is a circuit that codes analog signals into digital signals and decodes digital signals into analog signal according to various coding and decoding algorithms. The error correction bits [cyclic redundancy check (CRC) and redundant bits] are then appended and data interleaving is performed.

An application of data interleaving is as follows—a user talking on a mobile phone feels uneasy when there are abrupt periods of silence due to abrupt losses of connectivity. A technique to provide a relief is to interleave an induced background noise. The background noise can be home or office background noise or roadside traffic noise. When a voice activity detector is not detecting any voice traffic from the other end due to the user not talking or some processing related delays or losses of connectivity, interleaving at the listener's end helps the user, who does not feel uneasy due to abrupt intervals of silence.

Three types of voice traffic were mentioned in Fig. 3.1—voice at full 13 kbps, half data-rate speech, and enhanced full-rate speech. User data or subscriber data is also called traffic data. The different types of voice traffic are as follows:

- TCH/FS (traffic channel/full rate set for transmission)—Voice is coded with a codec (coder–decoder) at 13 kbps. As the additional bits are appended after coding, the data rate is enhanced to 22.8 kbps when transmitting at full speed.
- TCH/HS (traffic channel/half rate set for transmission)—Voice is coded with a codec at 5.6 kbps and, after the error correction bits, the data rate is enhanced to 11.4 kbps and transmission takes place at half speed. The available data rate is 22.8 kbps. The advantage of this sort of transmission is that double voice signals can now be transmitted. However this sort of voice-data results in degradation of voice quality.
- TCH/EFR (traffic channel/enhanced full rate set for transmission)—Voice is coded with another enhanced coding technique employing a codec. EFR gives an enhanced voice quality but has limited error correction bits because the data rate is limited to 12.8 kbps. It is advantageous since the voice quality is upgraded in those cases where the transmission error rate is small. A codec may function in automatic mode and code the voice as TCH/FS or TCH/EFR depending on the transmission error rate detected in the bursts.
- TCH/F14.4, TCH/F9.6, and TCH/F4.8 (traffic channel/full rate at 14.4. 9.6, and 4.8 kbps)—Due to large number of subscribers at a base station, the GSM specifications provide for the traffic rates of 14.4 kbps, 9.6 kbps and 4.8 kbps also.

3.2.5.2 Control Data Channels

The 184-bit packet from data link layer (Section 3.3) is formatted for the data burst bits. The 184-bits are added with 40 parity bits, four tail bits, and 224 half-convolution coding bits to result in the 456-bit packet. Section 3.2.4 discussed the use of control data bursts. There are four formats for the control data bursts in the data frames. The control data bursts can also be formatted as 456-bit packets with different coding procedure than in a traffic multiframe (Section 3.2.4). The various types of control channels (CCH) are as follows:

- DCCH (dedicated control channel)—An MS sends TCH traffic only after a call setup. A bi-directional communication channel is present between the BTS and MS before the TCH traffic starts. It is called standalone DCCH (SDCCH). It is used for the registration, authentication, and other requirements. Total 782 bits are sent as dedicated control channel data in 1 s in case of slow associated standalone DCCH (SADCCH). (950 bps can be sent as a control data slot in a traffic multiframe.) When more than 782 bits are to be sent per second, then the TCH part of the data bursts can be used. Then DCCH is called FACCH (fast associated control channel). Only FACCH and SACCH (slow associated control channel) data is transmitted in a traffic multiframe. The traffic channels transmit only after the call set up and transmit the user/subscriber data through the BTS. A separate multiframe is required for the control channels (other than FACCH and SACCH). That multiframe is called *control multiframe*. The latter has 51 data frames. One data frame is of 4.615 ms each (Section 3.2.2), therefore, a control multiframe transmits for 51×4.615 ms = 235.4 ms.
- BCCH (broadcast control channel)—A BTS needs to broadcast the frequency and cell identity. BCCH is used for that. A BTS needs to broadcast the information regarding frequencies and sequence options for hopping that can be assigned to the MSs in the cell to all the MSs. This enables an MS to get an available radio-carrier frequency channel and transmit with different frequencies on different hops and synchronize with the BTS. The synchronization and frequency correction bursts (Section 3.2.4) also use the BCCH.
- CCCH (common control channel)—A BTS (when granting access to an MS so that MS can use either SDCCH or TCH) uses a channel called AGCH (access grant channel). After the access is granted, the call setup or call forwarding can take place. The control channel used for such purposes is called a CCCH. When call setup requirements are transmitted from the MS, the CCCH is called RACH (random access channel). RACH data burst format is as follows—during 577 μs in place of the sequence (*H*, user data, *S*, *TR*, *S*, user data, and *T*), a 145-bit sequence is modified as eight *H* bits, 41 synchronization bits, 36 bits user data and 3 *T* bits (total 88 bits). The guard-space time intervals are now equal to $(68.25 \times 3.692)/2 = 126$ μs before *H* and after *T* bits. When call-forwarding information is transmitted from the

BTS, the CCCH is called PCH (paging channel). An example of paging is transmission of information to a select target MS, for example, the identity of the caller of an incoming call to the MS to which the call is to be forwarded. When access granting information is transmitted from the BTS, the CCCH is called AGCH (access grant channel).

3.2.5.3 Super and Hyper Frames

Section 3.2.5.2 mentioned that a control multiframe has 51 data frames. A control multiframe, therefore, transmits for 51×4.615 ms = 235.4 ms. A traffic multiframe transmits in 120 ms. After formatting, a super frame can have 51 traffic multiframes of 120 ms each or 26 control multiframes of 235.4 ms each. A *superframe*, therefore, transmits in 26×235.4 ms or 51×120 ms, both equal 6.12 s.

A *hyperframe* consists of 2048 superframes. A hyperframe transmits in 2048×6.12 s = 12533.76 s ≈ 3½ hours. In GSM systems, a sequence (count) number [from 0 to $(2048 \times 26 \times 51) - 1$ slots in a hyperframe] is assigned to each 4.615 ms data frame. The sequence number is encrypted along with the data so that after decryption, the original frame number can be recovered for sequential arrangement of data. Since the frames are sequentially transmitted through the eight TDMA time-slots, the sequence number also helps in identifying the original time-slot of a given data burst.

3.3 Protocols

Various protocols are used at different layers in a communication network. Each layer has specific protocols that govern its communication with adjacent layers. The layers defined in the open system interconnection (OSI) model are physical (layer 1), data link (layer 2), network (layer 3), transport (layer 4), session (layer 5), presentation (layer 6), and application (layer 7). When a transceiver receives signals, they are processed at the different layers arranged in order from layer 1 to layer 7. The OSI model specifies that when the transceiver transmits the signals, they are processed in reverse order of layers, i.e., from layer 7 to layer 1. Each layer adds header bits in a specific format so that these layer header bits for each layer can be stripped by the transceiver at the receiving end and various operations can be performed on the received data as per the received bits. However, for an actually used protocol, for example, TCP/IP or GSM, a transceiver need not define protocols for all seven layers as some layers perform the functions of neighbouring layer(s). The MS, BTS, BSC, and MSC, for example, have just three layers—physical, data link, and network. Transport and session layer functions are taken care of by network layer protocols. The tasks of the presentation layer are performed by other layers. TE (user) application at either end (caller and connected ends) controls the application layer protocols (Figs 3.5 and 3.8). The following subsections describe the signalling protocols between MS and BTS, BTS and BSC, and BSC and MSC

3.3.1 Mobile Station–Base Transceiver Signalling Protocols

The *physical layer* between the MS and the BTS is called *radio* and performs the functions shown in Fig. 3.11(a). Figure 3.11(b) shows all protocol layers between the MS and the BTS. The details of these radio interface functions are discussed in Section 3.2.

The *data link layer* controls the flow of packets to and from the network layer and provides access to the various services discussed in Section 3.1.1. The data link layer protocol between an MS and a BTS is LAPD$_m$ (link access protocol *D*-channel modified) for U_m. The protocol prescribes the standard procedure in GSM for accessing the *D*-channel link. The protocol is a modified version of the LAPD protocol for the *D*-channel of ISDN (integrated services digital network). The

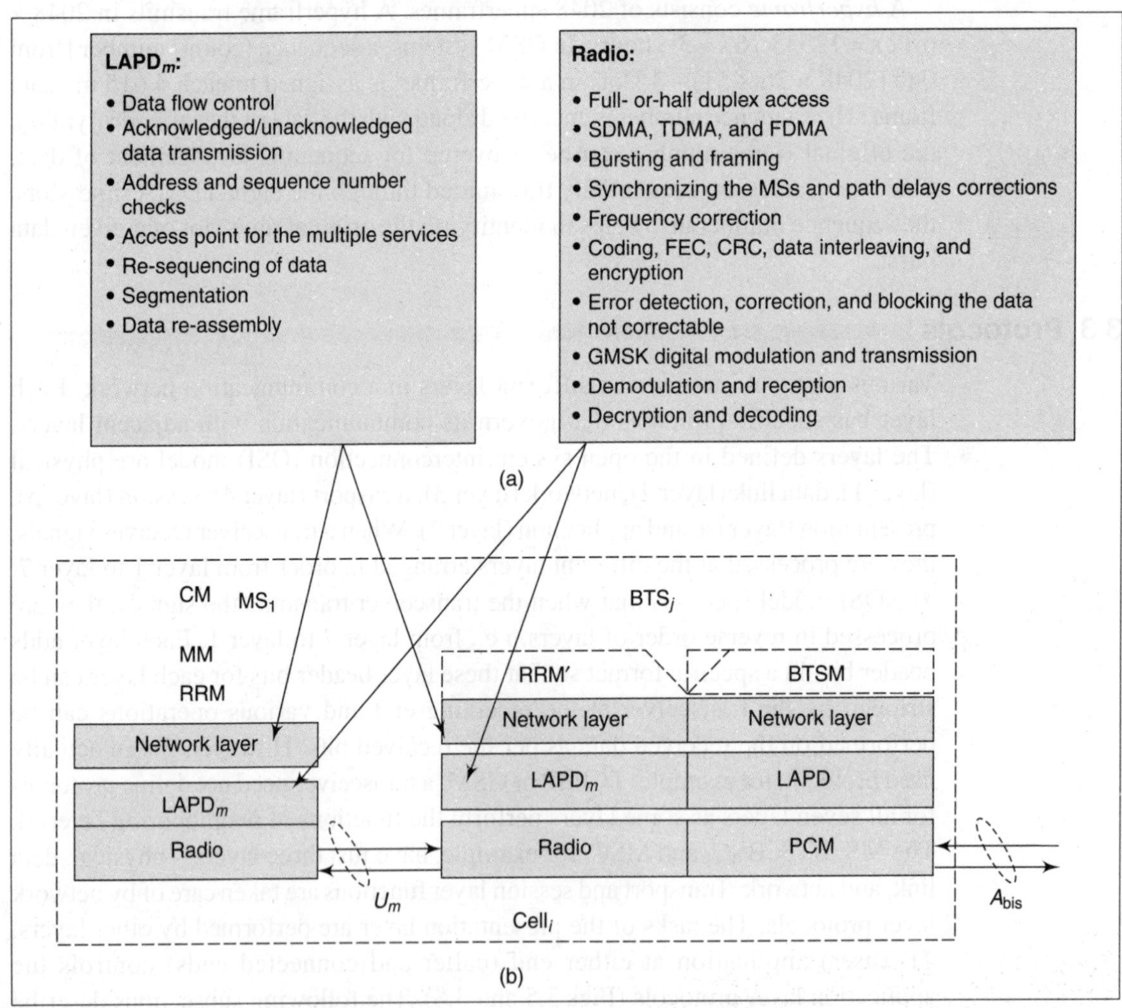

Fig. 3.11 (a) Functions of radio and LAPD$_m$ (b) All protocol layers between the MS and the BTS

functions of $LAPD_m$ are shown in Fig. 3.11(a). There is no need of appending and stripping of synchronization bits, *S* flag, and error correction bits to and from the layer in $LAPD_m$ because the radio interface (U_m) performs these functions at the physical layer itself. The $LAPD_m$ communicates wirelessly across the radio interface as opposed to the guided transmission of ISDN signals in case of the LAPD. The 184-bit $LAPD_m$ format differs from the LAPD in the following respects:

- Eight-bit address field (optional) [Address field identifies the service access point (SAP) to enable different services (Fig. 3.1).]
- Eight-bit control field (control field identifies the frame type, command frame, or response frame.)
- Eight-bit length field (length field identifies the length of information bits.)
- Information bits of variable length (information bits identify the operation required for a network layer message.)
- Remaining bits as 1s (filler bits). [This field is also used for acknowledgement in asynchronous communication (Section 3.1.1.7) and when $LAPD_m$ is used for that then there is no communication for layer 3 (no SAP address specification).] There are 16 filler bits to make a 184-bit LAPD, when there is 144-bit information (e.g., SACCH) along with 24-bit address, control, and length fields. When there is 160-bit information (e.g., FACCH) along with the 24-bit fields, then there are no fillers.

The *network layer* has three sublayers—call (connection) management (CM), mobility management (MM), and radio resource management (RRM). Figure 3.12 shows the operations in the CM, MM, and RRM sublayers.

The CM sublayer protocol supports call establishment, maintenance, and termination. The CM sublayer also controls and supports the functioning of the SMS and supplementary services. The CM also supports DTMF (dual tone multiple frequency) signalling. The MM layer controls issues regarding mobility management when an MS moves into another cell (location area). The RRM manages the radio resources. The BTS implements only RRM′(a part of RRM) as the BSC handles the handover.

The network layer performs the following functions:

- Defines protocols for implementation of addressed messages received from the data link layer
- Defines addresses of the messages
- Transmits the logical channels' (traffic and control data channels) data and information bits to the data link layer from an address (SAP)
- Receives the logical channels' data and information bits (e.g., FACCH, SACCH bits) from the data link layer for the addressed SAP

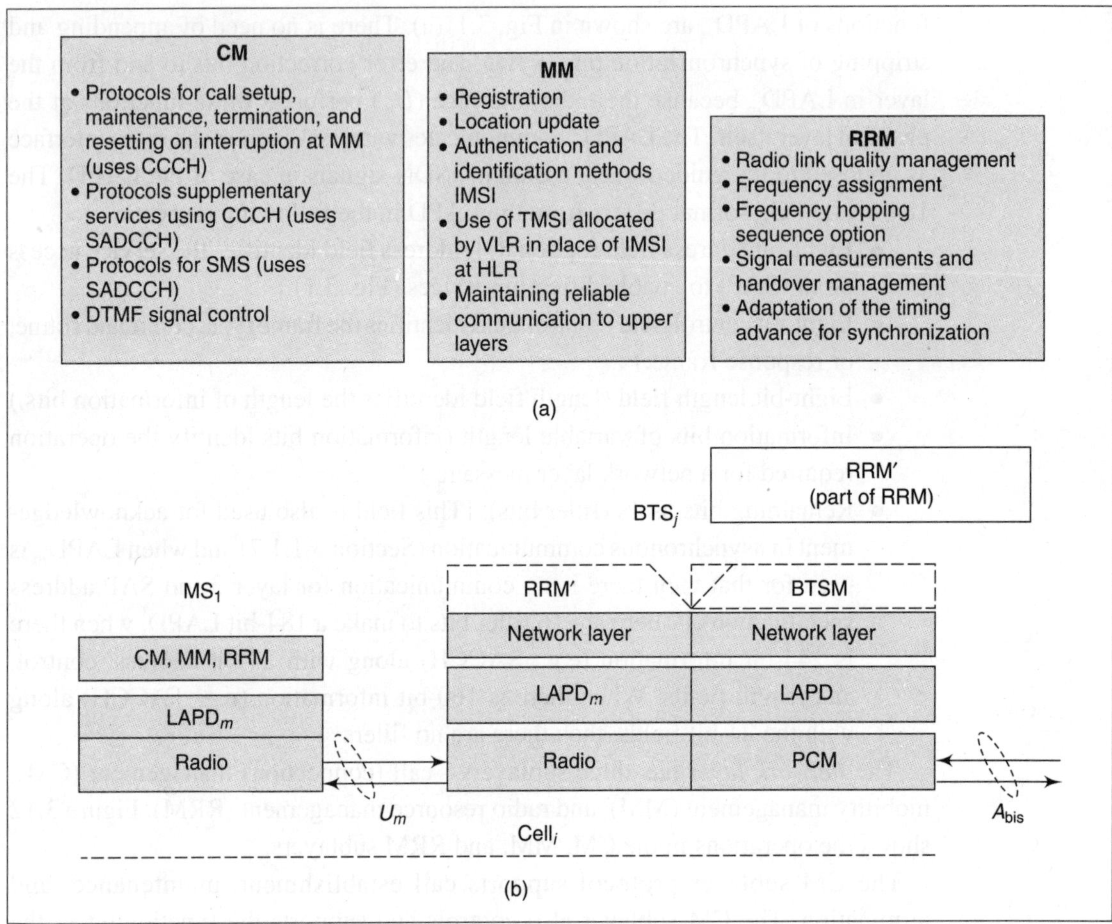

Fig. 3.12 (a) Functions and (b) interfaces of the network sublayers

All channel communications (with a few exceptions, such as data acknowledgements in asynchronous data transmission) take place between the network layers of the MS and the BTS. The network layer message format is as follows:

1. TI (transaction identifier) field so that the protocol to send sequence of messages is identified.
2. Eight-bit control field (control field identifies the frame type, command frame or response frame).
3. PD (protocol discriminator) field to identify the protocol operation. For example, whether the operation belongs to the test, call control, call management, or service category.
4. MT (message type) field to identify the type of message for protocol operation.
5. IE (information element) field for providing optional information for the operation.
6. IEI (information element identifier) field to identify the IE field information.

3.3.2 Base Transceiver–Base Station Controller Signalling Protocols

Figure 3.13 shows all the protocol layers in the BTS and BSC. The *physical layer* between the BTS and the BSC is the A_{bis} interface (of the PSTN, ISDN, or PSPDN networks). The connection between the BTS and the BSC is through a wired network (PSTN, ISDN, or PSPDN). Voice is coded in the 64 kbps PCM (pulse code modulation) format in a PSTN network. The A_{bis} interface between a BTS and a BSC, therefore, uses the 64 kbps PCM (or four multiplexed 16 kbps channels) format. PCM coding techniques are different from the 22.8 kbps TCH (Section 3.2.5) radio interface U_m (between MS and BTS). Translation between these coding formats is performed by recoding the TCH bits received from the caller MS to 64 kbps PCM and from PCM to TCH for the receiver MS. This translation and retranslation from one coding format to another may affect voice quality. Therefore,

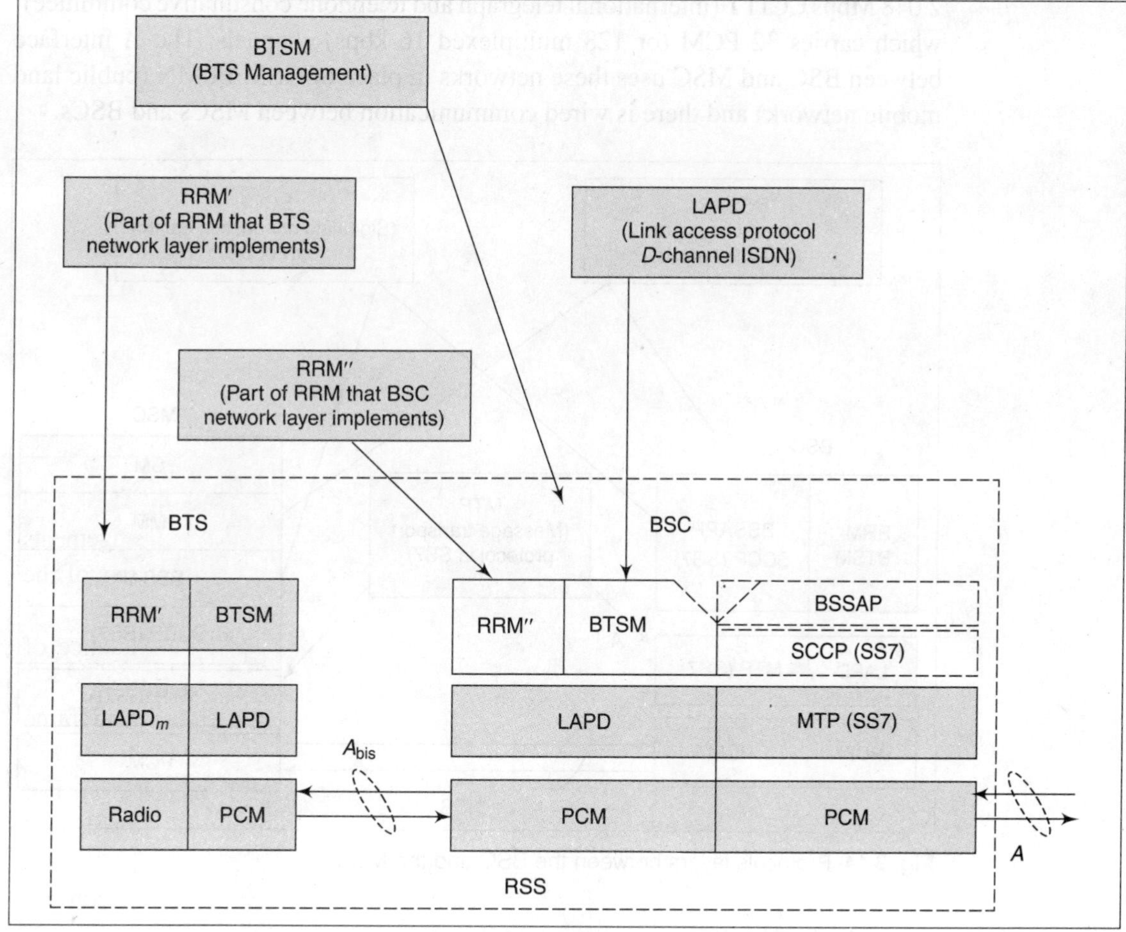

Fig. 3.13 Protocols layers between the BTS and the BSC

a procedure called TFO (tandem free operation) can be adopted at the BTSs, BSCs, and MSCs. (TFO means operate without performing translation and retranslation processes repeatedly.)

The *data link layer* protocol between BTS and BSC is LAPD (link access protocol *D*-channel) for A_{bis}. The protocol prescribes the standard procedure for the *D*-channel of ISDN (integrated services digital network). The *network layer* protocol between BTS and BSC is called BTSM (BTS management).

3.3.3 Base Station Controller–Mobile Services Switching Centre Signalling Protocols

Figure 3.14 shows the various protocol layers between the BSC and the MSC. The *physical layer* between the BSC and MSC is PCM′ (PCM multiplexed). The MSC connects to PSTN, ISDN, and PSPDN networks which employ the 64 kbps PCM or 2.048 Mbps CCITT (international telegraph and telephone consultative committee)[1] which carries 32 PCM (or 128 multiplexed 16 kbps) channels. The *A* interface between BSC and MSC uses these networks in place of GSM PLMN (public land mobile network) and there is wired communication between MSCs and BSCs.

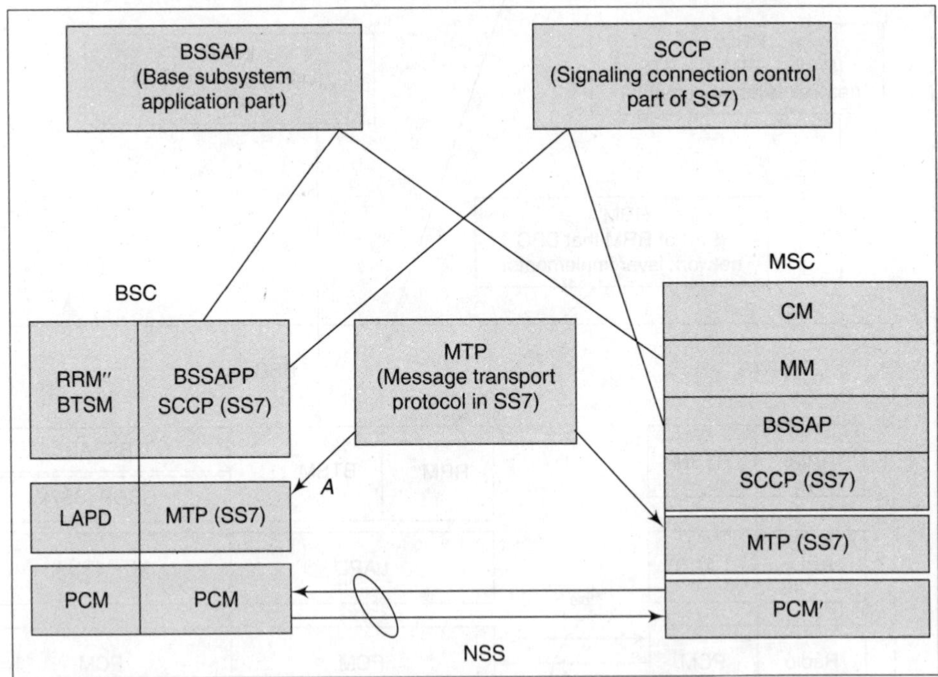

Fig. 3.14 Protocols layers between the BSC and the MSC

[1]Translated name of the *Comité Consultatif International Téléphonique et Télégraphique* which is a sattelite organization of the international telecommunication union.

Data link layer protocols between the BSC and the MSC are MTP (message transfer protocol) and SCCP (signalling connection control protocol). MTP and SCCP are parts of the SS7 (signalling system No. 7) used by interface *A*. The layer protocol prescribes a standard procedure for the MTP and SCCP for SS7 transmission and reception in a 2 Mbps CCITT PSTN/ISDN/PSPDN network.

Network layer protocol at the BSC is BSSAP (base subsystem application part). Figure 3.12 shows the CM and MM sublayers at the BTS. Network layer protocol sublayers at the MSC are CM, MM, and BSSAP.

3.4 Localization

Localization is a process by which a mobile station is identified, authenticated, and provided service by a mobile switching centre through the base station controller and base transceiver either at the home location of the MS or at a visiting location. Users want instantaneous connection setup for a call and service-on-demand even while they are on the move. The mobile service providers, on the other hand, will provide service(s) to the user only after identifying the mobile station (MS) of the user and verifying the services subscribed to by the user or the services presently allowed to that MS. Localization mechanism of the GSM system fulfils both the requirements.

The NSS (network subsystem) of GSM architecture periodically updates the location of those MSs which are not switched off and are not struck off (or blocked) from the list of subscribers to the given mobile service. The SIM in a mobile station MS_i stores *location-area identification* (LAI). LAI is location information which is updated by the MSC which covers the MS's current location area. The SIM also saves a temporary mobile subscriber identity (TMSI) assigned by the VLR associated to the current MSC. The location update is recorded at the VLR (visitor location register) and the LAI is updated at the SIM card in MS_i via the MSC, BSC, and BTS covering its current location (interfaces *j*, *7b*, *7a*, and *8a* in Figs 3.5 and 3.8).

The VLR updating process is as follows—Each MSC is associated to a VLR (visitor location register) and an HLR (home location register). An MSC_A has an HLR_A at which a mobile station is originally registered on subscription to a mobile service. The HLR_A copies the HLR information of an MS (via MSC_A and another MSC_B) to VLR_B when the MS moves out to the area covered by MSC_B. The copying saves time as the MSC_B gets instantaneous information regarding MS_i from VLR_B instead of having to fetch it from HLR_A. The time that would otherwise have been spent in the searching and fetching of information for the relocated MS_i from HLR_A is saved. The storing of LAI into the SIM from the VLR through the MSC_B, BSC, and BTS saves the time that otherwise would have been spent in identifying the BTS, BSC, and MSC_B. These processes, therefore, help the MS in getting service from MSC_B instantaneously.

The main functions of an HLR are as follows:

- Registration of information regarding IMSI (international mobile subscriber identity), MSISDN (mobile station international subscriber ISDN number), roaming restrictions, call forwarding, mobile subscriber roaming number (MSRN), present VLR, and present MSC. (MSISDN has the international country code, followed by the destination area code in a country and the subscriber number. The identical coding scheme for address is used in the ISDN network employing a fixed wire or fiber line.) The present VLR and MSC information can change when the user MS moves into another location area but the HLR which stores this information remains the same.
- Registration of information regarding all associated MSs which have the given MSC as their home location area. There is no deregistration of the information when an MS moves from its home location area into another location area. It connects to the AuC for authenticating an MS and granting service permissions. Only after the service permission access and authentication, does the MS registered at a given HLR get service though the MSC covering its location area.

The VLR stores the information regarding an MS currently in its area, which helps in saving time taken for searching and fetching the information of that MS from a remote home location register (HLR) at which the MS is registered. The information stored in a VLR is updated periodically. The main functions of a VLR are as follows:

- Registration of information pertaining to currently associated MSs. The information is about their HLR, IMSI, and MSISDN.
- Storing information of the MSs which are in its location area and to which the MSC (associated with the given VLR) is currently providing network services.
- Registration of any new MS that moves into the VLR's location area. It copies the information from the HLR of that MS.
- De-registration of an MS, if the MS dissociates from the MSC associated with the given VLR and moves out to another location area.

3.5 Calling

Figure 3.1 shows communication between a mobile station TE and another TE. The other TE could be a mobile station TE or other TE (such as a PSTN phone). The figure showed the caller TE to be an MS communicating to the other TE via the path 1–2–3–4–5–6–7–8. The caller TE can also be a PSTN phone. Different methods and protocols are used for establishing connection and maintaining communication in calling to and from mobile devices in a GSM PLMN network. The various types of calls handled by a GSM network are: calls originating from a mobile TE to a PSTN destination TE (Mobile→PSTN calls), calls originating from a mobile TE to a mobile destination TE (Mobile→Mobile calls), calls originating from a PSTN TE to a mobile destination TE (PSTN→Mobile calls), and message exchanges between the mobile station and the base transceiver (Mobile station ↔ Base transceiver message exchanges). The various types of calls and their respective procedures are discussed in the following subsections.

3.5.1 Mobile→PSTN Calls

Mobile→PSTN calls are calls originating from a mobile terminal and ending at a PSTN destination. Let us take an example to clarify how the connection is established and maintained in such a call. When a mobile station, MS_i, calls and communicates with a PSTN phone, TE_j, the connection is established through switching centres, MSC_i and MSC_j. MS_i establishes (switches) a connection (channel) between MS_i and TE_j in the following manner:

1. MS_i connects to BTS_i, then to BSC_i, and then to MSC_i (through interface paths 1*a*, 2*a*, 2*b*, 7*b*, 7*a*, and 8*a* in Fig. 3.5)
2. MSC_i verifies and authenticates MS_i using the VLR through *k* and *l* in Fig. 3.8. It also discovers available paths to the PSTN phone TE_j through MSC_j and $GMSC_j$.
3. MSC_i switches to MSC_j, then to $GMSC_j$, then to TE_j (through interface paths 2*n*, 3*a*, and 4*a* in Fig. 3.8).
4. TE_j transmits back to $GMSC_j$ and MSC_j (through interface paths 5*a*, 6*a*, 7*n*, and 7*m* in Fig. 3.8).
5. MSC_j switches to MSC_i, which transmits back to BSC_i, BTS_i, and MS_i (through interface paths 7*b*, 7*a*, and 8*a* in Fig. 3.5).

3.5.2 Mobile→Mobile Calls

Mobile→Mobile calls are calls originating from a mobile terminal and ending at a mobile destination. Let us take an example to clarify how the connection is established and maintained in such a call. When a mobile terminal MS_i calls and communicates with another mobile phone MS_j, the communication between them is routed through switching centres MSC_i and MSC_j. MS_i establishes (switches) a connection (channel) between MS_i and MS_j in the following manner:

1. MS_i connects to BTS_i, then to BSC_i, and then to MSC_i (through interface paths 1*a*, 2*a*, 2*b*, 7*b*, 7*a*, and 8*a* in Fig. 3.5).
2. MSC_i verifies and authenticates MS_i using the VLR through *i* and *j* in Fig. 3.8. It also discovers available paths to the mobile phone MS_j through MSC_j.
3. MSC_i switches to the MSC_j and verifies and authenticates MS_j using the VLR through *k* and *l*.
4. MSC_j connects to BSC_j, BTS_j, and MS_j (through interface paths 7*m*, 7*o*, and 8*p* in Fig. 3.5).
5. MS_j transmits back to MSC_j (through interface paths 1*p*, 2*o*, and 2*m* in Fig. 3.5).

6. MSC_j switches to MSC_i, which transmits back to BSC_i, BTS_i, and MS_i (through interface paths *7b*, *7a*, and *8a* in Fig. 3.3).

3.5.3 PSTN→Mobile Calls

PSTN→Mobile calls are calls originating from a PSTN phone and terminating in a mobile destination employ the GMSC. For example, when making a call to a mobile terminal MS_j, a PSTN terminal TE_i connects to $GMSC_i$ which, in turn, requests HLR_i to discover MSC_j. MSC_j uses VLR_j to verify and authenticate the MS_j and then MSC_j directs the call to MS_j through BSC_j and BTS_j.

3.5.4 Mobile Station↔Base Transceiver Message Exchanges

If a mobile station MS_i is setting up a call to another terminal TE_j, then the following message sequences are exchanged between BTS_i and MS_i before the voice or data exchange through the base transceiver can begin.

1. The MS requests the BTS to grant it a channel for communication and the BTS responds immediately by assigning a channel to the MS.
2. The MS sends a request to the BTS for service and the BTS replies with a request for authentication to the MS.
3. The BTS transmits its response for authenticating the MS along with a command (with a ciphering number) for ciphering at the MS. A ciphering number is generated using a random number sent by the AuC and a cipher key available at the BTS for accessing the MS.
4. The MS runs an algorithm on the ciphering number sent by the BTS and the cipher key stored in the SIM. This algorithm generates an encryption key. Upon completion of the mathematical algorithm, the MS transmits the information to the BTS. Only the message of completion of the algorithm is transmitted from the MS to the BTS and not the generated encryption key. The BTS must independently run the algorithm to get this key for deciphering the encrypted traffic coming in from the MS.
5. The call is set up using the CM (call management) protocol (Section 3.3.1) by the MS and the BTS management (BTSM) protocol by the BTS.
6. Call setup is confirmed by the BTS to the MS.
7. Assignment commands are sent by the BTS to the MS and assignment completion messages are sent by the MS to the BTS.
8. An alert message is sent from the BTS to the MS before the connection.
9. A connection establishment message is sent from the BTS to the MS and connection acknowledgement is sent from the MS to the BTS.
10. Voice or data interchange starts.

When a mobile station MS_i (bottom TE in Fig. 3.1) is setting up for a call from another terminal TE_j (top TE in Fig. 3.1), the followings message exchanges take place between the base transceiver BTS_i and MS_j before the voice or data exchange begins:

1. The BTS sends a request for paging to the MS and the MS requests the BTS for channel allocation.
2. A channel is immediately assigned by the BTS to the MS and the MS responds to the page by the BTS.
3. A request for authenticating the MS is sent by the BTS and a response for authentication is sent by the MS to the BTS.
4. The BTS sends a command for ciphering to the MS and the MS replies with the ciphering completion message to BTS.
5. Call is set up employing the CM protocol at the MS and the BTSM protocol at the BTS and call setup is confirmed by the MS to the BTS.
6. Assignment commands are sent by the BTS to the MS and assignment completion messages are sent by the MS to the BTS.
7. Before the connection established, an alert message is sent by the BTS to the MS.
8. Connection established message is sent by the BTS to the MS and connection acknowledgement message is sent by the MS to the BTS.
9. Voice or data interchange begins.

3.6 Handover

Handover (also known as handoff as handover to one is handoff from another), to the neighbouring cell, is defined as a mechanism to hand over the control of a mobile device to the neighbouring cell (Section 1.7). This is, however, too simplistic a picture of the handoff–handover process. Handover is technically the process of transferring a call (or data transfer) in progress from one channel to another. The core network may perform handovers at various levels of the system architecture or it may hand over the call to another network altogether. There are two main reasons for handover in cellular networks—(a) if the mobile device moves out of the range of one cell (base station) and a different base station can provide it with a stronger signal, or (b) if all channels of one base station are busy, then a nearby base station can provide service to the device. The handover process is an important one in any cellular network. Also, the handover must be completed efficiently and without any inconvenience to the user. Different networks use different types of handover techniques. The following subsections describe the various types of handovers and the handover processes within GSM networks.

3.6.1 Types of Handover

Different cellular systems follow different regulations for handoff–handover processes. With the development of 3G standards and technology, it is possible for several mobile phones to use the same channel and for neighbouring cells to use the same frequency bands. As a result of this, new handover methods have also evolved and are used in addition to the older techniques. The two main types of handover are *hard handover* and *soft handover*. These are discussed below.

3.6.1.1 Hard Handover

A hard handover is one where the existing radio link must be dropped for a small period of time before it can be taken over by another base station. In this type of handover, a call in progress is redirected not only from a base station to another base station but also from its current transmit–receive frequency pair to another frequency pair. An ongoing call cannot exchange data or voice for this duration. This break in call transmission is called call drop or call cut-off. Handover takes place in a few ms (at best in 60 ms) and the interruption is hardly discernible by the user. Handover to another cell is required when the signal strength is low and error rate is high. GSM systems perform hard handovers.

3.6.1.2 Soft Handover

3G CDMA systems have soft handover. Soft handover means an MS at the boundary of two adjacent cells does not suffer call drops due to handover in the boundary region. It gives seamless connectivity to an MS (Section 4.2.2.1). A method of soft handover that employs an offset to pseudo noise code is described in Section 4.2.2.2. Soft handover does not require breaking of the radio link for cell-to-cell transfer of a call. A mobile device can be connected to several base stations simultaneously.

3.6.2 Handovers in GSM

Figure 3.15 shows the different kinds of handover when a mobile station MS_1 moves from one cell to another or when the traffic through a specific stage becomes very high. These types of handover are described in the following subsections.

3.6.2.1 Inter-cell Handover

SDMA by multiple antennae at the same BTS aligned in different directions results in the formation of multiple cells. Section 3.3 described how the signal measurements are continuously performed at the RRM (Figs 3.12 and 3.13) sublayers in the MS, BTS, and BSC. The RRM is also responsible for handover management. When the signal strength goes weak due to several reasons (for example, the MS moving away from the cell in which it is presently localized to the boundary region of another cell), there is handover from a cell to another. This process is called inter-cell handover. The signal strength changes inversely with the square of the distance from the transmitter. Therefore, as distance of the MS from a BTS in an ith cell increases from s to $2s$, the signal strength decreases by factor of $(2s/s)^2 = 4$. Now, if the distance of the jth cell decreases from $2s$ to s, the signal strength increases by a factor of 4. There is a boundary region where the signal quality improves and error rates decrease on handover.

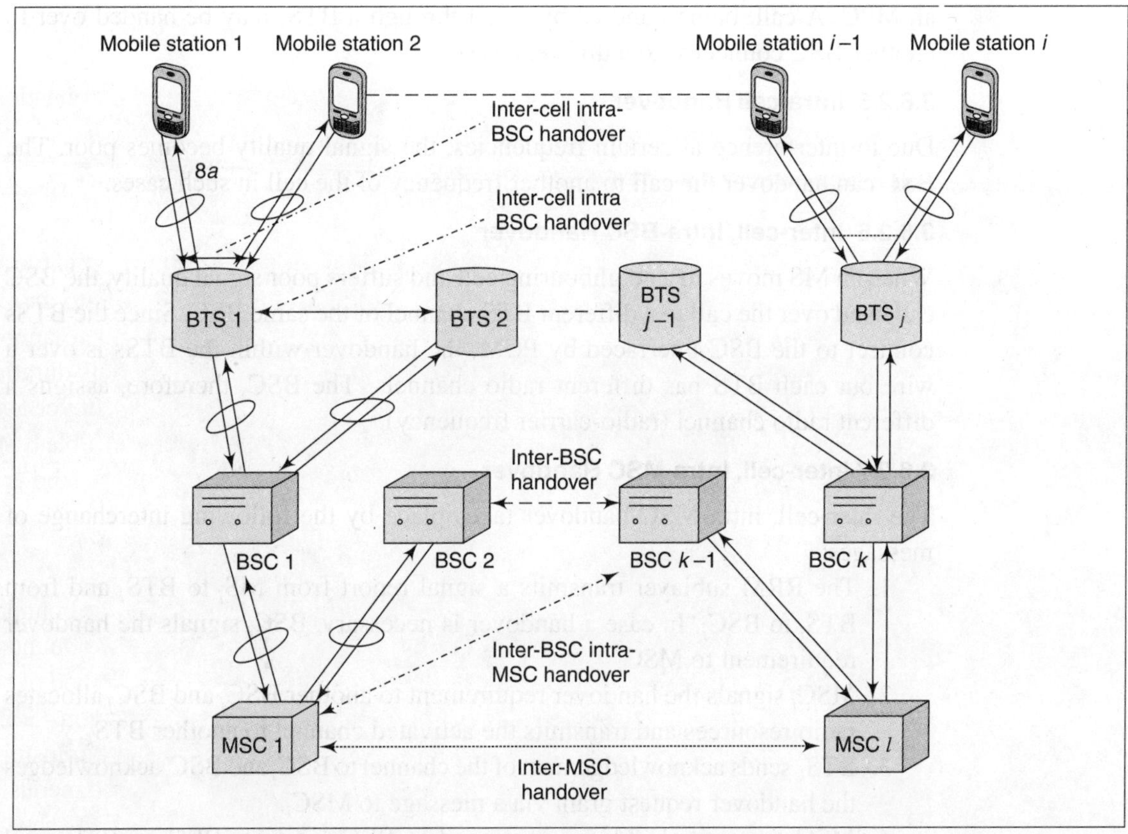

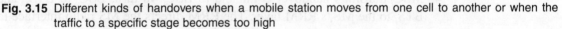

Fig. 3.15 Different kinds of handovers when a mobile station moves from one cell to another or when the traffic to a specific stage becomes too high

3.6.2.2 Inter-MSC Handover

Handover also takes place for load balancing when the traffic from the cells and BSCs is high. An ongoing call, which is being handled by a cell, may be handed over to another MSC. Since the two MSCs are interfaced through PCM (Fig. 3.14), the handover is performed over a wired line.

3.6.2.3 Inter-BSC Handover

Handover also takes place for load balancing when the traffic from the cells and BTSs is high. The BSCs connect to an MSC. A call, which is ongoing in a cell through a BTS, may be handed over to another BSC connected to the same MSC. Since the BSCs connect to the MSC interfaces by PCM, the handover is over a wired line.

3.6.2.4 Inter-BSC, Inter-MSC Handover

Handover also takes place for load balancing when the traffic from the cells and BTSs as well as the BSCs is high. The BTSs connect to a BSC and BSCs connect to an MSC. A call, being handled by a cell through a BTS, may be handed over to another BSC connected to a different MSC.

3.6.2.5 Intra-cell Handover

Due to interference at certain frequencies, the signal quality becomes poor. The BSC can handover the call to another frequency of the cell in such cases.

3.6.2.6 Inter-cell, Intra-BSC Handover

When an MS moves to a neighbouring cell and suffers poor signal quality, the BSC can hand over the call to a different BTS channel of the same BSC. Since the BTSs connect to the BSC interfaced by PCM, the handover within the BTSs is over a wire but each BTS has different radio channels. The BSC, therefore, assigns a different radio channel (radio-carrier frequency).

3.6.2.7 Inter-cell, Intra-MSC Handover

The inter-cell, intra-MSC handover takes place by the following interchange of messages:

1. The RRM sublayer transmits a signal report from MS_i to BTS_i and from BTS_i to BSC_i. In case a handover is necessary, BSC_i signals the handover requirement to MSC_i.
2. MSC_i signals the handover requirement to another BSC_j and BSC_j allocates radio resources and transmits the activated channel to another BTS_k.
3. BTS_k sends acknowledgement of the channel to BSC_j and BSC_j acknowledges the handover request grant via a message to MSC_i.
4. MSC_i transmits handover command to BSC_i, in turn, BSC_i to BTS_i, and BTS_i to the MS_i's RRM layer. The RRM directs the MS radio interface to operate at another channel linked to BTS_k.

There is handover of the call to BTS_k and voice or data interchange starts through BTS_k.

New generation (2.5G) networks ensure mobility by handover not only among the BTSs, BSCs, or MSCs but also among the in-between LANs. This ensures seamless (uninterrupted) connectivity to the user.

3.7 Security

Being a wireless radio-based network system, GSM is quite sensitive to unauthorized use of resources. GSM networks employ various security features. Some of these security features are designed to protect subscriber privacy and certain other features are used to secure the network against misuse of resources by unregistered users. For example, to control access to the network, the MS is required to use a PIN before it can access the network through U_m interface (Section 3.1.2.1). This section provides a detailed discussion of GSM security features.

3.7.1 Authentication

As discussed in Section 3.1.2.3, the operation and maintenance subsystem of the GSM network has an AuC (authentication centre) for authenticating an MS. The AuC first authenticates the subscriber MS and only then does the MSC provide the switching service [this service enables the MS to switch (communicate) to another terminal TE, which is also authenticated in case it is an MS]. Authentication algorithms use a random number sent by the AuC during the connection setup and an authentication key which is already saved in the SIM. Authentication algorithms used can differ for different mobile service providers.

3.7.2 TMSI

When an MS moves to a new location area, the VLR (visitor location register) assigns a TMSI which is stored in the SIM of the MS. Therefore, the identification of the subscriber during communication is done using not the IMSI but the TMSI. This ensures anonymous call number identity transmission over the radio channels. (Caller line identification provision is a supplementary service. The VLR assigned TMSI generates that ID.) This protects the MS against eavesdropping from external sources. (IMSI of the MS is its public identity. TMSI is the identity granted on moving to a particular location.)

3.7.3 Encryption

The BTS and the MS have to perform ciphering before call initiation or before connecting for receiving a call (Section 3.5.4). The MS uses a cipher (encryption key) for encryption. The cipher is a result of performing mathematical operations on (a) the cipher key saved in the SIM, and (b) the cipher number received from the BTS when the call setup is initiated. The BTS transmits the cipher number before a call is set up or transmitted. The mathematical operations performed are in accordance with the algorithms employed by mobile service providers. Only encrypted voice and data traffic and control channel data is transmitted to the BTS. This makes wireless communication secure (confidential) between the MS and the BTS. The encryption algorithm is identical for all mobile service providers. This ensures compatibility between the BTS, BSC, and MSC units made by different manufacturers. The BTS deciphers the voice and data channel by running a deciphering algorithm before communicating over the wired PCM (pulse code modulation) lines.

The random numbers used in authentication and ciphering processes are also known as *challenge* to the mobile station to generate the results (responses) of the algorithms and only if these results are correct, do the BTS and other units grant access to the *challenged* MS.

3.8 New Data Services

The GSM system provides data rates of TCH/13.4, TCH/HS11.4, TCH/12.8, TCH/F14.4, TCH/F9.6, and TCH/F4.8. These rates are good for transmission of voice-data but too low for high-speed data transfer. Speed enhancement is required for a GSM system to be able to provide data services such as transfer of large files and Internet access. New data services such as GPRS and HSCSD use different coding and multiplexing techniques to provide high transfer speeds to GSM users. The three major approaches to enhance transmission speed are as follows:

- Combining several slots in a packet-switched network. GPRS (general packet radio service) is an example this type of speed enhancement (Section 3.9).
- Combining several slots in a circuit-switched network. For example, HSCSD (high speed circuit switched data) is an improvement on GSM as it combines several time-slots for high-speed transmission of circuit-switched data (Section 3.10).
- Use of other technology such as DECT (digital enhanced cordless telecommunications system) which is used for short range communication (Section 3.11).

These new data services are explained in detail along with the techniques used for speed enhancement in the following sections.

3.9 General Packet Radio Service

General packet radio service (GPRS) is a speed enhanced data transmission service designed for GSM systems. Speed enhanced data transmission takes place by packetizing of data and simultaneous transmission of packets over different channels. The GPRS standard is defined by the European Telecommunications Standards Institute (ETSI). Section 1.1.2.6 described two switching modes—(a) *Circuit switching* in which a connection is first set up and then the entire data is transmitted through the path that has been set up during the connection, and (b) *Packet switching* in which packets of data at any given instant can take multiple time-slots, channels, paths, or routes depending on the idle slots available at that instant and the receiver assembles the packets into the original sequence in the data. GPRS is a packet-oriented service for mobile stations' data transmission and their access to the Internet. It uses the unused slots and channels in TDMA mode of a GSM network for packetized transmission from a mobile station. Data packets of a single mobile station are transmitted through a number of time-slots. GPRS deploys SGSNs (serving GPRS support nodes). An SGSN interfaces to BSCs (base station controllers) on one hand and to other SGSNs on the other hand. GPRS also uses GGSN (gateway GPRS support nodes) to interface to the SGSN on one hand and to a packet data network like the Internet on other hand. The BSCs also connect to the MSCs (mobile services switching centres) as in case of the GSM system. The GSM system described earlier can, therefore, be considered as a subsystem of a GPRS system. GPRS employs the GSM physical layer (Section 3.3) so that it connects the mobile stations to the Internet and packet data networks at higher data rates and also connects mobile stations for voice-data transmission.

Section 3.2.5 described the voice-data traffic channel rates of 22.4 kbps, 14.4 kbps, 11.4 kbps, 9.6 kbps, and 4.8 kbps in GSM. GPRS has four types of encoding schemes which allow transmissions at 21.4 kbps, 15.8 kbps, 13.4 kbps, and 9.05 kbps, named as coding schemes 4, 3, 2, and 1, respectively, as per increasing rate of possible transmission errors. There are eight TDMA time-slots. In case the data from a given TE (terminal) is transmitted in time slots 0, 1, 2,…, 7 simultaneously, the data rates are 21.4 kbps, 42.8 kbps, …, 171.2 kbps for coding scheme 4, and are 9.05 kbps, 18.2 kbps, …, 72.4 kbps for coding scheme 1.

Section 3.2.3 described half-duplex transmission in which there can be downlink at one instant and uplink at another instant after each 3.577 μs in GSM. All the eight available slots may not be available for a single mobile station for practical reasons. For the GPRS transmission, out of the eight slots, there can be k time slots used for receiving the bits and k' for transmitting in the eight-slot data frame. A maximum of k'' slots out of the eight can be used at an instant (k , k', and k'' are all less than or equal to eight). For example, a mobile station of class 10 has k, k', and k'' = 4, 2, and 5, respectively.

3.9.1 GPRS System Architecture

Figure 3.16 shows the GPRS system architecture. It shows four subsystems—RSS (radio subsystem), BSS (base subsystem), NSS (network subsystem), and GSS (gateway subsystem) consisting of a number of BSCs, MSCs, SGSNs, and GGSNs (gateway GPRS support nodes). The architectural elements are explained below:

- The RSS consists of a number of (MSs), (BTSs), and (BSCs), BSC_1, BSC_2, …, BSC_k.
- An MS having GPRS capability stores a CKSN (cipher key sequence number) similar to the cipher key stored in the SIM in GSM. It also stores a TLLI (temporary logical link identity) similar to the TMSI in the SIM.
- The NSS consists of a number of serving GPRS support nodes (SGSNs) $SGSN_1$, $SGSN_2$, …, $SGSN_l$, and mobile services (MSCs) MSC_1, MSC_2, …, MSC_j.
- The GPRS system creates a GPRS context which is stored in the MS as well as in the SGSN. The context has information of the status of MS, data compression flag, identifiers for the cell and channel for the packet data, and routing area information.
- An EIR (GPRS equipment identity register) stores the equipment data through the SGSN. EIR helps in the authentication, operation, and maintenance subsystems.

- Each SGSN and MSC in the NSS layer is connected to a number of BSCs at the RSS layer. The SGSNs use the frame relay protocol for this purpose.
- There are m home location-cum-GPRS registers (HLR/GRs) and n (VLRs). The GPRS register (GR) part in a HLR/GR stores the GPRS data and HLR part stores the information regarding mobile stations. The GR stores the information of the mobile station locations like the VLR in GSM. VLRs are connected to the MSCs as in the GSM subsystem.
- The GSS consists of the SGSNs and GGSNs and provides GPRS connections to the Internet and other PDNs (public data networks).

When a mobile station communicates voice-data in the GSM subsystem of the GPRS system, then the MSs, MS_1 to MS_i, connect to the transceivers, BTS_1 to BTS_j. Assume that the qth BSC interfaces to a number of BTSs and the rth MSC interfaces to a number of BSCs as shown in Fig. 3.16. When a mobile station communicates GPRS voice-data or data only packets in a GPRS system, then the mobile stations, MS_1 to MS_i, connect to transceivers, BTS_1 to BTS_j. Moreover, the kth SGSN interfaces to a number of BSCs. The kth SGSN is also connected to a

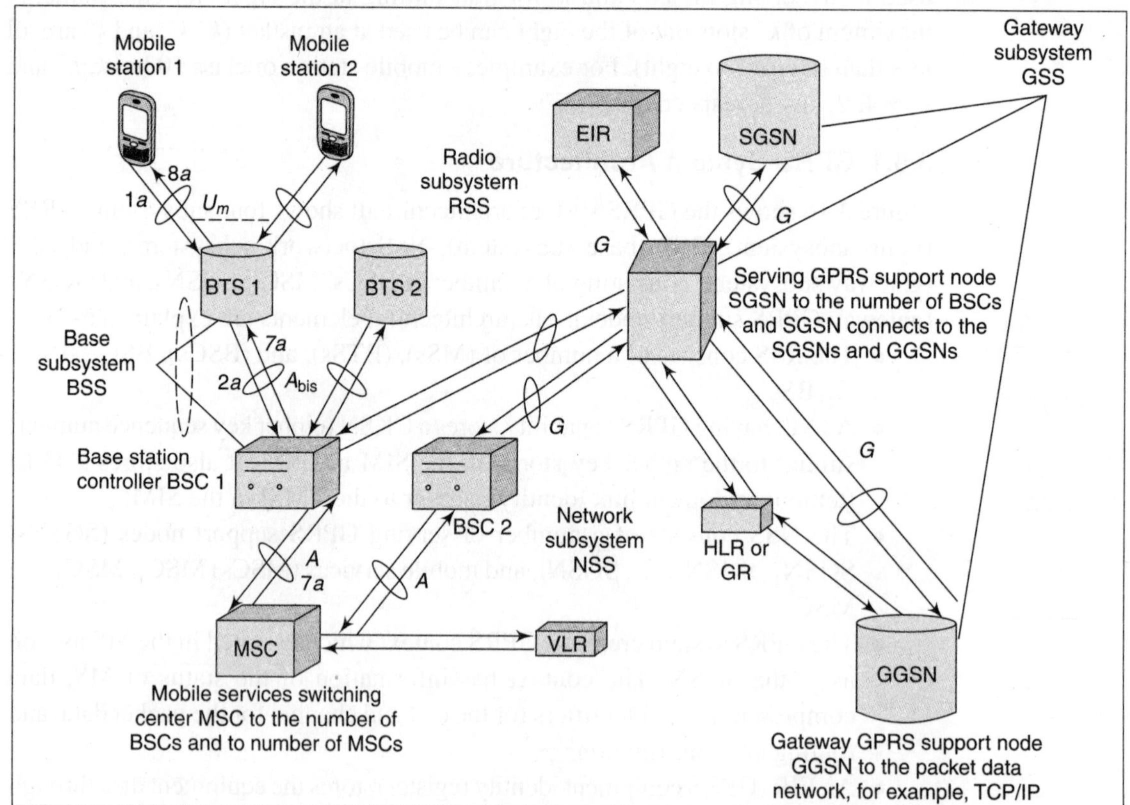

Fig. 3.16 GPRS system architecture—RSS (radio subsystem), BSS (base subsystem), NSS (network subsystem), and GSS (gateway subsystem)

number of SGSNs and GGSNs. The GGSNs interfaces to a packet data network, for example, a TCP/IP network.

3.9.2 GPRS Protocol Layers

As discussed in Section 3.3, a transceiver, subsystem, or node need not have all the seven OSI layers. They may also have some layers performing the function of the neighbouring layer(s). The GPRS protocol layers are similar to the GSM protocol layers as shown below:

- The MS has four layers—physical, data link, network, and application. Session presentation and transport layer issues are taken care of by the lower layers.
- The BSS has just three layers—physical, data link, and network. Transport and session layer functions are taken care of by network layer protocols.
- The SGSN and GGSN have four layers—physical, data link, network and transport. Presentation layer functions are performed by the lower layers.

The following subsections describe the signalling protocols in the layers between the MS and BSS, BSS and SGSN, and SGSN and GGSN (Figs 3.17, 3.18, and 3.19).

3.9.2.1 Mobile Station and Base Station Subsystem Signalling Protocols

Figure 3.17 shows all protocol layers between the MS and BSS. The *physical layer* between the MS and BTS is called the *radio* like in the GSM system and has similar functions.

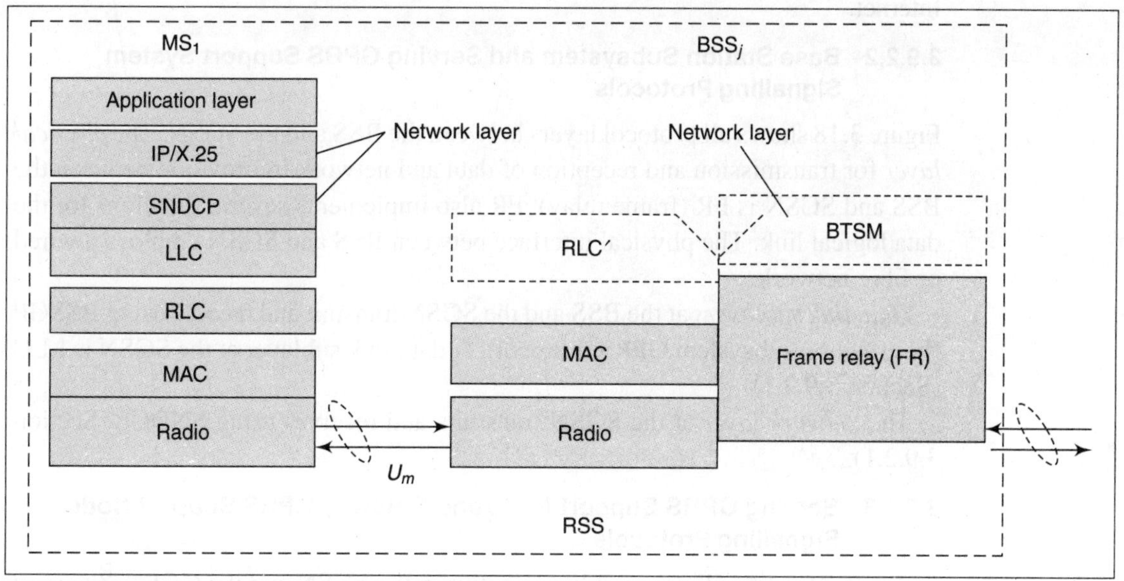

Fig. 3.17 All protocols layers of the MS and BSS

Data link layer protocols between MS and BSS link through the physical layer which has the interface U_m, as in the case of GSM. Data link layer controls the flow of the packets to and from the network layer and provides access to multiple services. GPRS data link layer at the MS has three sublayers—MAC (media access control), RLC (radio link control), and LLC (logical link control). A special LLC provides the FEC and ARQ (automatic repeat request) protocol. Data link layer at the BSS has two sublayers, MAC and RLC2. The RLC manages radio link resources issues. The MS can transmit maximum eight PDTCHs (packet data traffic channels). TCHs in GSM are formatted as packets and transmitted in PDTCHs. The BSS implements only RLC′ (a part of RLC) as the BSC and BSSGP (Section 3.9.2.2) handle the handover.

Network layer at the MS has two sublayers—IP/X.25 and SNDCP (sub-network dependent convergence protocol). These two sublayers are below a network sublayer for the convergence of IP/X.25 network sublayer above SNDCP and three sublayers at data link layer. Here, convergence means multiple ways of connecting network and data link layers through a common protocol (IP/X.25 in the present case). IP and X.25 are the packet-formatting protocols for the transmission and reception of packetized data. There are two services, voice-data (as in GSM) and the data only terminal service as in the Internet. MAC, LLC, and IP are used for the Internet service and RLC is used, for the voice-data link service. SNDCP is convergence protocol for both services in a GPRS system so that both can be delivered through IP/X.25.

The *application layer* at the MS provides end-to-end applications like voice and Internet.

3.9.2.2 Base Station Subsystem and Serving GPRS Support System Signalling Protocols

Figure 3.18 shows all protocol layers between the BSS and the SGSN. The *physical layer* for transmission and reception of data and network information between the BSS and SGSN is FR (frame relay). FR also implements several functions for the data logical link. The physical interface between BSS and SGSN employs a wired or fibre network.

Data link sublayers at the BSS and the SGSN transmit and receive using BSSGP (base station subsystem GPRS protocol). A data link sublayer at the SGSN is LLC (Section 3.9.2.1).

The *network layer* at the SGSN transmits and receives using SNDCP (Section 3.9.2.1).

3.9.2.3 Serving GPRS Support Node and Gateway GPRS Support Node Signalling Protocols

Figure 3.19 shows all protocol layers between the SGSN and the GGSN. *Physical layers* between the SGSN and GGSN are layer 1 (L1) protocols of the Internet or other PDN (PSTN, ISDN, and PSPDN). *Data link layer* protocol layers between

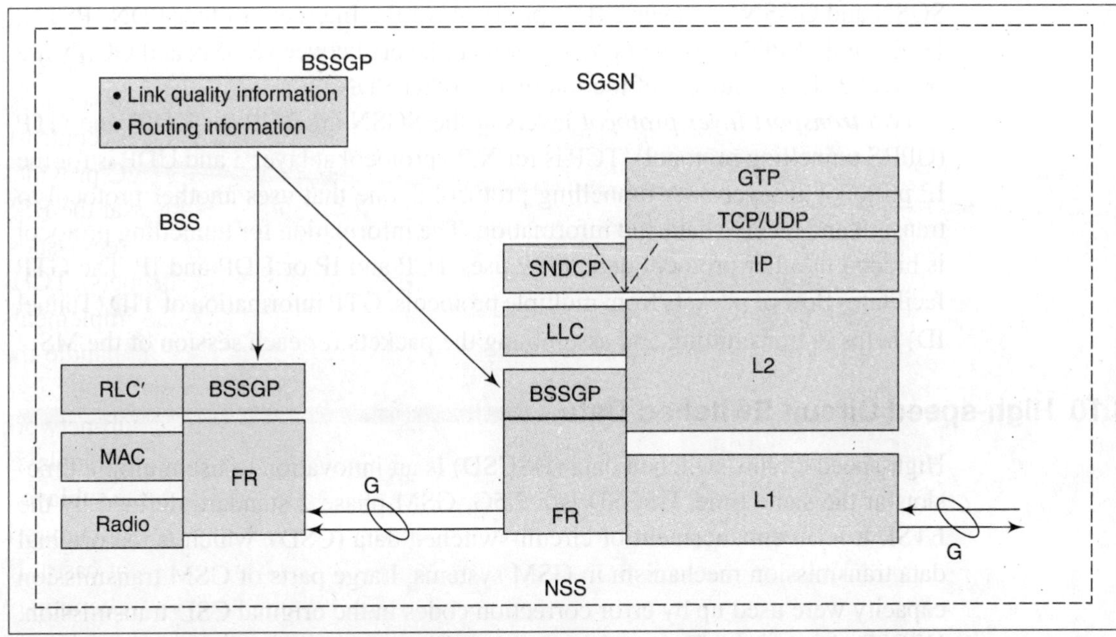

Fig. 3.18 All protocol layers between the BSS and SGSN

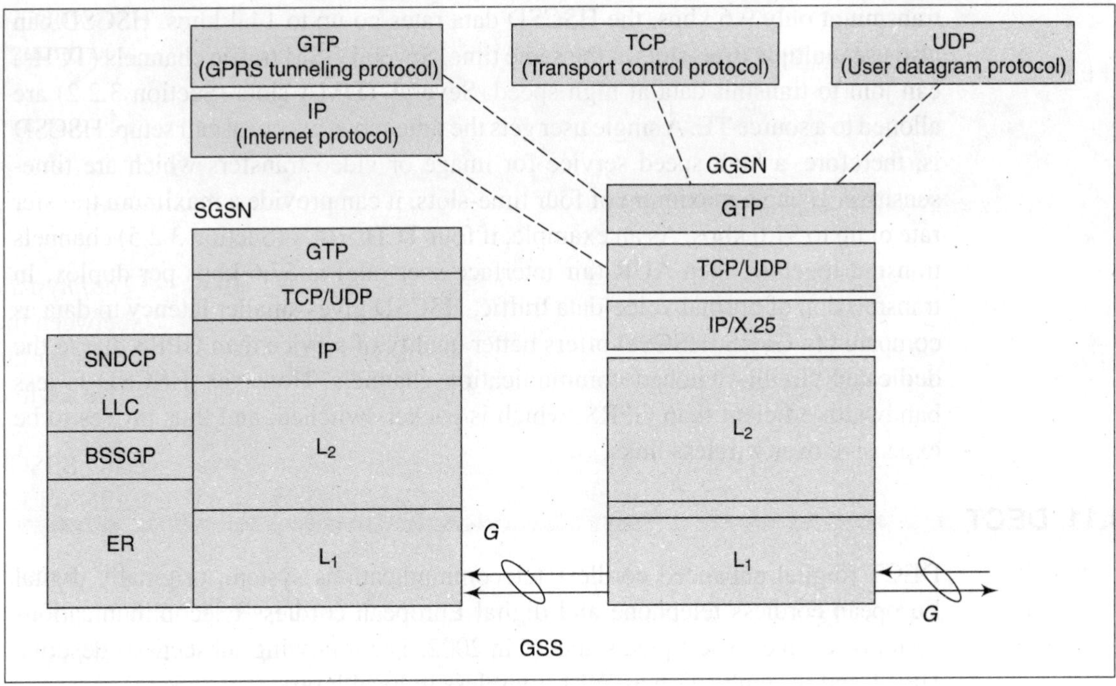

Fig. 3.19 All protocols layers between the SGSN and the GGSN

SGSN and GGSN are layer 2 (L2) protocols of the Internet or other PDN (PSTN, ISDN, and PSPDN). *Network layer protocol* layers between SGSN and GGSN are IP layer 3 (L3) protocols of the Internet or other PDN.

Two *transport layer protocol* layers at the SGSN are TCP [or UDP] and GTP (GPRS tunnelling protocol). TCP is for X.25 protocol at layer 3 and UDP is for the IP protocol at layer 3. A tunnelling protocol is one that uses another protocol to transmit and receive data and information. The information for tunnelling protocol is hidden in other protocol data. GTP uses TCP and IP or UDP and IP. The GTP facilitates flow of packets from multiple protocols. GTP information of TID (Tunnel ID) helps in transmitting and assembling the packets for each session of the MS.

3.10 High-speed Circuit Switched Data

High-speed circuit-switched data (HSCSD) is an innovation to use multiple time-slots at the same time. HSCSD is a 2.5G, GSM phase 2 standard defined by the ETSI. It is an enhancement of circuit-switched data (CSD), which is the original data transmission mechanism in GSM systems. Large parts of GSM transmission capacity were used up by error correction codes in the original CSD transmission. HSCSD, however, offers various levels of error correction that can be used in accordance with the quality of the radio link. As a result, so where CSD could transmit at only 9.6 kbps, the HSCSD data rates go up to 14.4 kbps. HSCSD can also use multiple time-slots at the same time. Several GSM traffic channels (TCHs) can join to transmit data at high speed. Several TDMA slots (Section 3.2.2) are allotted to a source TE. A single user gets the time-slots, except at call setup. HSCSD is, therefore, a high speed service for image or video transfers which are time-sensitive. Using a maximum of four time-slots, it can provide a maximum transfer rate of up to 57.6 kbps. As an example, if four TCH/F14.4 (Section 3.2.5) channels transmit together, then AUR (air interface user rate) is 57.6 kbps per duplex. In transmission of normal voice-data traffic, HSCSD gives smaller latency to data as compared to GPRS. HSCSD offers better quality of service than GPRS due to the dedicated circuit-switched communication channels. However, HSCSD is less bandwidth efficient than GPRS, which is packet-switched, and thus proves to be expensive over wireless links.

3.11 DECT

DECT (digital enhanced cordless telecommunications system, originally digital European cordless telephone and digital European cordless telecommunications system) became an accepted standard in 2002. The following subsections describe DECT and its application as WLL (wireless in local loop).

3.11.1 Frequencies and Functioning

DECT radio-carrier frequencies are 1880–1990 MHz. DECT is used in short-range communication. The distance between the transmitter and receiver is small. Same frequency in different time-slots is, therefore, used for uplink and downlink radio carriers. There are 10 radio-carrier frequency bands in this frequency region. Each link provides 120 channels for uplink and 120 channels for downlink. This is because each radio-carrier frequency has TDD (time division duplex) frame with 12 uplinks and 12 downlinks. Time division duplex means that the uplink and downlink instants are in separate time-slots. Each TDD time-slot is of 417 μs. The TDD frame duration is $(12 + 12) \times 416.7$ μs = 10 ms for each of the 10 radio carriers. DECT, like GSM, uses GMSK for transmitting 1s and 0s (Section 1.1.2.4).

A 1.728 MHz frequency band is used for each channel. Each channel during each successive 4 ms has access to 24 TDMA channels for each radio-carrier band. The channel bit rates for DECT are 1.152 Mbps (million bits per second). Speech coding is ADPCM (adaptive differential pulse code modulation) that gives a voice-data traffic rate of 32 kbps.

The two major differences between the DECT and GSM multiple access techniques are—(a) use of the same radio-carrier frequency for uplink and downlink and (b) use of TDD-TDMA slots. TDD of DECT differs from the half-duplex transmission between MS and BTS (Section 3.2.2). The carrier frequency bands are different (45 MHz more for downlink) but the time-slot is just 3.577 μs more for uplink (less than 1-bit interval of 3.692 μs). TDD of DECT also differs from GPRS (Section 3.9). During transmission by class-10 MS, there can be four receiving time slots and two transmitting time-slots in the data frame of the same frequency channel. A maximum of five slots can be used at an instant out of eight.

DECT 1900 operates at 1880–1990 MHz for the uplink and downlink frequencies for full-duplex channels of 10 radio carriers. The frequency ranges in which a carrier can operate are 1890.0 MHz ± 864 kHz, 1891 MHz ± 864 kHz, 1892 MHz ± 864 kHz, …, 1898 MHz ± 864 kHz, and 1899 MHz ± 864 kHz. Each radio carrier has 12 downlink time-slots (SL_0 to SL_{11}) and 12 uplink time-slots (SL_{12} to SL_{23}), a total 24 time-slots. Hence, the number of channels are $(12 + 12) \times 10 = 240$, 120 for uplink and 120 for downlink.

Each data burst is of 416.7 μs. A guard space at the beginning and the end, each of interval 26 μs (equal to 30-bit transmission time interval) is reserved to account for the delays in signals and computational time. The effective time for the data bits is, therefore, $(416.7 - 26 - 26) \approx 364$ μs. 420 bits are transmitted in the 364 μs interval. The data transmission rate is $24 \times 480/10\ \text{ms}^{-1} = 1152$ kbps. A GMSK signal is modulated and transmitted at 1152 kbps (= 0.868 μs/bit). The distribution of bits when transmitting the traffic is as follows:

- 30 bits guard space, 32 bits preamble in synchronizing field (*S*), 388 bits data, and 30 bits guard space in other fields, or
- 30 bits guard space, 32 bits preamble in synchronizing field (*S*), 64 network control bits, 320 bits user channel data in a 32 kbps mode and 4 parity bits (or transmission quality), and 30 bits guard space in other fields. 320-bit

data in a 25.6 kbps mode (also called protected mode) has four subsets of 80-bits each. Each subset has a 64-bit data segment plus a 16-bit CRC checksum.

3.11.2 Architecture, Services, and Protocols

Figure 3.20 shows the DECT architecture. It shows teleservices and supplementary services supported by the DECT systems. There are 12 bearer channels (each in one SL) per carrier. The terminal may be a portable wireless telephone (PWT) or a

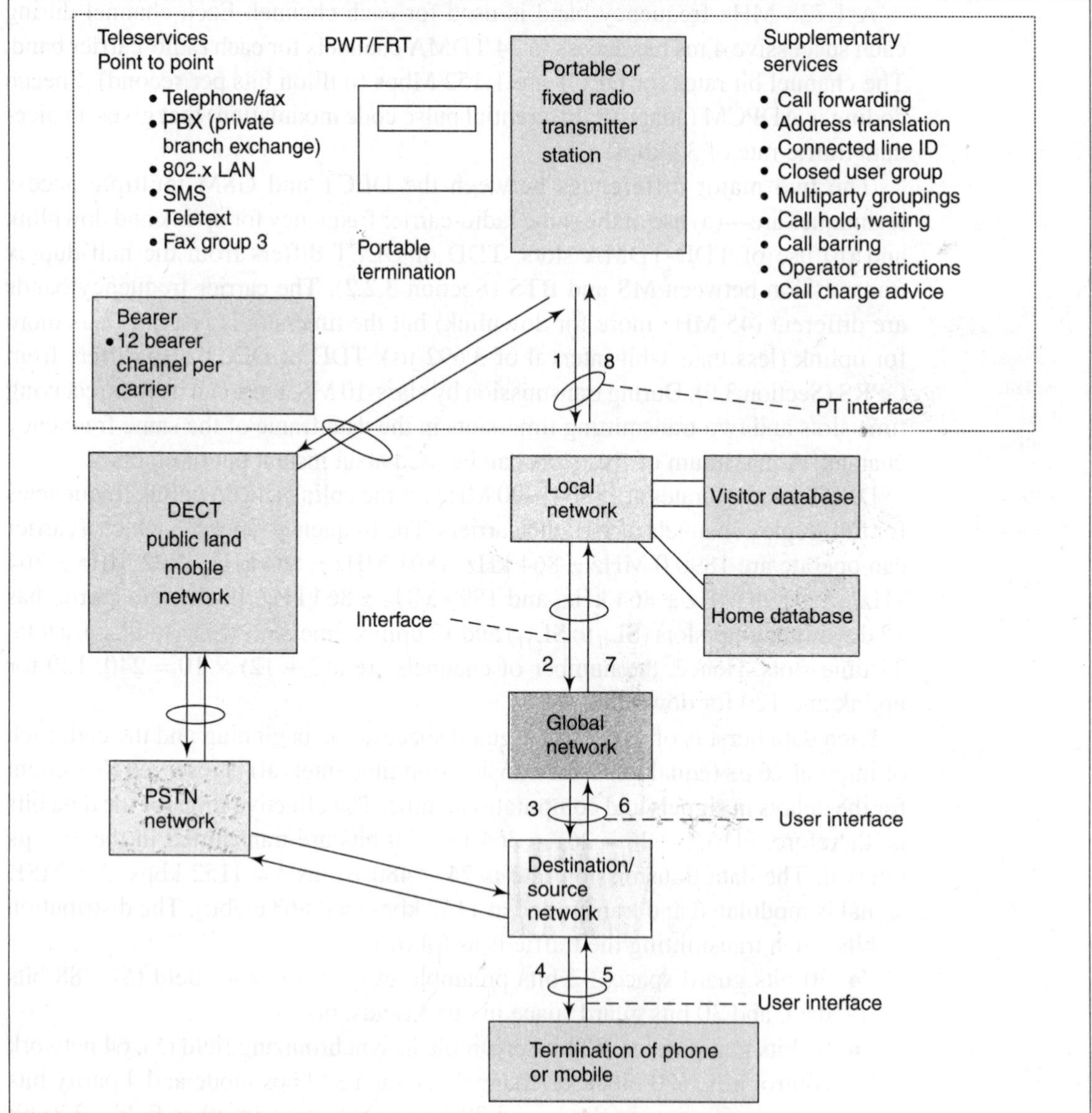

Fig. 3.20 DECT system architecture and services

fixed phone with radio interface (FRT). The PWT or FRT connects to a public land mobile network for calling to a mobile or to a local network. The local network has a visitor database (similar to VLR in the MSC) and home database (similar to HLR in the MSC). The local network can interface to a global network or an ISDN network. The latter connects to a destination–source network. This network acts as an interface to the destination terminal (PWT, FRT, or mobile).

Figure 3.21 shows the protocol layers in DECT. There are two planes—control plane (C-plane) and user plane (U-plane). The figure also shows the MAC layer functions. There is an additional network layer in the control plane as compared to the user applications plane. It transmits the DLC layer data directly to U-plane from the C-plane.

3.11.3 WLL Application

WLL (wireless local loop) also called FRA (fixed-radio access) or RITL (radio in the loop) connects a user to PSTN networks or broadband Internet using radio signals. WLL includes fixed and cellular systems, cordless access systems, and

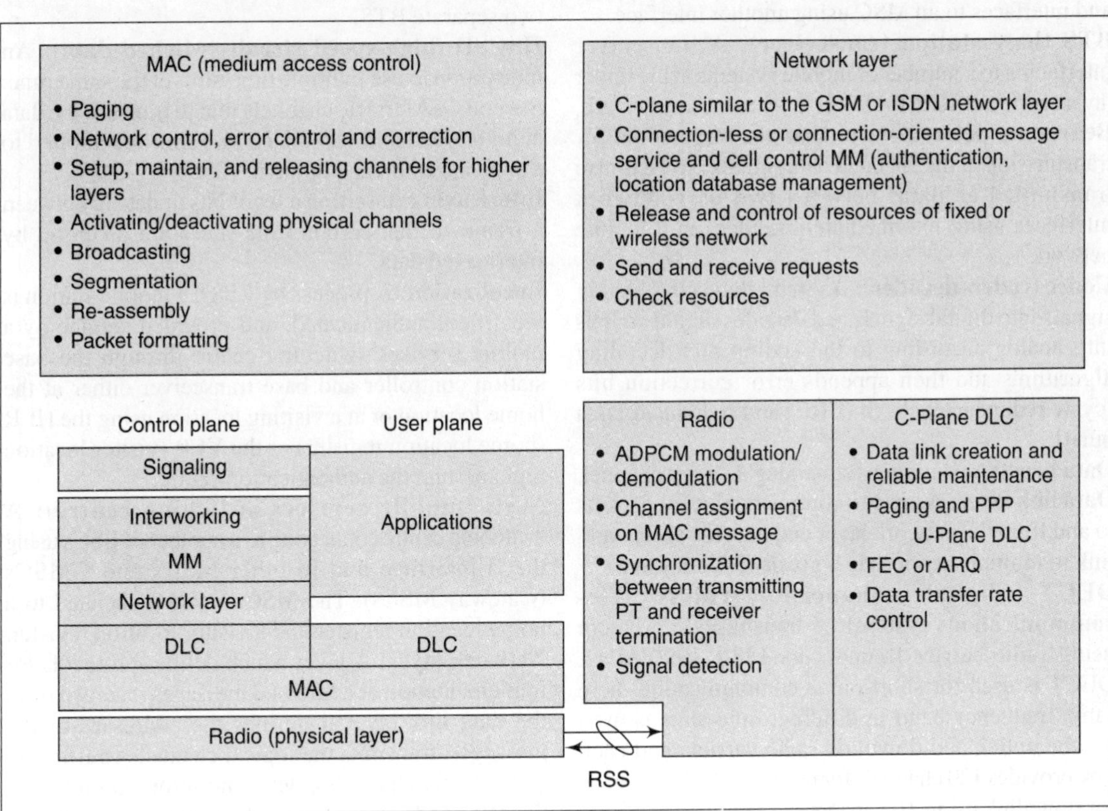

Fig. 3.21 Protocol layers in DECT

proprietary fixed radio access systems. WLL is implemented over DECT or other technologies to provide the link between two terminals (PWT or FRT), including CDMA, TDMA, GSM, and UMTS 3G. WLL, in addition to being an alternate system, helps in providing telecommunication and broadband services where wired or fibre lines do not exist.

Keywords

Asynchronous data transfer: Data transmission at variables rates, not maintaining constant time intervals between the bursts or frames. Asynchronous data transfer supports handshaking or acknowledgement of data. However, even if there is no acknowledgement, the data flow is maintained by using FEC plus buffers, which causes variable delays in reception.

BSC (base station controller): A controller that connects to a number of BTSs using the A_{bis} interface and interfaces to an MSC using another interface.

BTS (base station transceiver): A transceiver interfacing to a number of mobile systems at the upper layer and to a BSC at the lower layer using an interface.

Bearer services: Services that are responsible for transmission of the encoded data signals (voice is also transmitted as data) between two user-network interfaces using intermediate interfaces in a mobile network.

Codec (coder–decoder): A circuit that codes analog signals into digital signals and decodes digital signals into analog according to the coding and decoding algorithms and then appends error correction bits [cycle redundancy check (CRC) and redundant Data burst].

Data brust: A continuous streaming data in a channel.

Data link layer: A layer to control the flow of packets to and from the network layer and provide access and link to multiple services in a given network route.

DECT (digital enhanced cordless tele-communications system): A transmission protocol using radio-carrier frequencies 1880–1990 MHz. DECT is used for short-range communication; here same frequency band in different time-slots is used for the uplink and downlink radio carriers and each link provides 120 channels for uplink and 120 channels for downlink using 10 carriers.

FEC (forward error correction): The bits that insert redundant data, along with the data to be transmitted, after suitable coding. FEC helps detect errors during the transmission at the other end of the interface, as a result, the receiver selects only that portion which contains no recognizable errors. FEC transmission reduces data rate but helps in broadcasting without handshaking.

Handover: A mechanism to hand over the mobile device to the neighbouring cell when it crosses the boundary between two cells, which are serviced by two separate BTS.

HSCSD (high-speed circuit-switched data): An innovation to use multiple time-slots at the same time. Several GSM traffic channels join to transmit the data at high speed and several TDMA slots are allotted to a source terminal equipment.

Interleaving: Inserting a set of bits or data in between a frame so that certain time-slices are taken up by interleaved data.

Localization: A process by which a mobile station is identified, authenticated, and provided service by a mobile servises switching centre through the base station controller and base transceiver either at the home location or at a visiting location using the HLR (home location register) or the VLR (visitor location register) and the authentication centre.

MSC (mobile services switching centre): A switching centre connecting to a number of BSCs using the *A* interface and to other MSCs and GMSCs (Gateway MSCs). The MSC is also associated to a home location register and a visitor location register.

Network layer: A layer which defines protocols for implementation of addressed messages received from the data link layer it defines the addresses of the messages, transmits the logical channels (traffic and control data channels), data, and information bits to the data link layer from an address (SAP), and receives the logical channels' data and information bits from the data link layer for the addressed SAP.

Non-transparent data transfer: Data transfer through an interface using the physical, data link, and flow control layer protocols.
OSI (open systems interconnection) model: A model deploying protocols for the seven layers—physical (layer 1), data link (layer 2), network (layer 3), transport (layer 4), session (layer 5), presentation (layer 6), and application (layer 7). When the transceiver receives the signals, they are in order of layer 1 to layer 7, and in reverse order when transmitting.
Physical layer: A layer which transmits or receives through wireless, wired, fibre, or microwave networks after formatting or multiplexing. The protocol for the physical layer in GSM bearer service provides for FEC (forward error correction) also.
SIM: A card-like circuit, which personalizes the MS to the user and stores the following information, which does not change when the MS moves into another location—(i) international mobile subscriber identity (IMSI), (ii) card serial number and type and temporary information including a temporary mobile (dynamic) cipher key for encryption, temporary subscriber identity (TMSI), and location area identification (LAI).
Supplementary services: Services offered to the caller and destination devices in addition to teleservices, for example, caller line forwarding (redirection), caller line identification, additional services such as connected line identification to the caller, closed user group formation, multiparty groupings, call holding, call waiting, call barring to and from specified numbers or groups, operator restrictions on provisioning of certain features to a caller (for example, Internet access or games).
Synchronous data transmission: Data transmission from a transceiver at a fixed rate maintaining constant phase differences (and thus time intervals) between the data bursts or frames.
Transparent data transfer: Data transfer using an interface for service when using only the physical layer protocol.
WLL (wireless local loop): A protocol to connect a user to PSTN networks or broadband Internet using radio signals, includes fixed and cellular systems, cordless access systems, and proprietary fixed radio access systems. It is implemented over DECT, CDMA, TDMA, GSM, UMTS 3G, or other technologies to provide a link between two terminals.

Review Questions

1. What are the services provided in a GSM system? Explain how a mobile station connects to and talks with another mobile station? How will the in-between interfaces differ when a mobile station connects to a PSTN destination?
2. What features of a GSM system are provided with the help of the SIM card in a mobile station?
3. Describe transparent and non-transparent data transmission. How does the FEC help in reducing the BER?
4. Explain using schematic diagrams the synchronous, asynchronous, and synchronous packet types of data transfer.
5. Show the various subsystems and units in the GSM system architecture. How do these subsystems and units differ from those in GPRS?
6. Explain how interconnected mobile services switching centres enable a mobile station to communicate to another over long distances. How does the network subsystem change when the communication is to a PSTN terminal?
7. How does a single base transceiver provide access to many mobile stations in a GSM system?
8. How are the frequency channels and time-slots accessed by a mobile station?
9. Describe the format of a data burst. How is the data from a mobile station arriving by multiple paths synchronized? How is the data arriving from multiple mobile stations synchronized before granting access to them in different time-slots?
10. What is the need for defining a hyperframe of about 3.5 hours?
11. How is a traffic multiframe formatted?
12. Discuss the format of a channel data multiframe. How do the channel data multiframe and traffic multiframe interleave?

13. Describe the format of a superframe.
14. Describe data interleaving among different channels. Describe data interleaving for background noise.
15. Explain TCH/13.4, TCH/HS11.4, TCH/12.8, TCH/F14.4, TCH/F9.6, and TCH/F4.8 traffic.
16. Describe four kinds of control data bursts. Explain application of the control channel data.
17. How is an SMS message transferred?
18. Describe the protocols between the different units in GSM system architecture. Why is LAPD$_m$ used for GSM radio interface and not LAPD?
19. How is a mobile station localized to a new location?
20. How do the various types of calls (mobile-originated calls to PSTN destination, mobile-originated calls to mobile, and mobile-terminated PSTN calls) use the RSS and NSS units and interfaces?
21. Explain the message exchanges between the mobile station and the base transceiver.
22. Describe the process of call handover when a mobile station moves.
23. How is confidentiality maintained during a call using ciphering and encryption processes?
24. What is the need of anonymity during call transmission over the air? How is the anonymity maintained?
25. Describe authentication and access grant processes in GSM.
26. Describe the functions, architecture, and protocols of a GPRS system.
27. When is HSCSD used? How does the high-speed transmission occur in HSCSD? How is HSCSD data transfer different from that in GPRS?
28. Describe DECT system functions, architecture, and protocols.
29. What are the various protocols used in WLL? What are the services provided by WLL?

Objective Type Questions

Pick the correct or most appropriate statement from among the choices given.

1. GSM mobile stations and transceivers transmit and receive
   (a) full-duplex or half-duplex synchronous, asynchronous, or synchronous packet data by circuit switching.
   (b) full-duplex or half-duplex synchronous or asynchronous circuit-switched data.
   (c) full-duplex synchronous, asynchronous, or synchronous packet-switched data.
   (d) full- or half-duplex synchronous voice-data and synchronous packet SMS data.
2. Transparent data transfer rates in GSM are higher than non-transparent rates due to
   (a) use of forward error correction (FEC) for error correction.
   (b) use of TDMA slots.
   (c) use of acknowledgements.
   (d) use of forward error correction (FEC) for error correction and no waiting time spent for the acknowledgements.
3. Let *p* PIN, *q* PIN unblocking key, *c* cipher key, *a* authentication key, *t* TMSI, and *i* IMSI. A SIM card at a GSM mobile station saves
   (a) *p*, *c*, *t*, and *i*.
   (b) all of these.
   (c) all of these except *q* and *c*.
   (d) *p*, *q*, *c*, and *a*.
4. At a mobile station in a GSM system
   (a) three algorithms execute—one for authentication, second for ciphering, and third for encryption of the traffic data. The authentication and ciphering take place after two random numbers are sent by the base transceiver but before the mobile station starts transmitting voice-data.

(b) two algorithms execute—one for authentication and second for ciphering after two random numbers are sent by the base transceiver and during the period the mobile station transmits voice-data.
(c) two algorithms execute—one for authentication and second for ciphering using authentication and cipher keys stored at the SIM before the mobile station starts transmitting voice-data.
(d) authentication and encryption algorithms execute before the voice-data transmission starts.

5. In GSM system architecture, several mobile stations
(a) interface to a base station controller, several base station controllers to a base transceiver, and several base transceivers to a mobile services switching centre.
(b) interface to a base transceiver, several base transceivers to a base station controller, and several base station controllers to a mobile services switching centre.
(c) interface to a mobile services switching centre, several switching centres to a base station controller, and several base station controllers to a base transceiver.
(d) interface to a mobile services switching centre and several switching centres to base station controllers.

6. GSM900 uses uplink frequency-slices from the 890.1 MHz ± 50 kHz to 914.9 ± 50 kHz. Each channel has eight time-slots. Assume two channels are reserved for the system and ±50 kHz is the guard space. The number of mobile stations which can access a transceiver and transmit voice traffic channels are
(a) 11×8.
(b) 24×8.
(c) 22×8.
(d) $50 \times 22 \times 8$.

7. Data burst format in a GSM system is as follows:
(a) guard space, three trailing bits, 57 bit traffic data, one signalling bit, 26 training bits, one signalling bit, 57 bit traffic data, three trailing bits, and guard space.
(b) guard space, four trailing bits, 58 bit traffic data, one signalling bit, 26 training bits, 58 bit traffic data, four trailing bits, and guard space.
(c) guard space, four trailing bits, 57 bit traffic data, 64 training bits, 57 bit traffic data, four trailing bits, and guard space.
(d) guard space, 64 bit traffic data, 26 training bits, 64 bit traffic data, four trailing bits, and guard space.

8. Consider a GSM system. Enhanced full rate traffic channel gives
(a) low voice quality at full 22.8 kbps but transmission errors are very low.
(b) high voice quality at full 22.8 kbps data rate.
(c) high voice quality at full 22.8 kbps data rate and very low bit error rate.
(d) high voice quality when transmission errors are low.

9. Consider a GSM system. A *timing advance* is required
(a) for synchronization of the reflected signal reaching by longer path than the reflected signal reaching by shorter path.
(b) for synchronization of the transmitter and receiver clocks in synchronous data transmission.
(c) for synchronization when a base transceiver receives a signal from a far off mobile station compared to the short distance mobile station.
(d) for compensating for frequency drifts.

10. SMSs are transmitted from a GSM mobile station
(a) along with a spare capacity in the voice traffic channel data transmitting with half rate of 11.4 kbps and have maximum 128 characters.
(b) along with a spare capacity in signalling channel data transmitting with 782 bits as dedicated control channel data in 1 s having maximum 160 characters.
(c) as signalling channel data transmitting with 950 bits as dedicated control channel data in 1 s having maximum 256 characters.
(d) in idle time-slots of voice-data traffic channel.

11. Consider a GSM system
    (a) A super frame can have 52 traffic multiframes or two sets of 26 control multiframes and a hyperframe has 2048 multiframes.
    (b) A super frame can have 51 interleaved traffic multiframes and control multiframes and a hyperframe has 1024 multiframes.
    (c) Super multiframe transmits in 6.12 s and multiframe in ~3 hour 29 minutes.
    (d) A hyperframe can have 26 traffic multiframes.
12. In the GSM system
    (a) mobile station, base transceiver, base station controller, and mobile services switching centre protocols have four layers—application, physical, data link, and network.
    (b) mobile station has application, network, data link, and physical layer, and base transceiver, base station controller, and mobile services switching centre protocols have just two layers—physical and network.
    (c) mobile station, base transceiver, base station controller, and mobile services switching centre protocols have two layers—physical and data link.
    (d) mobile station, base transceiver, base station controller, and mobile services switching centre protocols have just three layers—physical, data link and network.
13. Consider the following GSM mobile service functions—(i) f1—Full- or half-duplex access, (ii) f2—SDMA, TDMA, and FDMA, (iii) f3—Bursting and framing, (iv) f4—Synchronizing the mobile stations and mobile station path delays corrections, (v) f5—Frequency correction, (vi) f6—Coding, FEC, CRC, data interleaving, and encryption, (vii) f7—Error detection, correction, and blocking the data not correctable, (viii) f8—GMSK digital modulation and transmission, (ix) f9—Demodulation and reception, (x) f10—Decryption, (xi) f11—Decoding
    (a) Physical layer at the mobile station performs all the above functions.
    (b) Physical layer at the mobile station performs all the above functions except f6, f7, f10, and f11 and these functions are handled by the data link layer.
    (c) Physical layer at the mobile station performs all the above functions except f6 and data flow control are the functions of data link layer.
    (d) Physical layer at the mobile station and base transceiver perform all the above functions.
14. A GSM service visiting location register registers information of the currently associated mobile stations.
    (a) The information is about their TMSI, IMSI, and MSISDN.
    (b) The information is about their HLR and IMSI.
    (c) The information is about their HLR, IMSI, and MSISDN.
    (d) The information is about their HLR and TMSI.
15. In a GSM system, the following message sequence takes place at base transceiver before the voice or data exchanges
    (a) channel grant, cipher command, and call setup.
    (b) call setup, channel grant, authentication command, and cipher command.
    (c) channel grant, authentication command, cipher command, and call setup.
    (d) channel grant, cipher command, encryption command, and call setup.
16. In a GSM system, GSM inter-cell intra-mobile services switching centre handover takes place by interchange of messages in the following sequence
    (a) weak signal report from mobile station to mobile services switching centre and the centre signalling to handover to another base station controller.
    (b) weak signal report from mobile station to base station controller, handover requirement to mobile services signal centre, and the centre signaling handover request to another base station controller.
    (c) weak signal report from mobile station to base station controller, handover request to another base station controller.

   (d) weak signal report from base station controller and handover request to another base station controller through a base transceiver.

17. Assignment of temporary mobile subscriber identity by a visiting location register into the GSM mobile station SIM card
   (a) ensures confidential call number identity transmission over the radio channels.
   (b) ensures anonymous call number identity transmission over the radio channels.
   (c) ensures service by nearest base transceiver.
   (d) ensures authentication by the authentication centre in which the mobile station registers on a visit to another location area.
18. GPRS (general packet radio service) is a
   (a) circuit switched-cum-packet-oriented service for mobile stations' data transmission and their access to the Internet and it deploys the unused slots and channels in TDMA mode of a GSM network for packet transmission from a mobile station.
   (b) packet-oriented service for mobile stations' data transmission and their access to the Internet and it deploys the unused slots and channels in TDMA mode of a GSM network for packet transmission from a mobile station.
   (c) synchronous packet-oriented service for mobile stations' data transmission and their access to the Internet and it deploys the time-slots sequentially in channels in TDMA mode of a GSM network for packet transmission from a mobile station.
   (d) asynchronous packet-oriented service for mobile stations' data transmission and their access to the Internet and it deploys the unused channels in FDMA mode of a GSM network for packet transmission from a mobile station.
19. DECT system frequency bandwidth, TDMA frame interval, number of TDMA slots, and channel access bit rates are
   (a) 270 kHz, 577 μs, 8, and 1.152 Mbps, respectively.
   (b) 1.728 MHz, 577 μs, 24, and 1.152 Mbps, respectively.
   (c) 1.728 MHz, 4 ms, 24, and 1.152 Mbps, respectively.
   (d) 22.4 kHz, 4 ms, 24, and 22.4 kbps, respectively.
20. HSCSD offers high speed because
   (a) several traffic channels can join to transmit the data at high speed. Several TDMA slots are allotted to a single source terminal.
   (b) several control and traffic channels can join to transmit the data at high speed. Several FDMA slots are allotted to a single source terminal.
   (c) several control and traffic channels can join to transmit the data at high speed. Several TDMA slots are allotted to multiple source terminals. It uses 8-PSK communication to achieve the higher speeds.
   (d) several packet-switched traffic channels can join to transmit the data at high speed. Several TDMA slots are allotted to a single source terminal.

Chapter 4

Wireless Medium Access Control and CDMA-based Communication

This chapter describes the problems faced during access to the transmission medium by multiple wireless systems for a given instant and frequency band. CDMA is a big step forward for medium access control (MAC). This chapter will also describe the features of CDMA systems for the transmission of multi-rate and multi-carrier signals including voice, data, picture, and video signals. This chapter also explains various CDMA-based standards such as IS-95, cdmaOne, IMT-2000, WCDMA, CDMA2000, i-mode, and OFDMA. This chapter describes the following concepts in detail.

- Modulation methods for medium access
- Multiplexing methods for enabling multiple accesses to the medium
- Medium access control (MAC) for transmission without collision (interference) in case of multiple terminals attempting to access a network
- Medium access control problem of receiving distinctly, the signals from exposed and hidden terminals, and the problem of drowning of far terminal signals by near terminal signals
- CDMA technique for a large number of channels and terminals accessing the medium at the same time using the same frequency band
- Direct sequence spread spectrum (DSSS) and frequency hopping spread spectrum (FHSS) techniques
- CDMA communication of multi-rate and multi-carrier signals of voice, data, picture, and video
- CDMA 2G+ IS-95 (also called cdmaOne), CDMA2000, IMT-2000, WCDMA, i-mode, and OFDMA communication systems

4.1 Medium Access Control

This section describes the access to transmission media in wireless networks through various modulation methods and medium access control by multiplexing the modulated signals. The following subsections will discuss the basic equations for amplitude of the modulated signals as a function of their frequencies and phase angles. When a number of signal sources attempt to access a wireless medium simultaneously, networks encounter the problem of receiving signals from each radio carrier distinctly. This is because the signals (which are actually electro-magnetic radiations) tend to interfere with each other when they are simultaneously transmitted through a medium. Also, networks encounter the problems of signals from hidden and exposed terminals as well as near and far off terminals, as explained in the following subsections. To overcome these problems, communication system receivers extract distinct signals from various terminals in the presence of signals divided into different cells, time-slots, frequencies, and codes (SDMA, TDMA, FDMA, and CDMA signals). These problems and processes are detailed in the following subsections.

4.1.1 Medium Access by Modulation

Voice-data or data signals propagate through the medium after modulation (Section 1.1.2.4) with a radio-carrier frequency. A wireless station accesses a medium by modulation of a radio carrier with the signal symbols (i.e., digitized form of the analog signal. Symbols are bits prepared for transmission after encoding of data and insertion of the error control and other bits).

Equation (1.3) gives the instantaneous value of signal amplitude, $s(t)$, at an instant t. A classical sinusoidal equation for $s(t)$ is

$$s(t) = S \times s_0 \sin\left(2\pi \times \frac{c}{\lambda_c} \times t + \varphi_{t0}\right) \text{ or}$$

$$s(t) = S \times s_0 \sin(2\pi \times f_c \times t + \varphi_{t0}) \tag{4.1}$$

where S is the symbol to be transmitted, i.e., 1 or 0.

Information is transmitted as sequences of symbols, 0 and 1. When ASK (Section 1.1.2.4) is used for modulation with a radio carrier of frequency, f_c, then $s_0 = A_0$ or $s_0 = A_1$ for transmitting the symbol, 0 or 1, respectively. When FSK (Section 1.1.2.4) is used for modulation with the radio carrier of frequency, f_{c0}, the transmitted frequencies are, $f_c' = f_{c0} - f_m/2$ and $f_c'' = f_{c0} + f_m/2 = f_c' + f_m$, depending on whether it is transmitting 0 or 1 at any given instant. Here, f_m is the modulating frequency. The FSK technique is also called BFSK (binary FSK). When $f_m \le f_{c0}$, then the transmission is said to be a narrow band transmission. Assuming that the transmitted power is equally distributed in frequencies f_c' and $(f_c' + f_m)$, the transmitted signals may be represented by the following equations:

$$s_0(t) = \frac{s_0}{\sqrt{2}} \sin(2\pi \times f_c' \times t + \varphi_{t0}) \text{ when } S = 0 \quad (4.2a)$$

$$s_1(t) = \frac{s_0}{\sqrt{2}} \sin[2\pi \times (f_c' + f_m) \times t + \varphi_{t0}] \text{ when } S = 1 \quad (4.2b)$$

The $(1/\sqrt{2})$ factor indicates that the signal power reduces by 1/2 and signal peak amplitude reduces by $1/\sqrt{2}$ for each signal component. Other modulation methods, such as BPSK, GMSK, QPSK, and 8-PSK, are also used to control medium access. Bandwidth requirements can also be reduced by using GMSK, QPSK, OQPSK, and $\pi/4$-QPSK. This allows access to a greater number of users for a given bandwidth.

Bandwidth of the BFSK signals is greater than f_m, due to the presence of higher harmonics. [Second harmonic is f_m, third is $2f_m$, and so on. A harmonic is a sinusoidal component of a frequency or its integral multiple as shown in Eqn (1.4).] Using the minimizing technique and a DSP-based Gaussian low pass filter, the higher harmonic components are removed in GMSK modulation (Section 1.1.2.4). The bandwidth required in GMSK is $2f_m$ on the assumption that higher harmonics are filtered out before transmission and $\pm f_m/2$ is the guard band for protection from transmitter drifts. GSM, MobiTex (mobile text), and CDPD (cellular digital packet data) systems use GMSK which greatly reduces the interference between adjacent channels.

Section 1.1.2.4 described BPSK, QPSK, and 8-PSK by which 2, 4, and 8 bits can be grouped and transmitted by phase shift keying [Eqns (1.14) to (1.21)]. When a number of bits are grouped together, the bandwidth use becomes more efficient. PSK modulation facilitates transmission of multiple bits (2, 4, or 8) in the same time-slots in a carrier frequency. More bits can be placed in a given carrier time period by PSK modulation. QAM modulation facilitates transmission of multiple bits (16 or 64) in the same time and amplitude slots in a carrier frequency.

In QPSK modulation, one of the four possible distinct sequences is transmitted using a specific phase angle of the radio-carrier frequency. Two symbols (00, 01, 10, or 11) represent a sequence. Offset QPSK (OQPSK) is a type of QPSK modulation in which the second of the two symbols is *offset* by 90° when applied to a QPSK modulator. Each alternate symbol is in the next quadrature. If 90° are added to the phase angle, the second symbol shifts to the next quadrature during transmission. An OQPSK receiver subtracts the phase angle by 90° and, therefore, receives the signal in the original quadrature and, therefore, it receives the second symbol which was originally present, and so on. The advantage of this is that the in-phase and quadrature signals overlap since they are now in the same phase quadrant. Thus, the number of sharp transitions in the signals is reduced to half of its original value. An OQPSK transmitted envelope is smoother as compared to one transmitted through QPSK. Therefore, the utilization-efficiency of the bandwidth allotted to a mobile service improves. {The coefficient of a harmonic in the Fourier equation [Eqn (1.4)] represents the peak amplitude of that harmonic. Now, the sharper the phase transition, the greater the number of Fourier coefficients which are of significance in the higher harmonics. This means that the bandwidth required will be greater and this greater bandwidth requirement reduces the total number of channels in the allotted bandwidth.} As an example, assume that a symbol transmission sequence is 10 00 11 01. After each successive time interval of T, the phase angles of the transmitted signal, s(t) [as per Eqns (1.17) to (1.20)], are $3\pi/4$ (135°), $-3\pi/4$(225°), $-\pi/4$(315°), $+\pi/4$ (+45°) in case of QPSK. (Quadrants are second, third, fourth, and first). These become $3\pi/4$, $-\pi/4$, $-\pi/4$, $+3\pi/4$ in case of OQPSK modulation. [Quadrants are second (135°), fourth (315°), fourth (315°), and second (135°)]. The OQPSK demodulator subtracts $\pi/2$ at each second bit in the pair (quadrants are second, third, fourth, and first). The original QPSK angles $3\pi/4$ (135°), $-3\pi/4$ (225°), $-\pi/4$ (315°), $+\pi/4$ (45°) are thus obtained from the demodulator. The symbol sequence recovered is 10 00 11 01, which is same as the one transmitted. The advantage of OQPSK is apparent from this example. The number of in-phase symbols is two out of four after OQPSK modulation as compared to four out of four in the QPSK case and the second and the third symbols are now in the same phase and the first and fourth are also in-phase with each other.

$\pi/4$-QPSK is a form of QPSK modulation in which the signal phase shifts by 45° after every two symbols. The advantage of this is that now there are no sharp transitions of π in the phase angle. The number of sharp transitions of the signals is halved. The transmitted envelope is smoother in $\pi/4$-QPSK as compared to that in QPSK. Therefore, bandwidth utilization efficiency is improved in $\pi/4$-QPSK (lesser sharp transitions imply a lower number of significant higher harmonics, lesser bandwidth requirement per channel, and an increased utilization of the allotted bandwidth to a wireless service provider). For example, assume that the symbol sequence 10 00 11 01 is to be transmitted after QPSK modulation. After each successive time interval of T, the phase angles of the transmitted signal, $s(t)$, which are $3\pi/4$, $-3\pi/4$, $-\pi/4$, $+\pi/4$ [as per Eqns (1.16) to (1.19)] become $3\pi/4$, $-\pi/2$, 0°, $\pi/2$ in $\pi/4$-QPSK. The $\pi/4$-QPSK demodulator at the receiving end subtracts $\pi/4$ after each successive bit pair, the original QPSK angles $3\pi/4$, $-3\pi/4$, $-\pi/4$, $+\pi/4$, are found and the bits are recovered as 10 00 11 01.

Equations (1.22) to (1.24) showed the 4-bit per symbol in the 16-QAM method. One of the 16 possible distinct sequences is transmitted by a specific phase angle of the radio-carrier frequency at a specific amplitude (one of the three values of amplitude s_0). Four symbols represent a sequence. Section 1.1.2.4 described the QAM modulation technique by which 16 bits can be grouped and transmitted by phase shift keying. 64-QAM employs two most significant bits for QPSK while reserving the remaining four for the 16-QAM signals. 64-QAM thus transmits six symbols (bits) in a sequence. When the bits are transmitted after 64-QAM, the spectrum bandwidth requirement greatly reduces. For example, assume that a

64-QAM modulated signal is generated and transmitted at 19.2 ksymbol/s. One of the 64 possible distinct sequences is transmitted at a specific phase angle, frequency, and amplitude. Six symbols represent a sequence. The bit transmission rate is 6 × 19.2 ksymbol/s = 115.2 kbps when 64-QAM is transmitted at 19.2 ksymbol/s, because each symbol is actually a sequence of 6 bits. The bandwidth requirement, in this case, is thus reduced by a factor of 1/6.

4.1.2 Controlling Medium Access

The biggest challenge regarding medium access control is that of controlling medium access so that wireless stations (WSs) can transmit at any instant without collision (interference) with signals from other WSs. A wireless station can be a mobile terminal (TE) at a mobile station (MS) a base transceiver system (BTS), or a wireless LAN node.

4.1.2.1 SDMA, TDMA, and FDMA

Different methods of multiplexing were described in Section 4.1.1 as ways to eliminate collisions (interference) between the signals from different WSs at any instant, *t*. SDMA, TDMA, FDMA, and CDMA facilitate access to the medium by multiple sources or channels of same source when each one is using a distinct set of physical space, time, frequency, and code.

Wireless stations, which are distantly located, access the medium by transmitting at the same f_{c0} as well as in the same time-slot SL ($t' \leq t \leq t''$) in different spaces (cells) only. Wireless stations located at suitable distances from each other are then said to transmit using SDMA. When the WSs ($\leq m$) are located in the same space (cell *c*), then the WSs can access the medium in *m* different time-slots, SL_0 to SL_{m-1}, when there are *m* slots in a TDMA communication system. DECT systems (Section 3.11) use TDMA for controlling medium access. Half of the TDMA slots are used for uplink and half for downlink. Each of the *m* stations can transmit with a maximum delay interval equal to the frame interval $m \times (t'' - t')$. As an example, the DECT system data bursts transmit in time-slots of 417 μs. The transmitting WS channels are allotted a fixed pattern by the BTS. There are total 12 uplink and 12 downlink channels in 24 slots in a total duration of 10 ms. After each successive 10 ms interval, the slots in a frame are repeated. Uplink and downlink frequencies can now be kept identical as the time-slots used for them are distinct.

Another scheme of operation is that the uplink and downlink accesses of the WSs in the medium are in different time-slots or in the same slots (shifted by a constant delay), SL_0 to SL_{m-1}, but the uplink and downlink frequencies of the radio carrier, f_c, are distinct (e.g., f_c and f_c + 45 MHz) for FDD (frequency division duplex) access (Section 3.2.3). Also different uplink–downlink frequency pairs are assigned distinct f_cs (out of the *n* values from f_{c0} to f_{cn-1}) in a cell. This is called FDMA. The access to the medium is by distinct f_c at any given instant t, when there are many WSs ($n > m$) accessing the medium simultaneously. GSM, GPRS, and HSCSD systems use TDMA-FDD-FDMA for medium access control.

When the WSs are using the same space (cell), time-slot, and frequency f_c, then there is the CDMA alternative to access the medium. Each WS uses a distinct code between C_0 and C_{p-1} (among p values from C_0 to C_{p-1}) when accessing the medium. CDMA-based systems use this scheme (Section 4.2).

4.1.2.2 ALOHA and CSMA

One medium access controlling mechanism is the division of wireless stations (WSs) on the basis of SDMA, TDMA, FDMA, or CDMA. It lets many WSs get simultaneous access to the medium. However, in spite of employing the above multiplexing schemes, there may be collisions between signals from different sources. The problem, then, is to control (restrict, coordinate, or synchronize) the access to the medium. Collisions may lead to interference of signals from different sources, which in turn, causes loss of information. Therefore, it is a must that collision (interference) between the signals from different WSs is eliminated, or at least reduced.

One simple method for reducing collisions is to employ the basic ALOHAnet protocol for a point-to-point or broadcast network [ALOHAnet was developed in 1970 at the University of Hawaii. Whenever a WS has any voice-data or data to transmit, it just transmits that data. In case there is interference or collision, the WS retransmits the data at a later instant. $LAPD_m$ and RLC protocols (variants of ALOHAnet) in GSM and GPRS systems use the retransmission scheme for non-transparent transmission.] Now, the question that arises is when to retransmit. One method is to randomly select the time. But the chance of success in such a case is small (the average is about 18%). An improved version of ALOHAnet is slotted ALOHA in which a WS retransmits in a discrete time-slot instead of transmitting at a random time. A WS cannot transmit at any time, but just at the beginning of that time-slot. Thus, the chances of collisions are reduced. The chance of success is improved (the average is about 36%) in slotted ALOHA. Another improved form of ALOHA is reservation ALOHA, in which the slots for the WSs are reserved as per the current demand of the WSs. Variants of the ALOHA protocol (such as slotted ALOHA) are used in MobiTex (a protocol for transmitting text on mobile). Hubbed Ethernet and non-switching Ethernet also use a variant of ALOHA. Wi-Fi systems use the IEEE 802.11b protocol, which is an ALOHA-based system for multiple access.

ALOHA-based multiple access has low success rates in cases where there are large numbers of users or transmission of data bursts. The problem of when to transmit or retransmit is solved by the following method—each WS first listens (senses) whether the carrier f_c is already present in the channel to be used and transmits only when there is no carrier present because only then will there be a negligible chance of interference or collision. This method is called CSMA (carrier sense multiple access). All nodes in the CSMA scheme attempt to sense at the same time and one of the nodes transmits as soon as it finds no carrier. CSMA is a variant of ALOHAnet in the sense that a WS *listens* to sense if the medium is busy and if it does not find the medium busy, then it transmits. Another scheme is the CSMA/CA (CSMA/collision avoidance). Each node waits (backs off) for a certain period of time after sensing the carrier after which the carrier is sensed again. A priority-assignments-based scheme can also be used in which the higher priority nodes have shorter back-off periods.

CSMA assumes that each WS can sense the presence or absence of each carrier on the network. On a wired network, it is easy for each transmitter to sense the presence or absence of a carrier on the network. However, in a wireless medium, listening (sensing) in CSMA and CSMA/CA also means consuming energy at the terminal station. It also means spending time in sensing. Cellular digital packet data (CDPD) uses a variant of ALOHAnet and CSMA/CD protocols. CSMA/CD (CSMA/collision detect) is a protocol in which it is checked whether a collision is detected at the transceiver before transmitting. Ethernet LAN uses the CSMA/CD protocol. When the channel is busy and medium access is not feasible, the BTS transmits the *transmission inhibit information* (TI). A data burst is sent to the MS for the TI. The MS senses only the TI data burst and backs off from transmission. An acknowledgement (ACK) data burst is sent to the MS. If ACK is not received by the MS, it means that a collision has been detected. The protocol is called DSMA (digital sense multiple access).

Another method of trasmission is ARCNET (attached resource computer network). Here, a token having an address is passed, first to the nearest node, then to the next nearest, and so on, in the form of a token-passing ring. The addressed node receives the signals.

4.1.3 Medium Access Control for Exposed and Hidden Terminals

Figure 4.1 shows a cell c with four radio-carrier channels ch_0, ch_1, ch_2, and ch_3, using the same radio-carrier frequency, f_c, in the same time-slot and using CSMA for medium access control. Each channel has uplink as well as downlink sub-channels. There can be a total of six channels, ch_0, ch_1, …, ch_5, between four WSs.

Exposed terminal problem in CSMA When ch_2 is active, then ch_0 cannot be used by WS_3 for transmitting to WS_0 even though there is no interference between ch_0 and ch_2. This is because WS_3 senses that the radio carrier f_c is being used by WS_2 and backs off. WS_3 is thus *exposed* to the WS_2 carrier.

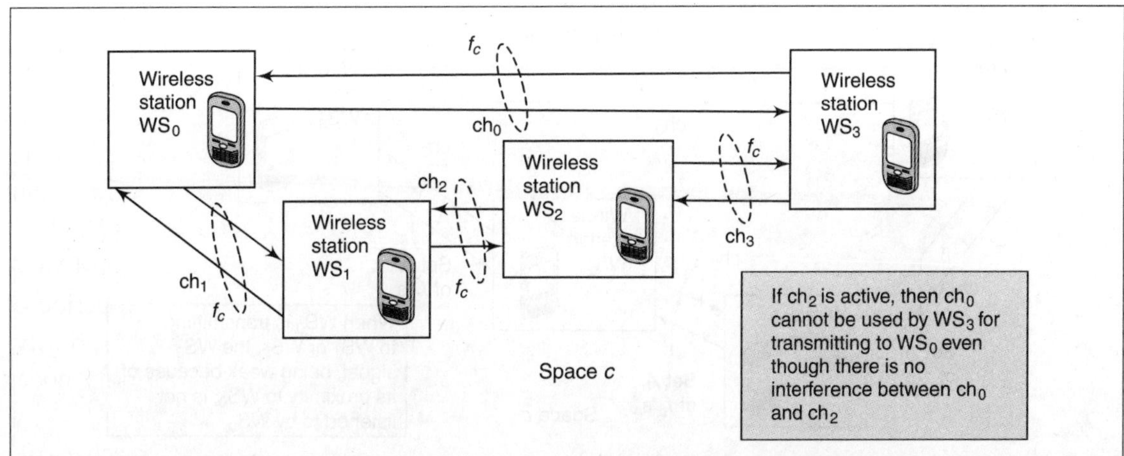

Fig. 4.1 Exposed terminal problem: space *c* with four radio-carrier channels, ch_0, ch_1, ch_2, and ch_3, of four wireless stations, WS_0 to WS_3, using the same radio-carrier frequency f_c and using CSMA for medium access control

Hidden terminal problem in CSMA There could also be cases where WS_3 cannot sense the ch_0 signals from WS_0. This is because the signal strength decreases as the inverse of the square of the distance between the two terminals. When WS_0 transmits to WS_1 or WS_2, since WS_3 does not sense that the radio carrier f_c is being used by WS_0, WS_3 also starts transmission to WS_1 or WS_2. The radio carriers from WS_0 and WS_3 interfere (collide) in the region near WS_1 and WS_2. The collisions of the signals from WS_3 with signals from WS_0 are not detected by WS_0 in CSMA (but they can be detected in CSMA/CD). This is because WS_0 is *hidden* to the WS_3 carrier.

4.1.4 Medium Access Control for Near and Far Terminals

Figure 4.2 shows space *c* with four radio-carrier channels ch_0, ch_1, ch_2, and ch_3, of four wireless stations, WS_0, WS_1, WS_2, and WS_3, using the same radio-carrier frequency f_c or the same set of f_c's. Now, let us assume that each WS transmits with a set of frequencies coded with a distinct code. Assume that WS_3 sends signals via ch_0 for WS_0. The signal strength is weak along the ch_0 region near WS_0. This is because the signal strength decreases as inverse of the square of the distance between the two terminals. When WS_0 is transmitting to WS_1 or WS_2, the WS_3 signal, being weak because of its proxinity to WS_0, is not *listened* to by WS_0. This is because the ch_1 signal strengths are higher near WS_0 as compared to the ch_0 signal strengths. The strong ch_1 signals superimpose on the weak ch_0 signals at WS_0. WS_3 is the *far terminal* and WS_1 or WS_2 are the *near terminals*. The radio carriers from both WS_3 and WS_1 will be listened to if the transmission power is raised in ch_0 or decreased in ch_1. Power control is, therefore, required for the far and near terminals to avoid

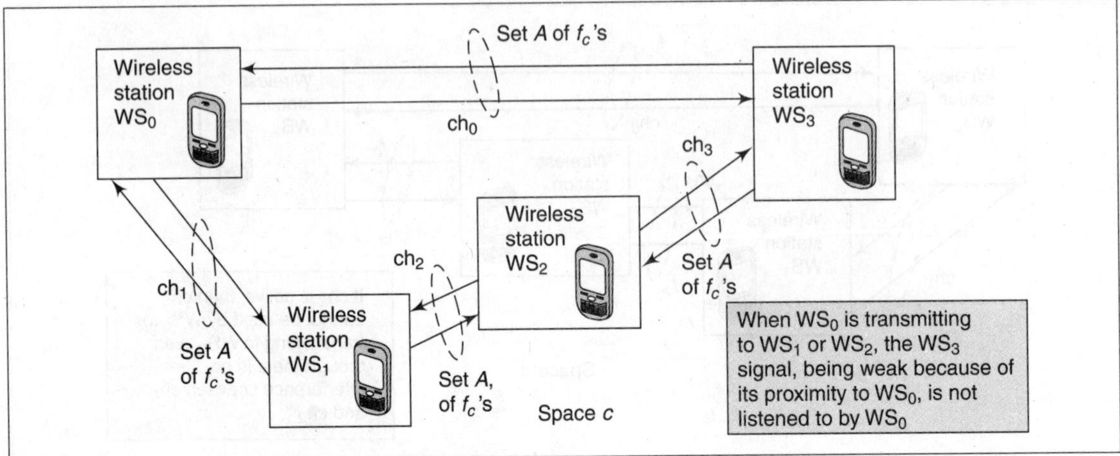

Fig. 4.2 Near and far terminal problem: space *c* with four radio-carrier channels, ch_0, ch_1, ch_2, and ch_3, of four wireless stations, WS_0 to WS_3, using the same set *A* of radio-carrier frequencies (f_c's) and using CSMA for medium access control

drowning of the far terminal signals in presence of signals from the near terminals. When a GSM system BTS transmits to an MS during CCH data bursts (Section 3.2.5), the required power transmission level from that MS is decided by measurements of the signal strengths from the MS. The RRM layer (Section 3.3) performs the signal measurement and power control tasks. GSM defines five levels of power transmission. CDPD transceivers transmit the power-received level during the CSI (channel stream identification) data bursts for an MS by measurements of the signal strengths at the RRM layer.

GSM systems use closed loop power control. Closed loop means that the MS and the BTS also measure the signal strength and the MS transmits information regarding the signal quality to the BTS. The MS adjusts its power level to minimize the transmitted power and still maintains an acceptable quality of signals. Section 4.5 describes the CDMA IS-95 power control open-loop mechanism for near and far terminals.

4.1.5 SDMA, TDMA, and FDMA Control

Section 1.1.2.5 described various forms of multiple access using space, time, and frequency division and Section 3.2.1 explained them in the context of GSM systems. In this section we shall revisit these as methods of medium access control.

Figure 4.3 shows four wireless stations, in four distinct cells, simultaneously transmitting with the same radio-carrier frequency f_c using SDMA. When the WSs are in different areas of space, there is no interference (collision). However, there is a limit to the number of cells which can be formed among the WSs using directed antennae. Hence SDMA does not completely resolve the interference issue in medium access control.

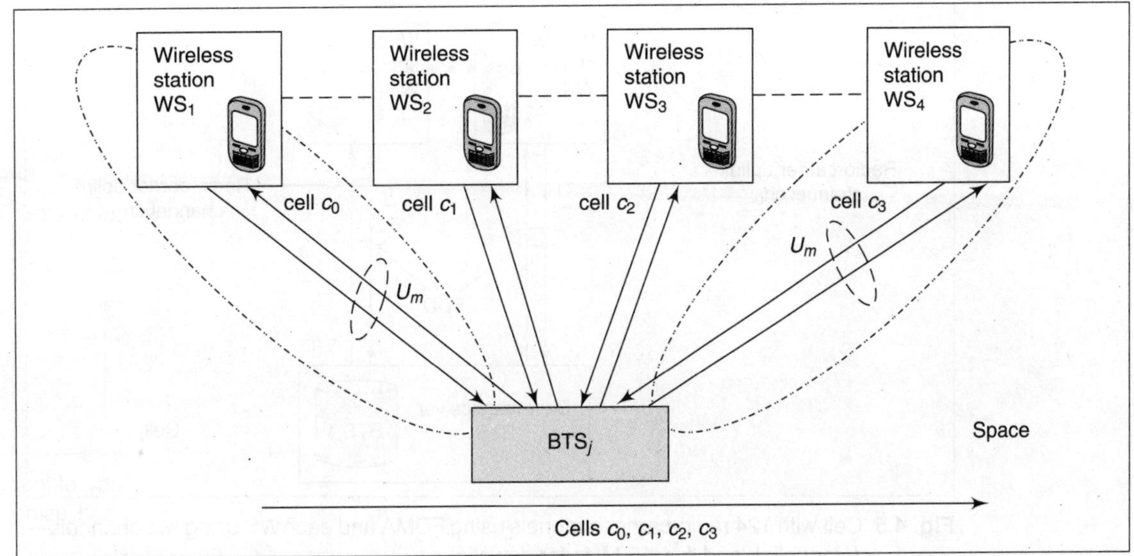

Fig. 4.3 WSs in four distinct cells simultaneously transmitting with same radio-carrier frequency, f_c, using SDMA

Figure 4.4 shows eight WSs transmitting in distinct time-slots, SL_0, SL_1, ..., SL_7, using the same radio-carrier frequency f_c using TDMA. When the WSs transmit in different slots, interference (collision) is avoided. But there is a limit to the number

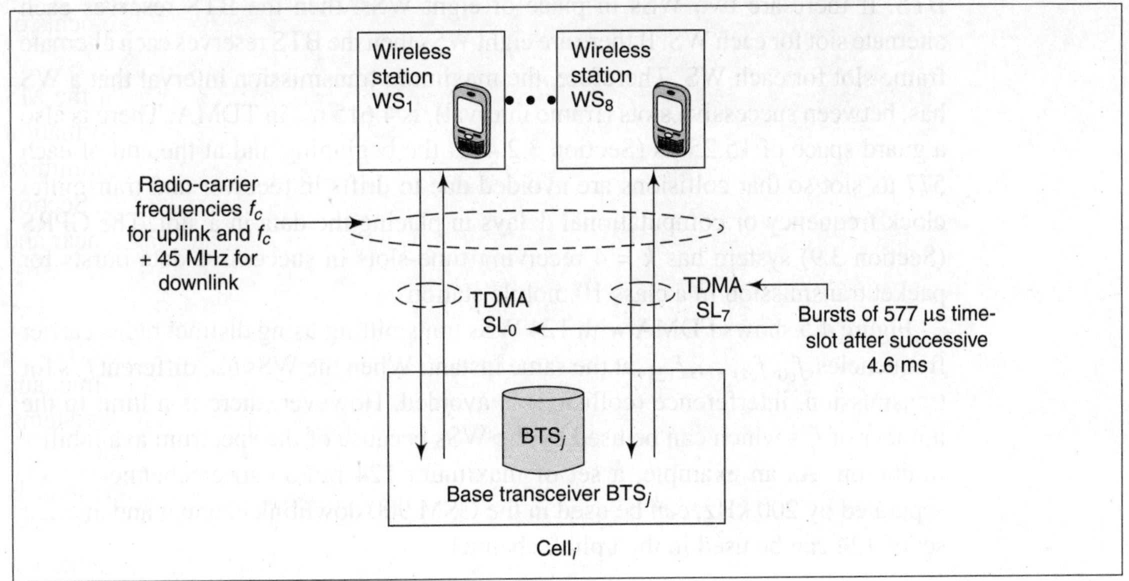

Fig. 4.4 Cell$_i$ with eight time-slots in a radio-carrier channel with frequency f_c for uplink and f_c + 45 MHz for downlink

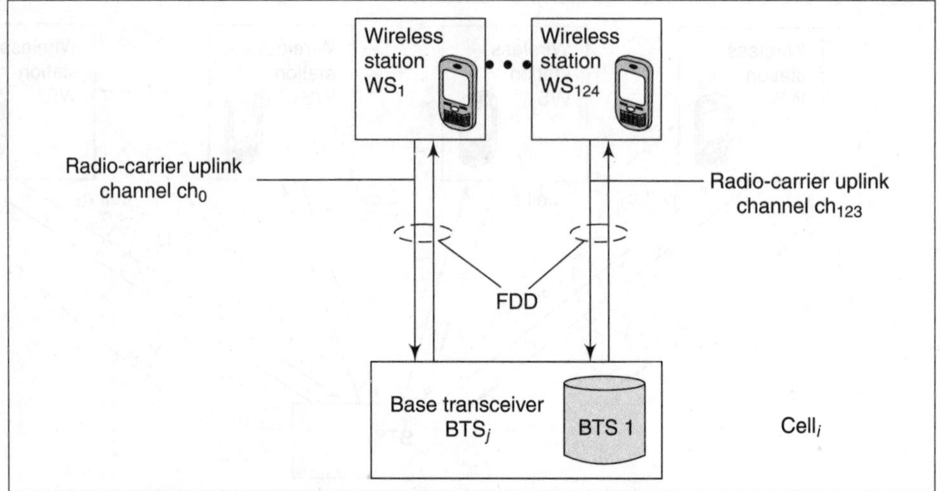

Fig. 4.5 Cell$_i$ with 124 radio-carrier channels using FDMA and each WS using two channels—f_c for uplink and f_c + 45 MHz for downlink

of wireless stations that can be served using different slots. This is because the transmission slots for a WS are repeated after small intervals (called frame intervals), otherwise the total data throughput from each WS will become too small. The GSM system, for example, provides for eight time-slots. Each time-slot is of 577 μs (Section 3.2.2) and the transmitting WS channels are allotted a fixed pattern by the BTS. If there are two WSs in place of eight WSs, then the BTS reserves each alternate slot for each WS. If there are eight WSs then the BTS reserves each alternate frame slot for each WS. Therefore, the maximum transmission interval that a WS has, between successive slots (frame interval), is 4.615 ms in TDMA. There is also a guard space of 15.25 μs (Section 3.2.4) at the beginning and at the end of each 577 μs slot so that collisions are avoided due to drifts in receiver and transmitter clock frequency or computational delays in placing the data in a slot. The GPRS (Section 3.9) system has k = 4 receiving time-slots in successive data bursts for packet transmission in a class 10 mobile station.

Figure 4.5 shows FDMA with 124 WSs transmitting using distinct radio-carrier frequencies, $f_{c0}, f_{c1}, \ldots, f_{123}$, at the same instant. When the WSs use different f_c's for transmission, interference (collision) is avoided. However, there is a limit to the number of f_c's which can be used by the WSs because of the spectrum availability limitation. As an example, a set of maximum 124 radio-carrier channels, each separated by 200 kHz, can be used in the GSM 900 downlink channel and another set of 124 can be used in the uplink channel.

4.2 Introduction to CDMA-based Systems

Code division multiple access (CDMA) is used as a multiplexing method in many mobile telephony systems. Such systems are called CDMA systems. The first CDMA standard was the IS-95 (interim standard-95) defined by Qualcomm, USA. Unlike GSM standards, which regulate all aspects of the GSM network infrastructure, the CDMA-based standards only govern the radio interface. The important features of CDMA systems are as follows:

- CDMA is more robust for multi-path delays and provides higher immunity towards frequency selective fading (Section 4.3).
- Signals from each mobile station or base transceiver are coded with two or three codes. An n-symbol code is assigned to each mobile or base transceiver wireless station (WS) and each channel of that user. Another m-symbol code can also be assigned to each user channel, for example, synchronization, paging, traffic, and pilot channels. Another k-symbol code can also be assigned for each carrier of a user. CDMA systems use a *good set* of codes. A good set of CDMA codes is one in which all the codes are orthogonal to each other and which results in autocorrelation at the receiver.
- A set of n equally spaced frequencies, known as chipping or hopping frequencies, is used for transmission either in direct sequence or by frequency hopping. The n chipping or hopping frequency signals are given as

$$s_0(t) = S_0 \times \left(\frac{s_0}{\sqrt{n}}\right) \sin(2\pi \times f_{c0} \times t + \varphi_{t0})$$

$$s_1(t) = S_1 \times \left(\frac{s_0}{\sqrt{n}}\right) \sin(2\pi \times (f_{c0} + f_s) \times t + \varphi_{t0})$$

$$\vdots$$

$$s_{(n-1)}(t) = S_{(n-1)} \times \left(\frac{s_0}{\sqrt{n}}\right) \sin\{2\pi \times (f_{c0} + (n-1) f_s) \times t + \varphi_{t0}\} \quad (4.2)$$

  The DSSS (direct sequence spread spectrum) technique uses all the frequencies defined by Eqn (4.2), simultaneously, in direct sequence. The frequency spread is between f_{c0} and $f_{c0} + (n-1)f_s$.
- Beside the above point, there may be hopping among n equally-spaced frequencies defined by Eqn (4.2) in a particular sequence. In the FHSS (frequency hopping spread spectrum) technique, only one channel frequency is used during a given interval and the channel frequency hops among n frequencies in a certain coded order of sequence.
- A symbol is transmitted in DSSS using the code as such for chipping when transmitting the symbol 0 and using the code's complement when transmitting 1. This means there is XORing between the user-signal symbols and chips. The chips are used as per the code.
- Alternatively, a symbol in a hop interval is transmitted using FHSS. The channel frequency used for transmission at a given hop interval is as per the hop sequence defined by the code.

- A CDMA DSSS receiver (Fig. 4.6) XORs the received signal with the chipping code, which was used for transmission, thus giving back the original symbol in user data. If XORing gives the result 0, then the transmitted symbol is 0, and if it results in a finite large value, then the transmitted symbol is 1. In a given hop interval, a CDMA FHSS receiver (Fig. 4.7) demodulates the signals of that channel frequency which corresponds to the code symbol for that hop.
- CDMA systems give signals of higher voice and data quality and small bit-error rates.
- CDMA systems have *soft handover*. Soft handover means that an MS at the boundary of two adjacent cells does not have to drop calls due to signal break during handover at the boundary region. CDMA systems provide seamless connectivity to the MS.
- CDMA systems perform power control by open loop or close loop methods to solve the problem of drowning of signals from far mobile terminals by those from the near terminals.

CDMA systems employ spread spectrum techniques (DSSS and FHSS) for medium access control. These techniques are described in the following sections along with codes used in CDMA systems and the structure and working of various CDMA-based systems.

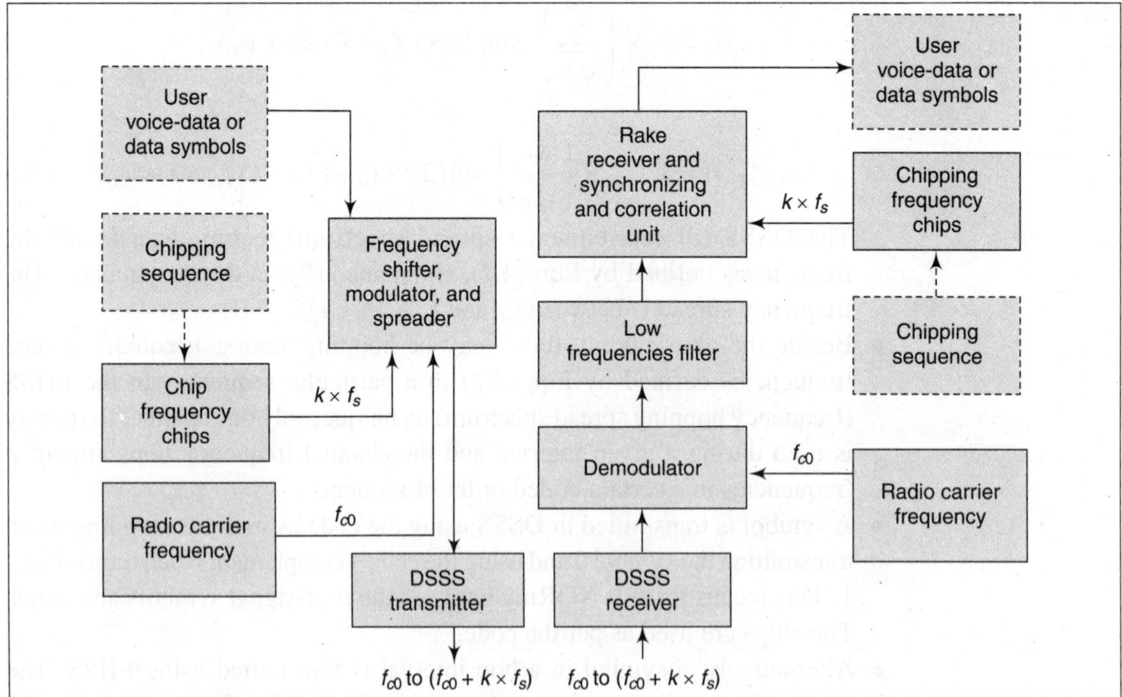

Fig. 4.6 DSSS tra nsmitter and receiver

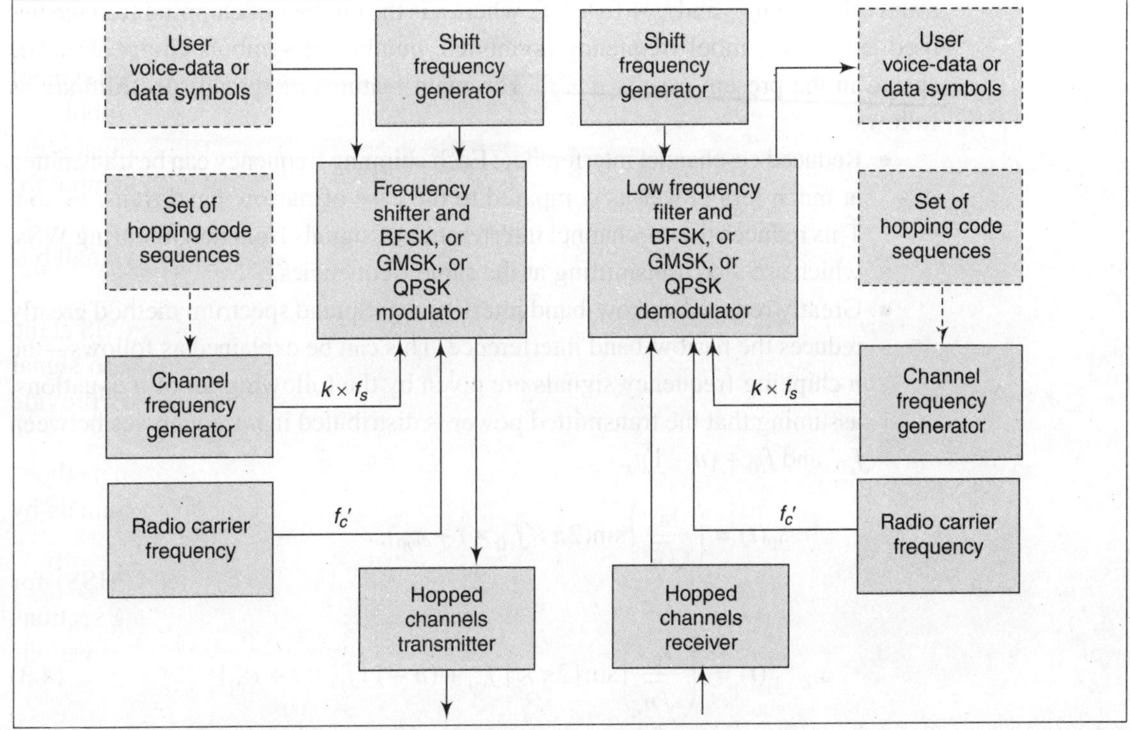

Fig. 4.7 FHSS transmitter and receiver

4.3 Spread Spectrum in CDMA Systems

Narrow band transmission uses a frequency band equal to f_m [Eqns (4.1) and (4.2)] for the transmission of 1s and 0s (assuming that higher harmonics are filtered). A major disadvantage of narrow band transmission is that there may be frequency-selective fading along a path (Fig. 1.5) between the MS and the BTS. The signals having frequencies close to f_{c0} may also be exposed to interference from mobile stations operating at the same frequency in a nearby cell (Fig. 1.14). This is called co-channel interference. One solution to this problem is that the data-frame channel frequency, assigned to each wireless (or mobile) station by the BTS, can be changed (hop) at select frequencies at a certain rate according to a predetermined sequence. The GSM hopping rate is 207.4 hops/s. (Use of hopping is optional for a GSM system.)

Spread spectrum is a transmission technique that provides a novel solution to the interference problem. The following subsections describe the spread spectrum techniques of direct sequence spread spectrum (DSSS) and frequency hopping spread spectrum (FHSS). CDMA systems use DS spread spectrum. As discussed in Section 1.1.2.5, we can transmit signals at frequencies, $f_{c0}, f_{c0} + f_s, f_{c0} + 2f_s, \ldots, f_{c0} + (n-2)f_s$, $f_{c0} + (n-1)f_s$, for example, in case of CDMA. The spectrum widens by a factor of

n and is between f_{c0} and $f_{c0} + (n-1)f_s$, where n is the number of chipping frequencies used and f_s is symbol frequency (symbol/s, number of symbols chipped/s). The spread in the present case is $n \times f_s$. The main features of spread spectrum are as follows:

- Reduced co-channel interference: Each chipping frequency can be transmitted at much less power as compared to the case of narrow-band transmission. This reduces the co-channel interference of signals from neighbouring WSs, which are also transmitting at the same frequencies.
- Greatly reduced narrow-band interference: Spread spectrum method greatly reduces the narrow-band interference. This can be explained as follows—the n chipping frequency signals are given by the following set of n equations, assuming that the transmitted power is distributed in n frequencies between f_{c0}' and $f_{c0} + (n-1)f_s$.

$$s_0(t) = \left(\frac{s_0}{\sqrt{n}}\right) \sin(2\pi \times f_{c0} \times t + \varphi_{t0}),$$

$$\vdots$$

$$s_{(n-1)}(t) = \left(\frac{s_0}{\sqrt{n}}\right) \sin\{2\pi \times [f_{c0} + (n-1) f_s] \times t + \varphi_{t0}\} \tag{4.3}$$

Spreading is a process of transmitting the signal using chipping frequencies given by Eqn (4.3). *De-spreading* is a process of retrieving back the signal, given by Eqn (4.1), from the received signals.

Example 4.1 A transmitter transmits a spread signal with 16 chipping frequencies. The eighth component of the spread spectrum, i.e., $s_8(t)$ faces narrow-band interference from an external signal $s_8'(t)$ with frequency, $f_{c0} + (n/2) f_s$. The strength of the interfering signal is 1.5 times that of the transmitted user signal and the de-spread signal at the receiving end is amplified by $10 \times \sqrt{n}$ times. (a) Write the equation representing $s_8(t)$. (b) Will the signal be interpreted correctly at the receiving end? Explain. (c) How can the interference from other chipping frequencies be removed from this signal?

Solution: (a) $n = 16$, therefore, the equation for $s_8(t)$ is given as

$$s_8(t) = \left(\frac{s_0}{\sqrt{n}}\right) \sin\{2\pi \times (f_{c0} + 8f_s) \times t + \varphi_{t0}\}$$

(b) the frequency of $s_8'(t)$ is $f_{c0} + (n/2)f_s = f_{c0} + 8f_s$

Also, $s_8'(t) = 1.5 \times s_8(t)$. Now, the interfering signal $s_8'(t)$ is from another source and is not spread as the transmitter spreads only the user data signal. The $s_8(t)$ peak amplitude will be affected by $s_8'(t)$ only on reaching the receiver. At the receiving end, the transmitted signal will be de-spread back and it is given that the signal is amplified by $10 \times \sqrt{n}$ times, which means the peak amplitude goes up to $10s_0$. The effect of the narrow-band interference will be $1.5 \times s_0$ which is just 15% of $10s_0$. Therefore, the receiver will correctly interpret 1 or 0 as per the transmitted symbol.

(c) The interference at other chipping frequencies can be removed by band pass filtering.

- Multiple access by greater number of channels: The chipping frequencies are coded distinctly for different wireless stations (WSs) and thus a greater number of channels can access the medium simultaneously. Let us assume that p WSs are coded in p ways out of 2^n possible ways of coding when using n chipping frequencies in the spread spectrum. Each of the p WSs can transmit at the same spread frequencies as given by Eqn (4.3). The following set of n equations gives the coded-X signal strengths at the chipping frequencies

$$s_0(t) = S_0 \times \left(\frac{s_0}{\sqrt{n}}\right) \sin(2\pi \times f_{c0} \times t + \varphi_{t0})$$

$$\vdots$$

$$s_{(n-1)}(t) = S_{(n-1)} \times \left(\frac{s_0}{\sqrt{n}}\right) \sin\{2\pi \times (f_{c0} + (n-1)f_s) \times t + \varphi_{t0}\} \tag{4.4}$$

where the code is a sequence of chipping symbols represented by $S_0, S_1, \ldots, S_{n-2}, S_{n-1}$. S stands for a symbol or bit (1 or 0). Each equation gives one chip, transmitted at a distinct frequency. There are n chips for an n-symbol code. The frequency spread and signal strengths for an exemplary code 1110000111100001 consisting of 16 chips are depicted in Fig. 1.8. Spread-spectrum transmission helps in reducing bandwidth requirements. For example, if six WSs transmit using FSK (Eqns (4.2.1) and (4.2.2)), the frequency band required is $12 \times f_m$ (assuming that there is a guard band of $\pm f_m/2$ at the low and high ends of each channel). Higher harmonics are assumed to be filtered and thus neglected. If we assume that each WS is coded using four chipping frequencies, then the required frequency band will be $4f_m + f_m/2 + f_m/2 = 5f_m$ (the $f_m/2$ factors represent the $\pm f_m/2$ guard bands). Now, let us assume that WS_0 is assigned the four-symbol chipping codes 0001 and 1110 for transmitting user data symbols, 0 and 1, respectively. Similarly, the remaining five WSs, WS_1 to WS_5, can be assigned the pairs of four-symbol codes, (0010, 1101), (0011, 1100), (0100, 1011), (0101, 1010), and (0110, 1001), for transmitting user data bits 0 and 1, respectively. Each individual WS thus shares the spread bandwidth. We find that the bandwidth required per WS is less when we use spread spectrum and multi-symbol coding for simultaneous transmission by a large number of stations.

DSSS and FHSS spread-spectrum methods mitigate the effects of narrow-band interference and effects of co-channel interference also when using different spreading codes for different channels. A spread-spectrum receiver uses complex computational processing with the help of digital signal processors (DSPs). The disadvantage of spread spectrum is the use of a large frequency band. Frequencies of the band can interfere with other stations and transceiver signals in the wireless medium. An appropriate power control strategy is thus required to control the near and far terminal effects. Also, all the codes formed by combination of bits are not usable. Selection of an appropriate code is important (Section 4.4).

4.3.1 Direct Sequence Spread Spectrum

Direct sequence spread spectrum (DSSS) is a method of transmitting data using the spread-spectrum technique. The data bit, *B*, is modulated (XORed) at successive intervals with a chipping frequency and the following set of *n* equations gives the transmitted DSSS signals at a given instant *t*.

$$s_0(t) = (B \text{ XOR } S_0) \times (s_0/\sqrt{n})\sin\{2\pi(f_{c0} + f_{\text{chip}})t + \varphi_{t0}\} \text{ in time interval } t = 0 \text{ to } t_{\text{chip}}$$

$$\vdots$$

$$s_{(n-1)}(t) = (B \text{ XOR } S_{(n-1)}) \times (s_0/\sqrt{n})\sin\{2\pi(f_{c0} + f_{\text{chip}})t + \varphi_{t0}\} \text{ in time interval } t = (n-2) \times t_s \text{ to } (n-1) \times t_{\text{chip}} \quad (4.5)$$

S_i $(0 < i < n - 1)$ is the ith chipping-symbol in the sequence (code), f_s is symbol frequency, and f_{chip} is the number of user symbols chipped per second.

$$t_{\text{chip}} = f_s^{-1} \text{ and } f_{\text{chip}} = (n-1) \times f_s$$

B is XORed with each of the *n* symbols. XORing means that if $B = 1$ and $S = 1$ or $B = 0$ and $S = 0$, then the amplitude is 0, else it is 1. The second term after the multiplication sign $[(s_0/\sqrt{n})\sin\{2\pi(f_{c0} + f_{\text{chip}})t + \varphi_{t0}\}]$ in each of the above equations is called chip. The term before the multiplication sign [$(B \text{ XOR } S_{(i-1)})$ for the ith chip] represents the operation performed at the spreader. There are *n* chips when there are *n* chipping intervals for spreading. The signals given by Eqn (4.5) are transmitted after modulation with a carrier frequency, f_{c0}.

The spread factor, D_{dsss}, for the bandwidth in DSSS is defined as the ratio of the period of a bit, t_{signal}, in the data to be transmitted, to the chipping time interval. Chipping time interval, t_{chip}, is the reciprocal of the frequency band used by one symbol for chipping (f_{chip}^{-1}). Therefore, $D_{\text{dsss}} = t_{\text{signal}}/t_{\text{chip}}$ and the bandwidth of DSSS data is $D_{\text{dsss}} \times f_s$.

Example 4.2 Data from a wireless station is transmitted using DSSS at the rate of 0.0192 Msymbol/s. The chipping rate is 1.2288 Mchip/s. Calculate the bandwidth of the DS spread spectrum.

Solution: The period of each transmitted bit, $t_{\text{signal}} = \dfrac{1}{(0.0192 \text{ Msymbol/s})} = 52\ \mu\text{s}$

Chipping frequency interval, $t_{\text{chip}} = \dfrac{1}{(1.2288 \text{ Mchip/s})} = 0.814\ \mu\text{s}$

$$\text{Spread factor, } D_{\text{dsss}} = \frac{52\ \mu\text{s}}{0.814\ \mu\text{s}}$$
$$= 64$$
$$\text{Bandwidth of the DSSS spectrum} = 64 \times 0.0192\ \text{Msymbol/s}$$
$$= 1.2288\ \text{MHz}$$

Figure 4.6 shows a DSSS transmitter and receiver. A chipping sequence for specifying the WS or BTS channel (paging, traffic, synchronizing, or pilot) is fed to generate chipped signals after XORing with the spreading frequency chips in a CDMA system. The chips and carrier frequency are inputs to the modulator and spreader unit. The output of the modulator is sent to the DSSS transmitter. The input of the DSSS receiver is demodulated and given to a low frequency filter to separate the f_{c0} carrier. A rake receiver with synchronizing and correlation units gets the input chips as per the chipping sequence used by the transmitter. The output from the rake receiver is the user voice-data or data symbols. The receiver might be receiving the transmitted signals through multiple paths. A rake receiver is a specialized receiver, which selects the signals of the strongest paths and reconstructs the signal by accounting for the variable delays in these paths and then correlates with the chip sequences used for transmission. The user voice-data or data symbols are thus retrieved in DSSS even in the presence of multiple path delays and narrow band and co-channel interferences.

A correlation unit performs the XOR operation between the demodulator output and the filtered output using the same chip sequence as used by the transmitter. When the result of XORing is either 0 or a value expected from the chipping code, then there is correlation between the transmitted and received data and the user symbol is correctly interpreted as 0 or 1. Even if the correlation is not perfect the user symbol can be correctly found for a certain range of errors. This happens when the code chosen is such that it results in autocorrelation, even when there is some chipping interval delay between the received signal and the receiver-generated chipping sequence.

The XOR operation for de-spreading at the receiver correlation unit can be represented by the following set of n equations for n chipping frequencies of an n-chip sequence.

$$b_0 = (S'_0 \text{ XOR } S_0)$$
$$\vdots$$
$$b_{n-1} = (S'_{n-1} \text{ XOR } S_{n-1}) \tag{4.6}$$

Here S'_i is the ith received symbol and S_i is the ith chip in the code. The value of b_0, b_1, ..., b_{n-1} is 0 if the transmitted user symbol is 0, and is some finite value as per the code if the transmitted user symbol is 1.

4.3.2 Frequency Hopping Spread Spectrum

Section 3.2.3 discussed how GSM systems provide for frequency hopping. However, hopping in a GSM system is optional. The period of transmission at one of the hopping frequency channels is 4.6 ms, which is also the period of one data frame of eight slots. If each frequency channel has a bandwidth of 100 kHz and a guard band of ±50 kHz, then each frequency channel is separated by 200 kHz. If there are n channels among which the channel frequency can hop, then spread-spectrum bandwidth is $n \times 200$ kHz for frequency hopping spread spectrum (FHSS).

FHSS is a method of transmitting user data using one of the carrier-frequency channels in a given interval of time from among multiple channels and then hop the channel frequency to another channel in the next interval of time. The channel-carrier frequency keeps hopping at a certain rate and hops in a certain specific sequence or set of sequences. Sequences are as per the code being used and are repeated in the next hop cycle. The Bluetooth radio interface (Section 12.5.1) uses FHSS. FHSS differs from DSSS in one aspect. The FHSS bandwidth during transmission at each given instant of time is just equal to the inter-channel separation. The DSSS bandwidth for transmission at each instant is equal to the full assigned spread spectrum. This means that the signal radio-carrier band is a narrow band, but the frequencies span over the spread spectrum during a complete sequence of hopping.

Each frequency channel is separated by a guard space. A hopping sequence is randomly designed. An example of a hopping sequence is that of the IEEE standard 802.11wireless-LAN—the transmitter transmits a set of three sequences. Hopping sequences are not repeated in the three sequences. Each sequence consists of 26 channels out of a total of 78 channels. The LAN specification is that frequency channel separations, f_s, be 1 MHz and basic radio-carrier frequency, f_{c0}, be 2.4 GHz. If the hop cycle frequency is f_h, then the hopping interval, $t_h = f_h^{-1}$. For a set of three sequences with 26 channels in each, there will be $26 \times 3 = 78$ equations representing the signals from the first to the 78th hop. For 78 channels, the spread spectrum bandwidth is 78 MHz when the f_s inter-channel separation is 1 MHz.

Let us take a simpler example. A signal is modulated using BFSK modulation. Radio-carrier signals after modulation are given by Eqn (4.2). The modulated frequencies in each channel are f_c' and $f_c' + f_m$. Let each channel be separated by frequency, f_s, where $f_s = f_m$ + lower guard band + upper guard band. An example of an eight-sequence code hopping sequence is (3, 4, 0, 1, 5, 3, 7, 2). Assume that the hopping sequence (code) for a transmitter is (i, j, k, l, m, n, o, p). Let the user data bit, B, be modulated with channel frequencies at different instants, then the following equations give the transmitted signals at successive hops as a function of time.

For the first hop interval

$$s_0(t) = (B \text{ XOR } S_0) \times s_0 \sin[2\pi \times \{f_c' + (i \times f_s) + B \times f_m\} \times t + \varphi_{t0}] \quad (4.7a)$$

For the second hop interval

$$s_1(t) = (B \text{ XOR } S_0) \times s_0 \sin[2\pi \times \{f_c' + (j \times f_s) + B \times f_m\} \times t + \varphi_{t0}] \quad (4.7b)$$

For the third hop interval

$$s_2(t) = (B \text{ XOR } S_0) \times s_0 \sin[2\pi \times \{f_c' + (k \times f_s) + B \times f_m\} \times t + \varphi_{t0}] \quad (4.7c)$$

For the fourth hop interval

$$s_3(t) = (B \text{ XOR } S_0) \times s_0 \sin[2\pi \times \{f_c' + (l \times f_s) + B \times f_m\} \times t + \varphi_{t0}] \quad (4.7d)$$

For the fifth hop interval

$$s_4(t) = (B \text{ XOR } S_0) \times s_0 \sin[2\pi \times \{f_c + (m \times f_s) + B \times f_m\} \times t + \varphi_{t0}] \quad (4.7e)$$

For the sixth hop interval

$$s_5(t) = (B \text{ XOR } S_0) \times s_0 \sin[2\pi \times \{f_c + (n \times f_s) + B \times f_m\} \times t + \varphi_{t0}] \quad (4.7f)$$

For the seventh hop interval

$$s_6(t) = (B \text{ XOR } S_0) \times s_0 \sin[2\pi \times \{f_c + (o \times f_s) + B \times f_m\} \times t + \varphi_{t0}] \quad (4.7g)$$

For the eighth hop interval

$$s_7(t) = (B \text{ XOR } S_0) \times s_0 \sin[2\pi \times \{f_c + (p \times f_s) + B \times f_m\} \times t + \varphi_{t0}] \quad (4.7h)$$

Here, $B = 1$ or $B = 0$, depending upon whether the user data symbol being transmitted is 1 or 0.

Slow FHSS FHSS spectrum in which the interval during a hop, $t_h \gg t_s$ (where $t_s = f_s^{-1}$), is called slow FHSS. A number of symbols get transmitted during a channel hop period. The advantage of this is that even if a channel frequency signal is faded at the receiver due to narrow band interference, the other symbols are received correctly. For example, let us assume that out of 78 channels, the 35th channel is affected by interference. Then the signals from the 35th channel are rejected and the transmitter later retransmits these symbols at another channel frequency. If FHSS is not used, then even retransmission does not help, because that channel will fade again.

Fast FHSS Fast FHSS is FHSS in which the interval during a hop, $t_h \ll t_s$. During a symbol period, a large number of frequency hops take place. The advantage of this is that even if a few channel frequencies are faded at the receiver due to narrow band interference, the symbol is received correctly. The synchronization of fast FHSS between the BTS and the WS is more complex than that of slow FHSS.

FHSS Transmitter and Receiver Figure 4.7 shows a FHSS transmitter and receiver. A hopping sequence for the WS or BTS channel is fed to generate frequency channels as per the hopped channels for the FHSS modulator. The channel frequency (hopped-frequency signal) and carrier frequency are inputs to the modulator. The output of the modulator is sent to the FHSS transmitter. The input of the FHSS receiver is demodulated and given to the low-frequencies filter to separate the $f_c' + (k \times f_s)$ carriers. The output of the filter unit is user voice-data or data symbols.

4.4 Coding Methods in CDMA

CDMA systems use distinctive spreading codes to spread the symbols before transmission. At the receiving end, a correlator is used to de-spread the original signal. Unwanted signals are not correlated and thus not de-spread. CDMA codes are carefully designed sequences of 1s and 0s. These sequences are produced from the codes as per the symbols at rates much higher than those of the symbols. The original data rates are called symbols per second and the code chipping rates are called chip rates. These codes uniquely distinguish sets of data from each other. CDMA codes enable unique identification of signals from different sources and allow different signals to be transmitted through the same space, time, and frequency slots without interference. It is, therefore, important that CDMA codes should not correlate to other codes or to time-shifted versions of themselves. The various types of codes and coding methods used in CDMA systems are discussed in the following subsections.

4.4.1 Autocorrelation Codes

Consider coded symbols of n bits between time interval t_0 and $t_0 + (n - 1)T$, where T is the period between successive bits. The sequence is transmitted and reaches the receiver in the interval between t_0' and $t_0' + (n - 1)T$ when $(t_0' - t_0)$ is the propagation delay. When the receiver attempts to correlate the received coded symbols with respect to any of the codes which it internally generates, it is not able to correlate even when it uses exactly the same code as the one used for transmission. Firstly, the reason can be that the receiver uses the code for extracting symbols within the interval t_0'' and $t_0'' + (n - 1)T$, and t_0'' and t_0' do not differ by $n \times T$. Secondly, the reason can be that the sequence of bits in the code itself is such that correlation is not possible by using a correlation method at the receiver.

The first reason is taken care of by the receiver by successively shifting the receiver generated sequence by a period = $m \times T$, which is an integral multiple of the time period, T [i.e., $m = 0, 1, 2, 3, \ldots, (n - 1)$]. The correlation method is to find the sum of products (SOP) to correlate the sequence of bits between $t_0'' + (m \times T)$ and $t_0'' + (n - 1)T + (m \times T)$ with the received coded signal, and vary m till the SOP is maximum. The second reason is taken care of by choosing an autocorrelation code. An autocorrelation code is a multi-bit code which, when used for coding the symbols before a transmission, enables the receiver to automatically correlate and extract the symbols using the correlation method. An example of an autocorrelation code is the Barker code, which is described in the following subsection.

4.4.1.1 Barker Code

Let us assume that the chipping signal symbol is 1 for +1 and 0 for –1. The Barker code is a sequence of n values (code-symbols) of +1 and –1. Assume that a code symbol is a_i where $i = 0, 1, \ldots, n - 2, n - 1$. SOP is given by sum of terms $a_i \times a_{i+k}$ from $i = 0$ to $n - k$ for all values of k between 1 and n. Also

$$|\Sigma a_i \times a_{i+k}| \leq 1 \text{ for } i = 1 \leq k < n \tag{4.8}$$

The IEEE 802.11 standard for the wireless LAN recommends Barker code of length $n = 11$. The code is {+1, +1, +1, −1, −1, −1, +1, −1, −1, +1, −1}. Consider a Barker code for $n = 13$. The code C_{13} is given by {+1, +1, +1, +1, +1, −1, −1, +1, +1, −1, +1, −1, +1}. For $k = 1$, the absolute value of sum of products = $|(+1 \times +1) + (+1 \times +1) + (+1 \times +1) + (+1 \times +1) + (+1 \times -1) + (-1 \times -1) + (-1 \times +1) + (+1 \times +1) + (+1 \times -1) + (-1 \times +1) + (+1 \times -1) + (-1 \times +1)| = 0$, and the condition of Eqn (4.5) stands verified for $k = 1$. Similarly, it can be found that the sum of products will always be less than or equal to 1 for all values of k between 1 and n. The code appears random and noise-like but is actually not random and is, therefore, known as a pseudo-noise (PN) sequence. $\bar{C}_{13}$ is the complement of {−1, −1, −1, −1, −1, +1, +1, −1, −1, +1, −1, +1, −1}, which is also a Barker code and is the complement of C_{13}.

$$C_{13} \text{ XOR } C_{13} = \bar{C}_{13}$$

During a DSSS transmission, when $B = 0$, the chipping frequencies are in the sequence {+1, +1, +1, +1, +1, 0, 0, +1, +1, 0, +1, 0, +1} and when $B = 1$, the successive chipping frequencies are in the sequence {0, 0, 0, 0, 0, +1, +1, 0, 0, +1, 0, +1, 0}.

To examine the autocorrelation characteristics of Barker codes let us find the SOP between {+1, +1, +1, +1, +1, −1, −1, +1, +1, −1, +1, −1, +1} and {+1, +1, +1, +1, +1, −1, −1, +1, +1, −1, +1, −1, +1}. This means that we must find the product of first symbol with first symbol, second with second, and so on up to the thirteenth symbol. The following equation gives the sum of products of the code symbols with themselves.

$$\Sigma a_i \times a_i \text{ for } 0 \leq i < n \tag{4.9}$$

The SOP is 13. Assuming that there is path delay at the receiver, equal to the on-chip interval, $t_s = f_s^{-1}$ (f_s is the chipping pulse frequency). The receiver then gets the spread-modulated signal with the one-bit right-shifted code {x, +1,+1, +1, +1, +1, −1, −1, +1, +1, −1, +1, −1}. The basic code {+1, +1, +1, +1, +1, −1, −1, +1, +1, −1, +1, −1, +1} is used by the receiver for de-spreading. Now the SOP is +1 assuming $x = 1$ and assuming the path delay to be $2 \times t_s$. Even then the SOP (= 1) is very low as compared to the case when the receiver chipping sequence matches with the

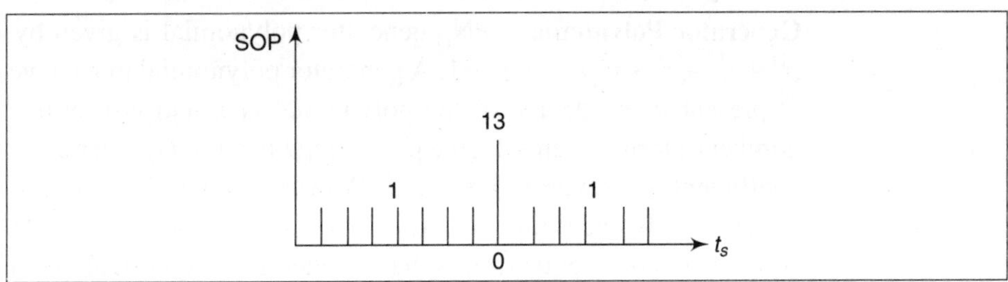

Fig. 4.8 Sum of the 13 products between the coded and received sequences for $n = 13$, Barker code as a function of the time interval

chipping sequence received. Figure 4.8 shows the autocorrelation property of Barker code with $n = 13$.

A good autocorrelation function is one, which gives a large value of sum $\Sigma e_i \times r_i$ when $e_i = r_i$, and a very low value when $e_i \neq r_i$ for $0 \le i < n$. Here, e_i is the expected chip symbol and r_i is the received chip symbol in an ith sequence at the receiver.

4.4.1.2 Pseudo-noise Codes

When a code appears random, like noise, but is actually not random, then it is known as pseudo-noise (PN) code. It is used to generate one or multiple sequences. PN codes are used for soft handover. A second BTS is added for the users on the edge of a cell. Edge signal quality improves and handover becomes robust. GSM systems have separate operating frequencies in adjacent cells. This is required to avoid inter-cell interference. At the edge of the cells, handover (Section 3.6) is performed. When a GSM MS is unable to handover to new frequencies, the traffic channels are discarded and the call is dropped. Adjacent cells of a CDMA system use the same set of carrier and chipping frequencies but different codes. When the cell changes, an offset is added to the pseudo-noise codes. Each cell has distinct pseudo-noise code offsets. Pseudo-noise code offset processing can be done easily. Only the offset value changes in case of handover when the signal of one cell becomes weak. The call is not dropped as the offset can be changed by the BTS depending on which cell has the stronger signal at the boundary of two adjacent cells.

M-sequences (maximum-length sequences) codes are codes generated by using *m* small-length shift registers. The feedback generates a large number of sets, each set having *m* sequences. For example, a set of 15 registers ($m = 15$) can be used to generate a set of $(2^{15} - 1)$ sequences. *M*-sequences PN codes find application in scrambling codes. A method of generating an *M*-sequences PN code is as follows—Consider an IS-95 cdmaOne quadrature component code sequence PN_Q. The term *quadrature component* refers to the 90° phase-shifted component and *in-phase* refers to the same phase component. Quadrature component is orthogonal to in-phase component. Figure 4.9 shows a circuit called linear feedback shift register (LFSR) for PN_Q.

Generator Polynomial PN_Q generator polynomial is given by $G_Q = z^{15} + z^{12} + z^{11} + z^{10} + z^6 + z^5 + z^4 + z^3 + 1$. A generator polynomial must have at least first term z^n present if the degree of the polynomial is n and last term $z^0 = 1$. Maximum number of terms in an n degree polynomial is $n + 1$. G_Q, therefore, has 16 terms, but coefficients of seven terms are 0s.We can write G_Q = {1001 1100 0111 1001}, which can also be written as (15, 12, 11, 10, 6, 5, 4, 3, 0). IS-95 CDMA in-phase component code sequence PN_I has a generator polynomial given by $G_I = z^{15} + z^{13} + z^9 + z^8 + z^7 + z^5 + 1$.

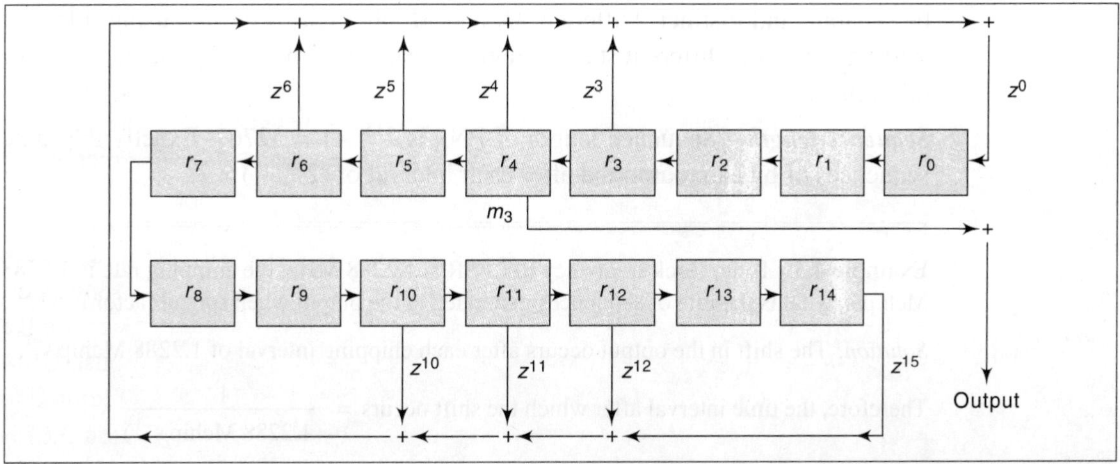

Fig. 4.9 Linear feedback shift register (LFSR) for PN_Q for generating multiple sequences using generator polynomial G_Q. (Here the mask vector is assumed to be ⟨000 0000 0001 0000⟩. An output bit is generated on each successive clock pulse. The + sign shows an XOR operation.)

Initial State Vector An initial state vector (also known as the reset vector) has n bits in case of an n degree polynomial. A set of n shift registers stores this vector on reset (at the start of the sequence generation). As an example, there is a set of 15 registers, $r_{14}, \ldots, r_1, r_0$, for G_Q. Assume that the initial vector is {000 1110 0011 1100}. It means that $r_{11}, r_{10}, r_9, r_5, r_4, r_3$, and r_2 store 1s and rest store 0s at the start of a PN sequence. Starting sequence should not be 0. Therefore, at least one of the registers should store 1 and at least one of the binary numbers in the vector is 1. After n sequences, the same sequence as the first one is used to generate the next output sequence of bits after $n \times T$, where T is the clock period. Since G_Q results in a different input to r_0, the sequence changes after each interval of $n \times T$.

Shift Parameter and Mask Vector There is a positive integer, called shift parameter, which defines how much should be shifted after each successive n sequences. If shift parameter is 3, then it means that each sequence starts from register r_3 in place of r_0 skipping r_0, r_1, and r_2. If shift parameter is 0, then it means that each sequence starts output from register r_0.

There is a vector called the mask vector with 16 elements ($m_{15}, m_{14}, m_{13}, \ldots, m_2, m_1$, and m_0). The mask vector specifies which register output is to be taken and is not masked, and which set of registers output is masked and is not the input in next sequence. Assume that the vector is {000 0000 0001 0000}. It means that shift parameter is 4 (because $m_4 = 1$), therefore, the next sequence will start after $n \times T$ from register r_4 in place of r_0. This is because this value m_4 is 1.

Gold codes WCDMA uses Gold codes[1]. A Gold code is created from two M-sequence codes, M_1 and M_2. M_1 and M_2 are added modulo 2. M_1 and M_2 should

[1]For more information on the use of Gold codes in WCDMA, see Andrew Richardson, 2005, *WCDMA Design Handbook*, Cambridge University Press, Cambridge.

be separate and distinct. Different M_1 and M_2 are created by just using different starting registers. Different starting registers can be set by setting the mask vector differently.

Sequence length Sequence length of PN_Q is $2^{15} - 1 = 32767$. Exactly the same sequences of bits are outputted after each interval of $(2^{15} - 1) \times T$.

Example 4.3 If the clock frequency to LFSR is 1.2288 MHz, the chipping rate is 1.2288 Mchip/s, what is the rate of sequences generated at the output when spread factor is 64?

Solution: The shift in the output occurs after each chipping interval of 1.2288 Mchip s^{-1}.

Therefore, the time interval after which the shift occurs $= \dfrac{1}{(1.2288 \text{ Mchip s}^{-1})}$

$$= 0.814\ \mu\text{s}$$

If the output sequence is divided by 64, then the output appears every $0.814 \times 64\ \mu$s

Therefore, the sequences appear at the rate $= \dfrac{1}{(0.814 \times 64\ \mu\text{s})}$

$$= 19.2 \text{ kchip/s}$$

4.4.1.3 Barker Codes vs Pseudo-noise Codes

Barker code C_{13} has nine 1s and four 0s, and the Barker code shows strong autocorrelation. Pseudo-noise (PN) codes have almost equal number of 0s and 1s. A PN code shows a strong peak with a few low, non-zero values (Fig. 4.8). This may result in interference with the other users using the same spread.

4.4.2 Other Important Codes

Two codes used for coding of two sets of symbols for transmission are said to be orthogonal when there is no effect of interference between the two sets of signals on the received output. Orthogonal codes require synchronization between the transmitter and the receiver as orthogonal codes do not show a strong autocorrelation property. Synchronization means that the instant of the received first bit of coded symbols and first bit of generated code for extracting symbols are in the same phase.

Best codes are optimized codes which enable significant correlation and do not cause significant interference among the different channels. The following subsections describe orthogonal codes and other codes that are used in CDMA.

4.4.2.1 Orthogonal Codes

Codes are said to be orthogonal, if they have zero cross-correlation. (Cross-correlation refers to the product of the ith symbol in two codes and the sum of products for all values of i.) When each transmitter adopts a unique orthogonal code, then there is no effect of interference on the received output because the fact that cross-correlation of two codes is 0 can be used to filter the other transmitter signals out. Two codes of n chips are orthogonal if the SOP of their components is 0, i.e.

$$\Sigma p_i \times q_i = 0 \tag{4.10}$$

When $p_i = q_i$ for $0 \leq i < n$, we find that SOP is 0. Here, p_i is the first code and q_i is the second code at the ith chip.

Best Codes Orthogonal codes require synchronization between the transmitter and the receiver as they do not exhibit a strong autocorrelation property. A WS can use PN sequences for uplink and the BTS can transmit using orthogonal codes for downlink, because the BTS has special synchronization units. Autocorrelation is important for synchronization. We may also use codes that are almost orthogonal, i.e., $\Sigma p_i \times q_i \sim 0$.

Orthogonal codes have almost zero cross-correlation and are used in identifying the user, user channel, and carrier. Long *M*-sequence PN codes have strong autocorrelation and are used in synchronizing and detecting the user channel signals. Short PN codes also have strong autocorrelation and are used for synchronizing and detecting the user carriers. All three coding schemes can be simultaneously used in a CDMA system.

4.4.2.2 Walsh Codes

The Walsh code used in IS-95 cdmaOne has a 64×64 matrix, with all pairs of rows orthogonal.[1] Walsh codes are generated from a matrix called the Hadamard matrix.

- W_0, zeroth row Walsh code = {0, 0, …, 0, 0}. IS-95 employs it for the pilot channel.
- W_{32}, 32nd row Walsh code = {0, 0, …, 0, 0} all 0s for first half column elements and {1, 1, ..., 1, 1} all ones for next half column elements. IS-95 employs it for the synchronization channel.
- W_1, 1st row Walsh code = {0, 0, …, 0, 0} all 0s for first half columns elements and {1, 1, ..., 1, 1} all ones for next half elements. IS-95 employs it for the paging channel.
- $W_{2-31,\ 33-63}$ can be used by traffic channels. If more than 1 and up to 7 paging channels are being used, then $W_{8-31,\ 33-63}$ are used for the traffic channels.

Let us examine how the zeroth row in the 8×8 matrix Walsh code W_0 = {0, 0, 0, 0, 0, 0, 0, 0} is orthogonal to the second row Walsh code W_2 = {0, 0, 1, 1, 0, 0, 1, 1}. The codes can be rewritten as {−1, −1, −1, −1, −1, −1, −1, −1} and {−1, −1, +1, +1, −1, −1, +1, +1}. Now, the cross-correlation SOP = $\Sigma p_i \times q_i$ when $p_i = q_i$ for $0 \leq i < n$. Here, p_i is the first Walsh code and q_i is the second Walsh code at the ith column sequence. The cross-correlation SOP is 0 between W_0 and W_2. Similar results are obtained between other pairs of Walsh codes.

[1]For all elements of the matrix, see Raymond Steele, Chin-Chun Lee, and Peter Gould, 2001, *GSM, cdmaOne and 3G Systems*, John Wiley, New York.

Variable Spread Factor Variable spread factor is obtained by using variable-length Walsh codes during multi-rate transmission. CDMA2000 uses variable length Walsh codes for transmitting data at variable rates. One procedure is that each user channel uses a distinct Walsh code, W_m. The receiver uses the same code W_m for identifying the data and for identifying that user and user channel. The chipping length of the Walsh code is varied. The code length depends upon the chipping rate and the data rate.

Example 4.4 A user signal is transmitted with a chipping interval of 814 ns. Find the code length which can be used for chipping this signal, if (a) it has a very low data rate of 4.8 kbps, (b) it has a low data rate of 19.2 kbps, and (c) it is used to transmit files of the CIF picture format at high data rates of 384 ksymbol/s with a rate matching reduction by a factor of 1.25.

Solution: A chipping interval of 814 ns means frequency $= \dfrac{1}{(814 \times 10^{-9})}$

$= 1.2288$ Mchip/s

(a) The code length for data rate of 4.8 kbps $= \dfrac{1.2288 \times 10^6}{4.8 \times 10^3} = 256$

(b) The code length for data rate of 19.2 kbps $= \dfrac{1.2288 \times 10^6}{19.2 \times 10^3} = 64$

(c) After data rate matching by reducing the rate by a factor of 1.25, the data rate 384 ksymbol/s becomes $= \dfrac{384}{1.25} = 307.2$ ksymbol/s

Therefore, the code length $= \dfrac{1.2288 \times 10^6}{307.2 \times 10^3} = 4$

4.4.2.3 Scrambling Codes

Scrambling codes have long sequence lengths. A transceiver supports large number of users and user channels. Long autocorrelation codes are thus required. A scrambling code can be a PN *M*-sequence code. Scrambling code must exhibit strong autocorrelation property. The long code generator polynomial, $G_I = z^{42} + z^{35} + z^{33} + z^{31} + z^{27} + z^{26} + z^{25} + z^{22} + z^{21} + z^{19} + z^{18} + z^{17} + z^{16} + z^{10} + z^{7} + z^{6} + z^{5} + z^{3} + z^{2} + z + 1$, used in cdmaOne is an example of a scrambling code. For uplink from an MS, a short code can also be used, for example, in WCDMA (Section 4.6.1).

4.4.2.4 Channelization Codes

A channelization code has a short length sequence and must exhibit the orthogonality property. Walsh codes are used for channelization due to their orthogonality property.

These are scrambled with long codes to achieve orthogonality as well as autocorrelation. As an example, in cdmaOne as well as CDMA2000 systems, a Walsh code performs the chipping of the signals after a PN M-sequence $(2^{42} - 1)$ long code scrambles the user channel symbols. In both cases, a processing unit performs XORing of the user symbols (scrambled with a PN long-sequence code) with the orthogonal coded chips.

4.4.2.5 Carrier Modulation Codes

A transceiver can support a limited number of carriers ($\ll 2^{14}$). Short autocorrelation codes can thus suffice. Orthogonal phase modulation (QPSK) is performed on the I- and Q-PN short code pilot waveforms XORed with the scrambled and then chipped signals. The modulated signals are transmitted using a carrier. (Remember that orthogonal phase modulation is in time-space and orthogonal code or PN code modulation (spreading) is in code-space. The purpose of the first is to synchronize the carriers of different base stations and the purpose of the second is to identify the multiple-user channels.) As an example, two short PN codes called PN_Q and PN_I form two pilots and are used for orthogonal phase modulation. PN_I means in-phase PN code and PN_Q refers to the quadrature component of the PN code. IS-95 cdmaOne as well as CDMA2000 employ orthogonal waveforms, which are first coded using a PN short code of $(2^{15} - 1)$ sequences.

4.5 IS-95 cdmaOne System

IS-95 was developed by QUALCOM, USA in 1991 and was accepted as a standard in 1997. It is now known as cdmaOne. It is a 2G+ technology, originally called IS-95 (interim standards 95). It uses 824–849 MHz and 869–894 MHz with multiple analog channels forming one digital carrier. The standard uses FDD for forward and reverse links, as in the case of GSM systems.

The forward link (called downlink in GSM) is between the BTS (base transceiver) and an MS (mobile station). Forward-link frequency is 870.000 MHz + $0.030i$, where i is the channel number. The channel numbers are between 1 and 777 when $1 \le i \le n_1$. Here $n_1 = 777$. Channel numbers are between 1013 and 1023 when $-10 \le i \le 0$

The reverse link (uplink in GSM) is from an MS (mobile station) to the BTS (base transceiver). Reverse-link frequency is 45 MHz less than the forward link and is 825.000 MHz + $0.030i$, where i is the channel number. The channel numbers are $1 \le i \le n_1$ between 1 and 777 where $n_1 = 777$ and $n_2 \le i \le n_3$. For n_2 in between -10 and $n_3 = 0$, the channels are numbered from 1013 to 1023. There are two types of systems. System A services are the ones using channels 1 to 311, 689 to 694, and 1013 to 1023. System B services are the ones using channels 356 to 644 and 739 to 777. There is guard band between the two services.

Pseudo-noise spreading codes PN_Q and PN_I with 15-degree generator polynomials G_Q and G_I and 32767 chip sequence generator using the LFSR-based circuit shown in Fig. 4.9 (Section 4.4.1.2), are used for spreading. Each BTS adds one of the 512 offsets to the PN codes, so that each BTS in a cell is identified. During the handover, the offset value changes when an MS moves from one cell to another. Quadrature and in-phase base-band signals are then modulated by QPSK in forward channels and OQPSK in reverse channels (Section 4.1.1).

4.5.1 Forward-link Channels

GSM systems have traffic and control data channels (Section 3.2.5). Similarly, there are four channels in forward link, called traffic, paging, synchronization, and pilot channels. The forward channel from the BTS uses orthogonal Walsh codes.

Power control Let TR be a BTS transmitter and *R* be a receiver MS. If TR transmits to *R* the data of the same signal strength that it received from *R*, then power control means that TR raises or lowers its power as per data obtained from *R*.

Close loop control means that both ends TR and *R* transmit the signal strengths received from the other ends and use it for power control. Open loop control means that only one end (TR) transmits the signal strength received from the *R* data.

Open loop power control is a type of control in which the input data of signal strength detected at the other end controls the power used by the transmitter. The forward channel uses open loop power control. A near terminal reduces the power level and a far off terminal raises the power level after detecting the strength of the data on a pilot channel from the BTS. Only the BTS measures the signal strength in IS-95 and the BTS transmits information regarding the signal quality to the MS. The MS adjusts its power level as per the received signal quality and minimizes the transmitted power and still maintains an acceptable quality of signals.

Pilot channel The pilot channel is the channel transmitted by the BTS to the MS before the data is transmitted. The MS transmits the data at reduced power if the pilot is strong and raises the power if the pilot is weak. The pilot channel provides a reference to all the MSs in a cell. A pilot channel transmits with all 0s in place of voice-data traffic through the in-phase and quadrature spreading units. The pilot channel uses the Walsh code W_0 and a circuit consisting of PN_I in-phase and PN_Q quadrature spreaders, base-band filters, and QPSK modulator. This part of the circuit is identical to the lower part of the processing unit described in Subsection 4.5.1.1.

4.5.1.1 Traffic Channel

Figure 4.10 shows the processing units used in the CDMA IS-95 standard for base-band transmission of traffic channel and power control messages. Figure 4.10 also shows the uses of long code and mask, PN_Q, PN_I, and Walsh coding units.

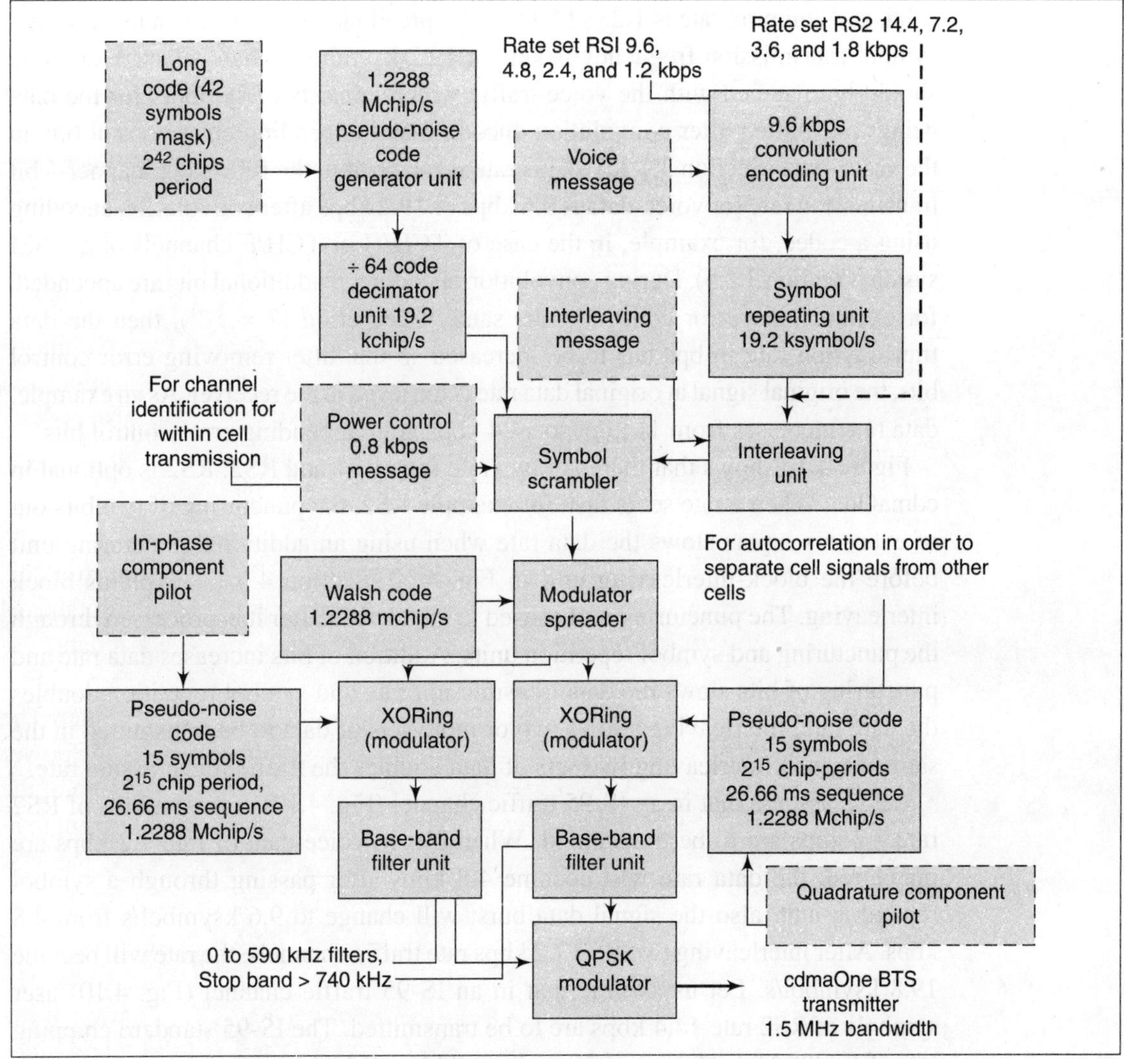

Fig. 4.10 Processing units used in CDMA IS-95 (cdmaOne) standard for base-band transmission of traffic channel and power control messages with use of long code and mask, PN_Q, PN_I, and Walsh codes

4.5.1.2 Data Rate Matching

Example 4.2 showed that the chipping frequency = spread factor × symbol transmission frequency. The spread factor is a constant positive integer (Eqns (4.2) and (4.3)). The code length per user symbol must be a positive integer. The spread factor is equal to the code length during the time in which one user symbol is transmitted. It must, therefore, be a positive integer. As an example, a symbol repetition unit doubles the data rates and thus matches the data rate with the set rate and set code of 9.6 kbps with 19.2 ksymbol/s when code length is 64, spread factor is 86, and chipping rate is 1.288 Mchip/s. If spread factor is kept constant at 64 and symbol transmission frequency is kept at 19.2 ksymbol/s, then selected data rate should be matched with the voice traffic which consists of symbols for the data being transmitted after convolution encoding and appending error control bits in the voice-data (Section 3.1.1.5). This rate is matched in the following manner—bit transmission rate for voice-data is 9.6 kbps or 19.2 kbps after convolution encoding using a codec, for example, in the case of TCH/H or TCH/F channels of a GSM system (Section 3.2.5). During convolution encoding, if additional bits are appended, for example, for error control in the same time period ($T = f^{-1}$), then the data transmission rate in bps has to be increased so that after removing error control bits, the original signal at original data rate is retrieved at the receiver. As an example, data rate increases from 11 kbps to 14.4 kbps after appending error control bits.

Figure 4.10 shows that there are two rate sets, RS1 and RS2. RS2 is optional in cdmaOne. When a rate set is fast, for example RS2, the puncturing of two bits out of every six inputs slows the data rate when using an additional puncturing unit before the block-interleaving unit in Fig. 4.10. Section 4.5.1.4 explains block interleaving. The puncturing unit is used to match data after it is processed through the puncturing and symbol repetition units. Addition of bits increases data rate and puncturing of bits slows the data transmission rate and symbol repetition doubles the data rate. Interleaving allows two or more sets of data to be transmitted in the same channel. Interleaving two sets of data doubles the traffic transmission rate.

Let us assume that in an IS-95 traffic channel (Fig. 4.10), user symbols of RS2 rate 7.2 kbps are to be transmitted. When bits in voice-data of rate 7.2 kbps are punctured, the data rate will become 4.8 kbps after passing through a symbol repetition unit, also the signal data burst will change to 9.6 ksymbol/s from 4.8 kbps. After interleaving two RS2 7.2 kbps rate traffic channels, the rate will become 19.8 ksymbol/s. Let us assume that in an IS-95 traffic channel (Fig. 4.10) user symbols of RS2 rate 14.4 kbps are to be transmitted. The IS-95 standard chipping rate of 1.2288 Mchip/s is used. Now, 1.2288 Mbps/14.4 kbps is a fractional number, but 1.2288 Mbps/19.2 kbps is an integer (64). This means that the Walsh code has a code length of 64. Therefore, the rate set for RS2 user symbols is matched with the RS1 rate of 9.6 kbps by puncturing two symbols for every six user symbols and then doubled by symbol repetition to 19.2 kbps. This facilitates the use of Walsh code of length 64 for transmitting each user symbol. Puncturing two symbols for every six and then repeating gives a signal of 19.2 ksymbol/s. At 14.4 ksymbol/s, a six-bit interval is 6×69.44 μs = 416.4 μs. At 9.6 ksymbol/s, a 4-bit interval is 4×104.1 μs = 416.4 μs. Now, two bits are punctured in this time interval of 416.4 μs. The symbol rate now becomes 4/6 times for the purpose of spreading. 14.4 kbps × 4/6 equals 9.6 kbps. The puncturing of two symbols out of every six still gives a voice signal of acceptable quality. Symbol repetition doubles the number of symbols in the 416.4 μs interval. The symbols for spreading (coding) appear at double data rates as 19.2 ksymbol/s. The RS2 signal of data rate 14.4 kbps is transmitted at a rate of 19.6 ksymbol/s. Each symbol is then coded and spread later by a 64-symbol long code.

4.5.1.3 Multi-encoded Transmission of Signals in a Traffic Channel

IS-95 cdmaOne user data has different data rates. Rate set RS1 is transmitted at 9.6, 4.8, 2.4, or 1.2 kbps. An optional rate set RS2 is transmitted at 14.4, 7.2, 3.6, or 1.8 kbps. It requires a different type of convolution encoding and error encoding depending upon considerations of service quality. Low or high service quality means high or low bit error rates, respectively. For example, GSM TCH/FS voice quality improves because the error control bits are appended and data rate is enhanced to 22.8 kbps from 13 kbps (Section 3.2.5). TCH/HS at 11.4 kbps has lower voice quality and transmits with double the data throughput rate when two sets of TCH/HS data interleave before transmission through a traffic channel.

Multi-encoded rate systems transmit signals at variable data rates after convolution coding. In IS-95, the RS1 and RS2 data rate symbols are matched, repeated, and interleaved so that before chipping, the symbol transmission rate is constant at 19.2 ksymbol/s. Data rate matching with a fixed rate for a channel is used to transmit multi-encoded signals using a constant spread factor and fixed-length codes. If a set of bits is repeated in the same time period (e.g., when matching a preset symbol rate by doubling it), then the symbol transmission rate is doubled. Here, the term 'symbol' refers to the symbol to be XORed with the code at a given chipping frequency and the term 'bit' is used for the user signal to be transmitted and extracted at the receiver.

4.5.1.4 Block Interleaving

Interleaving is a technique which enables the use of idle time-slots in the frames which have a pre-defined bit (or symbol) transmission rate in a channel. Block interleaving is when between successive blocks of a data set, another data set block is inserted. Block interleaving helps in sending two data sets simultaneously. After two or four data sets interleave at a block-interleaving unit, the data symbol rate is kept the same at 19.2 ksymbol/s. Interleaving is possible only when the interleaved signal frequencies are integral multiples of each other. As an example, an 800 bps signal can be interleaved with a 19.2 ksymbol/s signal because 19.2 ksymbol/s, when divided by 800 bps, gives 24 which is an integer. In 24 time-slots, one time-slot is usable by the 800 bps signal. Interleaving of 19.2 kbps with 19.2 kbps is possible, when 19.2 kbps signal time-slots in a frame are idle. Similarly, the signals of 9.6 kbps, 4.8 kbps, 2.4 kbps, and 1.2 kbps can be interleaved. Interleaving lets the signals of different rates simultaneously utilize the idle time in a data frame.

Data interleaving is used for addition of background noise (Section 3.2.5) when a voice activity detector does not detect any voice traffic due to the user not talking or some processing related delays. There are several idle time intervals in case of transmission of voice-data or other data. IS-95 block-interleaving unit interleaves the power control and other channel messages with user data. This kind of block interleaving permits utilization of those idle time intervals.

4.5.1.5 Long Code Scrambling

A long code is used for identifying a traffic channel. The chip generation rate (Section 4.2.1) is chosen as 19.2 kchip/s × 64 = 1.2288 Mchip/s. A long code decimator reduces the chipping signal rate by a factor of 64. The 19.2 ksymbol/s input, the 19.2 kchips/s input, and the 0.8 kbps power control input are interleaved and are coded at 1.2288 Mchip/s by a 64-bit Walsh code.

4.5.1.6 *I* and *Q* pilots for Carrier Identification and Synchronization

The Walsh-coded signal is spread after scrambling of user symbols with long PN codes. The pilots are modulated after the spread with a pseudo-noise code of 15 symbols with 2^{15} chip period sequences in the *I*-pilot and *Q*-pilot channels (see Section 4.4.2.5 for short autocorrelation codes PN_Q and PN_I). A base-band filter unit filters the spread chipping frequencies in each pilot channel. A QPSK modulator gets inputs from both the channels and then transmits with a bandwidth of 1.25 MHz for each carrier. The efficiency of transmission is given by (1.2288 Mcps/ 1.25 MHz) × 100 = 0.98 × 100 = 98%. Traffic channels and multiple-user data channels are spread by Walsh codes to enable their distinct identification at the receiver. Coded traffic channel data is transmitted by PN long code sequencing to synchronize multiple channels and by PN short code sequencing of *I* and *Q* pilots to synchronize the multiple transmitter carriers (the short code is used to identify the multiple transmitter carriers at the receiver).

4.5.1.7 Paging Channel Data

Figure 4.11 shows the processing units used in the CDMA IS-95 (cdmaOne) standard for base-band transmission of the paging messages and paging mask. The paging channel sends the TMSI (temporary mobile subscriber ID) information about the traffic channel, response of MS access request during call setup, and information about the adjacent cell base station and its PN offsets. One or more paging channels can be used. Each paging channel uses a distinct Walsh code. Bit transmission rate for voice-data is 9.6 kbps or 19.2 kbps after convolution encoding using a codec for TCH/H or TCH/F (Section 3.2.5). This rate is for transmission of a paging channel message. After passing through the paging symbol repetition unit, the signal data burst will become 19.2 ksymbol/s. After going through a block-interleaving unit, the rate remains the same at 19.2 ksymbol/s. A long code, known as paging channel code, is used. The chip generation rate is 19.2 × 64 = 1.2288 Mchip/s. A long code decimator reduces the chipping signal rate by a factor of 64.

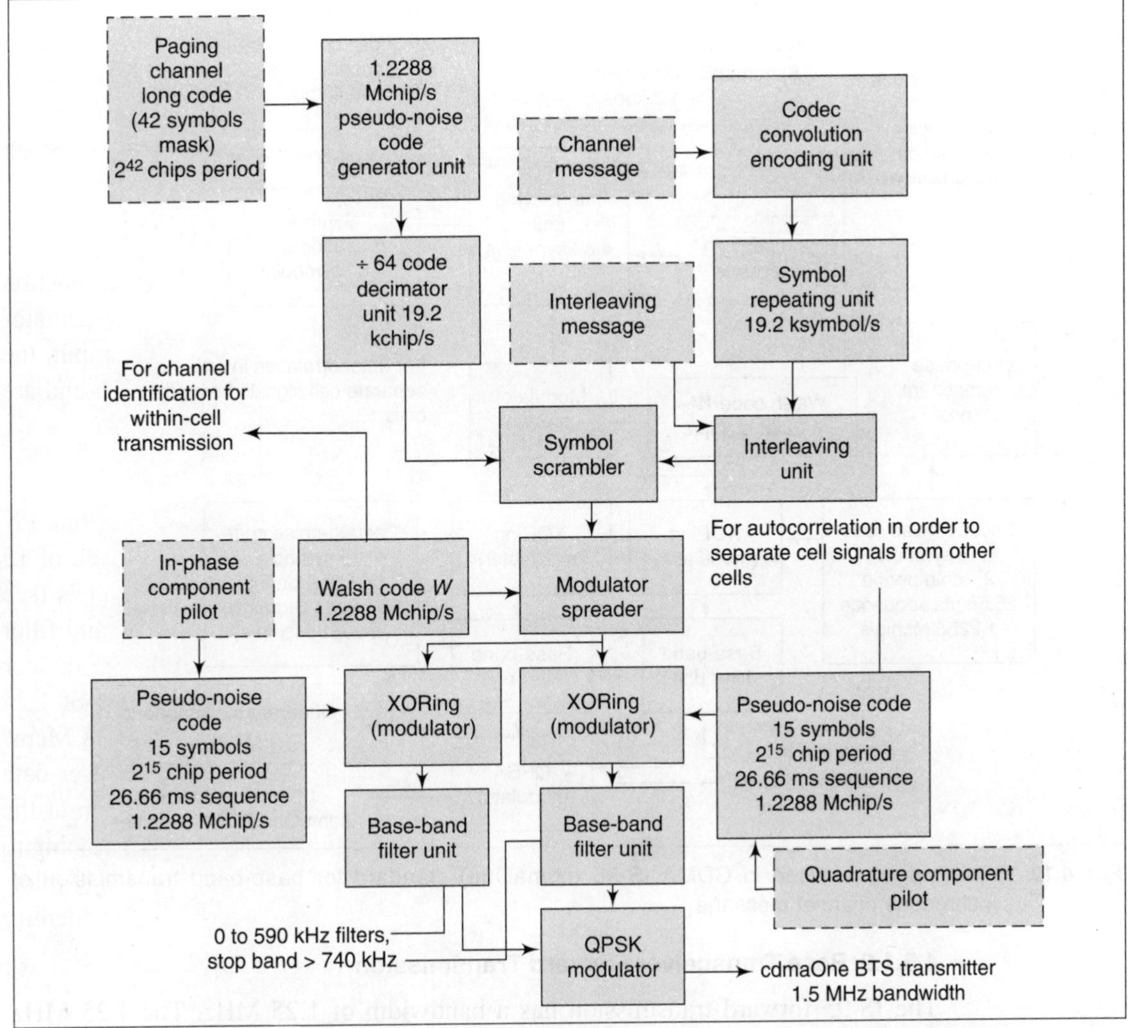

Fig. 4.11 Processing units used in CDMA IS-95 (cdmaOne) standard for base-band transmission of paging channel message and paging channel long code

4.5.1.8 Synchronous Channel Data

Figure 4.12 shows the processing units used in the CDMA IS-95 (cdmaOne) standard for base-band transmission of synchronous channel data at 1.2 kbps. The synchronous channel message is used for synchronization of chip sequences at the MS. Its messages include the system identification (SID), network identification (NID), system time, PN short sequence offset, and PN long sequence codes state. The synchronization channel uses Walsh code W_{32} and PN_Q and PN_I pilot channels like other channels. After passing through the synchronizing symbol repetition unit, the signal data burst will become 4.8 ksymbol/s. After going through a block-interleaving unit, the rate remains the same at 4.8 ksymbol/s. The 4.8 ksymbol/s are coded at a rate of 1.2288 Mchip/s by the Walsh code W_{32}.

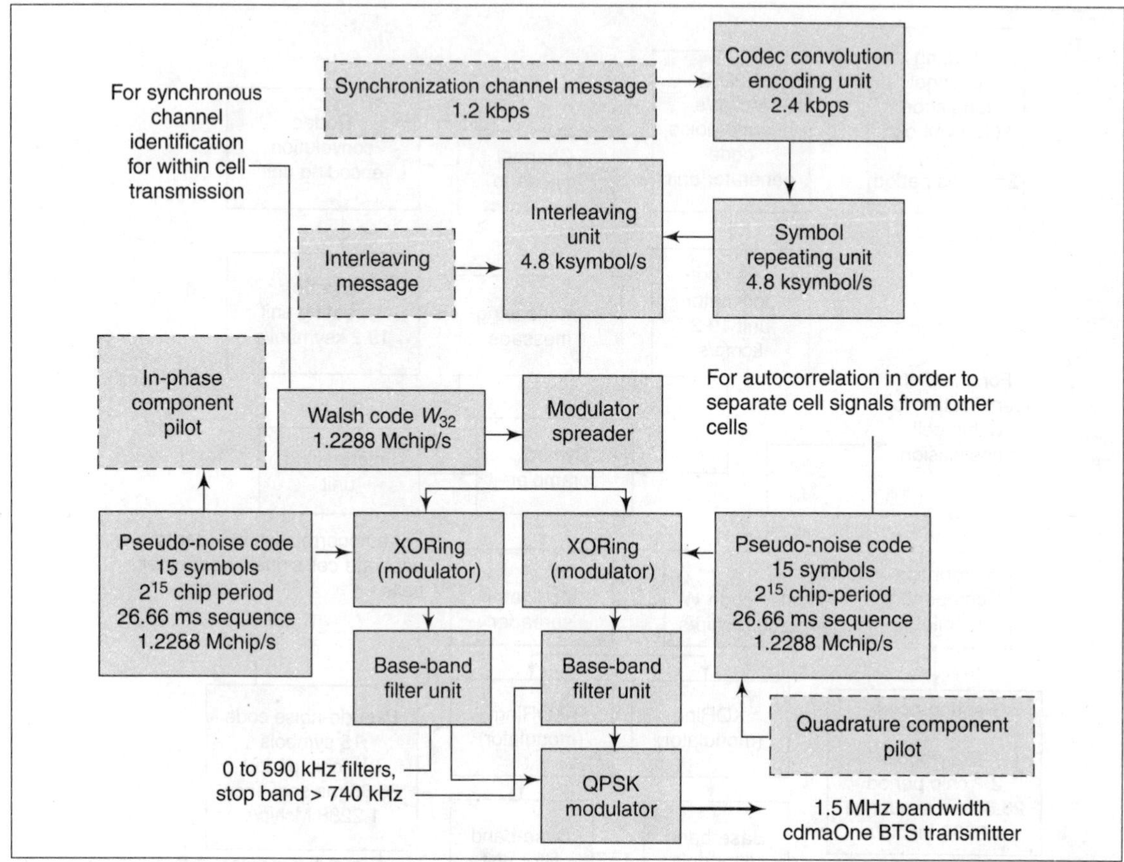

Fig. 4.12 Processing units used in CDMA IS-95 (cdmaOne) standard for base-band transmission of synchronous channel message

4.5.1.9 Base Transceiver Forward Transmission

The IS-95 forward transmission has a bandwidth of 1.25 MHz. The 1.25 MHz modulator output with radio carrier of forward-link channel has pilot (coded with Walsh code W_0), synchronizing channel (coded with Walsh code W_{32}), paging channels 1–7 (coded with Walsh codes W_{1-7}), and 55 traffic channels (each with a 0.8 kbps power control message) coded with the remaining Walsh codes (up to W_{63}). Traffic channel processing unit carries voice-data, fundamental code channel data, and supplementary data. A combining circuit is used to combine the PN_Q and PN_I base-band channel outputs from the total 64 traffic, synchronizing, paging, and pilot channels. The QPSK modulator (shown in Figs 4.10, 4.11, and 4.12) modulates the signals.

4.5.2 Reverse-link Channels

Figure 4.13 shows processing units used in the CDMA IS-95 (cdmaOne) standard for base-band transmission of reverse channels. These include fundamental mode

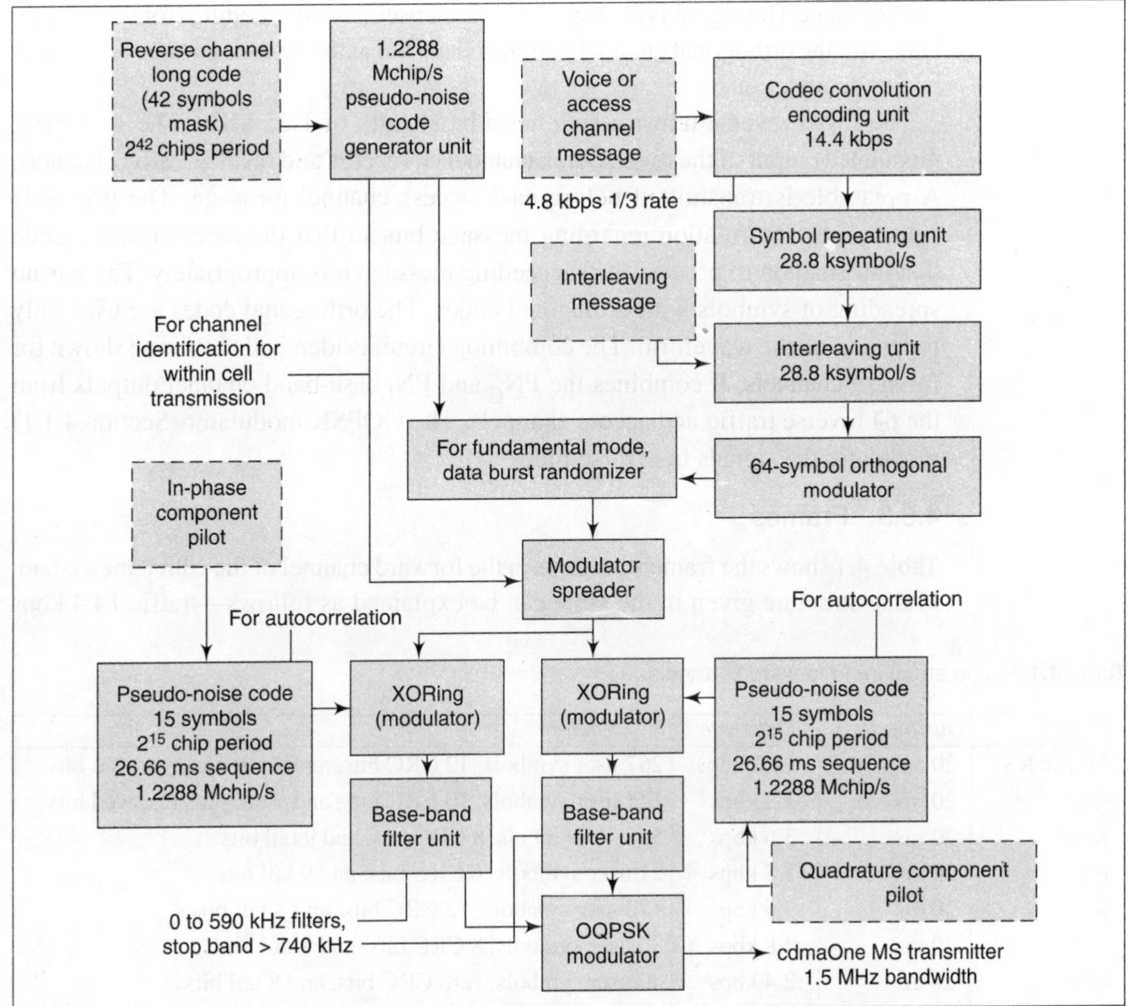

Fig. 4.13 Processing units used in CDMA IS-95 (cdmaOne) standard for base-band transmission of reverse channel access (when access message is used instead of voice) and fundamental mode traffic channels (64-symbol orthogonal modulator links directly to spreader in case of the access channel)

traffic channels and access channels (for access message transmission). A 64-symbol orthogonal spreader links directly to a PN long sequence spreader in case of the access channel. Processing units in forward channels shown in Figs 4.10, 4.11, and 4.12 differ from the reverse channel (Fig. 4.13) in the sense that in the latter 64-symbol orthogonal modulation is performed directly (not after the PN long *M*-sequence coding) on the user data (voice or access channel data). The user-signal waveform is mapped with the orthogonal Walsh codes [Eqn (4.5) shows how the mapping takes place]. The objective of user-signal waveform encoding is to reduce the bit error rates at the receiver. Orthogonal spreading modulation is used to identify the user data. The signal is also spread by long pseudo-noise coding (for scrambling later with the orthogonal channelization codes), but at the next stage after orthogonal coding modulation.

The IS-95 reverse transmission has a bandwidth of 1.25 MHz. The 1.25 MHz modulated output of the reverse-link channel has access and reverse traffic channels. A preamble is transmitted before each access channel message. The preamble contains the information regarding message bits so that the receiver can decode that information to process the succeeding message bits appropriately. There is no spreading of symbols with orthogonal codes. The orthogonal codes are used only for encoding the waveform. The combining circuit is identical to the one shown for forward channels. It combines the PN_Q and PN_I base-band channel outputs from the 64 reverse traffic and access channels. The OQPSK modulator (Section 4.1.1) modulates the signals in reverse link.

4.5.3 Frames

Table 4.1 shows the frame structures in the forward channel of the cdmaOne system. Frame structure given in the table can be explained as follows—traffic 14.4 kbps

Table 4.1 Frame structure in forward channels

Channel	*Duration*	*Data rate*	*Structure*
Traffic RS2	20 ms	14.4 kbps	267 user symbols, 12 CRC bits, and 9 tail plus reserved bits
RS2	20 ms	7.2 kbps	125 user symbols, 10 CRC bits, and 9 tail plus reserved bits
RS2	20 ms	3.6 kbps	55 user symbols, 8 CRC bits, and 9 tail bits
RS2	20 ms	1.8 kbps	21 user symbols, 6 CRC bits, and 9 tail bits
RS1	20 ms	9.6 kbps	172 user symbols, 12 CRC bits, and 8 tail bits
RS1	20 ms	4.8 kbps	80 user symbols, 8 CRC bits, and 8 tail bits
RS1	20 ms	2.4 kbps	40 user symbols, zero CRC bits, and 8 tail bits
RS1	20 ms	1.2 kbps	16 user symbols, zero CRC bits, and eight tail bits
Paging	10 ms (half frame)	9.6 kbps	Eight slots of 10 ms. 1-bit SCI (synchronization capsule indicator) preceding 95 paging bits in a 10 ms half frame. There can be eight slots in a paging message. SCI = 0 indicates the end of one message and the start of another message in a slot. SCI = 1 indicates continuation of the message.
Paging	10 ms (half frame)	4.6 kbps	8 slots of 10 ms with 47 bits after the SCI bit in each slot for the paging message.
Synchronous	20 ms	9.6 kbps	96 bits. A start of message (SOM) bit. SOM = 1 for first sub-frame of the message and SOM = 0 in successive extension of the message. The sequence is SOM bit, 31 data bits, SOM bit, 31 data bits, SOM bit, and 31 data bits. A frame is combined with other frames into a superframe. Each message consists of message length at the beginning, followed by data, error-checking codes, and padding bits.

RS2 frame consists of a sequence of 267 user symbols, 12 CRC bits, and 9 tail plus reserved bits. CRC means cyclic redundancy check (a standard way which is used for erroneous transmission check by comparing received frame CRC with expected CRC, for example, in Ethernet LAN). Paging channel half frame of 9.6 kbps consists of eight slots of 10 ms. One-bit SCI (synchronization capsule indicator) preceding 95 paging bits in a 10 ms half frame. There can be eight slots in a paging message. SCI = 0 indicates the end of one message and the start of another message in a slot. SCI = 1 indicates continuation of the message. Paging channel 4.6 kbps consists of eight slots of 10 ms with 47 bits after the SCI bit in each slot for the paging message. Synchronous channel 9.6 kbps consists of 96 bits. A start of message (SOM) bit is indicated as 1 for first sub-frame of the message and as 0 for the successive extension of the message. The sequence is SOM bit, 31 data bits, SOM bit, 31 data bits, SOM bit, and 31 data bits. A frame is combined with other frames into a superframe. Each message consists of message length at the beginning, followed by data, error-checking codes, and padding bits. Table 4.2 shows the frame structures in the reverse channel of the cdmaOne system.

Table 4.2 Frame structure in reverse channels

Channel	*Duration*	*Data rate*	*Structure*
Traffic	20 ms	14.4 kbps	267 user symbols, 12 CRC bits, and 9 tail plus reserved bits. At other data rates in RS1 and RS2, the structure is similar to the forward channel.
Traffic power control	20 ms	4.8 kbps	Sixteen groups of 1.25 ms, total frame duration remains 20 ms. When power control message is transmitted at 4.8 kbps, then eight out of 16 power control groups get masked. A data burst randomizer randomly performs masking. This reduces interference in the channel with voice-data at 9.6 kbps. Traffic channels can also carry voice and part of the data or power control data. This means that the traffic channel carries not only voice but also other data, for example SMS bits or MMS bits. It also carries power control data bits. These bits are transmitted after the measurement of the signal strength at the receiver to enable open loop power control during the CDMA communication.
Access	20 ms	4.8 kbps	88 bits plus eight bits as 0s. Preamble before message. Multiple frames which include with frames with all bits as 0s.

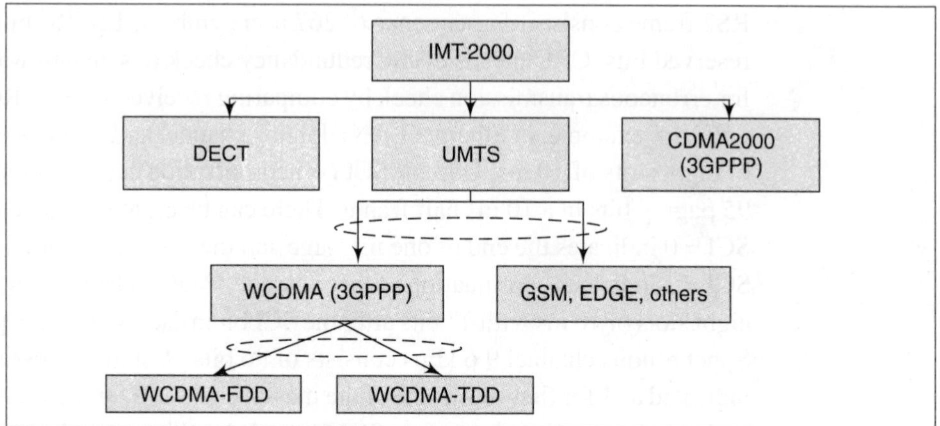

Fig. 4.14 3G technologies covered in the IMT-2000 global standards

4.6 IMT-2000

International mobile telecommunications-2000 (IMT-2000) is a 3G wireless communications standard defined by the recommendations of the International Telecommunication Union (ITU). Data rates lower than 153.6 kbps are considered below 3G. 3G technologies provide high quality of service and high data rate support for multimedia (audio, pictures, text, and video) transfer. The IMT-2000 standard represents a number of 3G technologies. Figure 4.14 shows the 3G technologies presently covered by the IMT-2000 global standard. These are DECT (Section 3.11), UMTS (universal mobile telecommunications system), and CDMA2000 standards. UMTS standardizes the mobile and transceiver station equipment, call setup session, security system, and radio access technologies. One set of technologies is based on WCDMA and the other on 3G evolution of GSM (EDGE) and others (Table 1.2).

4.6.1 WCDMA

WCDMA (wideband CDMA) supports data rates of 2 Mbps or higher for short distances and 384 kbps for long distances. It is also referred to as UTRA-FDD [universal (or sometimes UMTS) terrestrial radio access-frequency division duplex]. WCDMA access is either FDD or TDD (time division duplex). FDD separates reverse-link (known as uplink in GSM) and forward-link (known as downlink in GSM) frequencies. These are 1.920–1.980 GHz for uplink and 2.110–2.170 GHz for downlink and each uses a 5 MHz bandwidth. It is wider as compared to the 1.25 MHz of the IS-95 (2G CDMA) system. A brief description of WCDMA is given below.[1]

Figure 4.15(a) shows all the protocol layers in the WCDMA uplink terminal (MS) and downlink equipment (BTS). The physical layer between the MS and the

[1]For a detailed discussion on WCDMA see Andrew Richardson, 2005, *WCDMA Design Handbook,* Cambridge University Press, Cambridge.

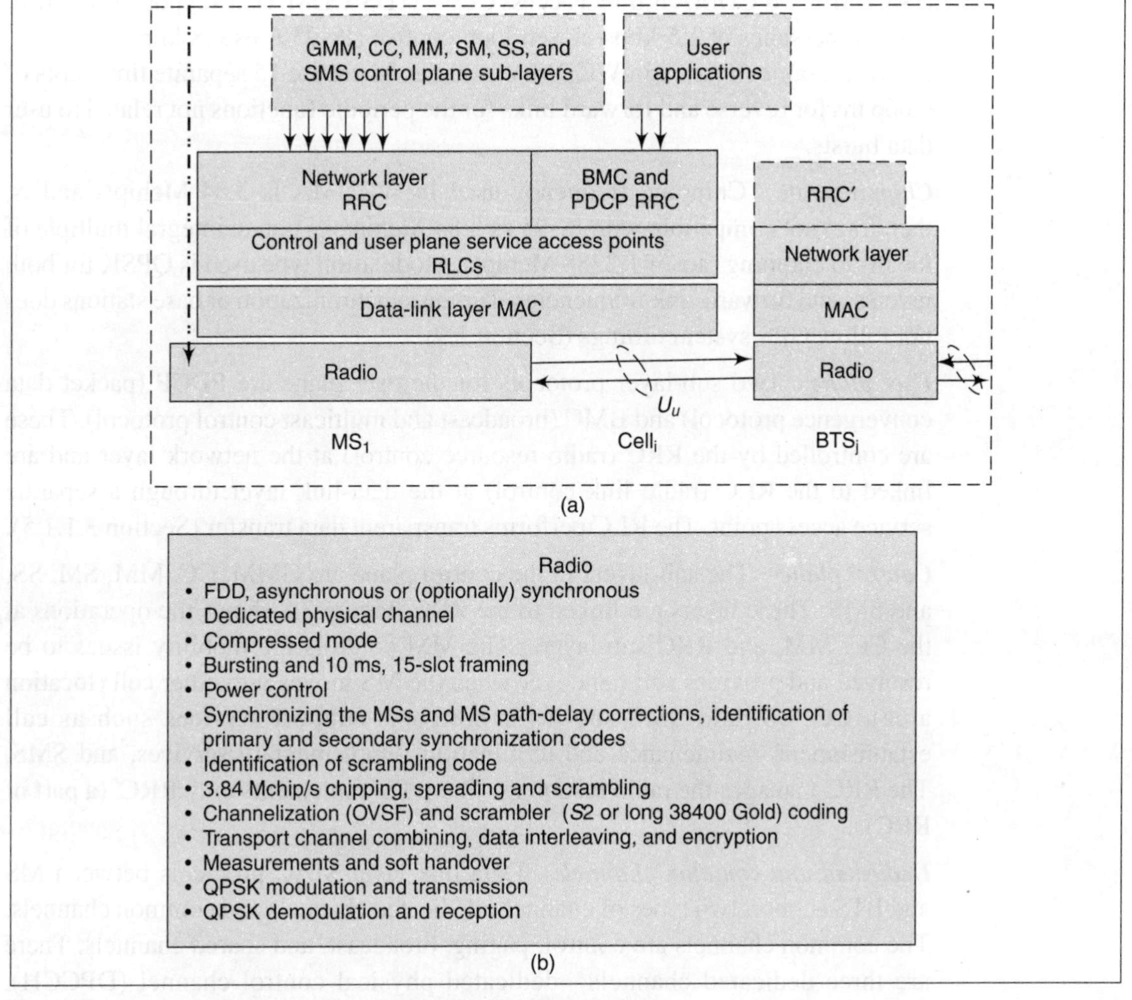

Fig. 4.15 Protocol layers between the WCDMA MS and BTS

BTS used for uplink and downlink is the U_u radio interface. Figure 4.15(b) shows the functions of the radio layer. The physical layer of WCDMA supports asynchronous transmission (Section 3.1.1.7). WCDMA(3GPPP) also supports synchronous transmission. Data link layer MAC controls the flow of packets to and from the network layer. The ciphering function is also assigned to the MAC layer protocol in WCDMA. It sends and receives data from control and user plane service access points at the radio link control (RLC) layer and sends it to the physical layer. The network layer above the RLCs provides access to multiple services such as BMC (broadcast and multicast control protocol), PDCP (packet data convergence protocol), GMM (GPRS mobility management), CC (call (connection) control), MM (mobility management), SM (session management), SS (supplementary service), and SMS (short message service).

DS-CDMA WCDMA is direct access CDMA (DS-CDMA). It supports fast power control messages at 1.5 kbps closed loop control (IS-95 uses 0.8 kbps open loop control). Frame duration in WCDMA is 10 ms. There are 15 separate time-slots of 0.666 ms for reverse and forward links for the periodic functions not related to user data bursts.

Chipping rate Chipping frequency used in WCDMA is 3.84 Mchip/s and is, therefore, not compatible with IS-95 as 3.84 Mchip/s is not an integral multiple of the IS-95 chipping rate of 1.2288 Mchip/s. Modulation type used is QPSK for both reverse- and forward-link frequencies. Timing synchronization of base stations does not follow GPS system timings (Section 2.9).

User plane Two sub-layer protocols for the user plane are PDCP (packet data convergence protocol) and BMC (broadcast and multicast control protocol). These are controlled by the RRC (radio resource control) at the network layer and are linked to the RLC (radio link control) at the data-link layer through a separate service access point. The RLC performs transparent data transfer (Section 3.1.1.5).

Control plane The sub-layers in the control plane are GMM, CC, MM, SM, SS, and SMS. These layers are linked to the RRC. Figure 3.8 shows the operations at the CC, MM, and RRC sub-layers. The MM controls the mobility issues to be resolved and provides soft handover when the MS moves into other cell (location area). CC, SS, and SMS sub-layer protocols support services such as call establishment, maintenance and termination, supplementary services, and SMS. The RRC manages the radio resources. The BTS implements only RRC′ (a part of RRC).

Dedicated and common channels Data link layer MAC protocols between MS and BTS support two types of channels, dedicated channels and common channels. The common channels are control, paging, broadcast, and shared channels. There are three dedicated channels—dedicated physical control channel (DPCCH), dedicated physical data channel (DPDCH), and dedicated physical channel (DPCH). Dedicated channels are assigned to the MSs for uplink. There is also a common channel for traffic. There are channels called transport channels (like access channel in cdmaOne). The channel code and structure is different for the uplink DPCCH and for the downlink DPDCH channel.

Short and long data packets Random access bursts are transmitted in 10 ms slots at fixed rates by a control mechanism and slotted ALOHA protocol is used for access. Short data packets are directly appended to the random access bursts in common (not dedicated) channel packet transmission. Longer packets are transmitted by dedicated channels at variable power, controlled by a power control message appended to the user symbols.

Different types of physical channels WCDMA supports several different types of physical channels. Certain physical channels are defined by special (distinct) waveforms only. Each pair of spreading and scrambling codes defines the remaining physical channels. WCDMA uplink employs channelization codes for controlling and synchronizing multiple data rate channels. UMTS terrestrial radio access network (UTRAN) employs channelization codes for synchronizing the multiple user terminals (MSs). (The functions of UTRAN are similar to those of the BTS in GSM systems.) WCDMA UTRAN employs scrambling codes for radio resource planning when downlinking to the MSs and uses channelization codes when downlinking the user channels to the MSs. WCDMA chipping rates are 3.84 Mchip/s.

DPDCH employs a pair of channelling and scrambling codes. UTRAN (UMTS terrestrial radio access network) performs code allocation functions for downlink channelization codes and uplink scrambling codes. The MS performs code allocation functions for uplink channelization code as per the data rate and the BTS performs the code allocation at the downlink radio-planning layer for the scrambling code.

Power control WCDMA reverse link (mobile terminal uplink) transmits pilot symbols, which are multiplexed with rate information as well as power control messages. Rate information facilitates coherent detection.

Multi-rate transmission A single code is used when transmitting small data rate signals multiplexed in time-space and multiple codes are used when transmitting large data rate signals multiplexed in code-space. WCDMA transmission uses a single code for small data rates and multiple codes for large data rates.

Use of variable rates by WCDMA processing units Certain types of data need to be transmitted at fast rates and some other types of data, for example, voice-data, power control data, and SMS text, require slow transmission rates. Variable rates are required in different types of services to form a system. Orthogonal coding for channelization is asymmetrical in uplink and downlink. WCDMA employs a constant chipping rate for spreading but variable spread factors, called OVSF (orthogonal variable spreading factors). The concept of variable spread factor was explained in Section 4.4.2.2 by taking an example of Walsh coding. WCDMA also uses variable spreading codes but ones that are different from the Walsh code. Use of variable spread factor controls the signals with multiple data rates. The code length used per symbol is four when the downlink user symbol data rate needed is 1.92 Mbps, and is 512, when the data rate needed is 15 kbps. Spreading codes of different lengths are thus used and the orthogonality of the codes is maintained. The source (MS) and channel for these symbols when using orthogonal code spreading then gets uniquely identified on de-spreading at the receiver.

Reverse channels use Gold and S(2) codes. The chipping rates for these codes are 38400 and 256 chips, respectively. Gold codes (Section 4.2.2) are used for the MS user symbols and S(2) codes identify the user at the receiver.

Spread factor control Spread factor controls the user data rate. When spread factor is 4, user data rate becomes one-fourth of the rate corresponding to a spread factor of 1. OVSF codes support both orthogonality as well as variable data rates for a physical channel. Uplink OVSF does not separate the users because of different delays expected from the near and far terminals (MSs).

Example 4.5 Consider a transmission with a code length of eight per symbol. The chipping rate is constant at 3.84 Mchip/s. Time interval between the chips is 1/3.84 μs. The time interval in which a symbol is chipped is 8/3.84 μs. What is the channel data rate? What is the effective transmission rate? Assuming that the code consists of 256 chips, what is the rate at which the symbols transmit?

Solution: The channel data rate will be the reciprocal of the time interval between each instance of chipping = 3.84/8 Msymbol/s

= 480 ksymbol/s

Two bits are paired during modulation in WCDMA transmission. Therefore, the effective transmission rate = 960 ksymbol/s

$$\text{If 256 chips are used, the transmission rate} = \frac{8}{256} \times 960 \text{ ksymbol/s}$$

$$= \frac{960}{32} \text{ ksymbol/s} = 40 \text{ ksymbol/s}$$

The key features of WCDMA processing units are as follows:

- Asynchronous base stations
- Employing the same direct sequencing FDD mode 1 transmission, same channelization codes (OVSF), same modulation (QPSK), and same carrier modulation (QPSK) for both uplink and downlink
- Chipping rate of 3.84 MHz
- Multi-rate transmission of signals by spread factor control
- Use of variable data rates

Compatibility with cdmaOne and CDMA2000 systems WCDMA systems can be made compatible with cdmaOne and CDMA2000 systems by use of synchronous base stations, employing multi-carrier mode with a chipping rate of 3.6864 Mchip/s (which is an integral multiple of the IS-95 chipping rate of 1.2288 Mchip/s), and adding a CDMA pilot to the direct spread mode data.

4.6.2 CDMA2000

CDMA2000 is a set of 3G communication standards. It is a registered trademark of the Telecommunications Industry Association (TIA). CDMA2000 can operate at 400 MHz, 800 MHz, 900 MHz, 1700 MHz, 1800 MHz, 1900 MHz, and 2100 MHz and is compatible with the cdmaOne standard. The CDMA2000 standards CDMA2000 1x, CDMA2000 1xEV-DV (evolution for high speed integrated data and voice), and CDMA2000 1xEV-DO (evolution for data optimized) are approved according to the IMT-2000 standard. CDMA2000 1xEV-DV transmission rates are up to 614 kbps. CDMA2000 1xEV-DO rates are up to 2.05 Mbps. 1xEV-DO is a 3G standard. CDMA2000 1xEV-DO and CDMA2000 1xEV-DV are enhancements of CDMA2000 1x, accepted as standards in 2004. CDMA2000 3x uses three 1.2288 Mchip/s channels. Channel bandwidths for CDMA channels are 1.25, 5, 10, 15, and 20 MHz.

4.6.2.1 External Synchronization Mechanism

CDMA2000 base stations (BTSs) use an external synchronization mechanism—satellite-based GPS. Adjacent cells can use the same frequency but must use distinct phase angles. Some period of time is required for synchronization among the adjacent base stations in asynchronous stations in WCDMA. Therefore, synchronizing processes in WCDMA stations have a certain latency period. The latency between base stations of adjacent cells causes call drops in WCDMA, even when a soft handover is performed. The external synchronization among stations in CDMA2000 reduces latency-related call drops.

4.6.2.2 Dedicated and Common Channels

Data-link layer MAC protocols between MS and BTS support two types of channels, dedicated channels and common channels for common control. Reverse access channel and reverse pilot channel are the only common control channels present during the access setup phase. The frame structure is 172 user data bits, 12 CRC bits, and 8 tail bits as 0s for a 4.8 kbps access channel. There is a preamble before a message (Section 4.5.2). Access is random and uses the slotted ALOHA protocol. User data is not transmitted through the access channel, but through dedicated channels with rate sets RS1 or RS2. On putting the user channel data in the 20 ms time-slots, the rate becomes 307.2 kbps so that after a four-chip Walsh code (spread factor of 4), the chipping rate becomes 1.2288 Mchip/s.

Low data rate services employ one fundamental mode and high data rate services and multimedia services employ a greater number of dedicated channels. A pilot signal is a reference signal. The MS uses the continuously transmitting, code-divided, dedicated pilot for uplink. The BTS uses the code-divided common pilot and dedicated or common auxiliary pilots for downlink. The code-divided dedicated pilot helps in coherent detection of the reference signal. The BTS pilot channel multiplexes the power control and a control bit called erasure indicator bit.

4.6.2.3 Short and Long Data Packets

Frame length modulation is QPSK for downlink and BPSK for uplink. Frame length modulation gives frames of 5, 10, 20, 40, and 80 ms duration as per the packet size. Short data bursts use the slotted ALOHA protocol and are transmitted at variable power. Power level is enhanced after an unsuccessful access.

4.6.2.4 Different Types of Physical Channels

CDMA2000 supports several different types of physical channels. Chip rates are $n \times 1.2288$ Mchip/s, where n is a positive integer. $n > 1$ facilitates multi-rate or single-rate data transmission at higher rates, which are n times 1.2288 Mbps. Supplementary channels for reverse link (uplink) use turbo codes. Other physical channels use 1/4 convolution encoding. The signals are scrambled with long code sequences before chipping with orthogonal codes. This is followed by short code spreading of the carrier channels. Long codes in CDMA2000 use the same length M-sequences but are used for channel data divided in two different phases (I and Q) for each channel before scrambling.

4.6.2.5 Multi-rate Transmission of Signals

Multi-rate transmission of signals can be explained as follows—A picture in CIF format has a resolution of 352 × 288 pixels. It is transmitted at a data rate of 384 kbps. Voice or other low data-rate services and power control messages are transmitted at RS1 data rates of 9.6, 4.8, 2.4, or 1.2 kbps or at RS2 data rates of 14.4, 7.2, 3.6, or 1.8 kbps. Low data-rate signals are used for voice and messages and high data-rate signals are used for pictures, videos, and large packets. A 3G system must support multi-rate transmission. An RS2 signal can be punctured to reduce data rates. Low data-rate signals of up to 19.2 kbps can be interleaved. For high data-rate signals of up to 307.2 kbps, variable length Walsh codes support both orthogonality as well as variable data rate for a physical channel. (Example 4.2 described an application of the concept of variable length spread factor). The spread factor can vary from four to 256 depending upon the data rates in CDMA2000. (IS-95 spread factor is constant at 64.) Downlink uses Walsh codes or quasi-orthogonal codes and uplink uses Walsh codes. (Quasi-orthogonal means not strongly orthogonal.) Orthogonal coding for channelization is asymmetrical in uplink and downlink.

4.6.2.6 Multi-rate Data Encoding in Traffic Channels

IS-95 cdmaOne user data has different data rates. Rate set RS1 transmits at 9.6, 4.8, 2.4, or 1.2 kbps. An optional rate set RS2 transmits at 14.4, 7.2, 3.6, or 1.8 kbps. It requires a different type of convolution encoding and error encoding depending on considerations of service quality. Low or high service quality means high or low bit error rates, respectively. As an example, multi-encoded rate signals are systems that transmit at variable data rates after convolution coding. In IS-95, RS1 and RS2 data rate symbols are matched, repeated, and interleaved so that before chipping, the symbol transmission rate is constant at 19.2 ksymbol/s.

4.6.2.7 Downlink and Uplink Modulations of Spreading Signals (Pilots)

Spreading signal (pilots) modulation is balanced QPSK modulation for downlink and dual channel QPSK modulation for uplink. The modulation of radio-carrier frequency is asymmetrical for uplink and downlink. Figure 4.16 shows a three-carrier system. The figure shows the processing units used in the CDMA2000 standard for power control messages and channels for data, sync, pilot, and traffic base-band transmission and use of a MUX/IQ unit, long code and mask, PN_Q, PN_I, and Walsh codes. In IS-95, the data channel is divided into the I (in-phase) and Q

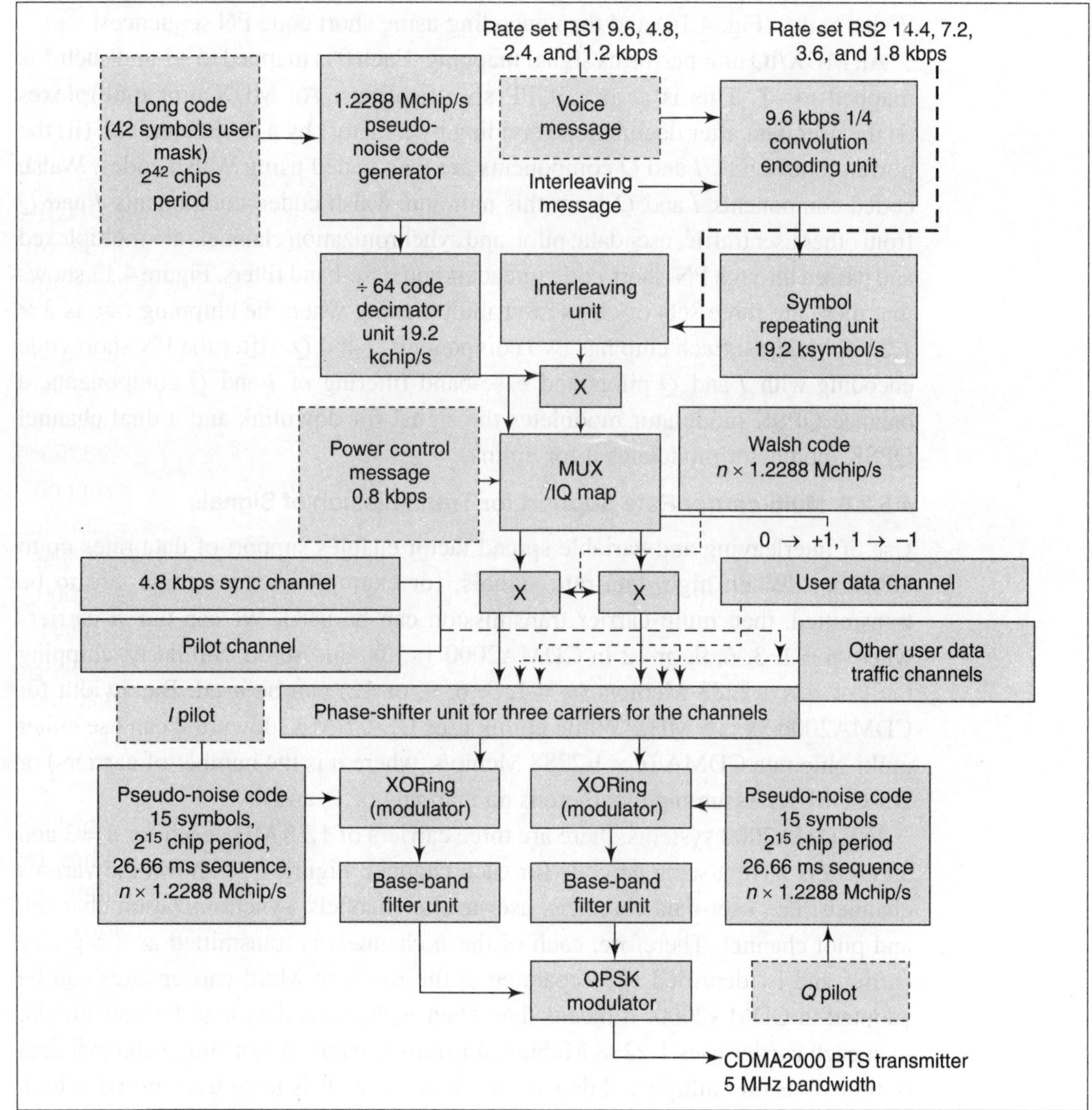

Fig. 4.16 Processing units used in the three-carrier CDMA2000 3X standard for the power control message and channels of data, sync, pilot, traffic base-band transmission (showing the use of a MUX/IQ unit, long code and mask, PN_Q, PN_I, and Walsh codes)

(quadrature) components after chipping with Walsh codes. In CDMA2000, data channel is divided before scrambling. The main difference with respect to the IS-95 processing unit is that in IS-95, the data channel divides into *I* (in-phase) and *Q* (quadrature) components after scrambling with Walsh codes (Fig. 4.10) and in CDMA2000, the MUX/IQ (multiplexer and signal mapping unit) first divides the signal into the *I* and *Q* components. This is followed by chipping of the signal using Walsh codes (Fig. 4.16) and then encoding using short code PN sequences.

An MUX/IQ unit performs signal mapping. Each 0 is mapped to +1 and each 1 is mapped to –1. This is as per 3GPP specifications. An MUX unit multiplexes (i) the user data after decimation (dividing by a factor) by a decimator, and (ii) the power control bits. *I* and *Q* components are then coded using Walsh codes. Walsh coded components *I* and *Q* from this unit and Walsh coded components *I* and *Q* from other user traffic, user data, pilot, and synchronization channels are multiplexed and passed through PN-short-code spreaders and base-band filters. Figure 4.17 shows that there are three sets of chips after multiplexing when the chipping rate is 3 × 1.2288 Mchip/s. Each chip has two components, *I* and *Q*. After the PN short code encoding with *I* and *Q* pilots and base-band filtering of *I* and *Q* components, a balance QPSK modulator modulates the signal for downlink and a dual channel QPSK modulator modulates it for uplink.

4.6.2.8 Multi-carrier Rate Support for Transmission of Signals

Use of interleaving and variable spread factor enables support of data rates up to 386 kbps. When high data-rate signals, for example video signals, are to be transmitted, then multi-carrier transmission can be used. We can use *n* carriers where n = 1, 3, 6, 9, or 12 in CDMA2000 1x, 3x, and so on. Similarly, chipping rates of $n \times 1.2288$ Mchip/s (n = 1, 3, 6, 9, or 12) can be used. Bandwidth for CDMA2000 3x is 5 MHz. While uplink uses DS-CDMA, downlink can use either multi-chip rate CDMA ($n \times 1.2288$ Mchip/s, where *n* is the number of carriers) or DS-CDMA. (Assuming that there is no mapping of 1s and 0s.)

In CDMA2000 systems, there are three carriers of 1.25 MHz each for $n = 3$ and all carriers have a separate code for each channel. Figure 4.18 shows the various channels, i.e., user-data channels, user-traffic channels, synchronization channel, and pilot channel. Therefore, each of the *n* channels is transmitted as a separate carrier and is identified and separated at the receiver. Multi-carrier rates can be adopted in CDMA2000. It means that when *n* channels data is to be transmitted, then each carrier uses 1.2288 Mchip/s for transmission. When single channel data or time division multiplexed data of different channels is to be transmitted at high data rates, then a single carrier uses $n \times 1.2288$ Mchip/s for chipping. One option for employing chipping rates of 1.2288 Mchip/s when transmission is at higher rates ($= n \times 1.2288$ Mbps) is as follows—one channel is transmitted in a time-slot by time division multiplexing and the same chipping rate is used for that channel. Another option is that only a single carrier be transmitted at an instant but at high transmission rates of $n \times 1.2288$ Mbps and using 1.2288 Mchip/s. WCDMA uses the second option (3.84 Mchip/s). Multiplexing in time- and code-space can be used for sending data from different channels and for sending data from different traffic channels. CDMA2000 uses the first option and three or more carriers (each with a chipping rate of 1.2288 Mchip/s) can be transmitted by multiplexing in time-space.

The phase-shifter unit for three carriers for the channels is shown in Fig. 4.16. Figure 4.17 shows the phase-shifted three-carrier transmission of carrier channels *A, B,* and *C* by multiplexing in time-space. (The code 0110101001 has been used as an example in Fig. 4.17. The actual code must be an orthogonal code and must have a code length per user symbol as per transmission channel requirements. Also 0s are transmitted as +1s and 1s as –1s using signal mapping as per 3GPP specifications.) All channel carriers are chipped using the same orthogonal code and scrambled using the same PN long codes, but the chipping instants of each carrier channel are shifted by $(1/3) \times 814$ ns (phase angle change of 120°), so that the combined chipping rate is $1.2288 \times 3 = 3.6864$ Mchip/s. Therefore, the spread-spectrum frequency bandwidth is 3.6864 MHz. The bandwidth of a CDMA2000 BTS is 5 MHz. For $n = 12$, 12 carrier channels are transmitted with a time-space shift of $(1/12) \times 814$ ns (phase angle change of 30°). Therefore, multi-carrier transmission in

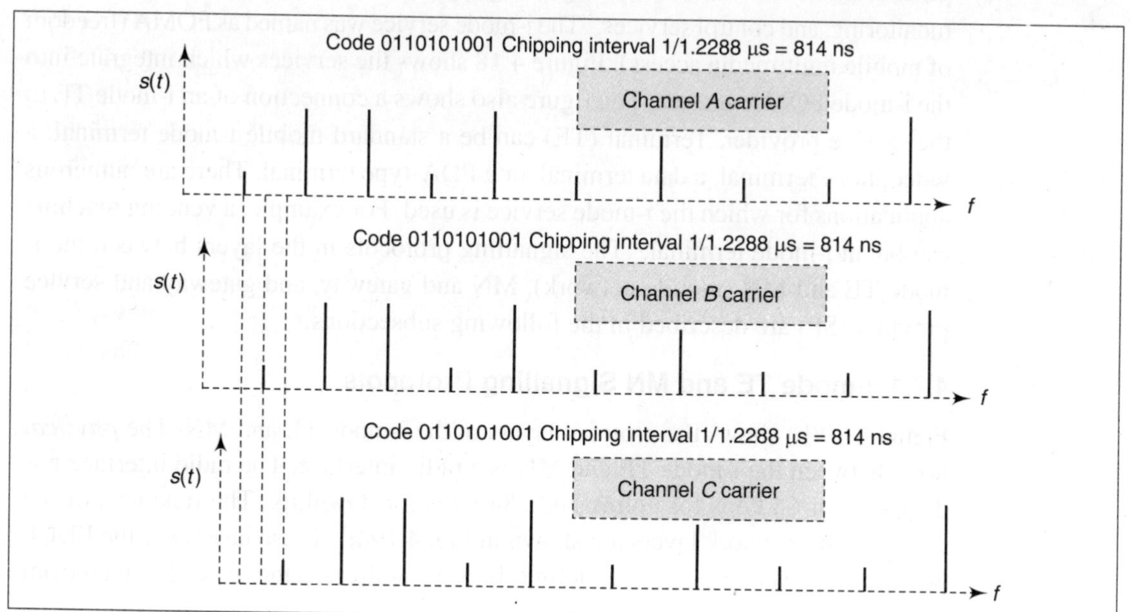

Fig. 4.17 Three-carrier signal frequencies with three chipping codes of 10 symbols (code 110101001 has been taken as an example. Dotted vertical lines show the phase shift between the three carriers)

CDMA2000 uses multiplexing in time-space. All channel carriers are chipped using the same orthogonal code and scrambled using the same PN long codes, but the chipping instants of each carrier channel are shifted by $(1/n) \times 814$ ns (phase angle change of $360°/n$) in CDMA2000.

The key features of CDMA2000 MS and BTS processing units are as follows:

- Synchronous base stations
- Interleaving of signals of low data rates (between 0.8 kbps and 4.8 kbps)
- Multi-rate transmission by variable spread factor (between 4 and 256) for data rates of 307.2 kbps to 4.8 kbps.
- Multi-carrier (MC) FDD mode-2 transmission by using chipping rate = $n \times 1.2288$ Mchip/s (where $n = 1, 3, 6, 9$, or 12 and all n carriers have unified power control)
- Employing distinct uplink and downlink multiple access, modulation, spreading, and channelization codes

4.7 i-mode

NTT DoCoMo, Japan evolved WCDMA based i-mode Internet services. It uses adaptive multi-rate encoding. i-mode is a cost-effective method for high-speed packet-switched data transfer. i-mode systems communicate user voice-data and provide Internet access. i-mode provides integrated services for voice, data, Internet, picture, music attachment to mail, gaming applications, ringtone downloads, remote monitoring, and control services.[1] The i-mode service was named as FOMA (freedom of mobile multimedia access). Figure 4.18 shows the services which integrate into the i-mode FOMA service. The figure also shows a connection of an i-mode TE to the service provider. Terminal (TE) can be a standard mobile i-mode terminal, a videophone terminal, a data terminal, or a PDA-type terminal. There are numerous applications for which the i-mode service is used. For example, a vending machine can be an i-mode terminal. The signalling protocols in the layers between the i-mode TE and MN (mobile network), MN and gateway, and gateway and service provider (SP) are described in the following subsections.

4.7.1 i-mode TE and MN Signalling Protocols

Figure 4.19(b) shows the protocol layers in the i-mode TE and MN. The *physical layer* between the i-mode TE and MN is a radio interface. The radio interface has data rates of 64 kbps for uplink and 384 kbps for downlink. The functions of the data link and network layers are shown in Fig. 4.19(a). These layers use the PDCP protocol. The *network layer* (a) defines how the addressed messages received from

[1]Detailed description of i-mode services and future perspectives towards 4G services are reviewed in Keiji Tachikawa 2003, "A perspective on the Evolution of Mobile Communication", *IEEE Communications*, vol. 41, no. 10, pp. 66–73.

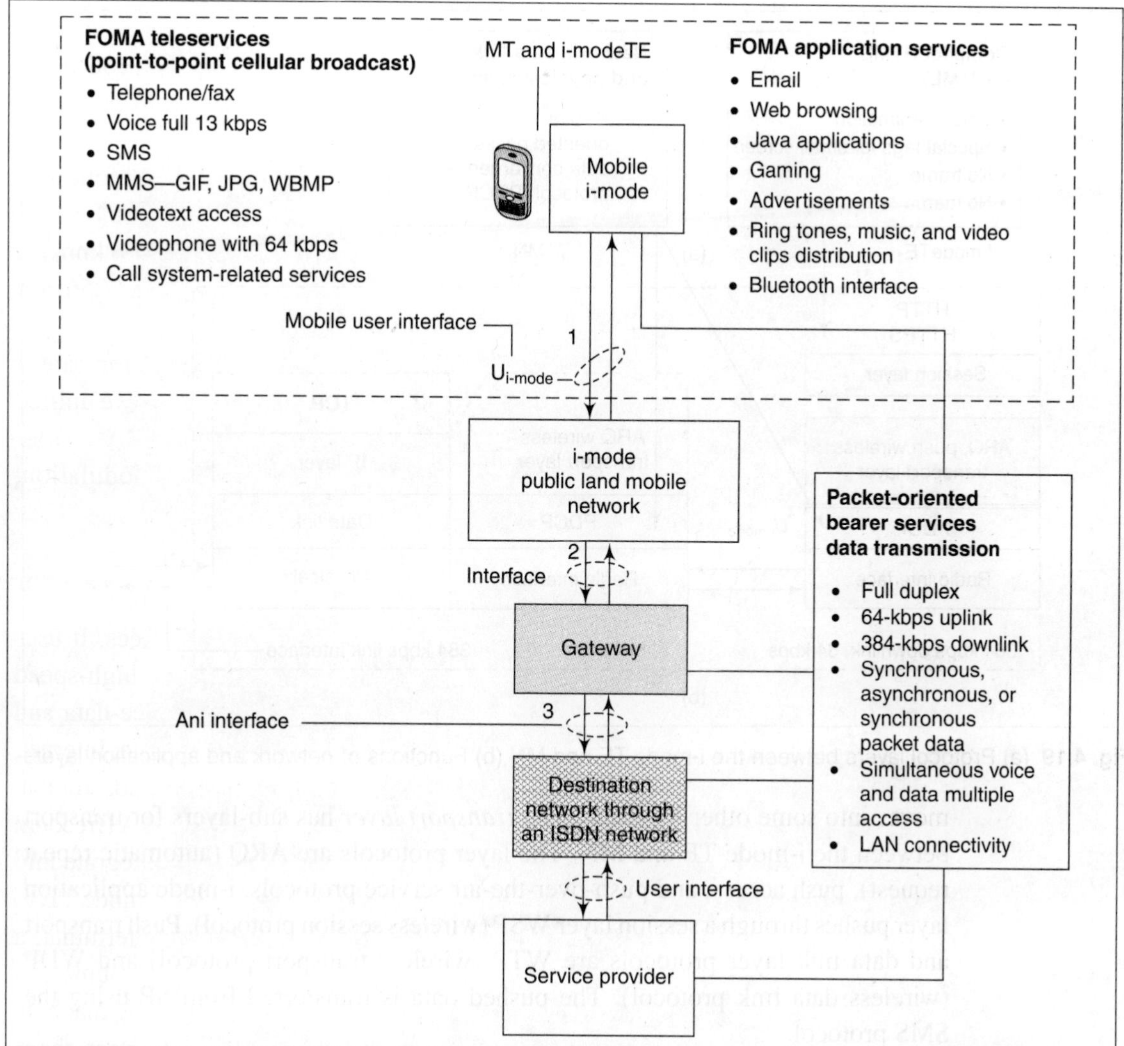

Fig. 4.18 FOMA telecommunication, application, and bearer services integration in an i-mode system

the data link layer are to be implemented by the operations of a protocol, (b) defines the addresses of the messages, (c) transmits the logical channel (FOMA service and control channels) data and information bits to the data-link layer from a service provider address (SPA), and (d) receives the logical channel data and information bits. It controls the flow of packets to and from the transport layer and provides access (through transport layer) to multiple FOMA services (Fig. 4.17) using the application layer HTTP (hyper-text transfer protocol) or HTTPS (hyper-text transfer protocol over SSL). SSL means secure socket sub-layer for HTTP.

Network sub-layer protocols support all services as packet-oriented services. The network layer also controls mobility management issues when the i-mode TE

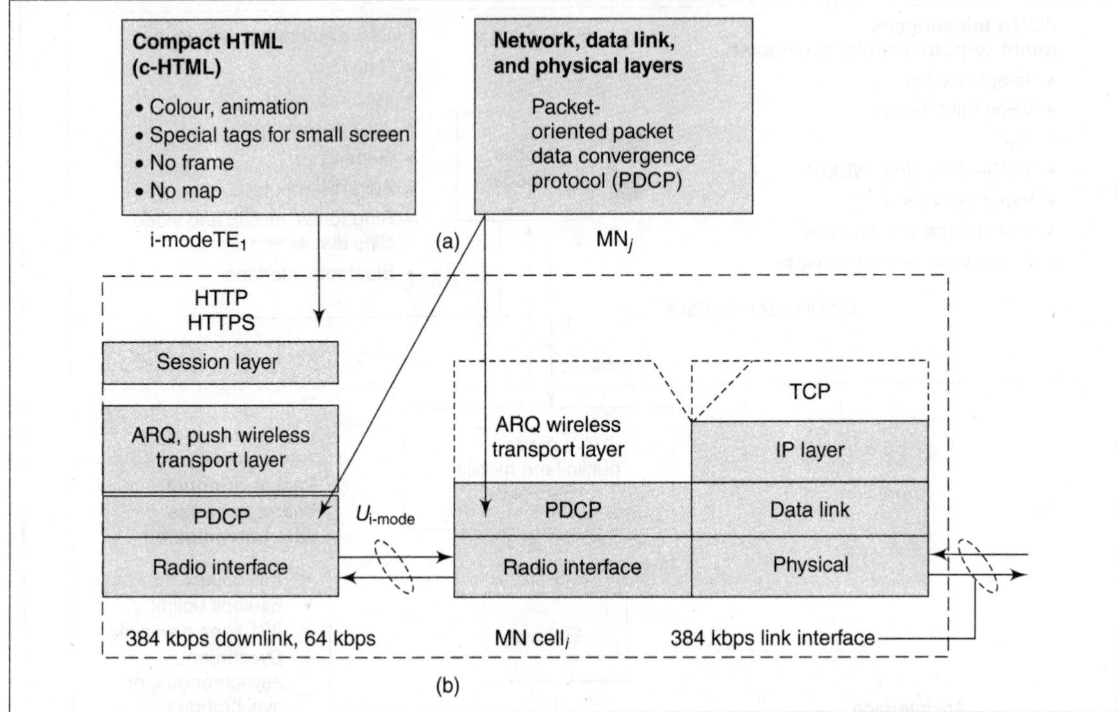

Fig. 4.19 (a) Protocol layers between the i-mode TE and MN (b) Functions of network and application layers

moves into some other MN area. The *transport layer* has sub-layers for transport between the i-mode TE and MN. The layer protocols are ARQ (automatic repeat request), push access, and push-over-the-air service protocols. i-mode application layer pushes through a session layer WSP (wireless session protocol). Push transport and data link layer protocols are WTP (wireless transport protocol) and WDP (wireless data link protocol). The pushed data is transferred from SP using the SMS protocol.

4.7.2 MN and Gateway Signalling Protocols

Figure 4.20 shows the protocol layers between the MN and the gateway. Voice is coded in 64 kbps format PCM (pulse code modulation) in a PSTN network. The *physical layer* between the MN and the gateway uses an ISDN or PSPDN network. The link operates at 384 kbps (six links of 64 kbps or 4×6 multiplexed 16 kbps channels). The interface between the MN and the gateway uses ISDN or PSPDN network and there is wired transmission and reception. The *data link layer* protocol between the MN and the gateway is LAPD (link access protocol *D*-channel) when using the ISDN A_{bis} interface or other L_2 layer protocol. The *network layer protocol* between the MN and the gateway is the globally used IP layer.

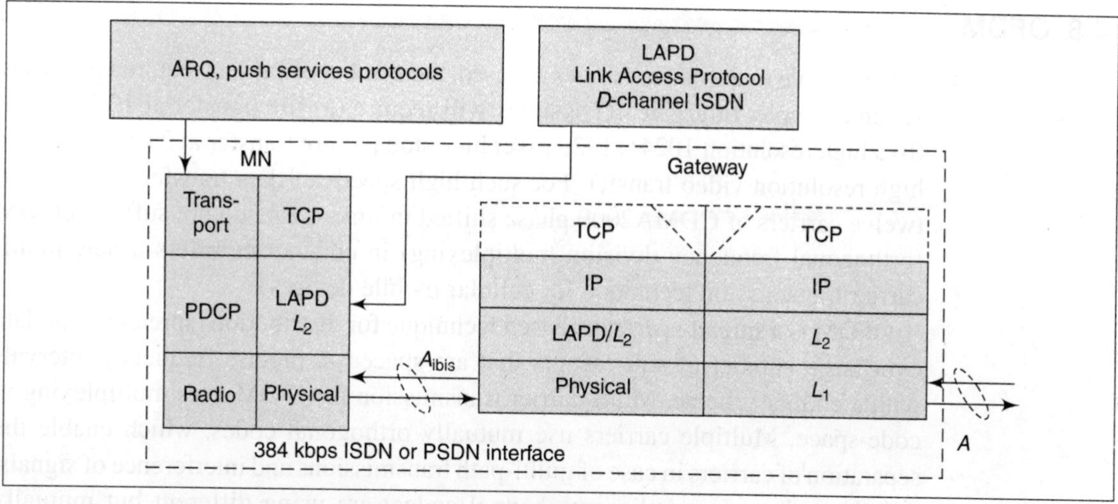

Fig. 4.20 Protocols layers between the MN and the gateway using ISDN or PSPDN

4.7.3 Gateway and Service Provider Signalling Protocols

Figure 4.21 shows the protocol layers between the gateway and the service provider. Physical to transport layers between the gateway and the SP use the same protocols as the global Internet service protocols. The protocols prescribe a standard procedure for the MTP (message transfer part) and SCCP (signalling connection control part) for SS7 (signalling system 7) transmission and reception in a 2 Mbps CCITT PSTN/ISDN/PSPDN network. (MTP is the part of the SS7. SCCP is also a part of SS7 which provides connection-less and connection-oriented network services above the MTP.) The *application layer* uses c-HTML and special tags. The layer employs HTTP or HTTPS for providing the services shown in Fig. 4.19.

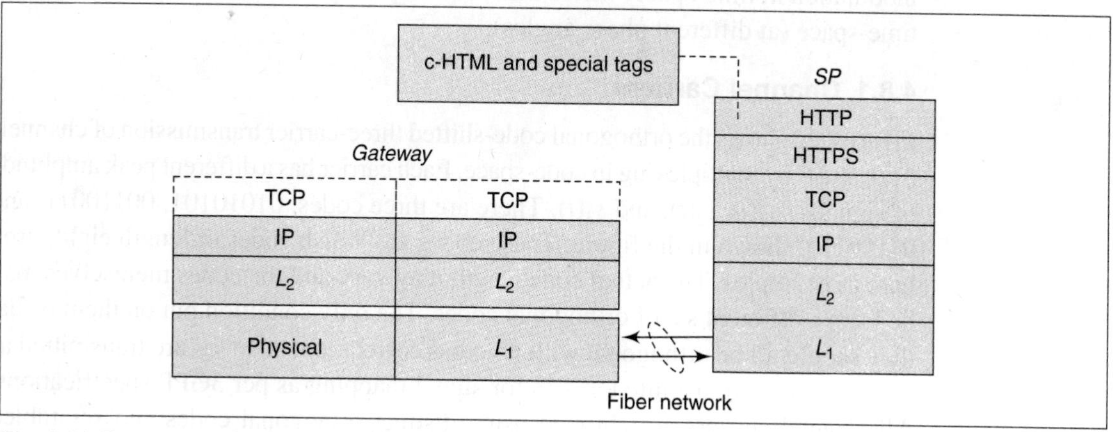

Fig. 4.21 Protocol layers in the gateway–SP interface

4.8 OFDM

When mobile multimedia services are used, the number of carriers required increases. As an example, futuristic 4G systems will require (a) file transfer at 10 Mbps, or (b) a high resolution 1024×1920 pixel hi-vision picture transfer at 24 Mbs, or (c) a high resolution video transfer. For such high speeds of data transfer, the three to twelve carriers of CDMA2000 phase shifted in time–space do not suffice. OFDM (orthogonal frequency division multiplexing) in code-space offers a new multi-carrier transmission technique for cellular mobile devices.

OFDM is a spread-spectrum-based technique for distribution (spreading) of data over large number of sub-carriers that are spaced at precise frequency intervals with a coding scheme. Multi-carrier transmission in OFDM uses multiplexing in code-space. Multiple carriers use mutually orthogonal codes, which enable the separation of carriers in case of multi-path transmissions and interference of signals. All channel carriers (adjacent channel codes) are using different but mutually orthogonal codes (Section 4.4.2.1). Each channel carrier has distinct amplitude (power level) and may have a time guard. The bandwidth remains equal to that in the single-carrier case. The three most important characteristics of OFDM are high spectral efficiency, strong resiliency to RF inter-symbol interferences, and lower multi-path distortions. As compared to a single-carrier system, the peak to average power ratio (PAR) is high in an OFDM system, but it gives a many-carrier system a bandwidth equal to that of a single-carrier system. Digital video broadcasting (Section 9.6) and HyperLAN-2 (Section 13.1.1) use OFDM systems.

OFDM is also called COFDM (code orthogonal frequency division multiplexing). OFDM is also called spread-spectrum based multi-carrier or discrete multi-tone modulation. The term COFDM is used in order to distinguish this modulation method from quadrature modulation QPSK or OQPSK. OQPSK is also an orthogonal modulation in time-space, where different frequency carriers transmits in different time-space (at different phase angles).

4.8.1 Channel Carriers

Figure 4.22 shows the orthogonal code-shifted three-carrier transmission of channels *A*, *B*, and *C* by multiplexing in code-space. Each carrier has a different peak amplitude of signals—$s_1(t)$, $s_2(t)$, and $s_3(t)$. There are three codes, 01010101, 00110011, and 01100110, shown in the figure. These codes are Walsh codes of length eight, used here as examples. The actual code length may vary and the codes themselves may be from a different set of orthogonal codes. The only condition put on them is that they should all be orthogonal with no cross correlation. Also, 0s are transmitted as +1s and 1s are transmitted as –1s for signal mapping as per 3GPP specifications. All channel carriers are chipped using distinct orthogonal codes and scrambled using the same or distinct PN long codes, but the chipping instants of each carrier

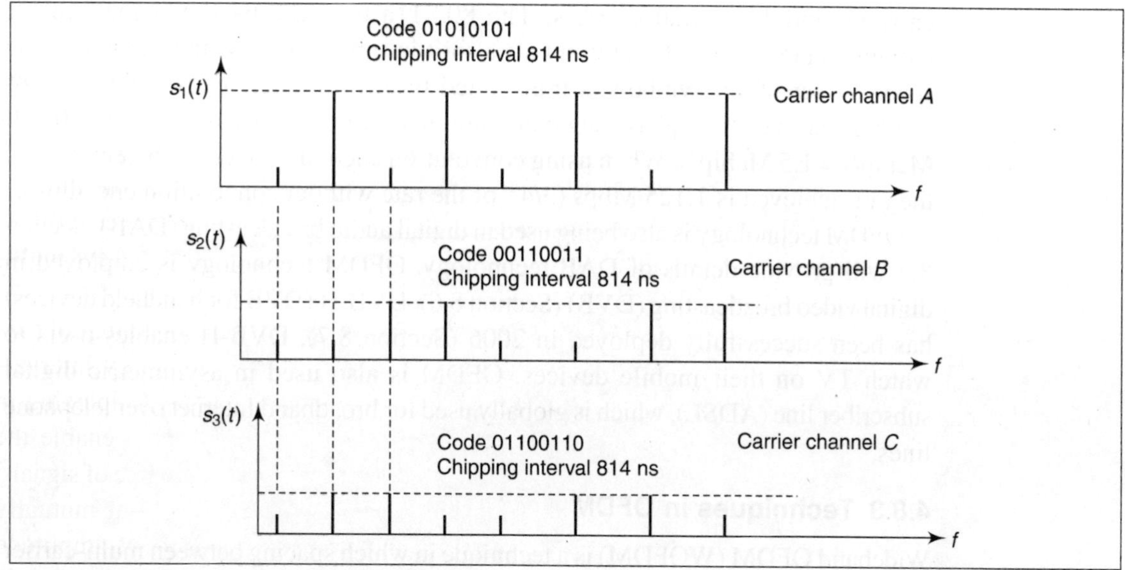

Fig. 4.22 Three-carrier frequency signals with three chipping codes of eight Walsh symbols. Dotted vertical lines show no phase shift (coding start time is the same for the three carriers). Dotted horizontal lines show three different power levels. Chipping rate is taken as 1.2288 Mchip/s in this example

channel use a different code (code-space divides into n when using n distinct orthogonal codes), so that the combined chipping rate remains 1.2288 Mchip/s with a bandwidth of 1.25 MHz. Here, the chipping rate is taken to be 1.2288 Mchip/s as an example. Actual chipping rate may be higher. Spread-spectrum frequency is 1.2288 MHz only when the bandwidth used is 1.25 MHz. For $n = 12$, 12 carrier channels transmit with n codes. Each carrier uses different power levels (amplitudes) so that the carriers can be separated at the receiver. A guard space in time between different OFDM carriers may also be used. (This will affect the effective data transmission rates. The wireless LAN (HiperLAN-2) IEEE 802.11a standard has 0.800 μs guard time.) Some carriers may be used by pilot channels for synchronizing and some may be redundant. Digital signal processors are used to perform modulation and demodulation in OFDM systems. Processing expressions for COFDM modulation are based on fast Fourier transforms and inverse Fourier transforms.[1]

4.8.2 Applications

OFDM technology is being used in wireless LANs for point-to-point transmission and for multicasting. Wireless LAN uses IEEE 802.11 protocol standards. In a supplement to the IEEE 802.11 standard, the IEEE 802.11 working group published the IEEE 802.11a standard which outlines the use of OFDM in the 5.8 GHz band. The 802.11a prescribes a chipping rate of 0.250 Mchip/s, 4 pilot carriers, 48 data

[1]For details see Jerry D. Gibson (Chief Editor) 1999, *Mobile Communications Handbook*, IEEE Press.

carriers, and 12 virtual carriers. The 802.11a uses the 64-QAM method for modulation (Section 4.1). This entails a shift in time as well as in peak amplitude space. A set of six symbols is transmitted through a single carrier. Each set of symbols has a distinct phase angle or amplitude. Bit transmission rate is 6×0.250 Mchip/s = 1.5 Mchip/s. When using convolution encoding to reduce bit error rates, the rate achieved is 1.125 Mbps (3/4th of the rate without convolution encoding).

OFDM technology is also being used in digital audio broadcasting (DAB). Section 8.5 will provide details of DAB technology. OFDM technology is employed in digital video broadcasting (DVB) (Section 8.6). DVB-H (DVB for handheld devices) has been successfully deployed in 2006 (Section 8.7). DVB-H enables users to watch TV on their mobile devices. OFDM is also used in asymmetric digital subscriber line (ADSL), which is globally used for broadband Internet over telephone lines.

4.8.3 Techniques in OFDM

Wideband OFDM (WOFDM) is a technique in which spacing between multi-carrier channels is made large. Therefore, any frequency errors between the transmitter and the receiver do not affect system performance.

Flash-OFDM (fast low-latency access with seamless handoff orthogonal frequency division multiplexing) is another technique based on the FHSS spectrum. It is also called fast-hopped OFDM.

MIMO-OFDM (multiple input, multiple output OFDM) is a technique in which multiple antennae are used for inputs and outputs. It provides broadband wireless access (BWA) and performs well in multiple non-line-of-sight multiple-path environments. VOFDM (vector OFDM) is another technique based on MIMO-OFDM. It has been developed by Cisco Systems, Inc.

Keywords

64-QAM: A modulation method which employs the two most significant bits for QPSK modulation and the remaining four bits for the 16-QAM signals. In this method, six symbols (bits) are transmitted for a sequence and thus the requirement of spectrum bandwidth is greatly reduced.

Autocorrelation code: A code of length n which gives a large value of sum $\Sigma e_i \times r_i$ when $e_i = r_i$ and a very low value when $e_i \neq r_i$, for $0 \leq i < n$. Here, e_i is the expected chip symbol and r_i is the received chip symbol in the ith sequence at the receiver.

cdmaOne: A transmission system, also known as IS-95, developed by QUALCOM, USA. It is a 2G+ technology using 824–849 MHz and 869–894 MHz, multiple analog channels interleaved in one digital carrier, and QPSK modulation. cdmaOne uses FDD (frequency division duplex) for forward and reverse links, *M*-sequence PN codes, Walsh codes, and PN short sequence codes for a carrier.

CDMA2000: A 3G transmission system using multiple rate channels interleaved in one digital carrier. The CDMA2000 system employs multiple n carriers, multi-chip rate ($n \times 1.2288$ Mchip/s), FDD (frequency division duplex) for forward and reverse links, *M*-sequence PN codes, variable length Walsh codes for data of variable rates, PN short sequence codes for the carrier, and QPSK modulation.

Channel: A transmission source operating in distinct space, time, frequency, or code. As an example the paging, traffic, synchronizing, or pilot channels in a user mobile device have different channelization codes defined by different Walsh codes.

Channelization code: A short length code showing the property of orthogonality, it is used for differentiating among different channels during CDMA transmission. Walsh codes are used for channelization due to their orthogonality property.

CSMA/CD (carrier sense multiple access/collision detect): A multiple access scheme in which all nodes attempt to sense the carrier at the same time, one of the nodes transmits as soon as it finds no carrier and after sensing the carrier each node waits (backs off) for a certain period of time in order to avoid collision. After the back off period the channel is sensed again for the carrier.

DSSS (direct sequence spread spectrum): A spread spectrum technique in which the spectrum is divided into output frequencies known as chipping frequencies. In DSSS, the spread is in direct sequence within a wide band so that the sequence of chipping frequencies is according to the result of the XOR operations between the *code* (consisting of a sequence of 1s and 0s) and the *symbol* to be transmitted. DSSS frequency changes from minimum to maximum value within a symbol interval.

Exposed terminal: A station which senses the carrier and backs off, even though that carrier is not interfering with the signal from the station.

Fast FHSS: A type of FHSS transmission in which the time interval for taking a hop is much less than the time interval of the symbol and a number of frequency hops take place during a symbol period.

FHSS (frequency hopping spread spectrum): A spread-spectrum technique in which the spectrum is divided into output frequencies known as hopping frequencies. In FHSS, the output frequencies hop within a wide band such that the hopping frequency value is according to a *code* consisting of a hopping sequence and the hopping frequency amplitude at an instant is 1 or 0 as per the *symbol* to be transmitted (1 or 0, respectively).

Hidden terminal: A station the signal from which is drowned in presence of a nearby strong signal.

i-mode: WCDMA-based services using adaptive multi-rate encoding for communicating user voice-data as well as Internet data. It provides integrated services for voice, data, Internet, picture, music attachment to mail, gaming applications, ringtone downloads, remote monitoring, and control services.

***M*-sequence code:** A PN code having maximum length sequences of codes, which are generated by using *m* small length shift registers.

OFDM (orthogonal frequency division multiplexing): A spread-spectrum-based multi-carrier or discrete multi-tone modulation technique with bandwidth equal to a single carrier system bandwidth because each carrier uses a code orthogonal to the others. This technique is used in digital audio broadcasting, digital video broadcasting, mobile TV, and HyperLAN-2.

Orthogonal codes: Two codes of n chips each are orthogonal if the cross correlation between them is zero, i.e., SOP $= \Sigma\, p_i \times q_i = 0$ when $p_i = q_i$ for $0 \le i < n$.

Paging channel: A channel standard for base-band transmission of paging messages and paging mask and for sending the TMSI (temporary mobile subscriber ID as in case of GSM), information about the traffic channel, response of a mobile station access request during call setup, information regarding base stations in adjacent cells, and PN offsets.

Pilot channel: A channel which provides a reference to all the stations of a cell channel. This channel is transmitted by the transceiver to a mobile station before the voice-data or data channel is transmitted. The mobile station transmits voice-data or data at reduced power if the pilot is strong and raises the power if the pilot is weak.

PN (pseudo-noise) code: A code having almost equal numbers of 0s and 1s and displaying good autocorrelation.

Power control: It is required for resolving the problem regarding far and near terminals. To avoid hiding of the signals from far terminals, in presence of transmissions from the near terminals, power is controlled using power information sent by the base transceiver. If power level indicated by the base station signal is high, the mobile device reduces the transmitted power level and if the level indicated is low, then the uplink power level is raised.

Rake receiver: A receiver consisting of synchronizing and correlation units, which receive the input chips as per the chipping sequence used by the transmitter and generate user voice-data or data symbol. The rake receiver also specializes in selecting the signals of the strongest paths and reconstructing them by accounting for variable delays in these paths and then correlating the signal with the code used for transmission.

Scrambling code: An autocorrelation code of long sequence length to enable support to large number of users and user channels and having a strong autocorrelation property.

Spread factor: The number of chipping or hopping frequencies within a period of symbols is called spread factor. It is equal to the number of XOR operations or the number of hops within the time interval of a user symbol in case of DSSS or FHSS, respectively.

Spread Spectrum: A transmission technique in which the carrier frequency is varied within a spread over a wide band and the output varies according to certain codes. 'Spread' here means change between minimum to maximum value of frequency within the band.

Synchronizing channel: A base-band channel data at 1.2 kbps giving message for synchroni-zation of chip sequences at the MS, system identification (SID), network identification (NID), system time, PN short sequence offset, and PN long codes sequence state.

Traffic channel: A channel for transmission of traffic and power control data using long codes and mask, PN_Q, PN_I, and Walsh coding units in case of cdmaOne systems.

Walsh code: A type of code used in IS-95 cdmaOne and CDMA2000. Walsh codes are generated from Hadamard matrix and are represented as a 64 × 64 matrix in which all pairs of rows are orthogonal.

WCDMA (wideband CDMA): A CDMA technology supporting 2 Mbps or higher rates for short distances and 384 kbps long distance trans-mission. Its main features are—3.84 Mchip/s chipping, spreading and scrambling, channeliza-tion, orthogonal variable spread factor (OVSF), and scrambler (*S*2 or long 38400 Gold code)

Review Questions

1. Explain GMSK and QPSK modulation methods. (a) What are the methods by which 2, 4, 8, 16, and 64 bits can be transmitted at the same frequency without affecting the carrier bandwidth? (b) Write the expressions for the signals to show the difference between QPSK, OQPSK, and $\pi/4$-QPSK.
2. Explain how space, time, frequency, and code division methods control the simultaneous access to the medium by multiple source or channels of mobile terminals and base transceivers.
3. Describe the medium access control problem of receiving distinctly, the signals from exposed and hidden terminals and from near and far mobile terminals. Also describe the medium access control problem of drowning of far terminal signals by near mobile terminals.
4. What are the protocols used to provide medium access to a terminal without collision (interference) in case of a large number of terminals attempting to access the medium simultaneously?
5. Draw graphically, eight coding signals when the code chips are 0000 0000, 0101 0101, 0011 0011, 0110 0110, 0000 1111, 0101 1010, 0011 1100, and 0110 1001 assume that the chipping interval is 1 μs. Also, represent graphically, all the eight signals after chipping a user signal with symbols 0110 with a 0.25 μs interval between the symbols.
6. Show that any two of the three codes 0011 0011, 0110 0110, and 0000 1111 are mutually orthogonal. Two orthogonal codes may or may not have autocorrelation. Why? Find the autocorrelation between 0110 0110 and 0000 1111.
7. What are the uses of long *M*-sequence PN codes, orthogonal Walsh codes, and short PN codes in a CDMA system? How is a long sequence code generated using the generator polynomial, the initial state vector, and the mask vector?
8. What are narrow band and co-channel interferences? Describe the DSSS technique. How does it mitigate narrow band interference?
9. How does the receiver extract the signals in presence of multiple path delays? How does the receiver extract the user signal without any co-channel interference and narrow band interference among the large number of user signals?

10. What are the functions of a rake receiver?
11. Describe FHSS frequency-hopping technique. How does it help in receiving signals in the presence of frequency-selective fading of the signals?
12. List the basic features of CDMA systems. Explain soft handover.
13. Describe the characteristics of the IS-95 (cdmaOne) system.
14. Explain the functions of IS-95 processing units for convolution, symbol repetition, interleaving, long code sequence generator, decimator, scrambling, DS Walsh code chipping, and *I*- and *Q*-pilots generators. What are the functions of base-band filters and QPSK modulators?
15. What are traffic, paging, pilot, and synchronization channels in a cdmaOne forward link (down link)? What are the channels in the reverse link (uplink)?
16. What are the functions of using in-phase and quadrature pilots?
17. How are transmitted signals reaching a receiver in a synchronized state when there are the multiple path, and thus time delays in a CDMA system?
18. Explain the use of orthogonal codes for coding the multiple user channels before transmitting. How does the receiver distinctly identify user channel after decoding?
19. Describe forward- and reverse-link structure and frames in IS-95.
20. Describe the processing units in CDMA2000 systems. How do these units differ from those in a cdmaOne system?
21. Why are the functions of using in-phase and quadrature carriers?
22. How do the EV-DV and EV-DO systems differ? How do CDMA2000 1x EV-DV and 3x EV-DV differ?
23. List the frequency bands and number of sub-carriers or carriers in a CDMA2000 system.
24. Describe WCDMA. How and why is the variable spread factor used in WCDMA?
25. How does the synchronization of a WCDMA base transceiver (UTRAN) differ from CDMA2000 transceivers? What are the options added by the 3GPP to make WCDMA compatible with CDMA2000?
26. Consider transmission of signals of many different data rates at a given chipping rate. How are multi-rate signals and multi-carrier signals processed by the system? How can data rate be matched to enable the use of a constant spread factor and the same chipping frequency? How are multiple carriers transmitted by time shifting (phase shifting) in CDMA2000?
27. How and why is the variable spread factor used in CDMA2000? Explain the differences between multiplexing of signals by time division and code division. How are the variable code-length divisions employed when using a carrier of fixed chipping frequency in a multi-rate system?
28. How is the power control message transmitted in CDMA2000? Why is it necessary to control the power in CDMA systems?
29. How does using puncturing and symbol repetition enable the transmission of signals of different rates at a fixed rate of chipping with a fixed code length per symbol in CDMA2000?
30. What are the recommended standards in IMT-2000?
31. Describe the main features of i-mode integrated services. How does PDCP enable the use of many different types of services in i-mode? What is *c*-HTML?
32. Explain the use of multiple carriers by orthogonal coding. How does an OFDMA system differ from a CDMA system? Where are the OFDM systems presently used?

Objective Type Questions

Pick the correct or most appropriate statement among the choices given:

1. A QPSK signal for the symbols 10 00 11 01 is transmitted at a frequency, which changes phase angles as $-\pi/4$, $+\pi/4$, $3\pi/4$, $-3\pi/4$. When BPSK and OQPSK modulation is used, then the sequences of phase angles of the waves will be
   (a) $+\pi$, $+3\pi/2$, 0, $-\pi/4$ and $+\pi/4$, $3\pi/4$, $-3\pi/4$, respectively.

(b) $+\pi/2, +\pi/2, +\pi/2, +\pi/2$ and $+3\pi/4, 3\pi/4, -\pi/4, -\pi/4$, respectively.
(c) $-\pi/2, +\pi/2, +\pi/2, +\pi/2, -\pi/2, -\pi/2, \pi/2, -\pi/2$ and $-\pi/4, +3\pi/4, 3\pi/4, -\pi/4$, respectively.
(d) $+\pi/2, +\pi, +3\pi/2, 0, -\pi/4, -3\pi/2, \pi, 0$ and $-\pi/4, -3\pi/4, 3\pi/4, -\pi/4$, respectively.

2. Assume that a 64-QAM modulated signal is generated and transmitted at 19.2 ksymbol/s. For QPSK modulation, the symbol rate will be
(a) 28.8 ksymbol/s.
(b) 115.2 9 ksymbol/s.
(c) 14.4 ksymbol/s.
(d) 64×19.2 ksymbol/s.

3. When accessing a medium for transmission without collision (interference), mobile systems use
(a) ALOHA or slotted ALOHA depending on whether it is a 2G or 3G application.
(b) slotted ALOHA because it assumes that each wireless node can sense the presence (or absence) of a carrier on the network in a given time-slot.
(c) ALOHA, slotted ALOHA, CSMA/CA, or CSMA/CD depending on whether the application is 2G or 3G.
(d) ALOHA or slotted ALOHA because CSMA assumes that each wireless node can sense the presence (or absence) of a carrier on the network and this assumption is not true.

4. A CDMA system may not be able to get the signals from a far terminal because
(a) far channel signals undergo interference and multi-path delays.
(b) far channel signals are not listened to at an intermediate station due to the fact that the strengths of signals from near terminals are high and that strong signals superimpose on the weak signals from far terminal, and thus drown the far terminal signals.
(c) far terminal signals have very low signal strengths due to the inverse square law.
(d) near channel signals do not get listened at an intermediate station due to very high signal strength which interferes strongly with all incoming signals.

5. A modulating signal of 19.2 ksymbol/s in IS-95 cdmaOne can transmit four signals of 2.4 kbps, one signal each of 4.8 kbps and 9.6 kbps by
(a) convolution encoding.
(b) rate setting.
(c) symbol repetitions.
(d) interleaving.

6. A processing unit transmits three modulating signals of 9.6 ksymbol/s, 19.2 ksymbol/s, and 153.6 ksymbol/s with a chipping frequency of 1.2288 Mchip/s by
(a) interleaving 128, 64, and 8 symbols/s.
(b) variable spread factor coding and interleaving 128, 64, and 8 ksymbol/s, respectively.
(c) variable spread factor coding by 128, 64, and 8 chip/symbol.
(d) variable spread factor coding by 8, 64, and 128 chip/symbol, respectively.

7. When using QPSK, an IEEE 802.11 standard wireless LAN uses a bandwidth of 26 MHz when chipping rates are 22 Mchip/s. The efficiency of bandwidth in chip/s/Hz is
(a) 0.85.
(b) 1.176.
(c) $0.85 \times$ spread factor.
(d) $1.176 \times$ spread factor.

8. Codes 0000 0000 0000 00000 and 0110 1001 1001 01100 are checked for orthogonality and autocorrelation by finding two cross correlation sums of products between the nearest cross correlation and next nearest cross-correlation. The codes are
(a) orthogonal and auto-correlated.
(b) orthogonal and cross-correlated.

(c) orthogonal only.
(d) auto-correlated only.

9. Consider a CDMA2000 system, (a) Orthogonal codes have non-zero cross-correlation and are used in identifying the user, user channel and carrier. Long M-Sequence PN codes have orthogonality and are used in synchronizing and detecting the user channel signals. (b) Orthogonal codes have almost zero cross-correlation and are used in identifying the user, user channel and carrier. The short PN codes have strong cross-correlation as well as orthogonality and are used in synchronizing and detecting the user channel signals. (c) Orthogonal codes have almost zero cross-correlation and are used in identifying the user, user channel and carrier. Long M-Sequence PN codes have strong autocorrelation and are used in synchronizing and detecting the user channel signals. (d) Short PN codes have almost zero orthogonality and are thus used in identifying the user, user channel and carrier. Long M-Sequence PN codes have strong autocorrelation and are used in synchronizing and detecting the user channel signals.
10. Assume that there is a guard band of 2.4 kHz at the lower and higher ends of each hopped frequency when a signal of data rate 19.2 ksymbol/s is transmitted with a spread factor of 64 using an FHSS transmitter. If basic carrier frequency in FHSS is 5 MHz. What are the frequencies at which hop can occur? [Assume that frequency hops in a linear sequence (0, 1, 2, 3, …, 62, 63)]
    (a) 5.0192 MHz, 5.0384 MHz, …, 6.2288 MHz.
    (b) 5.025 MHz, 5.050 MHz, …, 6.600 MHz.
    (c) 5.0216 MHz, 5.0432 MHz, …, 6.3824 MHz.
    (d) 5 MHz, 5.0216 MHz, 5.432 MHz, …, 6.3608 MHz.
11. There are three options to transmit multi-encoded signals of 76.8 ksymbol/s, 38.4 ksymbol/s, and 19.2 ksymbol/s—(i) interleaving symbols in successive time-slots of 0.208 ms each and transmitting with a carrier frequency of 1.2288 MHz, (ii) chipping by with a signal chipping frequency of 1.2288 Mchip/s with code of lengths 64 and transmitting with a carrier frequency of 1.2288 MHz, (iii) chipping each signal with a chipping frequency of 1.2288 Mchip/s with each signal chipped with variable code lengths of 64, 32, and 16 bits. Which of the above options requires the minimum energy expenditure in the processing units at the transmitter and the receiver?
    (a) (ii) and (iii).
    (b) (i) and (ii).
    (c) (iii).
    (d) (i).
12. CDMA systems exhibit soft handover due to
    (a) autocorrelation codes used in each cell transceiver.
    (b) each cell using same spread frequency spectrum.
    (c) negligible narrow band interference and co-channel interference of the signal.
    (d) each cell having a distinct pseudo-noise code offset, so that the handover to the adjacent cell is simply by adding the offset to the mobile terminal pseudo-noise code.
13. The function of the puncturing unit is to set the rates such that
    (a) it increases the data transmission rate.
    (b) symbols are spread with a positive integral number of chips for a given chipping frequency.
    (c) the redundant bits are removed.
    (d) the codes are made compatible between the forward and reverse channels.
14. (i) Data channel is divided into I (in-phase) and Q (quadrature) components after scrambling with Walsh codes. (ii) Channel signal is first divided into the I and Q components and then chipped using Walsh codes in CDMA processing units for transmission.
    (a) (i) is true for IS-95 cdmaOne and (ii) is true for CDMA2000.
    (b) (ii) is true for IS-95 cdmaOne and (i) is true for CDMA2000.
    (c) (i) and (ii) are true for forward and reverse channels, respectively.
    (d) both (i) and (ii) are incorrect. Walsh codes are used just before modulation.

15. OFDM is used for
    (a) multi-rate, multi-carrier transmission with signals of different rates coded by distinct orthogonal codes; it requires small power levels for each carrier.
    (b) multi-rate transmission with each carrier coded by a distinct orthogonal code; it needs greater power levels for each carrier.
    (c) multi-carrier transmission with each carrier coded by a distinct orthogonal code and needs greater power levels.
    (d) multi-carrier transmission with each carrier shifted by a phase angle of $360°/n$ coded by the same orthogonal code, which is distinct from that of another multi-carrier transmitter; it needs smaller power levels. (Here n is the number of carriers in multi-carrier transmission.)
16. The most important features of i-mode that facilitate many services (voice, data, Internet, picture, music attachment to mail, gaming applications, ringtone downloads, remote monitoring, and control services) on a mobile terminal are
    (a) PDCP (packet data convergence protocol) and c-HTML.
    (b) use of FOMA for Internet and asynchronous packet data transmission.
    (c) separate protocols for all integrated services.
    (d) gateways between the service provider and the mobile network.
17. (i) Open loop power control in which the pilot channel signal strength from the base transceiver is measured by a mobile station, the latter raises its power in case the signal strength is low and reduces the power if the signal strength is high. (ii) Closed loop control, base transceiver, and mobile station measure the power and adjust the power to acceptable levels.
    (a) (i) is true for GSM and (ii) is true for cdmaOne.
    (b) (i) is true for cdmaOne and (ii) is true for GSM.
    (c) (i) is true for both GSM and CDMA systems.
    (d) (ii) is true for both GSM and cdmaOne systems.
18. Long PN M-sequences are
    (a) autocorrelation as well as orthogonal codes that identify the carrier channel.
    (b) orthogonal codes used to identify a user and user channel distinctly at the receiver.
    (c) autocorrelation codes for chipping multi-rate data with variable spread factors.
    (d) autocorrelation codes used to identify a user and user channel distinctly at the receiver.
19. 3GPP recommends signal mapping of
    (a) 0 by –1 and 1 by +1.
    (b) 1 by –1 and 0 by +1.
    (c) 0 by 0 and 1 by –1.
    (d) 1 by 1 and 0 by 0.
20. (i) It is energy efficient to use a single code and multiplexing of signals in time-space. (ii) It is energy efficient to use multiple codes of short length and multiplexing of signals in code-space.
    (a) (ii) is true.
    (b) (ii) is true in WCDMA only.
    (c) (i) is true.
    (d) (i) is true for CDMA2000 only.

Chapter 5

Mobile IP Network Layer

The Internet network layer protocol (IP) is the most used network protocol today. This chapter describes the mobile IP network layer which facilitates Internet-based communication between mobile nodes (MNs). This chapter will discuss the following concepts.

- IP network protocols and methods used in the IP routing of packets
- Mobile IP network and delivery of data packets from one mobile node to another
- Functions of home and foreign agents in a mobile IP network including agent discovery, advertisement, and solicitation methods for management of mobility and handover
- Tunnelling and encapsulation methods, route optimization methods, and reverse tunnelling
- DHCP support to mobility when a node visits a new subnet on the Internet

5.1 IP and Mobile IP Network Layers

The OSI (open systems interconnection) defines seven layers for data communication between two ends (Section 1.3.6). The layers are—(i) physical (L_1) for sending and receiving signals wirelessly (e.g., TDMA or CDMA coding with FEC) or over wire or fibre, (ii) data link (L_2) (e.g., for linking to the destination computer by a MAC address), (iii) network (L_3) (e.g., for routing through a chain of routers), (iv) transport (L_4) (e.g., for defining sequencing and for repeat transmission, if required), (v) session (L_5) (e.g., for defining the transaction and session protocol to establish end-to-end connectivity), (vi) presentation (L_6) (e.g., for defining the data-encoding format), and (vii) application (L_7) (e.g., for running a web browser, mail transfer, or mobile e-business application). The first two layers, L_1 and L_2, are usually associated with the communication and physical network radio and switching infrastructure, for example, in GSM or CDMA systems. The third layer, L_3, is for networking (using a path chosen among the large number of paths available) through a chain of in-between routers. The mobile IP protocol discussed in this chapter operates in this layer. Section 5.1.1 introduces the Internet protocol and the mobile Internet protocol is explained in Section 5.1.2.

5.1.1 Internet Protocol

The Internet protocol (IP) is the basic protocol at L_3 which is used for transmission over the Internet. It is designed for use by networks which employ packet-switched data communication (Section 1.1.2.6). IP provisions the transmission of data packets. Each packet is treated independently in IP, which implies that every packet must contain complete destination addressing information. The following subsections describe the various aspects of the Internet protocol.

5.1.1.1 OSI Layer Functions

The OSI seven-layer model was introduced in Section 1.3.6 and Section 3.3 discussed the concept of protocols and protocol layers with examples of protocol layers in a GSM system. An application or service point transmits data or establishes interconnection. A physical network transmits data of one application to another application through the various intermediate layers. At a layer, data received from the upper layer can be suitably divided (e.g., packetized at the L_3 IP layer). At each layer, data or each section of data is suitably encoded (e.g., by adding a header) for transmission to the lower layer. At the receiving end of the network, the bits in the header fields are decoded (e.g., by extracting the header) upon reception before passing the data to the upper layer. Header fields carry information which is used by the successive layers at the transmitting end and by the corresponding layers at the receiving end. At a layer at the receiving end, data received from the lower layer can be suitably assembled (e.g., packetized IP layer data from L_2 is assembled at L_3). The L_4 to L_7 layers in the OSI model are as follows:

1. Transport layer, L_4: The protocol header fields of this layer define the sequences and other required fields for data transfer to and from a port in a network. Data transfers after a session are established through the session layer (L_5) between two ports on the network. The transport layer header, actions, and functions for the TCP/IP model are described in Chapter 6. The transport layer in the TCP/IP model (Section 5.1.1.2) has the additional function of session establishment.
2. Session layer, L_5: This layer defines how various sessions are established (e.g., a call setup session). The TCP/IP model does not provide for a separate session layer.

3. Presentation layer, L_6: It defines how the data from the port is to be presented or formatted. The TCP/IP model does not provide for a separate presentation layer and the application layer provides the functions of L_6.
4. Application layer, L_7: L_7 can support multiple applications. This layer defines a protocol for transferring application contents and data of a port. Each port supports a distinct service (application). Each port defines a service access point at the L_7 layer. Each port can deploy a protocol for data transfer between the L_7 layers at the transmitter and the receiver. An output port (software/service access point) at the L_7 layer transmits the service (application) contents and data employing a distinct protocol, for example, HTTP (hypertext transfer protocol). At the corresponding data input port (software) at the receiver L_7 layer, the same protocol is used to retrieve the service (application) contents and data. A port number can be defined for each application. A port can be numbered in order to define and distinguish each application's data easily.
   For example, the HTTP port in the TCP/IP model, through which all the HTML files are transferred through the network, is numbered 80.

The network layer L_3 in the OSI model is the IP layer (Section 5.1.1.3) in the TCP/IP model.

5.1.1.2 TCP/IP

Networking for the Internet follows a suite of protocols called the TCP/IP protocol suite or the Internet protocol suite. The set contains many other protocols besides the transmission control protocol (TCP) at L_4 and the Internet protocol (IP) at L_3. The suite was originally designed to have four layers but has evolved to a five-layer format. The functions of L_5 (session layer) are incorporated into L_4 (transport layer) and those of L_6 (presentation layer) are performed by L_7 (application layer) in TCP/IP. The five layers are, thus, L_7, L_4, L_3, L_2, and L_1.

The network layer, L_3, functions to facilitate transmission of data from one system with a common address for the ports (like service access points in a mobile system) to another system with a common address for the ports. The address is called the IP address in Internet terminology. The main network-layer protocol in the TCP/IP suite is called the Internet protocol (IP). It provides the connection to the router for transmission. A communication between two addresses on a physical network is carried out through routers. Connections to the Internet employ the IP protocol. The IP protocol specifies a header known as the IP header. The header encapsulates data from the upper layers, for example, the L_3 header encapsulates L_4 data after formatting it into packets.

5.1.1.3 Packet Formation in IP

Data from L_4 (transport layer) is divided into packets having a maximum size of 2^{16} bytes (2^{14} words) so that a packet-switched network can be used for transmission.

Data packets can hop on different routes to reach a destination so that more packets can be sent simultaneously through the network, unlike a circuit-switched network, where only one data frame can be transmitted at an instant (e.g., in GSM or HSCSD). An IP packet can be transmitted in fragments. This is because MTU (maximum transferable units per effort) may be much less than 2^{16} bytes in a source–destination path or sub-path in the network.

5.1.1.4 Header, Source, and Destination IP Addressing

The IP specifies certain header fields (a field is a set of bits placed in a word for a specific action, condition, or purpose) for encoding data from the transport layer at the transmitter and decoding the data received from the data link layer before passing it to the transport layer at the receiver. The header fields are as follows:

1. A 32-bit word to specify IP version (IPv4 or IPv6 for Internet or broadband Internet), length of the IP header, precedence of the IP packet, and total packet length.
2. A 32-bit word to specify the ID for the packet, flags, and fragment offset for fragments of the same ID.
3. A 32-bit word to specify time-to-live (not in seconds but in number of attempts to hop before expiry of packets in the network), type of protocol, and checksum of the header (for finding transmission errors, if any).
4. A 32-bit word (four decimal numbers, each separated by dots and each lesser than 256) to specify the IP address of the source.
5. A 32-bit word (four decimal numbers each separated by dots and each lesser than 256) to specify the IP address of the destination. For example, assume that the source IP address for routing is ($ns_1 . ns_2 . ns_3 . ns_4$) and the destination IP address is ($nd_1 . nd_2 . nd_3 . nd_4$). A packet transmitted from the source IP address reaches from the source to the destination by hopping among the various routers on a path. Paths can be different for different packets from the same source. The path for the routing of a packet depends on the paths and sub-paths which are available in the network at a given instant.

5.1.1.5 Routing Between Two IP Addresses

A router receives a packet from a source or a previous stage router, gets the destination address from the IP header, and forwards the packet to the next router or the destination router for that destination address. Each router maintains a table. This table, called the routing table, is maintained and updated by the router. A routing table has a large number of rows depending upon the maximum number of entries possible in it. Routing tables contain, in each row, the address of the destination router and the next router, so that the packets for that destination hop to that particular router. A router table is regularly updated.

5.1.1.6 Subnet

Imagine a global postal network. It is not possible for each sorting and routing station in such a system to maintain a database of information regarding each addressed person on the globe. Similarly, it is not possible for a router to hold routing table entries for all the IP destination addresses on the Internet and to store information about a large number of source and destination systems each having a distinct IP address. Each router has a 32-bit IP address. However, a router can connect to a maximum of 2^7, 2^8, or 2^{21} other routers depending upon the subnet in which the router is placed. Figure 5.1(a) shows an IP address and its structure.

The concept of subnets is used on the Internet. A subnet is a sub-network using standard specifications and protocols when connecting to the Internet on one end and to the host on the other end. A router can belong to a class *A*, *B*, or *C* subnet. The Internet consists of class *A*, *B*, and *C* subnets that are connected to the hosts (computers, nodes, and service terminals). Each subnet consists of a large number of connected local subnets or hosts. Figure 5.1(a) shows an IP address and how a router uses the 32-bit IP address which consists of four parts—(i) msbs (maximum significant bits) to specify the class, (ii) bits to specify the net ID in the network, (iii) bits to specify the subnet ID, and (iv) bits to identify the host. The two msbs

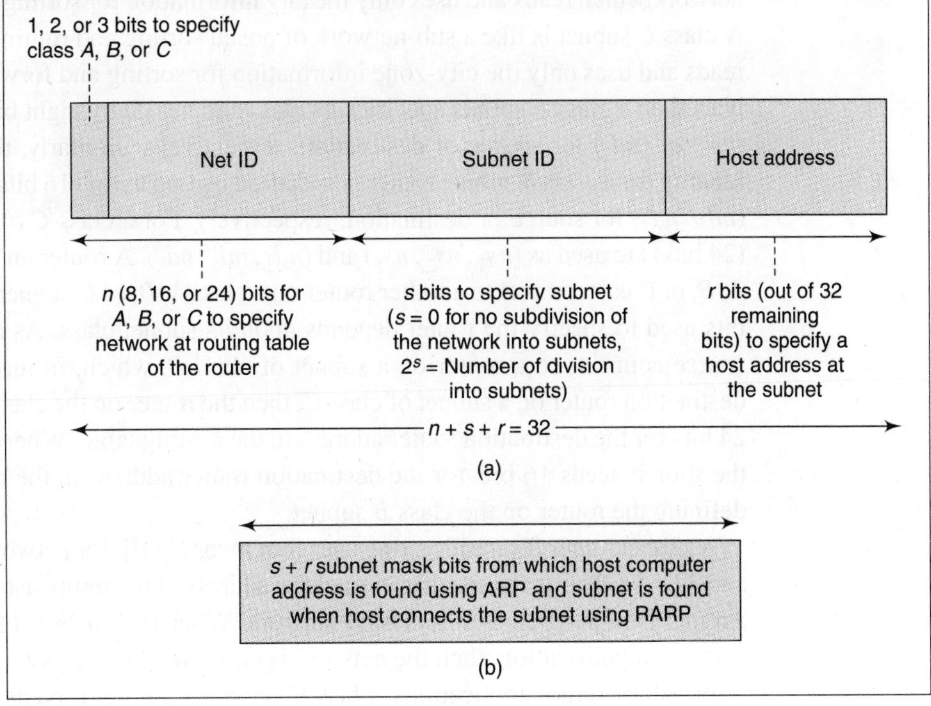

Fig. 5.1 (a) 32-bit IP address (b) Subnet mask

and the remaining n bits of the net ID help in routing from one network to another. The remaining bits after the n-bit net ID define a subnet ID of s bits and the bits remaining after the net ID and subnet ID bits define the r-bit host identity (Fig. 5.1). A router's subnet address (net ID) at other routers is defined by 7, 14, and 21 bits for classes A, B, and C, respectively. This is because one, two, or three bits (msbs) among the n network-ID-specifying bits are for identifying the class of subnet for that router as A, B, or C, respectively. The msbs used to identify A, B, and C type of networks are 0, 10, and 110, respectively. Therefore, the remaining $8 - 1$, $16 - 2$, or $24 - 3$ bits (7, 14, or 21), therefore, suffice as the router net ID address bits for a source. The remaining 24, 16, and 8 bits (subnet ID plus host address) in the IP address are required to be taken into consideration at the destined network only. While routing, the routers use 8, 16, or 24 bits only for the routing addresses (using net IDs) of other routers for the next hop. At the end network, the router placed on a subnet uses the remaining 24, 16, or 8 bits among the 32 bits of the IP address. The router selects the subnet using s bits and the host on that subnet, using r bits.

The class A, B, and C subnets are like sub-networks of postal sorting and routing stations. A class A subnet is like a postal sub-network which reads and uses the only country name for sorting and forwarding. A class B subnet is like a postal sub-network which reads and uses only the city information for sorting and forwarding. A class C subnet is like a sub-network of postal sorting and routing stations which reads and uses only the city-zone information for sorting and forwarding. A router placed on a class A subnet specifies its class and net ID by eight bits, i.e., one byte (ns_1) or (nd_1) for source or destination, respectively. Similarly, the class and net identity for a class B subnet router is specified by two bytes (16 bits) as $(ns_1 . ns_2)$ or $(nd_1 . nd_2)$ for source or destination, respectively. For a class C router, three bytes (24 bits) are used as $(ns_1 . ns_2 . ns_3)$ and $(nd_1 . nd_2 . nd_3)$. A router on a subnet of class A, B, or C can connect to another router on a class A, B, or C subnet. The number of bits used to specify the router depends upon its subnet class. As an example, if a source connects to a router on a subnet of class B, which, in turn, connects to a destination router on a subnet of class C, then the router on the class B subnet needs 24 bits for the destination router address in the routing table. Whereas, the router at the source needs 16 bits for the destination router address in the routing table for defining the router on the class B subnet.

A subnet router for multicasting uses four msbs (1110) for network identification and $32 - 4 = 28$ bits for specification of the address of the multicasting subnet. Such a router is said to belong to a class D network. When five msbs (11110) are used for network identification, then the network is classified as a class E network and it is reserved for future applications. Class E networks are not used as of now but will be used when a new type of routing or subnet mechanism is defined. Let us assume that the source routing address is $(ns_1 . ns_2 . ns_3 . ns_4)$ and the destination address is $(nd_1 . nd_2 . nd_3 . nd_4)$. The router maintains and updates the table in the following manner—each row has one column of 8, 16, or 24 bits $[(nd_1), (nd_1 . nd_2)$, or $(nd_1 . nd_2 . nd_3)]$ among the destination address bits. These bits specify the destination router's net ID and its subnet class. The next column has 8, 16, or 24 bits $[(nr_1)$, $(nr_1 . nr_2)$, or $(nr_1 . nr_2 . nr_2)]$ which define the address of the next router net ID to which the packets hop as and when they are received. Using this address, the packets can be forwarded to the next router, and so on till they reach the destination router's net ID address. At this router, the remaining bits are used to identify the addressed subnet and host.

5.1.1.7 Destination Router

Let us assume that an entry in a row for a destination address at the destination router consists of $(nd_1 . nd_2 . nd_3)$ and a null entry. A null entry indicates that there is no other router for the packet to hop to. When there is a null entry, then the router searches the destination address $(nd_1 . nd_2 . nd_3 . nd_4)$ from the received packet header. It uses a mask called the subnet mask. Figure 5.1(b) shows a subnet mask. The subnet mask unmasks the r bits (nd_4 when $r = 8$; the subnet finds nd_4 of the computer) and finds the addressed host's IP address $(nd_1 . nd_2 . nd_3 . nd_4)$. The address resolution protocol (ARP) then finds the destination computer MAC address on the LAN and forwards the packet to the destined computer. (An ARP cache stores the MAC address to enable ARP to translate the host IP address into the MAC address.) An Ethernet LAN computer has a 48-bit MAC (medium access control) address. ARP maps r-bits to the 48-bit MAC address.

Subnet masking and ARP mapping is like a process in a network of postal sorting and routing stations, which reads and uses the name and street information only for sorting and forwarding at a city zone and sends letters to their destinations. From the source computer, the packet is transmitted to a source router, using RARP (reverse ARP) and subnet mask. The latter masks the extra bits (nd_4 when $r = 8$). (An RARP cache saves the IP address and the computer address to enable the RARP to reverse translate the MAC address on LAN and get the IP address for sending the packets on the subnet at the Internet.) The packet is forwarded to the source router on the subnet with an n-bit source router net ID [e.g., $(ns_1 . ns_2 . ns_3)$ on class C]. RARP and subnet masking is like a network of postal sorting and routing stations, which reads and uses the city information only for sorting and forwarding a letter to the routing station for the destined postal network.

5.1.1.8 Point-to-Point, Multicast, and Broadcast

When a message or packet is transmitted to one destined IP address only, then it is called *unicast* or *point-to-point* transfer. When a message or packet is transmitted to a group of IP addresses, then the message or packet is said to be *multicast*. The IP protocol specifies the use of a class D subnet for multicasting. A subnet for multicasting has four msbs (1110) in the net ID part of the IP address for defining subnet as *multicast* network and 32 – 4 = 28 bits specify the address for the multicasting net ID and the subnet and host addresses. The address (224 . 0 . 0 .1) is used to multicast to all hosts in the links of a router. [$ns_1 = 224$ (in decimal system) and $ns_1 = 11100000$ (in binary system)]

The term 'multicast tree' refers to a multicasting source (root) multicasting to select multicast nodes (subnets) at level 1. Each level 1 node, then, transmits to multicast nodes (subnets) at level 2, and so on. A hierarchy of nodes is present in a multicast tree. Multicast tree nodes at one level can transmit to multicast nodes at another level simultaneously, via multiple paths. Thus, the time taken in multicasting a message is greatly reduced. Multicasting can be used for flooding a UDP (user datagram protocol) datagram on the network. [Flooding is required for advertisement (Section 5.3.1.1). Flooding means sending information along many paths. Some protocols block nodes, which have already received the relevant information, during flooding. One such protocol is the spanning tree protocol.]

When a message or packet is transmitted to all the IP addresses which are set for listening, then this type of transmission is known as *broadcasting*. The IP protocol specifies an address where all 32 bits are 1s (255 . 255 . 255 . 255) for broadcasting. It is used when broadcasting to all hosts and links of a router.

5.1.1.9 User Datagram Protocol

A datagram provides independent information. A datagram is stateless meaning that a datagram is not a sequential successor of a previous one or a predecessor of the next. Datagram data is sent using a connection-less protocol. The user datagram protocol (UDP) is one such protocol used to send the datagram using a connection-less protocol. A connection-less protocol means that there is no session establishment before the data transfer begins. For example, some phones have hotlines where one can just speak without dialling and waiting. A UDP datagram includes a header consisting of the following—a source port, destination port, source IP address, destination IP address, length of data, and checksum bytes for the header (to check erroneous receipt of header). UDP datagram consists of a maximum of 2^{16} bytes, transmitted as sequences of words, each of 32-bits (4 bytes).

5.1.1.10 Internet Control Message Protocol

Internet control message protocol (ICMP) is another connection-less protocol, which is a part of the IP network protocol suite. ICMP uses a datagram, which has an ICMP header, when (a) sending the messages for querying to find information, (b) reporting errors, and (c) making route address advertisement (Section 5.3.1.1) and for a router seeking (soliciting) messages (Section 5.3.1.2) to get the IP addresses of the linked subnets.

The ICMP protocol specifies the following words and fields for headers which encode data from the upper transport layer at a transmitter (subnet, router, or agent), and which decode the data received from the data link layer for passing it to the transport layer at the receiver.

1. A 32-bit word, to specify a byte for type of message, a byte for the code, and a two-byte checksum.
2. A 32-bit word, which specifies the number of addresses for advertising along with the address field size and the lifetime of message validity.
3. A set of pairs of words, each pair specifying the router address and preference (The pairs are arranged in sequence for level 1, level 2, and so on in a tree. The router of higher preference gets the messages earlier than the others.).
4. Extended words in headers are called options. First byte = 16 means that options are being used. One example of this use is the mobile IP protocol extension when an agent advertises (Section 5.3.1.1).

5.1.2 Mobile Internet Protocol

The mobile Internet protocol (mobile IP) is a protocol developed to allow inter-network mobility for wireless nodes without them having to change their IP addresses. Mobile IP is defined by the Internet Engineering Task Force (IETF) and is described in the IETF RFC 3344. The following subsections explain the requirements for the evolution of the new mobile IP protocol from the existing IP protocol and describe the working of the mobile IP protocol.

5.1.2.1 Evolution of Mobile IP

When a user computer (for example, a laptop) moves from one place to another, it is assigned a new IP address at the new hosting subnet to which the service-providing router provides connectivity to the Internet (Section 5.7). A mobile node can be considered to be a computer on the move from one area to another. The reasons for the use of a new mobile IP protocol instead of the existing IP protocol by the MNs are as follows:

Need for Enhancing IP Network Capacity Use of the existing IP protocol by a large number of MNs will lead to a decrease in the network throughput unless the capacity of existing IP network is scaled up, to cater to such large number of users.

Need for Upgrading Capacity of Routers and Data Link and Physical Layers of IP Networks For mobile nodes to move from one place to another while using the existing IP protocol, new protocols are required at lower layers—the data-link and physical layers. For example, the IP network protocols, as discussed in Section 5.1.1, support 48-bit MAC addresses. But when the number of MNs is large, then other interfaces and lower level protocols have to be added to the existing IP infrastructure. When an MN moves from one service area to another then the use of the existing IP protocol and assignments of new IP addresses is impractical due to the following reasons.

Security needs In Chapter 3, we discussed how a TMSI (temporary mobile subscriber identity) is assigned to a mobile node in a GSM system, when it moves to a new location area. The TMSI is used in place of the IMSI during the connection to secure the identity of the MN from eavesdroppers over the air. The mobility of the called MN is thus hidden from the calling MN. When a new IP address is allocated at the new hosting subnet of the existing IP-based infrastructure, the identity of the mobile node is not hidden from another host. The MN is, thus, exposed and lacks security when using the existing IP protocol.

Need for non-transparency from higher layers The transport layer establishes a connection between a given port at a given IP address (called socket) with another port at another IP address. The connection, established by the transport layers between the sockets, is broken as soon as the new address is assigned. The problems faced in the use of this technique are—(a) re-establishment of the connection takes time which means loss of data during that interval, (b) the re-establishment process must share the same network and the given transmission rate, and (c) any movement of the MN is transparent to the TCP and to L_7 in case the TCP layer re-establishes the connection when the IP protocol is followed by the MN. There is, therefore, a need for non-transparency of the MN to distant ports.

Another problem with the use of IP protocol is that of non-transparency in the routing table. Let us suppose that a distant router is sending data packets for an IP address, presently assigned to a mobile terminal using another router. When the terminal moves from one service area to another, the routing tables on the route need to be updated. Unless this is done, the packets will not reach their new destination. In this case, (a) the reconfiguration messages for updating the routing tables must share the same network and the given transmission rate, (b) re-establishment of the connection takes time and this means loss of data during that interval, and (c) any movement on the part of the MN is transparent and thus, not secure from the distant hosts on the network of distant routers. There is a need for non-transparency of the MN to the distant ports.

5.1.2.2 Working of Mobile IP

Figure 5.2 shows the architecture for a mobile IP network. It shows how a mobile IP network employs home and foreign agents. A mobile node (MN) connects to a mobile services switching centre (MSC) in the radio subsystem (Section 3.1.2.1). An MN can access Internet services using the mobile IP protocol. The MN can change its service router when visiting another location (which is serviced by a different router). A router has a home agent (HA) for a set of home networked MNs as well as a foreign agent (FA) for the visiting MNs. An agent is software employed at a router or the host serviced by a router. The same software can function as both the HA and the FA at different instants of time. An MN can also have software

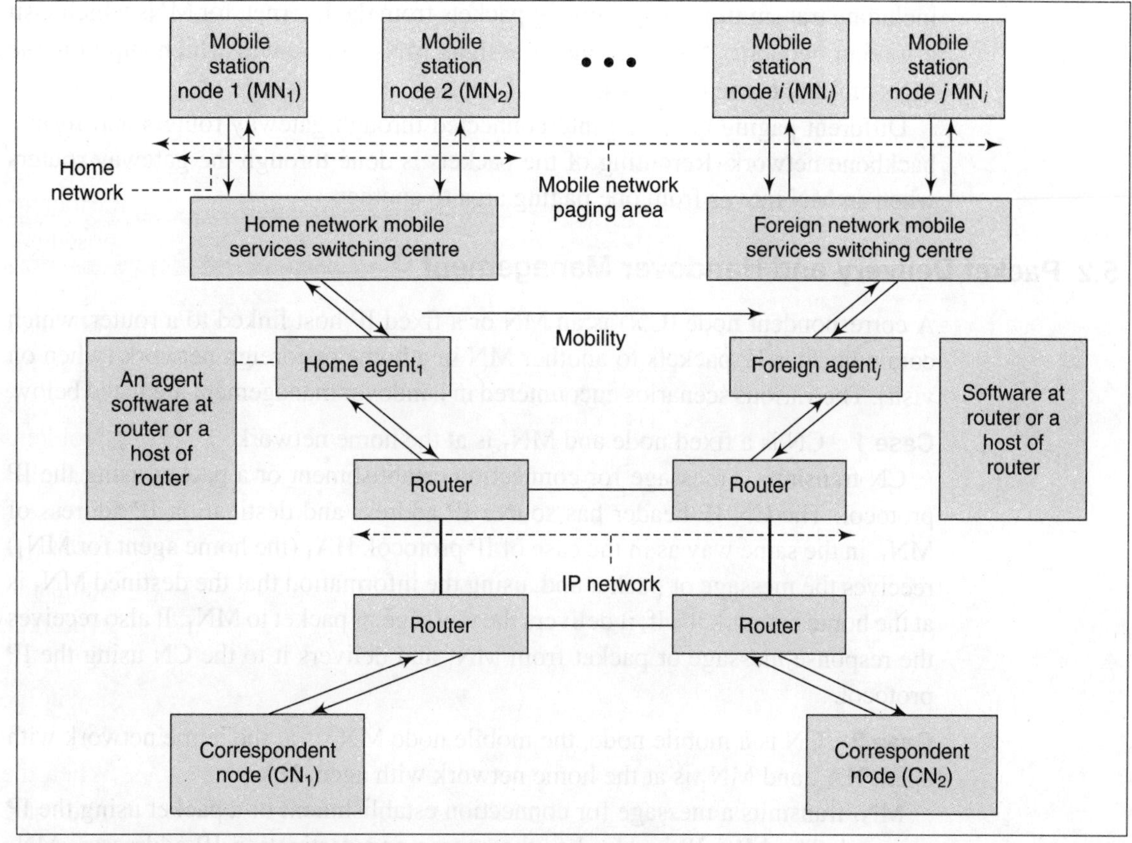

Fig. 5.2 Mobile IP network employing home and foreign agents

which functions as an FA instead of the FA at the router. The HA and the FA play a location management role similar to that of the HLR and the VLR in a GSM system, respectively.

A switching centre has a home agent (HA) in a home network. The home agent provides services to an MN at the registered home network. These services include transmitting and receiving packets from the Internet. A home agent assigns MNs to routers which support the MNs. A home network is a mobile radio subsystem's network within an area known as paging area. The home network is like a subnet. Similar to a subnet, which has a number of IP hosts, a home network has the MNs. The paging area is the area in which the MNs of home as well as foreign networks can be approached through a single MSC or a set of MSCs. Routing of packets through the routers is performed when an MN moves within one paging area. A switching centre has an FA for a foreign network of visiting MNs. Foreign network means another mobile radio subsystem network, which the MNs of home network visit, within the paging area. Foreign agent provides IP address and services, including transmitting and receiving packets from the Internet, for MNs which visit a foreign network. A foreign agent assigns MNs to a router, which supports the MNs of other home networks.

Different paging areas are interconnected through gateway routers and form a backbone network. Rerouting of the packets is done through the gateway routers when an MN moves from one paging area to another.

5.2 Packet Delivery and Handover Management

A correspondent node (CN) is an MN or a fixed IP host linked to a router, which communicates IP packets to another MN in a home or foreign network (when on visit). The various scenarios encountered in handover management are listed below.

Case 1 CN is a fixed node and MN_1 is at the home network.

CN transmits a message for connection establishment or a packet using the IP protocol. The CN IP header has source IP address and destination IP address of MN_1, in the same way as in the case of IP protocol. HA_1 (the home agent for MN_1) receives the message or packet and, using the information that the destined MN_1 is at the home network itself, it delivers the message or packet to MN_1. It also receives the response message or packet from MN_1 and delivers it to the CN using the IP protocol.

Case 2 CN is a mobile node, the mobile node MN_k is at the home network with agent HA_k, and MN_1 is at the home network with agent HA_1.

MN_k transmits a message for connection establishment or a packet using the IP protocol. The MN_k IP header has the source and destination IP addresses. MN_1 sends the message through HA_k. HA_k uses the same IP address of MN_k as in the IP header and forwards the packet on the Internet in the same way as in the case of the IP protocol and as in case 1. The packet is delivered to HA_1 and then to MN_1. MN_1 transmits back to MN_k like in case 1. Now, HA_k and HA_1 deliver the packets from one end to another and vice versa, instead of the CN directly performing the role of HA_k as in case 1 Both HA_k and HA_1 deliver by just forwarding the packets to their respective MNs using the IP protocol.

The cases when the destined or correspondent MN is visiting a foreign network, and the packets are handed over by the home agent to the foreign agent by tunnelling and encapsulation using the mobile IP protocol, are listed below.

Case 3 CN is a fixed node and MN_1 is at a foreign network.

CN transmits a message for connection establishment or a packet using the IP protocol. The CN IP header has source IP address and destination (MN_1) IP address, in the same way as in case of the IP protocol and as in case 1. HA_1 receives the packets and has the information that the destined mobile node MN_1 is not at the home network and is visiting a foreign network and is reachable via a foreign agent

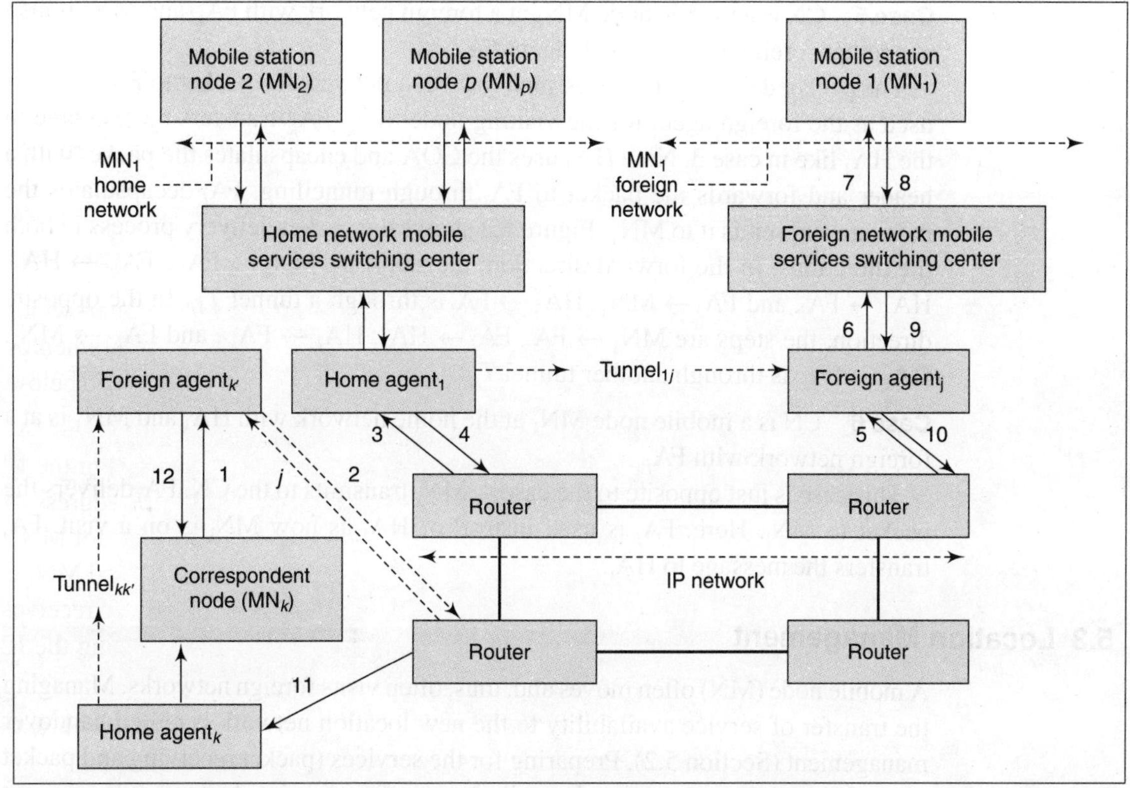

Fig. 5.3 Mobile IP network employing home and foreign agents and packet delivery to and from a mobile node MN_k at a foreign network with FA_k and the other mobile node MN_1 at the foreign network with FA_j

FA_j. It encapsulates the received IP packet using a new header. The new header over the IP packet has a care-of address (COA). The packet encapsulated with the new header is transmitted to FA_j by tunnelling. The FA_j reads the COA and decapsulates the IP packet. It then reads the destination IP address and transfers the packet to MN_1. When MN_1 sends the response message or IP packet with CN as the destination address, FA_j transfers the packet to CN as would have been done by HA_1 had MN_1 been at the home network. The mobility of MN_1 is secure from the CN as any movement on the part of MN_1 is known only to HA_1 and FA_j. Section 5.5 explains the concepts of protocol tunnelling and encapsulation.

Case 4 CN is a mobile node, MN_k, at a foreign network with FA_k and MN_1 is at the home network with agent HA_1.

The packet delivery process is similar to the step in case 3 when MN_1 transmits to CN. MN_k delivers the packet to FA_k. Here, FA_k is used instead of HA_k as now MN_k is on a visit. FA_k transfers the message to HA_1 like in case 1 where CN transfers the message to HA_1.

Case 5 CN is a mobile node MN_k at a foreign network with $FA_{k'}$ and MN_1 is also at another foreign network with agent FA_j.

The packet delivery process is such that MN_k delivers the packet to $FA_{k'}$. $FA_{k'}$ is used as the foreign agent for the visiting node MN_k. $FA_{k'}$ transfers the message to the HA_1 like in case 3. Now HA_1 uses the COA and encapsulates the packet with a header and forwards the packet to FA_j through tunnelling. FA_j decapsulates the message and sends it to MN_1. Figure 5.3 shows the packet delivery process in both the directions. In the forward direction, the steps are $MN_k \rightarrow FA_{k'}$, $FA_{k'} \rightarrow HA_1$, $HA_1 \rightarrow FA_j$, and $FA_j \rightarrow MN_1$. $HA_1 \rightarrow FA_j$ is through a tunnel T_{1j}. In the opposite direction, the steps are $MN_1 \rightarrow FA_j$, $FA_j \rightarrow HA_k$, $HA_k \rightarrow FA_{k'}$, and $FA_{k'} \rightarrow MN_k$. $HA_k \rightarrow FA_{k'}$ is through another tunnel $T_{kk'}$.

Case 6 CN is a mobile node MN_k at the home network with HA_k and MN_1 is at a foreign network with FA_j.

This case is just opposite to the case 4. MN_k transmits to the CN. FA_j delivers the packet to MN_1. Here, FA_j is used instead of HA_1 as now MN_1 is on a visit. FA_j transfers the message to HA_1.

5.3 Location Management

A mobile node (MN) often moves and, thus, often visits foreign networks. Managing the transfer of service availability to the new location network is called handover management (Section 5.2). Preparing for the services (packet receiving and packet transmitting) at the new network entails the use of appropriate location management protocols by the network for management of the MNs location. Agent discovery through agent advertisement and agent solicitation are two protocols used for this.

5.3.1 Agent Discovery

When visiting a foreign network, a mobile node (MN) must discover (find) a foreign agent (FA). Figure 5.4 shows the method of agent discovery by a mobile node MN_1 on receiving the COA during agent advertisement or by agent solicitation in case the COA is not discovered. The steps in the protocol for discovering an agent are as follows:

1. Listen to an advertisement (ICMP message) from an agent.
2. Proceed to step 3 if the advertisement is found, else solicit the agent from the routers. If agent found then proceed to step 3, else repeat the step.
3. If the COA discovered from the message is found to be the same as the previous COA, go back to step 1, else proceed to step 4.
4. If the discovered COA is the same as the home network, de-register at this network and go back to step 1, else if the current COA is a new COA, then register with the new COA.

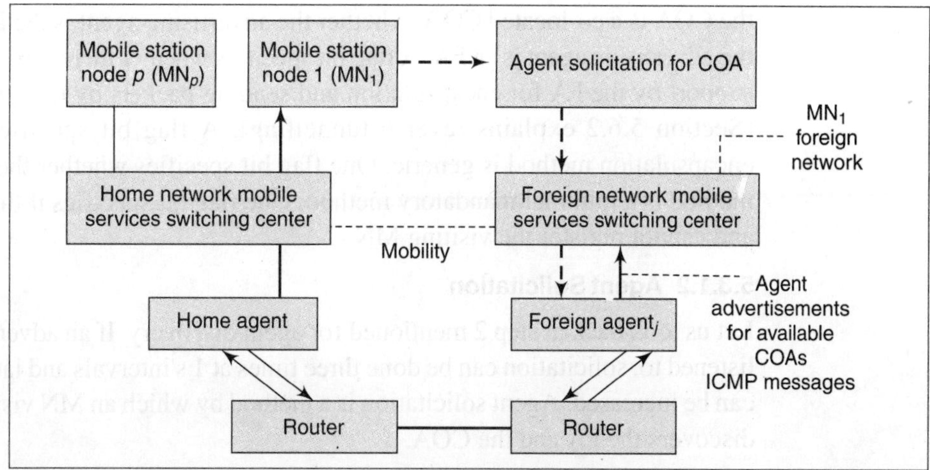

Fig. 5.4 Agent discovery by mobile node MN_1 on receiving COA during agent advertisement or by agent solicitation in case COA is not discovered

Discovery of an FA while on visit to a foreign network and then reverting back to the HA on moving back to the home network is done by the MN using one of the two protocol steps—agent advertisement and agent solicitation (registration) mentioned in steps 1 and 2 above. These are described in the following subsections.

5.3.1.1 Agent Advertisement

Agent advertisements are used by MNs to discover home and foreign agents while moving from one network area to another. Agent advertisements are essentially ICMP messages (Section 5.1.1.10) which are sent to a number of addresses. ICMP message uses the following options and words which are added for the mobility extension of ICMP header.

1. A 32-bit word, with first byte = 00010000 and second byte for length (= 2 words plus number of COAs specified in the extension to which the ICMP message is to be sent) and two bytes for the 16-bit sequence number (for the ICMP message advertised)
2. A 32-bit word has a two-byte specification by the agent for registration lifetime during which the MN can register with the new COA (step 4 in agent discovery). (Lifetime specification is in seconds). It has 8 bits for flags. The remaining byte is not used presently. It is reserved for any future requirements of modifications or specification expansion in ICMP.
3. A set of 32-bit words for the COA addresses for the MN at that agent

Second word has eight flag bits. A COA is said to be a co-located COA if the MN temporarily acquires an additional IP address while on visit to a new network; otherwise the COA is the same IP address for that MN while on visit and when at home. The FA obtains the co-located COA using the dynamic host configuration protocol (DHCP) (Section 5.7). A flag each in the second word specifies whether the COA is a co-located COA, whether the advertising agent is the HA, or whether the advertising agent is an FA. A flag bit specifies whether there is reverse tunnelling support by the FA for encapsulation and sending packets by tunnelling to the HA (Section 5.6.2 explains reverse tunnelling). A flag bit specifies whether the encapsulation method is generic. One flag bit specifies whether the encapsulation method is a minimal mandatory method. One flag bit specifies if the agent is busy and cannot register the visiting MN.

5.3.1.2 Agent Solicitation

Let us look back at step 2 mentioned for agent discovery. If an advertisement is not listened to, solicitation can be done three times at 1 s intervals and later this interval can be increased. Agent solicitation is a method by which an MN visiting a network discovers the FA and the COA.

5.4 Registration

When an MN discovers an agent for service and finds a COA from it, then the MN needs to register itself (for the service of receiving and transmitting of IP packets) with the new agent FA. The MN must also de-register (for the service of receiving and transmitting of IP packets) with the HA (step 4 of agent discovery in the protocol). The function of HA now is to encapsulate the IP packets and transmit them to the discovered FA (through tunnelling), whenever a CN communicates with the MN. Figure 5.5 shows a mobile node MN_k at a foreign network, after agent discovery of FA_j, seeking registration for creating a tunnel between HA_1 and FA_j.

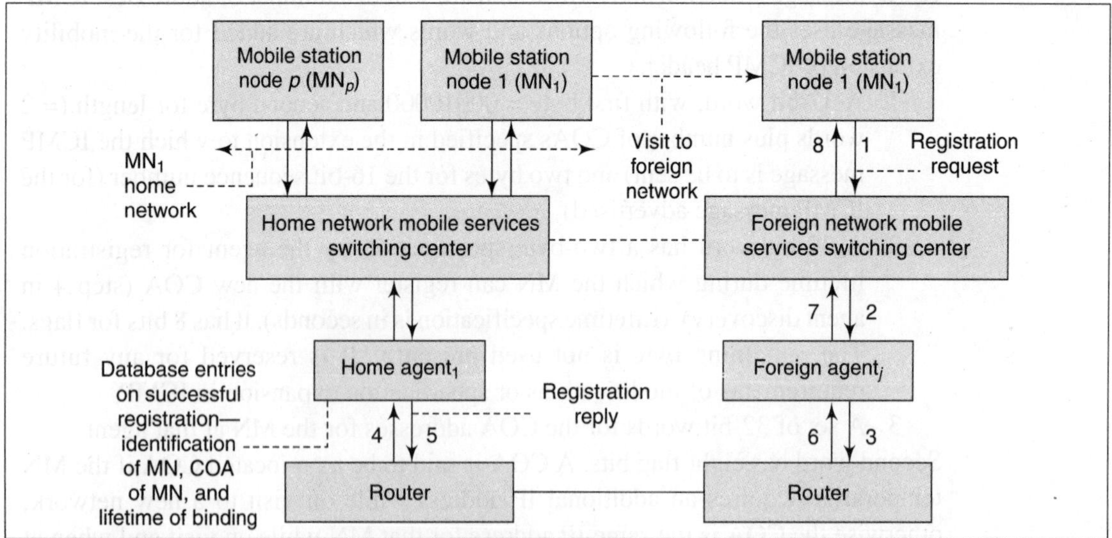

Fig. 5.5 Mobile node MN_k at a foreign network after agent discovery of FA_j in a mobile IP network seeking registration for creating tunnel between HA_1 and FA_j

Requests and replies are made by the MN, FA, and HA using a UDP datagram. Let us assume that the MN has IP address of the HA. If not, then the MN broadcasts the registration request to a paging area. The HAs then send the registration replies. The MN requests one of the HAs (out of those which reply) for registration.

The steps for registration at an agent are as follows:

1. The MN sends a registration request to FA. FA sends that request to the HA. When the COA is a co-located COA, then the request is sent directly to the HA.
2. The HA binds itself for mobility (binds itself for encapsulating and tunnelling the packets to the MN through a new FA). The binding period equals the lifetime of the COA. (It may be recalled that there is a lifetime field in the second word of the mobility extension words in the ICMP message sent by the FA.)
3. The MN registers again before the binding period expires, when it moves to another foreign network, or when it returns back to the home network.
4. The HA sends a registration reply to the FA and the FA to the MN. The MN checks whether the reply shows successful registration. This means that mobility binding now exists from the HA to the FA. It is possible that the reply shows that the registration was not successful. This is when there are too many tunnels created at the HA and the HA does not have the resources to handle new requests or there is an authentication failure or the HA is not reachable to the FA.

Registration Request Fields for Sending Mobile IP registration is by a UDP datagram, which sends the request using the following words after the UDP header (Section 5.1.1):

1. A 32-bit word with first byte = 00000001, eight bits for flags, and two bytes for the lifetime (in seconds)
2. A 32-bit word for the home IP address of the MN
3. A 32-bit word for the home agent IP address of the MN
4. A 32-bit word for the COA of the MN at the new agent
5. A 32-bit word for the identification of the MN
6. A set of words for extensions

The first word has eight flag bits. A flag each in first word specifies whether the COA is a co-located COA, whether the advertising agent is the HA, and whether the advertising agent is the FA. One flag bit specifies whether the MN requests previous mobility binding to be retained. This permits both new as well as the previous mobility bindings. A flag bit specifies whether the encapsulation method is generic, whether the encapsulation method is a minimal mandatory method, and whether the MN wishes to receive broadcast (multicast) messages, which the HA receives for tunnelling to the new FA. If not, then the broadcast messages are filtered at the HA. A flag bit is used to specify if there is reverse tunnelling support from the FA.

Registration Reply Fields for Sending UDP datagram sends the reply using the following words after the header:

1. A 32-bit word with first byte = 00000011, eight bits for a code specifying the result of registration, and two bytes for the lifetime (in seconds)
2. A 32-bit word for the home IP address of the MN
3. A 32-bit word for the home agent IP address of the MN
4. A 32-bit word for identification of the MN
5. A set of words for extensions

The result of registration (at the code sent in the first word of the reply) indicates (a) successful registration and whether the previous mobility binding still exists, (b) FA's rejection with one of the five reasons for it, or (c) HA's rejection with one of six reasons for it.

New Database Entry Fields after Registration at the HA The new database entry fields after registration at the HA consist of the following—*identification* of MN, COA of the MN, and *lifetime* of binding to tunnel the packets to the MN's COA. Figure 5.5 shows the entry fields for the MN registered at HA_i. The COA defined at the COA field facilitates the tunnelling of the packets to the registered new COA (of the MN on a foreign network) after identifying that MN using the identification field. The lifetime field specifies how long the tunnel binds and can be used by the HA for forwarding the packets to the new COA (at FA). (When the binding life expires, the tunnel is not forwarded from the HA to the FA using the COA.)

Database Entries in the Fields at the FA after Registration at the HA Following are the entries for the MNs visiting at the FA: (a) MN identification field, (b) home IP address of the MN, (c) IP address of the HA, (d) MN-link layer address for sending and receiving packets and messages to and from the MN, (e) UDP source port of the registration request, (f) received identification of the MN, (g) COA of the MN and lifetime of binding to tunnel the packets to the MN's COA (The binding of lifetime field specifies how long the tunnel can deliver the forwarded messages to new MN after registration at the FA. When the binding life expires the tunnel does not deliver to the FA using the COA), and (h) remaining lifetime.

5.5 Tunnelling and Encapsulation

When a mobile node (MN) is visiting a foreign network supported by a foreign agent (FA), then the FA has the COA (care-of address) of the MN. The FA receives the IP packets that were received at the HA through a tunnel from the HA to the FA (from HA IP address to the COA IP address at the FA). Tunnelling refers to establishing of a pipe. Pipe is a term used to specify a data stream between two connected ends. The data stream is inserted from one end and is retrieved as FIFO (first in first out) words from the other end. Packets received at the HA are transmitted through the tunnel after encapsulation.

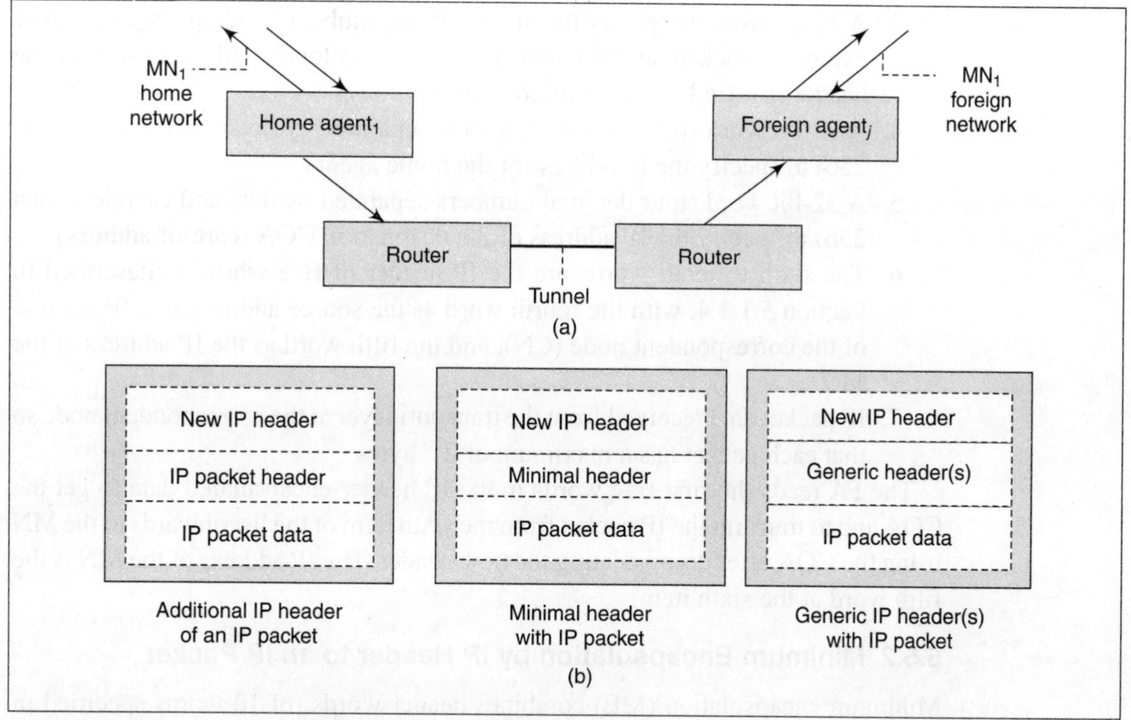

Fig. 5.6 Mobile node MN_k at a foreign network after agent discovery of FA_j in a mobile IP network seeking registration for creating a tunnel between HA_1 and FA_j

Figure 5.3 shows two tunnels (T_{1j} and $T_{kk'}$). Figure 5.6(a) shows a tunnel between the HA and the FA to carry the encapsulated packet. Figure 5.6(b) shows three ways of encapsulation. These ways are described in the following subsections.

5.5.1 Encapsulation by Additional IP Header of an IP Packet

Section 5.1.1 describes the packet formation in IP protocol and IP header fields. One way of encapsulating is as follows—the IP packet received at the HA has an IP header and data with a maximum 2^{16}-byte IP packet. Another IP header (new IP header) is placed over this IP packet. Now, the new IP header has the IP address of the HA as the source and the IP address of the FA as the destination. The data encapsulated with the new IP header has the following fields:

1. A 32-bit word, to specify the IP version (IPv4 or IPv6 for Internet or broadband Internet), length of the IP header (five words), precedence of the IP packet, and total packet length (which is now five words more than that of IP packet received at the HA)
2. A 32-bit word, to specify the ID for the packet, flags, and fragment offset for the same packet ID (a packet can be transmitted in fragments.)

3. A 32-bit word, to specify the time-to-live (number of attempts to hop before expiry of packets at the network), type of protocol, and checksum of the header (for finding transmission errors, if any)
4. A 32-bit word (four decimal numbers separated by dots and each less than 256) to specify the IP address of the home agent
5. A 32-bit word (four decimal numbers separated by dots and each less than 256) to specify the IP address of the destination COA (care-of address)
6. The sixth to tenth words are the IP header of five words as described in Section 5.1.1.4, with the fourth word as the source address, i.e., IP address of the correspondent node (CN), and the fifth word as the IP address of the MN
7. IP packet data received from the transport layer at the correspondent node so that each packet has a maximum of 2^{16} bytes

The FA reads the first five words in the IP header-encapsulated data to get the COA and to transmit the IP packet from the sixth item of the list onwards to the MN using the COA after decapsulating the new header. The IP address of the MN is the fifth word at the sixth item.

5.5.2 Minimum Encapsulation by IP Header to an IP Packet

Minimum encapsulation (ME) combines header words (of 10 words specified in Section 5.5.1) into seven or eight words in the following manner—the first words in the new IP header (of five words) and the IP packet header (of five words) are the same and are duplicating in case of IP-in-IP encapsulation (see Fig. 5.6).

1. The sixth and seventh words in the sixth item of the new IP header are not present in minimum encapsulation (ME) as both words are mere repetitions.
2. The eighth word in the sixth item is changed and now specifies the type of protocol, a one-bit flag, seven reserved bits, and a 16-bit checksum of the modified three-word IP header (from the original five) for finding transmission error, if any.
3. The ninth word in the sixth item is changed and now specifies (instead of the CN IP address) the MN IP address (which was earlier specified by the tenth word).
4. The tenth word in the sixth item is changed and now specifies (instead of the MN IP address) the CN IP address in case the flag bit is set to 1, and the tenth word in the sixth item is removed in case the flag bit is set to 0.

The FA reads the first five words in ME and transmits the packet to the MN using the COA. The MN IP address is specified by the seventh or the eighth word, depending upon the flag bit.

5.5.3 Generic Routing Encapsulation by IP Header to IP Packet

IP header-in-IP header encapsulation (Section 5.5.1) and minimal encapsulation

(Section 5.5.2) have the following deficiencies—(a) routing information for tunnelling is not given, (b) there is no provision for recursive encapsulations, and (c) there is no provision for a key that can be used for authentication or encryption. Recursive encapsulations are needed when the tunnel transmits multiple pieces of information for the MN. Each piece of information is encapsulated in one protocol. There may be one or more GRE headers depending on the number of recursions required to send multiple pieces of information. The three components in the GRE encapsulation method are described below.

New IP Header It is same as the first to fifth words in the first to fifth items of the new IP header. Time-to-live however, is, set as 1. This results in a once-only forwarding to the FA by the HA. The tunnel does not need an extra hop (attempt) and the tunnel does not get blocked like routers due to external IP address transmissions. It has fixed source and destination end points.

Generic Routing Encapsulation (GRE) Header(s) There is one GRE header for each protocol for encapsulation. The GRE header can be sent recursively after the new IP header (bottom right in Fig. 5.6).

1. The sixth word in encapsulation and the first word of the first GRE header has a total of 16 bits for flags—bits to define the number of recursions, reserve bits, and version bits, and the next 16 bits specify the protocol for encapsulating the information sent with the GRE header.
2. The seventh word specifies a 16-bit checksum and a 16-bit offset. Both are optional as indicated by the flag bits used to define these options. (Checksum of the GRE header enables locating of transmission errors, if any, at the receiver).
3. The eighth word specifies a 32-bit key. It is optional as indicated by the flag bit to used define the key option. (The key at the GRE header enables authentication or encryption at the FA.)
4. The ninth word specifies a 32-bit sequence number information. It is optional as indicated by the flag bit used to define the sequencing option. (Sequencing at the GRE header enables the FA to rearrange the packets sent by the HA.)
5. The tenth word specifies a 32-bit routing information. It is optional as indicated by the flag bit used to define the routing option. (Routing at the GRE header enables authentication or encryption at the FA.)
6. From the eleventh word, if number of recursions are defined in the first word of the GRE header, then the next GRE header is inserted before the IP header and IP data sent by the HA. There will be eleventh and twelfth words in case of one recursion. If the number of recursions specified in the eleventh word in the GRE header is two, then the next two GRE headers are also inserted before the IP header and IP data sent by the HA (Fig. 5.6).

IP Header and IP Packet Data This part remains the same as that in the un-encapsulated IP header and the data received from the CN (correspondent node) IP packet at the HA. The first word has five flag bits and three recursion-number-defining bits. The five flag bits are—checksum option flag, sequence number field option flag, key option flag, and source-routing option flag. Source routing is a method in which the source of a packet provides the route information also. Source-routing-method-based routers use the routing information word for routing a packet.

5.6 Route Optimization

Consider Fig. 5.3. Let us assume that MN_1 is visiting a foreign network which happens to be the home network of CN_2 and CN_2 is very close to CN_1 (Fig. 5.2). Consider the mobile IP network employing home and foreign agents and packet delivery to and from a mobile node MN_k at a foreign network with FA_k and MN_1 at the foreign network with FA_j. Figure 5.7 shows a mobile IP network employing a triangular route without mobility binding between COA_j and CN_k. It marks the path numbered in sequence as l, 2, 3, 4, and 5 to MN_1. Packets make a triangular trip to reach from CN_k to MN_1. It is also possible that FA_k and FA_j are identical. Optimization of the triangular route can be carried out in case the MN_1 opts to make its mobility known. Section 5.6.1 explains how the route is optimized.

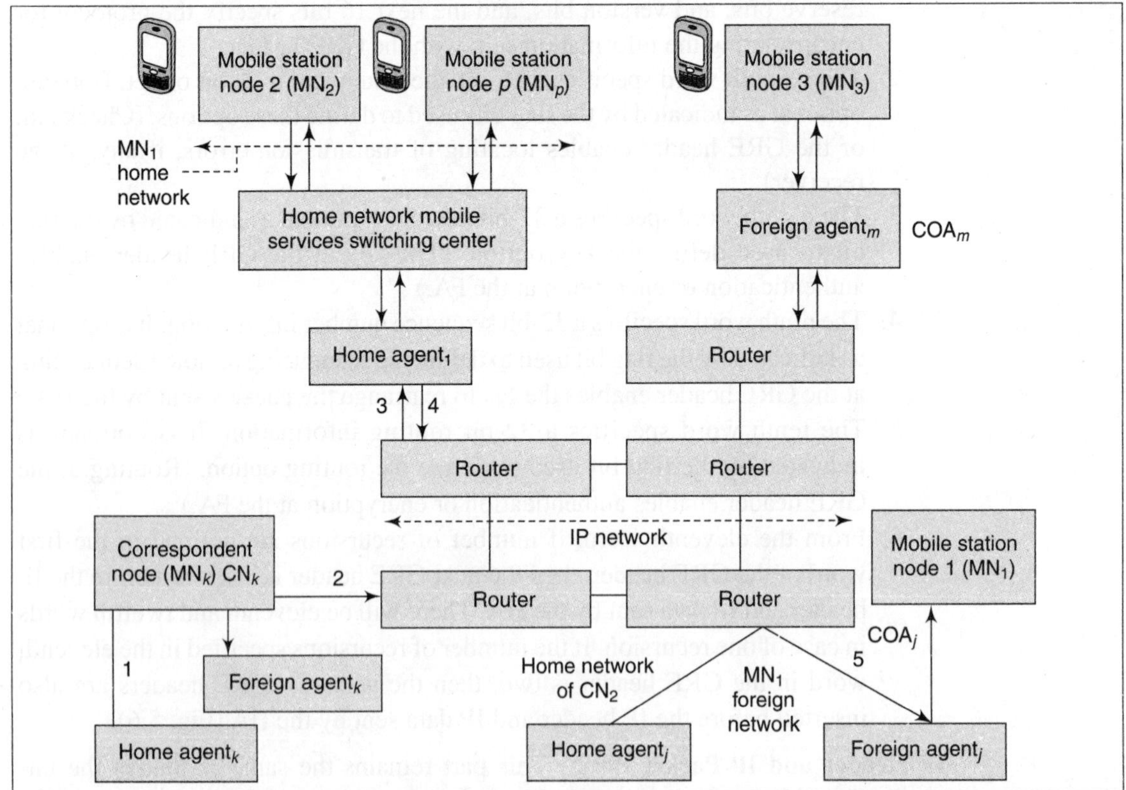

Fig. 5.7 Mobile IP network employing a triangular route without mobility binding between COA_j and CN_k

5.6.1 Mobility Binding

Triangular route optimization to a direct route is performed as follows—if MN_1 (to which CN sends the IP packet) permits the identification of its mobility when it visits a foreign network, then CN can use a mobility-binding cache. Mobility-binding cache is a cache that stores the current COA of the called MN. The CN can directly send the IP packets to MN_1, instead of triangular routing through HA_1, and then to the COA through a tunnel. Figure 5.8 shows the route optimization and path 1, 2, 3, 4, 5 after mobility binding of MN_1 at COA_j with CN_k. Figure 5.8 also marks the steps for establishing route optimization. The steps in the binding protocol at the calling network are as follows:

1. CN_k (fixed) or MN_k (mobile) network sends a mobility-binding request to HA_1.
2. HA_1 detects whether MN_1 (for which binding request is made) has blocked external mobility binding requests. (The word 'external' here does not include

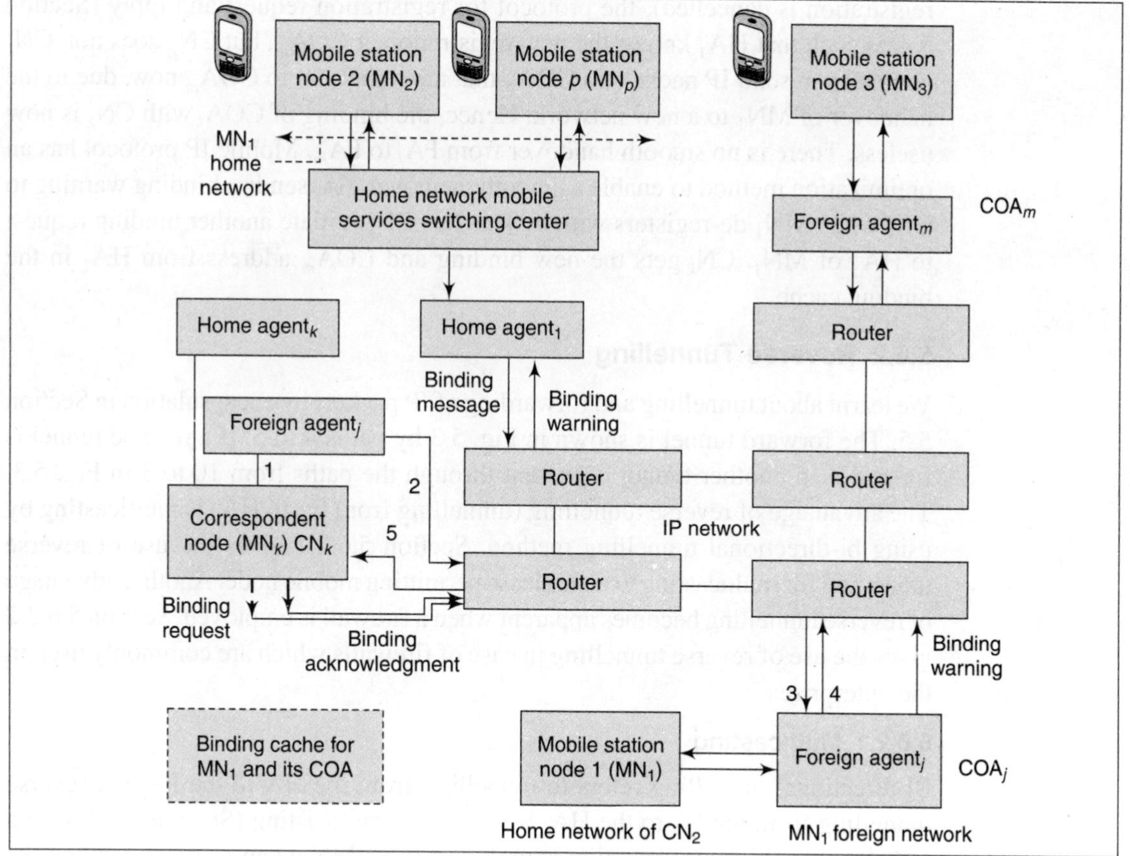

Fig. 5.8 Mobile IP network employing route-optimized path after mobility binding between COA_j and CN_k

the visiting network FA.) If not, then it sends the update for the mobility-binding message to the CN_k network.

3. Mobility-binding message has the IP address of MN_1 and the present COA (COA_j) of MN_1 when on visit to a foreign network and registered with FA_j.
4. CN_k issues an acknowledgement to HA_1 on receiving the binding message.
5. The CN_2 network decapsulates the IP packet (this decapsulation would have been performed by FA_j through HA_1 if MN_1 had blocked external binding requests) and sends a warning for binding. The warning is sent to HA_1 of MN_1. It serves an important purpose, i.e., HA_1 sends the binding update to CN_k when MN_1 moves to visit another foreign network or when it returns to the home network. [Warning for binding is, in fact, a message to the effect that the new IP addresses of MN_1 and CN_2 will decapsulate the encapsulated IP packets (from the moment that the warning is aired) instead of FA_j.]

Optimization for Smooth Handover When MN_1 makes a visit to a foreign network and FA_m is the new foreign agent which re-registers MN_1 (and the earlier COA_j registration is cancelled), the protocol for registration request and reply (Section 5.4) is such that HA_1 knows the new registration at COA_m, but CN_k does not. CN_k continues to send IP packets to COA_j, that are to be sent to COA_m now, due to the handover of MN_1 to a new network. Hence, the binding of COA_j with CN_k is now useless. There is no smooth handover from FA_j to FA_m. Mobile IP protocol has an optimization method to enable a smooth handover. FA_j sends a binding warning to CN_k when MN_1 de-registers with it. This lets CN_k initiate another binding request to HA_1 of MN_1. CN_k gets the new binding and COA_m address from HA_1 in the binding cache.

5.6.2 Reverse Tunnelling

We learnt about tunnelling and forwarding of IP packets by encapsulation in Section 5.5. The forward tunnel is shown in Fig. 5.3 by paths 4 to 5. If a reverse tunnel is formed then another tunnel is present through the paths from 10 to 3 in Fig. 5.3. The advantage of reverse tunnelling (tunnelling from FA to HA) is multicasting by using bi-directional tunnelling method. Section 5.6.2.1 gives the use of reverse tunnelling for multicasting to a multicast-permitting mobile node. Another advantage of reverse tunnelling becomes apparent when a firewall is employed. Section 5.6.2.2 gives the use of reverse tunnelling in case of firewalls which are commonly used in the enterprises.

5.6.2.1 Multicasting

Bi-directional tunnelling refers to tunnelling from the HA to the FA and reverse tunnelling from the FA to the HA. It facilitates multicasting (Section 5.1.1.8) to a mobile node (MN). Information is multicast to an MN when it sets the option for multicast listening (Section 5.4). Bi-directional tunnelling method over a mobile IP network works in the following manner—let us assume that a mobile node MN_i visits a foreign network with foreign agent FA_j. A multicast tree multicasts a packet to HA_j. HA_j forwards the multicast IP packets to MN_j. This is done after registration (Section 5.4). HA_j establishes a bi-directional tunnel (Section 5.5) between HA_i and FA_j. FA_j transmits the received multicast message or packet to MN_j. Now, suppose MN_i visits another foreign network with foreign agent FA_k. MN_i requests FA_k and FA_k forwards the transmit request for the multicast to HA_j. The disadvantages of this approach are as follows:

- Duplication of multicast IP packets takes place when multiple MNs of HA_j and other HAs visit the same FA. Since several HAs create several bi-directional tunnels, through which they transmit multicast packets multiple times, in case the built bi-directional tunnels do not converge into one, the packets maybe duplicated. Mobile multicast (MoM) protocol leads to convergence of the tunnels by defining an HA as the DMSP (designated multicast provider). Only the DMSP can build bi-direction tunnels.
- IP packets reach by short and long paths when there is no DMSP. When DMSP provides the multicast service, the IP packets reach by the longer path as the DMSP route length may not be the shortest.

The advantage of multicasting is that there is no reconfiguration (updating of the routing tables) of routers at the multicast tree.

Another multicast approach without the reverse tunnelling, called remote subscription approach, is described below:

1. Let us assume that MN_i visits a foreign network with foreign agent FA_j. FA_j transmits a 'join' request in case it is not presently registered for multicast at the multicast tree. The advantages of this are as follows:
   - There is no duplication of multicast IP packets.
   - IP packets reach through an optimal (shortest) path.
2. When MN_i moves to the next foreign agent FA_k, it again transmits a 'join' request and the previous subscription is cancelled.

5.6.2.2 Firewall Security

A firewall filters the packets assigned to an IP address received from another IP address. IP address of the MN is at the home agent. When it moves to a foreign agent, the MN sends the IP packets using the care-of address (COA) assigned at the foreign agent (Section 5.2). In case the firewall permits another IP address (at the FA) assigned to an IP address (at the HA), there is a security risk. This is because when there is reverse tunnelling of the packet from the FA COA to the home agent and the home agent transmits it to the correspondent node (CN), then the firewall gets the packets from the same IP address as the IP address registered at the firewall and it does not filter these packets. When the firewall at the CN sends the IP packets to MN_1, then the path followed (using the sequence numberings as given in

Fig. 5.3) will be 1, 2, 3, 4, 5, 6, and 7. The sub-paths 4 and 5 go across the forward tunnel. When the firewall at the CN receives the IP packets sent to MN_1, then the path followed will be 8, 9, 10, 3, 4, and *i*. The sub-paths 10 and 3 are across the reverse tunnel. The CN firewall continues to use the same IP address for the MN when transmitting and receiving packets.

5.6.2.3 Time-to-live

Time-to-live defines the number of attempts to hop before the expiry of packets at the network. GRE header encapsulation during tunnelling sets time-to-live as one, so the packets are forwarded only once. (The tunnel does not need extra hops. It has fixed end points.) This results in once-only forwarding through the tunnel from the home agent (HA) to the foreign agent (FA) when the mobile node (MN) visits a foreign network. At the foreign agent, the time-to-live setting might be too low. Therefore, when the MN sends the response to the correspondent network (CN), then the time-to-live set at the FA may not be sufficient. When the COA is used to send the response to the CN without reverse tunnelling, then a very low setting of time-to-live blocks the packets after a very small number of hops (attempts) to the CN.

Reverse tunnelling also sets the time-to-live as one because IP packets need to be sent only once. The tunnel does not need an extra hop, it has fixed source and destination end points. Reverse tunnelling facilitates guaranteed transmission of the IP packet responses through the tunnel to the HA. Now, the HA transmits the response to the CN. A low value of time-to-live at the FA does not lead to packet expiries.

5.7 Dynamic Host Configuration Protocol

When a mobile node visits a foreign network then it gets a new IP address known as care-of address (COA) by agent discovery process and advertisement of the COAs by the foreign agent (Section 5.3.1). Also, a co-located COA is obtained by the dynamic host configuration protocol (DHCP). Let us assume that a mobile computer (laptop or other) visits another network (which has a separate domain name server identity on the network and functions as a subnet on the Internet). The computer gets a new IP address in this case too. The server provides a dynamic IP address, subnet mask, and ARP and RARP caches to enable the computer to transmit and receive the IP packets at the new IP address from the Internet via the subnet. The server (subnet) has its own IP address to provide connectivity to the Internet. Dynamic host configuration protocol (DHCP) is a protocol to dynamically provide new IP addresses and set subnet masks for the visiting computer so that it can use the server and subnet router at the place being visited.

Any software in an agent (e.g., a foreign agent visiting a mobile node) or device software for connecting to the network can have a software component called the

DHCP client. The DHCP client protocol communicates with a server. The steps in the DHCP protocol for dynamically configuring the IP address and other networks are as follows:

1. The DHCP client in an agent, device, or computer broadcasts a discover request known as DHCPDISCOVER directly or through a DHCP relay agent to the servers. A subnet may have a number of DHCP servers. For this reason, the request is broadcasted to several servers. (A DHCP server may be part of the operating system of the computer seeking connection to the network. The server has software for allocation of network addresses to the computer.)
2. Each server listening to the discover request finds the configuration, which can be offered to the client. Server(s) send(s) the configuration parameters, including an IP address not presently in use, at the subnet. The configuration parameters are in the DHCPOFFER for the offered configuration.
3. The DHCP client can reject the offer from a server or servers. When DHCP offers from all the servers are rejected, the client repeats the steps from step 1, else it proceeds to step 4.
4. The client replies to the servers through a DHCPREQUEST to each server. The option 'reject' is set in each reply to those DHCP servers to which the client reply is 'reject'. The option 'select' is set for those servers to which the client reply is 'select'.
5. The selected DHCP server creates and manages bindings. (A binding is a collection of configuration parameters, including at least one IP address, which is associated with and binds to the DHCP client.) DHCP server also sets a time interval during which the offered IP address will be valid for the DHCP client computer. The required interval can vary depending upon the likely Internet connection interval at a particular Internet-serving network. The binding may periodically provide new IP addresses.
6. The DHCP server confirms the binding through a message. It sends DHCPACK after creating the binding.
7. When the DHCP client computer leaves the subnet, it sends DHCPRELEASE message. In case the client does not send DHCPRELEASE within a specified time interval, the server frees the created binding.
8. The server and client also use the authentication protocols before considering the DHCPDISCOVER from a client and before accepting a DHCPOFFER, respectively.

The DHCP protocol guarantees that any assigned network address, at a given instant, is in use by either one DHCP client or none.

Keywords

Agent: Software which (i) migrates and functions at a host *H* or system *B*, (ii) receives data or packet from network, and (iii) performs the actions and computations. Agent functions are the ones which are delegated by a computing system *A*. The agent transmits data or packets for *B* or reports to *A* and thereby takes the load of computations, transmission, and reception for *A* at *H* or *B*. The agent functions at a host *H* which is connected to *A* or functions at *B* to which the agent can migrate. (*B* is a server, router, or distributed computing system.)
Agent advertisement: A set of ICMP messages containing one or more IP addresses available at an instant at a router on the Internet. The router is connected to a foreign network (FN) and these addresses are multicast in the FN at periodic intervals so that a foreign agent (FA) can listen and obtain an IP address known as the COA (care-of address). The FA then provides Internet services to the mobile node entering into the FN using the COA as the IP address.
Agent discovery: A process by which a mobile node (MN) visiting a foreign network discovers a foreign agent (FA). The FA discovers a COA during either agent-advertisement or agent-solicitation processes.
Agent solicitation: A process at a foreign network by which a foreign agent (FA) seeks a COA on its own in case the COA is not listened at the FA, through a process of advertisements of COAs for the agents.
COA: An IP address assigned to an agent of a mobile node when it enters into a foreign network (FN) so that the foreign agent (FA) can use it for receiving and transmitting packets through the Internet.
Decapsulation: A process of recovering the original packet from *A* after using the header, which has encapsulated the data received at *B*. (Here, *A* and *B* are two computing systems, for example, the home and foreign agents.)
DHCP (dynamic host configuration protocol): A protocol to dynamically provide new IP address and set subnet mask for a visiting computer so that it can use the server and the subnet router at the place visited.
Encapsulation: A process in which a computing system receiving the packets on the Internet, uses a new protocol header placed over the packet and thus encapsulates the protocol used earlier on the network as well as data.
Firewall security: A security process in which a firewall (software for preventing unauthorized data communication) is used. An example of such a process is as follows—consider a home agent (HA) employing a firewall for receiving IP packets for a mobile node (MN). When the MN moves and visits a foreign network, the HA transmits to the foreign agent (FA) by tunnelling. The MN's FA transmits responses and packets through reverse tunnelling to the HA first, and then the HA transmits through the firewall before the router on the Internet. The FA does not directly transmit the responses and packets.
Foreign agent: An agent present either at an interconnected host or at a router of packets on the Internet. The foreign agent takes the load of computation, transmission, and reception for a mobile node's foreign network mobile services switching centre.
Foreign network: A network providing services to a mobile terminal equipment, station, or node and having a mobile switching centre. The centre does not register the node in a home location register for the nodes but registers at a visiting location register when the mobile node enters into the network.
Home agent: An agent either at an interconnected host or present at a router of packets on the Internet. It takes the load of computation, transmission, and reception for a home network mobile services switching centre.
Home network: A network providing services to a mobile terminal equipment, station, or node and having a mobile switching centre. The centre registers the node in a home location register for the nodes.
Internet protocol (IP): A protocol in which data from L_4 (transport layer) divides into packets such that a packet-switched network (such as the Internet) can be used. Using the IP protocol, the data packets, after appending an IP header, can hop on different routes to reach a destination so that more packets can simultaneously be sent through the network unlike in a circuit-switched network, where one data frame transmits at one instant (e.g., in GSM or HSCSD).
IP address: An address of 32 bits assigned to a node such that IP packets can be transmitted or received by the node on the Internet.
IP header: A set of 32-bit words transmitted along with a packet after appending them as per the IP protocol.

IP node: A station, terminal equipment, device, or computing system, which has a distinct IP address assigned to it at an instant. The IP nodes are interconnected through the routers and subnets on the Internet using the IP protocol.

IP packet: A part of data called a packet, which consists of a maximum of 2^{16} bytes and transmits and hops through different routes at different instants on the Internet using the IP protocol.

Mobility binding: A binding created for providing mobility to a mobile node after registration at a foreign network.

Mobile IP protocol: A protocol used by mobile services switching centres in the mobile network employing the home and foreign agents. Mobile IP provides Internet connectivity to a mobile station or node (MN) and a router employs a home agent for a set of home-networked MNs as well as a foreign agent for the visiting MNs.

Multicasting: A process in which a router transmits packets, datagram, or ICMP messages to the subnets and a number of hosts on a subnet. The selected nodes or their agents can get the data concurrently through multiple paths. It differs from broadcasting, as broadcasting is carried out to all nodes and multicasting is limited to a number of subnets (home and foreign networks). It differs from unicasting, as unicasting is carried out to one or those node(s) or devices(s) which request or subscribe to a service through one path or route at an instant.

Registration: A process by which a node, after getting a COA, registers itself for the service of transmitting IP packets in future with a new agent at a foreign network after de-registering (for the service of transmitting IP packets) with the home agent. After the de-registering, the mobile node receives the IP packets, which reach the home agent, by tunnelling to the new agent.

Reverse tunnelling: A process of forming another tunnel for tunnelling packets from a foreign agent to the home agent in order to facilitate multicasting by using the bi-directional tunnelling method and firewall security.

Time-to-live: A set of bits specifying the number of attempts to hop before expiry of packets at the network (it is not measured in seconds but in the number of hops).

Tunnelling: A process by which a tunnel (path or channel) is established for transmitting data streams from a computing system *A* to another computing system *B*, where *A* transmits to *B* through the tunnel after encapsulating the data using a new protocol header.

Review Questions

1. What are the five layers in the IP protocol? Explain the functions of each layer. What are stateless and state-oriented connections? Explain the difference between the connection-less and connection-oriented protocols. Give examples.
2. What is the difference between a datagram and a packet? What are the uses of datagram in the mobile IP protocol?
3. Explain the fields of the header in ICMP messages. What are the uses of ICMP messages on the Internet?
4. How does a routing table help in routing packets? Why do we use subnets for routing on the Internet? What do you mean by net ID, subnet, and host bits in IP addresses of class A, B, and C subnets?
5. How is a packet delivered to a computer on LAN after reaching a destination router on a subnet? Explain subnet mask, ARP, and RARP.
6. What do you mean by point-to-point, multicast, and broadcast communication on a network? What is multicast tree? How does it enhance the multicasting efficiency?
7. Describe the mobile IP protocol. Explain, with a diagram, how a correspondent mobile node on a visit sends and receives IP packets to and from another MN also on a visit at another foreign network. How does the packet delivery mechanism in the mobile IP protocol differ from that in the IP protocol?
8. What are the functions of home and foreign agents in the mobile IP protocol? How does an agent discover COA(s) when a mobile station node visits a foreign network? When and how does a mobile node solicit an agent? How does agent advertisement differ from agent solicitation?

9. What is the difference between the care-of address and the co-located care-of address?
10. Describe the registration of a visiting mobile node on handover. How is the binding between the home agent and the foreign agent created?
11. Describe the function of each header in the IP protocol. What do you mean by tunnelling, encapsulation, and decapsulation? How does a tunnel differ from a route? Why is the time-to-live set to one when IP packet tunnels through from the home to the foreign agent?
12. Explain each field in all the headers of an encapsulated IP packet with IP-in-IP, minimal, and GRE encapsulations. How is it possible to encapsulate with multiple protocol information in GRE?
13. When and how is a mobility binding created between the correspondent and mobile nodes? How does mobility binding optimize the route?
14. What do you mean by reverse tunnelling and bi-directional tunnelling? How does a reverse tunnel differ from a forward tunnel in the mobile IP protocol? How does the reverse tunnel help when the time-to-live for the packets at a foreign agent is small?
15. Describe multicasting in the mobile IP protocol. Show, using diagrams, the difference between the bi-direction tunnel multicast method and the remote subscription multicast method.
16. Explain the protocol used for finding a co-located care-of address. When is the DHCP used? Explain the DHCP protocol. How does a DHCP server bind a mobile node with an IP address?

Objective Type Questions

Pick the correct or most appropriate statement among the choices given:

1. A TCP/IP network has the OSI
   (a) session layer functions incorporated at the application layer and the presentation layer functions at the transport layer and there are a total of seven layers.
   (b) session layer functions incorporated at the transport layer and the presentation layer functions at the application layer and there are a total of five layers.
   (c) network layer functions incorporated at the transport layer and the presentation layer functions at the application layer and there are a total of five layers.
   (d) data link layer functions incorporated at the network layer and the presentation layer functions at the transport layer and there are a total of five layers.
2. A datagram is
   (a) data with a protocol header in a connection-oriented protocol.
   (b) less than 2^{16} bytes of data in an IP packet.
   (c) less than 2^{16} bytes of data with a protocol header in a connectionless protocol.
   (d) data with a protocol header for sending messages.
3. The third 32-bit word in the IP header specifies
   (a) time-to-live (which defines the number of attempts to hop before expiry of packets at the network), upper layer protocol type, and checksum of the header.
   (b) time-to-live in seconds, network type, and checksum of the header.
   (c) IP address of the source.
   (d) time-to-live (which defines the number of attempts to hop before expiry of packets at the network), lower layer type, and checksum of the packet.
4. A router's subnet address at each routing table is defined by
   (a) 2^{32}, 2^{24}, and 2^{16} bits for class A, B, and C, respectively.
   (b) 2^{16}, 2^{24}, and 2^{32} bits for class A, B, and C, respectively.
   (c) 2^{25}, 2^{18}, and 2^{11} bits for class A, B, and C, respectively.
   (d) 2^{7}, 2^{14}, and 2^{21} bits for class A, B, and C, respectively.

5. Consider a mobile IP network. A correspondent mobile node (CN_k) on a foreign network with foreign agent (FA_k) and home agent (HA_k) communicates IP packets to a mobile node (MN_j) on a visit to another foreign network with FA_j and the MN_j home network has HA_j. CN packet delivery and response from MN is through tunnels
   (a) from FA_j to HA_j and from FA_k and HA_k.
   (b) from FA_k to HA_k and from FA_j and HA_j, respectively.
   (c) from HA_j to FA_j and from HA_k and FA_k.
   (d) from HA_k to FA_k and from HA_j and FA_j, respectively.
6. Agent discovery (finding a care-of address) protocol entails
   (a) registering with a foreign agent and waiting for registration reply from the home agent.
   (b) requesting foreign agent advertisements three times in 1 s.
   (c) first listening to foreign agent advertisements for COAs and if COAs are not found, then agent solicitation first at the rate of three times at 1 s interval and then at increased solicitation intervals.
   (d) first agent solicitation and then (if COA is not found) listening to foreign agent advertisements for COAs at the rate of three in 1 s.
7. Agent advertisement is by
   (a) UDP datagram, which are sent to a number of addresses, UDP datagram message uses options and the words, which are added for the mobility extension of ICMP.
   (b) IP multicast messages, which are sent to a number of addresses as ICMP messages.
   (c) IP packets with mobility extension that are sent to a number of addresses, using options and words, which are added for the mobility extension of ICMP.
   (d) ICMP messages, which are sent to a number of addresses. ICMP message uses options and words, which are added for the mobility extension of ICMP.
8. First step in the registration of a mobile node on a visit is sending a registration request to
   (a) the HA, the HA sends that to the FA, when the COA is a co-located COA, then the MN sends the request directly to the FA.
   (b) the FA, the FA then sends that to the HA, when the COA is a co-located COA, then the MN sends the request directly to the HA.
   (c) the FA, the FA sends that to the HA, when the COA is a co-located COA, then the request is sent indirectly to the HA.
   (d) the FA, the FA sends that to a co-located COA by reverse tunnel.
9. After registration of a mobile node (MN), the database entry at the HA consists of
   (a) identification of MN, COA (care-of address) of the MN, and lifetime of binding to tunnel the packets sent to the MN's COA.
   (b) identification of the MN, IP address of the MN, and the COA of the FA.
   (c) identification of the MN, IP address of the FA, and the COA of the HA.
   (d) identifications of the FA and the MN, IP address of the MN, and the COA of the FA.
10. When a UDP datagram sends a registration request using the words after the header, the registration request field contains flags to specify
    (a) whether the datagram is for bi-directional tunnelling and whether the MN requests previous mobility binding to be retained.
    (b) whether the datagram is for multicasting and whether the MN requests previous mobility binding to be retained.
    (c) whether the COA (care-of-address) is sent by reverse tunnel to the HA and whether the advertising agent is the HA.
    (d) whether the COA (care-of address) is a co-located COA, whether the advertising agent is the HA, and whether the MN requests previous mobility binding to be retained.

11. During tunnelling from the home agent (HA) to the foreign agent (FA), minimum encapsulation (ME) combines encapsulation header and IP header words into
    (a) seven or eight words.
    (b) five words.
    (c) six words and checksum of the header.
    (d) six words and checksum of the packet through the tunnel.
12. Reverse tunnelling of packets is
    (a) from the HA (home agent) to the FA (foreign agent), and is used when multicasting or when there is a firewall at the correspondent node (CN) or when the time-to-live at the foreign agent (FA) is too short.
    (b) from the HA to the FA and is used only when multicasting.
    (c) from the FA to the HA and is used only when there is a firewall at the correspondent node (CN).
    (d) from the mobile node (MN) to the CN, and is used when multicasting or when there is a firewall at the CN or when the time-to-live at the foreign agent (FA) is too short.
13. The DHCP protocol specifies that after the client sends
    (a) DHCPDISCOVER to the selected DHCP server, the server creates and manages a binding and the binding is a collection of at least one IP address.
    (b) DHCPDISCOVER to the selected DHCP subnet, the subnet creates and manages a binding for the mobile node to the Internet.
    (c) DHCPREQUEST to the selected DHCP server, the server creates and manages a binding and the binding is a collection of configuration parameters, including at least one IP address, associated with or binding the DHCP client to a co-located address for the mobile node.
    (d) none of the above because mobile IP protocol, and DHCP protocol, is not used for binding of the IP address with the mobile node client.
14. When a mobile node (MN) visits a foreign network, then
    (a) the home agent seeks a co-located COA by the IP protocol when the COA for the MN is available.
    (b) the foreign agent (FA) seeks a co-located COA by the DHCP protocol when no COA for the MN is currently available.
    (c) the FA seeks the COA by the DHCP protocol when no co-located COA for the MN is currently available.
    (d) the FA seeks the COA by the mobile IP protocol when no co-located COA for the MN is currently available.
15. If a mobile node visiting a foreign network has sent a registration request, then which of the following is correct?
    (a) The FA sends a registration reply to the HA and the HA to the MN. The MN checks whether the reply shows successful registration indicating that mobility binding now exists from the FA.
    (b) The HA sends a registration reply to the MN. The MN checks whether the reply has the COA (care-of-address) for receiving and sending IP packets.
    (c) The HA sends a registration reply to the FA, and the FA to the MN. The MN checks whether the reply shows successful registration indicating that mobility binding now exists from the HA.
    (d) The MN does not need a registration reply for registration from the home agent since the registration at the FA suffices.
16. The mobile IP network employs
    (a) a triangular route when the mobility binding between COA_j and CN_k does not exist and the packets have short time-to-live.
    (b) direct IP protocol route when the mobility binding between COA_j and CN_k exists and the mobile node conceals the visit to a foreign network.
    (c) a triangular route when the tunnel between the home agent and the foreign agent does not exist.
    (d) a triangular route when the mobility binding between COA_j and CN_k does not exist and the mobile node conceals the visit to a foreign network.

Chapter 6

Mobile Transport Layer

In Chapter 1 it was discussed that mobile stations offer novel services such as email, web browsing, and enterprise solutions. The global network of the Internet facilitates these services all over the world. Chapter 5 described the mobile IP network layer used for access to the Internet. Mobile IP facilitates transmission of IP packets between a fixed or mobile node (MN) via the Internet. The chapter also introduced us to the transport layer above the IP network layer in the TCP/IP protocol suite for the Internet. The main function of the transport layer is sequential transfer of data from port to port (a port is an application or service access point). This chapter describes conventional TCP and then the additional TCP protocol sub-layers and methods for facilitating wireless TCP transport between mobile nodes. In this chapter, we will discuss the following concepts:

- Conventional TCP network, protocol header, TCP data streams, connection states, and packet delivery
- Mechanisms for flow control and congestion control
- Problems with employing the conventional TCP in a network of mobile nodes on the Internet
- Modifications made in the TCP for use over wireless and mobile networks using split TCP, indirect TCP, and mobile TCP methods
- TCP over wireless using TCP-aware link-layer solutions, use of data link layer for the TCP functions, and the snooping TCP method
- Fast retransmit/fast recovery and timeout freezing transmission methods
- Explicit notification method
- Selective retransmission and transaction-oriented TCP methods
- TCP over 2.5G/3G mobile networks

6.1 Conventional TCP/IP Transport Layer Protocols

The two main protocols for the TCP/IP transport layer, i.e., user datagram protocol (UDP) and transport control protocol (TCP) were introduced in Section 5.1.1. The following subsections describe these.

6.1.1 User Datagram Protocol

The user datagram protocol (UDP) is a connection-less protocol. There is no session establishment, data flow and congestion control, and session termination in UDP. It transmits like a person using a phone who just speaks without waiting, irrespective of whether the receiver at the other end is listening or not, replying or not.

A port is a service access point (software) for data input and output through which a service (application) is rendered by a node. Examples of services through the Internet are email transmission, email reception, and web browsing. An example of a port is port number 80 which specifies the www (HTTP) application in the TCP/IP protocol suite. The function of the transport layer is to transport the port data. UDP specifies the ports. The UDP header is used at the subsequent layers (from the transmitter transport layer up to the receiver transport layer) during transmission of port data (application layer data) to the receiver. The UDP header fields specify the following:

- First-word bits b_0–b_{15} for source port (optional) and b_{16}–b_{31} for destination port
- Second-word bits b_0–b_{15} for length of the datagram and b_{16}–b_{31} for the header checksum

A pseudo header is used as prefix. It has the first and the second words for the source and destination IP addresses, respectively, and the third word for the protocol and length. The bits b_0–b_7 are all zeros, b_8–b_{15} specify the network layer protocol, and b_{16}–b_{31} specify the UDP length. The UDP protocol is useful in transmitting datagrams (Section 5.4) such as those for multicasting, registration request, registration reply, etc. A pseudo header enables the identification of source and destination IP addresses of the ports, for example, when a datagram is sent for registration request or registration reply. It enables identification of the protocol to be used to route the datagram. For example, 'protocol = 17' shows the use of the IP protocol by the network layer. It also enables specification of the length up to which the UDP header and data extends so that the remaining part of datagram may be used for conveying other information.

6.1.2 Conventional Transport Control Protocol

Application data is first encoded using the application layer protocol header words by prefixing them over the data, following which the encoded data from the application layer is encoded again using the transport layer protocol header words by prefixing them over the previously encoded data. At the receiver end, the reverse process of decoding at each layer to retrieve back the application data takes place. The data is transported from the transport layer to next layer (L_3) using TCP (or UDP in the case of datagram). TCP is a connection-oriented protocol. TCP is a transport layer protocol for the Internet. The main features of TCP are as follows:

- *Transmission as data streams:* A segment is transmitted as a data stream of bytes, also known as packets due to the use of the IP protocol during transmission by the next layer. (It may be recalled that UDP transmits just one datagram of length $\leq 2^{16}$ bytes at an instant.) Transmission as streams indicates that octets are transmitted sequentially and are received by the transport layer at the other end in the same sequence as they are sent. An octet refers to a sequence of eight bits. A sequence number is first assigned to each byte before a TCP connection transfers the data. A set of 32 bits of a word (four bytes) has four sequences. The application layer data is transmitted and received as a stream consisting of sequences. The TCP header includes the sequence number of the first byte (not of each byte) to be transmitted through the stream.
- *Buffering and retransmission:* The TCP transmitter buffers the segment(s). The receiver acknowledges a data sequence. The transmitter empties the acknowledged bytes from the buffer. It retransmits from the byte next to the data sequence number of the last successfully transmitted byte. It also retransmits in case no acknowledgement is received within a timeout period.
- *Session-start, data transfer, and session-finish fully acknowledged end-to-end:* Handshaking of packets is used for acknowledgement and all transactions (connection start, establishment, data streaming, and finish) are acknowledged.
- *In-order delivery:* The TCP transport layer delivers the segments in a sequential order. When the nth segment has been delivered, only then is the $(n + 1)^{\text{th}}$ delivered to the application layer and not before that.
- *Congestion control and avoidance:* Lack of acknowledgement within the timeout period in a step or receipt of DACK (duplicate acknowledgement) signals congestion. (Refer to Section 6.1.4 for the meaning of duplicate acknowledgement.) One method of controlling congestion is to gradually increase, exponentially, the number of bytes in the stream in successive steps. The number of transmitted bytes in a step is reduced to half on encountering congestion. This reduces congestion (Section 6.1.5). Various congestion avoidance methods are also commonly used. A congestion control method for TCP over wireless is ECN (explicit congestion notification). Section 6.5.5 will explain this.

The following subsections discuss the fields at the TCP header and data streaming in TCP. Sections 6.1.3 and 6.1.4 explain TCP data delivery (transport) and data flow control methods, respectively. Section 6.1.5 explains congestion control methods. Section 6.1.6 gives the methods for wireless and mobile networks.

6.1.2.1 TCP Header

The TCP header is used at the subsequent layers during transmission of data from ports (application layer data) (after encoding as per application layer protocol) up to the transport layer at the receiver. The 32-bit words and the fields in the words of a TCP header are as follows:

1. *First word:* The function of the transport layer is to transport the port data. Therefore, the first 16 bits (b_0–b_{15}) are for the source port number and the next 16 bits (b_{16}–b_{31}) are for the destination port number.
2. *Second word:* This 32-bit field defines a 32-bit sequence number. The sequence number is for the first byte (octet) in a segment. The segment transmits by using the data streams from the application layer. The segment has the application data octets (after encoding at the application layer). The sequence number field takes into account the TCP data for synchronization request sequence. The sequence number field is reset to zero in case there are no more octets left to transmit from the segment. Using the sequence numbers (data byte number sent in the sequence), the packets, which reach non-sequentially, are reassembled and the receiver sets the acknowledgement number field as per the sequences successfully received by it.
3. *Third word:* It gives a 32-bit acknowledgement number (interpreted when the *A*-flag is set). The acknowledgement number is the value of the next sequence number (byte number), which the sender of the segment is expecting to receive from the receiver in case the sequences of bytes has been received successfully. This helps the receiver in knowing how much of the segment data was successfully transported. It also helps the transmitter in knowing the sequence number from which the data is to be sent to the receiver. When the connection is established, both the transmitter and the receiver use the field to inform each other about the upcoming sequences.
4. *Fourth word:* First four bits (b_0–b_3) are for the data offset and the next six bits (b_4–b_9) are reserved. The reserved bits facilitate the addition of more flag bits or provide a provision for extension of the present functions and features of TCP in any subsequent changes in the recommendations for TCP. The next six bits (b_{10}–b_{15}) are flags (*F*, *S*, *R*, *P*, *A*, and *U*) and the next 16 bits (b_{16}–b_{31}) are for setting window size, which is used for congestion control during transport. [The offset field specifies the word from where the application layer data octets will begin (after the header, options, and padding). The window-size field specifies the number of bytes the sender is willing to receive, starting from the acknowledgement field value.]
5. *Fifth word:* The first 16 bits (b_0–b_{15}) are for checksum of the header and the data and the next 16 bits (b_{16}–b_{31}) are for the urgent field. [Urgent field

communicates an offset value to be added to get a sequence number from the sequence number of the present segment. The value then points to the sequence number of the octet (in application layer data) following the urgent data. This helps in data flow control by specifying the urgent part of the data octets for transporting to the other end. The field is interpreted when the *U*-flag has been set during segment transport.]

6. Sixth and subsequent header words: These are used for options (each of eight bits) and padding.

Flags are control bits and are used as follows—*U* (URG) is set means urgent pointer field is being used for control of data flow. *A* (ACK) is set means acknowledgement number word field has significance and can be used to control flow and congestion. *P* (PSH) is set means push the data. *R* (RST) is set means reset the connection (sequence from the beginning). *S* (SYN) is set means synchronize the sequence number. *F* (FIN) is set means data is finished (no more data for sending at present).

6.1.2.2 TCP Data Streams

TCP provides a delivery mechanism for data streams. Each stream consists of bytes. Data streams are delivered using a virtual connection between sockets. Each socket has a port number and an IP address for rendering service to an application. The number of bytes in a stream is equal to the transport protocol data units (TPDU) that are placed in a memory buffer and transmitted by the TCP protocol. A TCP segment defines the TPDUs in a TCP data stream. TPDU depends on MTU (maximum transfer unit) presentable in a given network state. The protocol employs timers to synchronize the data stream between both ends.

RTT (round trip time) is the time interval between the start of the transmission of a data stream of the given TPDUs to the receiver layer and the receipt of acknowledgement of successful transmission. RTT is important in order to set the time interval during which the transmitter expects the receiver to send a TCP stream with the acknowledgement field set to indicate the successfully sent bytes. This is carried out so that the transmitter can retransmit from the acknowledge field sequence number onwards.

6.1.3 TCP Data Delivery

In TCP there is acknowledgement of sequences sent from one end to the other. (The acknowledgement field specifies the sequenced byte number from which the receiver end expects the sender end to transmit in the next sequence.) The bytes that do not reach the receiver successfully within a timeout period are retransmitted.

The checksum field detects errors. The field takes into account the header as well as the data. The receiver discards any erroneous packets in a data stream. Let us assume that after sending an acknowledgement, a receiver gets the packets in the stream which were delayed and acknowledges these too, but in the interim, the transmitter started retransmitting as per the previous acknowledgement and thus it retransmitted the delayed packets. In such cases the receiver discards the duplicate packets. Retransmission in case of no acknowledgement ensures reliable, guaranteed, and error-free data transfer. (Compare this with UDP which transmits with no buffering, no acknowledgement, and no retransmission.) *No retransmission* and *retransmission* mechanisms are similar to the transparent and non-transparent data transmission (Sections 3.1.1.5 and 3.1.1.6), respectively, in the radio layer of a GSM system. The steps for delivering the data are as follows:

Step 1: Connection establishment during four phases—LISTEN, SYN_SENT, SYN_RECEIVED, and ESTABLISHED. Three data streams are transferred in the last three phases. (The third one is with the data octets.)

Step 2: Data streams of application layer octets are transferred after establishment of the connection and before termination of the connection on completion of the transmission data octets.

Step 3: Connection *Finish* and then *Close*. During the flow of data, a TCP connection between the application stream transmitter (TCP_A) at sender end, *A*, and receiver (TCP_B) at end, *B*, may be in any of the following states:

(a) LISTEN: TCP_B receiver is waiting for listening to a connection request from the transmitter (TCP_A) transport layer.

(b) SYN_SENT: A SYN data stream sent by the transmitter TCP_A with the *S* (SYN) flag set to one and waiting for the receiver (TCP_B) data stream with SYN and ACK.

(c) SYN-RECEIVED: A SYN_ACK_B sent by receiver data stream TCP_B for TCP_A with the *S* (SYN) and *A* (ACK) flags set to one. TCP_B uses the acknowledgement field at the packet and sends an acknowledgement to TCP_A. For example, if SYN sequence number sent in step (b) is n, then TCP_B transmits an acknowledgement with the sequence number in the acknowledgement field set as $n + 1$ and the *A*-flag set to one in step (c). TCP_A expects the sequence number request, $n + 1$, from TCP_B. TCP_B sequence number field has sequence number m in it when TCP_A sends a sequence m to TCP_B.

(d) ESTABLISHED: An ESTABLISHED state is set at the transmitter data stream TCP_A for TCP_B. TCP_B sets the acknowledgement field (for TCP_A) accordingly in the next data stream and the *A*-flag is set to one and the *S*-flag is reset to zero in order to enable the port to send and receive the data stream from that point onwards. When the receiver receives the data stream from TCP_A, an ESTABLISHED state is set at the receiver end (TCP_B) for TCP_A. TCP_B sends and receives port data streams from that point onwards. For example, if acknowledgement field sequence number in TCP_B packet in step (c) is $n + 1$, then TCP_A transmits the acknowledgement with $n + 1$ in the sequence number field and the *A*-flag set to one in step (d). The acknowledgement number field in the TCP_A stream now is $m + 1$.

(e) DATA _TRANSFER: A TCP_A data stream always has the sequence number field for the byte starting from which the data of a segment is being transferred through the stream, and the acknowledgement number field for the byte starting from which the data of a segment is expected by TCP_B. Similarly, a TCP_B data stream always has the sequence number field for the byte starting from which the data of a segment is being transferred through the stream, and the acknowledgement number field for the byte starting from which the data of a segment is expected in a TCP_A data stream. TCP_B acknowledges a data stream by setting the acknowledgement field value to the sequence number of the last byte received successfully (from other end) plus one. This is so because TCP_A is expected to send the new data stream starting from this (last successfully received byte plus one) position. For example, a segment of 4020 bytes with sequence numbers 1025–5044 is to be transmitted. If the TCP_A layer begins transmitting 4020 bytes with a sequence number field at 1025 and all the bytes are transmitted without loss and without error, then the receiver sends a TCP_B stream with an acknowledgement number field at 5045. This is because the bytes of the next segment to be transmitted start from sequence number 5045. Now, if only the bytes up to sequence number 2047 are received successfully by the TCP_B layer in a specified time interval (i.e., RTT), then the receiver will send the TCP_B stream with an acknowledgement number field at 2048, as the receiver expects the transmitter to send bytes starting from sequence number 2048. Acknowledgements for data sent to the receiver, or the lack of acknowledgements from the receiver within a timeout period, are used by the sender to judge the present network conditions between the TCP sender and receiver, i.e., whether the network is congested or not and the data flow is controlled accordingly.

(f) FIN_WAIT_1: TCP_A sends a data stream with the *F*-flag (FIN) set to one and expects a FIN_ACK from TCP_B within a wait period. TIME_WAIT state is used to wait for FIN_ACK (acknowledgement of FIN) from the other end. After that the state changes to FIN_WAIT_2, else if the FIN_ACK is received from TCP_B, then the receiver TCP_B state changes to CLOSING.

(g) FIN_WAIT_2: TCP_A sends a data stream with the *F*-flag (FIN) set to one and expects a FIN_ACK from TCP_B during a wait period. TIME_WAIT state is used to wait for FIN_ACK (acknowledgement of FIN) from the other end and after that if no FIN_ACK is received, then the state changes (after timeout) to CLOSED. If FIN_ACK is received, then the TCP_B state changes to CLOSING.

(h) CLOSE_WAIT: TCP_A goes to CLOSE_WAIT state on receipt of the FIN_ACK in either step (f) or step (g), LAST_ACK is sent after CLOSE_WAIT to finally close the connection.

(i) CLOSING: TCP_B sends a CLOSING ACK to TCP_A, waits for timeout, and goes to the CLOSED state.

(j) CLOSED: A state in which there is no data transfer from TCP_A and TCP_B. [A wait period of maximum four minutes is recommended for a set time during which the acknowledgement for the connection finish request (FIN_ACK) is expected.]

6.1.4 TCP Data Flow Control

The octets in a segment transmit as data streams, which are transmitted in packets. A few packets of a data stream may reach the other end with an error, may be lost, or may not reach at the expected time. They need retransmission starting from the octet succeeding the last successfully received one. A controlled data flow prevents the need of large number of retransmissions. The important data flow control methods are listed below:

- *Window size adjustment (TCP tuning):* Window size is adjusted and throughput depends on RTT (round trip time) interval for the acknowledgement. The transmitter transmits all bytes received up to a sequence number specified by w at the window-size field of the other end, this indicates that it transmits from i to $i + w$ in case the next sequence number (in the acknowledgement field from receiver) to be sent from the transmitter is i. Timeout at the transmitter is set equal to the RTT for acknowledgement. If there is no acknowledgement in the timeout period, then the bytes sent using the window are considered as lost and are retransmitted. (This method known as *adjusting the window-size field method* is described in detail later in this section.)
- *Cumulative acknowledgement:* The receiver acknowledges all the bytes received up to a sequence number defined in the acknowledgement field. Cumulative acknowledgement from the TCP_B end can be called partial acknowledgement (PACK) if the acknowledgement field shows a lesser value than expected after transmission of all the bytes from the TCP_A end. PACK thus shows lost packets or data during transmission to the TCP_B end.
- *Reverse packet acknowledgement:* Reverse packet from the receiver piggybacks the acknowledgement.
- *Duplicate acknowledgement (DACK):* One of the packets of a segment may reach after a delay in comparison to the packet succeeding it. Acknowledgement from the receiver is duplicated by retransmission of an earlier acknowledgement without delay. DACK thus shows recovery of lost packets or data received after a delay at the TCP_B end.

- Delayed acknowledgement: Acknowledgement from the receiver is delayed if the receiver responds by including, after the TCP header, a large number of octets of the receiver-end segment. This is due to packetization time at the receiver network layer and the packets tracking different paths and hops to the transmitter. Acknowledgement is also delayed in case the receiver has started receiving another transmitted data stream from the next sequence number at the transmitter. This is because the receiver is now assigned a dual role at the same time. Delay at the transmitter can be set at 100–250 ms in order to reduce the number of delayed acknowledgements.

Adjusting the Window-size Field Method One method of data flow control is to adjust the window-size field. In this method, the window-size field is varied for congestion control during data flow (Section 6.1.5). Window size, w, specifies the number of bytes the receiver (TCP_A or TCP_B) is ready to receive from the sender (TCP_B or TCP_A, respectively). This method of specifying the window size is called TCP tuning. The bytes (from the other end) start from the acknowledgement field value, i, and are transmitted up to the $i + w$ value of the sequence number in a TCP segment. The window-size field is a 16-bit field and thus can be set to 2^{16} bytes.

Example 6.1 Assume that a segment of 4020 bytes has the sequence numbers of 1025–5044. Assume that w is set to 1000 B in the data stream received from other end, because the receiver data link layer has MTU (maximum transferable units = 1000 B). How does the TCP_A layer transmit?

Solution: The TCP_A layer transmits only 1000 bytes with a sequence number field = 1025 and bytes from sequence number 1025–2024, even though it has 4020 B in the buffer.

Example 6.2 Assume that w is set to 1000 B, the RTT delay in acknowledgement is 15,625 μs in Example 6.1. Calculate the throughput.

Solution: The throughput (data transmission rate finally achieved from end to end)

$$= \frac{1000 \text{ B}}{0.015625}$$
$$= 64{,}000 \text{ B/s}$$
$$= 512 \text{ kbps}$$

Example 6.3 Ideal window size, w_{ideal}, with which the TCP data stream should be sent depends on the link capacity, which is the bandwidth available at the given instance between the TCP_A and TCP_B layers. Assuming that the link capacity is 1024 kbps, how will you calculate w_{ideal}?

Solution: w_{ideal} = link capacity × RTT (i.e., the product of bandwidth and delay)
= 1024 kbps × 0.015625 s
= 2000 B

Window Scaling Method During high-speed data transfer, the window scaling method is used. The TCP tuning method scales up the widow size to 2^{30} bytes in case of high-speed networks. Number of bits, which left shifts the bits w at the window-size field, is set during a three-way handshake session (such as steps (b), (c), and (d) at Step 3 in Section 6.1.3). A left shift of 1 multiplies w by 2, a left shift of 2 multiplies w by 4, and so on. Number of shifts can be between zero and 14 and the window size is 2^{30+s}, where s is the number of shifts. Window scaling factor, s_w is represented by 2^s.

Example 6.4 Assume that a high-speed data transfer segment of 10,004,020 B has the sequence numbers from 1025–100,005,044. Assume that the window-size field is specified by the other end as 10,000 B. Since, the transmitter supports high-speed data transfer, the window is scaled up by a shifting number of 8. How is the window scaled? How does the TCP_A transmit?

Solution: The window scales up to $2^8 \times 10000$ B = 2,560,000 B.
The TCP_A layer transmits 2,560,000 B with a sequence number field of 1025. Bytes transmit from sequence numbers 1025–2,561,024.

Sliding Window Method Another method of data flow control is the sliding window method. In this method, a window is specified from a sequence number to another sequence number. S_{A0} and S_{A1} are the two sequence numbers which define the lower and upper boundaries of the window. $S_{A0} - S_{A1} = S_W$ (sliding window size). TCP_A can send only S_W bytes, then it must wait for an acknowledgement and a window update from the receiving end, TCP_B. As an example, let transmitter, TCP_A, transmit as per its present sliding window setting. The setting is such that the TCP_A data stream is transmitted between S_{A0} and S_{A1} before an acknowledgement is expected from the receiving end. TCP_A sets the sliding window to the next set of sequence numbers, S_{A1}–S_{A2}. When the receiver's acknowledgement is not received in a specified interval, then the window slides back to the original and the sequences between S_{A0} and S_{A1} are retransmitted. When the receiver's acknowledgement is received in the specified interval and it equals S'_{A0}, then the window slides back and sequences between S'_{A0} and $S'_{A0} + S_W$ is retransmitted. Sliding window size, S_W, also defines an acknowledgement delay period. Let us assume that one sequence is transmitted and acknowledged in time, RTT_0, then the next acknowledgement is expected after $RTT_0 \times S_W$, when the sliding window size is S_W. S_W is specified in the TCP_A data stream at the window-size field. The receiver TCP_B receives the stream and sends the acknowledgement number field after setting the acknowledging time interval as per the S_W value received from TCP_A.

6.1.5 Congestion Control

A TCP data stream receiver sets a window. The data from TCP_A (transmitter) is to be sent from sequence number S_0 and S_4. Let us assume that the data from a transmitter is received and sent to an application or a service access point (SAP) up to sequence number, S_0. Assume that at an instant, the situation is as follows—transmitted TCP_A octets are received and acknowledged by TCP_B receiver up to the sequence number, S_1. However, data is still being sent to the receiver SAP (application layer). TCP_A octets from sequence number $S_1 + 1$ transmit. Let us now assume that at the next instant, the situation is as follows—the data from TCP_A is received at TCP_B up to sequence number S_2 but it has not yet been acknowledged and it is yet to be sent to SAP. TCP_A octets from sequence number $S_1 + 1$ retransmit after the timeout. The data from the transmitter can be received up to a maximum of sequence number, S_3, when the window of the receiver extends from S_1 to S_3 and the window-size field bits in the TCP_A data stream header equal $(S_3 - S_1) \div s_w$, and s_w is one. Congestion network window size, cnwd is $S_3 - S_1$, and the window is for $(S_3 - S_1)$ octets and when s_w is two, the window is for $[(S_3 - S_1) \div 2]$ octets. The window scaling factor is set at the transmitter. By adjusting s_w, the congestion can be controlled. The following are methods that are usually employed for congestion control:

- Slow start and congestion avoidance
- Fast recovery (in place of slow start) after packet loss
- Fast retransmit and fast recovery (Section 6.5.1)
- Selective acknowledgement (Section 6.5.3)
- Explicit congestion notification (Section 6.5.5)

Of these, the first two methods are described in the following subsections.

6.1.5.1 Slow Start Method

The slow start method entails starting from a very small window and increasing the window size exponentially up to a threshold value and then linearly till congestion sets in. Once congestion sets in, the window is slow started again with the new threshold value set to one-half of the window size at congestion. Figure 6.1 shows the window sizes on successive round trips in three phases of the slow start method. The three phases are described below.

The *first phase* is known as exponential growth phase or slow start phase. The congestion network window size (cnwd) is equal to one at the start of a new data stream. After each RTT (after which the acknowledgement field is used to send the next data stream), for each subsequent stream, the window is doubled till a window threshold is reached. Now, cnwd = cnwd × 2 when 1 × cnwd ≤ cnwd_{th}, where cnwd_{th} is the threshold window after which the linear growth starts. This threshold value is defined to control further exponential growth in order to avoid congestion. As an example, let us assume that the exponential phase consists of up to the tenth

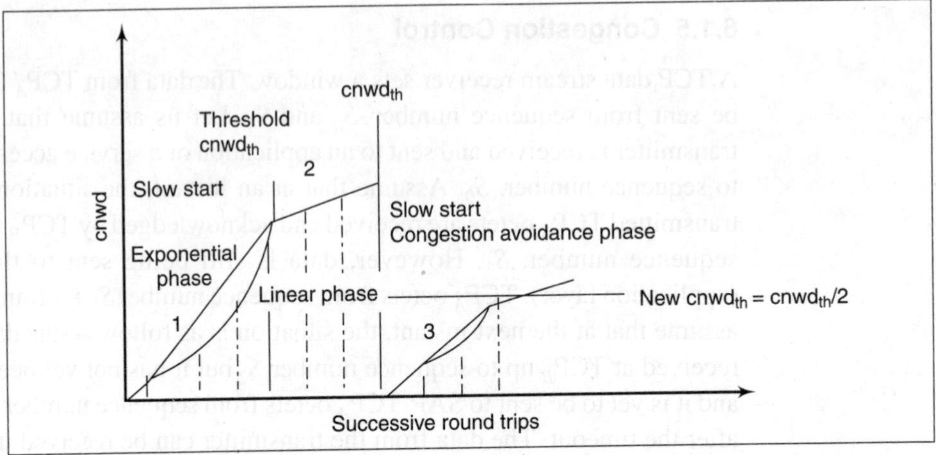

Fig. 6.1 Window sizes on successive round trips in exponential, linear, and congestion avoidance phases of the slow start method

round trip. Therefore, threshold window, $cnwd_{th} = 2^9$, and the number of segments transmitted is $1 + 2 + 4 + \ldots + 256 + 512 = 1023$.

The *second phase* is called linear growth phase. It is also called congestion avoidance phase. The congestion window size for new data stream, cnwd = 1 at the start. In subsequent streams, cnwd = cnwd + 1 between $cnwd_{th}$ and $cnwd_{cntrl}$, where $cnwd_{cntrl}$ is the window size when the data stream acknowledgement is missing in a timeout period and after which the congestion phase begins. As an example, if the exponential phase consists of up to tenth round trip and then the linear phase starts and congestion is signalled in the sixteenth round trip as there was no acknowledgement in the twelfth RTT period, then the window size in the eleventh round trip is 513 and that in the twelfth is 514. The number of segments transmitted up to the eleventh round trip is $1 + 2 + 4 + \ldots + 256 + 512 + 513 = 1536$. Timeout set in the twelfth trip is 514. Timeout set in the sixteenth trip is 518. RTT is $518 \times RTT_0$, where RTT_0 is the time when cnwd is 1.

Congestion avoidance phase is a phase that starts after linear growth causes congestion and this congestion needs to be controlled by two actions—resetting cnwd to cnwd ÷ cnwd, i.e., one and reducing the threshold window size to one-half of its present value ($cnwd_{cntrl}$) after the timeout period of the last trip indicating congestion. Now, the new $cnwd_{th} = (cnwd_{cntrl} \div 2)$ or 2, whichever is greater. In the congestion avoidance phase, the exponential phase (slow start phase) step starts again and when cnwd reaches $cnwd_{th}$, the linear phase starts. Let us take another look at the above example, cnwd at the start of congestion phase in the seventeenth RTT is set to 518/2. Therefore, new $cnwd_{th} = 259$. In the eighteenth RTT, cnwd = 1, in the nineteenth, cnwd = 2, and so on till cnwd becomes 256 in 26th RTT and $cnwd_{th} = 259$ in the 27th RTT. During the slow start phase, $1 \leq cnwd \leq$ new $cnwd_{th}$. During the linear phase cnwd will be incremented by 1 after each RTT, starting from the 28th RTT.

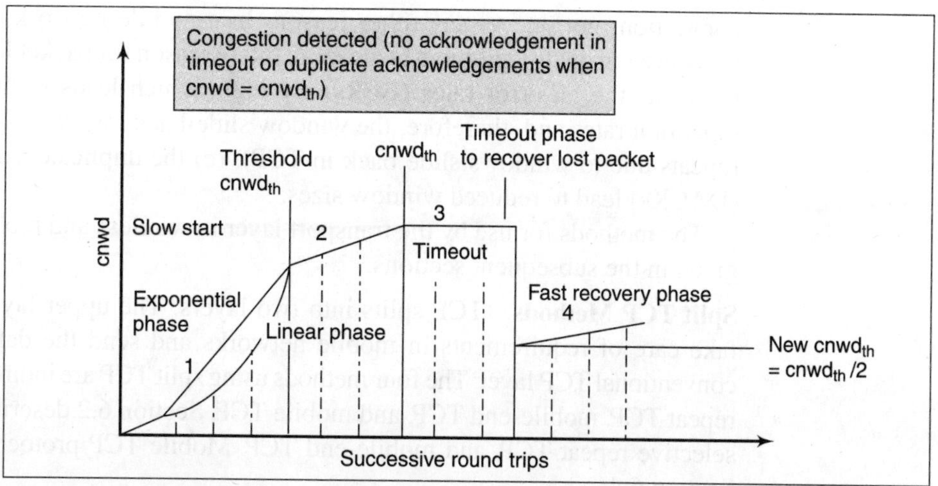

Fig. 6.2 Modification of slow start by fast recovery congestion avoidance phase after a timeout

6.1.5.2 Fast Recovery

Slow start congestion avoidance phase in Fig. 6.1 starts from cnwd at one. The congestion network window size (cnwd) is equal to one at the start of a new data stream. After each RTT (after which the acknowledgement field is used to send the next data stream), for each subsequent stream, the window is doubled till a window threshold is reached. Now, cnwd equals cnwd × 2 at $1 \leq \text{cnwd} \leq \text{cnwd}_{th}$, where cnwd_{th} is the threshold window after which the linear growth starts. Figure 6.2 shows first phase in region 1 and is called exponential growth phase. It is also called slow start phase. Figure 6.2 shows second phase in region 2 and the second phase is called linear growth phase. The congestion window size for a new data stream, $\text{cnwd} = \text{cnwd}_{th}$ at the start. In subsequent streams, cnwd = cnwd + 1 between cnwd_{th} and cnwd_{ca}, where cnwd_{ca} is the window size when the region 3 starts, which is known as congestion avoidance phase. Figure 6.2 shows the third phase in region 3. Figure 6.2 shows that in region 3, a modification of slow start by congestion avoidance (region 3) for a period equalling timeout, $\text{cnwd} = \text{cnwd}_{ca}$. Figure 6.2 shows that after timeout, there is fast recovery shown by region 4. The fast recovery (region 4) sets in after the timeout in the congestion avoidance phase. The new $\text{cnwd}_{cntrl} = \text{new cnwd}_{ca} \div 2$ in place of cnwd = 1 in case of Fig. 6.1. In other words, the slow start phase is not followed for congestion avoidance and the window returns to new cnwd_{cntrl} at new $\text{cnwd}_{ca}/2$ after the RTT (timeout) in which the congestion avoidance phase starts.

6.1.6 Methods for Wireless and Mobile Networks

The problems faced while employing the conventional TCP on a mobile network are as follows—(a) the slow start method presumes that a packet is lost due to congestion, not due to any other reason. In a mobile network, the data-linking transmission quality problem is the more likely reason for packet loss, (b) in mobile networks the bit error rates (BERs) are high, which leads to high transmission repetition rates and, therefore, the window slide-back rates are high (transmission repeats due to windows slide back in TCP), (c) the duplicate acknowledgements (DACKs) lead to reduced window sizes.

The methods for use by the transport layer in wireless and mobile networks are given in the subsequent sections.

Split TCP Methods TCP splits into two layers. The upper layer is designed to take care of requirements in mobile networks and send the data streams to the conventional TCP layer. The four methods using split TCP are indirect TCP, selective repeat TCP, mobile-end TCP, and mobile TCP. Section 6.2 describes indirect TCP, selective repeat TCP, and mobile-end TCP. Mobile TCP protocol is described in Section 6.4.

TCP-aware Link-layer Methods The word 'snoop' means secretly looking into or examining something. The data link layer snoops into the TCP layer data. These methods are based on TCP-aware link layer. The three TCP-aware link layer protocols are snooping TCP, WTCP, and delayed duplicate acknowledgement protocol. Section 6.3 will describe these.

Link-layer (L2) Modification Methods Data link layer in the mobile nodes (MNs) uses the following methods (in place of methods like L_4-TCP window sliding method) for error control:

- FEC: Section 3.1.1.5 discussed how forward error correction (FEC) recovers the bit errors during wireless transmission in GSM systems. The number of user bits effectively transmitted per second (throughput) is, however, reduced due to the need to carry additional FEC bits in the same time-slot. When there are no errors, the additional FEC bits unnecessarily reduce the throughput. When using adaptive techniques, FEC code lengths and frame sizes can be varied depending on the bit error rate (BER).
- ARQ (automatic repeat request): ARQ was discussed in Section 3.9.2.1 in the context of GPRS systems. When errors are detected, a repeat request is generated. The throughput remains unaffected in case there are no errors. However, in case of errors, the throughput, round trip time (RTT), and congestion in the network are affected due to repeated retransmission of data streams. When using adaptive techniques, retransmission can be limited and can be varied depending on the bit error rate (BER). As an example, in voice communication, a certain BER is tolerable. Hence, retransmission can be skipped.

For TCP layer transmission, in certain protocols, the link layer is modified to provide the additional features so that the TCP layer is not modified. An example of this is the timeout freezing transmission method (Section 6.5.2).

Explicit Notification Methods Section 6.5.5 will describe these methods.

6.2 Indirect TCP

Indirect TCP suggests splitting of the TCP layer into two TCP sub-layers. Figure 6.3(a) shows the indirect TCP sub-layer between the base transceiver and the fixed node and conventional TCP between the fixed nodes.

One connection is between the MN and the base transceiver (BTS) and the other between the BTS and a FN. The BTS has an access point at an agent TCP_M for TCP connection. TCP_M sends and receives the packets to and from the MN through the BTS. Indirect TCP functions in the following manner:

1. TCP_M sends and receives the packets to and from the TCP_F layer at the fixed node. The transfer mechanism is simple as there is only one hop. Retransmission delay between TCP_M and TCP_F is very small, unlike that between the fixed nodes.
2. TCP_F layer at the fixed node sends and receives the packets to and from another fixed node TCP'_F. The transfer mechanism is standard using multiple hops through the routers.

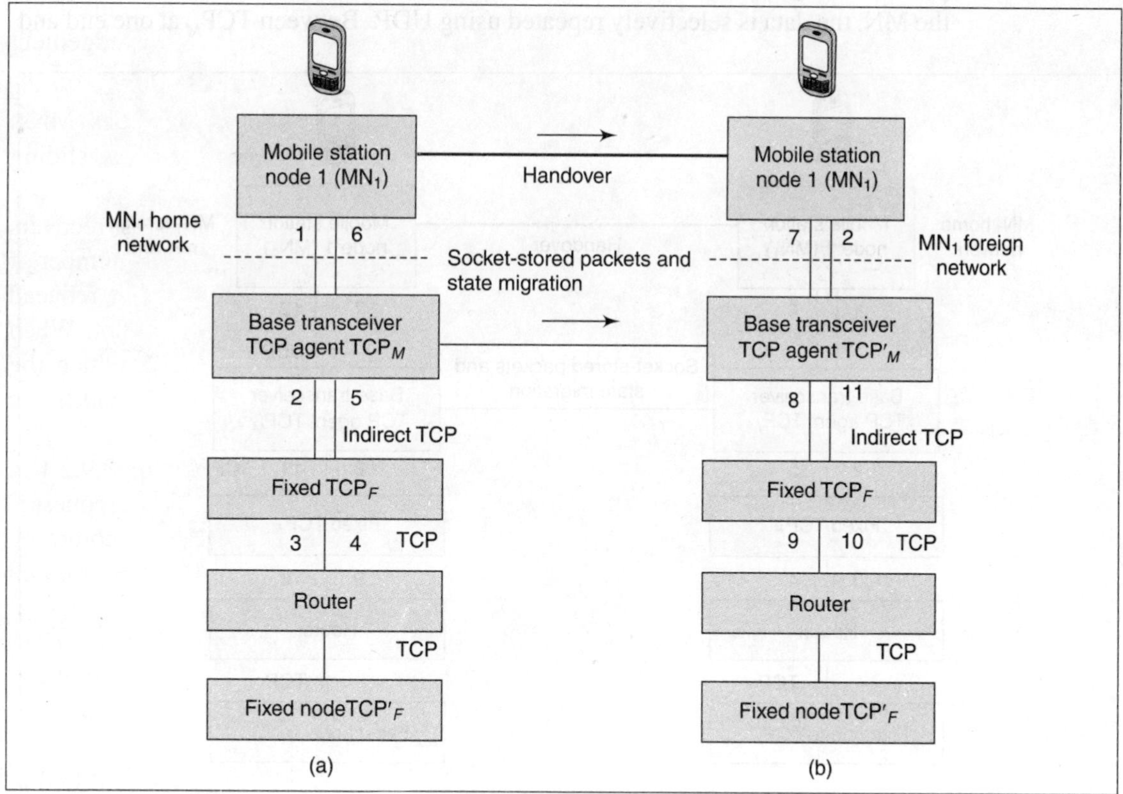

Fig. 6.3 (a) Indirect TCP sub-layer between the base transceiver and the fixed, node, and conventional TCP between the fixed nodes (b) Handover mechanism

The data streams are received from the service access point (application) at the MN and buffered at TCP_M. Figure 6.3(b) shows the handover mechanism. When there is handover when the MN visits a foreign network, the packets for transmission, buffered at TCP_M, are transferred to TCP'_M. On handover, the socket (port and IP address) and its present state migrate from TCP_M to TCP'_M. The transfer from TCP_M to TCP'_M has a latency period. (The states of a socket during a TCP connection are described in Section 6.1.3.) The advantage of indirect TCP is that the mobile part of the network is isolated from the conventional network and thus there is no change in the existing TCP network. The disadvantages of indirect TCP are the high latency period during handover of packets, possible loss of data at the base, and deviation from the end-to-end connection feature of conventional TCP, which guarantees reliable packet delivery. As an example, an acknowledgement to a sender may be lost during handover latency.

Selective Repeat Protocol A modification of the indirect TCP is the selective repeat protocol. Figure 6.4 shows a modification in indirect TCP in selective repeat protocol using UDP between the BTS and the MN. Between TCP_M at the BTS and the MN, the data is selectively repeated using UDP. Between TCP_M at one end and

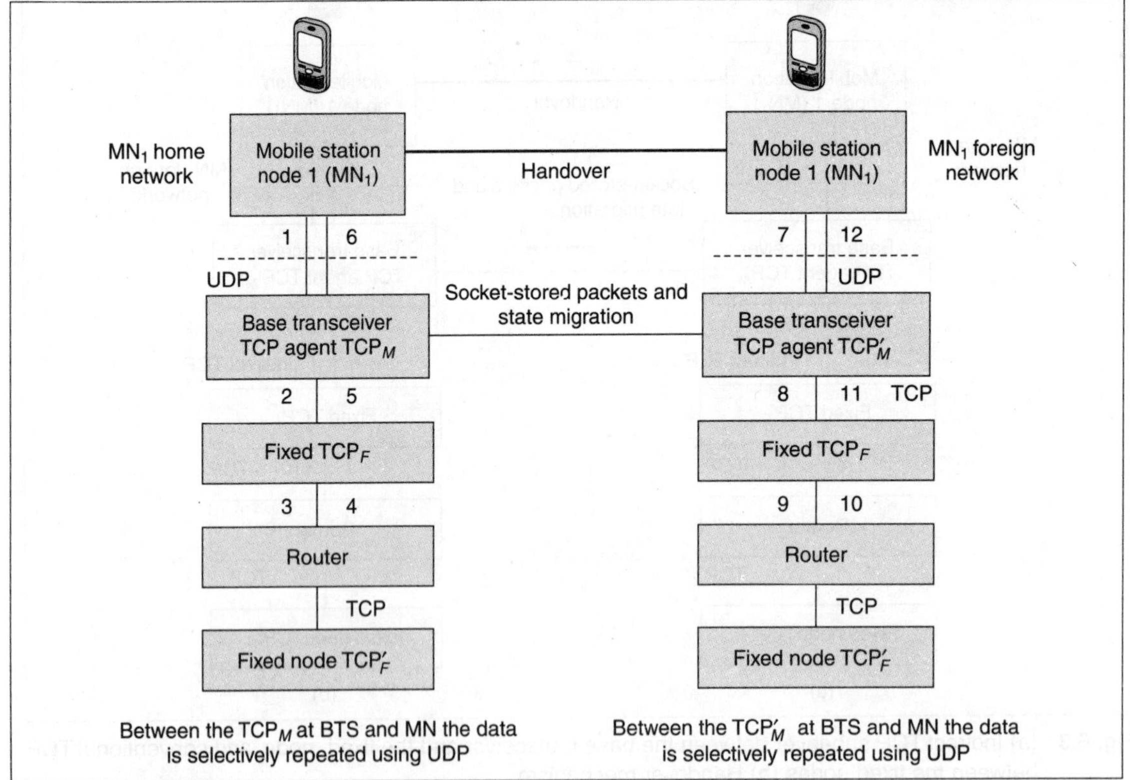

Fig. 6.4 Modification in indirect TCP in selective repeat protocol using UDP between the BTS and the MN

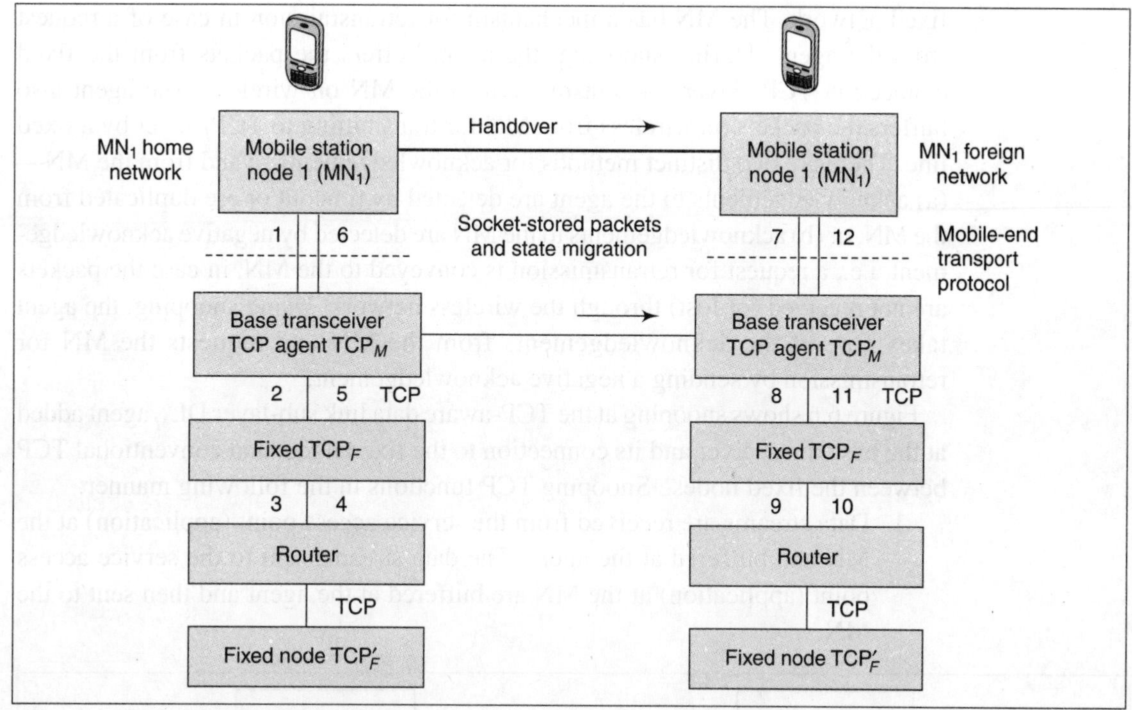

Fig. 6.5 Modification in indirect TCP by using mobile-end transport protocol

TCP'_F and TCP'_M at the other end, the data stream is transferred, as in case of conventional fixed-end TCP. UDP is a connection-less protocol, therefore, the selective repeat protocol does not guarantee the in-order delivery between the BTS and the MN, unlike TCP.

Mobile-end Transport Protocol Another modification of indirect TCP is the mobile-end transport protocol. Figure 6.5 shows the mobile-end transport protocol. Mobile-end transport protocol guarantees the in-order delivery between the BTS and the MN, like TCP. Between TCP_M at the BTS and the MN, the data is transferred by using the mobile-end transport protocol. Between TCP_M at one end and TCP_F and TCP'_M at the other end, the data stream is transferred, as in case of conventional fixed-end TCP.

6.3 Snooping TCP

The word 'snoop' means looking into or examining something secretly. A TCP connection splits into two. One connection is between the MN and the BTS and other between the BTS and the FN. Some changes are made in the BTS and some at the MN. The BTS has a TCP-aware data link sub-layer, DL_M. The sub-layer, DL_M, functions as an agent for snooping and buffering the TCP connection. The agent ensures the delivery of packets to the MN in their incoming sequence from the fixed network. The MN has a mechanism for retransmission in case of a request from the agent. During snooping, the agent buffers the packets from the fixed connection TCP_F layer for transmission to the MN on wireless. The agent also buffers the packets on wireless from MN for transmitting to TCP_F layer by a fixed line. There are two distinct methods for acknowledgements to and from the MN—(a) acknowledgements to the agent are detected by timeout or are duplicated from the MN, or (b) acknowledgements to the MN are detected by negative acknowledgement, i.e., a request for retransmission is conveyed to the MN, in case the packets are not received (or lost) through the wireless network. Using snooping, the agent takes note of the acknowledgements from the MN and requests the MN for retransmission by sending a negative acknowledgement.

Figure 6.6 shows snooping at the TCP-aware data link sub-layer DL_M agent added at the base transceiver and its connection to the fixed node, and conventional TCP between the fixed nodes. Snooping TCP functions in the following manner:

1. Data streams are received from the service access point (application) at the MN and buffered at the agent. The data streams sent to the service access point (application) at the MN are buffered at the agent and then sent to the MN.

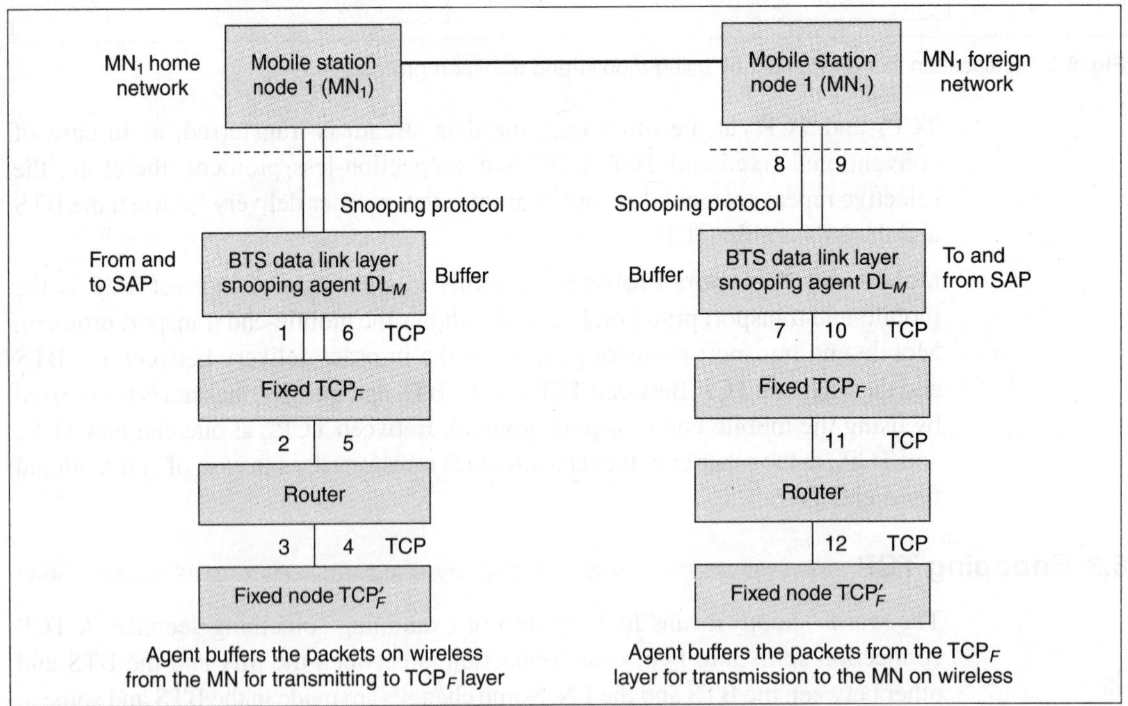

Fig. 6.6 Snooping at the TCP-aware data link sub-layer, DL_M (known as agent) added at the base transceiver and its connection to the fixed node (with conventional TCP between the fixed nodes)

2. The snooping agent DL_M at the BTS also buffers the data and sends and receives the packets to and from the TCP_F layer at the fixed node. The connection between the agent and TCP_F is transparent. DL_M layer agent at the BTS sends and receives the packets to and from TCP_F and then from another fixed node TCP'_F. The transfer mechanism is standard—using multiple hops through the routers.
3. DL_M agent at the BTS snoops into the packets (data stream) when sending and receiving to and from the MN.
4. The agent and the MN control the acknowledgements, lack of acknowledgements, and DACKs of the received data stream from the MN and to the agent, respectively, as described in the following steps.
5. Packet loss is discovered at the agent from the timeout or DACKs from the MN. The DL_M retransmits, if needed, to the MN in case of lost packets that are not acknowledged by the MN in the timeout period or if there are DACKs. Packet loss is discovered at the MN from the retransmission request from the agent to the MN. The MN retransmits to DL_M in case of lost packets for which a negative acknowledgement is sent by DL_M.
6. The buffer empties to the fixed network through TCP_F on receipt of an acknowledgement at the agent from the MN. The buffer empties to the MN on receipt of acknowledgement at the agent from the TCP_F.
7. The data stream transfer mechanism between the MN and DL_M is simple, such as in wireless networks. There is only one hop. In retransmission from DL_M to MN, the delay is very small, unlike that between the fixed nodes.

When there is handover of the MN while visiting a foreign network, the packets for transmission are snooped into at DL'_M at the other end. On handover, neither the socket (port and IP address) nor its present state migrates from DL_M to DL'_M. The advantage of snooping TCP is that the TCP connection is transparent end-to-end (without any transport layer changes), and the mobile part of the network between the base and the mobile node has very limited isolation and is completely isolated from the base and the conventional fixed TCP connection. There is no change in the existing TCP network, only a snooping sub-layer is added at the base. The disadvantages of snooping TCP are the security risks involved in snooping, difficulties in case of encrypted segment transmission, and insufficient isolation between the fixed-node transport layer and the snooping layer in case of an asymmetric path. (An asymmetric path is a fixed node connecting to the mobile node at the other end. The node at the other end does not have a snooping layer for retransmission and acknowledgement.)

Wireless TCP A modification of snooping TCP is wireless TCP (WTCP). WTCP modifies the time stamp on the packets while returning acknowledgements to compensate for the increased RTT. The advantage of WTCP is that the timeout periods compensate for the increased RTT between the agent at the base and the mobile node and is useful when retransmission intervals are greater than the timeout period. The disadvantage of WTCP is that it cannot be used on shared LANs.

Delayed-Duplicate-Acknowledgement Protocol for Delayed-Congestion Response Delayed-duplicate-acknowledgement protocol for delayed-congestion response (DCR) is a modification of snooping TCP. It is also called TCP-DCR. Figure 6.7 shows delayed-DACK protocol, a modification of snooping TCP. The DL_M agent is not TCP aware. The difference between TCP aware and unaware can be explained as follows—retransmission takes place in TCP-aware DL_M, whenever there is timeout or DAs (delayed duplicate acknowledgement), as both of these indicate loss of packets. When a packet reaches, the receiver out of order (TCP connection, for example, the TCP layer in the mobile node) sends the DA (Section 6.1.4). Further, the window is adjusted to a lower level for each instance of detection of packet loss, whether due to timeout or DA.

Packet losses are due to two reasons—congestion and channel transmission errors. Channel errors are due to interference in the medium. In TCP-unaware DL_M, the retransmission from the DL_M takes place when there is a timeout or acknowledgement

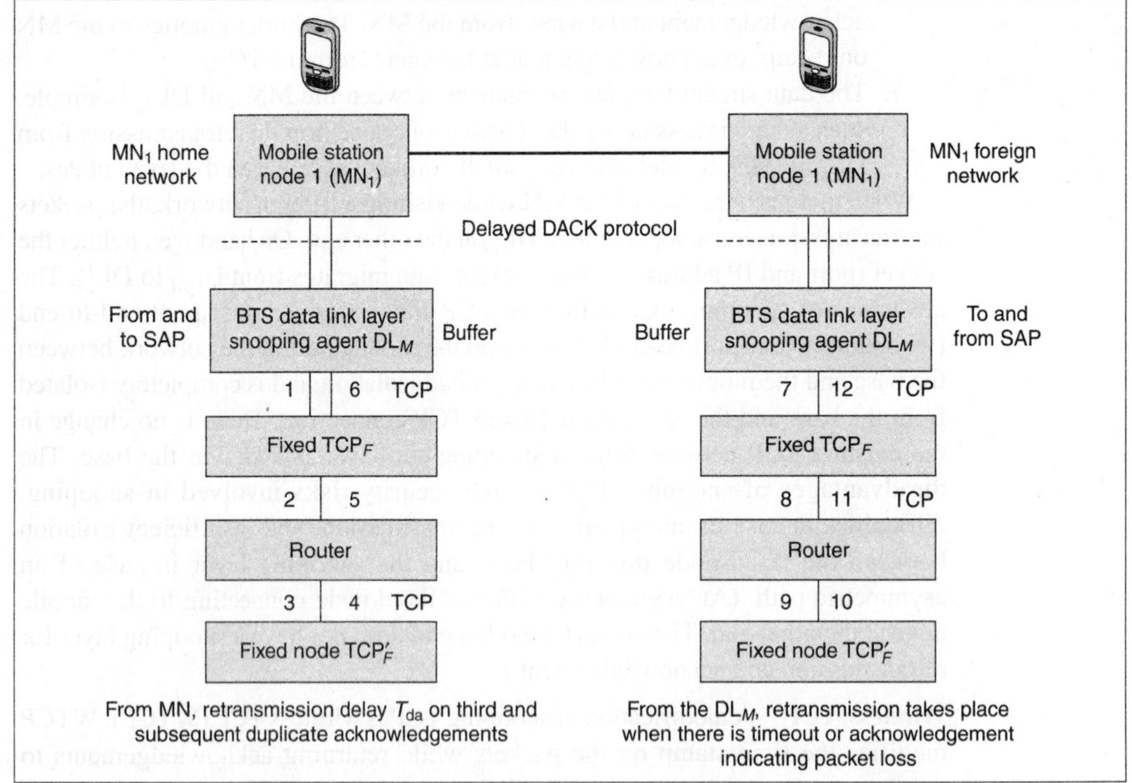

Fig. 6.7 Delayed duplicate acknowledgements protocol, a modification in snooping TCP

from the MN, but the MN delays the DAs so that the channel can recover from the channel errors. The MN attempts to reduce interference between data stream bits of the MN and retransmitted bits from DL_M in the following way—it delays the third and subsequent DACKs by a time interval, T_{da}. When the first consecutive out of order packets (COP) reach, the MN responds by a DA. When further consecutive out of order packets reach, each DA is delayed by a period, T_{da}. The MN no longer needs to send delayed DAs if the next packet is in sequence within T_{da}, i.e., the channel has recovered from channel errors. During T_{da}, the DL_M–MN channel recovers from channel errors. The MN may permit out-of-order packet delivery from the agent and it may also look into the TCP header and reassemble the packets itself in order of sequence. The data stream transfers between DL_M at one end and TCP_F and TCP'_F at the other, as in the case of conventional fixed-end TCP

The delayed duplicate acknowledgement protocol is advantageous since the TCP headers can be encrypted, as the agent is not TCP aware. The disadvantage of this protocol is that since the duplicate acknowledgements are delayed, the retransmission of the packets lost due to congestion is also delayed.

6.4 Mobile TCP

Mobile TCP (M-TCP) suggests the splitting of the TCP layer into two TCP sub-layers and a mechanism to reduce the window size to zero. The TCP split is asymmetric. The window is set to zero to prevent the transmission from the TCP transport layer at the MN or the fixed node, when disconnection is noticed. Disconnection is detected when the split connection does not get packets within a timeout interval. The window opens again on getting the packet. The M-TCP host at the base does not use slow start as it presumes that the packet loss is due to disconnection and not due to congestion or interference. Data flow control on the wireless part of the network is like an on–off control. The window size field is used for congestion control during transport. The window size field specifies the number of bytes the sender is willing to receive, starting from the acknowledgement field value. In slow start (Section 6.1.5), the window size is set to one at the beginning and on congestion detection. In M-TCP, it is set to zero on detection of packet loss or out of reach from the timeout or DACK from the other end.

Figure 6.8(a) shows the M-TCP supervisory host (SH_M) agent sub-layer between the base transceiver and the fixed node and conventional TCP between the fixed nodes.

One connection is between the MN and the BTS and the other between the BTS and the fixed node (FN). The BTS has an access point at an agent, SH_M for the TCP connection. SH_M sends and receives the packets to and from the MN through the BTS. M-TCP functions in the following manner:

1. SH_M sends and receives the packets to and from TCP_F layer at the fixed node. The transfer mechanism is simple. There is only one hop.

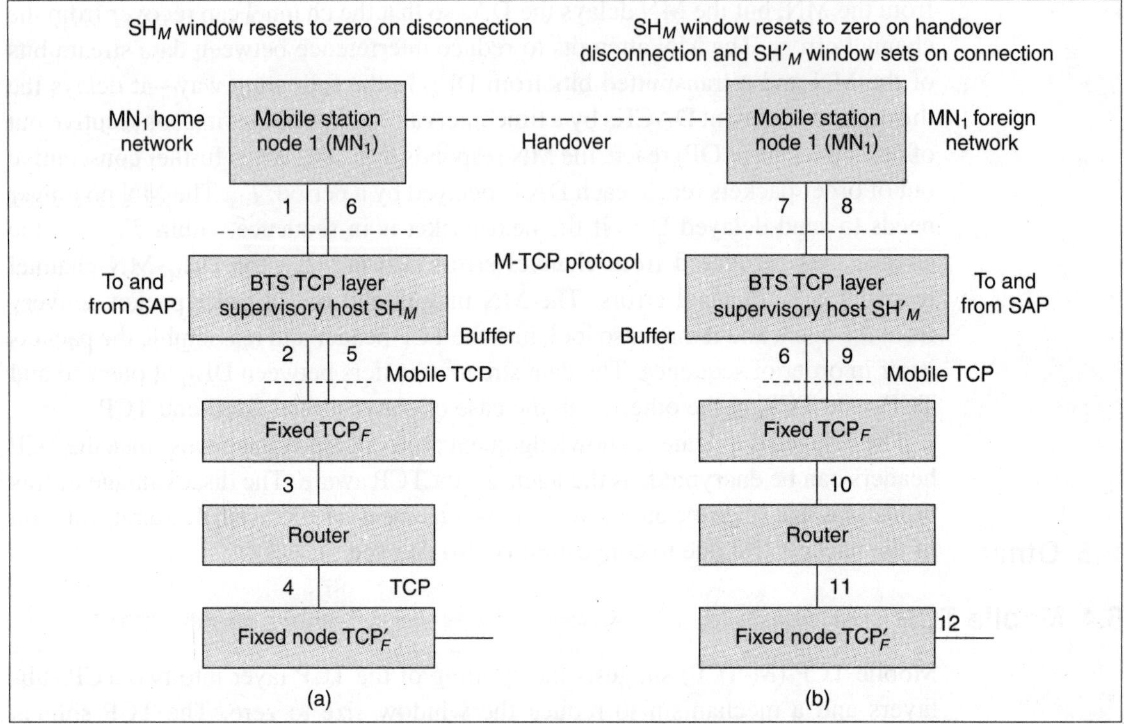

Fig. 6.8 (a) Mobile TCP supervisory host (SH_M) sub-layer between the base transceiver and the fixed node (with conventional TCP between the fixed nodes) (b) Handover mechanism

Retransmission of packets due to DAs between SH_M and TCP_F does not takes place, unlike the retransmission of packets between the fixed TCP nodes (Section 6.1.2). SH_M sets the window size to zero in case of timeout, as it presumes disconnection of the MN. The MN or TCP_F will also not retransmit as each of them finds that SH_M is not receiving packets within the timeout and has set the window to zero. When SH_M finds that the MN has sent the packet, it presumes that the connectivity is alive again and sets the window to its old value, i.e., the value when it last received the packets.

2. TCP_F layer at a fixed node sends and receives the packets to and from another fixed node TCP'_F. The transfer mechanism is standard, carried out by multiple hops through the routers.
3. Error detection and correction is done at the data link or physical layer at the BTS and the MN, not at SH_M.
4. The TCP header can be compressed during transmission between SH_M and the MN.

From the service access point (application) at the MN, data streams are received but not buffered at SH_M and, therefore, there is no retransmission from SH_M to the MN or to the fixed TCP_F layer. Figure 6.8(b) shows the handover mechanism.

When there is handover, the packets for transmission are not buffered at SH_M and no packet needs to be transferred to SH'_M. On handover, the socket (port and IP address) and its present state do not migrate from SH_M to SH'_M. The SH_M-to-SH'_M transfer has no latency period. SH_M sets the window size to zero at the beginning of the handover. SH'_M sets to a new window size at the new foreign network.

The advantages of mobile TCP are—(a) it maintains end-to-end connection between the base and the TCP layer at the other end, which guarantees reliable packet delivery and (b) it takes into account frequent disconnections of the mobile node in a wireless network as the most important factor for data loss. The disadvantages of mobile TCP are—(a) the mobile part of the network is not isolated from the conventional because, although there is no change in the existing TCP network, the bandwidth changes are frequent due to frequent settings of window size to zero, (b) security risks from the added supervisory hosts, and (c) presumption of low bit error rates in the wireless network.

6.5 Other Methods of TCP-layer Transmission for Mobile Networks

Consider that N packets are to be transmitted. The window size is equal to $(n_2 - n_1)$ when the window size is set such that the packets are transmitted from sequence number, n_1, to sequence number, n_2. The receiver does not acknowledge each and every packet starting from sequence n_1. After a set timeout, t, the receiver acknowledges the packets by sending the highest sequence number, n'_2, among the packets that the receiver has received during the period t. It is then assumed that all the packets between n_1 and n'_2 have been successfully received. Expected acknowledgement after t may or may not reach the transmitter. When the acknowledgement received at the transmitter is for a packet of sequence number n_2 then it means that there is no congestion. When the acknowledgement received at the transmitter is for packet n'_2, where $n'_2 < n_2$, it means that there is congestion and that the new window will start from $n'_2 + 1$ instead of $n_2 + 1$. When the transmitter finds no acknowledgement, it retransmits from the first sequence number, n_1. No acknowledgement also indicates congestion.

Let us assume that there is an acknowledgement after period t for n''_2, where $n''_2 < n_2$. The second cycle of transmission starts from sequence number $n''_2 + 1$ and the window resets for sequence numbers between $n''_2 + 1$ and $n''_2 + 1 +$ cnwd. One of the packets with sequence number less than n''_2 [$(n''_2 - 1)$, $(n''_2 - 2)$, or so on] may reach after a delay in comparison to packet n''_2, which succeeds these. The acknowledgement received at the transmitter from the receiver is then duplicated because the receiver resends an earlier acknowledgement for n''_2 (which was received without delay). The transmitter must note that the packets are being sent in a network with few congested paths which cause a lower sequence number packet to get delayed more than the period t. The number of duplicate acknowledgements, n_{DACK}, indicates the number of packets of sequence numbers less than n''_2 that got delayed in the network. Duplicate acknowledgements (DACKs) show the recovery of n_{DACK} packets (known as lost packets) after a period greater than t.

Section 6.1.5.1 explained a method called slow start. Once congestion is detected at the transmitter (from no acknowledgement after t or from DACKs after t), the transmitter must reduce the window size and transmit a lesser (than what it was before detecting congestion) number of packets during t. The slow start method entails starting again from window size equal to one and then increasing the window size and resetting the threshold to half of its earlier value.

Section 6.1.5.2 explained the fast recovery (FR) method. The first step of the FR method is to set another timeout for waiting, when congestion is detected from no acknowledgement in a timeout period. This is carried out for recovering the lost packets. The packets might reach the receiver during the additional timeout and are thus considered as recovered at the receiver and the receiver can now send the acknowledgement. There can be DACKs in a timeout. The second step of the FR method is that when there is no acknowledgement or when there is duplicate acknowledgement in period, t + timeout, then instead of setting the window size as one (slow start method), the transmitter sets the window size to half of the window size when congestion was detected. It is assumed that the network has recovered from congestion during the additional timeout and the transmitter can straightaway set the window to half of the window for congestion avoidance. Section 6.5.1 describes the fast retransmission/fast recovery (FRR) method for congestion control. An alternative method for the mobile transmitters and receivers network is to let the timeout continue till the data link layer detects the connection between the transmitter and the receiver. Another method is that the receiver sends selective acknowledgements and the transmitter retransmits the lost packets only. Timeout freezing transmission and selective retransmission methods are explained in Sections 6.5.2 and 6.5.3.

Packets may be used for short messages. Transaction-oriented TCP protocol (T-TCP) is a method used in such situations. Section 6.5.4 describes this new TCP mechanism called transaction-oriented TCP (T-TCP). Instead of sending the duplicate acknowledgement for n_2'' every time a packet of a lower sequence number n_2''' $(n_2''' < n_2'' < n_2)$ reaches it, the receiver makes an explicit notification method for the packet which was lost and recovered. Section 6.5.5 describes the explicit notification method.

6.5.1 Fast Retransmit/Fast Recovery Transmission

The slow start method (Section 6.1.5.1) presumes that congestion is the only reason for packet loss. In wireless networks, the causes are disconnection, handover, and channel errors due to interference. When a MN moves from the home network to a foreign network, there is a lack of acknowledgement from the other end. This is natural because it is the handover phase. Since lack of acknowledgement is natural,

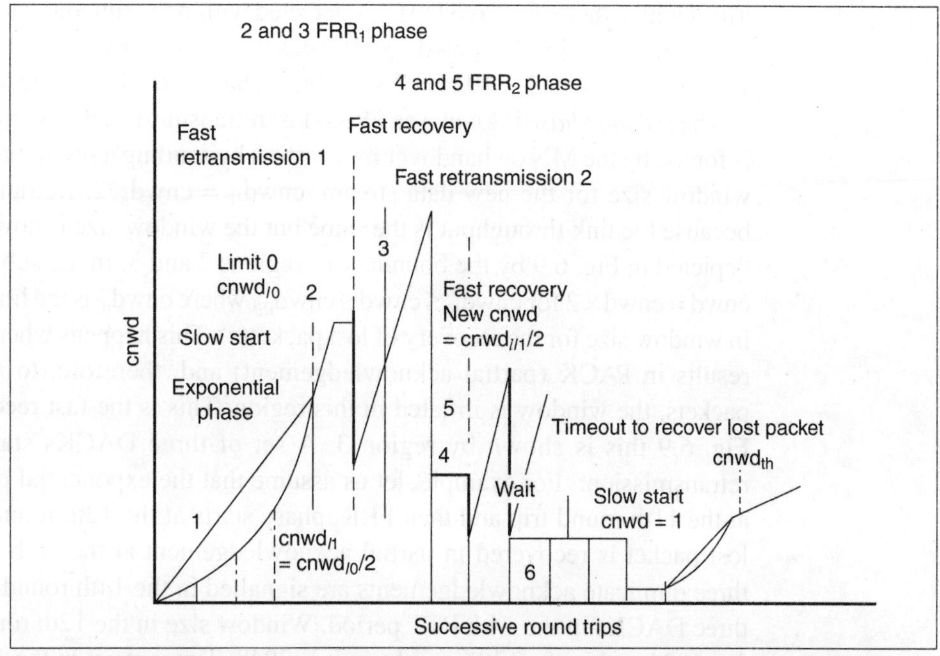

Fig. 6.9 Four phases of fast retransmit and fast recovery method—exponential, fast retransmit/recovery phase 1 (FRR1) on three duplicate-acknowledgements, fast retransmit/fast recovery phase 2 (FRR2), and wait (constant timeout and window size)

cnwd can be set to cnwd/2 and should recover fast to the value at which packet loss was detected. There is a method for recovering lost data called fast retransmit/fast recovery (FRR). The four phases of the fast retransmit and fast recovery method are—beginning (exponential), fast retransmit/recovery phase 1 (FRR_1) on three duplicate acknowledgements, fast retransmit/fast recovery phase 2 (FRR_2), and wait (constant timeout and window size). Figure 6.9 shows the four phases. FRR_1 can be forced on handover. The MN does this immediately after registering at a new foreign agent. The MN sends three DACKs. The details of the actions in the four phases are given below.

The *first phase* is known as beginning (exponential growth phase). Figure 6.9 shows this phase in regions 1 and 2. It can be called the beginning phase when the data streams start from a connection. The congestion window size for new data stream, cnwd is equal to one at the beginning. After each RTT (after which the acknowledgement field is used to send the next data stream), the window is doubled till there are three DACKs within the set timeout interval. Let us assume that cnwd reaches a limit $cnwd_{l0}$. Now, $cnwd = cnwd \times 2$ for $1 \leq cnwd \leq cnwd_{l0}$, where $cnwd_{l0}$ is the window at the limit zero, after which the FRR_1 phase starts. The threshold is defined to control further exponential growth in order to avoid congestion. For example, let us assume that the exponential phase consists of up to the 11th round trip at which there are three DACKs detected within the timeout interval. Therefore, the limit zero window, $cnwd_{l0} = 2^{10}$ and the number of segments transmitted is $1 + 2 + 4 + \ldots + 512 + 1024 = 2047$. FRR_1 phase sets in after the 11th round trip.

The *second phase* is known as FRR_1 (fast retransmit/fast recovery phase 1), which is forced by the MN on handover registration by sending a set of three DACKs. The window size for the new data stream, $cnwd_{l1} = cnwd_{l0}/2$. Retransmission is fast because the link throughput is the same but the window size is now halved. This is depicted in Fig. 6.9 by the boundary of regions 2 and 3. In the subsequent streams $cnwd = cnwd \times 2$ for $cnwd_{l0} \leq cnwd \leq cnwd_{th}$, where $cnwd_{th}$ is the limit of increments in window size for the recovery of lost packet(s). This happens when the data stream results in PACK (partial acknowledgement) and, therefore, to recover the lost packets, the window is inflated in this region. This is the fast recovery region. In Fig. 6.9 this is shown by region 3. A set of three DACKs starts another fast retransmission. For example, let us assume that the exponential phase terminates at the 11th round trip and then FRR_1 phase starts at the 12th round trip. Now, if a lost packet is recovered in partial acknowledgement at the 13th round trip, then three duplicate acknowledgements are signalled in the 14th round trip as there are three DACKs in the 14th RTT period. Window size in the 12th round trip is 2048/2 = 1024 and in the 13th round trip it is 2048. Timeout set in the 12th trip is 1024 RTT_0. Timeout set in the 13th trip is 2048. The number of segments transmitted from the beginning up to the 14th round trip is $1 + 2 + 4 + \ldots + 256 + 1024 + 0 + 1024 + 2048 = 5199$. As the lost packet is recovered, the window size is no longer incremented and RTT remains at $2048 \times RTT_0$, where RTT_0 is the time when cnwd is 1. Timeout set in the 14th trip is 2048 and in that there are three DACKs, hence, FRR_1 ends at the 14th trip and FRR_2 phase starts at the 15th trip.

The *third phase* is known as FRR_2 (fast retransmit/fast recovery phase 2). The cnwd remains constant (region 4 in Fig. 6.9) equals $cnwd_{l1} = cnwd_{l0}/2$ before the beginning of FRR_2. After this, exists the fast recovery (region 5). Fast retransmission starts by halving the window size and setting new $cnwd = cnwd_{l1}/2$. The boundary of regions 4 and 5 shows this drop in Fig. 6.9. Retransmission is now fast because for the same throughput, the window size is halved. In the subsequent RTT streams, $cnwd = cnwd \times 2$ for $cnwd_{l1}/2 \leq cnwd \leq cnwd_{th1}$, where $cnwd_{th1}$ is the limit for fast recovery when the lost packet is recovered by retransmission. Due to PACK(s) (Section 6.1.4), the fast recovery region sets in. Region 5 in Fig. 6.9 shows the fast recovery in FRR_2. There can be repeated successions of regions 4 and 5 due to additional FRRs. Three DACKs at any time in FRR_2 will trigger another FRR_j. ($j = 3$ in case it is only one additional FRR. $j = 3$ for the first additional FRR and $j = 4$ for the second additional FRR.) After the FRRs, the wait phase in region 6 starts. Considering the case discussed in the previous example, the cnwd at the start of FRR_2 in the 15th round trip is set to 2048/2. Therefore, cnwd = 1024. In the 16th round trip, cnwd = 1024/2 = 512 for fast retransmission 2. In the 17th round trip cnwd becomes 1024 and in the 18th, cnwd becomes 2048 in the fast recovery phase. However, in the 18th round trip, there is a lost packet recovery as a result of retransmission after a PACK. Hence the constant window size 512 sets in and the wait phase sets in with cnwd = 512.

The *fourth phase* is called wait/constant window-size phase. Figure 6.9 shows this phase as region 6. The window size for the new data stream, cnwd = $cnwd_{l1}/2$. The window size remains constant in order to recover lost packets. In subsequent streams, cnwd remains at $cnwd_{l1}/2$. If there is a missing acknowledgement and consecutive DACKs, the slow start phase sets in. If there are three DACKs, again the FRR_j phase sets in. Considering the case discussed in the previous example, cnwd at the start of the constant window-size/wait phase in the 19th round trip is set to 512. Therefore, $cnwd_{wait} = cnwd_{l1}/2 = 512$. In the 20th to 24th trips, cnwd remains at 512. Any lost packet is recovered and if a DACK is found in the 25th trip, then the slow start phase starts again on finding consecutive DACKs. If three DACKs are received in the antenna RTT, then FRR_j starts once again.

The fast retransmit/fast recovery method is efficient in a mobile network when there are disconnections and handovers. The disadvantage of the FRR method is that the wireless part is TCP mixes with that of the fixed network and is not sufficiently isolated. The fast retransmitted packets can force congestion and not reach the destination over the fixed network part.

6.5.1.1 TCP Reno

The FRR phase sets in after three DACKs. The TCP Reno protocol is a method of fast recovery by first halving the window size (cnwd) for fast retransmission followed by fast recovery after each fast retransmission.

When an acknowledgement is received before the DACK and retransmission starts, then it is called partial acknowledgement (PACK). After the PACK, when the lost packet is found by an acknowledgement during the fast recovery phase (fast recovery is by increasing timeout or window size), the fast recovery is stopped and the wait constant window size phase sets in or slow start sets in, as shown by region 4 in Fig. 6.9 after FRR_1 and region 6 after FRR_2. The TCP Reno method is used in the fast retransmissions and fast recoveries in FRR_1 and FRR_2. The window is inflated for extending size after slashing by a factor of two. Fast recovery ensures that the TCP connection has the data stream to transmit. Figure 6.9 shows that in FRR_2, there is an extension of the timeout period (region 4) for recovery in place of setting the window for slow start immediately on receiving the DACKs. Extension (inflation) of the window size refers to extending the timeout (RTT) by inflating the window in region 5 to recover the lost packet. After the recovery, there is an extended timeout in region 6 (RTT).

The advantage of TCP Reno is that the increasing of the window size after fast retransmission sets the larger RTT and recovers the lost packet. The lost packet may occur, for example, during handover of the mobile node. The disadvantage of Reno is that the increase in RTT and window size may lead to congestion at the fixed part of the TCP connection.

6.5.1.2 New TCP Reno

Figure 6.10 shows a modification of fast retransmit/fast recovery called new TCP Reno. It shows three phases of fast retransmit/fast recovery in new TCP Reno. There can be *m* phases in case of *m* PACKs. A PACK shows a set of lost packets. New Reno refines the fast retransmit/recovery by considering the PACKs. TCP Reno stops the extension of window size further for faster recovery after the PACK. The new TCP Reno method presumes that a single PACK shows the possibility of another lost packet. Therefore, the TCP transmitter (on getting a PACK) retransmits the first unacknowledged byte and subsequent sequence numbers. The fast recovery is not stopped, only the increment in window size is stopped. Figure 6.10 shows this in the regions 4 to 7. After each round trip and acknowledgement, the transmitter retransmits the unacknowledged segment. This is continued till all unacknowledged packets are obtained by retransmission. The RTT remains large and constant in the recovery phase in regions 4 to 7 (Fig. 6.10) until all the lost packets are obtained by retransmission. This enables the avoidance of multiple fast retransmissions from a single window of data when the window size is set to one in case of the slow start method. The new Reno permits FRR_1, FRR_2, ..., FRR_n phases. It suggests that the window size be kept high and constant till all *n* of the multiple lost packets are recovered. When *m* packets are lost, these are recovered in *m* RTTs.

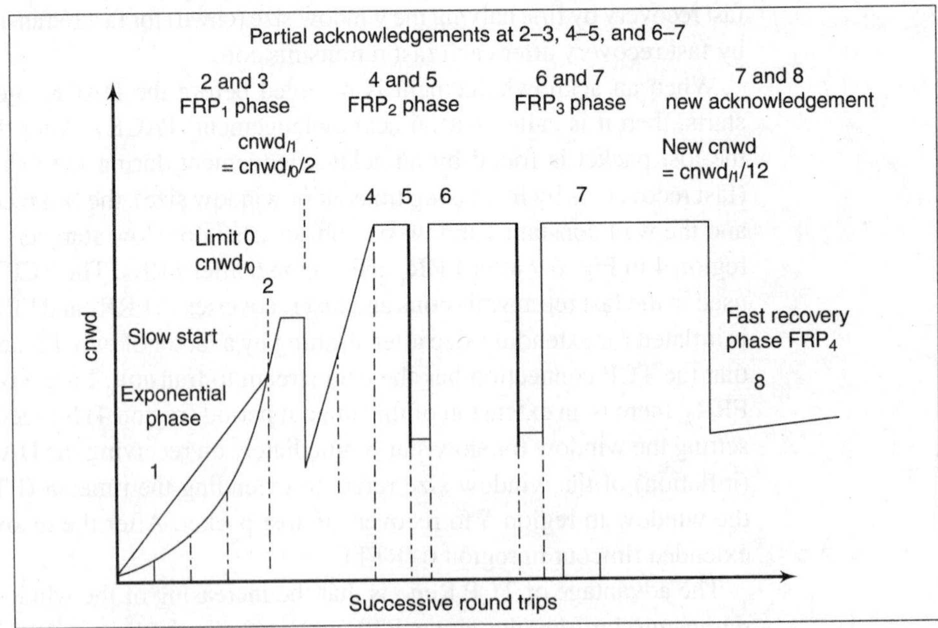

Fig. 6.10 Three phases of fast retransmit/fast recovery in new TCP Reno

The advantage of new Reno is that increasing the number of fast retransmissions and period settings for the RTT leads to recovery of multiple packets, which may be lost, for example, during handover of the mobile node. The disadvantage of new Reno is that increasing the number of retransmissions and period settings for the RTT by continuously extending the window size cannot go on endlessly.

6.5.2 Timeout Freezing of Transmission

Timeout freezing of transmission (TFT) is also used in situations where the MN faces long durations of disconnection. Figure 6.11(a) shows TCP connections in the timeout freezing of transmission protocol. Figure 6.11(b) shows the TCP connection timeout freezing. The timeout freezing transmission functions in the following manner:

1. A MAC sub-layer data link known as DL_{TF} agent senses the disconnection a little earlier than the TCP layer at the MN agent at data link MAC sub-layer

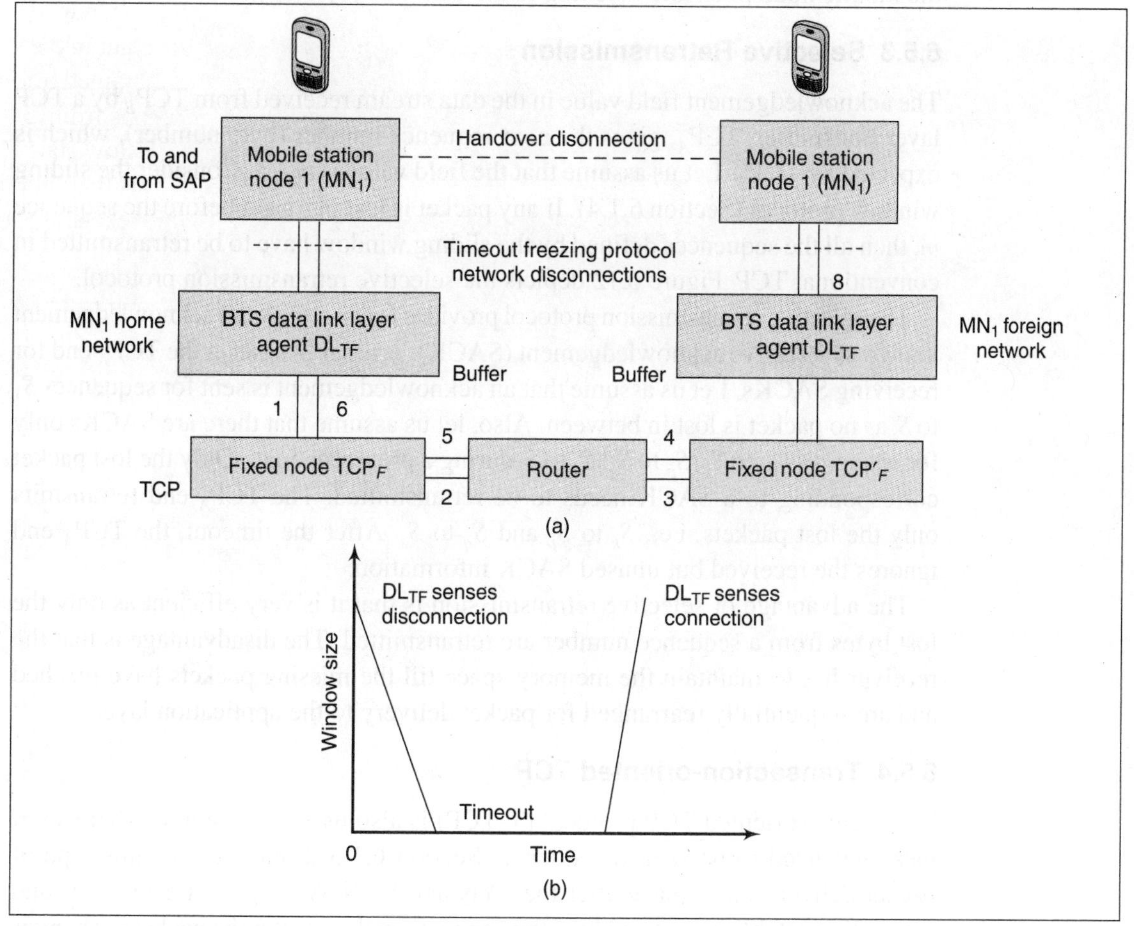

Fig. 6.11 (a) TCP connections in timeout freezing of transmission protocol (b) TCP window size timeout freezing at disconnection

known as DL_{TF} agent senses the disconnection a little earlier than the TCP layer at the MN. The TCP layer stops sending packets when disconnection is sensed by the DL_{TF} agent. The TCP layer presumes some congestion and, therefore, after a timeout the TCP layer freezes completely. During the timeout period, the MN may get some data sequences. After timeout, the TCP transmission freezes.

2. The TCP SYN and ACK data streams can still be received and transmitted through the lower layer with suitable encoding by the header.
3. When the DL_{TF} agent senses the establishment of connection, it activates the TCP transmission.

The advantage of timeout freezing is that long interruptions of mobile node are accounted for and are independent of the data stream contents. The disadvantage is that the mobile node data link layer needs to be modified by adding an agent to sense the loss and gain of connection and for making changes in the TCP layer at the mobile node to enable listening to the data link layer agent.

6.5.3 Selective Retransmission

The acknowledgement field value in the data stream received from TCP_B by a TCP layer transmitter, TCP_A, gives the next sequence number (byte number), which is expected by TCP_B. Let us assume that the field value is $m + 1$. Consider the sliding window protocol (Section 6.1.4). If any packet is lost in transit before the sequence m, then all the sequences defined by the sliding window have to be retransmitted in conventional TCP. Figure 6.12 depicts the selective retransmission protocol.

The selective retransmission protocol provides for an additional acknowledgement known as selective acknowledgement (SACK). A timeout is set at the TCP_A end for receiving SACKs. Let us assume that an acknowledgement is sent for sequences S_1 to S_j as no packet is lost in between. Also, let us assume that there are SACKs only for sequences S_j to S_k, S_p to S_q, S_s to S_t during a preset timeout. Only the lost packet corresponding to a SACK needs to be retransmitted. The TCP_A end retransmits only the lost packets, i.e., S_k to S_p and S_q to S_s. After the timeout, the TCP_A end ignores the received but unused SACK information.

The advantage of selective retransmission is that it is very efficient as only the lost bytes from a sequence number are retransmitted. The disadvantage is that the receiver has to maintain the memory space till the missing packets have reached and are sequentially rearranged for packet delivery to the application layer.

6.5.4 Transaction-oriented TCP

Transaction-oriented TCP protocol (T-TCP) is also used in situations where short messages need to be sent in sequence. Section 6.1.3 discussed how this type of packet delivery takes place after the SYN and SYN_ACK packet exchanges and connection establishment and how the connection closes after the packet exchanges

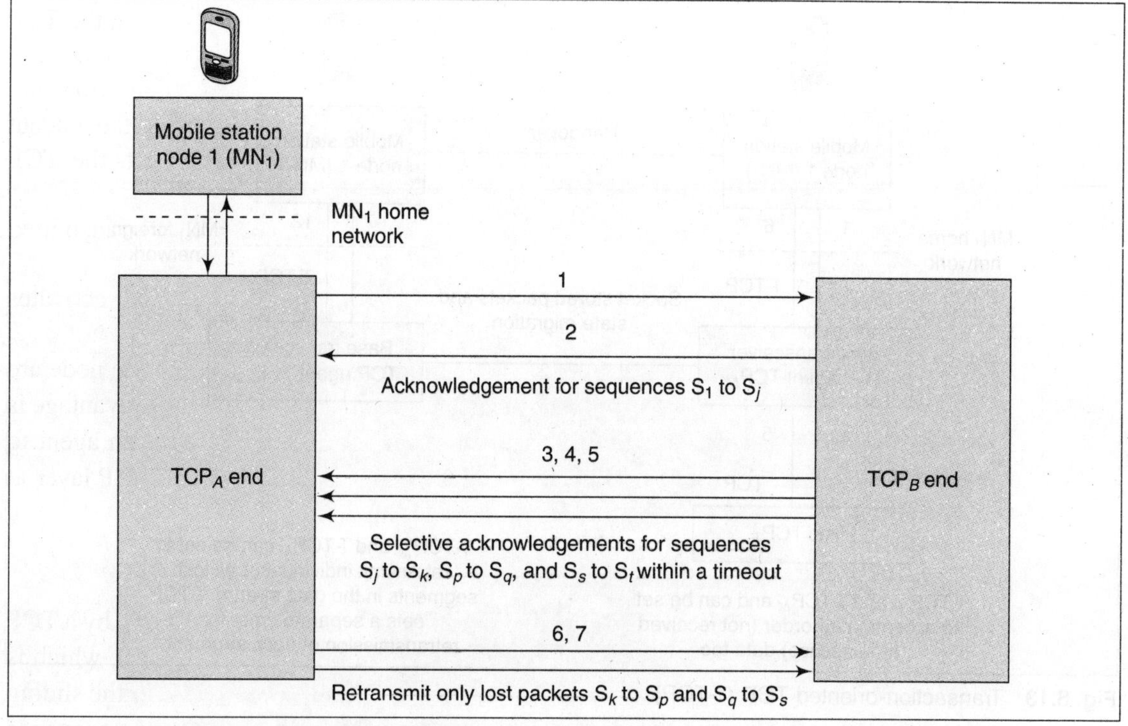

Fig. 6.12 Selective retransmission protocol

of FIN and FIN_ACK, CLOSING, and LAST_ACK. T-TCP combines the three functions—connection establishment, data transfer, and connection finish (close) for TCP data delivery such that only two or three packets are required for these functions. Figure 6.13 shows transaction-oriented protocol functions. The T-TCP functions efficiently due to the following reasons:

1. There are no multi-way handshakes, for example, SYN, SYN_ACK, and ESTABLISHED (Section 6.1.3) between the $T\text{-}TCP_M$ and $T\text{-}TCP'_M$ ends. Connection messages (SYN, SYN_ACK, ESTABLISHED, FIN, and FIN_ACK) are sent along with the data stream and not separately.
2. Sequence numbers are for two paired bytes in T-TCP instead of one as in the case of TCP.
3. Only immediate acknowledgements (such as SYN and SYN_ACK in TCP) and cumulative acknowledgements (no piggyback acknowledgements sent) are sent in T-TCP. (Piggyback acknowledgement is an acknowledgement sent by setting the acknowledgement field along with the data stream octets.) T-TCP employs caches for each state and RTTs at the transmitter and the receiver.
4. A parameter called connection count (cc) is used for keeping a count of the number of bytes. The header has the sequence number of the first byte of

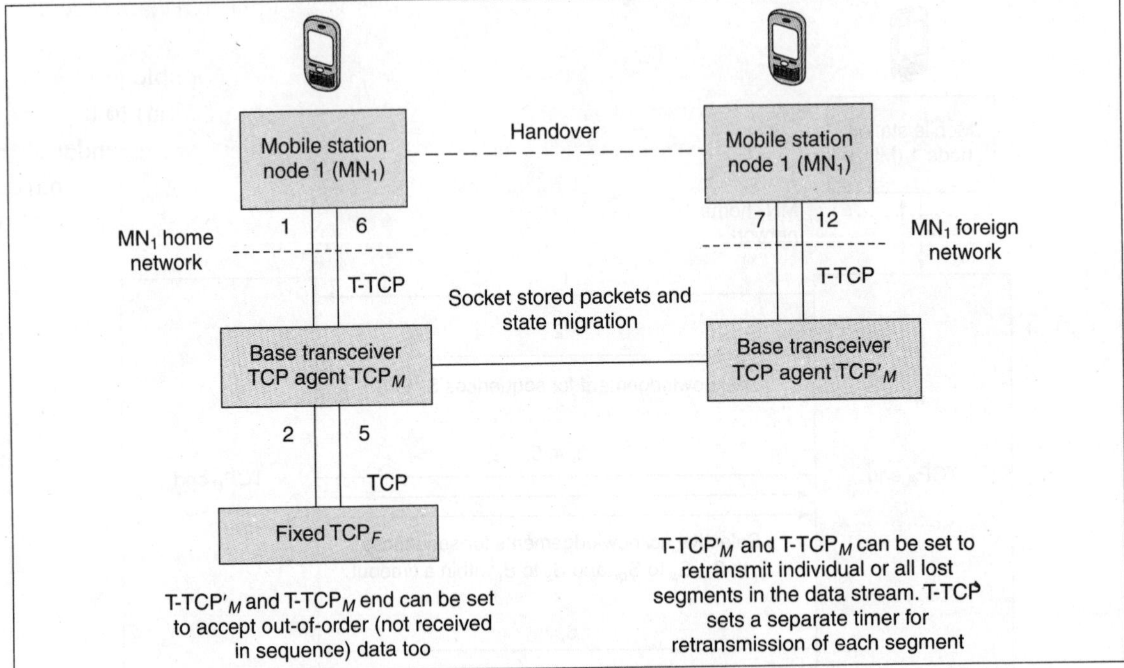

Fig. 6.13 Transaction-oriented TCP (T-TCP)

octets in a data stream. T-TCP server caches the cc for each connection at each end to restrict duplicate requests and replies.

5. The data stream is pushed during a preset timeout or sent as urgent by setting an urgent pointer (Section 6.1.2.1). The timeout is set to a small value, for example, $8 \times RTT_0$, where RTT_0 is the time for one round trip with minimum window. T-TCP thus becomes efficient for short data streams.

The advantage of T-TCP is the increase in efficiency by reducing overhead of packets for connection establishment, data transfer, and connection close. The disadvantage is that the mobile node TCP layer needs to be modified to enable the implementation of T-TCP.

6.5.5 Explicit Notification

Let a BTS detect the missing packets in the incoming TCP data stream from a MN. This is when the BTS gets a DACK (DA) from the MN. Explicit notification method (ELN method) works so that the BTS sets an ELN (explicit loss notification) bit when the DA is sent to the MN. When the MN gets the DA with the ELN bit, it retransmits. This is because the MN interprets the ELN bit from the BTS as signalling of packets loss. An alternative to this is the ELN2 method. ELN2 functions as follows—only the sequence numbers are buffered (in the cache) at the BTS in place of the packet. Whenever a DA corresponds to a buffered missing sequence number, the ELN bit is set with the DA. The MN retransmits on receipt of DA with the ELN.

Another alternative is that when the BTS receiver end is not able to transmit a packet to the MN, it sends an EBSN (explicit bad state notification) to the sender (fixed line) at the other end. The BTS can receive a packet from a sender at the other end, but may not be able to send it to the MN. Then, the partial acknowledgement protocol provides for a notification. The sender retransmits on receiving the notification.

6.6 TCP Over 2.5G/3G Mobile Networks

The characteristics of wireless transmission over a 2.5G/3G network are asymmetrical data rates for uplink and downlink and large data rates on downlink. The signals reach the other end with variable delays; delays also occur due to adaptive handling of data at the data link and physical levels. This is a result of the use of FEC and ARQ (Section 6.1.6) at the data link layer and of interleaving (Sections 3.2.4 and 4.5.1.4) at the physical layer. Therefore, mobile systems such as GSM, GPRS, and CDMA have high round-trip periods (up to 2 s) as compared to fixed lines.

In TCP over mobile 2.5/3G networks header compression is not recommended. Timestamps on the packets help in taking into account whether the delays are due to the propagation path or due to other reasons; also the sequences can be arranged as transmitted. SACK, ELN, EBSN, and FRR protocols can be used. Large windows are required to have a greater throughput. MTUs in the fixed part of the TCP connection need to be enlarged to decrease the RTT per unit window size.

Keywords

Congestion control: A method used to reduce congestion in the network. An example of this is the method of control in which the number of bytes in the stream starts slowly, increasing exponentially in successive steps. The number of transmitted bytes in a step is reduced to half or less on finding congestion and thus congestion is reduced.

Connection-oriented protocol: A protocol which entails session establishment, data flow, congestion control, and, in the end, session termination.

Explicit notification: A method of congestion control by explicit notification of congestion, for example, when a base transceiver at the receiver end is not able to transmit a packet to the mobile node then it sends an EBSN (explicit bad state notification) to the sender (on fixed line) at the other end.

Fast recovery: A method in which, if a congest-ion is detected after the slow start phase and doubling of data stream lengths (window) for transmission, the transmission window size is set to half of the window size when the congestion was detected (instead of zero).

Fast retransmit/fast recovery transmission: A method in which there are four or more phases of fast retransmit and fast recovery—first phase as slow start and beginning (exponential), then fast retransmit/recovery phase 1 (FRR_1) on three duplicate-acknowledgements, fast retransmit/fast recovery phase 2 (FRR_2), and wait (constant timeout and window size).

Indirect TCP: A split TCP method in which the TCP layer splits into two TCP sub-layers—one indirect TCP sub-layer between the mobile base transceiver and the fixed node and a conventional TCP sub-layer between the fixed nodes.

Mobile TCP: A method of splitting the TCP layer into two TCP sub-layers using a mechanism that reduces window size to zero. The split is asymmetric; the window is set to zero to prevent the transmission from the TCP transport layer at the mobile node (MN) or at the fixed node when disconnection is noticed. The window opens again on getting the packet, there is no slow start by the base transceiver and it is presumed that packet loss is due to disconnection and not due to congestion or interference.

RTT (round trip time): The time interval between the start of transmission of a data stream [of given transport protocol data units (TPDUs)] to the receiver layer and receiving back the acknowledgement of successful transmission of TPDUs (along with a parameter which enables the setting of the time interval during which the transmitter expects the receiver to send a TCP stream with the acknowledgement field set as per the successfully sent bytes, enabling the transmitter to retransmit or transmit from the acknowledgement field sequence number onwards).

Selective retransmission: A method in which there is an additional acknowledgement, known as selective acknowledgement (SACK); a timeout is set at the transmitting end for receiving SACKs. Only the lost packet corresponding to a SACK needs to be retransmitted.

Slow start: A method in which the window size is initially one transport protocol data unit and then the window size increases exponentially on successive round trips till congestion is noticed or a threshold size is reached.

Snooping TCP: A protocol in which an agent buffers the packets from the fixed connection layer for transmission to the mobile node on a wireless transceiver; the agent also buffers the packets on the wireless transceiver from the node for transmitting to a layer at a fixed line. The agent snoops at the transmission and reception in place of acknowledgement- or timeout-based TCP method in the mobile part of the network.

TCP: A globally used connection-oriented transport layer protocol using which the transmitter transmits data only after establishing a connection with the receiver. It entails data buffering and retransmission of unacknowledged parts, session-start, data transfer, and session-finish. The data transmission is fully acknow-ledged end-to-end in TCP and there is in-order delivery to the application running at the destination. Congestion control and congestion avoidance methods are also deployed in TCP.

TCP data delivery: A process of delivering data in which there is an acknowledgement of sequences sent from an end and there is retransmission of bytes that have not reached successfully within a timeout period.

TCP data flow control: A method of flow control which includes data flow control by specifying the urgent part of the data octets for transporting to the other end.

TCP data stream: A stream which consists of sequential bytes transmitted by the TCP protocol and later divided into packets, which are routed through routers on the Internet, using the IP protocol of the network layer. [The number of bytes in a stream is equal to the transport protocol data units (TPDUs) that are placed in a memory buffer.]

TCP header: A header used in the TCP protocol; it consists of fields in five 32-bit words followed by words for the option fields and padding.

TCP/IP protocol: A suite of protocol standards for the application, transport, and network layers when transmitting and receiving data through the Internet.

Timeout freezing transmission: A method used when a mobile node (MN) has long durations of disconnection and a data link layer (MAC) agent senses a disconnection a little earlier than the TCP layer at the node and from that point onwards, the TCP layer stops sending packets, presumes some congestion, and therefore, the TCP layer, after a timeout, freezes completely till the data link layer senses the connection again.

Transaction-oriented TCP: A protocol which is efficient and is used in situations where short messages are to be sent in sequence and a packet is delivered after the SYN and SYN_ACK packet exchanges and the connection closes after the packet exchanges of FIN, FIN_ACK, and CLOSING.

UDP (user datagram protocol): A protocol with a limited size header having source and destination IP addresses, port numbers, and header checksum. It transmits limited size data ($< 2^{16}$ bytes) and is a connection-less protocol, which means that there is no session establishment, data flow and congestion control, and session termination in UDP, and that the data transmits so that it is not in sequence or relation to any previous data transmitted.

Review Questions

1. What are the fields in the 32-bit words in the TCP protocol? Explain the function of each field.
2. How are the sequence number, window-size, and acknowledgement number fields used in data flow control at the TCP transport layer at the transmitter and receiver ends?
3. Describe the states of a TCP connection. How does the change of state from LISTEN to CLOSE take place at the transmitter and receiver ends?
4. Describe slow start of congestion control. How can fast recovery take place in the congestion avoidance phase?
5. Why is the presumption that congestion is the major factor limiting the data flow not valid for mobile and wireless networks? What are the differences in data flow control in mobile and fixed-line networks? List the deficiencies in conventional TCP on fixed-line networks that warrant modifications for the mobile networks connected to the Internet.
6. Describe indirect TCP. Explain the modifications of indirect TCP as the selective repeat protocol and mobile-end transport protocol. What are the advantages and disadvantages of indirect TCP?
7. What are the functions of snooping sub-layer in the snooping TCP protocol? How do the TCP packets transfer from a mobile node to the receiver end?
8. What can be the modifications in snooping sub-layer functions? List the advantages and disadvantages of snooping TCP.
9. Explain mobile TCP. How does a supervisory host send TCP packets to the mobile node and to a fixed TCP connection? Give the advantages and disadvantages of mobile TCP.
10. Describe with a diagram the fast retransmit/fast recovery transmission in mobile networks. Why are three DACKs used to force retransmission when a mobile node is handed over to a foreign network?
11. When are fast transmission and fast recovery triggered? What are TCP Reno and new TCP Reno modifications in the fast retransmit/fast recovery method? List the advantages and disadvantages of fast retransmit/fast recovery in TCP transmission.
12. Why is timeout freezing required in case of mobile nodes? What are the modifications made in data link and TCP layers to enforce timeout freezing?
13. How does selective transmission improve the transmission efficiency? What are the modifications required in the TCP receiver to implement the selective retransmission protocol?
14. Describe transaction-oriented TCP. How does the integration of connection establishment, data transfer, and close functions into one help in transmitting and receiving at the TCP nodes?
15. Describe the explicit notification schemes. What are the changes required at the receiver and the transmitter to enable explicit notifications?
16. What are the special requirements in transport layer protocols in case of 2.5G/3G mobile networks?

Objective Type Questions

Pick the correct or most appropriate statement among the choices given:

1. The TCP_A end transmits data segment octets of 20,000 bytes with sequence number (SN_A) specified as one. The TCP_B end receives the data and acknowledges to TCP_A, which sends four data octets from segment at *B* after TCP_B sets the SN field at $SN_B = 10$, acknowledgement number (AN) field at $AN_B = 1001$. Which of the following is true for the data octets sent by the TCP_A end in the next trip?
   (a) TCP_A must retransmit from sequence numbers 1001 with new $SN_A = 1001$ and $AN_A = 26$, and only when $AN_B = 20{,}001$, does it transmit the octets of the next segment.

(b) TCP_A transmits the next octets from sequence number 20,001 with $SN_A = 20001$ and $AN_A = 4$, and when there is a DACK, it retransmits from the previous segment.
(c) TCP_A must retransmit from sequence number 1001 with new $SN_A = 1001$ and $AN_A = 14$, and only when $AN_B = 20{,}000$, does it transmit the octets of the next segment.
(d) TCP_A must retransmit from sequence number 1001 with $SN_A = 251$ and $AN_A = 4$, and only when $AN_B = 20{,}001$, can it transmit the next segment.

2. Duplicate acknowledgement (DACK) is an acknowledgement
(a) from the receiver after transmission of an earlier acknowledgement (without delay) when a packet of a segment reaches after a delay in comparison to the packet next in sequence.
(b) from the receiver to request fast retransmission.
(c) from the receiver due to the congestion of previous acknowledgement at the link.
(d) from the transmitter to request acknowledgement when the acknowledgement from the receiver does not reach.

3. The TCP_A end transmits data segment octets of 20,000 bytes with sequence number (SN_A) specified as 1001, offset field $O_A = 10$, and window size $w_A = 2000$, after receiving data octets from the other end, TCP_B with $w_B = 4000$. Which of the following does TCP_A transmit?
(a) Bytes between sequence numbers 1001 and 3001 and the TCP_A header and options are of five 32-bit words.
(b) Bytes between sequence numbers 1001 and 20,001 and the TCP_A header and options are of ten 32-bit words.
(c) Bytes between sequence numbers 1001 and 5001 and the TCP_A header and options are of ten 32-bit words.
(d) Bytes between sequence numbers 1 and 20,000 and the TCP_A header and options are of 40 bytes.

4. Assume that at $w = 2000$ B, the RTT delay in acknowledgement is 15,625 µs. Then the throughput at the link is
(a) 1024 kbps
(b) 128 kbps
(c) 64 bytes per s
(d) none of the above. It cannot be calculated as throughput is independent of RTT delay.

5. Assuming that the slow start method is used for congestion control, the window size doubles upto fourth round trip and increases by one from that point onwards. Threshold window, $cnwd_{th}$, and the number of segments, n, transmitted are
(a) 2^8 and 15.
(b) 2^4 and 4.
(c) 4^4 and 15.
(d) 8 and 15.

6. In a slow start, the window size cnwd doubles up to the fourth round trip and $cnwd_{th} = 4$. It increases by one from that point onwards. A packet remains unacknowledged and is lost in the eighth round trip when $cnwd = cnwd_{cntrl}$. After a timeout period, in the fast recovery method the cnwd sets at
(a) 2^8.
(b) 6.
(c) 2^4.
(d) $\frac{cnwd_{th}}{2}$.

7. In indirect TCP, there is a split in the TCP layer between (i) the fixed node at the mobile node TCP_F end and the other end and TCP_F' (ii) the BTS TCP_M agent and the fixed node TCP_F. The retransmission delay between TCP_M to TCP_F is T_{MF}. On handover, the socket (port and IP address) and its present state migrate

from TCP_M to TCP'_M and the transfer from TCP_M to TCP'_F has a latency period equal to T_L. If the TCP_F to TCP'_F delay is T_{FF}, then which of the following is true?
(a) T_{FF} is very small and T_{MF} and T_L are high.
(b) T_L is 0 and T_{MF} and T_{FF} are high.
(c) T_{MF} is very small and T_L and T_{FF} are high.
(d) T_{MF} and T_L are very small and T_{FF} is high.

8. Consider the snooping TCP protocol. The snooping agent and the fixed node TCP_F connection and TCP connection between fixed nodes TCP_F and TCP'_F are
(a) transparent and non-transparent, respectively.
(b) both transparent.
(c) non-transparent and transparent, respectively.
(d) must be transparent and need not be non-transparent, respectively.

9. Consider the delayed duplicate acknowledgements (delayed DACKs) protocol. If MN is the mobile node and DL_M is the data link layer agent at the BTS for mobile node and for snooping TCP packets, then which of the following is true?
(a) The MN attempts to reduce congestion between the MN and retransmissions from DL_M by delaying all DACKs by a time interval, T_{da}.
(b) The MN attempts to reduce latency on handover between the MN and retransmissions from DL_M by delaying the third and subsequent DACKs by a time interval, T_{da}.
(c) The MN attempts to reduce congestion on handover between the MN and retransmissions from DL_M by delaying only the third DACK by a time interval, T_{da}.
(d) The MN attempts to reduce interference between the MN and retransmissions from DL_M by delaying the third and subsequent DACKs by a time interval, T_{da}.

10. The advantages of mobile TCP are that it
(a) need not maintain transparent packet delivery and it takes into account frequent disconnections of the mobile node by setting window size to a constant.
(b) must maintain end-to-end connection between the base and the other end TCP layer, which guarantees reliable packet delivery and there is no latency on handover; frequent disconnections are avoided because of the window size set to one (needed for fast recovery).
(c) maintains mobile-to-end connection between the base and the fixed-end TCP layer, which need not guarantee reliable packet delivery, but takes into account frequent disconnections of the mobile node by setting window size to zero.
(d) maintains end-to-end connection between the base and the other end TCP layer, which guarantees reliable packet delivery and takes into account the frequent disconnections of the mobile node by setting window size to zero.

11. Consider the fast retransmit/fast recovery transmission protocol. The mobile node
(a) sends three DACKs to force fast retransmission and performs fast recovery after dropping the window size to half.
(b) performs fast retransmission on handover and fast recovery after dropping the window size to one and then doubling after each round trip time.
(c) sends three DACKs to force fast retransmission after each round trip time and performs fast recovery without dropping the window size.
(d) sends three DACKs two times to force fast retransmission on handover and performs fast recovery without dropping the window size.

12. The advantage of TCP Reno is that
(a) by decreasing the window size for fast retransmission, it sets a larger RTT without waiting for the lost packet.
(b) the lost packet is recovered in retransmission.

(c) it increases the window size after fast retransmission, sets a larger RTT, and the lost packet is recovered.
(d) by keeping the window size constant after fast retransmission, it sets a larger RTT.

13. The advantage of new TCP Reno is that
(a) an increase in the number of fast retransmissions and period settings for the RTT causes recovery of multiple packets lost during handover of the mobile node.
(b) a decrease in the number of fast retransmissions and period settings for the RTT causes recovery of a single lost packet during handover of the mobile node.
(c) an increase in the number of fast retransmissions and period settings for the RTT causes recovery of unacknowledged packets during the TCP connection between the fixed nodes.
(d) the slow start causes recovery of multiple lost packets during handover of the mobile node.

14. The timeout freezing protocol provides
(a) sensing of disconnection by TCP layer, TCP_M, after a timeout freezes the data link layer, DL_M, and discontinues TCP, SYN, and ACK.
(b) sensing of disconnection by data link layer, DL_M, to let the TCP connection at the mobile node freeze completely and DL_M continue with TCP, SYN, and ACK.
(c) freezing on each timeout in which there is no acknowledgement and continues with TCP SYN only.
(d) sensing of disconnection by data link layer, DL_M, to let the TCP connection at the mobile node set the window to zero.

15. Explicit loss notification (ELN) is sent
(a) by the transmitter expecting an acknowledgement from the receiver by setting the ELN bit in the retransmitted packet.
(b) by the receiver expecting a new segment from the transmitter by setting the ELN bit in the acknowledged packet.
(c) by the receiver expecting a new packet from transmitter by sending request for ELN.
(d) by the receiver, which sets the ELN bit in the DACK to the transmitter.

16. The network characteristics to be taken care of while establishing TCP connection and data transfer are (i) variable delay signals, transparent transmission from the data link layer, interleaving, and error correction codes (ii) variable delay signals, transparent transmission from the radio link layer, and use of interleaving and multiplexing in time, frequency, or code division (iii) variable delay signals, non-transparent transmission from the data link layer, and very high RTT due to interleaving and error correction codes (iv) variable delay signals, non-transparent transmission from the data link layer, very low RTT, and use of interleaving and error correction codes. The TCP mobile network connection uses (v) large windows (vi) fast retransmit/fast recovery mechanism (vii) timestamps (viii) large MTU (maximum transfer unit) network. Which of the above are true for 2.5G/3G mobile networks?
(a) (iii), (iv), (v), (vi), (vii), and (viii).
(b) (iv), (vi), (vii), and (viii).
(c) (iii), (v), (vi), (vii), and (viii).
(d) (i), (iv), and (viii).

Chapter 7

Databases

Access to databases is made during mobile computing. This chapter focuses on database-related issues such as data caching, hoarding of database records, and accessing of the records in client–server architecture (with two, three, or more tiers). It also discusses database caches, access in power- and context-aware computing environments, queries, transactions, and recovery models. This chapter describes the following concepts.

- Hoarding and caching of data from databases
- Cache-invalidation mechanisms and mobile-data cache and web cache maintenance
- Client–server computing with *N*-tier architecture and adaptation mechanisms
- Power- and context-aware computing methods
- Transaction models
- Query processing
- Data recovery process
- Quality of service issues

7.1 Database Hoarding Techniques

Section 7.1.1 describes the concept of databases. Computations require data. It is better to organise this data in a *database* which enables raising queries, data transactions, and the retrieval of the required section of data during a computation. For example, consider contact data (with fields such as name, work-place phone number, fax number, mobile number, and address). The contact data of a number of people and companies can be organized as a database *Contacts*. Queries can be raised to the database *Contacts* during computing and the required information, such as a select mobile number, can be retrieved. Transactions can be carried out with the database *Contacts* for modifications in or additions to the information stored in it, such as a transaction to modify a select mobile number.

Section 7.1.2 describes hoarding (caching) of databases in mobile devices. A mobile device cannot store a large database due to memory constraints. Also, retrieving the required data from a database server during every computation is also impractical due to time constraints. Therefore, large databases are available on servers, remote computing systems, or networks. A mobile device is not always connected to the server or network, neither does the device retrieve data from a server or a network for each computation. Rather, the device caches some specific data, which may be required for future computations, during the interval in which the device is connected to the server or network. *Caching* entails saving a copy of select data or a part of a database from a connected system with a large database. The cached data is *hoarded* in the mobile device database. Hoarding of the cached data in the database ensures that even when the device is not connected to a network or server (disconnected mode), the data required from the database is available for computing. For example, consider the train schedules in a railway timetable. The schedule of specific trains to and from the device-user location is stored in the huge databases of the servers and network of the company operating the trains. The user device caches and hoards some specific information from the databases and the hoarded device-database is used during computations for retrieving the data for a specific set of train schedules.

7.1.1 Databases

A database is a collection of systematically stored records or information. It is not just arbitrarily stored data without any logic. Databases store data in a particular logical manner, for example, as lookup tables. A lookup table is a database which stores information in tabular form. The table has the following structure—the first column in a row has a reference for looking into the data and the subsequent column or columns contain the data. The reference is called key to the data-values. Another logical structure is a database which stores data using *tags*. A tag is also known as the key to the data-values. For example, "*contact*: 1 John, 2 Lucy." and "*address*: 1 ABC Street, 2 DEF Street." Here *contact* and *address* are the tags in the database. Using these tags during computation, the data for John's address can be fetched. Computational actions, such as connecting to a database and using the database for querying for a record, deleting a specific set of records, modifications of records, insertions into the records, and appending of the records, may be considered as an act of business (transactions) between the application software and the database. The command which is sent for retrieving the data from the database, embodies the logic used for obtaining (and storing) the data.

As discussed above, data stored in databases follows logic. The term 'business logic' indicates the logical way in which transactions (business) are carried out between two ends, for example, between database-client (application) and database-server or between an API (application program interface) and a database. Examples of transactions involving databases are—(a) establishing connection between API and database, (b) updating data records by inserting, adding, replacing, or deleting, (c) querying for records, and (d) terminating the connection between the API and the database. The API is a section of a program used to run an application (software). The API may run instructions to retrieve a queried record from a database. The API may also issue outputs or queries and commands to another program and receive the inputs from another program during a program-run.

7.1.1.1 Implicit Business Logic

As discussed above, the term 'business logic' indicates the logical manner, flow, or pattern, in which business (or transactions) may be carried out with a database. The structure and components of the database itself define *implicit business logic*, which is used in retrieving (or modifying) data from the database. The logic of transactions (business logic) is *implicit* when it comes from within the database. That is to say that no external definition is required for the business logic to function. For example, the components of data can be structured in such a way that business logic is implicit—assume that there is a directory in which the first word of each line is structured alphabetically. In such a situation, business logic is implicit in the structure of the database because the transactions can be carried out in the alphabetical order which is an inherent part of the given database. The telephone directory or phonebook with names and telephone numbers arranged alphabetically shows implicit business logic. Names and telephone numbers are structured in rows with each row having a name and the corresponding address and telephone number. When *business* is carried out between the API and the database, the database is scanned for the required information through an alphabetical search algorithm that searches the database, beginning with items starting with 'A', 'Aa', and so on. The key to guide the API commands up to the row which contains the required data is implicit. The row has one key (in this case *name*). The key enables the retrieval of the telephone number using two records, i.e., the address and the corresponding numbers. Whenever a telephone number is queried from the database, the DBMS (database management system) first searches for the first occurrence of the first character of the name, the details of which are to be recovered from the database, then a search is made for the occurrence of the first two characters together, then the first three, and so on. Finally, a match is found and the record that gives the telephone number is retrieved. Thus a telephone directory database operates with implicit business logic.

Another example of the implementation of implicit business logic is a search directory arranged alphabetically in an XML database. A database designed using

XML uses a tag as a key. The key enables business (transaction for retrieving, deleting, inserting, or modifying data). Ending of a tagged data-record is specified by an end-tag. An end-tag is a tag with a slash sign before the tag. Between the tag and corresponding end-tag, there are data values or text for the database record. For example, if `<mobile_number>` is used as a tag, then the corresponding end-tag is `</mobile_number>`, and the mobile number is between the tag and the end-tag. A tag can also have associated attribute(s) in XML. A database entity designed employing XML uses a tag, its associate attributes, and the corresponding end-tag. An end-tag separates one set or subset of data from another set or subset. Consider the following tag and its attributes—`<number` *`spec`* `= "mobile"` *`usage`* `= "default">`. The corresponding end-tag is `</number>`. Between the tag and the end-tag, there are digits for the number. The attribute *`spec`* of the tag specifies that the number is the mobile number and the attribute *`usage`* specifies that it is the number used as the default number. The following sample code clarifies this.

Sample Code 7.1 This code is for a database `Contacts` designed in XML. Here we are using `search`, `Allnames`, `name_record`, `address`, and `telnumber` as tags. The tags and attributes function as primary and secondary keys for the given XML database. Let us use the attribute *`first_character`* with the tag `<Allnames>`. All names with `first_character = R` are grouped together between the tag `Allnames` and the end-tag for it `</Allnames>`.

```
<search>
<Allnames first_character = "R">
<name_record>
Raj Kamal
<address> ABC Street, …. </address>
<telnumber> 9876543210 </telnumber>
</name_record>
</Allnames>
<Allnames first_character = "S">
.
.
</search>
```

Within the group, there are data records for names. `name_record` is the tag used for a contact name and related data for address and telephone number. Within the name record, there is the name and the additional data. `Allnames` can be used as the primary key and `first_character` as the secondary key to get the name and related records. `name_record` can be used as the primary key and `address` as the secondary key to get a name and its address.

XML databases incorporate implicit business logic. The key used to search for a record is a tag and the record is present between the tag and the end-tag. The primary key is a tag and secondary key is an attribute or another in-between tag.

7.1.1.2 Explicit Business Logic

Stored queries and procedures define business logic explicitly. Assume that a database stores the records of structure, language, flight number, flight origin, scheduled arrival time, on-time or delayed, and expected arrival time for all the incoming flights in a day at the airport. Some examples of queries and procedures in such a database are as follows:

A transaction (business) between the API and the database uses an explicitly defined *query*. A query can be as follows:

```
If Structure = "most recently added entry list" Content_Type = "English", source = "English_Records", flight origin = "Frankfurt", airline = "Lufthansa", present time = "0800 hrs" and status = "Not arrived" then Get_Records
```

Another query for business can be

```
If flight origin = "Frankfurt", airline = "Lufthansa", present time = "0800 hrs" and status = "arrived" then Delete_Records
```

`English_Records` is a section of the database having English contents. `Get_Records` and `Delete_Records` are two procedures that carry the transactions.

Sample Code 7.2 gives an example of how to specify business logic explicitly in a call history database in a mobile phone.

Sample Code 7.2 Let us specify explicit business logic for a database in a mobile phone which stores the history of called numbers in the database `call_history`. The *content type*, *structure*, *methods*, and the *connectivity protocol* can be explicitly specified as follows:

```
<!- - The following are the XML database explicit specifications of business logic to be used for transactions—It must be in English. It must be structured as most-recently-added entry first. The querying method for the database must be a method called GetRecord. The database record insertion method must be PutRecord and deletion method must be DelRecord. - ->
 Content_Type = "English "
 Structure = " Most recently added entry first" [or = "Most often used entries first" or "Alphabetical ascending" or = "Alphabetical descending order" or = "Auto-delete entries added before last 7 days"]
 Query_Method = "GetRecord"
 Insert_Method = "PutRecord"
 Delete_Method = "DelRecord"
 Database connectivity protocol = "SOAP"
 <!- - The following is the XML database call history for which the explicit specifications to be used for transactions are as above.- - >
 .
 .
```

7.1.1.3 Connectivity Protocol

A connectivity protocol is an API that has predefined methods to handle the various data access functions. The API defines ways to connect to and access a database and methods for sending queries and updating or retrieving database records. The API connects a client or server to the database. The connectivity protocol describes the set of permitted commands, transaction methods, and the order in which commands are interchanged between the API and the database at the server or the client. Using the connectivity protocol API, a program issues commands to access a database and query in order to select and retrieve queried record(s) from the database. Some examples of connectivity protocols are Java database connectivity (JDBC), open database connectivity (ODBC), and simple object access protocol (SOAP). These protocols connect the server to the database.

7.1.1.4 Relational Database

A relational database is defined as a database structured in accordance with the relational model. The relational model of data organization helps the database designer to create a consistent and logical representation of information. The relational model follows relational logic which means that it is assumed that all data can be represented as n-ary (binary means n = 2, tertiary means n = 3,...) relations. An n-ary relation is a subset of the Cartesian products of n-sets.

A relational database entails that it is always possible to mathematically model the relations between the data records and get the answers to the relational equations for the queries. The answers are as in two-valued predicate logic. Two-valued predicate logic means that there are only two possible results on evaluation, for each proposition—*true* or *false* and no third result, for example, 'null' or 'unknown', is possible.

7.1.2 Database Hoarding

Database hoarding may be done at the application tier itself. Figure 7.1(a) shows a simple architecture in which a mobile device API directly retrieves the data from a database. Figure 7.1(b) shows another simple architecture in which a mobile device API directly retrieves the data from a database through a program, for example, IBM DB2 Everyplace (DB2e). The two architectures shown in Fig. 7.1 belongs to the class of one-tier database architecture. This is so because the databases which are specific to a mobile device, i.e., not meant to be distributed to multiple devices and not synchronized with the new updates, are stored at the device itself. Some examples of database stored in one-tier architecture are downloaded ring tones, music, and video clips.

IBM DB2 Everyplace DB2 Everyplace (DB2e) is a relational database engine. It needs a memory of about 100 kB. It has been designed to reside at the device. It supports databases of sizes up to 120 MB. An enterprise server employing DB2 can

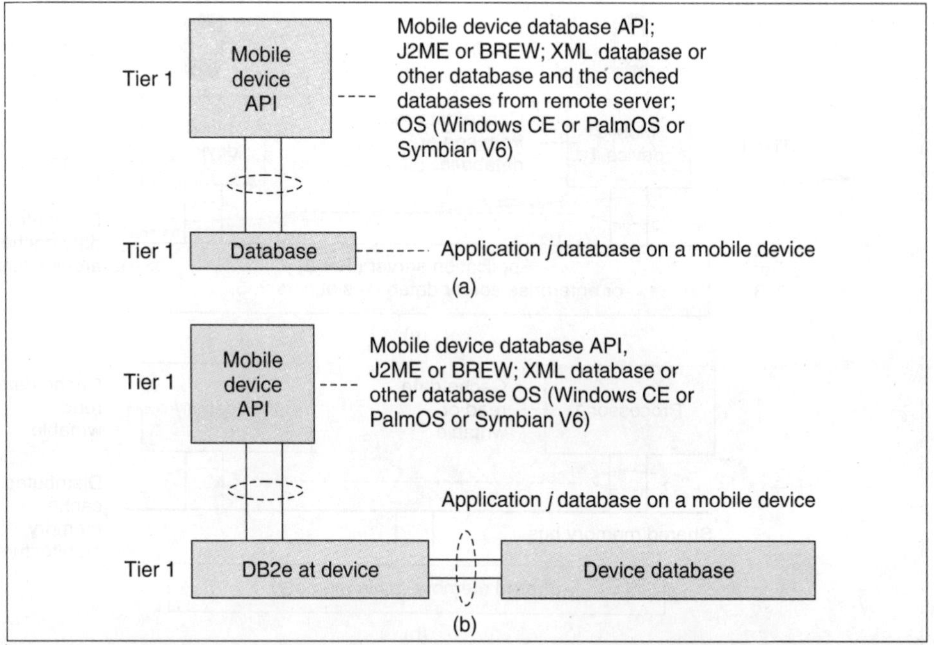

Fig. 7.1 (a) API at mobile device sending queries and retrieving data from local database (Tier 1) (b) API at mobile device retrieving data from database using DB2e (Tier 1)

deliver and synchronize the local copies of data contents at mobile devices. Synchronization means that if a data record is modified at the server then the copy of that record at the client device also changes accordingly.

DB2e supports J2ME and most mobile device operating systems, including Windows CE, PalmOS, and Symbian V6 and handheld computer (PDA) operating systems, including Win32, Linux, QNX (an RTOS based on Unix), and Neutrino. DB2e synchronizes with DB2 databases at the synchronization, application, or enterprise server.

Hoarding of Cached Data The database architecture shown in Fig. 7.2(a) is for two-tier or multi-tier databases. In such architecture the databases reside at the remote servers and the copies of these databases are cached at the client tiers. This type of architecture is known as client–server computing architecture. A cache is a list or database of items or records stored at the device. The device may select and save the item from a set of records broadcasted or pushed by a server or the device may access (by request, demand, or subscription) the item at the server and save it to the list or database. Databases are hoarded at the application or enterprise tier, where the database server uses business logic and connectivity for retrieving the data and then transmitting it to the device. The server provides and updates local copies of the database at each mobile device connected to it. The computing API at

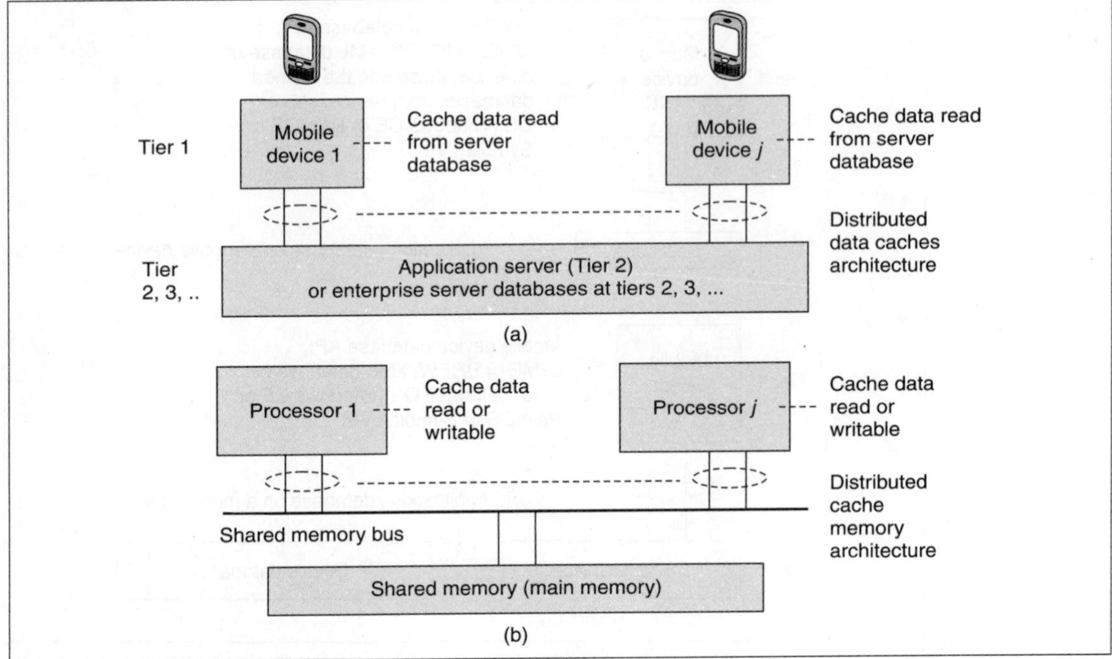

Fig. 7.2 (a) Distributed data caches in mobile devices (b) Similar architecture for a distributed cache memory in multiprocessor systems

the mobile device (first tier) uses the cached local copy. At the first tier (tier 1), the API uses the cached data record(s) using the computing architecture shown in Fig. 7.1. From tier 2 or tier 3, the server retrieves and transmits the data record(s) to tier 1 using business logic and sends and synchronizes the local copies at the device. These local copies function as device caches. This method of computing by an application is called client–server computing, as the devices are the clients of the server. Section 7.3 gives the details of this method.

The advantage of hoarding is that there is no access latency (delay in retrieving the queried record from the server over wireless mobile networks). The client device API has instantaneous data access to hoarded or cached data. After a device caches the data distributed by the server, the data is hoarded at the device. The disadvantage of hoarding is that the consistency of the cached data with the database at the server needs to be maintained.

7.2 Data Caching

Hoarded copies of the databases at the servers are distributed or transmitted to the mobile devices from the enterprise servers or application databases. The devices cache and hoard the required database records. Figure 7.2(a) shows the architecture of distributed data caches in mobile devices and a similar architecture of distributed cache memory in multiprocessor systems is shown in Fig. 7.2(b). The copies cached at the devices are equivalent to the cache memories at the processors in a multiprocessor system with a shared main memory and copies of the main memory data stored at different locations.

Cache Access Protocols A client device caches the pushed (disseminated) data records from a server (Section 8.2.1). Caching of the pushed data leads to a reduced access interval as compared to the pull (on-demand) mode of data fetching (Section 8.2.2). Further, it also reduces the dependence on pushing precedence at the server. Caching of data records can be based on pushed 'hot records' (the most needed database records at the client device). Also, caching can be based on the ratio of two parameters—access probability (at the device) and pushing rates (from the server) for each record. This method is called cost-based data replacement or caching. However, the access probability of each record and its comparison with probabilities of other records is often unpredictable in the wireless environment. An alternative method is that the least frequently pushed records and the pushed records having larger access time are placed in the database at the device. This access method, therefore, use the ratio of two parameters—average access time between two successive instances of access to the record and pushing rates for the record.

Pre-fetching Pre-fetching is another alternative to caching of disseminated data. The process of pre-fetching entails requesting for and pulling records that may be required later. The client device can pre-fetch instead of caching from the pushed records keeping future needs in view. Pre-fetching reduces server load (Section 8.2.2). Further, the cost of cache-misses can thus be reduced. The term 'cost of cache-misses' refers to the time taken in accessing a record at the server in case that record is not found in the device database when required by the device API.

7.2.1 Caching Invalidation Mechanisms

A cached record at the client device may be invalidated. This may be due to expiry or modification of the record at the database server. Cache invalidation is a process by which a cached data item or record becomes invalid and thus unusable because of modification, expiry, or invalidation at another computing system or server. Cache invalidation mechanisms are means by which the server conveys this information to client devices. Figure 7.2 depicts the similarity between the architecture of distribution of databases at the mobile device caches and the cache and shared memory data architecture in multiprocessor systems. This similarity between the two systems extends to the cache invalidation mechanism. Just as cache invalidation mechanisms are used to synchronize the data at other processors whenever the cache-data is written (modified) by a processor in a multiprocessor system, cache invalidation mechanisms are also active in the case of mobile devices having distributed copies from the server.

A cache consists of several records. Each record is called a cache-line, copies of which can be stored at other devices or servers. The cache at the mobile devices or server databases at any given time can be assigned one of four possible tags indicating its state—*modified* (after rewriting), *exclusive*, *shared*, and *invalidated* (after expiry or when new data becomes available) at any given instance. These four states are indicated by the letters M, E, S, and I, respectively (MESI). Cache-invalidation mechanisms under the MESI protocol entail that each record (line) in a cache has a tag to specify its state at any given instant and the tag is updated (modified) as soon as the state of the record changes. The tag always indicates one of the four states—*M, E, S,* or *I*—of the line at a device or server. The states indicated by the various tags are as follows:

(a) The *E* tag indicates the *exclusive* state which means that the data record is for internal use and cannot be used by any other device or server.

(b) The *S* tag indicates the *shared* state which indicates that the data record can be used by others.

(c) The *M* tag indicates the *modified* state which means that the device cache or server has rewritten a database record, which can be shared and used for computations and the modified state is to be noted on access initiated by the client.

(d) The *I* tag indicates the *invalidated* state which means that the server database no longer has a copy of the record which was shared and used for computations earlier. The invalidated state is reported by the server to the clients so that it can be noted by the clients.

Figure 7.3 shows the four possible states of a data record *i* at any instant in the server database and its copy at the cache of the mobile device *j*. The state of a data record *i* at the *j*th device can be any one of the following: E_{mi} (when it is exclusive at the device and the server does not modify it), S_{mi} (when it is shared with the server or other devices), M_{mi} (when it is modified at the device after new record copy of *i* is cached from server), or I_{mi} (when the server has sent invalidation report for *i* after invalidation report from the server for *i*. The state of a data record *i* at the server can be any one of the following: E_{si} (when exclusive at server, the device record does not affect it), S_{si} (when shared with other devices or servers), M_{si} (when modified at the server), or I_{si} (when it is invalidated at the server and invalidation report is sent to the devices).

Another important factor for cache maintenance in a mobile environment is *cache consistency* (also called *cache coherence*). This requires a mechanism to ensure that a database record is identical at the server as well as at the device caches and that only the valid cache records are used for computations. If the database server changes a data record (either read-only or writable), of which one or more client devices have copies (read-only), then the server-record is invalidated, forcing all client devices that currently have copies of the record to give up (not use further in computations) their cache-copies. Client devices can get new copies of the updated

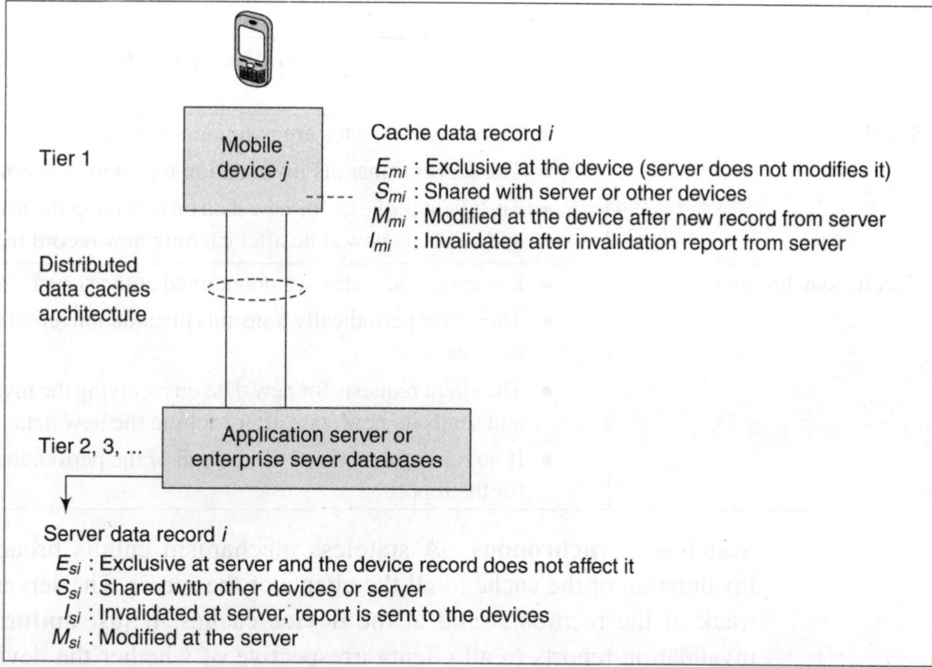

Fig. 7.3 Four possible states (*M*, *E*, *S*, or *I*) of a data record *i* at any instance at the server database and device *j* cache

record on request. Access in shared database systems entails that each device either send a request message to the server that contains a record to get a modified copy of the record or wait for the data-record pushes (during periodic broadcasts) from the server (Section 8.2.1). Cache invalidation mechanisms in mobile devices are triggered or initiated by the server. There are four possible invalidation mechanisms—stateless asynchronous, stateless synchronous, stateful asynchronous, and stateful synchronous. Table 7.1 gives the procedure used in each of these. These mechanisms are described in detail below.

Table 7.1 Cache invalidation mechanisms

Mechanism	*Procedure*
Stateless asynchronous	• Device cache states are not maintained at the server. • The server advertises invalidation report. • On receiving an invalidation report, the client requests for or caches the new data record.
Stateless synchronous	• Device cache states are not maintained at the server. • The server periodically advertises invalidation report. • The client requests for or caches the data on receiving the invalidation report.

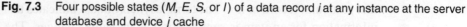

(*Contd*)

(*Contd*)

	• If no report is received till the end of the period, the client requests for the report.
Stateful asynchronous	• Device cache states are maintained at the server. • The server transmits invalidation report to concerned devices. • The client requests for new data on receiving the invalidation report and sends its new state after caching new record from the server.
Stateful synchronous	• Device cache states are maintained at the server. • The server periodically transmits invalidation report to the concerned devices. • The client requests for new data on receiving the invalidation report, and sends its new state after caching the new data. • If no report is received till the end of the period, the client requests for the report.

Stateless Asynchronous A stateless mechanism entails broadcasting of the invalidation of the cache to all the clients of the server. The server does not keep track of the records stored at the device caches. It just uniformly broadcasts invalidation reports to all clients irrespective of whether the device cache holds that particular record or not. The term 'asynchronous' indicates that the invalidation information for an item is sent as soon as its value changes. The server does not keep the information of the present state (whether E_{mi}, M_{mi}, S_{mi}, or I_{mi}) of a data-record in cache for broadcasting later. The server advertises the invalidation information only. The client can either request for a modified copy of the record or cache the relevant record when data is pushed from the server. The server advertises as and when the corresponding data-record at the server is invalidated and modified (deleted or replaced), for example, on $S_{si} \rightarrow I_{si}$ or $S_{si} \rightarrow M_{si}$ transition, the state of record i becomes invalid and the I_{mi} or M_{mi} information is transmitted from the server. Figure 7.4(a) shows the stateless asynchronous mechanism.

The advantage of the asynchronous approach is that there are no frequent, unnecessary transfers of data reports, thus making the mechanism more bandwidth efficient. The disadvantages of this approach are—(a) every client device gets an invalidation report, whether that client requires that copy or not and (b) client devices presume that as long as there is no invalidation report, the copy is valid for use in computations. Therefore, even when there is link failure, the devices may be using the invalidated data and the server is unaware of state changes at the clients after it sends the invalidation report.

Stateless Synchronous This is also a stateless mode, i.e., the server has no information regarding the present state of data records at the device caches and broadcasts to all client devices. However, unlike the asynchronous mechanism, here the server advertises invalidation information at periodic intervals as well as whenever the corresponding data-record at server is invalidated or modified. This method ensures synchronization because even if the in-between period report is not detected by the device due to a link failure, the device expects the period-end report of invalidation and if that is not received at the end of the period, then the device sends a request for the same (deleted or replaced). In case the client device does not get the periodic report due to link failure, it requests the server to send the report.

The advantage of the synchronous approach is that the client devices receive periodic information regarding invalidity (and thus validity) of the data caches. The periodic invalidation reports lead to greater reliability of cached data as update requests for invalid data can be sent to the server by the device-client. This also helps the server and devices maintain cache consistency through periodical exchanges. The disadvantages of this mode of cache invalidation are—(a) unnecessary transfers of data invalidation reports take place, (b) every client device gets an advertised invalidation report periodically, irrespective of whether that client has a copy of the invalidated data or not, and (c) during the period between two invalidation reports, the client devices assume that, as long as there is no invalidation report, the copy is valid for use in computations. Therefore, when there are link failures, the devices use data which has been invalidated in the in-between period and the server is unaware of state changes at the clients after it sends the invalidation report.

Stateful Asynchronous The stateful asynchronous mechanism is also referred to as the AS (asynchronous stateful) scheme. The term 'stateful' indicates that the cache invalidation reports are sent only to the affected client devices and not broadcasted to all. The server stores the information regarding the present state (a record *i* can have its state as E_{mi}, M_{mi}, S_{mi}, or I_{mi}) of each data-record at the client device caches. This state information is stored in the home location cache (HLC) at the server. The HLC is maintained by an HA (home agent) software. This is similar to the HLR (Section 3.1.2) at the MSC in a mobile network. The client device informs the HA of the state of each record to enable storage of the same at the HLC. The server transmits the invalidation information as and when the records are invalidated and it transmits only to the device-clients which are affected by the invalidation of data. Based on the invalidation information, these device-clients then request the server for new or modified data to replace the invalidated data. After the data records transmitted by the server modify the client device cache, the device sends information about the new state to the server so that the record of the cache-states at the server is also modified.

The advantage of the stateful asynchronous approach is that the server keeps track of the state of cached data at the client device. This enables the server to synchronize with the state of records at the device cache and keep the HLC updated. The stateful asynchronous mode is also advantageous in that only the affected clients receive the invalidation reports and other devices are not flooded with irrelevant reports. The disadvantage of the AS scheme is that the client devices presume that, as long as there is no invalidation report, the copy is valid for use in computations. Therefore, when there is a link failure, then the devices use invalidated data.

Stateful Synchronous As is clear from the name, the stateful synchronous stage is an amalgamation of the stateful and the synchronous modes of cache invalidation. This means that the cache invalidation reports are sent only to the affected clients (stateful) and that they are sent periodically (synchronous). The server keeps the information of the present state (E_{mi}, M_{mi}, S_{mi}, or I_{mi}) of data-records at the client-caches. The server stores the cache record state at the home location cache (HLC) using the home agent (HA). The server transmits the invalidation information at periodic intervals to the clients and whenever the data-record relevant to the client is invalidated or modified (deleted or replaced) at the server. This method ensures synchronization because even if the in-between period report is not detected by the device due to a link failure, the device expects the period-end report of invalidation and if it is not received at the end of the period, then the device requests for the same.

The advantage of the stateful synchronous approach is that there are reports identifying invalidity (and thus, indirectly, of validity) of data caches at periodic intervals and that the server also periodically updates the client-cache states stored in the HLC. This enables to synchronize with the client device when invalid data gets modified and becomes valid. Moreover, since the invalidation report is sent periodically, if a device does not receive an invalidation report after the specified period of time, it can request the server to send the report. Each client can thus be periodically updated of any modifications at the server. When the invalidation report is not received after the designated period and a link failure is found at the device, the device does not use the invalidated data. Instead it requests the server for an invalidation update. The disadvantage of the stateful synchronous approach is the high bandwidth requirement to enable periodic transmission of invalidation reports to each device and updating requests from each client device.

Use of Home Location Cache through Home Agent Figure 7.4 shows the four-step stateful asynchronous cache invalidation mechanism for database–cache consistency maintenance. The figure also shows the use of the home location cache (HLC) and the home agent (HA). In the stateless asynchronous mechanism, steps 1 to 3 take place and the server does not use the HLC and the HA. However, in the stateless synchronous mechanism, steps 1 to 4 take place at periodic intervals and the server uses the HLC and the HA at periodic intervals. In the stateful synchronous mechanism, steps 1 to 3 take place periodically, but the server does not maintain the HLC and uses the HA only.

Cache Invalidation Reports Each record at the server is assigned an identifier, which may be a number or code. Assume that the identifier for the data record *i* is

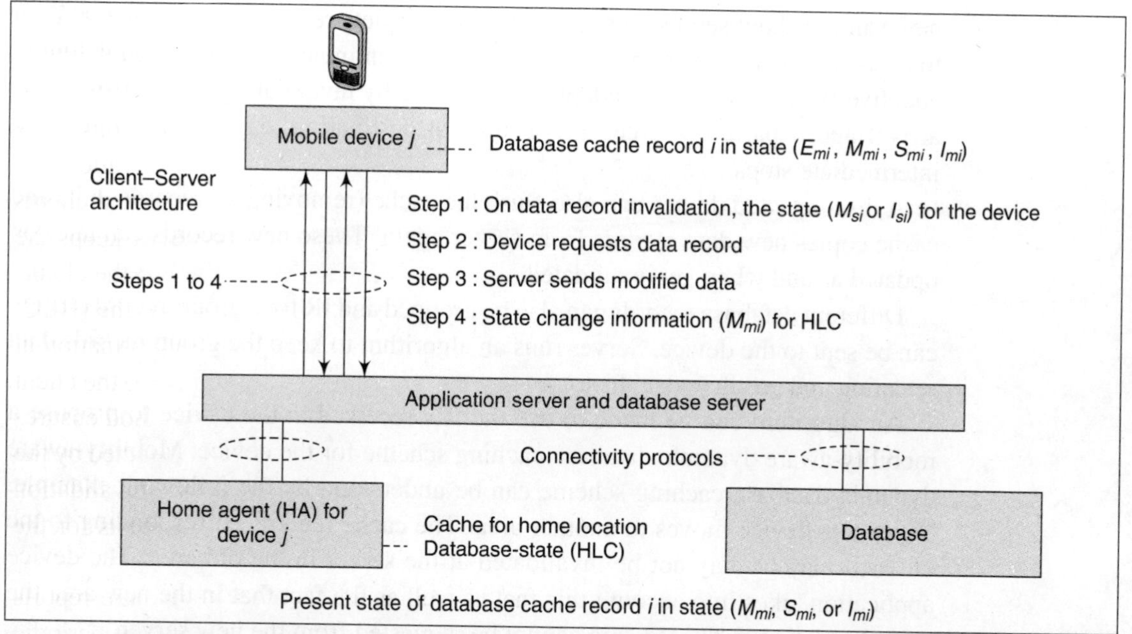

Fig. 7.4 Four steps asynchronous stateful cache invalidation mechanism for database-cache consistency maintenance and use of HLC and HA

id_i and that record i has been pushed to or pulled by the *j*th device. After a certain period, the record becomes invalid or modified. A report must therefore be sent by the server saying that it has been invalidated or modified at the server and therefore the cache should also modify its state as invalidated, take steps to cache it again, and modify it. Therefore, a cache invalidation report (IR) for record *i* is sent to the *j*th device (and to other devices). Cache IR consists of the identifier id_i of each record and its state as M or I (state is M when it modified or I when it invalidates). The server notifies the devices of IRs and the changes in the records. However, only the report is sent by the server and not the complete record *i.* This helps in serving the data records to the devices in case of limited bandwidth availability. Along with an IR, a brief-copy or incremental in the record can also be sent. Brief copy means a very short copy of the new record can be sent so that the device can choose whether the complete data record is to be invalidated and modified by request to the server or some changes at the device itself will suffice. The device may also decide to ignore the change. Incremental copy means only the increments or additional insertions in the record can be sent with the IR so that the device can take note of these and invalidate or modify them without requesting the server for a full copy. Alternatively, an adaptive IR can be sent by the server in response to on-demand queries (user device requests for a data record from the database). Adaptive IR means that the IR contains new record information in such a way that the information contained in the IR enables the device to adapt the previous copy to a new value without sending a request for the complete record. For example, a flight to a destination is discontinued. Only flight-discontinuation information is sent in adaptive IR and the devices adapt to the change by invalidating all information of arrival and departure source, arrival time at destination, days of the flight, and intermediate stops.

A disconnected device can also flush the cache (removing all records in it) and cache copies new data records from server again. These new records can later on updated as and when get invalidated.

Different database records can also be grouped and IR for a group-record change can be sent to the device. Server runs an algorithm to keep the group records data retention and group update history.

An algorithm can be made to use the IRs received at the device to propose a mobility-aware dynamic database-caching scheme for the device. Mobility-aware dynamic database-caching scheme can be understood by the following example. Suppose a device moves to another area. The cache records corresponding to the old area may or may not be invalidated at the server in the old area. The device application takes into account this fact as well as the fact that in the new area the records relating to the old area cannot be requested from the new server.

The cache invalidation scheme and appropriate use of IRs at the device improves the system performance greatly when scheme is compared to full data copying (replication) scheme for a device.

7.2.2 Data Cache Maintenance in Mobile Environments

Assume that a device needs a data-record during an application. A request must be sent to the server for the data record (this mechanism is called pulling). The time taken for the application software to access a particular record is known as *access latency*. Caching and hoarding the record at the device reduces access latency to zero. Therefore, data cache maintenance is necessary in a mobile environment to overcome access latency. *Data cache inconsistency* means that data records cached for applications are not invalidated at the device when modified at the server but not modified at the device. Data cache consistency can be maintained by the three methods given below:

(i) *Cache invalidation mechanism* (server-initiated case): the server sends invalidation reports on invalidation of records (asynchronous) or at regular intervals (synchronous), as described in Section 7.2.1.

(ii) *Polling mechanism* (client-initiated case): Polling means checking from the server, the state of data record whether the record is in the valid, invalid, modified, or exclusive state. Each cached record copy is polled whenever required by the application software during computation. The device connects to the server and finds out whether the cached data record copy at the device has become invalid or has been modified at the server. If the record is found to be modified or invalidated, then the device requests for the modified data and replaces the earlier cached record copy.

(iii) *Time-to-live mechanism* (client-initiated case): Each cached record is assigned a TTL (time-to-live). The TTL assignment is adaptive (adjustable) previous update intervals of that record. After the end of the TTL, the client device requests the server to check whether the cached data record is invalid or modified. If it is so, then the device requests the server to replace the invalid cached record with the modified data. When TTL is set to 0, the TTL mechanism is equivalent to the polling mechanism.

7.2.3 Web Cache Maintenance in Mobile Environments

The mobile devices or their servers can be connected to a web server (e.g., traffic information server or train information server). Web cache at the device stores the web server data and maintains it in a manner similar to the cache maintenance for server data described above (Section 7.2.2). If an application running at the device needs a data record from the web which is not at the web cache, then there is access latency. *Access latency* is the time required to access the data record from the remote system or server. Web cache maintenance is necessary in a mobile environment to overcome access latency in downloading from websites due to disconnections. Web cache consistency can be maintained by two methods. These are:

(i) Time-to-live (TTL) mechanism (client-initiated case): The method is identical to the one discussed for data cache maintenance in Section 7.2.2.

(ii) Power-aware computing mechanism (client-initiated case): Each web cache maintained at the device can also store the CRC (cyclic redundancy check) bits. Assume that there are N cached bits and n CRC bits. N is much greater than n. Similarly at the server, n CRC bits are stored. As long as there is consistency between the server and device records, the CRC bits at both are identical. Whenever any of the records cached at the server is modified, the corresponding CRC bits at the server are also modified. After the TTL expires or on-demand for the web cache records by the client API, the cached record CRC is polled and obtained from the website server. If the n CRC bits at server are found to be modified and the change is found to be much higher than a given threshold (i.e., a significant change), then the modified part of the website hypertext or database is retrieved by the client device for use by the API. However, if the change is minor, then the API uses the previous cache. Since $N >> n$, the power dissipated in the web cache maintenance method (in which invalidation reports and all invalidated record bits are transmitted) is much greater than that in the present method (in which the device polls for the significant change in the CRC bits at server and the records are transmitted only when there is a significant change in the CRC bits).

7.3 Client–Server Computing and Adaptation

Consider a network of distributed nodes (computers and computing devices). The network architecture can be designed such that a node is either a client or a server. A client node runs application software which depends on server node resources (files, databases, web pages, other resources, processor power, or other devices or computers connected or networked to it). The server node has larger resources and computing power than the client nodes. This architecture is known as *client–server computing architecture*. It is different from peer-to-peer architecture, where each node on the network has similar resources and the various nodes can depend on each other resources.

Mobile devices function as client nodes due to their resource constraints. A server networks to a number of devices. Client–server architecture is used for mobile computing. Section 7.3.1 describes the client–server computing and how the clients receive required database copies through the server(s). Section 7.3.2 describes the client–server computing with adaptation software.

7.3.1 Client–Server Computing

Client–server computing is a distributed computing architecture, in which there are two types of nodes, i.e., the clients and the servers. A client can request the server for data or responses, which the client then uses in computations. The client can either access the data records at the server or it can cache these records at the client device. The data can be accessed either on client request or through broadcasts or distribution from the server. The client and the server can be on the same computing system or on different computing systems. Client–server computing can have an N-tier architecture ($N = 1, 2, \ldots$). When the client and the server are on the same computing system (not on a network), then the number of tiers, $N = 1$. When the client and the server are on different computing systems on the network, then $N = 2$. Assume that the server networks or connects to other computing systems which provide additional resources to the server for the client then $N > 2$. $N > 1$ means that the client device at tier 1 connects to the server at tier 2 which, in turn, may connect to other tiers, 3, 4, and so on. The client device connects to or synchronizes with higher tiers for data. A command interchange protocol (e.g., HTTP) is used for obtaining the client requests at the server or the server responses at the client.

The following subsections describe client–server computing in 2, 3, or N-tier architectures. Each tier connects to the other with a connecting, synchronizing, data, or command interchange protocol. Examples of data or command interchange protocols are Java Remote Method Invocation (RMI) and C++-based remote procedure call (RPC).

7.3.1.1 Two-tier Client–Server Architecture

Figure 7.5 shows the application server at the second tier. The data records are retrieved using business logic and a synchronization server in the application server synchronizes with the local copies at the mobile devices. Figure 7.5 shows a structure for retrieving data records for computations and a database interchange mechanism. It shows two-tier architecture in a client–server computing environment. A client is a program, for example, an API to retrieve records from databases. A server is a program that connects to databases and sends the output (response) to the client. A server is defined as a computing system, which responds to requests from one or more clients. A client is defined as a computing system, which requests the server for a resource or for executing a task.

A specific application may require an application server to send a local copy of data to multiple devices. This enables running of the application independently on the devices without the need for a run-time retrieval. The various APIs synchronize with each other through a synchronization API. Synchronization means that when copies of records at the server-end are modified, the copies cached at the client devices should also be accordingly modified. The APIs are designed independent of hardware and software platforms as far as possible as different devices may have

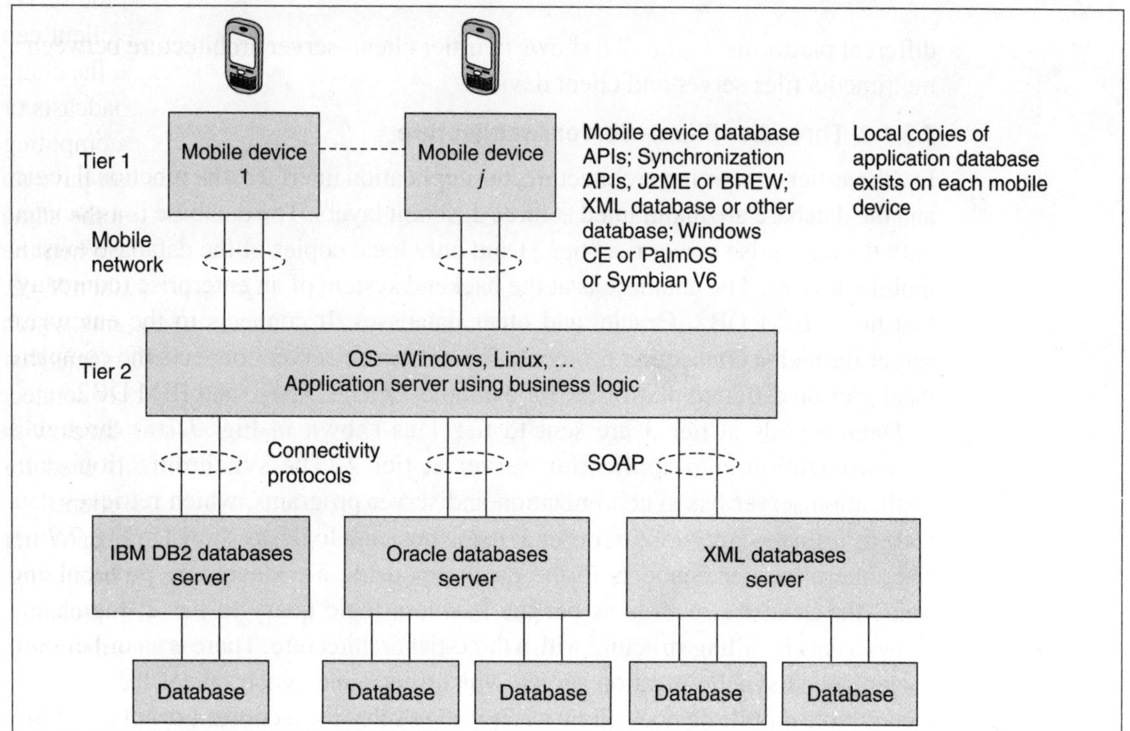

Fig. 7.5 Application server in two-tier client–server computing architecture (local copies 1 to *i* of database hoard at the mobile devices)

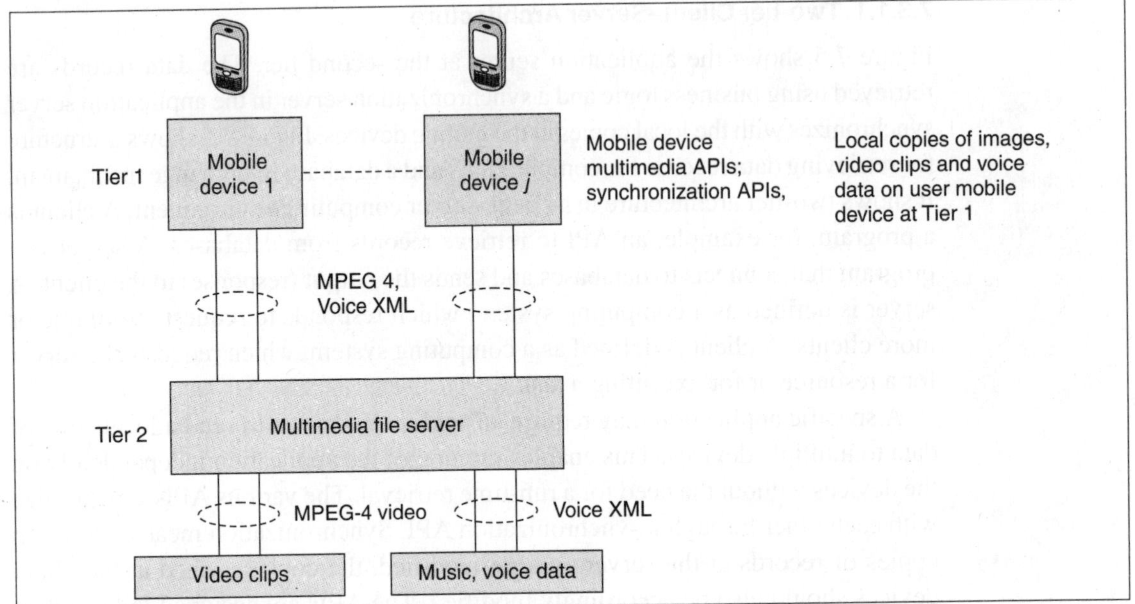

Fig. 7.6 Multimedia file server in two-tier client–server computing architecture (local copies 1 to *i* of image and voice hoarding at the mobile devices)

different platforms. Figure 7.6 shows two-tier client–server architecture between a multimedia files server and client devices.

7.3.1.2 Three-tier Client–Server Architecture

In a three-tier computing architecture, the application interface, the functional logic, and the database are maintained at three different layers. The database is associated with the enterprise server tier (tier 3) and only local copies of the database exist at mobile devices. The database is at the backend system of an enterprise (company) that holds IBM DB2, Oracle, and other databases. It connects to the enterprise server through a connecting protocol. The enterprise server connects the complete databases on different platforms, for example, Oracle, XML, and IBM DB2.

Data records at tier 3 are sent to tier 1 as shown in Fig. 7.7(a) through a synchronization-cum-application server at tier 2. The synchronization-cum-application server has synchronization and server programs, which retrieves data records from the enterprise tier (tier 3) using business logic, as shown in Fig. 7.7(a). The enterprise tier connects to the databases using a connectivity protocol and sends the database records as per the business logic query to tier 2. Figure 7.7 shows a data hoarding structure with a three-tier architecture. There is an in-between server, called synchronization server, which sends and synchronizes the copies at the multiple mobile devices. Figure 7.7(a) shows that local copies 1 to *i* of databases are hoarded at the mobile devices for the applications 1 to *j*. The figure also shows enterprise database connection to a synchronization server, which synchronizes the

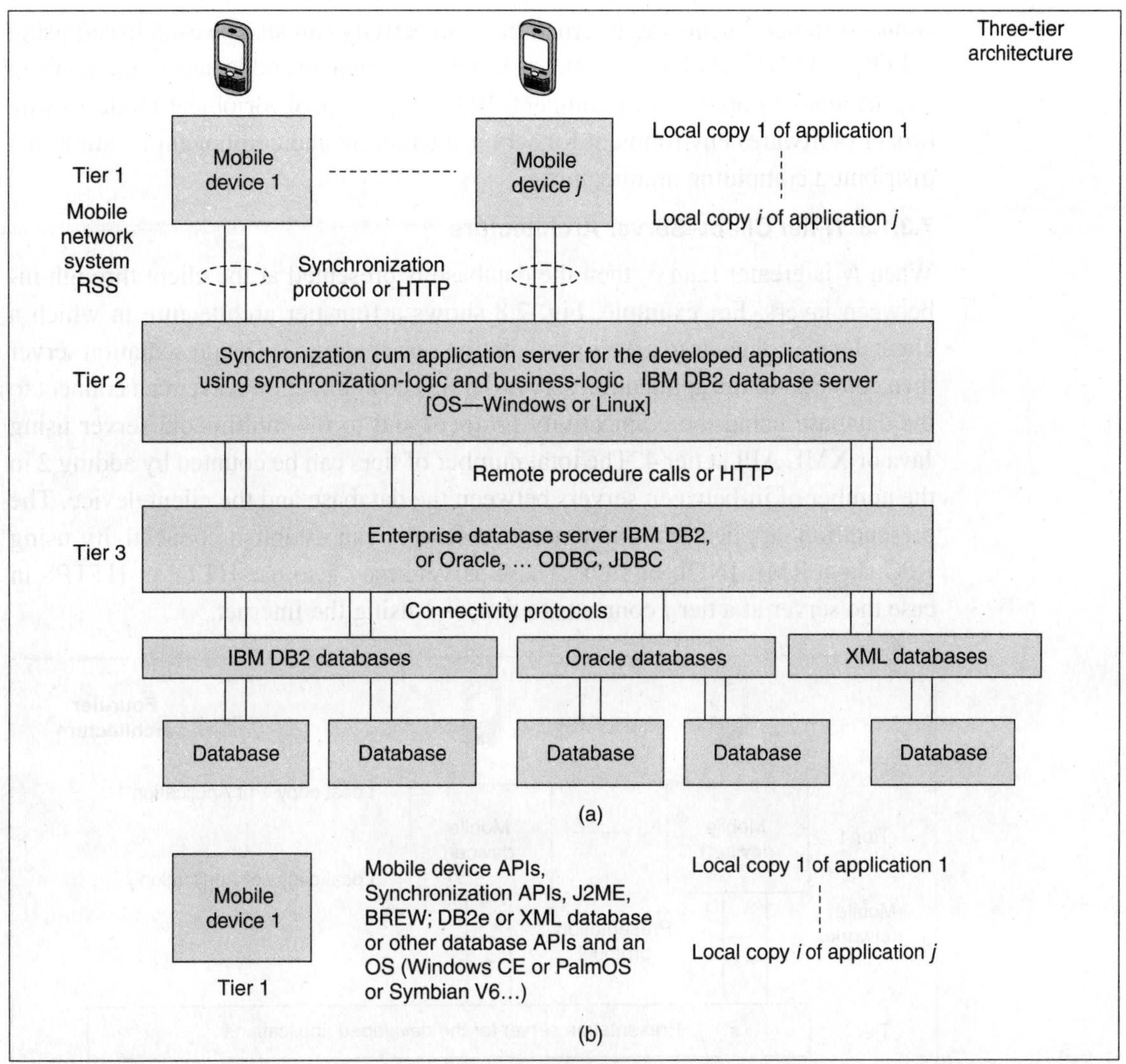

Fig. 7.7 (a) Local copies 1 to *i* of database hoarded at the mobile devices using an enterprise database connection to a synchronization server, which synchronizes the required local copies for application with the enterprise database server (b) Mobile device with J2ME or BREW platform, APIs an OS and database having local copies

local copies required by the applications with the enterprise database server. HTTP can also be used as synchronization protocol in case of a Web-client at tier 1.

Figure 7.7(b) shows the mobile devices with J2ME or BREW platforms, APIs for database applications, the OS, and local copies of the database. The connectivity of the synchronization-cum-application server to the enterprise server is by RPC, RMI, JNDI (Java naming and directory interface), or IIOP [Internet inter-ORB (object request broken) protocol] protocols. In case the application client at tier 1 connects to tier 2 using the Internet, the connectivity can also be established using HTTP or HTTPS. [JNDI is a protocol in Java to organize and locate components in a distributed computing environment. IIOP is a protocol for object request using broker (software) environment for server interaction and component location in a distributed computing architecture.]

7.3.1.3 *N*-tier Client–Server Architecture

When N is greater than 3, then the database is presented at the client through in-between layers. For example, Fig. 7.8 shows a four-tier architecture in which a client device connects to a data-presentation server at tier 2. The presentation server then connects to the application server tier 3. The application server can connect to the database using the connectivity protocol and to the multimedia server using Java or XML API at tier 4. The total number of tiers can be counted by adding 2 to the number of in-between servers between the database and the client device. The presentation, application, and enterprise servers can establish connectivity using RPC, Java RMI, JNDI, or IIOP. These servers may also use HTTP or HTTPS in case the server at a tier j connects to tier $j + 1$ using the Internet.

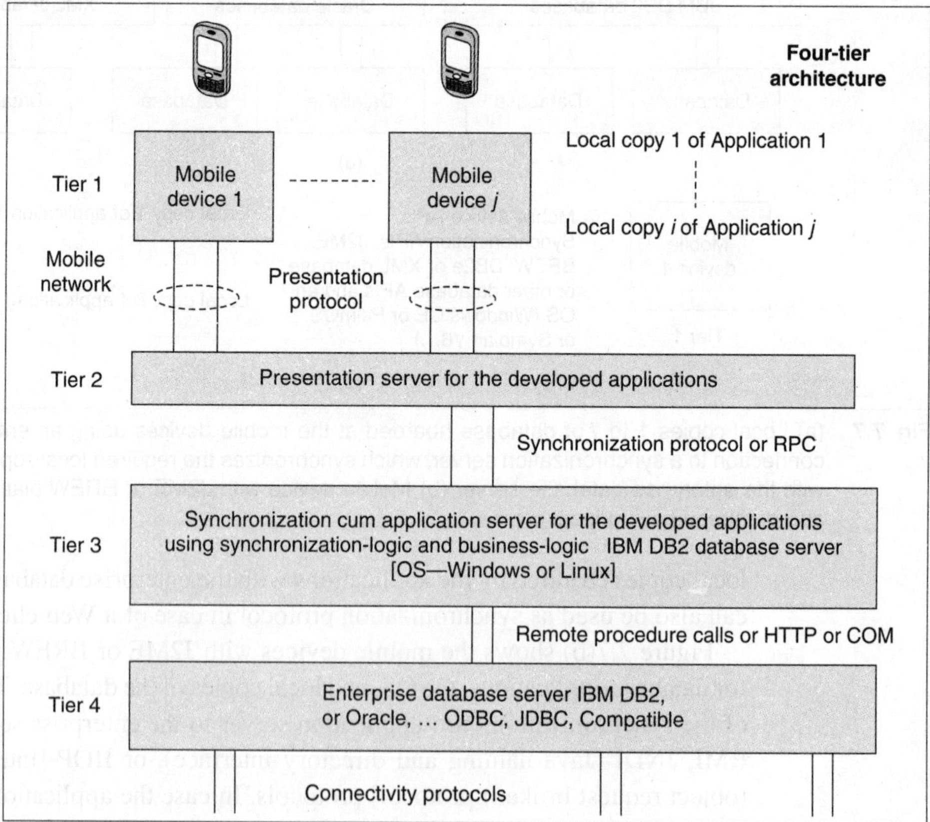

Fig. 7.8 4-tier architecture in which a client device connects to a data-presentation server

7.3.2 Client–Server Computing with Adaptation

As discussed in Section 2.1.2, a mobile device has a number of interfaces (APIs), for example, the PIM (personal information manager) interfaces for the *calendar* and *Contacts*, Microsoft Outlook 98, 2000, 2002, 2003, Microsoft Outlook Express/ Windows Address Book, Intellisync Wireless e-mail, Lotus Notes (5x and 6x), and Lotus Organizer (5x and 6x) and the APIs for IBM WebSphere Everyplace Access (WEA), BlackBerry Connect, Oracle Collaboration Suite, Secure Mobile Connections via VPN (virtual private network) Client and Symantec Client Security 3.0, Fujitsu mProcess Business Process Mobilizer. Since the data formats of data transmitted from the synchronization server and those required for the device database and device APIs are different in different cases, there are two adapters (adaptation software) at a mobile device—an adapter for standard data format for synchronization at the mobile device and another adapter for the backend database copy, which is in a different data format for the API at the mobile device. An *adapter* is a software to get data in one format or data governed by one protocol and covert it to another format or to data governed by another protocol.

Figure 7.9 shows an API, database, and adapters at a mobile device and the adapters at the synchronization, application, or enterprise servers. This is modified

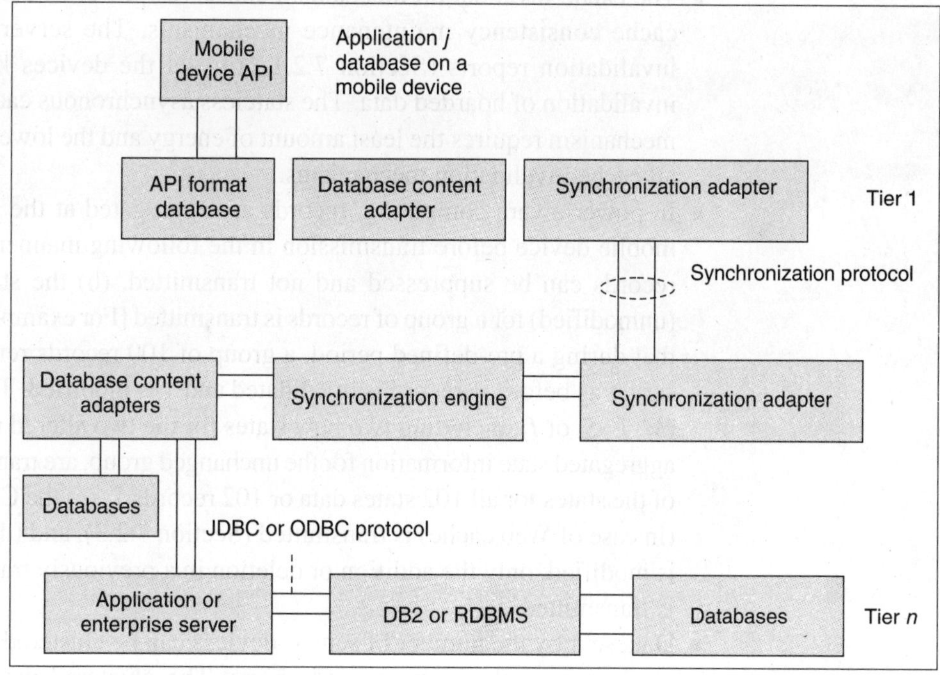

Fig. 7.9 APIs, database, and adapters at a mobile device and the adapters at the synchronization or application or enterprise server

database architecture. It is based on the architectures shown in Figs 7.1(a) and 7.5. Here the adapters are an addition used for interchange between standard data formats and data formats for the API. WEA provides adapters for synchronization objects (e.g., XML format synchronization objects) and the objects of API databases (e.g., for the PIM APIs).

7.3.3 Power-aware Computing

Mobile computing processes must be energy efficient as the power resources at mobile devices are limited due size constraints and mobility requirements. Devices run on batteries which need to be charged frequently. Power-aware computing takes into account these constraints and devises methods to cut down the energy requirements of computing processes in mobile devices. In certain cases, for example, in sensors or labels, the power is derived from the RF energy received from the signals. Power-aware computing is required to run mobile devices efficiently. The following methodologies are used in power-aware computing:

- Data caching at the devices conserves power as multiple requests (for data) made by uplinking need more energy. Moreover, the server's power resources are not limited like those of the client device. So the server can advertise the data records and the devices can caches the frequently required records.
- The cache invalidation mechanism conserves power as compared to other cache consistency maintenance mechanisms. The server advertises the invalidation reports (Section 7.2.1.2) to let the devices know about the invalidation of hoarded data. The stateless asynchronous cache invalidation mechanism requires the least amount of energy and the lowest bandwidth of all cache invalidation mechanisms.
- In power-aware computing, records are aggregated at the server or at the mobile device before transmission in the following manner—(a) duplicate records can be suppressed and not transmitted, (b) the state-information (unmodified) for a group of records is transmitted [For example, let us assume that during a pre-defined period, a group of 100 records remains same at a server as before, 1 record is invalidated and 1 is modified. Then three states (*M*, *E*, *S*, or *I*), including two new states for the two altered records and one aggregated state information for the unchanged group, are transmitted, instead of the states for all 102 states data or 102 records.] , (c) the CRC information (in case of Web cache) is transmitted (Section 7.2.3), and (d) when a record is modified, only the addition or deletion in a previously transmitted record is transmitted.
- Data sent by the number of sensor-devices can be clustered and aggregated at a server-node as mentioned above. The clustered data record server communicates the aggregated data to a base station. Aggregation reduces

the power requirements as it reduces the number of packets or packet size for transmission.

- Protocol optimization is used in power-aware computing. Optimized protocols use smaller size headers and need less frequent round trips than un-optimized protocols.

7.3.4 Context-aware Computing

The context of a mobile device represents the circumstances, situations, applications, or physical environment under which the device is being used. For example, the context is *student* when the device is used to download faculty lectures or PowerPoint slides. Let us assume that a mobile phone is operating in a busy, congested area. If the device is *aware* of the surrounding noises, then during the conversation, it can raise the speaker volume by itself and when the user leaves that area, the device can again reduce the volume. Also, if there is intermittent loss of connectivity during the conversation, the device can introduce background noises by itself so that the user does not feel discomfort due to intermittent periods of silence. This is one example in which the computing system is aware of the surrounding physical context in which the conversation is taking place.

A context-aware computing system is one which has user, device, and application interfaces such that, using these, the system remains aware of the past and present surrounding situations, circumstances, or actions such as the present mobile network, surrounding devices or systems, changes in the state of the connecting network, physical parameters such as present time of the day, presently remaining memory and battery power, presently available nearest connectivity, past sequence of actions of the device user, past sequence of application or applications, and previously cached data records, and takes these into account during computations. The following subsections describe meaning of context, context-aware computing, and context types used in context-aware computing.

7.3.4.1 Context

The dictionary meaning of the word 'context' is 'the circumstances that form the setting of an event, statement, or idea, and in terms of which it can be fully understood'. The term 'context' refers to the interrelated conditions in which a collection of elements, records, components, or entities exists or occurs. Each message, data record, element, or entity has a meaning. But when these are considered along with the conditions that relate them to each other and to the environment, then they have a wider meaning. Understanding of the context in which a device is meant to operate, results in better, more efficient computing strategies.

Structural Context To explain what is meant by structural context let us consider a few examples of records with structural arrangement. The fields *name*, *address*, *experience*, and *achievements* of a person have an individual meaning. However, when put together to form a résumé, these fields acquire a significance beyond their individual meanings. This significance comes from the fact that they are now arranged in a structure which indicates an interrelationship between them. The structure of the résumé includes the records and their interrelationship and thus defines a context for these records. Whereby, the records have a new meaning in the context of the résumé (which is a structure). Contexts such as the context of the résumé of an individual are called structural contexts. The context in such cases comes from the structure or format in which the records in a database are organized. Consider another example, this time that of a line in a telephone directory. It has a sequence of records including a name, an address, and a 10-digit number. Each record has an individual meaning. But a collection of these records shows an interrelationship and thus defines a context, i.e., a telephone directory.

Implicit and Explicit Contexts Context may be implicit or explicit. Implicit context provides for omissions by leaving out unimportant details, takes independent world-views, and performs alterations in order to cope with incompatible protocols, interfaces, or APIs by transparently changing the messages. Implicit context uses history to examine call history, to manage omissions, or to determine recipients and performs contextual message alterations. It provides for and manages transitions at the boundaries between world-views where contextual dispatches occur. Consider the context *'Contacts'* which has a set of contacts. Implicit context takes independent world-views and does not require components to have global view of the system. The name, e-mail ID, and telephone number are implicit in a *contact* in the context Contacts. When a computing device uses a contact to call a number using a name record, the system takes independent view and uses the telephone number implicitly and deploys CDMA or GSM protocols for connecting to the mobile network implicitly. Context CDMA is implicit in defining the records 'Contact'. When a computing system uses a contact to send an e-mail using a name record, the use of the e-mail ID record is implicit to the system and the use of SMTP (simple mail transfer protocol) or other mail sending protocol is also implicit. The context of the mobile service protocol, mail transfer protocol, and use of specific interfaces and software are implicit too. The inter-relationship between the fields (name, telephone number, e-mail ID), the protocols and the APIs needed to use the contacts is implicit and is not defined. There is no condition on structural positioning for maintaining the sequence name, then e-mail ID, and then telephone number in a contact or contacts in the structure 'Contacts'. Implicit context performs alterations and copes with incompatible interfaces, protocols, etc. by transparently changing the messages. For example, name gets automatically altered to e-mail ID when the context is sending of e-mail. The implicit context also copes with incompatible interfaces, for example, mail sending and receiving software handling data in different formats. Consider the context *document*. In *document* context, the contact or personal information is an extrinsic context. In context to processing of a document, the existence of document author contact information is extrinsic. The *contacts* context is imported into the *document* context to establish interrelationship between *document* and *contact*.

7.3.4.2 Context-aware Computing

Context-aware computing leads to application-aware computing. This is so because the APIs are part of the context (implicit or explicit contexts). For example, when using an e-mail ID, a mail receiving or mail sending application software is used for computing. An application can adapt itself to the context. For example, if context is a contact, the phone-talk application will adapt itself to use of the telephone number from the 'contact' and to the use of GSM or CDMA communication.

Context-aware computing leads to pervasive or ubiquitous computing. For example, consider the computing context during mobile device data-communication. Computing context includes the existence of the service discovery protocol, radio-interface, and corresponding protocol. Suppose service discovery protocol senses the context and finds that communication protocol is Bluetooth then the device uses Bluetooth to communicate. When it finds the protocol is 802.11 WiFi LAN, it uses the WiFi for communication.

Use of context in computing helps in reducing possibility of errors. It helps in reducing the ambiguity in the action(s). It helps in deciding the expected system response on computations. For example, if *name* is input in personal biodata context, then the *address*, *experience*, and *achievements*, which correspond to that name, are also required for computations. This is because all four are related and needed in biodata context. When *name* is input in telephone directory context, then the *address* and, which correspond to that name, are also required for computations. This is because all three are related in context to telephone directory. The *name* in two different contexts (personal biodata and telephone directory) during computations needs computations to perform different actions.

7.3.4.3 Context Types in Context-aware Computing

The five types of contexts that are important in context-aware computing are—physical context, computing context, user context, temporal context, and structural context.

Physical Context The context can be that of the physical environment. The parameters for defining a physical context are service disconnection, light level, noise level, and signal strength. For example, if there is service disconnection during a conversation, the mobile device can sense the change in the physical conditions and it interleaves background noise so that the listener does not feel the effects of the disconnection. Also, the mobile device can sense the light levels, so during daytime the display brightness is increased and during night time or in poor light conditions, the device display brightness is reduced. The physical context changes and the device display is adjusted accordingly.

Computing Context The context in a context-aware computing environment may also be computing context. Computing context is defined by interrelationships and conditions of the network connectivity protocol in use (Bluetooth, ZigBee, GSM, GPRS, or CDMA), bandwidth, and available resources. Examples of resources are keypad, display unit, printer, and cradle. A cradle is the unit on which the mobile device lies in order to connect to a computer in the vicinity. Consider a mobile device lying on a cradle. It discovers the computing context and uses ActiveSync (Section 2.4.1) to synchronize and download from the computer. When a mobile device lies in the vicinity of a computer with a Bluetooth interface, it discovers another computing context resource and uses wireless Bluetooth for connecting to the computer. When it functions independently and connects to a mobile network, it discovers another computing context and uses a GSM, CDMA, GPRS, or EDGE connection. The response of the system is as per the computing context, i.e., the network connectivity protocol.

User Context The user context is defined user location, user profiles, and persons near the user. Reza B'Far[1] defines user-interfaces context states as follows—'within the realm of user interfaces, we can define context as the sum of the relationships between the user interface components, the condition of the user, the primary intent of the system, and all of the other elements that allow users and computing systems to communicate.'

Temporal Context Temporal context defines the interrelation between time and the occurrence of an event or action. A group of interface components have an intrinsic or extrinsic temporal context. For example, assume that at an instant the user presses the switch for *dial* in a mobile device. At the next instant the device seeks a *number* as an input. Then user will consider it in the context of dialling and input the number to be dialled. Now, assume that at another time the user presses the switch to *add* a contact in the mobile device. The device again prompts the user to enter a *number* as an input. The user will consider it in context of the number to be added in the *contacts* and stored in the device for future use. The device then seeks the name of the contact as the input. Response of the system in such cases is as per the temporal context. The context for the VUI (voice user interface) elements also defines a temporal context (depending upon the instances and sequences in which these occur).

Structural Context Structural context defines a sequence and structure formed by the elements or records (Section 7.3.4.1). Graphic user interface (GUI) elements have structural context. Structural context may also be extrinsic for some other type of context. Interrelation among the GUI elements depends on structural positions on the display screen. When time is the context, then the hour and minute elements

[1]Reza B'Far 2005, *Mobile Computing Principles—Designing and Developing Mobile Applications with UML and XML*, Cambridge University Press, Cambridge.

structurally appear in such a manner that the hours are on the left of the minutes. In another context, the same elements may mean date or amount.

7.4 Transactional Models

A transaction is the execution of interrelated instructions in a sequence for a specific operation on a database. Database transaction models must maintain data integrity and must enforce a set of rules called ACID rules. These rules are as follows—

1. *Atomicity*: All operations of a transaction must be complete. In case, a transaction cannot be completed; it must be undone (rolled back). Operations in a transaction are assumed to be one indivisible unit (atomic unit). For example, if a transaction entails transfer of a balance (say, Rs 1000) from account *A* to account *B*, then the transfer from *A* to *B* should be completed in two operations, operation 1 subtracting the balance in *A* by 1000 and operation 2 adding in the balance Rs 1000 in *B* in totality. If both operations cannot be completed, then the transaction must be rolled back to the previous state of the accounts (i.e., the balance amounts in *A* and *B* should be rolled back to the state before the start of the transaction). Operations 1 and 2 form one atomic unit of the transaction.
2. *Consistency*: A transaction must be such that it preserves the integrity constraints and follows the declared consistency rules for a given database. Consider the transaction example discussed above. Consistency means the data is not in a contradictory state after the transaction. The amount transferred must be subtracted from account *A* and added into account *B*. Consistency means that the sum total of the balances in accounts *A* and *B* is the same as it was before the transaction.
3. *Isolation*: If two transactions are carried out simultaneously, there should not be any interference between the two. Further, any intermediate results in a transaction should be invisible to any other transaction. For example, consider two simultaneous transactions—transferring a balance from account *A* to account *B* and another from account *C* to account *A*. Then, account *B* gets the balance transfer from *A* and account *A* gets the balance transfer from *C*. Both transactions are isolated. If the two transactions were not isolated, then the balance from *C* as well as previous balance in *A* will be transferred to *B* and *A* will be left with nothing. *Serializable isolation* is a more robust form of isolation. When using serializable isolation, read and write operations appear as sequential. The read operation of a new state starts only after the write operation of the previous state is completed.
4. *Durability*: After a transaction is completed, it must persist and cannot be aborted or discarded. For example, in a transaction entailing transfer of a balance from account *A* to account *B*, once the transfer is completed and finished there should be no roll back.

A database management system (DBMS) transaction model may provide for selective relaxation in the four ACID rules. A concurrency control method ensures that the transactions are executed in a safe manner. The ACID rules permit concurrent running of multiple threads (program tasks). These multiple threads do not interfere with each other when running concurrently. A DBMS ensures that only serializable and recoverable sequences (schedules) are allowed for transactions and that no result of the committed transaction is lost after undoing an aborted transaction. Aborted transactions are caused by network failure or supply failure.

Consider a base class library included in Microsoft .NET. It has a set of computer software components called ADO.NET (ActiveX Data Objects in .NET). These can be used to access the data and data services including for access and modifying the data stored in relational database systems. The ADO.NET transaction model permits three transaction commands:

1. `BeginTransaction`: It is used to begin a transaction. Any operation after `BeginTransaction` is assumed to be a part of the transaction till the `CommitTransaction` command or the `RollbackTransaction` command. An example of a command is as follows:
   ```
   connectionA.open ( );
   transA = connectionA.BeginTransaction ( );
   ```
   Here `connectionA` and `transA` are two distinct objects.
2. `Commit`: It is used to commit the transaction operations that were carried out after the `BeginTransaction` command and up to this command. An example of this is
   ```
   transA.Commit ( );
   ```
   All statements between `BeginTransaction` and `commit` must execute atomically.
3. `Rollback`: It is used to rollback the transaction in case an exception is generated (caught) after the `BeginTransaction` command is executed. [Consider the statements `catch (e) {...};`

When a section of codes is executed and an exception condition denoted by *e* is discovered by a process, then another section of codes is executed and this process is known as *catching an exception*. Similarly, when the `BeginTransactions` command is executed and an exception is caught before the completion of the transaction, then the transaction rolls back using the statement `transA.Rollback ( ); ]`

```
transA.Rollback ( );
```

All statements after the `BeginTransaction` command are rolled back and undone. Whenever an error occurs, the database honours transactions when recovering. Honouring a transaction means that either the database returns to the state it was in just before the error occurred or it finishes the whole transaction. A DBMS may provide for an auto-commit mode. *Auto-commit mode* means that the transaction is finished automatically even if an error occurs in between. An example of the `autocommit` command is:

```
set autocommit = 1;
```

In this case, each transaction operation is automatically committed after the *begin* or *start* command. Executing the command below disables the auto-commit mode:

```
set autocommit=0;
```

Whenever a lock command is used, the database is locked (assigned) to the current transaction only. On using the unlock command, the other transactions can also access the database and perform operations.

7.5 Query Processing

During a transaction with a database, queries are sent to read and get the records from the database. Queries can be better understood by the following example. Assume that there are two sets of database records `Contacts` and `SavedNumbers`. `Contacts` stores the rows of records consisting of first character (`firstChar`) of name, contact-name (`cName`), and contact telephone number (`cTelNum`). A record in `Contacts` can be searched by `firstChar`, `cName`, or `cTelNum`. `DialledNumbers` stores the rows of records consisting of dialling sequence number (`seqNum`), time of call (`cTime`), and dialled telephone number (`dTelNum`). A record in `DialledNumbers` can be searched by `seqNum`, `cTime`, or `dTelNum`.

Assume that a query for processing is as follows—get the contact name and telephone number in which the first character of the contact name is R and contact telephone number is among the previously dialled numbers. If high-level language SQL is used for declarations, then the SQL query during a transaction is programmed as follows:

```
SELECT cName, cTelNum FROM Contacts, DialledNumbers WHERE
Contacts.firstChar = "R" AND Contacts.cTelNum =
DialledNumbers.dTelNum
```

Query processing means making a correct as well as efficient execution strategy by *query decomposition* and *query-optimization*. A relational-algebraic equation defines a set of operations needed during query processing. Either of the two equivalent relational-algebraic equations given below can be used.

$$\pi_{\texttt{cName, cTelNum}}\ (\sigma_{\texttt{Contacts.firstChar = "R"}}\ (\sigma_{\texttt{Contacts.cTelNum = DialledNumbers.dTelNum}}\ (\texttt{Contacts})\times \texttt{DialledNumbers})) \tag{7.1}$$

This means first select a column `Contacts.cTelNum` in a row in `Contacts` in which `Contacts.cTelNum` column equals a column `DialledNumbers.dTelNum` by crosschecking and matching the records of a column in `Contacts` with all the rows of `DialledNumbers`. Then in the second step `select` the row in which `Contacts.firstChar = "R"` and the selected `cTelNum` exists. Then in the third step project `cName` and `cTelNum`.

$$\pi_{\texttt{cName, cTelNum}}\ (\sigma_{\texttt{Contacts.firstChar = "R"}\ \wedge\ \texttt{Contacts.cTelNum = DialledNumbers.dTelNum}}\ (\texttt{Contacts}\times \texttt{DialledNumbers})) \tag{7.2}$$

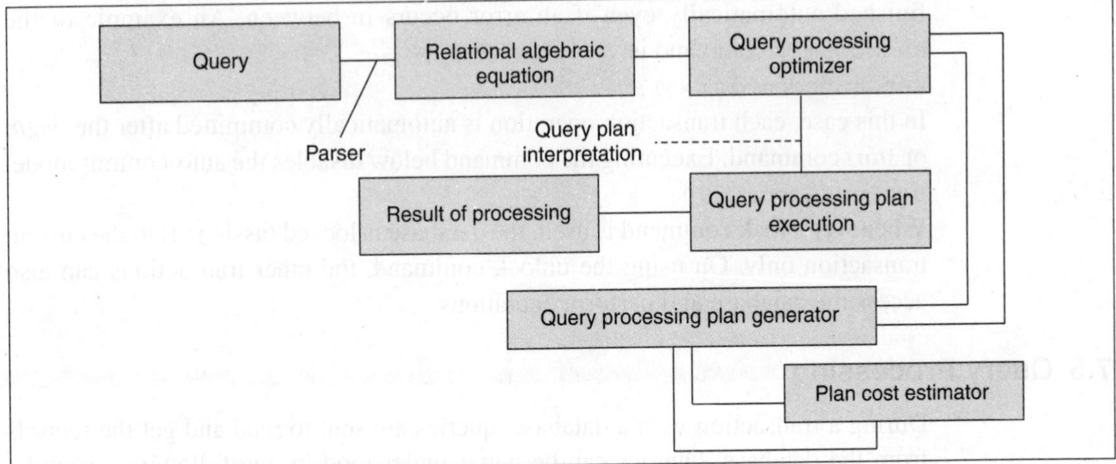

Fig. 7.10 Query processing architecture

This means that in first series of step, crosscheck all rows of `Contacts` and `DialledNumbers` and select, after AND operation, the rows in which `Contacts.firstChar = "R"` and `Contacts.cTelNum = DialledNumbers.dTelNum`. Then in the next step project `cName` and `cTelNum` form the selected records.

π represents the projection operation, σ, the *selection* operation, and $\wedge$, the AND operation. It is clear that the second set of operations in query processing is less efficient than the first. Query decomposition of the first set gives efficiency. Decomposition is done by (i) analysis, (ii) conjunctive and disjunctive normalization, and (iii) semantic analysis.

Efficient processing of queries needs optimization of steps for query processing. Optimization can be based on cost (number of micro-operations in processing) by evaluating the costs of sets of equivalent expressions. Optimization can also be based on a heuristic approach consisting of the following steps: perform the selection steps and projection steps as early as possible and eliminate duplicate operations. For example, in the above case, in first set of operations, the column `Contacts.cTelNum` was first selected and then the other operations were carried out.

Figure 7.10 shows a query processing architecture. The query optimizer employs (a) query processing plan generator and (b) query processing cost estimator to provide an efficient plan for query processing. Here, plan means a tree like structure of relational algebraic operations for the given equation.

7.6 Data Recovery Process

As discussed in Section 7.4, transactions are modelled according to ACID rules. Rule *A* is atomicity in each transaction and *D* is durability of each transaction.

There are number of reasons, which warrant database recovery. There might be media failure, system failure, or transaction abortion. There may be data destruction due to intentional external attack or due to unintentional (due to careless handling) user carelessness. Data may also be destroyed due to destruction of the physical media hoarding the data. There may also be logical program errors and a transaction may not materialize. Finally, there may be loss of main memory due to system errors (hardware or software).

Let us assume that transactions started at time t_0 and system crash or failure occurs at $t_0 + T$. Assuming that transactions T_0 to T_{n-1} are required to be completed in sequence $T_0, T_1, T_2, ..., T_{n-1}$, the following cases are possible:

Case 1: Last transactions incomplete

T_0 to T_{i-1} completed and stored at the secondary memory from the database buffer. However, T_{i-1} to T_{n-1} were not completed and are not yet committed. Therefore, T_{i-1} to T_{n-1} must be aborted.

Case 2: Initial and Last transactions incomplete

Let us assume that transactions started at t_0 and system crash or failure occurs at $t_0 + T$. Also, let transactions T_0 to T_{n-1} be required to be completed in sequence T_0, $T_1, T_2, ..., T_{n-1}$. Now, (a) T_0 to T_{i-1} and T_j to T_{n-1} are not completed and, therefore, must be aborted, (b) only T_i to T_{j-3} were completed, committed, and saved at the secondary storage, and (c) T_{j-2} and T_{j-1} were committed but not saved at the secondary memory from the database buffer.

In case 1, the atomicity rule of transaction comes into effect and the last transactions are automatically aborted. In case 2, the atomicity rule of transaction comes into effect and initial and last transactions are automatically aborted. But the in-between transactions T_{j-2} and T_{j-1} are completed and also committed but are not saved in the secondary storage up to the instant $t_0 + T$ when the crash occurs. The transactions T_{j-2} and T_{j-1} need to be recovered by a database transaction recovery mechanism.

Data is non-recoverable in case of media failure, intentional attack on the database and transactions logging data, or physical media destruction. However, data recovery is possible in other cases. Figure 7.11 shows recovery management architecture. It uses a recovery manager, which ensures atomicity and durability. Atomicity ensures that an uncommitted but started transaction aborts on failure and aborted transactions are logged in a log file. Durability ensures that a committed transaction is not affected by failure and is recovered. Stable state databases at the start and at the end of transactions reside in secondary storage. Transaction commands are sent to the recovery manager, which sends fetch commands to the database manager. The database manager processes the queries during the transaction and uses a database buffer. The recovery manager also sends the flush commands to transfer the committed transactions and database buffer data to the secondary storage. The recovery manager detects the results of operations. It recovers lost operations from

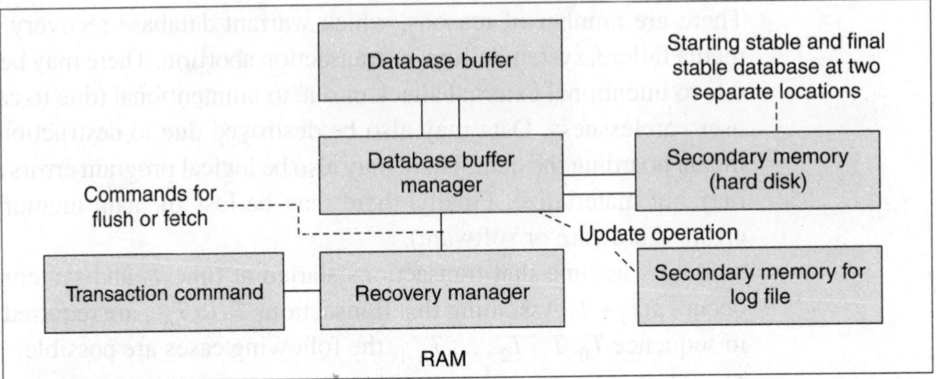

Fig. 7.11 Recovery management architecture

the secondary storage. Recovery is by detecting the data lost during the transaction. The recovery manager uses a log file, which logs actions in the following manner—

(a) Each instruction for a transaction for update (insertion, deletion, replacement, and addition) must be logged.
(b) Database read instructions are not logged.
(c) Log files are stored at a different storage medium.
(d) Log entries are flushed out after the final stable state database is stored.

Each logged entry contains the following fields:

(i) transaction type (begin, commit, or rollback transaction)
(ii) transaction ID
(iii) operation-type
(iv) object on which the operation is performed
(v) pre-operation and post-operation values of the object

The recovery manager recovers or aborts a transaction using the logged entries. One of the log protocols used by the recovery manager is the write-ahead log protocol which entails that (a) updating of a record must be first logged and then the updated record should be sent to disk and (b) all operations must be logged in an *entry* before the commit operation. [The entry has the list of all the operations to be completed before the commit command starts execution and the updated record is sent to the disk. In case of transaction failure, the recovery manager recovers the lost transaction operations from the logged operations.] Actions (a) and (b) in the protocol guarantee atomicity and durability, respectively.

Assume that a transaction consists of 1, …, N operations. A data recovery procedure is as follows—assume that a checkpoint is inserted on kth operation such that 1, …, k operations need not be backtracked, since recovering the lost data in transactions and data recovery is done from the $(k + 1)$th to Nth operations. A checkpoint-based data recovery procedure uses the checkpoints for operations on the data during a set of transactions. Recovery is always made by back-scanning the logged records. A checkpoint-based data recovery procedure defines the stage, up to which the back-scanning of logged operations in the secondary storage is to be done. For example, if there is a checkpoint at transaction operation T_i, which is in progress when the crash occurs, then back scanning must be performed up to the beginning of T_i.

A procedure called the Aries algorithm is also used for recovering lost data. The basic steps of the algorithm are:

1. Analyse from last checkpoint and identify all dirty records (written again after operation restarted) in the buffer.
2. Redo all buffered operations logged in the update log to finish and make final page.
3. Undo all write operations and restore pre-transaction values.

The recovery models used in data recovery processes are as follows:

(i) The *full recovery model* creates backup of the database and incremental backup of the changes. All transactions are logged from the last backup taken for the database.

(ii) The *bulk logged recovery model* entails logging and taking backup of bulk data record operations but not the full logging and backup. Size of bulk logging is kept to the minimum required. This improves performance. We can recover the database to the point of failure by restoring the database with the bulk transaction log file backup. This is unlike the full recovery model in which all operations are logged.

(iii) The *simple recovery model* prepares full backups but the incremental changes are not logged. We can recover the database to the most recent backup of the given database.

7.7 Issues Relating to Quality of Service

Quality of service (QoS) in a mobile network is affected due to network connectivity. QoS tools measure the bandwidth availability, which controls data transfer rates, connection reliability, and data-loss risks. GPRS supports user specification of QoS (data throughput and classes for delay, reliability, and precedence) and the radio interface allocates resources as per the QoS profile. Loss of connection affects QoS. Some network protocols, for example, WAP, take into account the connection loss issues while maintaining the QoS. Network tools, such as J2ME CLDC (Connection Limited Device Configuration), support emulation of QoS of the network during application development.

Mobile device applications adapt to the QoS. QoS is an important consideration in distributed multimedia streams. QoS is important for adapting of visual displays on small screen to a certain sustainable level in case of loss of connectivity and loss of frames.

An example of maintenance of QoS is adaptation of low quality voice to an acceptable threshold by the device voice API. The loss of connectivity is adapted to by providing background noise. This is done with the assumption that connectivity will be re-established and to avoid discomfort to the user due to abrupt intermittent periods of silence. The noise reduces in volume at an appropriate rate after a certain period of time. Another example of maintaining QoS in mobile networks is that of handoff of the mobile device when there is loss of QoS in one cell and the neighbouring cell can provide a higher QoS.

7.7.1 QoS Object Models

Consider an object-oriented programming language, such as C or C++. We define the objects in the language. An object has the data fields and the methods which operate on those fields. An object can be extended or modified. For example, consider an object's QoS capability. It defines the data for specifying the capability of a given mobile network in providing a specified QoS. Unified modelling language (UML) is a language used to model objects. An object can have an interface. The interface has unimplemented methods (functions) which are implemented in the interfaced objects.

QoS can also be modelled with the objects in UML. For example, `QoS_capability` is an object to model capability. A network can be specified to have QoS capability at 10 Mbps with 5% times needing retransmission. Following is a list of the objects which specify and model the QoS.:

(i) `QoS_capability` [An object to model the capability.]
(ii) `QoS_requirement` and two interfaces `QoS_obligation` and `QoS_expectation`
(iii) `QoS_contract` and two interfaces `QoS_obligation` and `QoS_expectation`
(iv) `QoS_offer`
(v) `QoS_constraint_container_type`
(vi) `QoS_constraint_container` and three interfaces `QoS_constraint`, `QoS_obligation`, and `QoS_expectation`
(vii) `QoS_value` and its extended objects `QoS characteristic_type`, `QoS_type`, `QoS_characteristic`, and interface `QoS_characteristic_feature`

`QoS_requirement` is an object which specifies the quality requirements from a network. It also specifies the obligatory quality and expected quality. The `QoS_requirement` object can also have the methods to estimate the requirements. An object can have an *interface* or set of interfaces. An interface is an object, the *specifications* of which are given in the interfacing object. The object `QoS_requirement` consists of two interfaces—`QoS_obligation` and `QoS_expectation`. `QoS_obligation` is an interface to interfacing object `QoS_requirement`. Hence `QoS_obligation` is specified and defined in `QoS_requirement`. `QoS_expectation` is another interface to `QoS_requirement`. Hence `QoS_expectation` is also specified and defined in `QoS_requirement`.

Keywords

Adapter: Software for converting data which is in one format (or by a given protocol) into another format (or by another protocol).

Asynchronous: When the transmission of bits or actions does not occur at predefined periods, intervals, or phase differences.

Atomic: An action or process which must be completed if initiated, which cannot occur in parts, and which, once completed, cannot be undone.

Atomic transaction: A transaction which cannot be done in parts, which cannot be undone once completed, and which must be undone (rolled back) if not completed.

Business logic: A logical way in which transactions are carried out between two ends, for example, between database-client (application) and database-server or between an API and a database.

Cache: A list or database consisting of saved items or records at the device, which the device saved for faster access at the later instants, rather than reselecting or re-tuning or re-fetching when needing them. The cached items are the ones which device either selects and tunes to from a set of broadcasted or pushed records by a server or which the device fetches by request, demand, or subscription to the server and saves in the list or database.

Cache invalidation: A mechanism by which a cached data item or record becomes invalid and, thus, unusable. The reason for this could be modification, expiry, or invalidation at another computing system or server.

Client: Software which requests and then gets the input responses from a server or set of computing systems after the server has performed the required computations.

Client–server computing: A computing structure in which a client requests for computations and data and then receives the requested data or responses after necessary computations from a server. It is a computing structure in which client caches or accesses to the data record(s) after computations at a server. The access may be either on client request or on broadcasts or distribution from the server. The client and the server can be on same computing system or on different computing systems.

Connectivity protocol: A protocol with API(s), which has predefined methods to handle various data access functions.

Context: A set of interrelated conditions on which the existence, action(s), and result of operations(s) depend.

Database: A set of data records stored with business logic. A record can be queried or updated through transaction only.

QoS (Quality of service): A measure of achieving the fulfilment of contracted, committed, or offered services within the contracted time frame.

Record: Stored data with a key using which transactions can be carried out.

Relational database: A database structured in accordance with the relational model, which helps the database designer to create a consistent, logical representation of information.

Server: Software at a computing system, which hoards data for distribution on requests from clients or data subscriptions. The server also broadcasts hoarded data to the clients.

Stateful: An entity or data transfer operation is considered stateful when its state is defined or considered and state can be defined as exclusive, modified, shared, or invalidated. The state can be a state succeeding some other state after a state transition. (State-transition is an action by which state changes on an event or input. For example, a time in a clock undergoes state transition on each tick.)

Stateless: An entity or data transfer operation is considered stateless when its state is not defined or considered or is not a successor of a state.

Synchronous: Indicates that transmissions or operations are performed at predefined periods, intervals, or phase differences

Temporal: An action, entity, process, or feature which changes over a period and has relationship with time.

Transaction: The execution of interrelated instructions in a sequence for a specific operation on a database according to business logic or rules.

VUI (voice user interface): Software used to recognize voice commands, words, or inputs received from the user. The VUI can interpret voice inputs and initiate appropriate action(s).

Review Questions

1. How does data stored in a file differ from a database? Explain the function of the database connectivity protocol.
2. What are the advantages of hoarding data at the mobile device?
3. Describe data caching architecture. Explain data cache maintenance in a mobile environment. How is the web cache maintained in a mobile environment?
4. Explain cache invalidation mechanisms. Explain the advantages and disadvantages of stateless and stateful cache invalidation. Explain the advantages and disadvantages of asynchronous and synchronous cache invalidation.
5. Show a client–server computing architecture in which the database is at the application tier. How does this architecture differ if the application server fetches the data from the enterprise server tier?
6. Draw and explain four-tier architecture. How do multimedia databases serve a mobile device in client–server architecture?
7. Explain the advantages of using an adaptation mechanism in client–server architecture.
8. Explain power-aware computing. What do you mean by context? Explain with examples. Describe context-aware computing. What are different context-types used in context-aware computing?
9. Explain the database transaction models and ACID rules.
10. Explain query-processing architecture for processing a query using distributed databases.
11. Explain the situations in which a database can crash. How does a database recover using recovery manager? What is the role of logged entries in updating transactions?
12. Why does a mobile device take quality of service issues into account while computing? List the object models for application adaptation for the QoS constraints.

Objective Type Questions

Pick the correct or most appropriate statement among the choices given:

1. The connectivity protocol is used for interchanges of data sets between the database-server (for sending data sets to clients) and database for the data sets. It describes
   (a) the network connection headers.
   (b) the relations between various data sets.
   (c) the transaction model for connecting various data sets.
   (d) an API with predefined methods to handle various data access functions.
2. The advantage of hoarding or caching is that it provides
   (a) protection from frequent disconnections of mobile device in a wireless network.
   (b) the facility to distribute new data records at periodic intervals.
   (c) no access latency and delay in retrieving the queried record from the server.
   (d) synchronization of server database with the mobile device database.
3. A cache invalidation mechanism is
   (a) a server-initiated mechanism for informing mobile devices that data sent previously is invalid.
   (b) a server-initiated mechanism of sending invalidation report(s) of a data record to the device in order to synchronize two records and maintain cache-consistency.
   (c) a client-initiated mechanism for checking data validity and rejecting invalid data.
   (d) a client-initiated mechanism which defines a cache-consistency protocol to discard invalid data.
4. The advantage of the asynchronous cache invalidation approach is that
   (a) the server transfers the cache invalidation report at periodic intervals and needs greater bandwidth than the synchronous approach.

(b) the client device transfers the cache invalidation report at variable intervals and needs smaller bandwidth than the synchronous approach.
(c) frequent unnecessary transfers of data reports from the server do not take place and, thus, the mechanism is more bandwidth efficient.
(d) the client device transfers the cache invalidation report and data requests at variable intervals as and when cached data become invalid and the bandwidth requirement is the same as that in the synchronous approach.

5. Client–server computing has
(a) *N*-tier architecture ($N = 2, 3, \ldots$) and at each tier the client device or server connects to another using a connecting protocol, a synchronizing protocol, or a data or command interchange protocol for obtaining the client requests or server responses.
(b) two-tier architecture in which the server distributes data for computations at the client.
(c) architecture in which the server distributes computed data for the client device API.
(d) architecture in which the client requests the server for distribution of computed data for the client device API.

6. Adapter is software
(a) for converting in one format (or by a given protocol) into another format (or another protocol).
(b) for adapting an application according to context.
(c) for synchronization with another protocol.
(d) adapting data from the server.

7. "Power-aware computing can use (i) data caching, (ii) cache invalidation in place of time-to-live updates, (iii) aggregation of records and transmitting incremental data only, and (iv) protocol optimization in order to save power." Which of the following is true for the above statement?
(a) Only (iii) is correct.
(b) All except (i) are correct.
(c) Only (iv) is not correct.
(d) All are correct.

8. Context-aware computing helps in
(a) eliminating computations on out-of-context data.
(b) defining the context of the computing problem better.
(c) application-aware computing which helps in reducing possibility of errors and eliminates ambiguity in action by considering the interrelated conditions, which decides according to the expected system response on computations, and which helps in ubiquitous computing.
(d) computing by considering the conditions at the mobile device.

9. Consider the following statements—(i) When not setting *autocommit*, a transaction will be committed when the user requests a lock-and-execute commit command. (ii) Isolation in transaction means operations transform the database from one consistent state to another. (iii) Atomic transaction means operations transform the database from one consistent state to another. (iv) Durable transaction means committed transactions should be permanent. (v) When using the read command the user may read a *dirty record* (reading after the data record has been modified by another command). (vi) Consistency in transactions entails that a transaction must preserve the integrity constraints and follow the declared consistency rules for a given database.
Which of following is true for the above set of statements?
(a) Only (ii) and (iii) are correct.
(b) Only (ii) and (iii) are not correct.
(c) All are correct.
(d) Only (v) is correct and (vi) is partially correct.

10. Efficient processing of queries needs
    (a) query-processing cost estimator.
    (b) binary relational-algebraic equations.
    (c) query optimizer employs plan generator and cost estimator for query-processing, where plan means a tree-like structure of relational algebraic operations for the given query processing equation.
    (d) a systematic plan to decompose the queries.
11. "The recovery manager uses a log file which logs the following actions—(i) Database read instructions are not logged. (ii) Log files are at different storage media. (iii) Logged entries are flushed out after the final stable state database is stored. (iv) Each logged entry contains the transaction type (begin or commit or rollback transaction). (v) Each logged entry contains the transaction ID. (vi) Each logged entry contains the operation-type. (vii) Each logged entry contains the object on which the operation is performed. (viii) Each logged entry contains pre-operation and post-operation values of the object. (ix) Recovery manager using logged entries recovers or aborts a transaction." Which of the following are true for the above statement?
    (a) All except (viii) are correct.
    (b) All are correct.
    (c) All except (vii) are correct.
    (d) The recovery manager does not use log entries but a database buffer, a database manager, and a database backup.
12. During transactions the recovery manager must ensure
    (a) the ACID rules.
    (b) atomicity and durability.
    (c) database consistency.
    (d) recovery on media failure.
13. Quality of service (QoS) in a mobile network is affected by
    (a) transmission errors and the quality of audio received is the measure of QoS.
    (b) atmospheric conditions and the number of mobile users connected to base station at a given instant is the measure of QoS.
    (c) network load and the effective measure of QoS is the data lost per second.
    (d) network connectivity and the effective bandwidth availability (which controls data transfer rates), connection reliability, and data-loss risks are a measure of QoS.
14. An alternative to cache invalidation is that each cached record is
    (a) verified for validity before use at the mobile device.
    (b) polled from the server.
    (c) assigned a TTL (time-to-live) and the TTL assignment is adaptive (adjusts) on previous update intervals of that record. TTL = 0 can be used in the polling mode.
    (d) downloaded and hoarded in the mobile device.
15. Two transactions
    (a) if performed simultaneously, must not have any interference between them.
    (b) cannot be performed simultaneously.
    (c) should be performed in sequence and the second must begin after the first is committed.
    (d) can be performed on different databases.
16. Stateful asynchronous communication from the server entails that
    (a) the client keeps the information of the present state of each data-record at the client-cache.
    (b) the server keeps the information of the present state of each data-record at the client-cache(s).
    (c) the client periodically checks the information of the present state of each data-record at the server.
    (d) the server and the client both must keep the information of the states of the data.

Chapter 8

Data Dissemination and Broadcasting Systems

In Chapter 7, we discussed how a database is cached at a mobile device. Cache invalidation mechanisms by which old, invalidated data records are replaced with new ones and transaction models for transactions with the databases were also described. It was presumed that an application running on a mobile device uses the hoarded and cached data and that there is no active participation from the server end in mobile computing, i.e., mobile computing for an application is restricted to the device only. This chapter describes how data is disseminated in a mobile network in an asymmetric communication environment so that it can be cached at the device. Data dissemination entails distributing and pushing data generated by a set of computing systems or broadcasting data from audio, video, and data services. The output data is sent to mobile devices. A mobile device can select, tune, and cache the required data items. The devices can cache items for application programs. There is active participation of the one or more distributed computing systems with the mobile devices. (Interconnected and networked computers, including the device and the server, are at distributed locations. A distributed computing system consists of two or more of such computers.) This chapter also covers the broadcast systems for digital audio and video broadcasts in a mobile network. The following concepts are discussed in this chapter:

- Asymmetric nature of communication between mobile devices and fixed computing systems
- Classification of data delivery mechanisms used for distribution to multiple mobile devices
- Push-based, pull-based, and hybrid mechanisms for data dissemination
- Broadcast disk models for cyclic repetition (pushes) of data
- Indexing techniques for selective tuning at the devices
- Digital audio and video broadcasting
- Mobile TV

8.1 Communication Asymmetry

Mobile communication between a mobile device and a static computer system is intrinsically asymmetric. A device is allocated a limited bandwidth. This is because a large number of devices access the network. Bandwidth in the downstream from the server to the device is much larger than the one in the upstream from the device to the server. This is because mobile devices have limited power resources and also due to the fact that faster data transmission rates for long intervals of time need greater power dissipation from the devices. In GSM networks data transmission rates go up to a maximum of 14.4 kbps for both uplink and downlink. The communication is symmetric and this symmetry can be maintained because GSM is only used for voice communication. The i-mode, on the other hand, is used for many applications including multimedia transmission, Internet access, voice communi-cation, etc. The i-mode communication base station provides downlink at 384 kbps but the uplink from the devices is restricted to 64 kbps and the communication, is thus, asymmetric. The characteristics of interference and time-dispersion in wireless signals result in signal distortion and transmission errors at the receiver end. These characteristics also lead to path loss and signal fading, which cause data loss. Also, there is greater access latency in wireless networks than that in wired networks. This is because data loss has to be taken care of by repeat transmissions and transmission errors have to be corrected and taken care of by appending additional bits, such as the forward error correction bits. Mobile devices also have low storage capacity (memory) and, therefore, cannot hoard large databases. Accessing the data online not only has a latency period (is not instantaneous) but also dissipates bandwidth resources of the device. Figure 8.1 shows communication asymmetry in uplink and downlink in a mobile network. The participation of device APIs and distributed computing systems in the running of an application is also shown in Fig. 8.1.

A special case of broadcasting corresponds to unidirectional (downlink from the server to the devices) unicast communication. Unicast means the transmission of data packets in a computer network such that a single destination receives the packets. This destination is generally the one which has subscribed to a broadcasting or application–distribution service. Mobile TV (commercially launched in early 2006) is an example of unidirectional unicast mode of broadcasting. Each device receives broadcast data packets from the service provider's application–distribution system. The application–distribution system broadcasts data of text, audio, or video services. Figure 8.2 shows a broadcasting architecture.

8.2 Classification of Data-Delivery Mechanisms

Data-delivery mechanisms can be classified into three categories, namely, push-based mechanisms (publish–subscribe mode), pull-based mechanisms (on-demand mode), and hybrid mechanisms (hybrid mode). These are described in the following subsections.

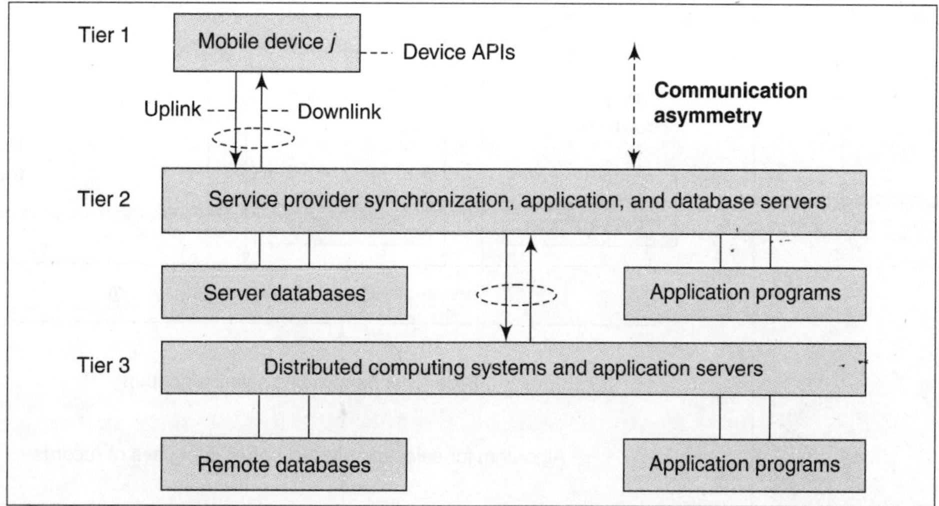

Fig. 8.1 Communication asymmetry in uplink and downlink and participation of device APIs and distributed computing systems when an application runs

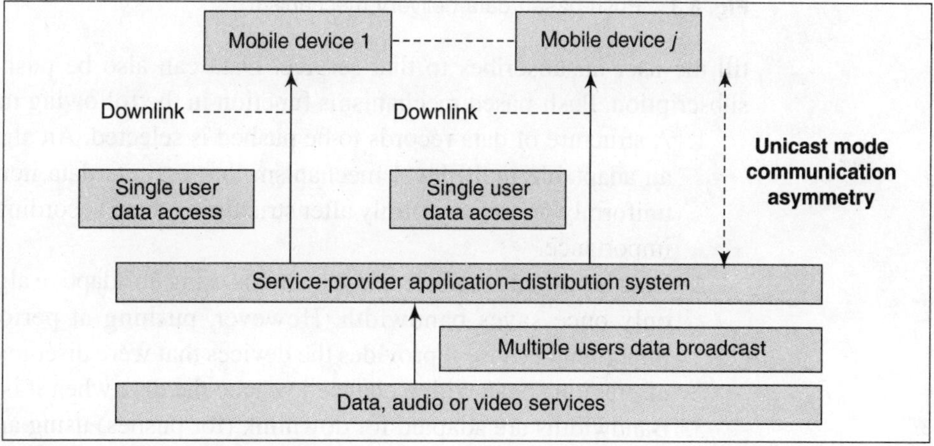

Fig. 8.2 Broadcasting architecture

8.2.1 Push-based Mechanisms

The server pushes data records from a set of distributed computing systems. Examples of distributed computing systems are advertisers or generators of traffic congestion, weather reports, stock quotes, and news reports. Figure 8.3 shows a push-based data-delivery mechanism in which a server or computing system pushes the data records from a set of distributed computing systems. The data records are pushed to mobile devices by broadcasting without any demand. The push mode is also known as publish–subscribe mode in which the data is pushed as per the subscription for a push service by a user. An example of subscription is subscription to stock quotes. The subscribed query for a data record is taken as perpetual query

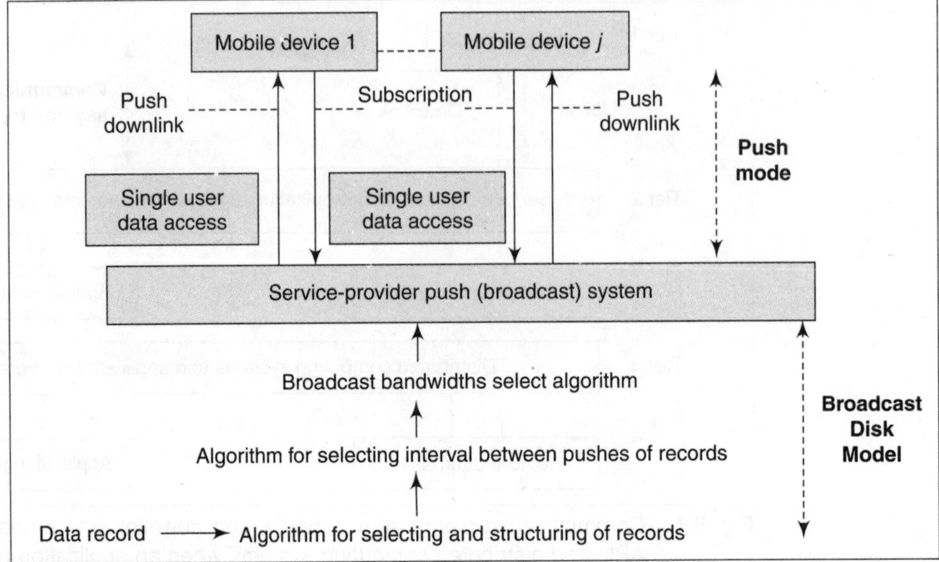

Fig. 8.3 Push-based data-delivery mechanism

till the user unsubscribes to that service. Data can also be pushed without user subscription. Push-based mechanisms function in the following manner:

1. A structure of data records to be pushed is selected. An algorithm provides an adaptable multi-level mechanism that permits data items to be pushed uniformly or non-uniformly after structuring them according to their relative importance.
2. Data is pushed at selected time intervals using an adaptive algorithm. Pushing only once saves bandwidth. However, pushing at periodic intervals is important because it provides the devices that were disconnected at the time of previous push with a chance to cache the data when it is pushed again.
3. Bandwidths are adapted for downlink (for pushes) using an algorithm. The same fixed periods can be used for pushing all records but usually higher bandwidth is allocated to records having higher number of subscribers or to those with higher access probabilities.
4. A mechanism is also adopted to stop pushes when a device is handed over to another cell.

The application–distribution system of the service provider uses these algorithms and adopts bandwidths as per the number of subscribers for the published data records. On the device handoff, the subscription cancels or may be passed on to new service provider system.

The client–device caches (Section 7.2) the required data from the pushed data records. An advantage of push-based mechanisms is that they enable broadcast of data services to multiple devices. Moreover, the server is not interrupted frequently by requests from the mobile devices. Push-based mechanisms are the best option for the server as they prevent server overload, which may occur due to flooding of device requests. Another advantage of pushing data is that a user gets even that data, that he would have otherwise ignored. Examples of such data are information about a forthcoming storm, traffic congestion, or the movement of the user to another territory. The disadvantage of push-based mechanisms is the dissemination of unsolicited, irrelevant, or out-of-context data. Also, the user may not be interested in the disseminated data and may be inconvenienced.

8.2.2 Pull-based Mechanisms

The user-device or computing system pulls the data records from the service provider's application database server or from a set of distributed computing systems. Examples of distributed computing systems are music album server, ring tones server, video clips server, or bank account activity server. Records are pulled by the mobile devices on demand followed by the selective response from the server. Selective response means that server transmits data packets as response selectively, for example, after client-authentication, verification, or subscription account check. The pull mode is also known as the on-demand mode. Figure 8.4 shows a pull-based data-delivery mechanism in which a device pulls (demands) from a server or computing system, the data records generated by a set of distributed computing systems.

Pull-based mechanisms function in the following manner:

1. The bandwidth used for the uplink channel depends upon the number of pull requests. Assume that an uplink channel uses a bandwidth of 19.2 kbps and service provider's application–distribution system server can accept 384 kbps. Then only 20 pull requests can be used at 19.2 kbps. If number of pull requests

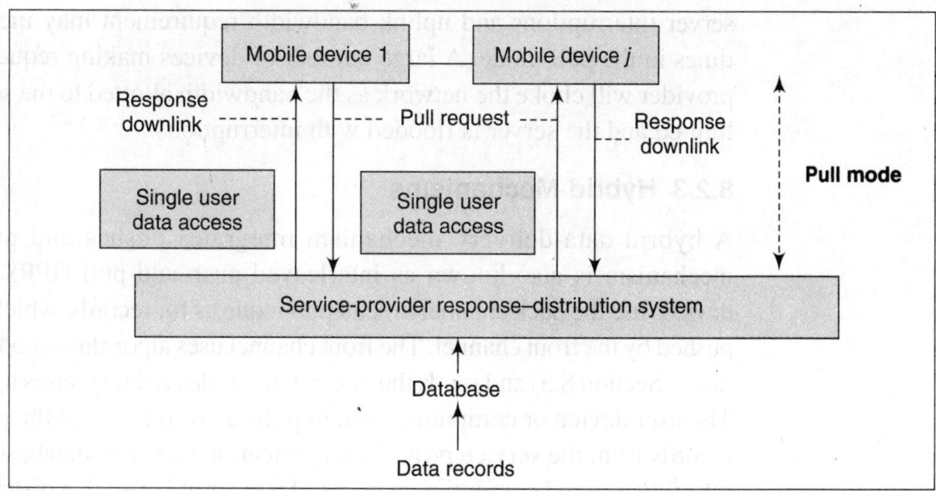

Fig. 8.4 Pull-based data-delivery mechanism

is larger, the uplink channel may send requests at 9.8 kbps or 4.8 kbps. Similarly, the service provider's application–distribution system uses a mechanism which consists of adapting to the bandwidth required for serving the requests (downlink) in case the server is unable to deliver the response in a reasonable period. For example, for downlink from server, if few pull requests are there, the server can use 384 kbps. If many pull requests are there, the server adapts to the higher bandwidths, for example, 1.536 Mbps in order to send the responses in a reasonable period.

2. A pull threshold is selected. This threshold limits the number of pull requests in a given period of time. This controls the number of server interruptions.
3. A mechanism is adopted to prevent the device from pulling from a cell, which has handed over the concerned device to another cell. On device handoff, the subscription is cancelled or passed on to the new service provider cell.

In pull-based mechanisms the user-device receives data records sent by server on demand only. The advantage of pull-based mechanisms is that no unsolicited or irrelevant data arrives at the device and the relevant data is disseminated only when the user asks for it. Pull-based mechanisms are the best option when the server has very little contention and is able to respond to many device requests within expected time intervals. The disadvantage of pull-based dissemination mechanisms is that the server faces frequent interruptions and queues of requests at the server may cause congestion in cases of sudden rise in demand for certain data record(s). For example, during a world cup match there may be multiple requests for score update. Another disadvantage of the on-demand mode is the energy and bandwidth required for sending the requests for hot items and temporal records (records changing with time). For example, the score of a world cup football match is a temporal data record in which thousands of users are interested at the same time. The number of server interruptions and uplink bandwidth requirement may increase a thousand times in the pull mode. A large number of devices making requests to the service provider will choke the network as the bandwidth allotted to the service provider is limited and the server is flooded with interruptions.

8.2.3 Hybrid Mechanisms

A hybrid data-delivery mechanism integrates pushes and pulls. The hybrid mechanism is also known as interleaved-push-and-pull (IPP) mechanism. The devices use the back channel to send pull requests for records, which are not regularly pushed by the front channel. The front channel uses algorithms modelled as broadcast disks (Section 8.3) and sends the generated interleaved responses to the pull requests. The user device or computing system pulls as well receives the pushes of the data records from the service provider's application server or database server or from a set of distributed computing systems. An example of such a distributed computing

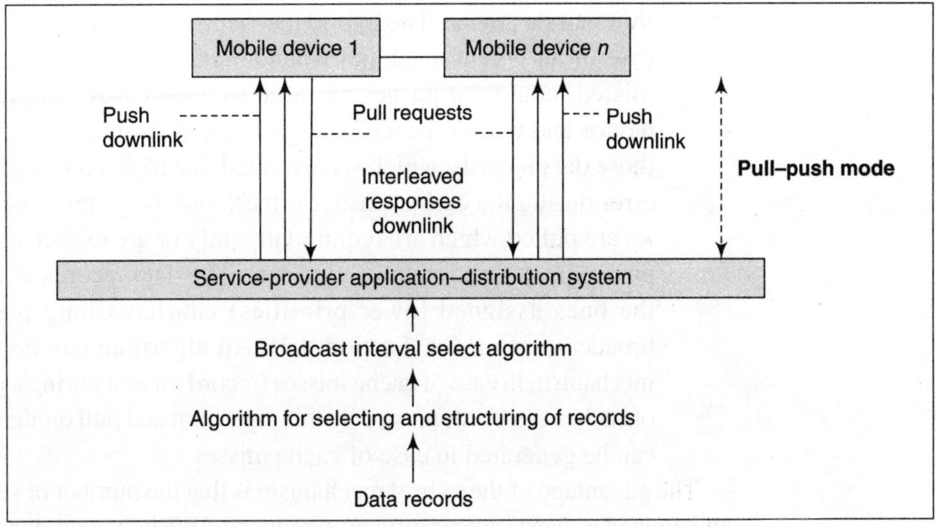

Fig. 8.5 Hybrid interleaved push–pull-based data-delivery mechanism

system is a system for advertising and selling music albums. The advertisements are pushed and the mobile devices pull for buying the album. Figure 8.5 shows a hybrid interleaved, push–pull-based data-delivery mechanism in which a device pulls (demands) from a server and the server interleaves the responses along with the pushes of the data records generated by a set of distributed computing systems. Hybrid mechanisms function in the following manner:

1. There are two channels, one for pushes by front channel and other for pulls by back channel.
2. Bandwidth is shared and adapted between the two channels. It is adapted in downlink and uplink channels depending upon the number of active devices receiving data from the server and the number of devices requesting data pulls from the server.
3. An algorithm can adaptively chop the slowest level of the scheduled pushes successively into larger number of pieces. Assume that at the slowest level, M data records each of length n bits are broadcast and pushed at successive interval of time, T_s each. The bandwidth used by these records is $M \times n \times (1/T_s)$ bps. When number of pull requests increase, the push records are adaptively chopped into M at an instant, then front-channel bandwidth will become $M \times n \times (1/T_s)$ bps. This increases the bandwidth available for pull and back-channel bandwidth increment is given by $(M - M') \times n \times (1/T_s)$. When number of pull requests decrease, the number of data records pushed at intervals T_s can again be increased to M. Assume that there are M records from R_0 to R_{M-1}. The records R_0 to R_{M-1} are pushed repeatedly after successive interval of time. This is because in wireless environment a device can have a disconnection at a given instant. Assume that a record R_j is needed urgently

by a mobile device. The hybrid mechanism provides for adaptation to pull in case of an urgent need and when a data record pushed from the server is missed, otherwise pushes are used in most cases. The R_j is pulled by the device and the server transmits it by back channel. User devices pull only those data records, which were missed due to disconnection or transmission error during the earlier push. Furthermore, only those records, for example R_j, are pulled which are required urgently or are expected to be available by push only after a long push interval. The data records at lower levels (e.g., the ones assigned lower priorities) can have long push intervals in a broadcasting model (Section 8.3). An algorithm can be used to adapt the mechanism in case of cache misses (record missed during a push). The content of caches can be monitored by this algorithm and pull on-demand data requests can be generated in case of cache misses.

The advantage of the hybrid mechanism is that the number of server interruptions and queued requests are significantly reduced. IPP, however, does not eliminate the typical server problems of too many interruptions and queued requests. Further, the adaptive chopping of the slowest level of scheduled pushes poses a disadvantage. Refer to the example of M data records. In spite of chopping M records at intervals of T_s into M' at an instant, there may still be insufficient back channel bandwidth for pulling requests and there may be considerable performance degradation. The devices are not able to pull the items that were not pushed or chopped.

8.3 Data Dissemination Broadcast Models

Wireless environment entails that a device may not always remain connected and networked to the data-server. It may experience several breaks in its connection. Therefore, the data records pushed from the server may be missed and not be cached. One strategy to overcome this problem is that the server can repeat pushing the records at successive intervals of time and adopt some specific algorithm or model for repeating each pushed record. In such a case when a device is connected at a certain instant with no breaks in its connection, it may tune to the selected record and cache it for its applications. A disk model in which it is presumed that all the n records to be broadcast are stored on a circular disk from 0° to 360° is called broadcast disk model. As the hypothetical disk revolves and the angle changes from 0° to 360°, the entire N bits in n records get pushed through a hypothetical reading-head over the disk. The head continuously reads each bit of a record just beneath it and broadcasts it instantaneously on the wireless network. During next revolution each bit in the records positioned from 0° to 360° is broadcast once again in the same sequence as in earlier revolution. If a device misses a record in first revolution, it can cache the same in next or any of the successive revolutions. Broadcast bandwidth $(t_s)^{-1}$ equals the number of bits (N) stored between 0° and 360° divided by T_s which is the time taken for one revolution of the disk. Here t_s is the time interval between successive bits transmitted from the disk. Each bit in each record is thus repeatedly broadcast at successive time intervals ($=N \times t_s = T_s$). Time interval between transmission of successive bits (T_s/N) [$= t_s$ = reciprocal of bandwidth].

There are number of adaptations and algorithms for broadcast. Figure 8.6 shows four algorithms and adaptations of the broadcast disk model. For example, we can assume that there are disks at multiple levels (Fig. 8.6(a)) and each level has identical bandwidth (t_s^{-1}) but different number of bits N or records n. The broadcast disk model is called flat disk model when all bits of all records are stored on and broadcast from one level only (Fig. 8.6(b)). An alternative model involves variable repetition rates as per record priority but same number of bits and transmission rate at each level. Records with a high priority are distributed at a number of levels, the number being proportional to priority (Fig. 8.6(c)). Another alternative model employs skewed distribution on a disk (Fig. 8.6(d)). Section 8.3.1 describes the algorithms for cyclic repetition of data records. Broadcast disk model is the basis of each algorithm.

8.3.1 Cyclic Repetition of Data—Broadcast Disks

The algorithms for pushes are based on a model for adaptive structuring of data records, periodicity of pushes, and bandwidth of records at different levels. The records at the highest level have highest priority for being pushed. There can be cyclic repetition of data during pushes analogous to retrieval of data records from a disk rotating at a uniform time interval t_s between successive bits. One possibility is that at the highest level in a disk (Fig. 8.6(a)), fewer records (smallest number n ($= i$) and thus N ($=N_i$)) are placed so that when the disk rotates keeping the t_s same, these records are retrieved at a greater repetition rate than those at the lowest level (largest n ($= j$) and thus N ($=N_j$)). In other words, larger number of records is placed on the lowest level and are repeated at lower bit transfer rates $(=T_s/N_j)^{-1}$ than those placed at the highest level.

Figure 8.6(a) shows a broadcast disk model, called circular multi-disk model, in which each block of records is pushed with a repetition rate proportional to its hierarchical level. This model of pushing data records is a multi-level mechanism. At the highest level on the disk, only a select few records (i) (hot items or items subscribed to by many devices) are placed and pushed (broadcast) at highest bandwidth by lowering T_s with smallest periods between two pushes (T_{si}). At the lowest level, the records with the lowest subscription rates are placed and are broadcast at lowest bandwidth by raising T_s with longest periods between two pushes T_{si}.

Assume that there are n records to be broadcast and distributed. The broadcast disk model pushes the records in the following manner—the total bandwidth for broadcast ($= N \div t_s$) for pushes of data records at various levels remains the same. Here N is the sum of bits at each level and t_s is time interval between successive bit.

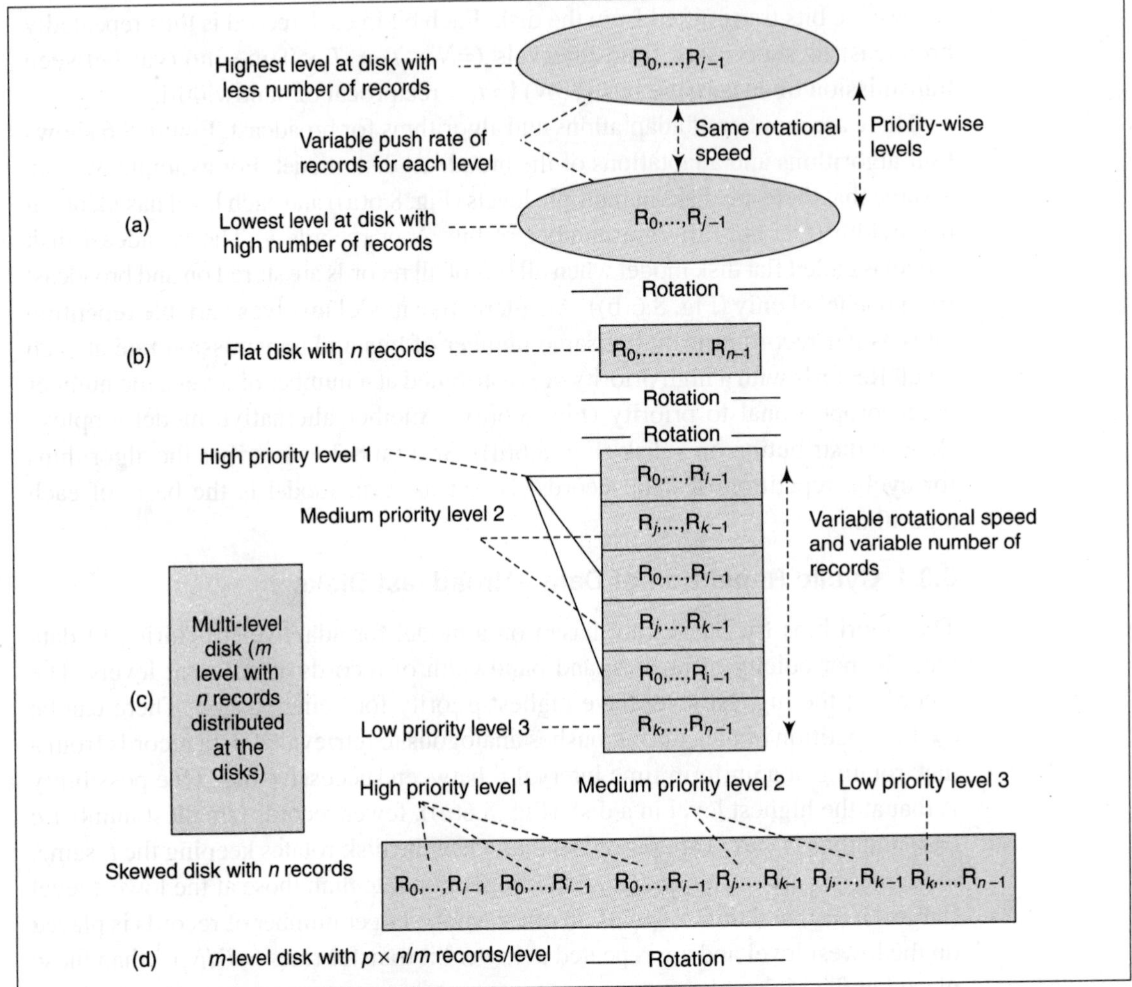

Fig. 8.6 (a) Circular multi-disk broadcast model (b) Flat disk model (c) Multi disk model (*m*-level disk with *n*/*m* records/level with repetition rate directly proportional to *p* (d) Skewed disk model (*m*-level disk with *p* × *n*/*m* records/level, where *p* is priority)

But the individual broadcast periods ($N_i \div T_s$) of different set of records can be varied using highest bandwidth at the highest level and lowest bandwidth at the lowest level. (Total bandwidth equals sum of individual bandwidths at each level.) To explain this, let us consider the example of a highway that is divided into three lanes of different widths and speed limits. Now, vehicles with high priority, such as ambulances, etc. are allowed the use of the widest lane and largest speed limit. Vehicles with medium priority are permitted to use the second widest lane and intermediate speed limit and low priority vehicles must travel in the narrowest of the three lanes and with lower speed limits. In this case wider the lane, faster the traffic. Similarly, in a broadcast disk model, greater the bandwidth, faster the delivery of the pushed records of that level. Now, the total width of the highway remains the same, therefore, to increase the width of one lane the width of the other lane(s) must be reduced. In a similar manner, increasing the broadcast rate (and hence the bandwidth at a level) of one set of records necessarily entails decreasing the broadcast rate (and hence the bandwidth at another level) of another set of records.[1] The root of this constraint is the limited total bandwidth allocated to each server.

Putting data records (or sets of records) at different levels represents an independent disk transmitting at a repetition rate inversely proportional to the number of records associated with the disk. The broadcast from the disk at the highest level adapts to the greatest bandwidth and broadcast from the lowest level adapts to the smallest bandwidth and the total sum of the bandwidths of transmission from the server remains fixed.

In all, the broadcast disk model for push-based mechanisms provides periodic dissemination of data and a multi-level mechanism in which the data records with different priorities are placed at multiple hierarchical levels thus leading to non-uniform pushing intervals.

8.3.1.1 Types of Broadcast Disk Models

As discussed above, the broadcast periods of the individual records are based on their priority ratings. The priority p of a broadcast is decided by the access probability p_a of that record by the users. Records required more often by the users or records that are frequently updated, need to be aired after short intervals of time. There are different methods for bandwidth allocation to records based on their priority assignment in the network. Based on these methods, broadcast disk models are of three types as follows.

Flat-disk Model Figure 8.6(b) shows a broadcast disk model, called flat-disk model, in which each block of records is pushed after equal intervals of time. Figure 8.7(a) shows a rotating disk cycle with record blocks R_0, R_1, R_2, R_3, R_0, R_1, R_2, R_3, transmitted in a single broadcast cycle. All record blocks have an equal priority level. Using the flat-disk model, the server broadcasts the data as per cyclic requests (subscriptions) without taking into account the number of devices that subscribe to a particular record.

Multi-disk Model Figure 8.7(c) shows the multi-disk model, in which each block of records at a level is pushed in equal time T_s (=$N \times t_s$), but a block of higher priority is repeated more often at more levels. The multi-disk model entails multiple levels of records on the broadcast disk. The number of levels assigned to a block of records depends on its priority for being pushed or the number of users subscribing to it. It is assumed that the transmission rate at each level is same but the repetition

[1]For more on broadcast disk models, access probabilities, and bandwidth allocation see Swarup Acharya, Rafael Alonso, Michael Franklin, and Stanley Zdonik 1995, "Broadcast Disks: Data Management for Asymmetric Communication Environments", *ACM SIGMOD International Conference on Management of Data*, San Jose. [Refer *http://www.cs.siena.edu/courses/csis355s01mobile.pdf*].

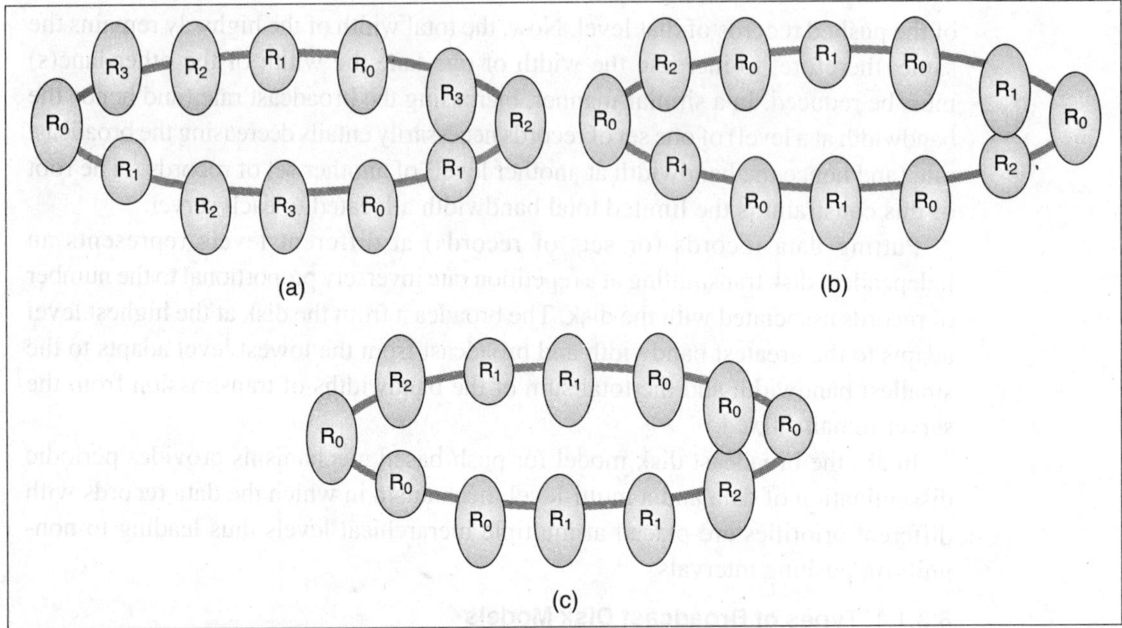

Fig. 8.7 (a) Flat-disk model, (b) Multi-disk model and (c) Skewed-disk model

rate of a block of records is proportional to the record's priority level. The multi-disk model has an m-level disk with n/m records per level and a repetition rate proportional to p, where p is the priority of a set of x records. [$x = j$ at highest level, $= k - j$ at intermediate priority level, and $= n - k$ at the lowest level in Fig. 8.6(c) and $x = 1$ at all levels in Fig. 8.7(b).] The pushing rate and number of disk levels for each record block depends on its subscription (or on access or caching probability).

It is assumed that there are multiple disks, each with same pushing rate but different number of records in them and data sets with the same pushing rates are modelled as belonging to the same disk. The high-priority records are pushed more often by the multiple disks than the low priority ones. For example, consider three records R_0, R_1, and R_2 which are to be transmitted in a given broadcast cycle. The priority order of these records is—R_0 has the highest priority, R_1 and then R_2 .In a multi-disk broadcast program these records are transmitted during one broadcast cycle in the order R_0, R_1, R_0, R_1, R_0, R_2, R_0, R_1, R_0, R_1, R_0, and R_2 . From this order it is evident that the most often pushed record is R_0, followed by R_1 and R_2. This order is repeated cyclically and the time between two successive pushes of a particular data record remains the same. This example illustrates a six-level multi-disk model but more levels can be incorporated as long as for a particular disk, the time between two consecutive cyclic pushes remains constant.

Skewed-disk Model Figure 8.7(c) shows a broadcast disk model, called skewed-disk model. In this model the block of records are repeated as per their priorities for pushing or as per number of subscribers of a given record. High priority record blocks are pushed more often than the low priority ones. However, unlike the multi-disk model, the skewed-disk model entails consecutive repeated transmissions of a record block, followed by consecutive repeated transmissions of another record block, and so on. The number of transmissions of a record block depends upon its priority. Also, in a particular broadcast cycle, the transmission of record blocks is in decreasing order of their priorities. For example, R_0, R_1, and R_2 are the three record blocks to be transmitted in a single broadcast cycle and R_0 has the highest priority, followed by R_1, and then R_2. In a skewed-disk model, these record blocks are transmitted in the order R_0, R_0, R_0, R_1, R_1, R_2, R_0, R_0, R_0, R_1, R_1, and R_2 in a disk rotation. This order is repeated cyclically and the time lapse between two consecutive transmissions of a record block varies. This is evident from the order listed above as the first three show that the record R_0 is repeated three times but the next transmission of the record R_0 occurs after three other records have been transmitted. Records are repeatedly transmitted several times in the skewed-disk model and the number of repetitions is as per their priorities for pushing or as per number of subscribers of a given record. The skewed-disk model has an *m*-level disk with $p \times n/m$ records per level, where *p* represents priority (*p* is proportional to the priority of a set of *i* records), along with the interleaving of the responses of devices from the server. Interleaving is a process in which other entity insertions are made at periodic intervals in between the entities being sent at periodic intervals. Different records are pushed with different frequencies. The time interval between two consecutive pushes of a record with higher subscription is matched to the requirement of the user devices.

8.3.1.2 Broadcast Disk Modelling for Special Cases

Section 8.2.3 described a special case in which there is push–pull-based hybrid mechanism. The server pushes the records through downlink and the devices send the requests to server through uplinks. Hybrid mechanism entails interleaving of responses to the pull requests when the server is pushing the records at periodic intervals. In case of real-time environment, the records need to be pushed in real time. The instants at which a record is pushed also matters. Two models for such cases are considered below:

Modelling for Hybrid Data-Delivery Mechanism A broadcast disk model for the interleaved, push–pull-based hybrid mechanism must provide caching on pushes and pre-fetching by pulls by interleaving of the responses during the broadcast of the disk records. Figure 8.8 shows a broadcast disk model based on the skewed-disk model. It assumes an *m*-level disk with $p \times n/m$ records per level, where *p* represents priority. The figure shows the interleaving of responses I_p, I_q, and I_r of requests from the devices *p*, *q*, and *r* along with the pushed records R_0 to R_{i-1} from the server.

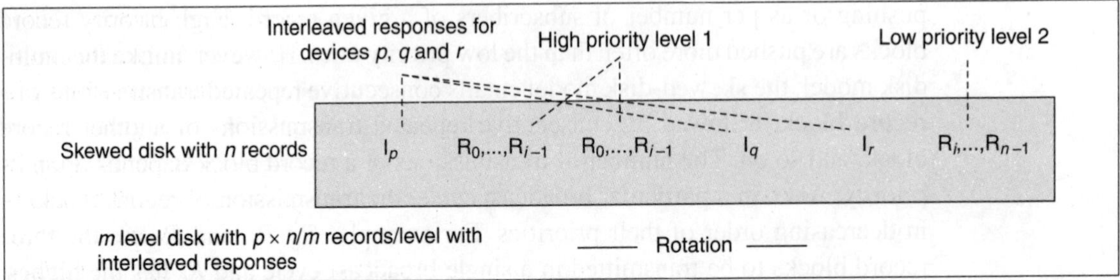

Fig. 8.8 Hybrid data delivery using broadcast disk model along with the interleaving of the responses of devices from the server

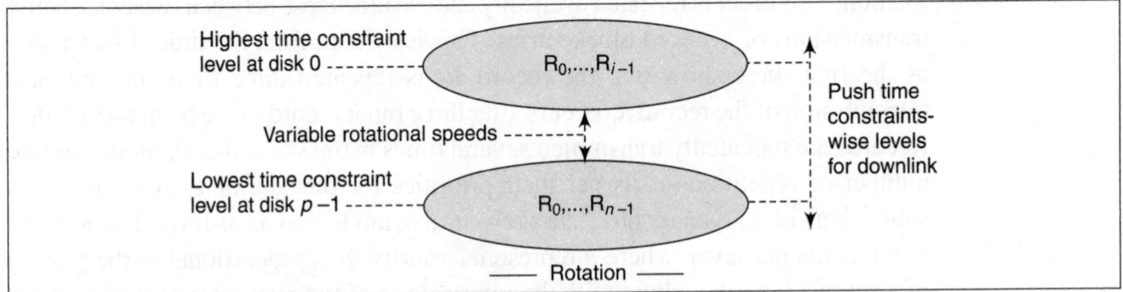

Fig. 8.9 Broadcast multi-disk model for real-time environment

Broadcast Disk Modelling for Real-time Environment Broadcast disk models described above assume that successive bits are transmitted at constant interval of time and total allocated bandwidth for transmission from the server remains fixed. However, in a real-time environment, for example, audio or video data services, there are real-time constraints. Broadcast disk models can be adapted to real-time environments. Consider the following model—each data record (or each set of data records) represents an independent disk rotating at a speed inversely proportional to the time constraint associated with the record. This facilitates the delivery of data records within a deadline in a real-time environment. Assume that data records between 0 and $n - 1$ are spread over p disks. Figure 8.9 shows multi-disk broadcast model adapted to real-time environment in which each disk (from 0 to $p - 1$) rotates at a speed inversely proportional to the time constraint associated with the records present on the disk.

8.3.2 Adaptive Information Dispersal Algorithm

One of the functions of an information dispersal algorithm (IDA) is to identify the duplicate or redundant parts of a file and to make out the additions, deletions, or repetitions of a part, which can reconstruct the original file. For example, the information consisting of record blocks R_0 and R_1 in a file is arranged as R_0, R_1, R_0, R_0, R_0, R_1, R_1, R_0, R_1, R_0, R_1, and R_0 in 12 parts. The file can be broken and AIDA algorithm can find two parts x and y which correspond to R_0 and R_1R_0. Now the file information can be transmitted as x, y, $2x$, $y - x$, $3y$ only in five record blocks on a broadcast disk. The device can later reconstruct all the information in the original form. An adaptive IDA (AIDA) algorithm adapts the parts used for transmitting big sized information depending upon the requirements. Assume that a file, data item, or record is too big. In this case a high time (T_s) is required for broadcast cycle. A wireless environment entails that a device can have a disconnection at several instants during a big broadcast cycle and the device may not be able to cache the record even in several cycles of broadcast. An adaptive information dispersal algorithm (AIDA) can be used in such a situation. A file F_i for data dissemination and push is partitioned into k parts. Then, j parts are selected at the server using an IDA such that from the pushed parts (dispersed, disseminated, or broadcasted parts), the mobile device can reconstruct all k parts and the file F_i is reconstructed at the client device using an algorithm based on the IDA or AIDA.

Directory-based Data Dissemination A directory can hold a group of data records which are of interest to a number of client devices. The records of each directory are then broadcast using a broadcast disk model. Assume that a server gets updates of temperature, weather bulletin, and other related information for dissemination to the mobile devices. All the records related to weather condition can be grouped into a directory. A device can select the directory and tune and cache the weather records. An advantage is that the time, which would otherwise be spent by a device in first selecting and then tuning to the individual records needed by it in case of specific interests, for example, weather records, is saved.

8.3.3 Bandwidth Allocation and Broadcast Scheduling

Bandwidth refers to the number of bits transmitted per second from a server disseminating data by successive pushes. Broadcast model bandwidth is t_s^{-1} where t_s is the time interval between successive bits transmitted from the disk. Assume that the ith record, R_i, on a disk has length l_{ii}, then $N = \Sigma l_{ii}$ where the sum extends over all records R_i with i varying from 0 to $(n - 1)$ for n records on a disk. Each bit in each record is thus repeatedly broadcast at successive time interval $N \times t_s = T_s$. Here T_s is the time taken for one revolution of the disk. Bandwidth also equals to the total number of bits ($N = \Sigma l_{ii}$) stored between 0° and 360° divided by time interval T_s between repeated transmissions in successive broadcast cycles. Now, bandwidth allocation means allocation of frequency $f\,(=(t_s)^{-1})$ to the records in case of multi disks or multi-level disk. It also means allocation of revolutionary speed $(T_s)^{-1}$ to given disk at a given level.

Bandwidth Allocation Strategy Consider a broadcast disk model in which each bit in each record at each level is transmitted at an identical rate. Now, out of the total available bandwidth, the bandwidth used for broadcasting each record is proportional to its length. This is one strategy for bandwidth allocation. There can be another strategy in which bandwidth allocation can be independent of length. In other words, the shorter length records are broadcast at high frequency (repetition rate) than the longer records.

Mathematically, the first bandwidth allocation strategy can be expressed as follows—a record is allotted bandwidth in proportion to its length and its frequency of pushes (push rate = record repetition rate $= f_i$). Assume that time interval between two pushes of record R_i is denoted by $t_s(i)$. Then f_i is $1/t_s(i)$. Bandwidth allotted to ith record (R_i) is proportional to $l_i/\, t_s(i)$, where l_i is the record length.

Broadcast Scheduling When there is no time interval between the successive pushes of record R_i, then it is scheduled to broadcast f_i $(= [t_s(i)]^{-1})$ times in 1 s, that is, after every f_i^{-1}s. However, if there are time intervals between the successive pushes of record R_i, $t_s(i)$ increases and f_i decreases. In such a case, a record broadcast is scheduled after a longer time. Calculation of a record broadcast in case of a given or adopted allocation strategy and rate of pushes is done as described below.

Consider ith record R_i of length l_i. This record is pushed periodically and the time lapse between two consecutive pushes of record R_i is $t_s(i)$. The time interval between the two instants, when the device reads last bit during a push and when it starts reading the first bit in the next push is $t_s(i)/2$. Let us assume that R_i is pushed in successive cycles after equal intervals of time and that there are a total of n records. If p_i represents the subscription probability and f_i, the push rate [reciprocal of $t_s(i)$], then the average wait for ith record (access latency), $t_{\text{lat}}(i)$ and average latency, $\bar{t}_{\text{lat}}$, for all records are given by the following equations:

$$t_{\text{lat}}(i) = t_s(i)/2 \tag{8.1}$$

$$\bar{t}_{\text{lat}} = (1/2)\ \Sigma t_s(i) \times p_i \tag{8.2}$$

where the summation goes from $i = 0$ to $n - 1$ for n records.

A rule, called *square root rule*, states that $\bar{t}_{\text{lat}}$ is minimum when $t_s(i)$ is proportional to $(l_i/p_i)^{1/2}$. This is because if p_i/l_i subscription probability per unit record-length (size in number of bits) of a record increases, then an adaptation algorithm is required to decrease the time spacing of the successive pushes of the record, not in direct proportion to p_i/l_i, but according to the square root rule. If (p_i/l_i) increases four times, $t_s(i)$ should decrease by a factor of two. This square root factor arises because $t_s(i) \times p_i$ is proportional to the fraction of bandwidth allotted which, in turn, is proportional to $l_i/t_s(i)$. The total bandwidth available for broadcast is fixed and must be distributed among all records (here, one may recall the example of the highway and its lanes discussed in Section 8.3.1). Therefore, $(t_s(i))^2$ is proportional to l_i/p_i. The product $(t_s(i))^2\ p_i/l_i$ is a constant and an algorithm must be used to minimize this constant to obtain minimum average access time, $\bar{t}_{\text{lat}}$.

Assume that present time is t_p and $t_s(i)$ is a previous instant when ith record R_i was broadcast. Consider a parameter $c(i)$ given by the equation

$$c(i) = [t_p - t_s(i)]^2\ p_i/l_i \tag{8.3}$$

The parameter $c(i)$ is such that it is proportional to the square of the time gap between present instant and previous instant when the R_i was scheduled for broadcast. This parameter is useful in deciding whether R_i should be scheduled at this instant or later with the help of computations. R_i should be scheduled if $c(i)$ is highest for R_i among n records. $c(i)$ is proportional to p_i because higher the subscription probability, higher is the chance of it being scheduled after a lower time gap $[t_p - t_s(i)]$. $c(i)$ is inversely proportional to l_i because higher the length l_i of the record R_i, lower is the chance of it being scheduled after a lower time gap $[t_p - t_s(i)]$.

Online Scheduling Algorithm In order to decide which of the n records is to be scheduled for broadcasting, the parameter $c(i)$ introduced in Eqn (8.3) can be used as follows—an online scheduling algorithm minimizes $[t_s(i)]^2\, p_i/l_i$ by carrying out online computations for $c(i)$ for each record and discovering which records have been waiting for long. The various steps of online scheduling algorithm for broadcasting are as follows:

1. Compute $c(i)$ for all records.
2. Compute c_{max}, where c_{max} is the maximum value among the various values of $c(i)$.
3. Select the record for which c is maximum and call it record q. When more than one records show maximum c, then select any one of them as q.
4. Broadcast using DVB-H service qth record.

Online Bucket Scheduling Algorithm A bucket is defined as the smallest logical unit of a broadcast. It is similar to a packet which is minimum logical unit for communication in a TCP/IP network. A record can also be considered to be consisting of buckets during a broadcast. When the number of records is large, the computation of $c(i)$ and c_{max}, that is, steps 1 and 2 of the online scheduling algorithm, takes a longer time. Computation time is proportional to n. Greater the number of records, the greater will be time taken for the computations involved. An approach used to minimize computation time is to partition each record into buckets, then schedule their broadcast and place the buckets in a circular queue for transmission. The number of times a bucket of a record is sent into the queue depends upon subscription probability and number of buckets in the record. The online scheduling algorithm minimizes $[t_s(i)]^2\, p_{bji}/l_{bji}$ by performing online computations for each bucket-record $c_b(j) = [t_p - t_s(i)]^2\, p_{bji}/l_{bji}$ and finding out which buckets have been waiting for long. Here, p_{bji} and l_{bji} are average subscription probability and average length of the jth bucket of the ith record, respectively. Computation time also reduces when the number of bits in bucket lengths are chosen to be uniform because then l_{bj} is constant, i.e., independent of j. Then computations of $c'(j) = c_b(j) \times l_{bj}$ and maximum of $c'(j)$ suffices for steps 1 and 2, respectively. The steps of the online bucket scheduling algorithm are as follows:

1. Compute $c_b(j)$ for all bucket records ($j = 0, 1, \ldots, k - 1$), where k is the number of buckets in the records.

2. Compute $c_{b\max}$, where $c_{b\max}$ is maximum among the values of $c_b(j)$.
3. Select the bucket record for which c_b is maximum and call it q. When more than one bucket record shows maximum c_b, then select any one of them as q.
4. Broadcast the qth bucket record.

The online bucket scheduling algorithm is efficient as it computes for only k bucket records in place of n records. The number k is adaptable to give minimum online computation time at the server. Large values of k will, again, increase the computation time.

Off-line Scheduling Algorithm Usually, the parameters, subscription probability p_i and length of record l_i, do not change significantly so as not to warrant modifications in schedules of each broadcast cycle. Thus, continuous online computations involved in steps 1 and 2 are not required to be carried out at the server. The initial stages of off-line scheduling algorithm require an online scheduling algorithm to compute the schedule following steps 1–4 for one broadcast cycle such that each record is broadcast at least once.

For one broadcast cycle the sequences, repetitions, and schedule of broadcasting of each of the n records, once computed, are used for subsequent off-line schedules. The computations are thus done only once. If the number of subscribing devices is modified, which means p_i is modified, or the data records are modified, which means l_i are modified, and the modifications are up to a significant extent, this warrants modifications in the schedules of each broadcast cycle. Rescheduling is then carried out by performing steps 1–4 of online scheduling once again. This fixes the schedules again for the off-line scheduling.

8.4 Selective Tuning and Indexing Techniques

The purpose of pushing and adapting to a broadcast model is to push records of greater interest with greater frequency in order to reduce access time or average access latency. A mobile device does not have sufficient energy to continuously cache the broadcast records and hoard them in its memory. A device has to dissipate more power if it gets each pushed item and caches it. Therefore, it should be activated for listening and caching only when it is going to receive the selected data records or buckets of interest. During remaining time intervals, that is, when the broadcast data buckets or records are not of its interest, it switches to idle or power down mode.

Selective tuning is a process by which client device selects only the required pushed buckets or records, tunes to them, and caches them. Tuning means getting ready for caching at those instants and intervals when a selected record of interest broadcasts. Broadcast data has a structure and overhead. Consider the following:

1. There are n records R_0 to R_{n-1} which are interleaved and broadcast as in a multi-disk model (Fig. 8.6(c)).

2. Only the records $R_{i'}$ and $R_{j'}$ are of interest and required by applications at a device.
3. The broadcast disk broadcasts $R_{i'}$ and $R_{j'}$ thrice and once, respectively, as the subscription probability of $R_{i'}$ is three times that of $R_{j'}$.
4. The record $R_{i'}$ is partitioned into k buckets, bi_0 to bi_{k-1}.
5. The record $R_{j'}$ is partitioned into k' buckets, bj_0 to $bj_{k'-1}$. Each bucket has equal length l_b, which means equal number of bits and the devices takes identical time $l_b \times t_s$ to cache each bucket data. Here t_s is the time interval between successive bits.
6. In addition to data, each broadcast cycle broadcasts a directory, hash-key, or index which is the overhead prefixed by server before the data. [The directory, hash-keying, and indexing are explained later.]

Selective tuning means that the device selects only the buckets of $R_{i'}$ and $R_{j'}$ which are of interest and receives the signals only during first, second, or third instances of $R_{i'}$ or during instances of $R_{j'}$, that is, during the intervals $\text{Ti}_0, \ldots, \text{Ti}_{k-1}$, $\text{Tj}_0, \ldots, \text{Tj}_{k'-1}$ of broadcasting of $bi_0 \ldots bi_{k-1}, bj_0 \ldots bj_{k'-1}$, respectively, in a broadcast disk cycle. In the remaining intervals, where either the other records which are not of interest are being broadcast or when record of interest is already cached in an earlier broadcast cycle, the device remains idle. During this period it does not dissipate power and hence saves energy. When there are cache-misses during the instants of broadcast of buckets or records of interest, the device tunes to the missed buckets or records either during instants of repeat broadcast of that record in same broadcast cycle or in subsequent broadcast cycle or cycles. The presence of directory key or index enables a device to select and tune to specific records before caching them by first listening only to the directory or key or index.

Data broadcast from server, which is organized into buckets, is interleaved. The server prefixes a directory, hash parameter (from which the device finds the key), or index to the buckets. These prefixes form the basis of different methods of selective tuning. Before proceeding with the description of these methods, access time (t_{access}) needs to be defined. It is the time interval between pull request from device and reception of response from broadcasting or data pushing or responding server. Two important factors affect t_{access}—(i) number and size of the records to be broadcast (greater the n and N, greater will be t_{access}) and (ii) directory- or cache-miss factor—if there is a miss then the response from the server can be received only in subsequent broadcast cycle or subsequent repeat broadcast in the cycle.

8.4.1 Directory Method

One of the methods for selective tuning involves broadcasting a directory as overhead at the beginning of each broadcast cycle. If interval between the start of the broadcast cycles is T, then directory broadcasts at each successive intervals of T. A directory can be provided which specifies when a specific record or data item appears in data being broadcasted. For example, a directory (at header of the cycle) consists of directory start sign, 10, 20, 52, directory end sign. It means that after the directory end sign, the 10th, 20th and 52nd buckets contain the data items in response to the device request. The device selectively tunes to these buckets from the broadcast data. If bucket length is l_b and t_s is the time interval between successive bits, then the device tunes to first bucket after time interval of $9 \times l_b \times t_s$ and tunes to the data for the interval $l_b \times t_s$. It again tunes to data after time interval of $9 \times l_b \times t_s$ for 20th bucket data. It again tunes to data after time interval of $31 \times l_b \times t_s$ for 52nd bucket data. In case the size of the data broadcast by server is constant, say, 100 bucket lengths ($100 \times l_b \times t_s$), the device again listens to the directory start sign after time interval of $48 \times l_b \times t_s$.

A device has to wait for directory consisting of start sign, pointers for locating buckets or records, and end sign. Then it has to wait for the required bucket or record before it can get tuned to it and, start caching it. Tuning time t_{tune} is the time taken by the device for selection of records. This includes the time lapse before the device starts receiving data from the server. In other words, it is the sum of three periods—time spent in listening to the directory signs and pointers for the record in order to select a bucket or record required by the device, waiting for the buckets of interest while actively listening (getting the incoming record wirelessly), and caching the broadcast data record or bucket.

The device selectively tunes to the broadcast data to download the records of interest. When a directory is broadcast along with the data records, it minimises t_{tune} and t_{access}. The device saves energy by remaining active just for the periods of caching the directory and the data buckets. For rest of the period (between directory end sign and start of the required bucket), it remains idle or performs application tasks. Without the use of directory for tuning, $t_{tune} = t_{access}$ and the device is not idle during any time interval.

Assume that the time period between two instants of repetiting the directory before the broadcast of the records is T. This also equals the time interval between two instances of a record item in the repeatedly pushed records if a record is pushed only once during a cycle. If first instance is missed, then second can be utilized. Broadcast data record can be missed if t_{tune} is large. Directory (header)- or cache-miss occurs if t_{tune} is longer than T.

8.4.2 Hash-Based Method

Hash is a result of operations on a pair of key and record. Advantage of broadcasting a hash is that it contains a fewer bits compared to key and record separately. The operations are done by a hashing function. From the server end the hash is broadcasted and from the device end a key is extracted by computations from the data in the record by operating the data with a function called hash function (algorithm). This key is called hash key. Hash-based method entails that the hash for the hashing parameter (hash key) is broadcasted. Each device receives it and tunes to the record as per the extracted key. In this method, the records that are of interest to a device or those required by it are cached from the broadcast cycle by first extracting and identifying the hash key which provides the location of the record. This helps in tuning of the device.

Hash-based method can be described as follows:

1. A separate directory is not broadcast as overhead with each broadcast cycle.
2. Each broadcast cycle has hash bits for the hash function H, a shift function S, and the data that it holds. The function S specifies the location of the record or remaining part of the record relative to the location of hash and, thus, the time interval for wait before the record can be tuned and cached.
3. Assume that a broadcast cycle pushes the hashing parameters H ($R_{i'}$) [H and S] and record $R_{i'}$. The functions H and S help in tuning to the H ($R_{i'}$) and hence to $R_{i'}$ as follows—H gives a key which in turn gives the location of H ($R_{i'}$) in the broadcast data. In case H generates a key that does not provide the location of H ($R_{i'}$) by itself, then the device computes the location from S after the location of H ($R_{i'}$). That location has the sequential records $R_{i'}$ and the devices tunes to the records from these locations.
4. In case the device misses the record in first cycle, it tunes and caches that in next or some other cycle.

8.4.3 Index-Based Method

When we search for a particular topic in a book, we search the index and find a number which tells the page number where that topic can be found. Similarly in a broadcast cycle, a number called index can be first sent. It specifies the location of the bucket or record. Consider a simple example. Let index be 20 at the beginning of a broadcast cycle. It specifies that 20th bucket is of interest and is sent to the device in response to previous subscription. Indexing is another method for selective tuning. Indexes temporally map the location of the buckets.

At each location, besides the bits for the bucket in record of interest data, an offset value may also be specified there. While an index maps to the absolute location from the beginning of a broadcast cycle, an offset index is a number which maps to the relative location after the end of present bucket of interest. Offset means a value to be used by the device along with the present location and calculate the wait period for tuning to the next bucket. All buckets have an offset to the beginning of the next indexed bucket or item. Following example explains the method.

Assume that a number I represents the index which maps to the location of the bucket of length l_b. Assume that 20th and 52nd buckets from the beginning of a broadcast cycle contain the records of interest to the device. The index I at the beginning of a broadcast cycle gives the location of the data bucket, record, or record block. For example, assume that first data bucket of interest is 20th from start of the broadcast cycle. Then let us assume that index I (=20) specifies the time interval $(I-1)\times l_b \times t_s = 19 \times l_b \times t_s$ after which the bucket of interest to be cached by the device will be starting. The device should tune to the broadcast data using I. Assume that a new offset I' to the I at the 20th bucket gives the location of the second data bucket, record, or record block of interest. Let $I' = 32$. Then let us assume that this index specifies the offset and the time interval $(I'-1)\times l_b \times t_s = 31 \times l_b \times t_s$ after which the bucket of interest (52nd bucket from beginning of the broadcast cycle) to be cached by the device will be starting. Assume that a new offset index I' at the 52nd bucket gives $I' = 48$ as location of the end of the broadcast cycle and beginning of new broadcast cycle.

Indexing is a technique in which each data bucket, record, or record block of interest is assigned an index at the previous data bucket, record, or record block of interest to enable the device to tune and cache the bucket after the wait as per the offset value. The server transmits this index at the beginning of a broadcast cycle as well as with each bucket corresponding to data of interest to the device. A disadvantage of using index is that it extends the broadcast cycle and hence increases t_{access}.

An index is not as simplistic as given in above example. The index I has several offsets and the bucket type and flag information. A typical index may consist of the following:

1. I_{offset} (1) which defines the offset to first bucket of nearest index (20th in above example).
2. Additional information about T_b which is the time required for caching the bucket bits in full after the device tunes to and starts caching the bucket. This enables transmission of buckets of variable lengths.
3. I_{offset} (next) which is the index offset of next bucket record of interest.
4. I_{offset} (end) which is the index offset for the end of broadcast cycle and the start of next cycle. This enables the device to look for next index I after the time interval as per I_{offset} (end). This also permits a broadcast cycle to consist of variable number of buckets.
5. b_{type} which provides the specification of the type of contents of next bucket to be tuned, that is, whether it has an index value or data.
6. A flag called dirty flag which contains the information whether the indexed buckets defined by I_{offset} (1) and I_{offset} (next) are dirty or not. An indexed bucket being dirty means that it has been rewritten at the server with new values. Therefore, the device should invalidate the previous caches of these buckets and update them by tuning to and caching them.

The advantage of having an index is that a device just reads it and selectively tunes to the data buckets or records of interest instead of reading all the data records and then discarding those which are not required by it. During the time intervals in which data which is not of interest is being broadcast, the device remains in idle or power down mode.

Every record or bucket has a prefixed index. For example, consider the example of broadcast cycle having I1, b1—I′1, b2—I′2, b3—I′3, that is, b1, b2, and b3 have I1 indexes and I′1, I′2, and I′3 offset indices, respectively. Indexing increases access latency. Assume that a record or bucket has 100 bits and index has 16 bits. The access latency of a record will increase by 16% when broadcast cycle has just one record and by 116% if the device misses *I* due to disconnection. So the total increase in access latency is much more.

Transmission of an index *I* only once with every broadcast cycle increases access latency of a record as follows: This is so because if an index is lost during a push due to transmission loss, then the device must wait for the next push of the same index–record pair. The data tuning time now increases by an interval equal to the time required for one broadcast cycle.

An index assignment strategy (*I*, *m*) is now described. (*I*, *m*) indexing means an index *I* is transmitted *m* times during each push of a record. An algorithm is used to adapt a value of *m* such that it minimizes access (caching) latency in a given wireless environment which may involve frequent or less frequent loss of index or data.

The index format is adapted to (*I*, *m*) with a suitable value of *m* chosen as per the wireless environment. This decreases the probability of missing *I* and hence the caching of the record of interest. If *m* is small then the power dissipated by device is less. If *m* decreases, the chances that the cache be missed go up and the data access latency increases. The value of *m* therefore needs to be optimized which can be done by employing an algorithm as stated earlier. One method is to assign an offset *j* to the next index segment *I* for each bucket of interest and only *j* is sent *m* times before the bits of an item, record, or bucket of interest are transmitted.

Indexing reduces the time taken for tuning by the client devices and thus conserves their power resources. Indexing increases access latency because the number of items pushed is more (equals *m* times index plus *n* records). An alternative technique is distributed indexing transmission. It is described as follows.

8.4.4 Distributed Index-based Method

When index *I* is repeated *m* times, the access latency increases significantly even though the cache-miss probability reduces drastically. The fact that indexing in (*I*, *m*) method infact increases access latency will become apparent from the example that follows. Assume that there is one record or bucket of interest having 100 bits of data and index *I* made of 16 bits with $m = 3$. Therefore, at the start of broadcast cycle, the bucket is prefixed by the index *I* three times. Further assume that the broadcast cycle has just two records ($n = 2$) out of which one is of interest. The access latency of a record will increase by (time interval for broadcast of 16 bits × $m \div n$) which comes out to be 24% in this case.

Distributed index-based method is an improvement on the (I, m) method. In this method, there is no need to repeat the complete index again and again. Instead of replicating the whole index m times, each index segment in a bucket describes only the offset I' of data items which immediately follow. Each index I is partitioned into two parts—I' and I''. I'' consists of unrepeated k levels (sub-indexes), which do not repeat and I' consists of top j repeated levels (sub-indexes).

For example, assume that the index I has 50 bits for offset I_{offset} (1). The I [= I_{offset} (1)] is partitioned into 40 bits (I'') which do not repeat again and 10 bits (I') which repeat m times in a broadcast cycle. This is a typical case, the length j of I'' being four times the length k of I'. The I' with shorter length is used for repeating the transmission of index m times instead of repeating all the bits of index I. This is done mainly because only I' in I_{offset} (1) is relevant in order to define the offset of the data (bucket or record item) which follows after m times repetition of I'.

In the present example, out of the 50 bits of I_{offset} (1), only 10 bits of I' are used by the device to tune to the bucket of interest. The 40 bits of I'' function as a primary key to tune to the distributed buckets of interest with index segment in each. Each bucket of interest has replication of only m times the 10 bits of I' with a new value of I' (corresponding to I_{offset} (next) or I_{offset} (end) for each bucket, chosen such that it defines the offset of next bucket to be tuned by the device or that of the end of the broadcast cycle and beginning of the new cycle. In other words, I'' need not be replicated with each bucket of interest. Therefore, m times 10 bits suffice in this case in contrast to m times 50 bits required in case of (I, m) method of undistributed indexing. Also access latency is reduced by a time-period equal to that required for broadcasting $(m - 1)$ times 40 bits at the beginning of broadcast cycle and m times 40 bits when each bucket having data of interest is broadcast.

Assume that a device misses I (includes I' and I' once) transmitted at the beginning of the broadcast cycle. As I' is repeated $m - 1$ times after this, it tunes to the pushes by using I'. The access latency is reduced as I' has lesser levels (one-fourth that of I'' in the typical case considered above). When the device interprets I' for selecting a data record, it also takes into account its push frequency. However, when the pushes are from skewed disk (Fig. 8.6(d)), then the effect of skewing is also considered.

8.4.5 Flexible Indexing Method

Assume that a broadcast cycle has number of data segments with each of the segments having a variable set of records. For example, let n records, R_0 to R_{n-1}, be present in four data segments, R_0 to R_{i-1}, R_i to R_{j-1}, R_j to R_{k-1}, and R_k to R_{n-1}. Some possible index parameters are (i) I_{seg}, having just 2 bits for the offset, to specify the location of a segment in a broadcast cycle, (ii) I_{rec}, having just 6 bits for the offset, to specify the location of a record of interest within a segment of the broadcast cycle, (iii) I_b, having just 4 bits for the offset, to specify the location of a bucket of interest within a record present in one of the segments of the broadcast cycle. Flexible indexing method provides dual use of the parameters (e.g., use of I_{seg} or I_{rec} in an index segment to tune to the record or buckets of interest) or multi-parameter indexing (e.g., use of I_{seg}, I_{rec}, or I_b in an index segment to tune to the bucket of interest).

Assume that broadcast cycle has m sets of records (called segments). A set of binary bits defines the index parameter I_{seg}. A local index is then assigned to the specific record (or bucket). Only local index (I_{rec} or I_b) is used in (I_{loc}, m) based data tuning which corresponds to the case of flexible indexing method being discussed. The number of bits in a local index is much smaller than that required when each record is assigned an index. Therefore, the flexible indexing method proves to be beneficial.

8.4.6 Alternative Methods

In the methods described above, address (offset), with respect to start of the broadcast cycle or with respect to presently tuned bucket, record, or record block, is used to tune to the bucket of interest. The offset is just like an address which can be in units of the bucket length l_b since bucket is the minimum logical unit of broadcasting. It is assumed that each bit is transmitted after a uniform time interval and hence the device can tune to the data bucket or record of interest. Some alternatives to the use of indexing are discussed as follows.

Temporal Addressing Temporal addressing is a technique used for pushing in which instead of repeating I several times, a temporal value is repeated before a data record is transmitted. When temporal information contained in this value is used instead of address, there can be effective synchronization of tuning and caching of the record of interest in case of non-uniform time intervals between the successive bits. The device remains idle and starts tuning by synchronizing as per the temporal (time)-information for the pushed record. Temporal information gives the time at which cache is scheduled. Assume that temporal address is 25675 and each address corresponds to wait of 1 ms, the device waits and starts synchronizing the record after 25675 ms.

Broadcast Addressing Broadcast addressing uses a broadcast address similar to IP or multicast address. Each device or group of devices can be assigned an address. The devices cache the records which have this address as the broadcasting address in a broadcast cycle. This address can be used along with the pushed record. A device uses broadcast address in place of the index I to select the data records or sets. Only the addressed device(s) caches the pushed record and other devices do not select and tune to the record. In place of repeating I several times, the broadcast address can be repeated before a data record is transmitted. The advantage of using this type of addressing is that the server addresses to specific device or specific group of devices.

Use of Headers A server can broadcast a data in multiple versions or ways. An index or address only specifies where the data is located for the purpose of tuning. It does not specify the details of data at the buckets. An alternative is to place a header or a header with an extension with a data object before broadcasting. Header is used along with the pushed record. The device uses header part in place of the index *I* and in case device finds from the header that the record is of interest, it selects the object and caches it. The header can be useful, for example it can give information about the type, version, and content modification data or application for which it is targeted.

8.5 Digital Audio Broadcasting

Analog radio uses amplitude modulation (AM) or frequency modulation (FM) for radio broadcasts all over the globe. The concept of digital radio broadcasting was developed in the late 1980s and was commercialized in the year 2000. Since early 2000, over 50 commercial stations have come up in the UK alone. Digital audio broadcasting (DAB) is used in terrestrial digital audio broadcast. Using a radio card, one can listen to DAB through a personal computer. Many countries in Europe have adopted the Eureka 147 protocol for DAB. Other protocols are IBAC and IBOC used in the USA and called HD-radio (high definition radio).

DAB supports reception of CD-quality audio on radio. It does not require re-tuning while roaming in a single frequency network (SFN) as the operational frequency band does not change. DAB functions on principles of OFDM (also called COFDM) (Section 4.8).

Section 8.5.1 describes digital audio broadcast system and Section 8.5.2, the multimedia object transfer protocol for broadcast.

8.5.1 Digital Audio Broadcast System

DAB systems are allocated frequency spectrum in VHF Band III and UHF L band. Since DAB functions on OFDM principles described in Section 4.8, it has multiple carriers with each carrier using mutually orthogonal codes. This enables separation of carriers in spite of the multi-path transmissions and interference of signals. An OFDM carrier uses a FHSS (frequency hopping spread spectrum) based technique (Section 4.3.2) for distribution (spreading) of data over large number of sub-carriers that are spaced at precise frequency intervals with the help of a coding scheme. Multi-carrier transmission and multiplexing in each carrier facilitates single frequency networks (SFN) for multiple applications. For example, multiplexing of stereo radios and radios for traffic reports. DAB uses differential QPSK (Section

4.1.1). Successive sub-carriers have a quadrature relationship and I and Q components (Section 4.5.1.6). Each transmitted frame has three time–space multiplexed channels for main service, synchronization, and fast information. DAB permits use of various audio-coding protocols. These protocols provide multiplexed text with audio for radio text real-time information. Radio text is the text displayed on LCD screen of the device which is rendering the audio. Radio text may have advertisement or information pertaining to the broadcast audio. Table 8.1 gives the characteristic features of DAB systems.

Table 8.1 Characteristics of DAB

Characteristic	*Description*
Frequency spectrum and bandwidth	Eureka 147 protocol, VHF Band III (174 to 240 MHz) and UHF L band (1.452 to 1.492 GHz), bandwidth 1.5 MHz block (192 to 1536 sub-carriers) per carrier multi-carrier transmission
Carriers and inter-symbol multiplexing	DAB uses OFDMA access for each carrier. It facilitates multiple radio stations and multiplexes (by orthogonal code division) multiple carriers for broadcast from a single radio station. For example, three short wave radios (SWR) of 192 kbps stereo, one 160 kbps stereo, two SWR services for traffic reports at 18 kbps use a SFN in Germany
Transmitter	Lower power needs with SFN of all stations in a country for same radio program network and using OFDM same frequency spread multiplexing
Digital modulation	Differential QPSK (DQPSK) also called four-phase modulation (QPSK phase) angle depends on associated pair of symbols. In DQPSK modulation, phase angle of modulated carrier depends on associated pair of symbols after finding difference in phase angles of present pair with respect to previous pair of symbols (Section 4.1.1). DQPSK is comparatively resilient for fading of carriers. Symbols use Gray codes (e.g., 00, 01, 11, and 10—only one bit changes between adjacent numbers)
Sub-carrier frequencies spacing	Within the time interval given by reciprocal of the active symbol period there are successive sub-carriers and they have a quadrature relationship with each other. The modulated sub-carrier spectra efficiently occupies the bandwidth with a degree of controlled overlapping and I-Q demodulation (Section 4.5.1.6). I-Q demodulation gives zero intermediate frequency (IF) and thus no hardware filters for receiving sub-carriers
Frequency–space interleaving in data streams	Frequency hopping FHSS (Section 4.3) of the sub-carriers frequencies [refer Eqns (4.7a) to (4.7h)]. FHSS gives better signal quality for most of the period as only the select frequency signals fade
Frame	A frame consists of main service channel (MSC), synchronization channel (SynC), and fast information channel (FIC)
Time–space interleaving	Data symbols are transmitted in different time slots in sequence other than that in which they are generated from the audio music or voice source (studio). Receiver rearranges the sequence and generates the original analog signals

(*Contd*)

(*Contd.*)

	Computation delay occurs due to this sequencing and re-sequencing. The data buffering and other processing also contribute to delay, typically of a few seconds, between the studio source and the receiver. The advantage is that error bursts resulting from multi-path interference in a case of moving vehicle are averaged out over the time. (Effect of delay is felt only in time reference signals.)
Common interleaved frame (CIF)	CIF is a frame, which transmits after the time–space interleaving of data fields in the frame
Capacity unit (CU)	CIF consists of CUs and each CU is addressable and has 64 bits
Audio coding	MP2 (MPEG-1 layer-2) audio codec[1] needs 192 kbps plus for good stereo audio, but at 192 kbps, only six DAB stations can multiplex [because a 1.5 MHz block of the frequency band is used and 1.5 MHz block multiplexes six 192 kbps stations by frequency multiplexing] UK adopted DAB 128 kbps codec and now dual rate codec (128 kbps and 192 kbps) protocols are being adopted. New DAB, DABv2 (DAB version 2, DAB-2), is being adopted in 2006. DABv2 supports MPEG2, AAC (advanced audio coding), MPEG4, BSAC (bit slice arithmetic error resilient coding), AAC+ (only in new DAB+ standard), SBR[2] (spectral band replication), and Windows Media audio codecs together with convolution coding and RS coding.
Multiplexing audio and text	Dynamic label segment (DLS) multiplexes text with audio to provide radio text real-time information such as weather report, stock-quotes, and traffic congestion reports. The text can be read on LCD screens
Guard space	Each symbol has a guard space
Bit-error correction	The Viterbi method is used for bit-error correction and is based on forward error correction (FEC) [Section 3.1.1.(5)] Very low BER (bit error rate) in FEC for the control of critical features in the receiver (e.g., synchronization channel) and normal BER in FEC for traffic channel
Multi-path interference	It is eliminated provided received signal-strength is sufficient and above a required threshold

[1]*CODEC is a way of coding and decoding. MPEG 2 and AAC+ (only in new DAB+ standard) are two examples of codecs. Coding is for data compression for reducing the number of bits (also called symbols) for broadcasting and caching by the mobile device. Decoding is retrieving back the full audio file data by decompression.*

[2]*SBR is a method which is used in combination with a codec for audio compression in order to provide replication of full spectral features on decompression at the receiver. Fourier series Eqn (1.4) shows that an audio signal will contain lower frequency as well as high frequency harmonics. The codec itself transmits the lower frequencies of the spectrum and the SBR synthesizes associated higher frequency harmonics of the lower frequencies in the audio signal and transmitted side information.*

Format of Data Frames DAB frame has three channels—main service channel (MSC), synchronization channel (SyncC) and fast information channel (FIC). MSC has audio encoded data from a radio service and data from data services (e.g., images, low-resolution video, traffic reports, news as text, weather report, and stock quotes) as inputs to the MSC multiplexer. Data traffic (including audio and multimedia services data) is transmitted along MSC. A frame transmits a null symbol plus n_s symbols. Symbols reflect the frame data. For each symbol, number of bits equal to spread factor are transmitted by a spreader (Section 4.3). The transmitter of each frame provides additional guard space. A guard space of time interval t_g is incorporated at the beginning of transmission of the symbols, Guard interval is the period provided before synchronization of the transmitted bits takes place in the receiver. Each symbol is of duration t_s. Hence, the total duration taken by a frame is $(1 + n_s) \times (t_g + t_s)$. This can be 96 ms, 48 ms, or 24 ms (European Standard). A 24 ms frame transmits data at 55296 bit/24 ms or 2.304 Mbps. In other words, a frame of 55296 bits is sent every 24 ms. Let us assume that n_h is the spread factor which defines the number of frequency values, at which the sub-carrier is to hop. n_h can take the values 1536, 768, 384, or 192 for a symbol.

Calculation of guard intervals in terms of the time taken by number of symbols transmitted is as follows: Each 24 ms frame transmits either (i) 76 symbols using spread factor $n_h = 768$ or (ii) 153 symbols using spread factor $n_h = 384$. The ratio of the number of frame bits and spread factor is equal to number of symbols for data and is 55296/768 = 72 for first case and 55296/ 384 = 144 for the second case. Therefore, the guard interval period is equal to time interval corresponding to 4 symbols for the first case and 9 symbols for the second case. This period is spent in guard intervals after each set of symbols.

The fast information channel (FIC) carries control information and has 240 bits plus a 16-bit checksum. It is interleaved in the frame along with the MSC and the synchronization channel. The synchronization channel is of the duration of the transmission time of two symbols $[2 \times (t_g + t_s)]$ and is also interleaved with the MSC and the FIC.

Figure 8.10 shows the architecture of a DAB transmission unit. Audio encoded data from a radio service and data from data services (e.g., traffic reports, news as text, weather report, and stock quotes) are inputs to the MSC multiplexer. Transmission multiplexer (interleaving) unit receives inputs from the MSC multiplexer, FIC, and SynC. An OFDM multiplexes two or more transmission multiplexers of multiple carriers (e.g., six carriers). OFDM output first divides into I and Q carriers and then it is DQPSK modulated. The DQPSK modulator output is broadcasted as DAB output.

8.5.2 DAB Objects and Object Transfer Protocol

An object consists of a collection of logically bonded data fields and properties which define the state of the object and methods (functions) which manipulate the state of the object. We can consider a DAB transmitter as transferring the DAB objects with each object consisting of the data fields and services. A device can be considered as receiving DAB objects in real-time environment and the server disseminating or broadcasting these objects to the devices.

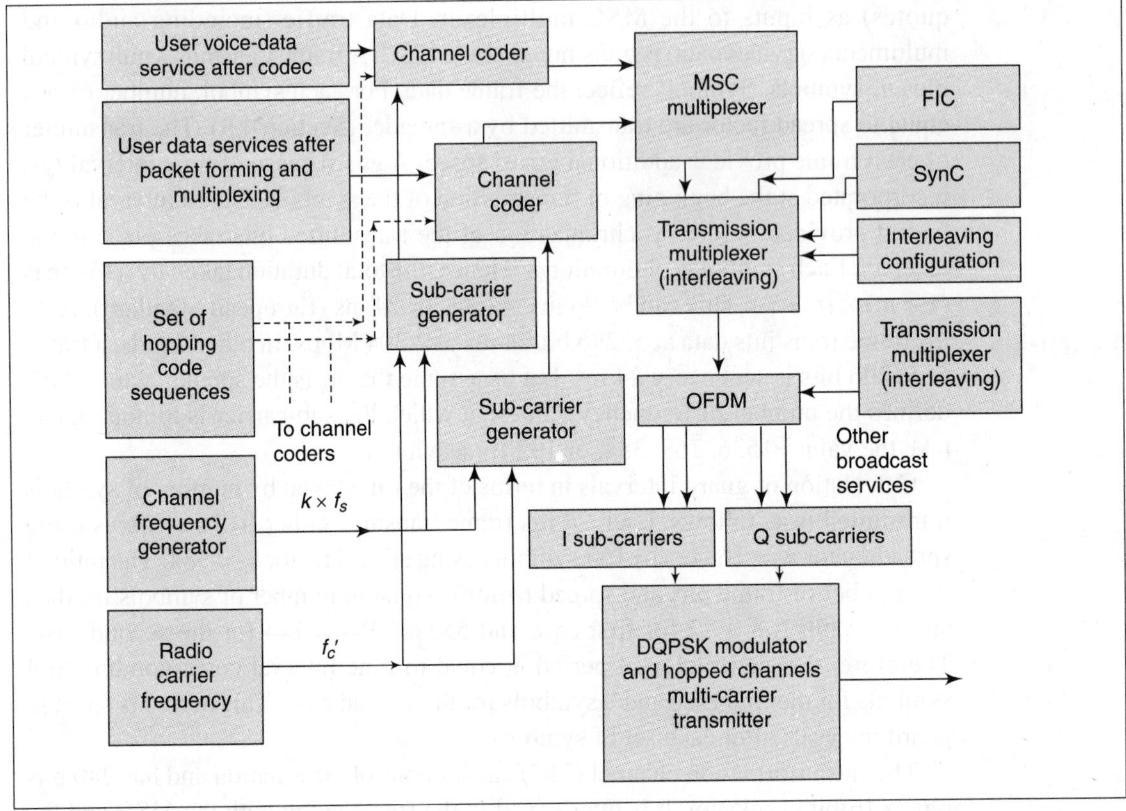

Fig. 8.10 DAB transmission unit architecture

The server (broadcaster) transmits the objects and the objects (e.g., traffic reports, news as text, weather report, and stock quotes) from the server can also be repeated at defined push frequencies (Section 8.3.1). Objects with high subscription (demand) can be repeated more often with defined push frequencies proportional to subscription probability (Sections 8.3.1 and 8.3.3). Broadcasting can occur with interleaved segments (Fig. 8.8) of different objects. The server can broadcast such that only object header is repeated (similar to index (*I*, *m*) technique described in Section 8.4 in which index *I* repeats *m* times). A protocol for sending the objects is multimedia object transfer protocol (MOT) which is discussed next.

Multimedia object transfer (MOT) protocol is a standard, which supports DAB and multimedia formats (e.g., JPEG, Java, HTML, GIF, BMP, ASCII, and HTTP). A DAB receiver can be of various types. For example, audio only, audio with colour graphic display, audio with multi-line text display, audio with single-line text display, audio with slide show, and audio with HTML web page. Different types of devices (receivers) can receive the broadcast data. The receiver should be able to identify the object as well as assess its resources. For example, whether the data object received is a JPEG file, a text in ASCII, or an audio codec output.

MOT protocol involves transmission of data with a core header, a header extension, and a body. The core has seven byte fields for header size, body size, and content type. The extension header fields present after the core provide additional information such as push frequency for a data file or priority of the data, segmentation, supporting caching mechanism (e.g., method to be used for selective tuning and caching of DAB objects or segments). The DAB objects or interleaved segments form the body of the transferred data. The body can have arbitrary data the content type of which is given in the header field.

8.6 Digital Video Broadcasting

Analog TV uses amplitude modulated (AM) transmission for the video component and frequency modulated (FM) transmission for the audio component. The concept of digital video broadcasting (DVB) was developed in the early 1990s and was commercialized in 2000. Since then, over 50 commercial stations have come up in the UK alone. Mobile TV has been commercialized since early 2006.

DVB systems broadcast data by cable (DVB-C), terrestrial TV (DVB-T), satellite (DVB-S), and terrestrial TV for handheld mobile devices (DVB-H). DVB-H is used in mobile TV and is described in Section 8.6.2. DVB-T and DVB-H services are primarily digital video broadcasts and include a few digital audio channels for utilization of spare bandwidth.

Section 8.6.1 describes DVB data broadcasting. Sections 8.6.2 and 8.6.3 describe the applications of DVB for the mobile TV and high speed Internet access. Section 8.6.4 describes the convergence aspect of broadcasting and mobile communication.

8.6.1 DVB System

Commercial DVB-T was deployed in late 1998. DVB systems are allocated frequency spectrum in the range of VHF (130–260 MHz) and UHF (430–882 MHz). DVB functions on principles of OFDM (Section 4.8) [OFDM is also called COFDM]. It employs multiple carriers with each carrier using mutually orthogonal codes. Each carrier has 8 MHz block in the multi-carrier transmission. DVB receiver downlink at user has data rate between 5 and 30 Mbps. Broadcaster data rate for downlink to high speed broadband Internet is between 6 and 38 Mbps. OFDM carrier uses a frequency hopping spread spectrum (FHSS)-based technique (Section 4.3) for distribution (spreading) of data over large number of sub-carriers that are spaced at precise frequency intervals with a coding scheme just as in case of DAB. Multi-carrier transmission and multiplexing in each carrier facilitates single frequency networks (SFN) for multiple applications, for example, multiplexing of TV broadcaster, Internet service provider, mobile service provider channels, and service information tables. DVB-C uses 64-QAM or uses 256-QAM and DVB-T uses 16-QAM or 64-QAM (or QPSK) (Section 4.1.1). DVB has following MPEG encoding formats: MPEG-2/DVB single channel HDTV (high definition TV) 1920 × 1080 pixel, MPEG-2/DVB multiple channels EDTV (enhanced definition TV), MPEG-2/DVB multiple channels SDTV (standard definition TV) 640 or 704 × 480 pixel, and MPEG-2/DVB multiple channels multimedia data broadcasting. Table 8.2 gives the characteristic features of DVB systems.

Table 8.2 Characteristics of DVB

Characteristic	*Description*
Frequency spectrum and bandwidth	VHF (130–260 MHz) and UHF (430–882 MHz). 8 MHz block per carrier multi carrier transmission
Carriers and inter-symbol multiplexing	DVB uses OFDMA access for each carrier. It facilitates multiple video or TV stations, Internet service provider, cable service provider, and mobile service provider station. DVB multiplexes (by orthogonal code division) multiple carriers for broadcast from a single station
Digital modulation	DVB-C 64-QAM or 256-QAM and DVB-T 16-QAM or 64-QAM (or QPSK)
Data rates	Downlink 5–30 Mbps data
Additional protocols	DVB-VBI (vertical blanking interval) for conversion from DVB to analog TV, Data connections DVB-DATA, return channels (e.g. GSM, DECT, PSTN, ISDN, Satellite) DVB-RC, Internet protocol DVB-IPI, DVB-NPI (Network protocol independent) and DVB-Text for text and DVB-SUB for sub-titling on the video frames
Video coding	DVB-MPEG (MPEG2 with additional constraints) for audio and video transmission
Multimedia home platform	DVB-MHP (multimedia home platform) is Java based for the development of consumer video system applications, interfaces for network card control, application download, and layered graphics
Service information	Specifies the MPEG container in four tables for the set top boxes—(1) event (current status) information table, (2) service description table with broadcast service names, parameters, and multiplexed channels information, (3) network information table, (4) time and date information table
Encryption	DVB-CA for conditional access system and DVB-CPCM for content protection and content management

8.6.2 Mobile TV—DVB-H

DVB-H stands for DVB for handheld devices (e.g., mobile phone or PDA). DVB-H broadcast downlink channel of mobile service provider using DVB-T/DAB/IP data broadcaster with high rates of data broadcast (5 Mbps) to the handheld devices. Mobile TV devices also network for uplink and interact with mobile service provider network, for example, enhanced telecommunication mobile GSM/GPRS network. Standard broadcast frequency spectra allocations are Band IV (554 MHz), VHF Band III (174–230 MHz) or a portion of it, UHF Band IV/V (470–830 MHz) or a portion of it, and Band L (1.452–1.492 GHz). DVB-H can coexist with DVB-T in the same multiplex DVB-H+ hybrid (satellite/terrestrial) architecture using S band.

The technical specification for bringing broadcast services to handheld mobile and PDA devices has been formally adopted as ETSI standard EN 302 304 since November 2004. DVB-H commercial broadcast for mobile TV started in Oxford in early 2006. Participating companies were Arqiva/O2 with content providers—BBC, ITV, Turner Broadcasting, Shorts International, Discovery, Eurosport, MTV, Channel 4, Five TV, and BSkyB. They broadcast for receiving the signals on mobile TV as well for 16 popular TV programs.

Table 8.3 gives the characteristic features of DVB-H broadcasts for handheld devices at Oxford in UK from early 2006. DVB-H/Mobile TV is allocated broadcast frequency spectrum Band IV (554 MHz) at Oxford. DVB multiplexes multiple carriers for broadcast from a single SFN video transmission station. Oxford DVB-H service multiplexes 16 popular TV programs in an SFN. These TV programs include CNN, Cartoon Network, Discovery Channel, and others. Oxford SFN uses H.263 video format.

Table 8.3 Characteristics of DVB-H from Oxford

Characteristic	*Description*
Frequency spectrum	Band IV 554 MHz (Oxford), VHF Band III (174–230 MHz) or a portion of it, UHF Band IV/V (470–830 MHz) or a portion of it, Band L (1.452–1.492 GHz). DVB-H can coexist with DVB-T in the same multiplex DVB-H+ hybrid (satellite/terrestrial) architecture using S band
Video codec	MPEG2 single channel HDTV and multiple channels SDTV [DAB-2 cannot be used for mobile TV because it does not include any video codecs]
Carriers and inter symbol multiplexing	DVB uses OFDMA access for each carrier. It facilitates multiple TV programs on SFN stations. Multiplexes multiple carriers for broadcast from a single SFN video transmission station. For example, 16-popular TV programs (CNN, Cartoon Network, Discovery Channel, and others) in Oxford multiplex in an SFN
Transmitter	9 Harris Semiconductor transmitters transmit and each employs SFN multiplexing. The transmitter covers Oxford, UK
Video format	H.263
Digital modulation	QPSK
Transmission technique	IP datacast
Time division multiplexing	10 Time-slices
Video-coding format	A new standard for temporally compressed distribution to mobile devices
Receiver device	Nokia 7710
Server system	Nokia "ServiceSystem" v2.2
Conditional access system	IPSec
Error correction	½ FEC
Development tool	Nokia N92 & Sagem myMobileTV DVB-H handsets and DVB-H ESG Simulator

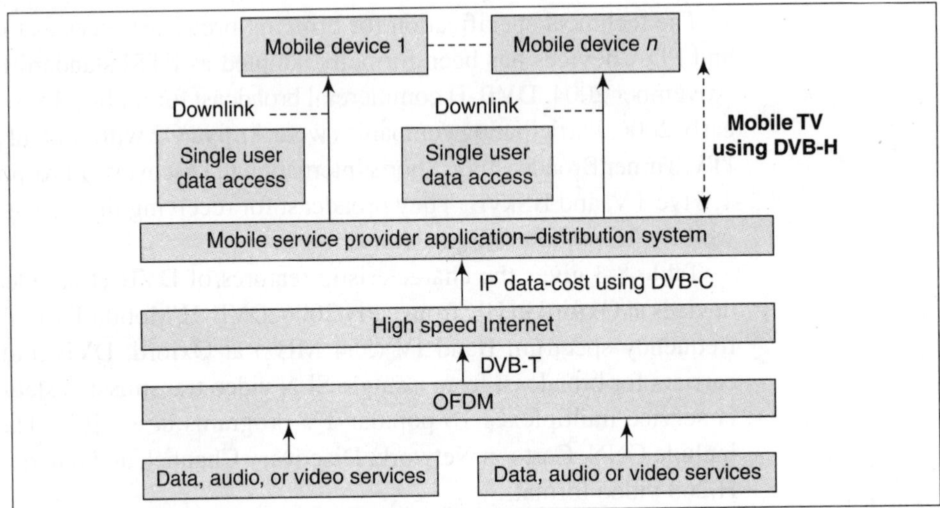

Fig. 8.11 Mobile TV/DVB-H broadcasting architecture for downlink to handheld devices

Figure 8.11 shows mobile TV broadcast architecture. At first stage, the data, audio, video, and TV services use OFDM DVB-T transmission standard. Multiplexed data at second stage is carried all over the globe using high speed Internet using IP datacast. At third stage, the IP datacast networks to a mobile service provider application distribution system (MSPADS). This system finally provides service to handheld devices which have the subscription using DVB-H service.

IP datagrams are transmitted as data bursts in small time slots by TDMA by the MSPADS. For example, there are 16 time slots, T_0, ..., T_{15}. In each time slot the data bursts for 16 different popular TV channels, Arqiva/O2 and content providers—BBC, ITV, ..., BSkyB are transmitted from MSPS. The handheld device at the receiver end selects only those time slots during which selected service (e.g., BBC or discovery channel) is on air. The front end of the receiver device switches 'on' only for those time intervals when the data burst of a selected BBC or discovery channel service is on air. Within a data burst slot, the transmission of data is at rate higher than the one used for display on the screen. Data bursts store in a device-buffer. From the device-buffer, the device either caches the application data or can play it live on its small LCD screen.

8.6.3 DVB for High-speed Internet Access

IP datacasting enables high speed Internet access through Internet service provider (ISP). IP datacast is then employed by DVB-H for connecting handheld devices. Assume that a device seeks access to high speed Internet and to a remote-networked video channel linked to Internet. The complete process consists of the following steps:

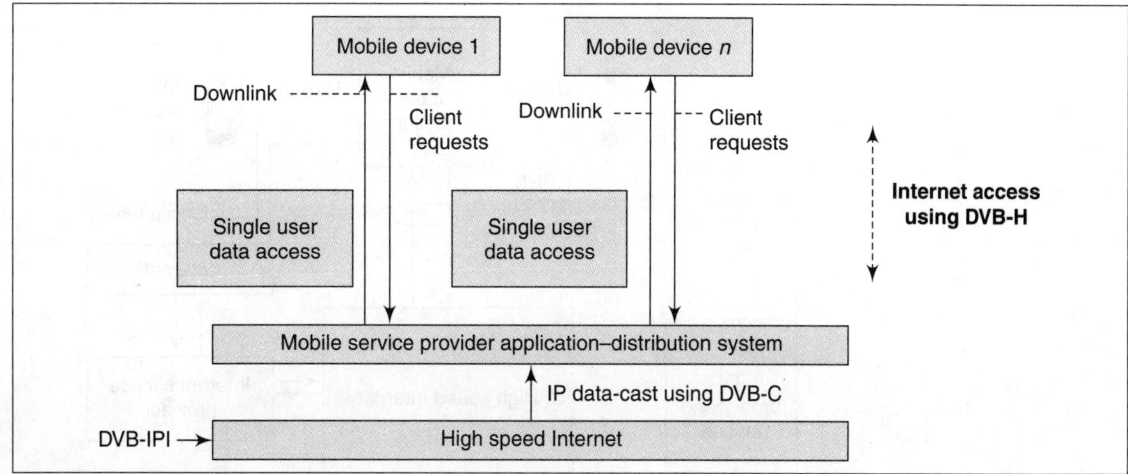

Fig. 8.12 High-speed Internet access architecture

1. A handheld device transmits the device request through uplink channel to MSPADS.
2. MSPADS passes the request to Internet through cable and to video channel through Internet.
3. Video channel sends the response using DVB-IPI to high speed Internet.
4. The Internet service datacasts the data to MSPADS using DVB-C through a cable connection.
5. MSPADS transmits the response through a downlink using DVB-H to the handheld device.

Figure 8.12 shows high speed Internet access architecture. The set of specifications for IP datacast (phase 1) was approved by DVB in October 2005.

8.6.4 Convergence of Broadcasting and Mobile Communication

Figure 8.13 shows DVB-H+ and convergence of mobile, Internet, and broadcasting network architecture. Assume that a device accesses Internet IPv4, high speed Internet IPv6, and a remote networked data service, TV broadcast service, or video channel linked to Internet. The complete process is as follows:

1. A handheld device transmits the device request through uplink channel to mobile base station.
2. The base station functions as a MSPADS and connects to IPv4 Internet through GPRS gateway which networks the device to the IPv4 service provider. The device is thus able to get Internet IPv4 service.
3. Base station passes the request for high speed Internet IPv6 access through cable using DVB-C.
4. IPv6 Internet connects to a multiplexer using DVB-IPI.
5. The multiplexer connects to video channel or TV service through DVB-T and to data service using DVB-data.

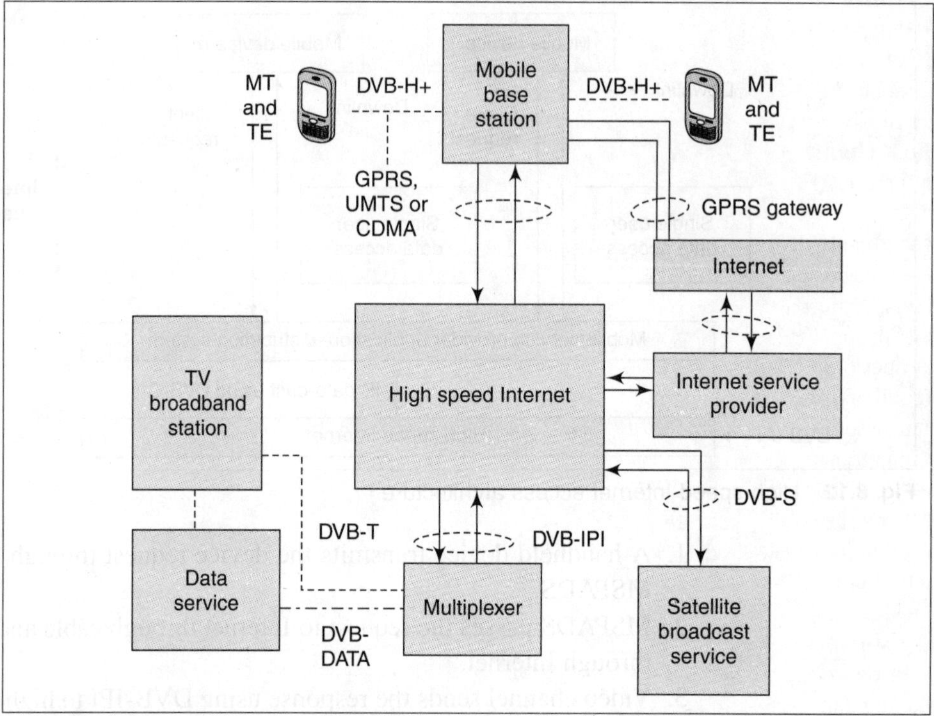

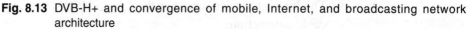

Fig. 8.13 DVB-H+ and convergence of mobile, Internet, and broadcasting network architecture

6. Data service, video channel, or TV station transmits the response to the multiplexer and multiplexer uses DVB-IPI protocol to communicate to high speed Internet using the IPv6 protocol.
7. The Internet IPv6 service datacasts the data to MSPADS using DVB-C through a cable connection.
8. MSPADS transmits the response to the device request through a downlink using DVB-H+ to the handheld device. Due to use of IP datacast, it has become feasible to use DVB-H+ for accessing Internet, mobile telephone, and TV, audio, and satellite broadcast services. DVB-H+ provides hybrid network.
9. The device receives DVB-H+ downlink using a mobile service provider GPRS, UMTS, or 3G system.

Keywords

Access latency (Access time): Time required for caching a data item. This interval includes the time for waiting for push of the data item, selecting and tuning to it. It is also defined as the time interval between pull request from the device and reception of server response at it.

Adaptive information dispersal algorithm (AIDA): A data dissemination algorithm, which entails the partition of pushed records into *k* parts. A limited number of these parts are selected and pushed so that the receiver can adapt and reconstruct all parts.

Back channel: A channel for devices to send their pull requests for the items and records, which are not pushed by the server regularly.

Broadcast disk model: A model in which a disk is assumed to store and sequentially send the records, items, or their parts and each record is transmitted at rate proportional to the disk rotational speed and number of times the record repeats on the disk.

Broadcasting: A special case of communication or data dissemination in which there is unidirectional data stream transmission.

Bucket: The smallest logical unit of a broadcast similar to a packet which is minimum logical unit for communication in a TCP/IP network. A record can also be considered to be consisting of buckets during a broadcast. Buckets are equal length and the records have variable lengths and variable number of buckets.

Data dissemination: A process of distributing and pushing data generated by a set of computing systems or broadcasting data from audio, video, and data services. The output is sent to many systems, so that a system can select and tune to the needed data item.

Digital audio broadcast: Broadcasting of digital data of audio and data services employing multi-carrier multi-tone modulation using COFDM and a single frequency network (SFN) as the operational frequency band does not change.

Digital video broadcast: Broadcasting of digital data of TV, a video, audio, and data service employing multi-carrier multi-tone modulation using COFDM and a single frequency network (SFN) as the operational frequency band does not change.

Distributed indexing: Dividing the index into the main and offset parts and repeating only the offset part after the main part when an item or record is transmitted.

DQPSK: A QPSK-based modulation technique, in which the phase angle of the modulated carrier depends on an associated pair of symbols after finding the difference in phase angles of the present pair with respect to the previous pair of symbols.

Front channel: A channel from a server or set of computing systems for pushing data items or records.

Hybrid data dissemination mechanism: A mechanism of data dissemination that incorporates two-way pushing and pulling from a server or set of computing systems which transmit after interleaving pulled items with the items to be pushed.

Index: An entry placed before a record or item so that the receiver can select and tune in order to cache the record.

Interleaved push and pull: During different time intervals, the pulled items are placed in between the pushed items (interleaved) so that both of them can be transmitted concurrently.

Mobile TV: A system to view a selected TV program on a mobile device employing the DVB-H protocol (DVB for handheld) after transmission of TV programs through COFDM-based DVB-T. The program is unicast first to high-speed Internet, then to the mobile service provider, and then to the devices using DVB-H.

Multi-disk model: A model in which records are placed on to multiple levels or disks and level assigned to a record depends on its priority for pushing or on its number of subscribers. It is assumed that the transmission rate at each level is proportional to the priority or the number of subscribers.

Online scheduling algorithm: An algorithm to schedule the transmission of a record or item after computing the bandwidth allocation for each record depending on the present subscription probability, length of the record, and how long the record has been waiting for transmission.

Pulling mechanism: A mechanism by which a user can receive a data item or record on demand. The item or record is generated or sent from database records by a server or set of computing systems and transmitted either directly or after interleaving with the other items to be pushed.

Pushing mechanism: A mechanism of data dissemination of database records (or items) once or many times. The records are sent either in anticipation of requirements or as per subscription probability and distributed to many devices or computing systems so that one or more receivers can access the needed record after selecting and tuning to it.

Skewed disk: A disk in which there is contiguous placement of records by repeating records several times and number of repetitions is as per their priorities for pushing or as per number of subscribers of a given record.

Tuning: A process for synchronizing received bits of an item with the bits transmitted from a server or set of computing systems. A receiver can access the bits after selection based on directory, index, address, or header information and tuning to the item.

Tuning time: Time taken by the device for selection of records. This includes the time lapse before the device starts receiving data from the server. In other words, it is time spent in listening to the directory, index, hash key, address, or header in order to select a bucket or record required by the device, waiting for the buckets of interest while actively listening (getting the incoming record wirelessly), and caching the broadcast data record or bucket.

Review Questions

1. Show architecture for data dissemination and broadcast. Explain the reasons for communication asymmetry in mobile network. Give examples of asymmetric communication architecture for data dissemination.
2. Describe push-based data-delivery mechanism. What are the advantages and disadvantages of push-based data dissemination?
3. Describe pull-based data-delivery mechanism. What are the advantages and disadvantages of pull-based data-delivery?
4. Describe hybrid data delivery mechanism. What are the situations in which pull-based and push-based mechanism are preferred? Show by diagram, how responses to the device requests interleave along with pushed data. What are the advantages of hybrid delivery?
5. Compare flat-disk, skewed-disk and multi-disk broadcast models. List the situations in which one is preferred over another.
6. Describe the online scheduling algorithm for scheduling the broadcast of data having various subscription probabilities and lengths.
7. Describe the online bucket scheduling algorithm for scheduling the broadcast of data having various subscription probabilities and lengths. What are the advantages of forming buckets for the data broadcast?
8. How does a mobile device tune to the broadcast data? Explain selective and distributive indexing. Show how access latency of a record reduces as a result of greater bandwidth requirement from server.
9. Describe digital audio broadcasting. Explain the DAB technology.
10. Describe digital video broadcasting. Show how it facilitates mobile TV, high-speed Internet access, and convergence among the networks.

Objective Type Questions

Pick the correct or most appropriate statement among the choices given:

1. Communication asymmetry arises due to
   (a) wireless transmission being asymmetric in nature as a result of transmission errors.
   (b) small screen sizes and low-power-battery operation.
   (c) devices having limited power, uplink bandwidth, and memory.
   (d) limited power, uplink bandwidth, and access latency.
2. Push-based delivery mechanism enables
   (a) broadcast of audio, video, and data services.
   (b) distribution of data when a set of computing systems generate and send data output to many mobile devices.
   (c) a device to cache data from a database server.
   (d) broadcast of audio, video, and data services and distribution of data when a set of computing systems generate and send data output to many mobile devices which cache the data.

3. The advantage(s) of push-based mechanism is (are)
   (a) periodic pushes helps in preventing cache misses at the mobile devices.
   (b) broadcast of data services, no frequent server interruptions due to requests for response from multiples devices, and periodic pushes help in preventing cache misses at the mobile devices.
   (c) database server data is pushed.
   (d) there is less load on server due to requests for response from multiple devices.
4. The advantage(s) of pull-based mechanism is (are)
   (a) there is very little server contention and server is able to deliver the response to many device requests within expected time interval.
   (b) there is no unsolicited or irrelevant data and dissemination is during user-interest period only.
   (c) device has less access latency.
   (d) device needs much less memory than when it caches pushed data.
5. Reason(s) for larger data-access latency for a device in the case of push-based data-delivery is (are) (i) wireless transmission errors needing repeat transmissions (ii) cache misses needing repetition of data (iii) that server has to push data of interest to many devices (iv) that server pushes the data which have greater demand with greater periodicity (v) temporal variations in data records (vi) communication asymmetry (vii) use of index before each data record (viii) repetition of index several times before each data record (ix) time taken by the device in tuning to a data record (x) path delays. Which of these are correct?
   (a) All are incorrect except (i), (viii) and (ix).
   (b) All are correct except (i), (v), (vi) and (ix).
   (c) Only (i), (iii), and (ix) are correct.
   (d) Only (viii) is correct.
6. Total allocated bandwidth in hybrid data-delivery mechanism
   (a) is separately allocated for uplink and downlink.
   (b) depends on number of devices sending the requests for data pulls.
   (c) depends on number of records repeatedly pushed.
   (d) is shared and adapted between two channels and is adapted in downlink and uplink channels as per the number of active devices receiving data from the server and requesting the data pull from server.
7. Digital audio and video broadcasting
   (a) uses one of the broadcast disk models.
   (b) uses flat-disk broadcast model.
   (c) does not use any of the broadcast disk models.
   (d) uses multilevel disk model.
8. A computation time efficient approach to schedule broadcasts in push-based delivery is
   (a) to schedule according to subscription probability.
   (b) to schedule according length of data records.
   (c) online scheduling algorithm.
   (d) to partition each record into buckets, then schedule their broadcast and place the buckets in a circular queue for transmission.
9. Skewed-disk broadcast model is that in which
   (a) a record having higher priority is repeated continuously more number of times compared to lower priority record.
   (b) a record having higher priority is repeated more often compared to lower priority record.
   (c) higher priority record is transmitted from disk of faster speed.
   (d) each record is transferred with same speed by rotating a disk at continuously incrementing or decrementing speeds.
10. Adaptive information dispersion algorithm
    (a) is based on a directory which holds a group of data-records of interest for a number of client-devices.
    (b) is that a file of records is partitioned into parts and then limited number of parts are pushed such that the mobile device reconstructs all parts from cached parts.
    (c) uses variable speed broadcast-disk model.
    (d) adapts bandwidth according to number of data records being pushed.

11. Selecting data records by index
    (a) helps in tuning to the needed record only.
    (b) helps in selecting the records according to priority of records for a device.
    (c) reduces device tuning time and power dissipation at device in tuning to the data.
    (d) reduces server load due to sending data repeatedly.
12. When a device tunes to a data with minimum tuning time (i) it needs repeat broadcast of data (ii) it needs repeat broadcast of index (iii) the index can be distributed in two parts, non-replicated and replicated (iv) it may employ temporal information in place of index (v) it may employ broadcast address in place of index (vi) it may use a header for selection. Which of these are correct?
    (a) All are incorrect except (iii).
    (b) All are correct except (i) to (iii).
    (c) (ii) and (iii) are correct.
    (d) All are correct except (i).
13. DAB employs (i) frequency–space interleaving in data streams (ii) time–space interleaving (iii) continuous allocation of time slots for consecutive parts of data (iv) consecutive allocation of frequencies to sub-carriers (v) $\pi/4$ QPSK (vi) COFDM for multiple main service, synchronization, and fast information channels (vii) multiplexing of multiple radio-station broadcasts. Which of these are correct?
    (a) All are correct except (iii), (iv), and (vii).
    (b) All are incorrect except (i), (ii), and (vii).
    (c) (iii) to (vi) are correct.
    (d) Only (i), (ii), and (vi) are correct.
14. Mobile TV employs
    (a) 1.452–1.492 GHz band, MPEG4 video coding, and OFDMA access for each channel of a TV program.
    (b) 800–959 MHz COFDMA access for each station of the TV programs and IP datacast.
    (c) 174–230 MHz, 470–830 MHz, or 1.452–1.492 GHz band, MPEG2 video coding, IP datacast, and OFDMA access for each carrier for facilitating multiple TV programs on a single frequency network of stations.
    (d) 174–230 MHz, 470–830 MHz, or 1.452–1.492 GHz band, CDMA2000, and IP datacast.
15. DVB employs
    (a) QPSK and 5 Mbps data transfer rates and cannot be used to receive TV programs on analog TV receiver.
    (b) 64- or 256-QAM for cable path, 16- or 64-QAM for terrestrial path, 5–30 Mbps data transfer rates, and using a protocol converts the data and outputs it to analog TV.
    (c) QPSK and 5 Mbps data transfer rates and video coding scheme cannot accept DAB system coded data.
    (d) VHF (200–260 MHz), UHF (750–882 MHz), 6 MHz block per carrier multi-carrier transmission, 5–30 Mbps data transfer rates, and using a protocol converts the data and outputs it to analog TV.
16. DAB
    (a) main service channel (MSC) transmits a null, data, audio-service, and multimedia-service symbols. Each symbol has a guard space in a frame-time, and interleaves synchronization and fast information channels.
    (b) main service channel (MSC) has each frame with a guard space in a frame-time. Synchronization and fast information channels transmit in separate TDM frames.
    (c) main service channel (MSC) transmits a null, data, audio-service, and multimedia-service symbols. Each symbol has a guard space in a frame-time and synchronization channel in separate frame interleaves with fast information channel.
    (d) main service channel (MSC) has a guard space in a frame-time as well as TDM multiplexing of symbols. Synchronization and fast information channels have FDM multiplexing.

Chapter 9

Data Synchronization in Mobile Computing Systems

In the preceding chapters we learnt how a server or a distributed mobile computing system disseminates data to mobile devices. We also discussed the mechanisms used for data dissemination. A mobile device selects, tunes, and caches the required item or database record. It then hoards the cached data for future computations. The hoarded data in the device database is then used for mobile computing directly. A device application can send the queries for data and perform transactions with the hoarded database.

The disseminated data at the service-provider server can be modified in future. The disseminated data must therefore be synchronized so that its hoarded copies at the devices are consistent, without any discrepancies or conflicts among the disseminated or distributed copies of data. Also, the hoarded data copies at a personal area computer, which may be associated with a mobile device, and data at the mobile device can also be modified. Therefore, the data hoarded at the device and computer must also be synchronized so that it become consistent, without any discrepancies or conflicts.

This chapter describes the synchronization of data hoarded at different computing ends in a mobile computing system. In this chapter we will learn how data generated and disseminated at the data source end is changed and how it is synchronized at other ends in the system. Any change at one end must be reflected at all other ends by the process of synchronization. The following concepts are covered in this chapter:

- Need of synchronization for mobile computing applications
- Domain-dependent specific rules for synchronization and conflict management strategies
- Software, protocols, and languages used for synchronization
- SyncML (Synchronization Markup Language)
- Sync4J (Funambol) which provides Java-based synchronization using SyncML
- SMIL (synchronized multimedia integration language)

9.1 Synchronization

Data is disseminated and replicated at either remote or local location(s). Data replication may entail copying of data at one place after copying from another (i.e., recopying), copying from one to many others or from many to many others. For example, videos of faculty lectures or music files get replicated at iPods. Figure 9.1 display how data gets replicated in one-to-one, one-to-many, or many-to-many environments. A replicated data copy can be a full copy or a partial copy.

Full copy from a source means that the full set of data records replicates according to certain domain-specific data format rules at the replicating devices or systems. Assume that a server has a set of 8 images of an event with resolution 640×640 pixels. In the domain of a mobile device, it can replicate and hoard with 160×160 pixels. When all 8 images are copied, though with the different resolution, then it is known as full copy replication.

Partial copying of data from the source means that a subset of the data set is copied according to certain domain-specific rules at the devices or systems. Assume that a server has a set of 24 temperature records with a resolution of 0.1°C. In

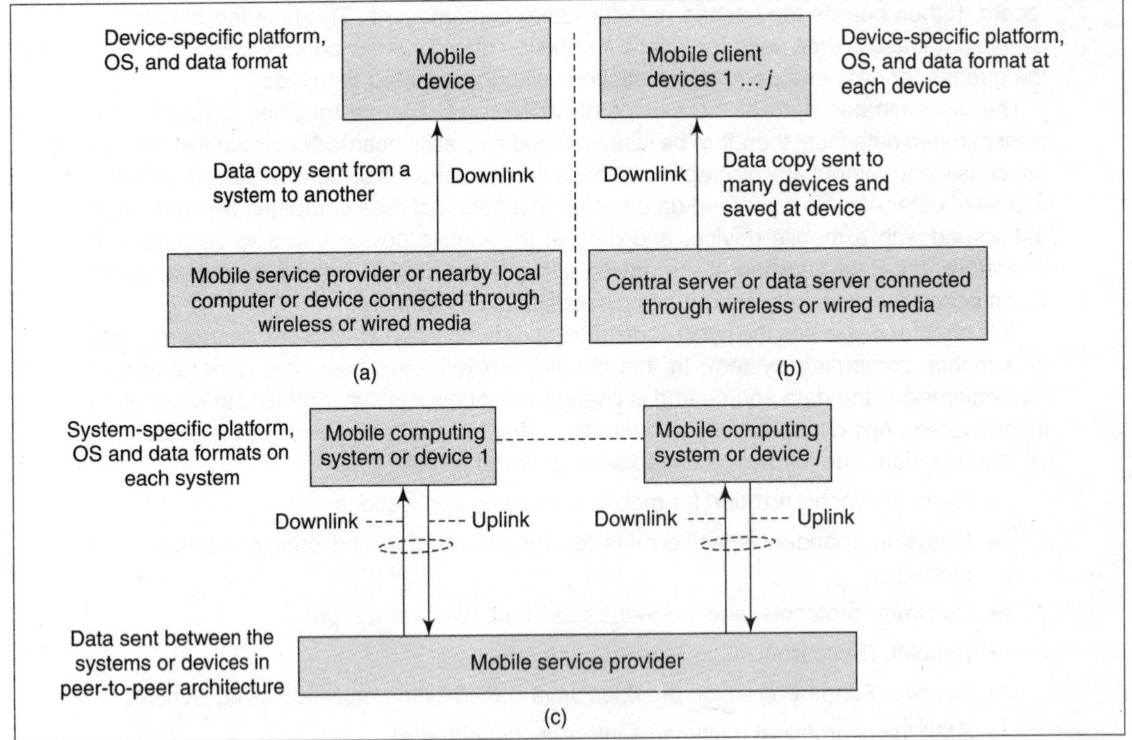

Fig. 9.1 (a) Data replication from data source and device (b) Data replication from data source server to many clients (devices) (c) Data replication among many data sources and devices in peer-to-peer architecture

mobile device domain, assume that it replicates and hoards some of the records being disseminated by the server every three hours with a resolution of 1 °C. This is the case of partial copy replication.

Data replication precedes data synchronization which refers to maintaining data consistency among the disseminated or distributed data. Data consistency means that if there is data modification at the server then that should reflect in the data with the device within a defined period.

In one-to-many synchronization, each system or device caches the data pushed from the server or sends a pull request to the central server and gets a response. This method of synchronization employs client–server architecture. The server maintains a synchronization server which may have separate accounts of many client subscribers for synchronizing data copies separately for each client. Many-to-many synchronization employs peer-to-peer architecture where each system is capable of sending pull requests and of pushing responses.

This section explains in detail synchronization within mobile computing systems. It is divided into five subsections which describe synchronization in the mobile computing systems, usage models for synchronization in mobile computing environment, domain-dependent specific rules followed for data synchronization, conflict resolution strategies, and the use of a synchronizer.

9.1.1 Synchronization in Mobile Computing Systems

Data synchronization is defined as the process of maintaining the availability of data generated from the source and *maintaining consistency* between the copies pushed from the data source and local cached or hoarded data at different computing systems *without discrepancies or conflicts* among the distributed data. A consistent copy of data is a copy which may not be identical to the present data record at the data-generating source, but must satisfy all the required functions and domain-dependent specific rules (Section 9.1.3). These rules can be in terms of resolution, precision, data format, and time interval permitted for replication. A consistent copy should not be in conflict with the data at the data-generating source.

Data-generating source for synchronization may be a server, a distributed mobile computing system, or a device. A device is a source when it sends device data, configuration information, or new subscriptions for the services required by the device.

Data synchronization helps mobile users in accessing data and using it for computing on mobile devices. When a device is not connected to a source or server, the user may employ data that is not in conflict with the present state of data at the source. Data synchronization helps mobile users in hoarding the device data at the personal area computer. It also helps mobile users in hoarding the personal area computer data. Data synchronization initiated at frequent intervals enhances device mobility and ensures that device applications use the latest updated data from the source, even when the device is disconnected. The process of data synchronization helps an enterprise server *store* large chunks of information for the many devices connected to it and *update* partial copies of data at frequent intervals.

Figure 9.2(a) shows synchronization between a server as the data-generating source and a device when the server sends full copy. Figure 9.2(b) shows synchronization between a device as the data-generating source and a server when the device sends full copy. Figure 9.2(c) shows synchronization between a server as the data-generating source and a device when partial but consistent copy is stored

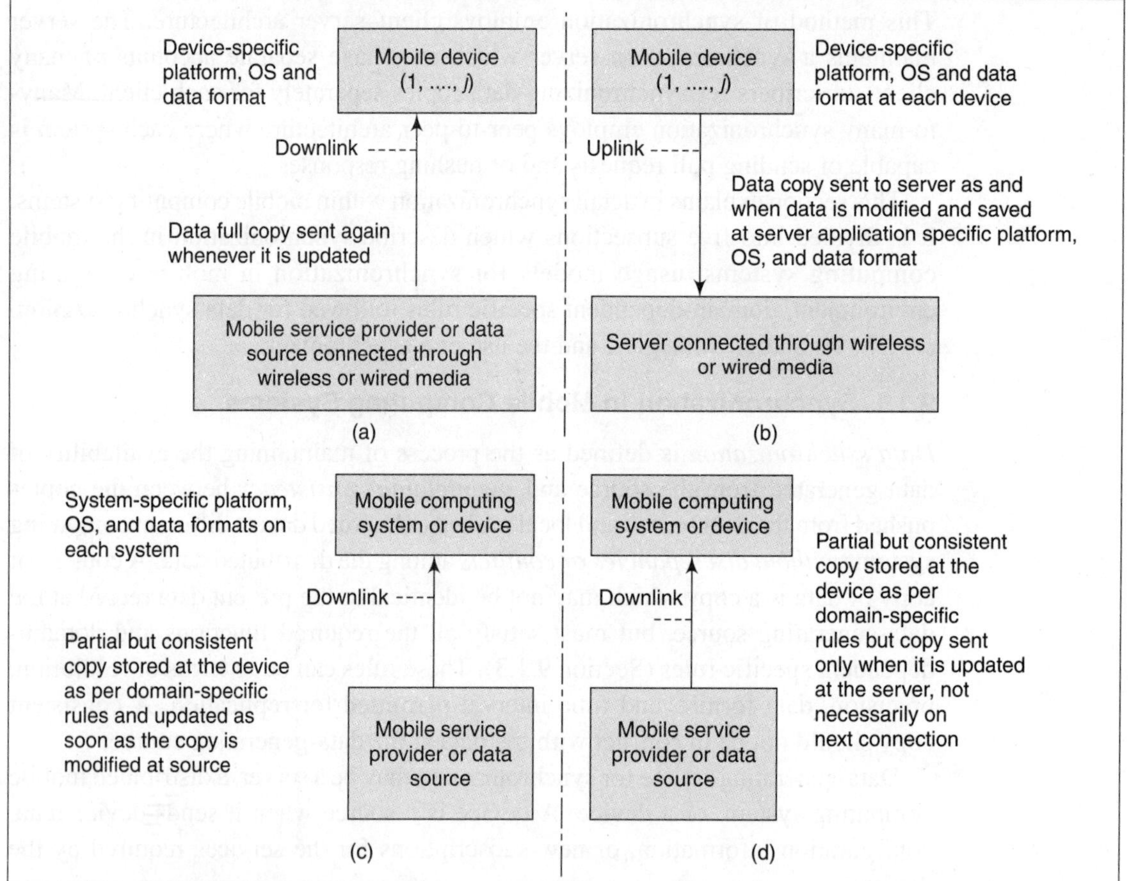

Fig. 9.2 (a) Synchronization between a server as the data-generating source and a device when the server sends full copy
(b) Synchronization between a device as the data-generating source and a server when the device sends full copy
(c) Synchronization between a server as the data-generating source and a device when partial but consistent copy stored at the device and updated as soon as the copy modifies
(d) Synchronization between server and device for partial but consistent copy stored at the device, but copy sent after delays by an interval when it updates at server, not necessarily on next connection

at the device and updated as soon as the copy is modified. Figure 9.2(d) shows synchronization between server and device for partial but consistent copy stored at the device, but copy sent only when it is updated at server, not necessarily on next connection. (The device is part of the mobile computing system.)

9.1.1.1 Types of Synchronization

Data synchronization is required between the mobile device and the service provider and between the device and the personal area computer, nearby wireless access point (in WiFi connection), or another nearby device (e.g., Bluetooth device). Synchronization of data may be categorized as listed below:

1. *Two-way synchronization* Partial or full copies of data are synchronized between the device and the server. For example, consider synchronization of the list of contacts and personal information manager data. The two-way synchronization between mobile device and personal area computer means that whenever the list of contacts and personal information manager data is modified at any of them, it is made consistent after synchronization.
2. *Server-alerted synchronization* When data is modified at the server (mobile service provider, PC, or remote mail server), the server alerts the client and the client synchronizes the modified data by pull request. For example, if a new email arrives at a server, it alerts the device as and when the device connects to the server. The device then sends a request to the server in order to pull the mail.
3. *One-way server-initiated synchronization* Server initiates synchronization of any new modification since communication of last modification and sends modified data copies to the client. For example, if a new email arrives at a server, it initiates the synchronization as and when the device connects to the server and sends the mail.
4. *Refresh synchronization at client* The client initiates synchronization with the server for refreshing its existing data copies and for refreshing the configuration parameters saved at the server for it. For example, a computer or mobile device initiates refreshing of the hoarded contacts and personal information data either at periodic intervals or as and when it connects to the server. Another example is that if the device configuration changes or a new device connects to a server, then the configuration parameters sent earlier refresh at the server so that in future data dissemination from the server and data refresh synchronization at the device is as per the new device configurations.
5. *Client-initiated synchronization* The client initiates synchronization with the server for sending its modifications, for example, device configuration for the services. For example, a client mobile device initiates synchronization of the mails or new ring tones or music files available at the server either at periodic intervals or as and when it connects to it.

6. *Refresh from client for backup and update synchronization* The client initiates synchronization and sends backup to the server for updating its data. For example, a computer or mobile device initiates refreshing of the hoarded contacts and personal information data either at periodic intervals or as and when it connects to the server.
7. *Slow (full data copy) synchronization* Client and server data are compared for each data field and are synchronized as per conflict resolution rules. For example, a mobile device and a device image at the server compare each data record and synchronize the data. The full copy synchronization usually takes place when the device is in idle state and not immediately on connecting to the server. That is why it is also called *slow* synchronization, though it actually means *thorough* synchronization.

9.1.1.2 Formats of Synchronized Data Copies

The data formats at a server and device can be different from each other. When the data at a source synchronizes with the data at other end, it does so as per the format specified at that end. Synchronized data copies at the device can be in the following formats:

1. *Database records* The records are indexed enabling search by querying using the indexes, for example, the relational database records. Consider the personal information management data at the device. Assume that it is indexed as per the first alphabet in the name and then each contact name data is indexed as per the information, name, telephone number, address, email address, and web address. Each telephone number is indexed as per the office, home, mobile, default, and fax numbers. The database record is retrieved by sending a query specifying the entries in these indexes. For example, find number when (name = 'xyz' and selection = 'mobile') or find numbers when (first alphabet = 'R').
2. *Flat files* Flat file means that data can be interpreted only if the file is read from beginning to end and that data cannot be picked from anywhere within the file. For example, an XML or html file at the server synchronizes with the file at the device which is in text format or is a binary file depending upon the information format. Information format in mobile computing is XML document format and for transmission it is WBXML (WAP Binary XML) content format. An example is address book data at a mobile device with the data transmitted in WBXML format.
3. *Device-specific storage* For example, AAC (Apple Audio Communication) files are used for audio communication with an Apple iPod. A file in AAC format synchronizes with music files in some other format at a computer or remote website serving the music files. At a mobile device the *Contacts* information is in vCard format, while calendar, tasks-to-do list, and journal information are in vCalendar, vToDo, and vJournal formats.

9.1.2 Usage Models for Synchronization in Mobile Applications

Consider a mobile computing system consisting of (i) mobile device, (ii) personal area computer connected by WLAN or Internet, (iii) nearby devices, for example, printer, and (iv) mobile service provider connected by wireless. The synchronization can be carried out in different ways. The four usage models employed for synchronization in mobile computing systems are shown in Fig. 9.3(a), (b), (c), and (d). These are discussed below:

1. *Synchronization between two APIs within a mobile computing system* The usage model for synchronization between two application programming interfaces (APIs) is such that the data generated by an application is synchronized and used in another application. An API running at the device synchronizes data with another application on the same or another device or computer. Figure 9.3(a) shows the synchronization between two APIs, *i* and

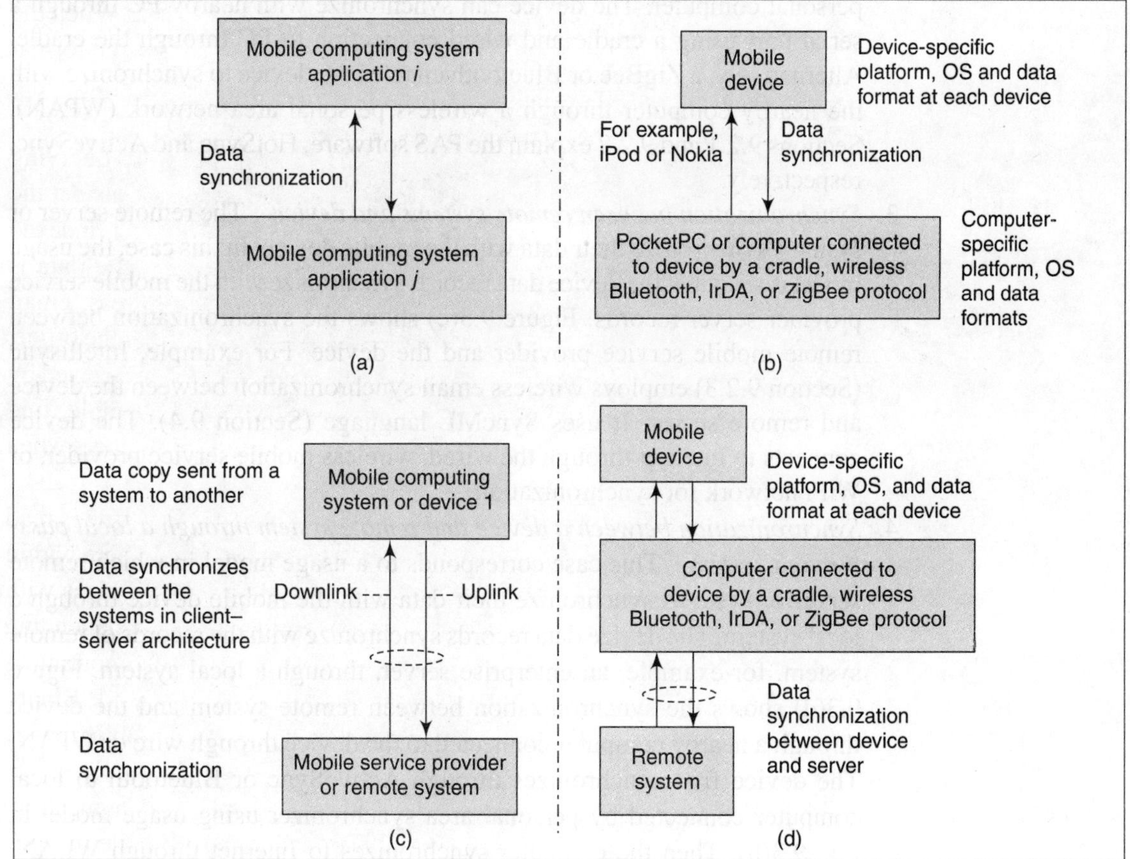

Fig. 9.3 (a) Synchronization between two APIs of the same mobile computing system (b) Synchronization between the device and local computer connected to wireless or wired network (c) Synchronization between remote system and device (d) Synchronization between a device and remote system through a local pass-through system

j. For example, data records at personal information manager (PIM) API are synchronized with the email API. When email from a new source is retrieved at the email API in the device, the name and email address data fields at the application are saved as new data record at PIM API. When an email is to be sent to the same person, the email API uses the same data record from the PIM API. As soon as user types few characters of that person's name, the email name field characters complete the name automatically from the PIM API and fill the email address automatically due to synchronization of data fields in the PIM and email APIs.

2. *Synchronization between the device and nearby local computer or device connected through a wireless or wired network* This type of synchronization corresponds to another usage model in which device and computer synchronize their data. This is also called personal area synchronization (PAS). Figure 9.3(b) shows the synchronization between device and nearby personal computer. The device can synchronize with nearby PC through a serial port using a cradle and wired connection to PC through the cradle. Alternatively, a ZigBee or Bluetooth enables the device to synchronize with the nearby computer through a wireless personal area network (WPAN). Sections 9.2.1 and 9.2.2 explain the PAS software, HotSync and ActiveSync, respectively.
3. *Synchronization between remote systems and device* The remote server or systems synchronize their data with the mobile device. In this case, the usage model is such that the device data records synchronize with the mobile service provider server records. Figure 9.3(c) shows the synchronization between remote mobile service provider and the device. For example, Intellisync (Section 9.2.3) employs wireless email synchronization between the device and remote server. It uses SyncML language (Section 9.4). The device connects to Internet through the wired, wireless mobile service provider, or WiFi network for synchronization.
4. *Synchronization between a device and remote system through a local pass-through system* This case corresponds to a usage model in which remote server or systems synchronize their data with the mobile device through a local system. The device data records synchronize with the records of remote system, for example, an enterprise server, through a local system. Figure 9.3(d) shows the synchronization between remote system and the device through a nearby computer connected to the device through wire or WPAN. The device first synchronizes through ActiveSync or Bluetooth to local computer connected by personal area synchronizer using usage model in Fig. 9.3(b). Then the computer synchronizes to Internet through WLAN, WiFi, or wired network using usage model of Fig. 9.3(c). ActiveSync, HotSync, and Intellisync software (Sections 9.2.1, 9.2.2 and 9.2.3) are personal area synchronizers which provide synchronization between the device and remote system through a local pass-through system (PC).

9.1.3 Domain-dependent Specific Rules for Data Synchronization

A mobile computing system consists of multiple domains between which data synchronization occurs. Data in mobile device domain has different rules for storage and usage as compared to the ones at the remote server, WPAN computer, or nearby device. For example, data in computer domain can be saved in a file identified by a name having alphabets up to maximum of 255 characters and data in mobile device domain can be saved in a file identified with an 8-bit number. There are domain-specific rules for existence of data. Domain-dependent specific rules for synchronization are explained below:

1. *Data synchronization in domain-specific platforms and data formats* Data synchronization can take place between data-generating domain and destined domain, both having different platform and data formats. For example, a copy of database record at the device may be in structured text or XML format and the device OS platform Symbian. The record may be synchronized with the database record at the server where it can be in DB2 or Oracle database format and the OS may be Windows. Another example of a domain-specific rule is that a data set at the device may be in text format and OS may be Window CE and at the synchronizing device it will be structured text and OS will be Symbian.
2. *Domain-specific data-property-dependent synchronization* Data synchronization can take place between one domain with one property of data and another domain having different property. For example, assume that, a data record at a device has an ID specified by a byte. It synchronizes with the record, which has an ID specified by 16-bit word at the server. Another example is that the device uses 8-bit ASCII characters for an ID while the server uses 16-bit unicode characters.
3. *Synchronization up to the last successful act of synchronization* One of the domain-specific rules is that data record is considered to be synchronized if it was updated at the last connection. Assume that a phonebook has the records of missed calls, dialled numbers, and received calls. A data record at a device is synchronized with the record in the phonebook and if it was updated at the last connection, then it will eventually update again on the next connection.
4. *Memory-infrastructure-dependent rule-based synchronization at two different domains (device and server)* A domain-specific rule can be that data records are synchronized up to the allotted memory. For example, assume that at a remote server full address book is maintained in the allotted memory of 8 MB and a device is allocated 128 kB for the address book. Thus only a part of database or items (e.g., only 100 new email addresses, which were used for sending and receiving emails) can be synchronized and saved in the device PIM (personal information manager) (Section 9.1.4). The addresses saved

by PIM are updated and maintained in text or a simple format (e.g., XML database format). PIM synchronizes addresses from a remote server which stores full address book for each device. Another example of domain-specific rule is that a maximum of 100 addresses and 200 phonebook entries may be synchronized at the device and arranged in order such that the last cached address or phone number is at top and least recently cached one is last. Another example of domain-specific rule is one in which only 100 latest music files may be synchronized at the device from the local computer and arranged in order such that last cached file is at top and least recently cached file is last.

5. *Synchronization with temporal properties of data* A domain-specific rule can be that data records are synchronized with data generated at source within specific time interval and at the time specified at the domain. For example, the data set may be synchronized every week (e.g., a flight time table) or once every day (e.g., weather report). Another example is that at a device, a data record (e.g., weather report) is updated and synchronized up to the last day and will eventually update on a day if it is available at the server. There may be periods of inconsistency when temporal properties of data are being used for synchronization. However, mobile applications will remain unaffected if there are no temporal conflicts and unaccountable discrepancies. Another example of domain-specific rule is that in which a critical and newly updated data record must be sought before using the data if the data is 2 seconds old (e.g., for live stock-quotes), otherwise the data is considered to be in temporal conflict.

9.1.4 Personal Information Manager

The functions of a manager include creation of copies, updation, replacement, addition, and deletion of data records. It is also responsible for sending a data record to an API, receiving it from another API, and sorting of the data records. Sorting can be configured according to first alphabet of the name, calendar (entries in a list for tasks-to-do at current date first followed by entries for next date), last name in the names at the list of Contacts (entry of last name with first alphabet as A first and as Z last), or priority. Smart mobile phone devices have the APIs for a software component called personal information manager (PIM). PIM manages the data record copies. PIM APIs are for managing data copies for device user. PIM data record copies are for the multiple selections and options provided by the PIM APIs, for example, calendar, address book, and tasks-to-do. Address book is a list of contacts with their office, home, mobile, and fax numbers, home and office addresses, web addresses, and email IDs. Tasks-to-do is a list of tasks planned by the device user, which also provides along with each task the time at which to do the task. The device reminds the user of the tasks as per their planned time. PIM data record copies can be synchronized with the copies on a personal area computer, WPAN computer, a remote system, or another mobile device. PIM options for synchronization are described next.

A mobile device may loose the PIM data records, for example, due to accidental deletion or battery power loss. There is an option for synchronization called *Restore* which helps in restoring the copies at the device. When the mobile device connects to a personal area computer, WPAN computer, a remote system, or another mobile device, the copies are synchronized with the copies at the latter and the computer copies the selected or whole data to the device.

A mobile device can have the backup of the PIM data records to protect the user due to accidental deletion or battery power loss. This option is called *Back Up*. When the mobile device connects to a personal area computer, WPAN computer, a remote system, or another mobile device, the copies are transferred completely with all data as a set to the PC or laptop in a selected directory. The backup enables restoration later.

Another option for synchronization is called *Reports*. The *Reports* option provides a device with the entries added, deleted, or modified at the other device or PC. It not only specifies time and date of last synchronization but also gives reports of synchronization failures.

The PIM also provides the configuration option and setting of preferences for synchronization. The configuration option can be set to *User Selection* or *Auto*. *User-Selection* configuration means synchronization is done only when user selects the option 'synchronize'. *Auto* configuration means synchronization is done automatically on connection between the device and the computer. An example of setting the preferences is setting of connection preference. It can be configured as Bluetooth, IrDA (infrared data association), USB, or COM port cable.

9.1.5 Synchronization and Conflict Resolution Strategies

A *conflict* in synchronization arises when an item (or data copy) has changed at one end but is not simultaneously modified at other ends. Therefore, the same data item at two ends, P and Q, will be in conflict during computation in the time interval between t_1 and t_2, where t_1 and t_2 are the instants when P and Q get the modified data copy. A conflict resolution strategy is adopted in such cases to resolve conflicts. This strategy specifies the rules that need to be applied for conflict resolution. Some examples of different types of conflict resolution rules are listed below:

1. *Priority-based resolution rule* A data-server can be specified as dominant higher priority entity for conflict resolution of synchronized data records. Assume that S is mobile service provider server having a list of missed, dialled, and received calls for the device D. D has a synchronized list of missed, dialled, and received calls. Assume that the list at D is in conflict with the list at S. Priority-based resolution rule specifies that the server priority is higher. This means that the list at the server is resolved as correct and acceptable in comparison to the list at the device.

2. *Time-based resolution rule* A data node P is specified as dominant entity when P always receives copies first from the server S. Assume that S is having the emails which are disseminated to the device D at an instant t_1. D connects to a personal area computer (PC) to which the device always synchronizes the mails at a later instant t_2. Time-based resolution rule specifies that the D is dominant because it receives the mails earlier than the PC.
3. *Information-based resolution rule* A data node can be specified as dominant entity when information specific to it is synchronized with other nodes. Assume that S is a server having the device configuration record which was disseminated from the device D. Information-based resolution rule specifies that since the information is for the device D hence D is dominant node. For device-specific information, the device data is accepted rather than the server data.
4. *Time-stamp-based resolution rule* Time-stamp-based resolution rule necessitates that a time-stamp must be used while sending a data copy and the copy found to be latest will resolve the conflict. Assume that S is a server having flight information which it disseminates at regular intervals with a time stamp over it to the device D and as well as to a PC. Time-stamp-based resolution rule specifies that the node with flight information with latest time stamp is dominant.
5. *User-interaction-based resolution rule* An API at a device allows a user to interact with the device and this interaction resolves the conflict arising out of the duplicate or multiple entries. User-interaction-based resolution rule means that the duplicate data entries are permitted at the node and a receiver API should resolve the conflict after interaction with user. For example, when there are two phone number entries found for same name and address, the device prompts user to resolve the conflict. The user resolves the conflict by opting for one of it.

9.1.6 Synchronizer (Synchronization Engine)

A synchronizer (also called a synchronization engine) is a software tool for synchronization. It resides at a host. If the synchronizer host is a server, then it is called synchronization server. A synchronizer

- is configured as per the specified domain-dependent and conflict resolution rules
- finds the data item or record appended at the data set(s)
- discovers modifications with respect to the last action of synchronization of the specified data copies which have been deleted, added, or modified and makes changes in the copies accordingly. In case of conflict, it resolves it as per resolution rule(s)

- disseminates (propagates) the changes to other devices or systems so that others devices also incorporate the changes in their data copies.

A method which can be adopted by a synchronizer for performing the task of synchronization is explained next. This method is more efficient compared to one in which the synchronizer directly retrieves the changed records. In this method, the synchronizer first logs in the information about the records which have been changed and then retrieves the changed records. The process of logging in the information about the changes can be fast. For example, it can be done by just setting or resetting a flag. This flag, also called *dirty* or *invalid* flag, indicates whether the data record associated with an entry remains unchanged or not. When it is in set state (=1), it means that the record *R*, with which this flag is associated, has been *modified* or *invalidated* since previous synchronization. When it is in reset state (=0), it means that *R* is not modified or invalidated since previous synchronization. This reduces the possibility of using conflicting entries at the server (or device) hosting the synchronizer. An example explaining the use of flags by synchronizer follows.

Consider a *Contact* in a PIM data of a mobile phone device D. Assume that a Contact with name N_i exists. A flag f_{nd} associated with N_i sets to 1 when N_i and associated data are entered for the first time at D. Another associated flag ef_{nd} resets to 0 because data fields of contact N_i are not empty now and the data is valid. An additional data record containing personal information about office, home, mobile, default, and fax telephone numbers, address, email address, and web address is also associated with N_i. Assume that fax number, email, and web fields in the records are empty. Further assume that a priority-based resolution is adopted as the conflict resolution rule for the PIM data. When the data at D and C are conflicting, the data at D has priority and is the dominant data. The synchronizer acts as follows:

1. *Synchronizer Action 1* Assume that synchronizers Sy_c and Sy_d at a WAP computer C and device D, respectively, both use the method of first logging in the information about the changed records. When the device is placed on a cradle, connected to C, PIM data at the computer and the device synchronize at an instant t_1. When Sy_c synchronizes the contact data for N_i at D, it looks for the flags f_{nd} and ef_{nd} at D. If f_{nd} is found to be 1 and ef_{nd} 0, the Sy_c synchronizes and replicates the N_i and associated data at C. On the other hand, if f_{nd} and ef_{nd} are both 0, the Sy_c ignores the synchronization of N_i data and does not copy it at C. After the synchronization is completed at C using Sy_c, the Sy_d at D resets f_{nd} to 0.
2. *Synchronizer Action 2* Now assume that device user at a later instant t_2 has no changes in the contact N_i. The N_i is not dirty now because it has not been modified. Therefore, the dirty flag f_{nd} and invalidation flag ef_{nd} at the device remain reset to 0. Further assume that when the device is placed on the cradle next time, it connects to C at t_2. The Sy_c at C synchronizes the contact

data for N_i at D. It looks for the flag f_{nd} and ef_{nd} at D. Since both are found to be 0, the Sy_c ignores the synchronization of N_i data.

3. *Synchronizer Action 3* Now assume that the device user at a later instant t_3 enters the email address at the contact N_i at D. The N_i is now called dirty because it has been modified. Therefore, the dirty flag f_{nd} at the device sets to 1. The ef_{nd} still remains in the reset state, as data is not invalidated. Again assume that when the device is placed on the cradle next time, it connects to C at the instant t_3. The Sy_c at C synchronizes the contact data for N_i at D. It looks for the flag f_{nd} at D. Since f_{nd} is found to be 1, the Sy_c synchronizes and copies the changes in N_i and associated data at C. The Sy_d at D resets f_{nd} to 0 after synchronization.
4. *Synchronizer Action 4* Now assume that the device user at a later instant t_4 deletes the contact N_i at D. The N_i becomes empty because the data has been now invalidated. Therefore, the invalid flag ef_{nd} sets to 1 at the device. When the device is placed on the cradle next time, assume that it connects to C at the instant t_4. The Sy_c at C synchronizes the contact data for N_i at D. It looks for the flag ef_{nd} at D. Since ef_{nd} is found to be 1, N_i and its associated data at C are deleted. In this case, the Sy_d at D does not change ef_{nd} to 0 after synchronization as long as contact N_i is not restored at D using the option *Restore*.

An advantage of the above method is that only the associated flags are retrieved by the synchronizers initially and, if necessary, the associated data record is retrieved later. Thus the synchronization actions take very little time.

9.1.7 Mobile Agents for Synchronization

Mobile agents can be deployed for synchronization. The synchronization-agent software recognizes inconsistency and conflict in data received at a client from the connected remote server, local server, or a device using resolution rules. The agent is present at a host which can be the client, an intermediate node, or the server.

A characteristic of an agent is that it can migrate from one node in a mobile computing system to another and if needed, to another host for data synchronization. The host is one which is connected to the server. The migrated mobile agent can function as proxy to a mobile device or computing system.

A synchronizer can use a method for synchronization, which employs a mobile agent. This method is explained below with the help of an example. Consider the example of Section 9.1.6 once again. Assume that A_c is the agent for the synchronizer Sy_c at a computer C. It hosts at the device D but does the work for Sy_c. Assume that a priority-based resolution is adopted as the conflict resolution rule for the PIM data. When the data at D and C are in conflict, the data at D has a higher priority and is dominant. The agent A_c hosted at D logs in the information about the records which have been changed so that the Sy_c can synchronize the changed records later whenever D connects to C. The device connects to C for synchronization. The A_c communicates the status of the modified and invalid flags to C for use by Sy_c. The agent acts as follows:

1. *Agent and Synchronizer Action 1* When the device is placed on a cradle, connected to WAP computer C, PIM data at the computer and the device synchronize at an instant t_1. When C connects to D, the flag f_{nd} is communicated as 1 and ef_{nd} as 0 using A_c. Sy_c synchronizes the contact data for N_i at D and copies data for N_i at C. Once the synchronization at C is complete, A_c at D resets the status of synchronization using Sy_c and thus resets f_{nd} to 0.
2. *Agent and Synchronizer Action 2* Now assume that device user has no changes at the contact N_i at a later instant t_2. When the device is placed on the cradle next time, there is no communication from A_c when the device again connects to C at t_2. The Sy_c at C does not initiate the synchronization.
3. *Agent and Synchronizer Action 3* Now assume that device user enters the email address at the contact N_i at D at a later instant t_3. When C connects to D, the flag f_{nd} is communicated as 1 and ef_{nd} as 0 using the A_c. The Sy_c synchronizes the contact data for N_i at D. Once the synchronization is completed at C, A_c at D resets the status of synchronization using Sy_c making f_{nd} 0.
4. *Agent and Synchronizer Action 4* Now assume that device user deletes the contact N_i at D at a later instant t_4. When the device is placed on the cradle next time, it again connects to C at the instant t_4. A_c at D communicates ef_{nd} as 1. Sy_d deletes the record of contact N_i at D. A_c at D does not change ef_{nd} to 0 after synchronization as long as contact N_i is not restored at D using the option *Restore*.

The agent A_c need not host at the server which is the device D in the above example. It can be present at an intermediate Bluetooth device placed between the computer C and the device D. It can also be present at the computer C itself which is the client in the present case. This fact is one of the advantages of an agent. In addition to this, the agent at C can periodically seek the status of the flags from D and Sy_c can initiate the synchronization as per data state of A_c. In other words, the synchronizers initiate action as per the data state communicated by the agent and take no action in case no communication is received from it. Moreover, an agent A_c need not send only the status as in the above example. It can communicate the changed record directly from the server to the client, that is, from D to C in the above example. Therefore, the use of agents for synchronization proves to be beneficial.

9.2 Synchronization Software for Mobile Devices

A synchronizer is used for data synchronization as per configuration, domain-specific rules, and conflict resolution strategies. HotSync (Section 9.2.1) is personal area synchronization (PAS) software. It is used in a mobile device when device operating system (OS) is PalmOS (Section 14.2). ActiveSync (Section 9.2.2) is used in a mobile device when device OS is Windows CE (Section 14.3). Intellisync (Section 2.4.2) is also PAS software when the device OS is PalmOS.

HotSync or ActiveSync synchronizer synchronizes the mobile device when it connects to a computer (PC) or a device in the personal area network or WPAN. In other words, it synchronizes the mobile data copies with data copies on PC.

A protocol is used for communication from one node to another in a mobile computing system. HotSync or ActiveSync synchronizer uses these protocols to connect a mobile device to the PC or another device. The connection is through a USB (universal serial bus) port, serial port, or infrared port at PC. The synchronizer can also connect to an Ethernet LAN.

A synchronizer is used for remote as well as local area data synchronization. Intellisync is a synchronizer (Section 9.2.3) which provides synchronization with remote system as well as PAS.

9.2.1 HotSync

HotSync is a software used for synchronizing handheld devices based on PalmOS with Windows- or Mac-based personal computers in a synchronizing architecture (Fig. 9.3(b)). It organizes data and most programs in the device based on PalmOS. Copies of the device data are also listed at the PC in a backup list, so that these can be restored to the device in case of a device overrun.

Each copy (file) has an attribute bit (similar to dirty, invalid, or modified flags) which sets when there is a change in the copy at the PC. A backup program at the device, using the attribute, copies those copies from the PC, which have their attribute bits set. After copying is complete for a copy, the attribute bit is reset at the PC.

HotSync has small programs called conduits. Each task which is synchronized uses a conduit. Auxiliary software for HotSync is installed in PCs. It synchronizes memos, tasks-to-do list, calendars, contacts, and data files. It can be used to install new applications on the device. The auxiliary software can also be used to transfer photos, music, and other multimedia objects to the handheld device. It enables saving of settings, preferences, favourites, call history, and speed-dialling. The data can also be tuned with Microsoft Outlook.

The PalmOS device connects to the PC through a cable called HotSync cable or using a wireless connection which could be WiFi, Bluetooth, or infrared IrDA. It also enables synchronization with remote systems using the architecture shown in Fig. 9.3(d).

9.2.2 ActiveSync

ActiveSync is a software from Microsoft for synchronizing mobile devices with Windows-based PCs (Fig. 9.3(b)). ActiveSync 4.1 synchronizes data at the PC with devices based on Windows Mobile 5.0 or Windows CE. ActiveSync is also available for the Symbian OS platform. ActiveSync synchronization between mobile devices and PCs is not restricted to communication through serial or USB ports or through the Bluetooth or IrDA protocols; a serial port may also be configured to support non-encrypted or encrypted (using secure connection configuration) communication.

The PC can be configured as a Bluetooth ActiveSync partner. Bluetooth communication sets a virtual COM port, which is utilized by ActiveSync for synchronizing with the PC, thus allowing the device to synchronize with the PC using Bluetooth.

ActiveSync synchronizes personal information (email, calendar, and contact information) on the mobile device even when the device is away from the PC. It also functions as personal sync server (PSS) which enables a mobile user to synchronize the PIM data set, which consists of address book, calendar, and tasks-to-do list, on the handset with PIM data on a PC or a laptop. The PSS can be integrated with Microsoft Outlook and Outlook Express on the PC, enabling email in Outlook to be transferred to the mobile device even when the user is on the move.

ActiveSync 4.x synchronizes mobile devices with Windows Media, videos, pictures, music, and MS Office files. It has a partnership wizard to help device users set up a synchronizing partnership easily.

ActiveSync resolves the conflicts between different versions of files during data exchange and uses multiple service providers and service managers. The PC connects to the Internet through a wired network or a wireless WiFi network in a synchronization architecture (Fig. 9.3(d)). The PC then stores copies of the data for synchronization with the device. For example, the PC downloads music files and videos and synchronizes them with the mobile device.

ActiveSync 4.x does not allow synchronization through direct Internet connection (TCP/IP connections using WiFi, WLAN, or dial-up telephone line). This is to prevent any external system from accessing the device and accessing the device data directly. Device and remote system synchronization through PC (Fig. 9.3(d)) is done using ActiveSync in conjunction with Microsoft Exchange Server. Microsoft Exchange 2003 Service Pack 2 with devices running the Messaging and Security Feature software for Windows Mobile 5.0 enables direct push email,[1] device wipe, and certificate-based authentication with the Exchange Server.

The PC is required to be ready for Internet, WiFi, and phone access for ActiveSync to work. For synchronizing with PCs running Windows Vista, the mobile device needs Windows Mobile Device Center in place of ActiveSync.

[1]Push email means that the service-provider server pushes email through the PC, which the device selects, tunes, caches, and hoards. There may be some period push latency of remote server.

9.2.3 Intellisync

Intellisync was developed at the Intellisync Corporation (now acquired by Nokia). It synchronizes PIM data between mobile devices and the Internet. It has an open architecture and can, therefore, be integrated with enterprise architecture. This enables enterprise connectivity features. Intellisync provides security—it supports end-to-end encryption and password protection.

Intellisync desktop synchronizer supports Microsoft Windows Mobile-based Smartphones. It supports most mobile device platforms and handheld devices based on PalmOS, Pocket PC, Symbian, BREW, and other mobile device operating systems. It supports database synchronization employing DataSync software. Intellisync supports the synchronization architectures of Fig 9.3(c) and (d). It supports email, PIM synchronization, email accelerator, device management and file distribution using Systems Management/File Sync software, and synchronization with Microsoft Exchange Server.

Intellisync synchronizes email, calendar, Microsoft Outlook, and Lotus Notes. Remote server Intellisync Wireless Email Express saves inbox, contacts, tasks-to-do list, calendars, personal travel information, weather forecasts, driving directions, etc. for each client, which can be updated while the mobile user is on the move. It also supports meeting requests and viewing attachments with emails. It also provides wireless email and push email features on a mobile device, just as in a Blackberry mobile. [Push email has push latency of 30 minutes.]

9.3 Synchronization Protocols

A protocol is used for communication of data between two nodes of a computing system. Also at each layer of communication, a distinct protocol can be used when communicating data between two layers of a computing node. For example, TCP/IP protocol is used for communication on the Internet and there are five protocol layers in the TCP/IP protocol suite. Similarly, a synchronizer can deploy different protocol layers for communicating to the different devices and computers. The synchronizer can also use the different protocol layers for communicating to the personal area and remote area devices and computers. Figure 9.4 shows synchronization protocols for the synchronization of the PIM, email and application data records with the personal and remote area devices and systems. Following are the popular synchronization protocols for synchronizing mobile applications at the mobile devices.

9.3.1 Bluetooth

The Bluetooth protocol is used for synchronization among mobile devices and Bluetooth-enabled PCs in a wireless personal area network. Figure 9.5 shows synchronization between a Bluetooth-enabled computer and a mobile device.

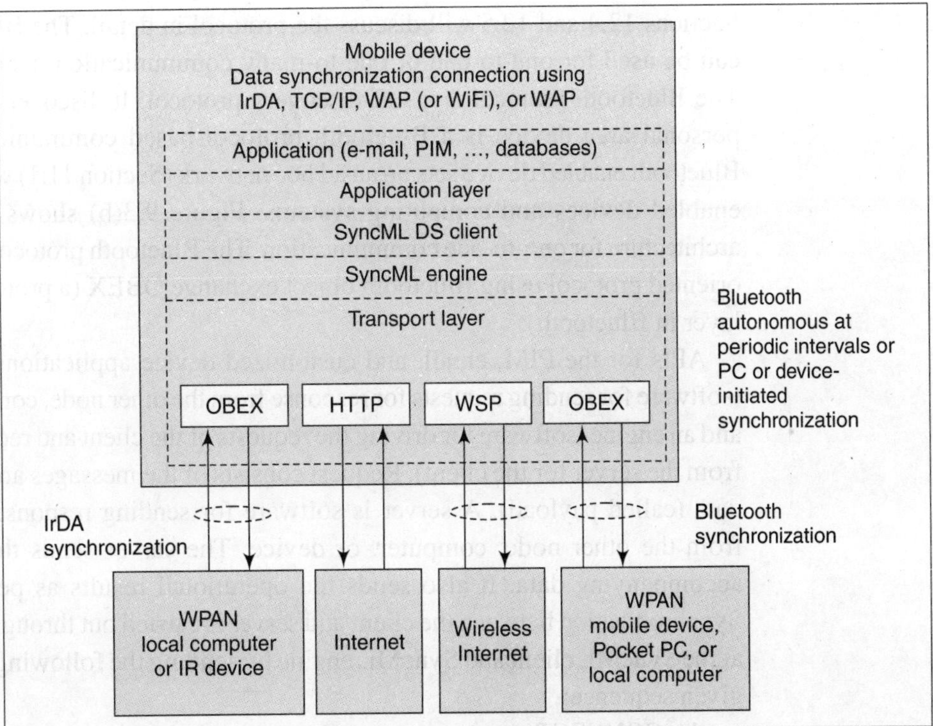

Fig. 9.4 Synchronization between infrared- and Bluetooth-enabled computer and mobile device in wireless personal area network (WPAN) and between Internet or Wireless Internet and mobile device having synchronization client and engine (e.g., SyncML DS client and SyncML engine)

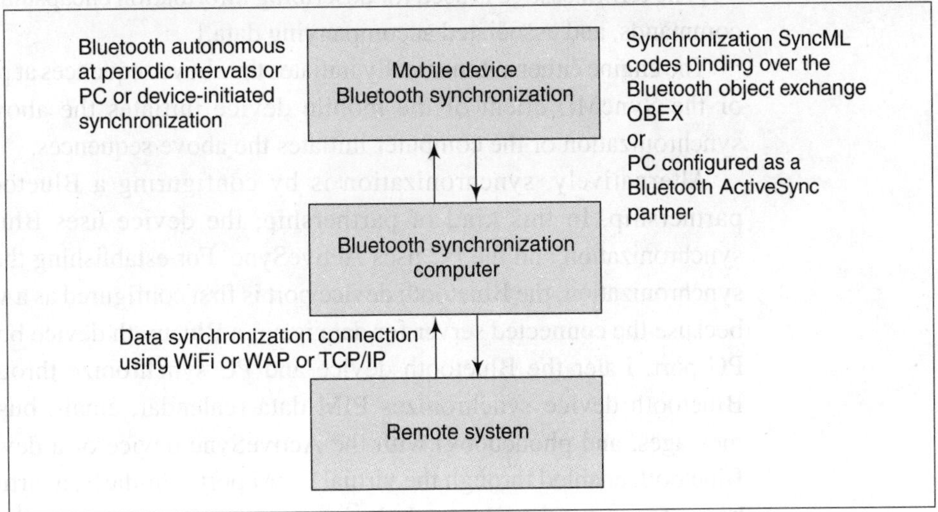

Fig. 9.5 Synchronization between Bluetooth-enabled computer and device

Sections 12.4 and 12.5 will discuss the protocol in detail. The Bluetooth protocol can be used for one-to-one or one-to-many communication over short distances. The Bluetooth protocol is a self-discovery protocol. It discovers whether nearby personal area device is a Bluetooth protocol-based communicating device. A Bluetooth enabled device sets up an ad hoc network (Section 11.1) with the Bluetooth enabled devices and computing systems. Figure 9.3(b) shows synchronization architecture for one-to-one communication. The Bluetooth protocol is a connection-oriented protocol using Bluetooth object exchange OBEX (a protocol for transport layer in Bluetooth).

APIs for the PIM, email, and customized device applications deploy a client (software for sending requests for response from the other node, computer, or device) and an engine (software for driving the requests of the client and receiving responses from the server for the client). Request consists of the messages and accompanying data (called payload). A server is software for sending response to the requests from the other node, computer, or device. The server sends the messages and accompanying data. It also sends the operational results as per the messages. Synchronization between the client and server is carried out through SyncML codes at the SyncML client and SyncML engine by sending the following messages in the given sequence:

1. *CONNECT*
2. *PUT* (to put the response to the request) or *GET* (to get the response to the request)
3. *ABORT* (to abort the connection)

[The term *message* is used for describing information encapsulated in the header, commands, and associated accompanying data.]

The engine either automatically initiates the above sequences at periodic intervals, or the SyncML client of the mobile device initiates the above sequence for synchronization or the computer initiates the above sequences.

Alternatively, synchronization is by configuring a Bluetooth ActiveSync partnership. In this kind of partnership, the device uses Bluetooth protocol synchronization and the PC uses ActiveSync. For establishing the partnership for synchronization, the Bluetooth device port is first configured as a virtual COM port because the connected server for data is not a Bluetooth device but an ActiveSync PC port. Later the Bluetooth device and PC synchronize through ActiveSync. Bluetooth device synchronizes PIM data (calendar, email, business card, text messages, and phonebook) with the ActiveSync device or a device which is not Bluetooth enabled through the virtual COM port. Similarly, a virtual USB port can be configured at the Bluetooth device.

9.3.2 IrDA

Infrared data association (IrDA) is an infrared protocol for infrared-based synchronization of mobile devices and computers within the same room. Section 12.7 describes IrDA in detail. IrDA specifications include connection-oriented or connectionless protocols. IrDA specifies five levels of communication—minimum, access, index, sync, and SyncML (levels 1 to 5). IrDA synchronization can be used to synchronize PIM data (calendar, email, business card, text messages, etc.).

9.3.3 WAP 2.0

The wireless application protocol (WAP) is a protocol for wireless synchronization of WAP client computers and WAP servers. Section 12.2 discusses WAP 2.0 in detail. Synchronization is through SyncML codes binding over the WAP application layer client or server. A WAP gateway connects WAP client to HTTP servers which serve Internet websites. The HTTP layer in TCP/IP protocol suite is an application layer protocol used when connecting to Internet in a wired network. Similarly, WSP (wireless session protocol) layer (Section 12.2.5) in WAP 2.0 is an application layer protocol when connecting to the Internet in a wireless network. SyncML codes bind with WSP for Internet connectivity.

9.4 SyncML—Synchronization Language for Mobile Computing

A mobile computing system consists of (i) mobile device, (ii) personal area computer in a WPAN, a computer connected by WLAN, or Internet, (iii) nearby devices, for example, printer, and (iv) mobile service provider connected by wireless. In general, each of these devices can use different platforms. For example, one device may use Symbian OS, another PalmOS, and the computer on WLAN Windows OS. Also, each one can use different languages. For example, a device may use Java and the computer C/C++.

The database connector for SyncML-based mobile application synchronizes data at the device with any relational database. SyncML-based software synchronizes data for PIM (email, calendar, memo, tasks-to-do list, or contacts list). A client framework, for example, JCF (Jataayu client framework) enables functioning of the software on different platforms. The SyncML framework is described in Section 9.4.1.

APIs for PIM, email, and customized device application can deploy SyncML client and SyncML engine. SyncML server is the software for sending response to the requests from the other node, computer, or device. It sends the messages which accompany the data and the operational results as per the messages.

SyncML is a data synchronization language based on XML. In other words, it is a markup language used for writing the codes for interfaces used for synchronization between the mobile devices and the server. In order to understand how SyncML is used for coding, the concept of XML is first described. SyncML protocols are described in Section 9.3.

XML Base of SyncML XML is a language for marking up a given text with tags and attributes and it is extensively used in mobile computing. Section 13.2 will give the details.

Sample Code 9.1 This is a sample XML code which documents a search list of contacts in a mobile phone with contact names.

```
<search_list>
…
<!- - Contact information about contacts identified by first
alphabet as R - - >
<alp_name first_character = "R">
<contact_name>
Raj Kamal
<address> ABC Street, …. </address>
<telnumber> 9876543210 </telnumber>
</contact_name>
<contact_name>
Raveena
<address> XYZ Street, …. </address>
<telnumber> 9848543210 </telnumber>
</contact_name>
</alp_name>
…
</search_list>
```

The code shows how to put the comments which improve the readability of the code. The comments are not used in parsing (processing) of the codes. XML comments start with <!- - (less than sign, ! sign and two dashes) and ends with - - > (two dashes before the greater than sign).

The sample code also shows how to use start and last tags for writing the text of a search list in between the tags. The start tag starts with '<' sign, followed by a tag name which is `search_list` in this case, and '>' sign. The last tag starts with <, followed by a slash sign, the same tag name as at the start, and '>' sign.

The document also demonstrates the use of inner tags. For example, for address, there is a start tag followed by text for the address, and an end tag. There is an inner tag with attributes also. For example, `alp_name` with attribute specification `first_character = "R"`. The coding format is '<' sign, tag name, tag attribute or attributes, '>' sign, followed by text associated with the given tag and attributes, '<' sign, slash sign, tag name, and '>' sign.

Other tags in Sample Code 9.1 are `contact_name` and `telnumber` to specify a name and telephone number.

For an XML document, there is a data type definitions (DTD) file which specifies the rules. For example, for the document given above, it specifies that `search_list` is the root element and that `contact_name` contains the address and telephone number.

XML is a platform-independent processing language. This implies that it can be used for processing not only by Java but also by any other language. Tags, attributes, and metadata formats in XML are standardized so that they can be universally interpreted on different platforms, machines, or networks. *SyncML* is an open standard based on XML. It has revolutionized mobile application-development, services, and devices. *VoiceXML* is another language based on XML and is used for communicating voice messages.

The information is exchanged between a client and server in form of an *envelope* representing a message. The term message describes information. A tag and its attributes define an *envelope* of a text message. Each *envelope* has a *header* followed by *body*. The header exists between header start and end tags. The *body* which includes the *commands* is present between body start and end tags. The *envelope* is exchanged between client and server by request and response messages. The accompanying data (payload) is placed between a start tag and the corresponding last tag. Therefore, XML can be used between client and server for sending messages and data and for synchronization. However, the XML tags used in Sample Code 9.1 are non-standard. SyncML provides the standard ways (Section 9.4.2).

Before embarking on the description of SyncML framework, some of the terms or concepts related to SyncML are first explained. This includes the definition of the term *Datastore* and a brief description of the method of parsing and data type definitions in SyncML.

DataStore *Data store* is a term used in SyncML codes for persistent data storage. It is also written as DataStore. This term refers to storage of data in a file system or database or in any other way in which persistency is maintained. In the context of mobile devices, persistency means that the data remains intact till deleted even if the device power is interrupted. It also implies that a change in data during an operation is simultaneously stored and reflected at all related files in a file system, database, or any other record. Suppose a telephone number is stored in a mobile device. The flash memory file stores it persistently and it is used in call operations. When user modifies the telephone number in *Contacts* at PIM using an API, it is also stored and modified at the flash storage and will be available any time on next call. When user deletes the telephone number in *Contacts*, only then the number looses persistency.

The APIs, for example, PIM or email, use a *Datastore* mechanism. For synchronizing data using different *Datastore* mechanisms at the nodes (client, database, file system, or server), with each one, in general, using different platforms and languages, is a difficult task. A common language for synchronization is therefore required. This language can be used for sending messages of the APIs and data using the *Datastore* mechanism at both the client and server ends. The use of a common and standard language enables interoperability. It also provides specifications for the protocols for sending messages from one node to another and representation of the messages.

Parsing A SyncML *message* from SyncML client or server is used after parsing the tags, attributes and the text. A SyncML *message* is parsed similar to an XML document. For example, the `search_list`—an XML document, in Sample Code 9.1 can be processed using a Java program for XML parsing. SyncML parser like an XML parser can be made in any language, Java, C, or in C++. The parser will be able to retrieve the entities in any XML text document. For example, in Sample Code 9.1, the entities are the contact name, telephone number, and address. The parser will also able to retrieve the data for the contact name, telephone number, and address. An API at the device or server can save the data using the *Data store* mechanism. An API can use the data for synchronization with another API in personal area network or at server. Assume that an API uses a parser and parses for contact names, then firstly the parser will parse for `alphabet tag`, then `first character "R"`. As a result, the parser gets the contact names with first character as the alphabet 'R'. A parser program associated with PIM API can be used to get a list of contact names. Using the document in Sample Code 9.1, an item in the list of items will be—first line: `Name Raj Kamal`, second line: `Address ABC Street`, ..., and third line `Phone 9876543210`.

SyncML Data Type Definitions For an XML document, there is another XML file without which the data types used in the document are undefined and the roots and daughters of the elements remain undefined. This XML file defines the data types and is called data type definitions (DTDs) file. Sample Code 9.1 DTD file will give the DTD specifications for the XML document.

SyncML, like XML, uses DTDs. These include device information DTD, service information DTD, meta information DTD (for metadata), and main SyncML DTD. Service information DTD represents the functional capabilities of the data objects supported by client or server. For example, if a client supports vCard version 2.1 for two-way synchronization, the server must also have this capability; otherwise the synchronization session for the vCard data object cannot proceed.

SyncML *messages* are SyncML-based XML documents. Similar to XML documents, the SyncML *message* has associated DTDs which define the data and metadata synchronization formats. SyncML DTDs have the specifications for the rules of SyncML *message*. SyncML DTDs include main SyncML *message* DTD, meta info DTD, and device info DTD. The DTD is also exchanged with the message. SyncML defines a universal data synchronization format DTD, which is exchanged with a SyncML message.

Metadata in Data Type Definitions Metadata is information regarding the stored data or it is information about information. Like XML document, the SyncML metadata is structured data that describes various characteristics of information-carrying elements. Examples of SyncML metadata are as follows—*meta info* has information about the *Data store info*, that is, whether *Data store* mechanism is database-based or file-based. *Data store info* is used in an API for synchronization for receiving or sending the data items. *meta info* also has the information about the data used for synchronization, that is, whether the data is in 64-bit or 32-bit binary form. It has information of number of commands, number of data chunks, device info, specific DataStore info for DataStore, and specific synchronization info. *Device info* has information about the device capability for synchronization. *Device info* consists of device type, model number, manufacturer, device hardware version specification, device software specification, and device capabilities for maintaining data copies and synchronization.

The metadata helps in discovering and locating data or in its assessment. The metadata data helps in the identification or management of data in the subsequent text data in the document. Metadata is considered as information about data or as information about information.

Open Standard The initiative for universal acceptability of SyncML is a project of the Open Mobile Alliance (OMA). The OMA SyncML initiative came from IBM, Motorola, Nokia, Palm, Ericsson, Lotus, Starfish, and Psion. Later many organizations joined the OMA, which has lead to development of a large number of mobile applications. *Sync4J* (Section 9.4), now called *Funambol*, provides an open source standard for Java-based processing of SyncML messages and synchronization formats.

Sections 9.3.1 to 9.3.4 describe the SyncML framework, protocol, phases, and packages required for using SyncML.

9.4.1 Synchronization Client and Server Framework

SyncML provides seven types of synchronization—(a) two-way synchronization, (b) server-alerted synchronization, (c) one-way server-initiated synchronization, (d) refresh synchronization at client, (e) client-initiated synchronization, (f) refresh synchronization from client for backup and update, and (g) slow (full data copy) synchronization. These were explained in Section 9.1.1.

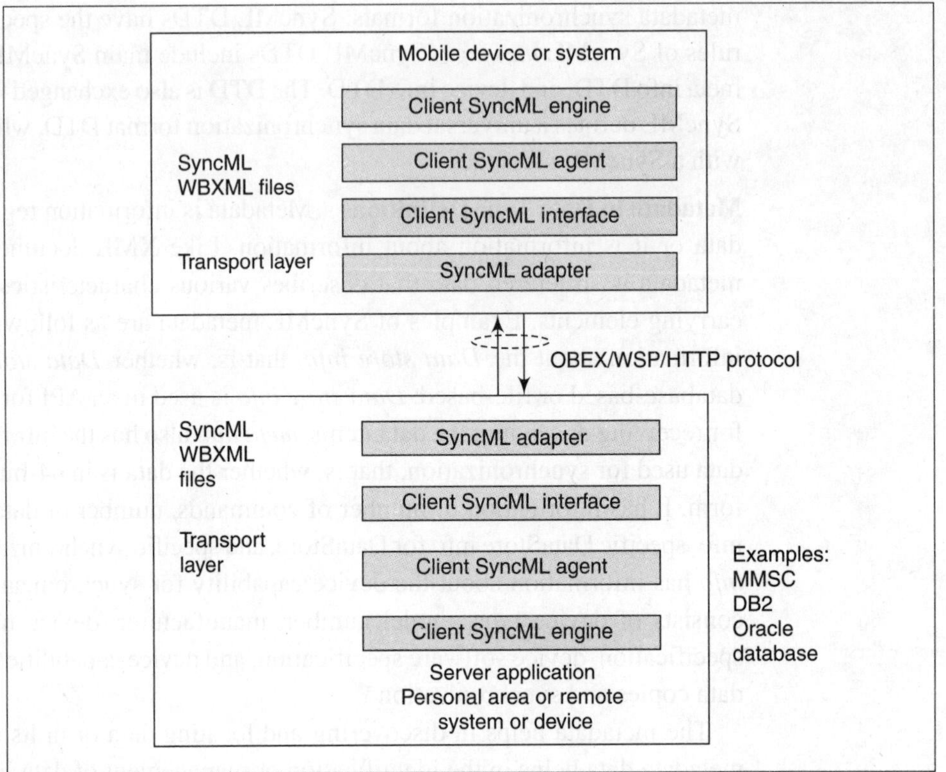

Fig. 9.6 SyncML client and server framework

Figure 9.6 shows the framework for client and server synchronization using SyncML. A SyncML client framework provides portability and interoperability. It may be ported on single or multiple platforms—J2ME, BREW, Windows Mobile, PalmOS, Symbian, and others. Framework has a SyncML agent which performs data synchronization in association with a SyncML engine. The agent employs a SyncML interface to send and receive SyncML WBXML format data to and from a SyncML adapter. WBXML makes the SyncML data compact. SyncML adapter is a piece of software that maintains transfer of SyncML data using OBEX, WSP, or HTTP protocols.

The use of adapter can be explained by the following examples. Consider an MMS server which provides SyncML interface for data synchronization servers. Consider that a device has an MMS and a database client. An MMS server provides a feature rich range of multimedia messaging and also provides essential core of the MMS environment for the MMS clients. A SyncML adapter provides the SyncML interface for the MMS server data. The interface provides the SyncML transport using OBEX, WSP, or HTTP protocol. The MMS client at device receives the MMS data from the SyncML adapter at the device.

DB2 and Oracle database servers provide interfaces for SyncML data synchronization servers. Interfaces synchronize the database records with database clients at the devices. A SyncML adapter enables SyncML transport over OBEX , WSP, or HTTP to the database clients.

Figure 9.7 shows synchronization of a mobile device in the SyncML protocol architecture. A device synchronizes with a personal area network computer or remote server using SyncML data synchronization (SyncML DS) client, SyncML engine, and binding with OBEX/WAP/HTTP.

The *application layer* executes the application and provides extensible adapters for different applications and multimedia interfaces for device user interactions. An application is an IMPS (instant messaging and presence service) client used for online chatting. Other applications include browsing, using HTML or WAP 2.0, MMS client, email, and Sync client. IMPS is a Wireless Village protocol compliant software used to provide an instant messaging client. IMPS chat client (in Nokia) for chatting employs the *MyPresence* protocol.

The *SyncML DS client* executes SyncML commands for applications, PIM, and email. It encodes and decodes PIM data. It encodes the PIM data that is received from the SyncML DS server and decodes the PIM data that is sent to the SyncML

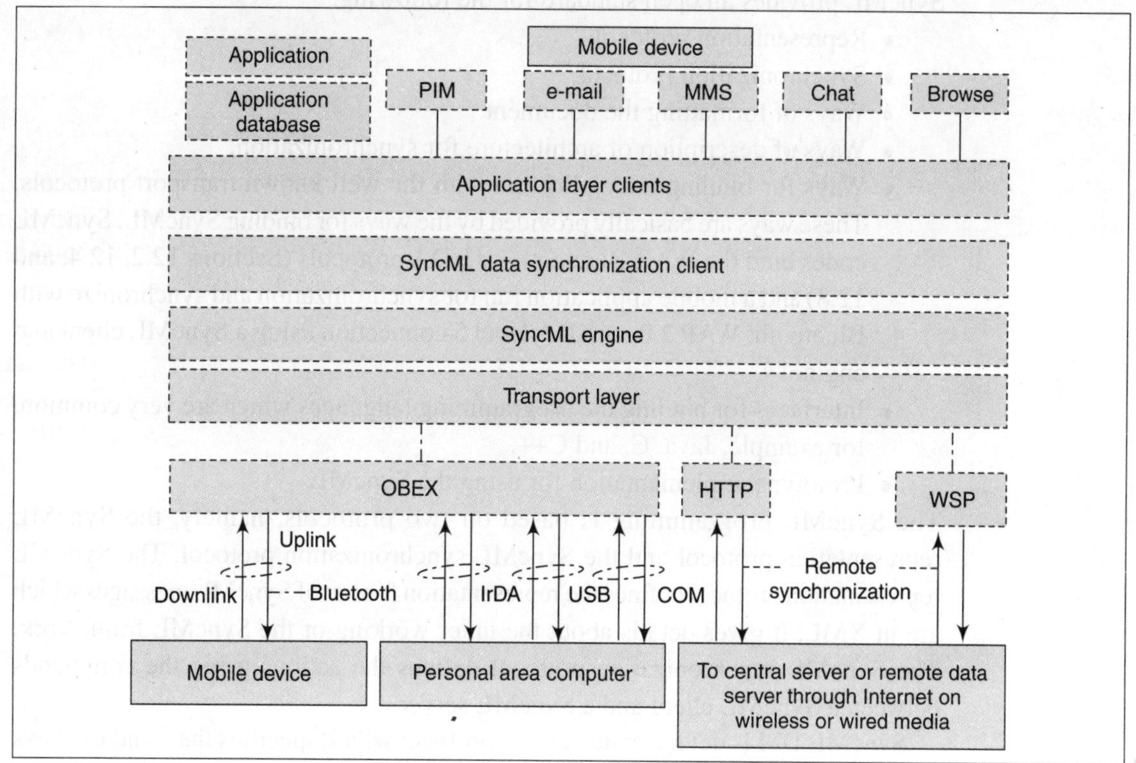

Fig. 9.7 Synchronization of mobile device in SyncML protocol architecture

DS server. It generates status codes (Section 9.4.2) and communicates these to the SyncML protocol engine.

The *SyncML data engine* performs SyncML code generation, parsing of received SyncML data, validation of DTD in WBXML and XML format data, base-64 encoding/decoding and notification message parsing, credential checks, security operations, and HMAC data integrity check. Credential checks mean authentication of the server before accepting SyncML data. HMAC (keyed hash message authentication code) is used for data integrity check along with the message authentication by calculating a hash function along with a secret key. HMAC employs an iterative function SHA-1 or MD5.

Application layer API uses HTTP and WSP. Transport layer functions are provisions of object-exchange adapters using OBEX for synchronization through Bluetooth, IrDA, COM ports, and USB ports. A transport layer function is a method of communication for remote synchronization. It receives notification buffers over OBEX or WAP PUSH receiving and sending SyncML Package data to and from the DS server (Fig. 9.2).

9.4.2 SyncML Protocol

SyncML provides an open standard for the following:

- Representation protocol.
- Synchronization protocol.
- Ways of formatting the document.
- Ways of description of architecture for synchronization.
- Ways for binding an application with the well known transport protocols. These ways are basically provided by the ways for binding SyncML. SyncML codes bind the WAP, Bluetooth, or IrDA protocols (Sections 12.2, 12.4, and 12.8) and a mobile application run for synchronization and synchronize with Bluetooth, WAP 2.0, or IrDA level 5 connection using a SyncML client and engine.
- Interfaces for binding the programming languages which are very common, for example, Java, C, and C++.
- Prototype implementation for using the SyncML.

The SyncML programming is based on two protocols, namely, the SyncML representation protocol and the SyncML synchronization protocol. The SyncML representation protocol defines the representation format of SyncML messages which are in XML. It gives details about the inner working of the SyncML framework. The SyncML synchronization protocol defines the actions using the commands between a SyncML client and a SyncML server.

SyncML DM is device management protocol which specifies the standard ways for device management and ways of configuring the device so that it can download the applications from the server. It also specifies ways to update the data at the device. A server carries out device management using SyncML DM. It requests the device for the client configuration and sets the data. It also updates the device data.

Following are the standard specified terms and methods in the SyncML protocol.

9.4.2.1 SyncML Header and Body

SyncML message has two components—(a) a header which starts with the start tag `<SyncHdr>` and ends with the end tag `</SyncHdr>` and (b) a body which starts with `<SyncBody>` and ends with `</SyncBody>`. Sample Codes 9.2 and 9.3 illustrate how to use `<SyncHdr>` and `<SyncBody>` tags, respectively.

The header consists of DTD, protocol, and data (message or session) identifications. The header also includes target and source information. The SyncHdr element carries initialization information which is required before synchronization. The initialization information enables device authentication and includes information on available device functions.

Sample Code 9.2 This is a sample code which documents header part of a SyncML message. It defines the SyncML DTD version, protocol version, session ID, message ID, target, and the source.

```
<SyncML>
<SyncHdr>
  <VerDTD>1.0</VerDTD>
  <VerProto>SyncML/1.0</VerProto>
  <SessionID>session01</SessionID>
  <MsgID>message09</MsgID>
  <Target><LocURI>http://www.syncml.org/sync-
server</LocURI></Target>
  <!- - LocURI means local URI. URI means Universal Resource
Identifier - - >
  <Source><LocURI>IMEI:49..........800</LocURI></Source> < - - IMEI is
International Mobile Equipment Identity represented by a number -
->
   ...
 </SyncHdr>
 <SyncBody>
 ...
 </SyncBody>
 </SyncML>
```

SyncML protocol specifies that a *message* has SyncHdr and SyncBody, with meta info as well as device info present within SyncHdr. Meta info is placed between the start and end tags <MetInf> and </MetInf> and device info between <DevInfo> and </DevInfo>.

9.4.2.2 Universal Resource Identifier

Universal resource identifier (URI) means a full path specification for a device resource. Assume that PIM data is structured like a tree. The parent PIM data object is the root object. At the branches of the root object, there are objects called daughter data record copies, these are *calendar*, *address_book*, and *tasks_to_do*. The address_book object has the branches having the contacts. Each of the contacts object has branches and daughter objects which may be leaves. The leaves are name, office, home, mobile and fax numbers, home and office addresses, web addresses, and email IDs.

Assume that the root object of device PIM data is dev_PIM_data and that of server PIM data is PIM_data. The root PIM_data has a hierarchy at the server—PIM_data → address_book → contacts → Raj Kamal → telNum_RK. This is because PIM data has the object address_book which has daughter object contacts, contacts have the object Raj Kamal. Raj Kamal object has a leaf, the resource telephone number, telNum_RK [telNum_RK contains the telephone number data in digits], where resource refers to an entity used in the computations or for synchronization. For an API, the resource is at leaf during computations using telNum_RK.

SyncML protocol specifies that the URI be used to identify any resource at the object. The URI for identifying a resource at the contacts at the server is PIM_data/address_book/contacts/Raj Kamal. The URI for the resource at the leaf telNum_RK is PIM_data/address_book/contacts/Raj Kamal/telNum-RK. If the object Raj Kamal has two branches, telNum_RK and address_RK, then the URI for the leaf object address_RK will be PIM_data/contacts/Raj Kamal/address_RK, where address_RK has the text of address.

Assuming that the device is yet to obtain the contact Raj Kamal from the server, the URI for the leaf at the device is dev_PIM_data/dev_contacts. The object dev_contacts has no branches because the object Raj Kamal is not yet replicated at the device. The replication will occur only after the synchronization of PIM_Data and device PIM data.

9.4.2.3 Commands

The SyncML protocol uses standard commands such as Create, Delete, Add, Update, Sync, Alert, Atomic, Search, Read, Get, Exec, Put, Copy, Results, Map, and Status. The *Alert* command is a special SyncML feature by which the server alerts the client. Table 9.1 gives the meanings of the SyncML commands. Sample Code 9.3 shows how to use these commands.

Table 9.1 SyncML commands for synchronization initiated from client or server

Command	*Description*
`<Delete>`	Permitted within `<sync>` command and is an instruction to request client or server for deletion of data item(s).
`<Add>`	Permitted within `<sync>` command and is an instruction to request client or server for addition of a data item.
`<Update>`	Permitted within `<sync>` command and is an instruction to request client or server for an update of the data item(s).
`<Sync>`	Precedes synchronization commands (e.g., `add` or `delete`) during data synchronization of two data objects.
`<Alert>`	Server- or client-initiated notification. For example, when data is modified (e.g., new email or newly downloaded music file) at the server (PC or remote mail server), then the server alerts the client and the client synchronizes with the modified data.
`<Atomic>`	start and end tags of the command define a transaction (data exchange) between server and client using a set of commands placed between the two tags. The transaction should be treated as a basic unit, which is not decomposable. It should either be completed in full or should be rolled back.
`<Search>`	Defines a query to a client or a server for an item at the DataStore.
`<Get>`	Defines a request to client or server for meta info about a data object and precedes the `<Sync>` command when meta info is required.
`<Exec>`	Defines a request to client or server for executing a program.
`<Put>`	Defines a request to send meta info about a data object to client or server and precedes the `<Sync>` command when meta info is sent.
`<Copy>`	Defines a request to copy a data object at client or server from server or client, respectively.
`<Results>`	Precedes `<Search>` or `<Get>` command(s).
`<Map>`	Defines a request to client or server for updating an identifier. The identifier maps the two data objects.
`<Status>`[1]	Precedes the request made by command(s) for client operations on server data item(s), DataStore, or object(s). As the result of this command, a status data is returned. Status codes of SyncML are similar to those of HTTP. For example, status code is 200 for successful operation and 212 for authentication for a session.

[1]A receiving end (client or server) uses a *Status* code for conveying the status of the request (through the commands) to the source. This code is usually an integer. The source is SyncML DS client and the receiving end is SyncML DS server or vice versa.

Sample Code 9.3 `<SyncBody>` for the `SyncHdr` in Sample Code 9.2 can be written as follows:

```
<SyncBody>
  <Status>
   <Cmd>SyncHdr</Cmd> <CmdRef>0</CmdRef>
<MsgRef>1</MsgRef>
   <TargetRef> IMEI:49..........800</TargetRef>
```

```
    <SourceRef> http://www.syncml.org/sync-server
</SourceRef>
   <Data>212</Data>
   </Status>
   <Status>
    <Cmd>Alert</Cmd> <CmdRef>1</CmdRef> <MsgRef>1</MsgRef>
    <!- - Assume that Device contact data object at the device is
dev_contacts, which was the target defined at the header and
assume that new contact added at the server is URI ./contacts/Raj
Kamal and Alert is sent to client device - - >
   <TargetRef>./dev_contacts</TargetRef>
   <SourceRef>./contacts/ Raj Kamal </SourceRef>
    <Data>200</Data>
    </Status>
</SyncBody>
```

9.4.2.4 Security

For secure synchronization of messages between source and destination, a message *M* sent from a source to destination needs verification of its integrity and authenticity of the source. One method is that *M* is sent along with a keyed hash message authentication code (*HMAC*). *HMAC*, which is sent along with *M*, is generated after computations at the source. At the destination, the *HMAC* is regenerated. If the *HMAC* computed at the destination matches with the received *HMAC*, the integrity of *M* stands verified and the source of *M* is authenticated.

The computations for *HMAC* are done using M, a secret key S and an iterative hash function. Iterative function is a function repeatedly operated on the operand. SHA-1 (secure hash algorithm-1) and MD-5 (message digest algorithm-5) are two of the several iterative hash functions that can be used. Hash or message-digest is the term used for a computed condensed digital representation of the full message. Interested reader can find the details of these in cryptography books. The key S is known only to the source and destination and M is used in computing HMAC, therefore, using HMAC, the destination verifies the integrity as well as authenticates the source.

The transmission of SyncML messages employing HMAC uses an interactive function, SHA-1 or MD5. Status code is returned to the message source after authentication of HMAC at destination. The code conveys to the source that the message reached the destination securely. For example, status code 212 is returned in case of secure transmission.

9.4.2.5 Handling Large Objects

During wireless transmission, there are interruptions. Therefore, a large data object should be broken into many chunks. For transmitting large object data in chunks, there is a tag called `<MoreData>`. Using this tag, a large object can be exchanged by dividing it into data chunks. Each chunk has a start tag `<MoreData>` and an end tag `</Moredata>`. This indicates that next chunk is expected in next session. The receiving objects join the chunks to recreate a large object.

9.4.2.6 Errors

Errors might occur during transmission. The type of error is usually conveyed in the form of an integer. SyncML protocol defines error codes in the event of data conflict, incomplete document, and other errors that occur during synchronization. Status codes between 500 and 599 are received for response from the receiver. For example, assume that a status code sent by the receiver of SyncML message is error code 501. This implies that the command could not be implemented at the receiver because the SyncML client or server did not support it for any of the resources. Another example is return of status code 511 which implies server failure.

9.4.3 Setup and Synchronization Phases

SyncML DM protocol specifies a standard method for set up and synchronization. Device management is one-way synchronization of the device from the server. The server sets and updates the data on the client device. It also requests for the device data, for example, data for enabling device authentication by it. Figure 9.8 shows the synchronization procedure for device management by the server.

The data exchange occurs in phases using packages 0–4. A *package* is a SyncML message containing XML document. Phases of data exchanges in packages 0–2 are called setup phases. Phases of data exchanges in packages 3 and 4 are called synchronization or management phases. A server can initiate synchronization using

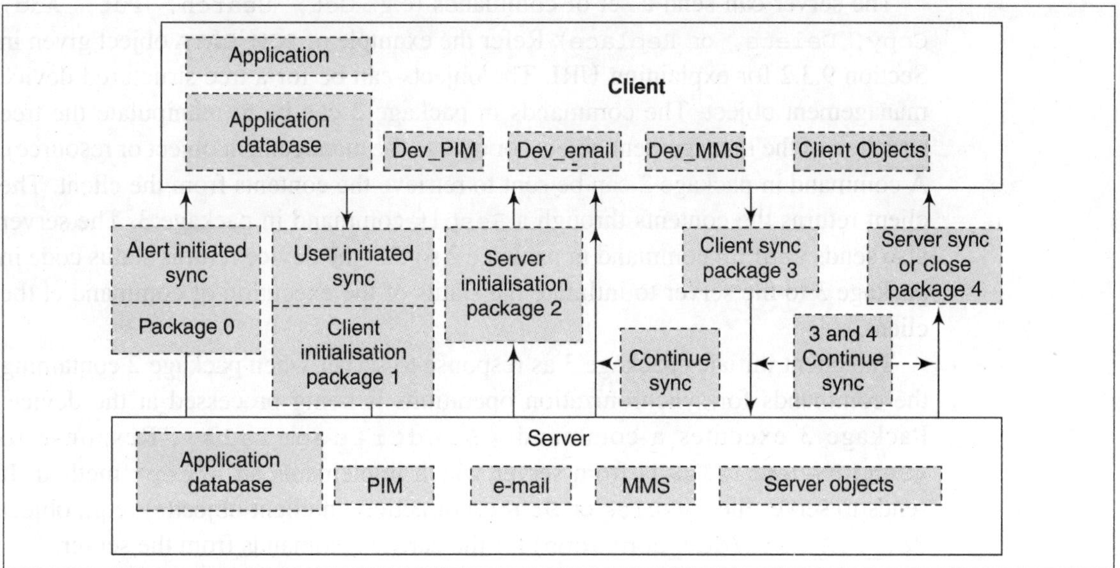

Fig. 9.8 Synchronization of mobile device in SyncML protocol architecture

package 0 which includes the Alert command. When a device listens to a notification (an uninvited message), it initiates connection back to the server using package 1.

A user interface at the client can also initiate synchronization using package 1. Package 1 is also sent when SIM card is inserted in a device for bootstrapping it. Bootstrapping means running initial set-up program before a device or computer can be used for operations. Package 1 from client MUST send the device info in the first message. The device executes a command 'SendCredentialsforAuthentication'. The package MAY specify requested database access and type of sync desired. The server identifies (authenticates) the device credentials when it receives it. The package MAY also let a server access the device credentials. It executes a command which requests for secure resources from the server. [MUST specifies the protocol entity as *must be sent*. MAY specifies the protocol entity as *may be sent*.]

Both client and server must be authenticated by each other using credentials or their information. Next is package 2 for server initialization with the server information to enable client identification of the server. The server implements this package after a command <Challenge> from the device in package 1. The challenge is authentication challenge. When a device or server sends some data object, for example, a header for an object exchange or synchronization, it also sends <Challenge> command. It is a request to the destination to authenticate during the next response from the receiver. The server MAY communicate the credentials in package 2 to the device. The client uses HMAC and status data in the SyncML authentication framework. Sending commands and data while package 2 is in process is optional. Another option is to send a response to the user interaction commands.

The server can send a set of commands (e.g., Get, Search, Put, Add, Copy, Delete, or Replace). Refer the example of PIM_DATA object given in Section 9.3.2 for explaining URI. The objects can be for a tree structured device management object. The commands in package 2 can be to manipulate the tree structure of the management objects or update the management object or resource). A command in package 2 can be sent to retrieve the contents from the client. The client returns the contents through a Result command in package 3. The server also sends Status command in package 2 so that the device returns status code in package 3 to the server to intimate the status of the execution of command at the client.

The client initiates package 3 as response to server when package 2 containing the commands for synchronization operations is being processed at the device. Package 3 executes a command <SendCredentials>, Response to <SendCredentials> is from server which implements an accept method. It sends to server the *Results* or *Status* of action on client object(s) (e.g., object dev_PIM_data/dev_contacts) for the earlier commands from the server.

A package 2 received at the device may initiate an interaction with the user. Package 3 may contain the results of user interaction in the response to the server.

The server initiates package 4 for the device. It may be for closing a session. It may be for further user interaction and operations on the device objects. If package 4 is not for closing then the server sends information regarding the operation needed and client restarts from package 3.

9.5 Sync4J (Funambol)

Section 9.4 described SyncML protocol and the commands which are exchanged for synchronization. These commands have to processed at the receiving and source ends. For example, a SyncML message contains a command for specifying the maximum display time (MAXDT). `<Item><Data>MAXDT = 15> </Data> </Item>`. A parser parses the SyncML messages and the parser program can be in Java. After parsing, the commands are to be executed. The parser parses MAXDT and its data = 15. A program then has to implement the parsed command. That program can also be in Java.

Initially named Sync4j, Funambol provides Java-based synchronization using SyncML messages. Funambol is an open-source mobile application server software. [Open source means openly available for use.] It is also a software development platform for mobile applications.

The codes are compiled as byte codes which are implemented on the Java Virtual Machine (JVM) for Java2 Standard Edition (J2SE) and Java2 Enterprise Edition (J2EE). First, a SyncML message containing XML document is parsed using either Java and DOM (document object model) (Section 9.5) or SAX modelled APIs. Then the appropriate commands are implemented (Section 13.2).

Funambol implements itemized commands—`Alert`, `Put`, `Get`, `Modification`, and `Response`. It also implements `Atomic`, `Map`, `Search`, and `Sequence` commands (Table 9.1).

A Funambol server for mobile application includes a suite of tools. The tools enable programmers to develop, deploy, and manage mobile applications. Funambol implements SyncML OMA data synchronization and device management commands, data synchronization server, open source DBConnector (database connector), push email and Microsoft Outlook functionality, iPod, Windows Mobile, Palm client and gateways for DataStore in databases, files, and email systems.

9.6 Synchronized Multimedia Markup Language (SMIL)

So far, data synchronization has been described. Mobile devices not only synchronize data but also synchronize multimedia (music, video clips, images, and slide shows). The transmitted text is shown on the display along with the video clip. The text is also rendered with the voice. The displayed text, images, and audio needs to be synchronized. A language is required to specify the synchronization messages in order to enable appropriate synchronization and for integration of multi-modal multimedia communication. *Multi-modal* means usage of different modes—text, image, video, or audio. Multi-modal communication integrates and synchronizes multimedia.

Synchronized Multimedia Markup Language (SMIL) is a language used for text, speech, or multimedia integration for multi-modal communication. Like SyncML, the SMIL is based on XML. SMIL version 2.1 enables coding of messages for interactive audio–visual presentations. The coding of SMIL messages needs a simple text editor. SMIL specifies the standard ways and tags which integrate text, images, and streaming audio and video. SMIL is a W3C council (World Wide Web Consortium) recommended language.

Assume that the coding for a text-to-speech synthesis (TTS) engine is required. TTS is deployed in automobiles (Section 2.7) and many other applications. If the driver of an automobile attempts to read a text message, his attention may get diverted. A TTS engine converts text messages to voice messages. SMIL is a language which can be used for coding the TTS engines. Just as SyncML refers to objects by URIs (Section 9.4.2), SMIL refers to multimedia objects by URLs (universal resource locators). Multimedia objects need to be shared between presentations and may be required to be stored on different servers for load balancing. SMIL provides the commands for these actions. Different media objects may be required to be transmitted at different bandwidths. SMIL provides commands for these actions also.

Each SMIL document has two components—(i) a header between start and end tags, `<header>` and `</header>` and (ii) a body between start and end tags, `<body>` and `</body>`. Due to multimedia synchronization, the `<body>` section provides the timing information.

Sample Code 9.4 This is a sample SMIL message to describe use of the `<SMIL>` tag and encapsulation of `<SyncHdr>` and `<Syncbody>` tags.

```
<smil>
<SyncHdr>
...
</SyncHdr>
<SyncBody>
...
</SyncBody>
</smil>
```

SMIL specifies the use of following four tags within the `<SyncBody>` tag for coding:

(i) `<layout>` within the header element to specify layout of SMIL document.

(ii) `<seq>` for sequential operations within the body element. The specified operations should be performed in sequence.

(iii) `<par>` for parallel operations within the body element. The specified operations should be carried out in parallel.

(iv) <switch> for a different set within the body element for presenting multimedia contents.

Sample Code 9.5 This is a sample SMIL message to describe use of the <seq> tag in Syncbody and to specify the source of the video clip file using which video clip is rendered by the device for 10 s such that the clip repeats five times in 10 s. It plays for 1 s. It also specifies the source of text and audio files. The text from the file is also rendered along with the video clip.

```
1. <SyncBody>
<seq>
<!- - Source of WBXML text file is my.wml,
maximum period 10 s, duration is 1 s, repetition is 5 times- - >
    <text src = "my.wml" max= "10s" dur = "1s" repeatCount ="5"/
>
<!- - Source of video file is my.mpg, identifier is my_vid and end
duration = 10s. - - >
    <video src = "my.mpg" id = "my_vid" end = 10s/>
</seq>
</SyncBody>
2. <SyncBody>
<!- - Assume that all operations image and audio begin in parallel
and begin at 2s" - - >
<par begin = "2s">
<!- - Source of image file is myimage.jpg, end is 2s and freeze on
display for 2s - - >
<img src = "myimage.jpg" begin = "2s" end myimage.end - 2s" fill =
"freeze"/>
<!- - Source of audio file is myaudio.wav, id is "my_id", begin at
t = 0s for 10s duration. - - >
<audio id = "my_id" src = "myaudio.wav" begin = "0s" dur = "10s" /
>
    </par>
</SyncBody>
```

The MMS, which is the video and picture equivalent of SMS, is a subset of SMIL. It can be implemented on handheld computers and mobile devices.

SMIL messages can be structured as in a document object model (DOM) tree. DOM tree specifies a tree-like structure of the tags. An example of tree-like structuring of objects was given in Section 9.4.2 in reference to URI specifications. In a DOM tree, a manager is the root and SMIL messages are the branches and leaves.

SMIL messages are scheduled using SMIL scheduler software. Just as an XML document or SyncML message, SMIL message needs to be parsed. A DOM interface specifies the parsing and layout.

Keywords

ActiveSync: Synchronizer of PC or device using Windows Operating System with a Windows Mobile 5.0 or Windows CE-based device through a serial port, USB port, Bluetooth, or IrDA protocol.
Agent: Software executing commands for a client at either the location of client device or at remote or intermediate location. A characteristic of the agent is its migratory behaviour.
Alert: A notification for an event.
Atomic: A command defining a transaction (data exchange) between server and client using a set of SyncML commands placed between the two tags. The transaction should be treated as a basic unit, which is not decomposable. It should either be completed in full or should be rolled back.
Back Up: Transfer the data set of device to PC or laptop (server to enable synchronization) in a selected directory.
Bluetooth: A wireless protocol to communicate and synchronize in personal area.
COM RS232C port: A port identical to IBM PC RS232C serial port in which data communicates in UART (Universal Asynchronous Receiver and Transmitter) format with a baud rate up to 9600 kbps.
Conflict: When a copy or item has changed at one end but modifies at other ends after different intervals.
Consistent copy: A copy which need not be identical to the copy at the data-generating source but satisfies all the required functions and domain-dependent specific rules and is such that it is not in conflict with and is not different from the copy at the source.
Data replication: Creation of data copies at one place after copying from another or copies from one to many others or many to many. The copies may be in different formats.
Data synchronization: Maintaining availability of source-generated data and maintaining consistency between the pushed copies from data source and local cached or hoarded data at different computing systems without discrepancies or conflicts among the distributed data. [Definition applicable for mobile computing environment.]
Flat file: Text or binary files as per the information format.
Funambol: Originally called Sync4J, provides an open source standard for Java-based processing of SyncML messages and synchronization formats.
HotSync: Software used for synchronizing handheld devices based on PalmOS with Windows- or Mac-based personal computers in a synchronizing architecture. Synchronization is through a serial port, USB port, Bluetooth, or IrDA protocol.
Intellisync: A synchronizer from Intellisync Corporation (now a Nokia-acquired company). It synchronizes PIM data between mobile devices and Internet employing an open architecture and it can therefore be integrated with an enterprise architecture. It enables enterprise connectivity features along with the security with end-to-end encryption and password protection.
IrDA: Infrared protocol to communicate or synchronize in a room with client and server in line of sight.
Map: A command defines a request to client or server for updating an identifier which maps one data object with another.
Mobile agent: Software to recognize data inconsistency and conflict in the data received at the server connected with the host device (client) using resolution rules.
Personal area synchronizer: A synchronizer to synchronize data copies at device with the copies at nearby PC, laptop, PDA, or device which connects directly to Internet by WAP or HTTP protocols. Nearby specifies the range for the distance between the device laid on a cradle and PC as 10–50 meter for wireless and 1 meter for cable.
Personal information manager (PIM): A manager for personal information, for example, calendar, address book and tasks-to-do. Tasks-to-do is a list of tasks planned by the device user. Address book has contacts, their addresses, website URL, email address, and telephone number(s).
Refresh synchronization: A client- or server-initiated synchronization with the server so that copies that are deleted at other end, get deleted at this end and copies that are modified are copied again at this end, so that client or server have non-conflicting and consistent copies.
Reports: An option which provides the device with the entries that are added/deleted/modified at server.
Restore: Copying selected or all data to the device.

Results: A command which precedes `<Search>` or `<Get>` command(s) used for requesting for the search of queried data item(s) or getting meta info, respectively.

Serial port: RS232C COM port is usually referred as a serial port.

Server alert: When data is modified (e.g., new email or newly downloaded music file) at the server (PC, or remote mail server), the server alerts the client and the client synchronizes the modified data by pull request.

Status: A command which precedes the request made by command(s) for client operations on server data item(s), DataStore, or object(s). As the result of this command, a status data is returned. Status codes of SyncML are similar to those of HTTP. For example, status code is 200 for successful operation and 212 for authentication for a session. The receiving end returns status code to the source to intimate the result of execution of the messages at the receiving end.

Slow synchronization: Synchronization in which client and server data are compared for each data field and are synchronized as per conflict resolution rules. It is actually thorough synchronization. The full copy synchronization usually takes place when the device is in idle state and not immediately on connecting to the server. That is why it is also called slow synchronization.

SMIL: XML-based language for text, speech, or multimedia integration for multi-modal communication.

Synchronization client: Software at a device or computer system, which initiates synchronization of its data with synchronization server.

SyncML: An XML-based OMA-initiated widely accepted language.

Synchronization engine: Same as synchronizer.

Synchronization server: Software at a device or computer system that responds to requests for data made by a synchronization client and also sends an alert to the synchronization client.

Synchronizer: A software tool for synchronization that resides at a host. If the synchronizer host is a server, then it is called synchronization server.

Sync4J: Same as Funambol.

Tasks-to-do: A list of the tasks to be performed by the user of a mobile device stored in chronological order.

Temporal property: A property of a record, item, entity, or file, which can vary with time. For example, time of the day or mailbox.

Two-way synchronization: Synchronization in which partial or full copies of data are synchronized between device and server.

WBXML: A binary format for communication on web in compact form. It is an alternative to text format of XML.

XML: Extensive markup language in which there are markings which include start and corresponding end tags that help in providing a meaning to the text within. The tags can be extended unlike in HTML where only the standard tags recommended by W3C can be used. XML is therefore said to be extensible by including tags and attributes. An example of a non-standard tag in XML is the address tag. It is used to make it apparent that the given characters within start and end tags represent the address.

Review Questions

1. Illustrate the architecture for data synchronization in mobile computing systems. Explain the terms replication, consistency, data conflict, and discrepancy in data in mobile computing systems. Give examples of data conflict and data discrepancy.
2. What are the different types of synchronization? Display the application of each.
3. Describe synchronization usage models in mobile applications.
4. Explain the need for domain-dependent specific rules. Explain the term WBXML.
5. What do you mean by PIM data synchronization? What are the PIM objects managed by PIM synchronization messages?
6. What are the differences between PIM server and Personal Area Synchronizer? Give an example of each.
7. What are the types of conflict resolution strategies used when there are conflicting data copies?

8. Compare the features of HotSync, ActiveSync, and Intellisync. List the synchronization protocols and their data synchronization features.
9. Explain with diagram client and server framework. Explain how SyncML synchronization client, agent, engine, interface, and adapter are used in a mobile application.
10. Describe the features of SyncML protocol.
11. Describe, with examples, setup and synchronization phases in data synchronization between mobile device and server.
12. Exemplify the application of SMIL.

Objective Type Questions

Pick the correct or most appropriate statement among the choices given.

1. (i) Data replication means creation of data copies at one place after copying from another or copies from one to many others or many to many. (ii) A replicated data copy can be full or partial. (iii) The videos of faculty lectures or music files can be replicated at iPods. Which of these are true?
   (a) (i), (ii), and (iii) are true.
   (b) (i) and (ii) are true.
   (c) Only (i) is true.
   (d) Data replication means data dissemination from one place to another without any change in data format.
2. Data synchronization between computer *A* and handheld device *B* is
   (a) maintaining the availability of source-generated data from *A* to *B*, from *B* to *A*, or both and *maintaining consistency* between the pushed copies from one data source and the local cached or hoarded data between the two *without discrepancies or conflicts* among the distributed data.
   (b) from *A* to *B* as well as from *B* to *A* and *maintaining consistency* between the two *without conflict.*
   (c) from *A* to *B* as well as from *B* to *A* at the same instant and *maintaining copies* between the two *without discrepancies.*
   (d) from *A* to *B* as well as from *B* to *A* at the same instant and *maintaining* consistency between the two *without discrepancies.*
3. A data synchronization domain-specific rule can be as follows: On the device a database record copy can be
   (a) in structured text or XML format and may be synchronized with device employing Symbian OS, while at the server, it can be in structured text, HTML, or WML format.
   (b) in structured text or XML format and may be synchronized with device employing Symbian OS, while at the server, it can be in DB2, Oracle, or another OS.
   (c) in structured text or XML format and may be synchronized with device employing Symbian OS, while the server deploys another OS.
   (d) in structured text or XML format and may be synchronized with device employing Symbian OS, while at the server, it can be in DB2 or Oracle but the operating system must be Symbian.
4. When synchronization is done with temporal properties of data,
   (a) there should be no inconsistency in case there is no temporal conflict and unaccountable discrepancy.
   (b) there may be consistency for a short period but that can affect the mobile applications for that period.
   (c) there may be inconsistency for a period but that does not affect the mobile applications in case there is no temporal conflict and unaccountable discrepancy.
   (d) there may be inconsistency and there is temporal conflict and accountable discrepancy for a period but that affects the mobile applications for a short period only.

5. What type of synchronization is provided by PIM between device and computer and what is the configuration for connection preference?
   (a) *Device selection* or *voice user interface* and connection preference is configured as IrDA or Bluetooth.
   (b) *Device selection* or *time-specific* and connection preference is configured as TCP/IP, WLAN, or Bluetooth.
   (c) *User-selection* or *time-specific* and connection preference is configured as TCP/IP, WLAN, or Bluetooth.
   (d) *User-selection* or *auto* and connection preference is configured as *Bluetooth, IrDA,* USB, or COM port cable.
6. Conflict resolution means that a
   (a) copy or item has changed at one end but should not be modified at other ends after different intervals and for data synchronization a conflict resolution strategy is not adopted.
   (b) copy or item has not been changed at one end and for data synchronization, a strategy specifies the rules.
   (c) copy or item never changes at one end but is modified at other ends after a fixed interval and for data synchronization, a strategy specifies the rules.
   (d) copy or item has changed at one end but is modified at other ends after different intervals and for data synchronization, a conflict resolution strategy is adopted and the strategy specifies the rules.
7. (i) Mobile agent is software to recognize data inconsistency and conflicts using resolution rules in the data received at the remote server connected with the host device-client. (ii) The agent can also migrate to another host connected to a server for data synchronization, if needed. (iii) The migrated mobile agent can function as proxy to a mobile device or computing system.
   Which of these are true?
   (a) (i) and (ii) are true.
   (b) (i), (ii), and (iii) are true.
   (c) Only (i) is true.
   (d) Mobile agent means software which functions as migrated proxy for data dissemination.
8. (a) ActiveSync 4.1 is synchronization software from Nokia. It synchronizes data at the PC with Windows Mobile 5.0 or Nokia device and ActiveSync for Symbian OS platform is not available.
   (b) ActiveSync 4.1 is synchronization software from Palm. It synchronizes data at the PC with Palm-based devices.
   (c) ActiveSync 4.1 is synchronization software from Microsoft. It synchronizes data at the PC with Windows Mobile 5.0- or Windows CE-based device and ActiveSync for Symbian OS platform is also available.
   (d) Symbian OS platform HotSync 3.1 is synchronization software from Symbian. It synchronizes data at the PC with Windows Mobile 5.0.
9. (a) Intellisync provides support for PIM synchronization, email accelerator, device management, and file distribution using Systems Management/File Sync software on Nokia devices. It also supports synchronization with Microsoft exchange server and wireless and push email features on a mobile device; push email has push latency of 30 minutes.
   (b) Intellisync provides Windows and Outlook email and instant push email feature on a mobile device, and supports PIM synchronization, email accelerator, device management, and file distribution using systems management.
   (c) HotSync provides Windows and Outlook email and instant push email feature on a mobile device from server.
   (d) ActiveSync 4.x allows synchronization through direct Internet connection (TCP/IP connections using WiFi, WLAN, or dial-up telephone line).
10. (i) WAP 2.0 is wireless protocol for synchronization of WAP client computers and WAP server. (ii) Synchronization is through SyncML codes binding over the WAP application layer client or server. (iii) A WAP gateway connects WAP client to HTTP servers of the web sites on the Internet.

(a) (i) and (ii) are true.
(b) (i), (ii), and (iii) are true.
(c) Only (i) is true.
(d) SyncML is not used in WAP client mobile devices.

11. (a) SyncML codes bind for synchronization using Bluetooth, WAP, or IrDA and a mobile application run and are synchronized with Bluetooth, WAP 2.0, or IrDA level 5 connection.
(b) SyncML codes bind for synchronization using Bluetooth, and a mobile application run and are synchronized with Bluetooth and Internet connection.
(c) SyncML codes bind for voice user interface and SyncML server running Java applications.
(d) SyncML is a language for PIM and voice data synchronization.

12. (a) SyncML-based mobile application database connector synchronizes data at the device with XML database.
(b) XML-based mobile application database synchronizes data at the SyncML server. The *Alert* command of SyncML defines a request for updating an identifier at client or server and the identifier maps the two data objects.
(c) SyncML-based database synchronizes data with the device database. The *Update* command of SyncML defines a request for updating an identifier at client or server and the identifier maps the two data objects.
(d) SyncML-based mobile application database connector synchronizes data at the device with any relational database and MAP Command of SyncML defines a request for updating an identifier to client or server and the identifier maps the two data objects.

13. (i) SyncML code generation (ii) parsing of received SyncML data (iii) validation of DTD in WBXML and XML data format (iv) base-64 encoding/decoding (v) notification message parsing (vi) credential checks (vii) security operations (viii) HMAC data-integrity check
Which of these are true in SyncML OMA standard?
(a) (i), (ii), and (iii) are true.
(b) All are true except (viii) and (iv).
(c) All are true.
(d) (i) to (v) are true.

14. URI means
(a) a relative path specification.
(b) a full path specification.
(c) directory, folder, and subfolder path.
(d) URL path specification for a device or server resource.

15. (a) SyncML provides Java-based synchronization using open-source mobile application server software and software development platform for mobile applications.
(b) Java provides SyncML-based synchronization.
(c) Funambol provides Java-based synchronization using SyncML, open-source mobile application server software, and software development platform for mobile applications.
(d) Funambol provides C++ and Visual C++ Java-based synchronization using SyncML, open-source mobile application server software, and software development platform for mobile applications.

16. SMIL is a language for
(a) text, speech, or multimedia integration for multi-modal communication and integrates streaming audio, video, and speech.
(b) text, speech, or multimedia integration for multi-modal communication and integrates text, image, and streaming audio.
(c) text, speech, or multimedia integration for multi-modal communication and integrates video with text.
(d) text, speech, or multimedia integration for multi-modal communication and integrates text, image, streaming audio, video, and speech.

Chapter 10

Mobile Devices: Server and Management

In the preceding chapters, we learnt how a server or a distributed mobile computing system disseminates data to mobile devices and how devices synchronize data with the system. The important concepts of mobile computing which are described in this chapter are as follows:

- Mobile agents and their applications
- Application servers and their applications
- Gateways and their applications
- Service discovery by a mobile computing system and its application in dynamic interactions and provisioning of services to client devices
- Device management by the Tivoli device support infrastructure and OMA DM
- Mobile file system—CODA
- Security problems and their solutions

10.1 Mobile Agent

Section 5.2 described how home and foreign agents manage mobility and handover in a mobile IP network. When a mobile node moves from a home network to a foreign network, it discovers a foreign agent, which enables it to get a care-of address (COA) by agent advertisement or solicitation. Section 7.3 described *N*-tier client–server architecture (Figs 7.5–7.9). In peer-to-peer architecture, each system or device can send a request to and get a response from the other and each device can have access to databases or database servers (Fig. 10.1).

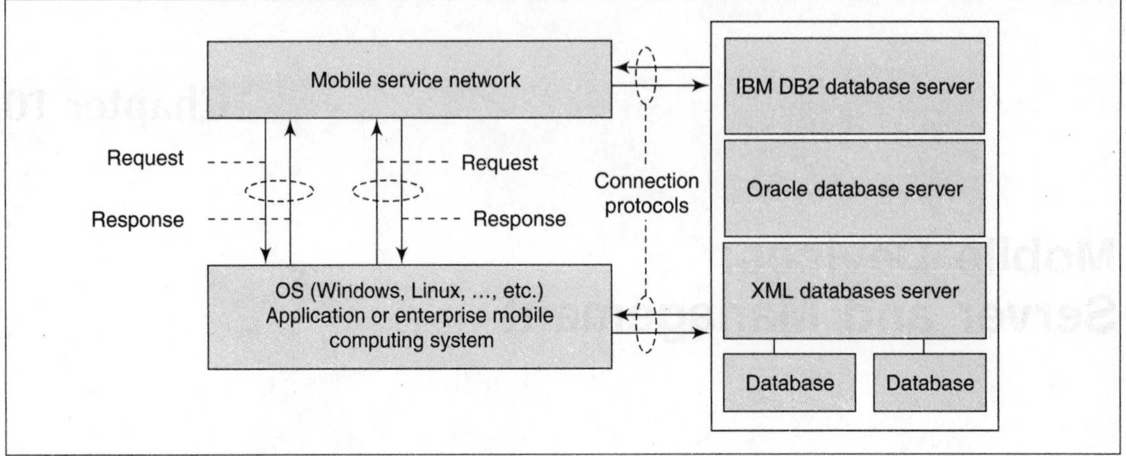

Fig. 10.1 Peer-to-peer computing architecture

A *mobile agent* consists of software and data, which can move from one computing system to another autonomously and functions for a device or system the host. A mobile agent can also be described as an autonomous software which runs on a host with some data and dynamically moves to another host (which has other required data) as and when required. Mobile-agent-based architecture provides for mobility of codes and data.

Figure 10.2 shows a mobile-agent-based architecture, in which the agent moves at instants T1, T2, and T3 to process a request, get email, and get records from a database, respectively. When a mobile agent moves at instant T1, T2, or T3, it saves its own state at the host and transmits this saved state to the next host in order to resume execution of the codes starting from the saved state.

The various characteristics of a mobile agent are:

- Mobility of code and data from one computing system (host) to another
- Ability to learn in order to adapt code and data to a host computing system
- Ability to clone, extend, or dispose itself after its role is over
- Compatibility to the hosts
- Ability to continuously and autonomously process requests and send responses and alerts (an alert is an unsolicited message, record, or information.)

Some of the advantages of a mobile agent are:

- Asynchronous running of codes on diversified heterogeneous hosts
- Reduced computational and data requirements on devices with limited resources
- Tolerance to connection failures
- Only the agent source (e.g., device middleware, which sends the agent) needs to be modified in order to redefine the functions expected from the agent.

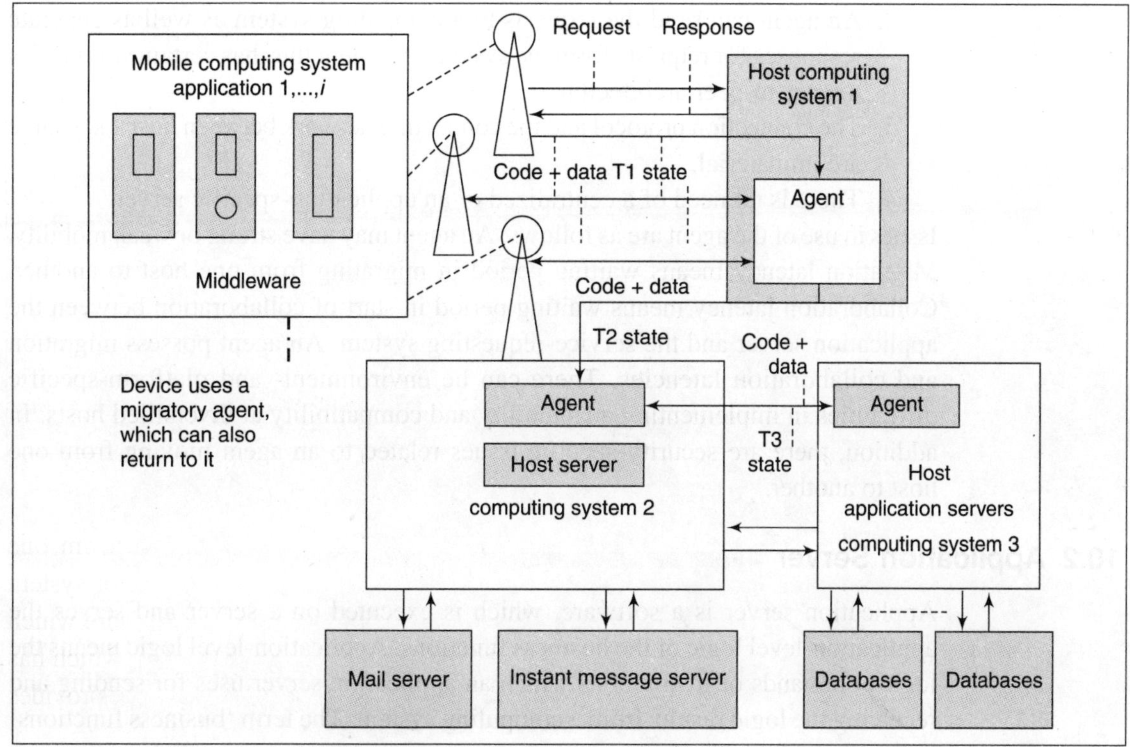

Fig. 10.2 Mobile-agent-based architecture

A mobile agent is a powerful tool for distributed applications and retrieval of remote host information. The agent gives dynamic software that runs on different hosts at different times. The agent makes available the resources of its host to the resource-scarce devices, discovers new resources, and monitors the distribution of resources. It also manages the network and the distributed computing systems. It is a very effective alternative to the use of application-specific server(s) and device middleware for retrieving information and messages.

Advantages when deploying an agent in a computing system are as follows:

1. Consider AgentOS, an operating system developed in 2006. It is one of the popular mobile agent technologies. It provides application virtualisation environment and allows automatic thread migration (ATM), the threads running on the host being independent of the OS. Application virtualisation environment means that an application instead of running in its own environment provided by the OS and the system hardware actually runs at a host OS and host system hardware of the mobile agent of that system. To a user, the application appears to be running at the system environment while it is actually running at the host environment, which is host to the agent.

2. An agent can send the requests to a computing system as well as generate responses for requests from the system. An agent thus has certain similarities to peer-to-peer architecture.
3. The connection protocol and the connecting network between host and source are immaterial.
4. There is no need of a centralized or an application-specific server.

Issues in use of the agent are as follows: An agent may have strong or weak mobility. Migration latency means waiting period in migrating from one host to another. Collaboration latency means waiting period in start of collaboration between the application server and the service-requesting system. An agent possess migration and collaboration latencies. There can be environment- and platform-specific difficulties in implementing adaptability and compatibility at diversified hosts. In addition, there are security-specific issues related to an agent moving from one host to another.

10.2 Application Server

Application server is a software, which is executed on a server and serves the application-level logic of the business functions. Application-level logic means the logic commands or instructions which an application server uses for sending and receiving the logic results from a computing system. The term 'business functions' indicates the logical way in which transactions (business) are carried out between server at one end and application at the other. Examples of transactions involving mail application server are—(a) establishing connection between mail APIs (application program interfaces) and mail server, (b) updating mails by inserting, adding, replacing, or deleting, (c) querying for the mails, and (d) terminating the connection between the API and the mail server.

An application server gets requests from the collaborating or independent mobile devices of an enterprise or from a distributed mobile computing system. It processes these requests and generates responses. The application server also handles presentation services to the devices or computing systems. It employs records at the server database(s) for this purpose. The server also integrates itself with the backend databases and systems, for example, an enterprise system.

Figure 10.3 shows an application-server-based N-tier architecture ($N \geq 3$) with requests processed at the application server using backend database(s) and systems. Assume that there are j clients which can request to the server. A client 1, …, or j sends the request from collaborating or independent mobile devices of an enterprise or from a distributed mobile computing system. The figure shows the stages of data transmission requests at instants from T0 to T4. Requests are processed through tiers 1–N. Responses are sent from the backend system to tier 1 at instants from T5 to T9.

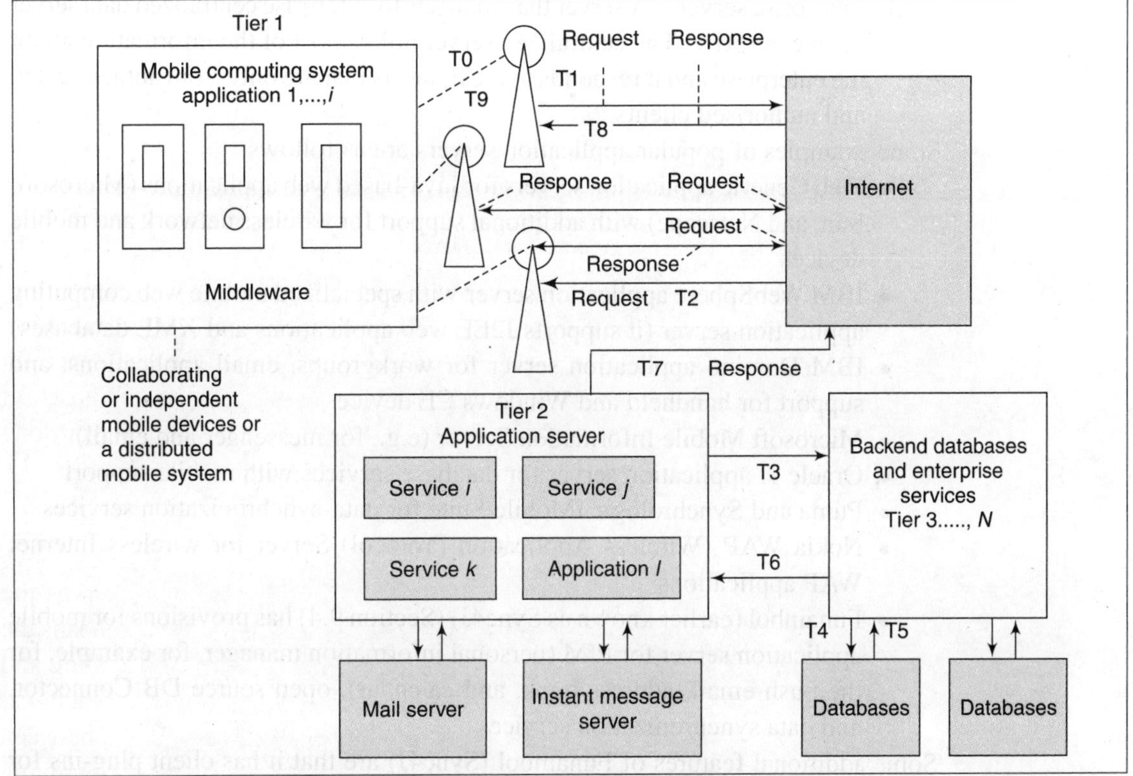

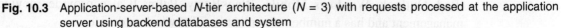

Fig. 10.3 Application-server-based *N*-tier architecture (*N* = 3) with requests processed at the application server using backend databases and system

Consider the application server at Tier 2. It provides the following services:

- Service *i*: application logic processing at the server
- Service *j*: presentation services for device responses and decoding the device requests (e.g., presentation service of a middleware application server for universal device access)
- Service *k*: transaction services with support to pervasive computing model of mobile applications
- Application *l*: system integration service for backend services and database at tiers 3, ..., *N*.

Consider the examples of web database and enterprise application servers at tiers 3, ..., *N*:

1. IBM DB2 database server—IBM DB2 is an RDBMS (Relational Database Management System) data server from IBM and its versions run on handheld devices and in enterprises the application logic processing is at the server.
2. Oracle 9i database server—RDBMS Oracle9i server has a large number of features and it supports XML documents and has an option for cluster database.

3. Enterprise server—A server that connects to enterprise centralized data server in an enterprise. The centralized server holds most of the information about the enterprise and it responds to the requests that are only from authenticated and authorised clients.

Some examples of popular application servers are as follows:

- Web Generic application servers for Java-based web applications (Microsoft, Sun, and Netscape) with additional support for wireless network and mobile devices
- IBM WebSphere application server with specialized mobile web computing application server (it supports J2EE web applications and XML databases)
- IBM Domino application server for workgroups, email applications, and support for handheld and Windows CE devices
- Microsoft Mobile Information Server (e.g., for messenger and email)
- Oracle 9i application server for database services with mobile support
- Puma and Synchrologic iMobile Suite for data-synchronization services
- Nokia WAP (Wireless Application Protocol) Server for wireless Internet WAP applications
- Funambol (earlier known as Sync4J) (Section 9.4) has provisions for mobile application server for PIM (personal information manager, for example, for the push email, address book, and calendar), open source DB Connector, and data synchronization services

Some additional features of Funambol (Sync4J) are that it has client plug-ins for iPod, Outlook, Windows Mobile, BlackBerry, Palm. It also provides device management and has a number of applications for wireless devices.

One of the popular enterprise servers is the BlackBerry Enterprise Server (BES). BES is a middleware software for BlackBerry wireless devices. A brief description of some of the application servers mentioned above are as follows.

10.2.1 Sun Java System Web Server 6

Sun Java System Web Server 6 is meant for large business applications and is compatible with a number of operating systems. It allows deployment of CGI (common gateway interface) scripts, PHP (pre hypertext processor), ColdFusion, Servlets, JSPs (Java server pages) and ASPs (active server pages). (CGI is a protocol. It is a language-neutral interface for processing. It can use the language Perl or PHP (Pre-Hypertext Processing). PHP is a language for processing the text and preparing it in hypertext format and enables dynamic web pages. ColdFusion software creates pages by joining different entities, elements, and databases before a web server sends the response to the client. Servlets are Java classes at server side for processing the web page. JSP is a web page which connects to the databases and deploys codes, JavaScript codes, Java components and directives, and generates the response in HTML for the requesting clients. ASP is a web page which connects to the databases and deploys the JavaScript and Visual Basic scripts, codes, and directives, and generates the response in HTML for the requesting clients.)

The server provides application services and runtime environment. Application services include email service, security, file system, and session management. An access manager provides a secure access to Web-based resources. Session management entails starting a session, carrying out transactions during the session, and closing the session. Runtime environment means a platform for running CGIs, Java servlets, or PHP.

The server has a content engine which is a software layer (in between software component) for publishing the Web application response. In other words, content engine is a program which uses the content sources at a server and drives them to the destination. Driving means sending, publishing, or generating response in a web application so that the client gets the required contents at the destination. The engine has three components—search engine, content management engine, and HTTP engine. The content engine is a layer for web content creation, addition to existing contents, and storage of content page using the HTML, JSPs, and ASPs. It maintains integrity of the client contents with the server. For example, a client has submitted a user ID, a password, and other client information to the server to enable it to get response contents. Integrity means the client information is used for client-specific application only.

10.2.2 IBM WebSphere MQe

Figure 10.4 gives an overview of IBM WebSphere application server for web applications on mobile devices with the feature of MQe (Messaging and Queuing Everyplace). IBM WebSphere application server runs on a mobile device. Application accelerator accelerates the running of an application so that results are obtained faster. MQe refers to an added set of features. MQe assures secure messaging, decouples from the application, and uses a queue manager. It messages onto a queue. It has adapters to adapt itself to device interfaces. It enables connection and remote connection to mobile devices, for example, sensors, phones, and PDAs. Everyplace means for mobile devices at all places including sensors, phones, PDAs. MQe also has IBM WebSphere Voice Server, text-to-speech (TTS) software, and IBM WebSphere RFID Premises Server.

MQe development kit shown in Fig. 10.4 is used for writing applications for messaging and queuing for mobile devices. WebSphere MQe development kit writes applications based on Java and C. The figure shows that there is the Presentation Everyplace suite. There are portal, transcoding, and personalization servers. Deployment and integration are two other functions of the WebSphere. Transcoding is an abbreviation for transformation-cum-coding. A web portal (Section 10.4) or server deploys data formats and codes which are different from the ones at a mobile device. For example, characters may be in ASCII format at device and in Unicode at the server. Database may be in XML format at the device. Transcoding means

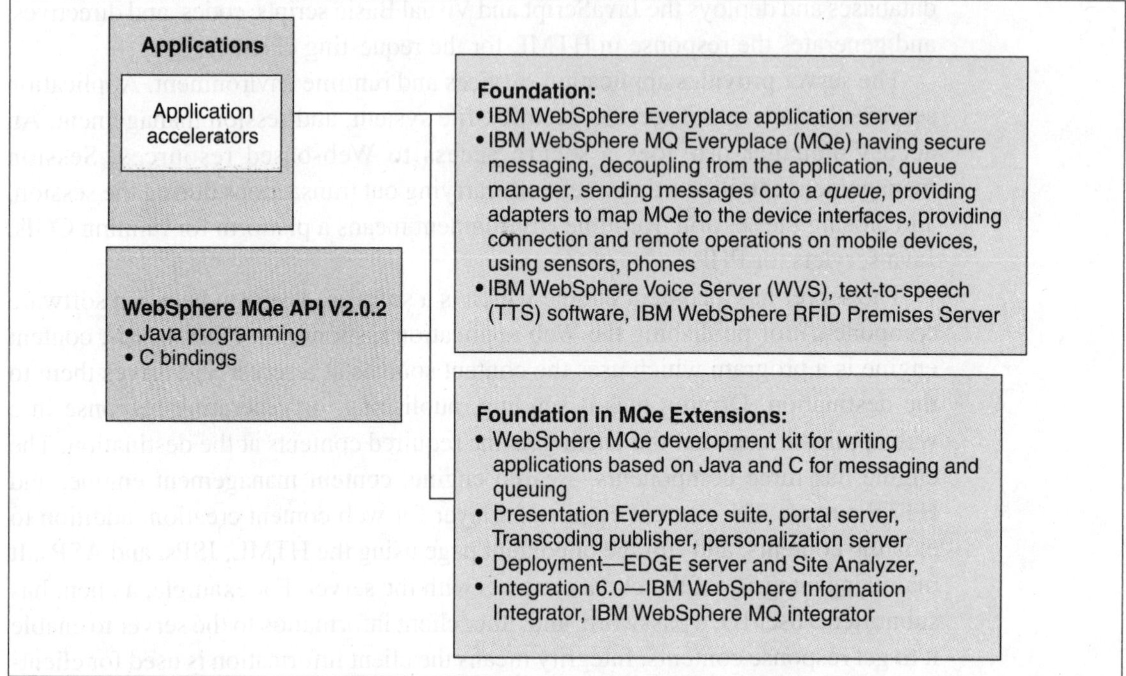

Fig. 10.4 IBM WebSphere application server for mobile devices

adaptation, conversions, and changes using software which renders the web responses, messages, or records in formats and representations required and acceptable at a device. Similarly, the device requests are adapted, converted, and changed into the required formats acceptable at the server by transcoding software.

Figure 10.5 gives an overview of IBM WebSphere Everyplace services and functions. The IBM WebSphere Everyplace has three *basic services—configuring service*, *directory service*, and *networking protocols*. A basic service of this software is configuring of the server and operations. Second basic service is directory services. It deploys JNDS (Java naming and directory service) and LDAP (lightweight directory access protocol). IBM WebSphere Everyplace also has an HTTP server which functions as website server. SyncML and WAP are used for website server and applications. Other services of IBM WebSphere Everyplace are as follows:

- device management and management of subscriptions for the applications. Everyplace provides the devices with connectivity protocols. These are TCP/IP, CDMA, GSM, and WAP.
- provides a synchronization engine which synchronizes the data available with a device having the records at the server. It maintains consistency without discrepancies or conflicts (Section 9.1.1).
- provides for device caching (Section 7.2).
- enables load balancing which means distribution of time between multiple client requests.

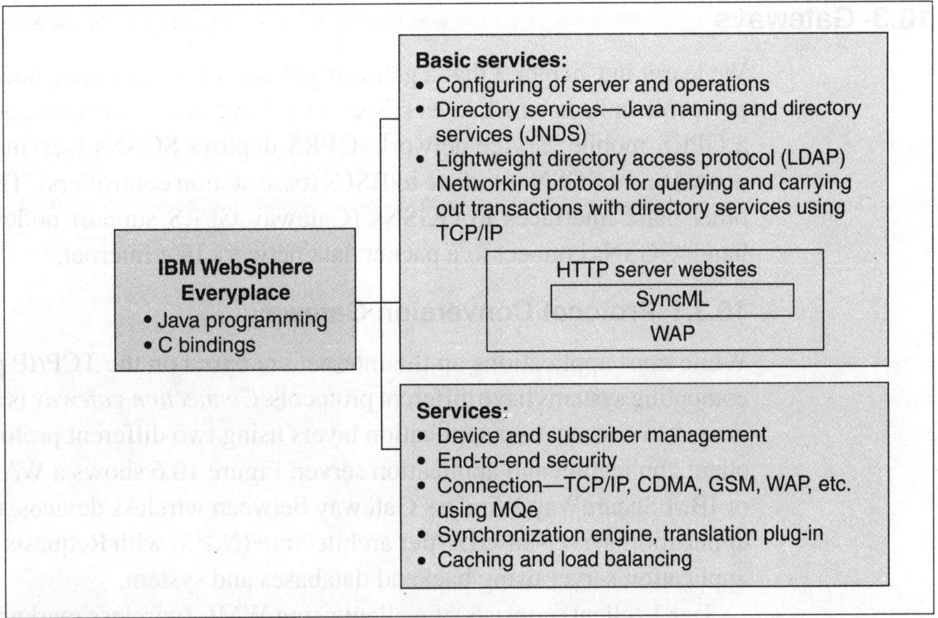

Fig. 10.5 Services and functions of IBM WebSphere Everyplace

10.2.3 Oracle Application Server

Oracle is the most popular RDBMS. Application server (AS) enables web-based transactions with the databases. Oracle Application Server 10g is the latest version. The last character g stands for grid computing feature. Grid computing means a number of computers forming a grid for computing. A brief description of Oracle Application Server 9i (9iAS) are as follows:

9iAS is used at Windows as well as UNIX platforms. It includes the Oracle XML Parser (version 2) and OracleJSP (Oracle Java Server Pages). 9iAS has two components—HTTP server and 9iAS portal with OC4J (Oracle AS containers for J2EE 1.4 specification EJBs). Section 10.4 will give the meaning of the term portal. The HTTP server responds to the SQL (simple query language and HTTP requests from the clients. A client sends requests and gets responses through an in between web cache called 9iAS Enterprise Web Cache. The HTTP server interfaces to 9iAS portal. The portal has a parallel page engine (PPE). It enables retrieving of pages in parallel at the portal. The SQL calls received at the HTTP server connect to database schema and database providers. 9iAS has OC4J J2EE EJBs (Java 2 Enterprise Edition Enterprise JavaBean) in a container. J2EE provides web applications. The EJBs establish the sessions with the client and connectivity with portal through which the pages are rendered using PPE. 9iAS integrates business logic, management, security, e-business, and portal functions.

10.3 Gateways

We learnt in Chapter 3 that a gateway connects two networks, each using different protocols in its network layers. Section 3.9 introduced the concept of a gateway in a GPRS mobile service network. GPRS deploys SGSNs (serving GPRS support nodes) and a SGSN interface to BSCs (base station controllers). The SGSNs on the other hand interfaces to GGSNs (Gateway GPRS support nodes). On the other hand, GGSNs connect to a packet data network like Internet.

10.3.1 Protocol Conversion Gateway

While most applications on the Internet are based on the TCP/IP protocol, mobile computing systems have different protocols. *Connection gateway* is a software which is used to connect two application layers using two different protocols. It connects client application and application server. Figure 10.6 shows a WAP (Section 12.2) or IBM SecureWay Wireless Gateway between wireless devices, the Internet, and application-server-based *N*-tier architecture ($N \geq 3$) with Requests processed at the application server using backend databases and system.

Tier 1 (client) consists of a client using WML (wireless markup language). The client sends the application service request and obtains the responses (Section

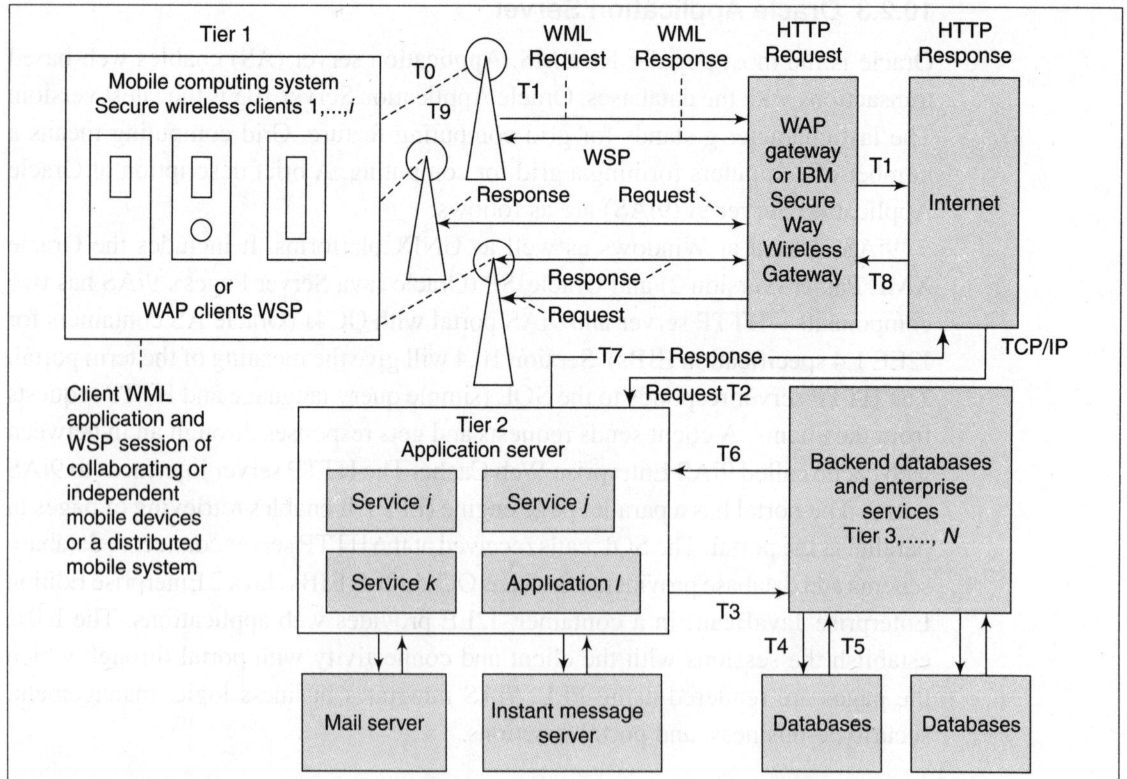

Fig. 10.6 WAP or IBM SecureWay Wireless Gateway

12.2.6). A WSP (wireless session protocol) establishes session between the application server (AS) at the Internet and the client(s). A client sends requests for the service required by collaborating or independent mobile devices or a mobile computing system.

Tire 2 AS connects to tier 1 through the Internet. The Internet uses TCP/IP, while the wireless devices use WAP or other protocols. When the protocol used is WAP, a gateway which connects tier 1 to Internet which has the AS is called WAP gateway. IBM SecureWay Wireless Gateway is a WAP gateway which connects these two tiers. The connection is established through WSP. Tiers 2, 3,..., *N* are the same as in Fig. 10.3.

When Funambol is used for developing the mobile computing applications, the connector objects provide the gateways. The gateways connect to the file systems, databases, email systems, and applications. They are used for two-way synchronization with existing data assets. The connector objects also provide gateways for datastores (databases, files, and email systems). IBM WebSphere consists of Partner GateWay. It is used for the connections and transcoding between the tiers.

10.3.2 Transcoding Gateway or Proxy

A gateway is called *transcoding gateway* when it adapts its response to the content format of the client device and its requests to the format required by the application running on the application server. *Transcoding proxy* has conversion, computational, and analyzing capabilities, while a gateway has conversion and computational capabilities only. A proxy can execute itself on the client system or application server.

Transcoding applications involve formats, data, and code conversion from one end to another when the multimedia data is transferred from a server to the mobile TV, Internet TV, WAP phone, and Smartphone as the client devices. Transcoding applications also involve filtering, compression, or decompression. Transcoding gateway is configured to provide authoring and deployment. Transcoding facilitates TTS (text-to-speech) conversion and STT (speech-to-text) conversion. It is used for converting XML style sheets to WML or HTML, video to image, image to hyperlink, and one image format to another. Some of its other functions are scaling, mapping, and colour adaptation.

10.3.3 Residential Gateway

A home has number of systems—Cable TV or set-top box, personal computer, laptop, digital camera, iPod, home theatre, iPhone or mobile phone, and Bluetooth-enabled appliances. These form a home network. Outside world typically consists of Internet, enterprise server, backend database servers, and application servers.

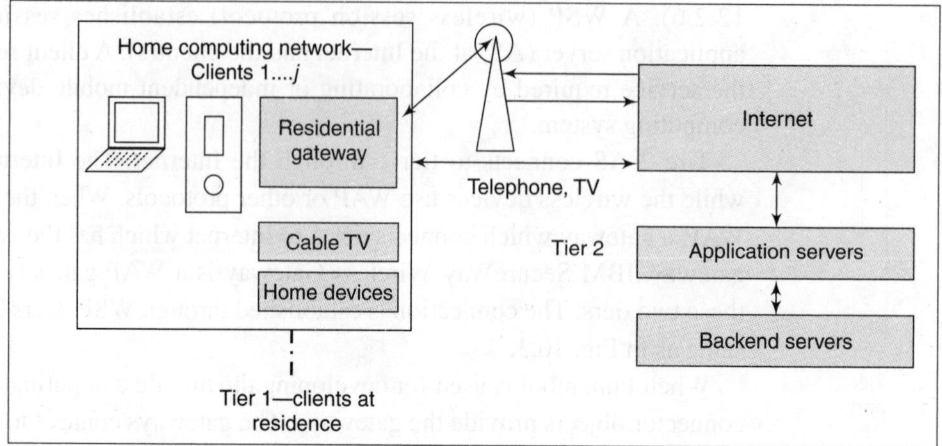

Fig. 10.7 Residential gateway architecture

The outside network typically consists of high speed Internet IPv6. A residential gateway is a wireless device which provides connectivity through a service provider. The network also connects to application servers and backend server through the Internet. Residential gateway is a gateway between a network of home devices and the outside world. Figure 10.7 shows residential gateway architecture.

10.4 Portals

A Web portal is a website that provides application-specific programs and personalized capabilities like business transactions to client device(s) or computing systems. It also provides a path to other content related to the portal personalization criteria and enables running of distributed or collaboration applications. (An example of a collaboration application is business transactions through an enterprise portal.) The portal has portlets (application-specific small portals), e.g., mail (POP3) portlet. A portal may have the PPE, for example, in 9iAS.

The portal provides services available from a number of websites. It enables them using diversified middleware. It provides services to PCs as well as mobile devices.

IBM WebSphere portal is one portal which provides a software architecture for the applications and business transactions. It has scalability of any size organization.

WSRP (Web services for remote portlets) is a software for content aggregators (e.g., Web portals). WSRP accesses and displays content sources which are at remote sites or portals. An IBM WebSphere portal software has an extension for voice-based applications. The extension deploys IBM voice tool kit. It gives voice-based access to the IBM WebSphere portal. IBM WebSphere portal interfaces with Voice Gateway which connects to a network of mobile service providers. The Java Portlet Specification defines a contract between the portlet container and portlets. Programmers deploy JSR (Java Specification Recommendation) 168.

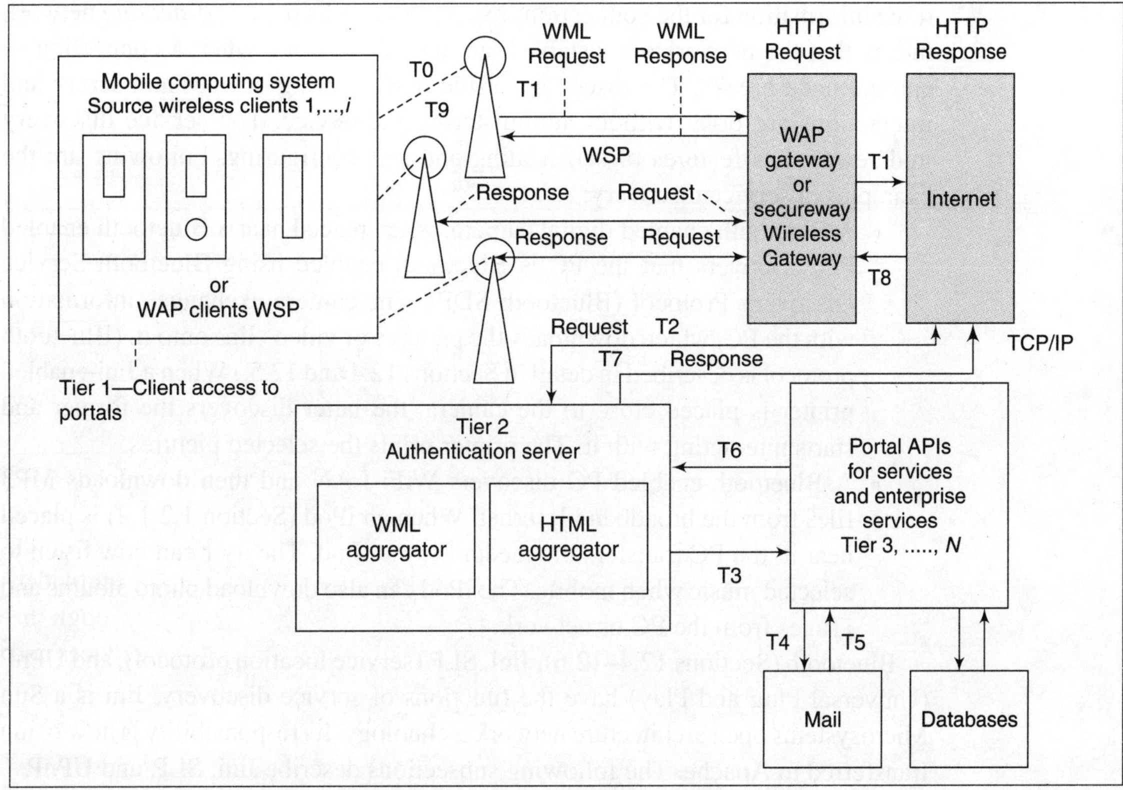

Fig. 10.8 Portal based client–server architecture for mobile clients

Figure 10.8 shows a portal-based client–server architecture. A portal includes an authentication server, a content aggregator, and APIs for services. Authentication of client is done before permitting access to the contents of the portal.

10.5 Service Discovery

Service discovery is an adaptable middleware in a device (or a mobile computing system) that dynamically discovers services by carrying out the following steps in chronological order:

1. letting nearby service network (or device or system) recognize that device
2. letting the nearby network know of device service(s)
3. searching and discovering a new service(s) at the network
4. interacting with nearby network using discovered service(s)

Self-administration software means software for starting the operating system, allocating network and system access addresses, initiating accesses, establishing and terminating the connections, and making secure connections by the provisions by a system on its own without using network administration software when connecting to a network. *Self-configuration* means establishing and modifying the route information for the connections by a system on its own. *Self-healing* network means that the network can establish an alternative route when a connection or enroute node breaks. The discovery of the neighbouring devices and server and interactions are done without administering the device. The service discovery middleware has features of self-healing and self-configuring. Following are the examples of service discovery:

- A Bluetooth enabled digital camera, when placed near a Bluetooth enabled PC, discovers that the PC is Bluetooth enabled using Bluetooth Service Discovery Protocol (Bluetooth SDP). The camera exchanges information with the PC, which downloads the pictures or video clips onto it. (Bluetooth protocol is described in detail in Sections 12.4 and 12.5.) When a Jini-enabled printer is placed close to the camera, the latter discovers the former and starts interacting with it. The printer prints the selected pictures.
- A Bluetooth-enabled PC discovers WiFi LAN and then downloads MP3 files from the broadband Internet. When an iPod (Section 1.2.1.3) is placed near it, the PC transfers the media files to iPod. The user can now listen to selected music when mobile. The iPod can also download photo albums and iTunes from the PC or network.

Bluetooth (Sections 12.4–12.6), Jini, SLP (service location protocol), and UPnP (Universal Plug and Play) have the functions of service discovery. Jini is a Sun Microsystems open architecture network technology. Its responsibility is now being transferred to Apache. The following subsections describe Jini, SLP, and UPnP.

Jini Jini enables the programming for the distributed computing system environment, for example, mobile devices and server, and for developing modular and cooperating services as it enables the download of classes for the service components. Jini provisions for not only service discovery but also for the lookup for the databases, RMIs (remote method invocations), and the joining (binding) of APIs and programs of a device with those of other devices discovered using the lookups.

Figure 10.9 shows Jini core protocols for service discovery, join, and lookup.

Jini includes JavaSpaces Technology, which is a simple and powerful high-level tool. It is used for programming collaborative and distributed applications. Its purpose is to share network-based object space(s). Object space is associated memory for the distributed objects on the network, which can be used both for object storage and as exchange area. Communicating nodes use this space indirectly. It creates a simple API for use by source and destination nodes.

Jini also includes extensible remote invocation (Jini ERI). (Remote invocation means invoking a method in a Java class on a remote system.) It enables programming and dynamic computing in a device and provides a platform to create adaptable, scalable, evolvable, and flexible network-centric services. It is highly adaptive to changes in software and hardware. It provides spontaneous interactions between devices and network.

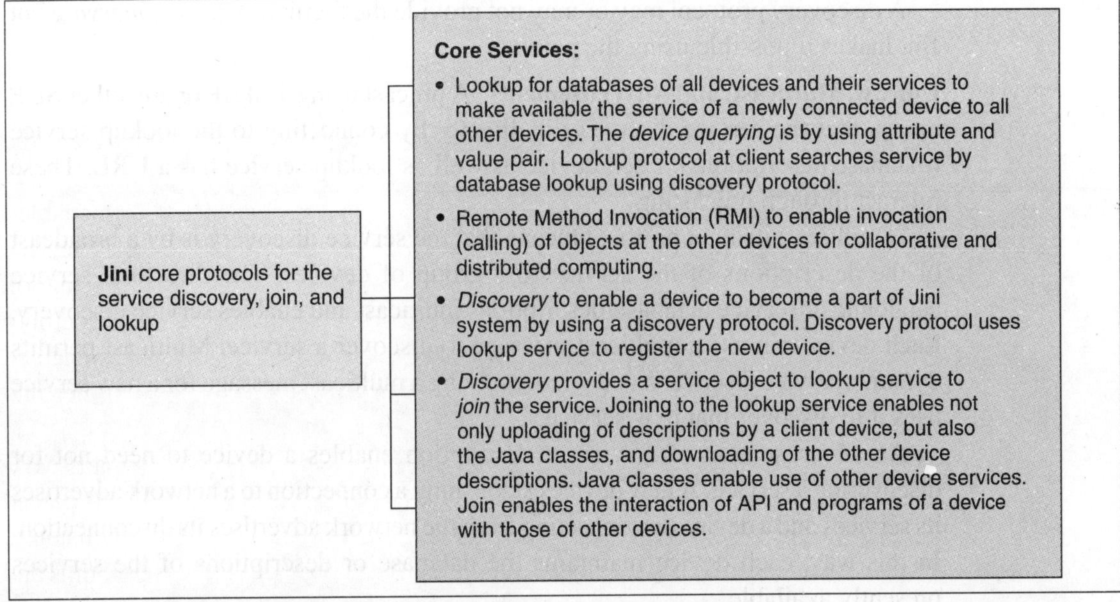

Jini core protocols for the service discovery, join, and lookup
Core Services:
Lookup for databases of all devices and their services to make available the service of a newly connected device to all other devices. The device querying is by using attribute and value pair. Lookup protocol at client searches service by database lookup using discovery protocol.
Remote Method Invocation (RMI) to enable invocation (calling) of objects at the other devices for collaborative and distributed computing.
Discovery to enable a device to become a part of Jini system by using a discovery protocol. Discovery protocol uses lookup service to register the new device.
Discovery provides a service object to lookup service to join the service. Joining to the lookup service enables not only uploading of descriptions by a client device, but also the Java classes, and downloading of the other device descriptions. Java classes enable use of other device services. Join enables the interaction of API and programs of a device with those of other devices.

Fig. 10.9 Jini core protocols for the service discovery, join, and lookup

Service Location Protocol (SLP) A client device using SLP dynamically discovers a service. SLP stack consists of SLP header and SLP service URL (universal resource locator). A URL is a universally accepted unique identifier for resource and thus a location. It enables other devices to use the service of the client device. It provides the location of the device in the network. It advertises its service needs in multicast mode and does not use directory service. (Unicast, multicast, and advertisement modes of broadcasting are described next.) Multicast transmission is to the multiple nodes having for getting to a service directory, service catalogue or service database/ descriptions. Client can use a multicast message to discover a service. Multicast permits discovery dynamically by getting a multicast message. A device receiving a multicast message for a new service uses it to discover that new service.

SLP is an alternative to lookup service database in Jini or a service catalogue. It provides Java classes (program codes) to a client device. SLP difffers from Jini as it just provides the lookup for the URLs and description of other devices during discovery phase and it does not provide for database *lookup*, *join*, and *spaces* as Jini.

A description specifies a service. Device textual description at a lookup database service or service catalogue enables a new device to use a service of a specific device. It is uploaded at the database when a new device joins a network and is downloaded to the new device to let the device discover the services. It can be in XML (Section 7.1.1, Sample Code 7.1, and Section 13.1). The XML tags with attribute–value pairs are used for the description of the service and device.

A discovery protocol may or may not provide the facility of *device querying* but Jini makes it possible using the pairs.

Unicast, Multicast, and Advertisement A unicast using TCP/IP or any other SDP means that the service discovery is directly by connecting to the lookup service (database/descriptions). Each device as well as lookup service has a URL. These interact through unicasting.

A multicast using a protocol means that the service discovery is by a broadcast of the descriptions of the service to a group of devices. The directory, service catalogue or service database/descriptions multicast and enables service discovery. Each device can use a multicast message to discover a service. Multicast permits dynamic service discovery. A device receiving a multicast message for a new service uses it to discover that new service.

An advertisement of service or description enables a device to need not for discovering a service. A new device establishing a connection to a network advertises its services and a device disconnecting from the network advertises its disconnection. In this way, each device maintains the database or descriptions of the services presently available.

UPnP Universal plug and play (UPnP) is a Microsoft solution for service discovery, service, and device descriptions, use of control points, control for registering the subscription events and withdrawal after subscribed interval, and presentation and event notifications (called eventing). Figure 10.10 shows UPnP core protocols for the service discovery, description, control, and eventing.

UPnP provides for control point(s). A control point is a registry for any device establishing connection with the network or disconnecting from the network. The registry provides descriptions and URLs of all devices presently connected to the

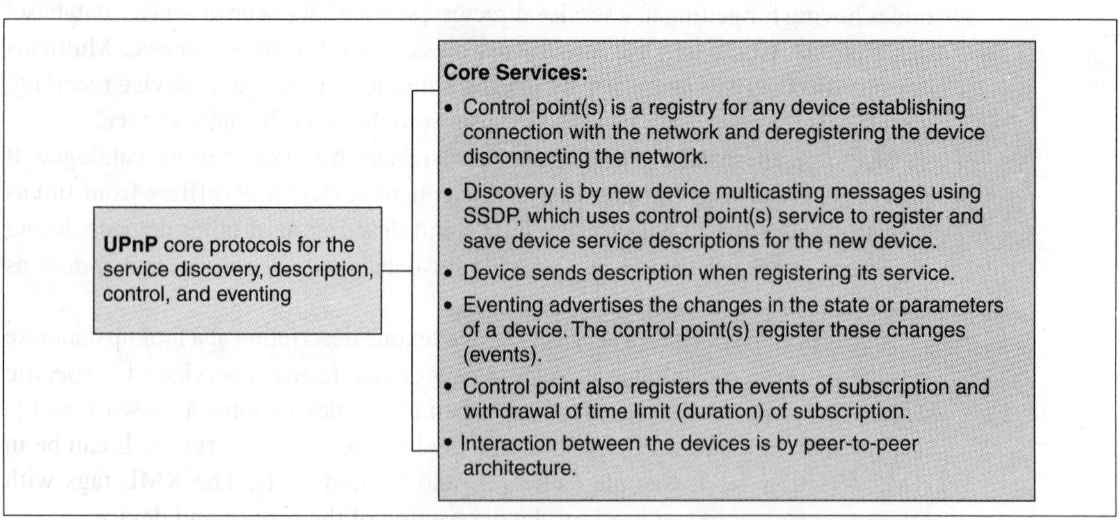

UPnP core protocols for the service discovery, description, control, and eventing
Core Services:
• Control point(s) is a registry for any device establishing connection with the network and deregistering the device disconnecting the network.
• Discovery is by new device multicasting messages using SSDP, which uses control point(s) service to register and save device service descriptions for the new device.
• Device sends description when registering its service.
• Eventing advertises the changes in the state or parameters of a device. The control point(s) register these changes (events).
• Control point also registers the events of subscription and withdrawal of time limit (duration) of subscription.
• Interaction between the devices is by peer-to-peer architecture.

Fig. 10.10 UPnP core protocols for the service discovery, description, control, and eventing

network and makes available the service of a newly connected device to all other devices. The registry does not allow *device querying*. It provides only *description* of device service. A device can itself have API(s) for querying, downloading, and uploading of classes as all this is outside the purview of UPnP protocol. A device itself has to be programmed for dynamic computing and adaptability using the descriptions and URLs provided by the control point(s).

UPnP includes description. UPnP clients are called control points. For example, a master device client functioning as digital audio player. Control point automatically discovers UPnP server on the network. Client gets the streaming media files from the server. A device sends description to the control point when registering its services.

Each *description* has a URL which the control point provides to other devices requiring the service. The descriptions do not include classes. UPnP provides for downloading of the description of device and services. An XML document for description is as per UPnP template.

UPnP devices do the multicasting of messages and services. *Discovery* by new device is from the messages and after discovery for its service(s) at other devices. The device becomes a part of UPnP system. Discovery is by using a simple service discovery protocol (SSDP). SSDP uses the control point(s) service to register and save device service descriptions for the new device.

10.6 Device Management

There are many types of devices in a mobile network. These devices are managed by a mobile service provider. Device management means *configuring at initialisation* (bootstrapping), *monitoring current configuration*, *processing maintenance requests*, and *taking care of location and handover* of each device. In pervasive computing environment, it means *managing the infrastructure of a large number of networks at the same time*. Each device can have applications downloaded from different sources in an enterprise. The service provider has to manage and serve the applications. In addition to this, each device may subscribe to different types of services for different durations. For example, a device may subscribe to specific gaming applications for a month. The account and authentication of each device is managed.

Each device is to be managed invisibly without system administrator. It means that the device is self-administered. It boots up, starts the operating system, initiates accesses, establishes and terminates the connections, and makes secure connections on its own without using a network administration software when connecting to a network. It runs through a setup phase and exchanges packages to get the allocated network and system access addresses. Each device in a mobile computing system interacts with the other and has features of self-healing and self-configuring network.

10.6.1 Device Support Infrastructure

Tivoli DSI (Device Support Infrastructure) is an IBM software. It is used for ATMs, handheld devices, set-top boxes, and cable modems. Figure 10.11 shows the Tivoli DSI architecture. A Device Gateway has a device management agent to connect devices at one end with the gateway at other end. The Gateway includes Tivoli Management Gateway. It connects to device management server of the service provider. Device manager at the device management server assigns a unique ID to a device and a local ID to the device which is supported by the support infrastructure. Device unique ID remains fixed and is assigned once. Local ID can be reassigned when the device moves from one personal area network to another. Device information is saved at the device manager. When a number of devices are of identical types, for example, Smartphones type, then a group object can be used by assigning the same type of devices to a group. A device group object then manages large number of devices of same type.

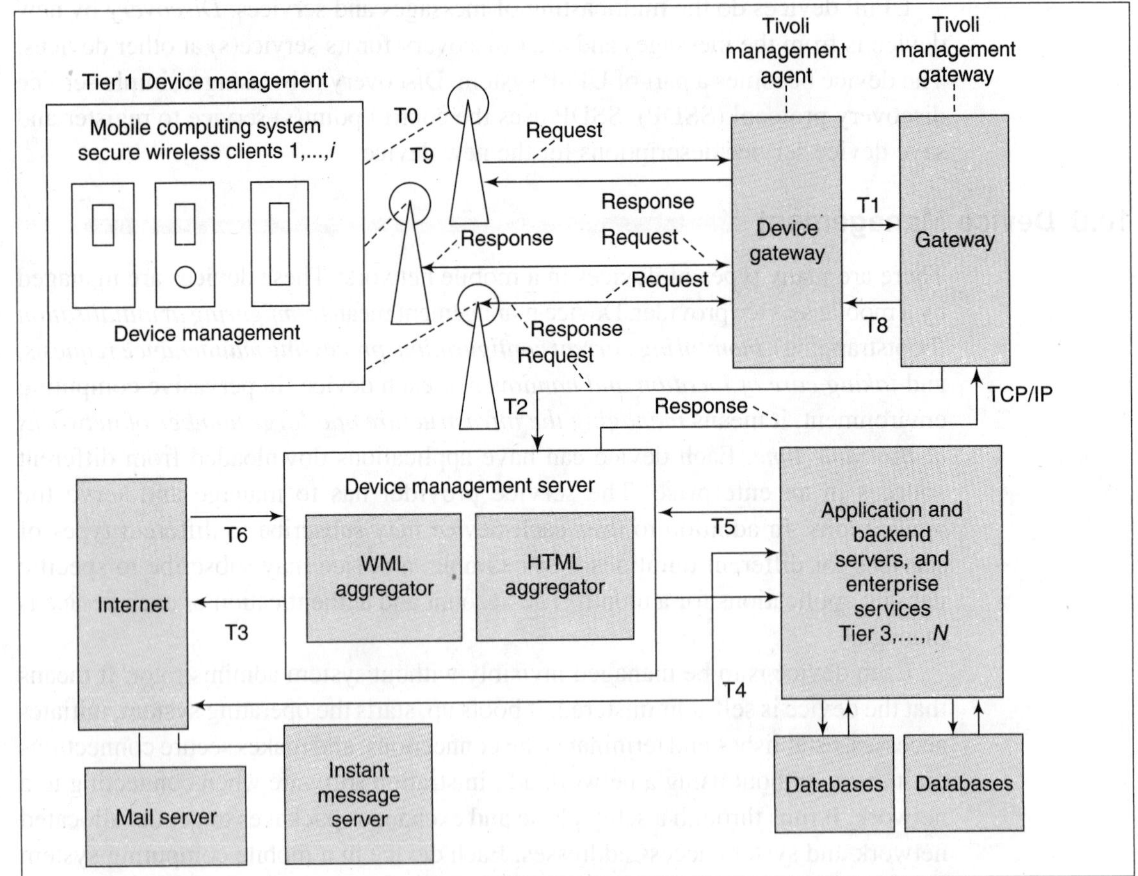

Fig. 10.11 IBM Tivoli device support infrastructure architecture

10.6.2 User, Device, and Network Profiles

Profiles provide a specification for the use of software such as Device manager or Device management server. Device management requires profiles for the user, device, and network. Mobile information device profile (MIDP) provides a specification for the mobile devices such as mobile phone to enable the use of Java microedition programming framework.

User profile is used as follows—Device management is facilitated by system access to user profile which consists of user password and ID. A user can also add PIM data, individual preferences, and security credentials to the profile. Device management needs system access to device and network profiles also. Device profile includes a unique ID, local ID, individual preferences, and available resources. Network profile specifies the current location address of the device and networked devices and the description of the network services.

Information given in profiles includes specification of groups, services, names, and objects. A profile may give descriptions of the types of the devices which can group and be managed concurrently. A profile may give descriptions of the services at the devices in the network. A profile may include the names (services and object names).

10.6.3 Directory Service

A service means a software or protocol for specifications and provisions for a set of operations with the given objects or entries. Directory service means a service protocol which specifies and provisions for the set of operations with the given objects or entries in a directory. Directory is an efficient way of storing and accessing data. It has a tree-like structure with entries at the tree-leaves and nodes representing the printers, documents, persons, organizational units, groups of persons, or anything else which may represent a given entry or multiple entries at the tree.

The tree has a root with number of nodes. A root object is a parent object which has the number of child objects (daughters). Each child object has number of child objects (daughters). At the end of hierarchy, there is the last child object (daughter) called leaf. A child object or leaf object is identified by a URI (universal resource identifier) in a function (method). For example, assume that *A* is root object and it has child objects *B1*, *B2*, and *B3*. B1 has children *C1* and *C2* and assume that *C1* is a leaf object. URI for *C1* is *A.B1.C1*. A URI specifies the hierarchy position of a node under consideration with respect to the root. URI is used when referring to an object in a function or method. Relative URI means specification relative to a present node not necessarily from the root. For example, when using the object *B1*, a relative URI is *B1*. Another relative URI can be used for reference to *C1* from *B1*. The relative URI of *C1* is *B1.C1* in that case.

An object is accessed by its name and attributes. *Name* is referred to as DN (distinguished name) for the object. Each attribute is also named. For example, consider a contact object. Name of the contact is its DN. For example, a DN is Raj_Kamal. The e-mail_ID and tel_Num (telephone number) are the named attributes that are associated with a DN. Each attribute, e-mail_ID, and tel_Num, has a value, which may be specified as mandatory or optional.

LDAP is an open source networking protocol for accessing, modifying, and querying TCP/IP directory services. Its current version is LDAPv3. Lightweight means that the protocol does not depend on OS and system resources. Accessing, querying, or modifying an object in LDAP can involve a tree of directory entries, each of which consists of a set of named attributes with values. Some attributes are mandatory and some optional. Most services use LDAP as a simple starting point for their database organization. The basic operations are *bind*, *start TLS* (transport layer security) protocol (Section 12.2), *add* entry, *delete* an entry, *modify* DN, *abandon* to abort an earlier request, *search*, *compare*, *extend*, and *unbind*. *Bind* is an operation (function or method) to link an accessing object with the acccessed object. Without successful bind operation, the directory entries (objects) cannot be accessed, modified, or queried. Bind operation may not succeed, for example, in case of unauthenticated object trying to access an entry (object). The *start* TLS is an operation to enable use of TLS when accessing, modifying, or quering. *Add* is an operation to append or insert an entry at destined directory. *Delete* is an operation to delete an entry at destined directory. *Modify* is an operation to modify a named entry identified by a DN. *Abandon* is an operation to abort a previous request. For example, a request made earlier is *add*. The abandon operation will unroll the previous operation. *Search* and *compare* are the operations for searching an entry or comparing an entry with another. *Extend* is an operation to extend a tree-like object by specifying child objects of a leaf and the child objects will now be the leaves. *Unbind* is a delinking operation which inhibits accessing the directory entries by the object which was bound earlier. After an unbind operation, a bind operation is required to access the entries again.

10.6.4 OMA DM (Open Mobile Alliance Device Management)

OMA DM objects are most used standard in mobile device computing system. DM defines a description framework and has hierarchical structure in which there is a management object tree.

DM is based on SyncML Data Synchronization (SyncML DS) specifications (Section 9.3). In a hierarchical structure for the management objects, the OMA DM protocol (standard) provides for specifying how many times (how many children of a parent) an object node can occur in the hierarchy. For example, clientUserName (user name of client which server recognizes) and clientPW (password using which server authenticates a client before providing the service) are specified in OMA

DM as having zero or one occurrences. Another example of parent node of a management object for data synchronization is DSAcc (data synchronization account) (one or more occurrences). The child objects (nodes) and leaves (objects which are last in the hierarchy of objects for DSAcc) of the parent object DSAcc with specifications for their occurrences and functions (methods) required for their accessibility are mentioned below:

1. *Addr* (address) which has one occurrence, must exist, and be configured. The node Addr is accessible by Add and Get functions.
2. Nodes *AddrType* (address type), *PortNbr* (Port number), *Name*, *DB* (database), *clientUserName*, *clientPW*, *ToNapID* (default physical reference by relative URI for connectivity information stored elsewhere in the management tree), and *ServerID* (Data synchronization server ID or DSS ID) each having zero or one occurrence.
3. DB node is a named parent (root) of the database objects. For example, *ContactsDB* is a parent of database for the contacts. CTType is object at the leaf, which is used to define the supported media contents by the database. LDBURI (local database relative URI) and RDBURI (remote database relative URI) are the leaves of a DB root object.

Figure 10.12 show the SyncML DM stack. Bootstrapping is either by SyncML message based on DTD or by WAP provisioning by an SMS push. (Section 9.3 described Bootstrap and setup phases and packages 0–4.)

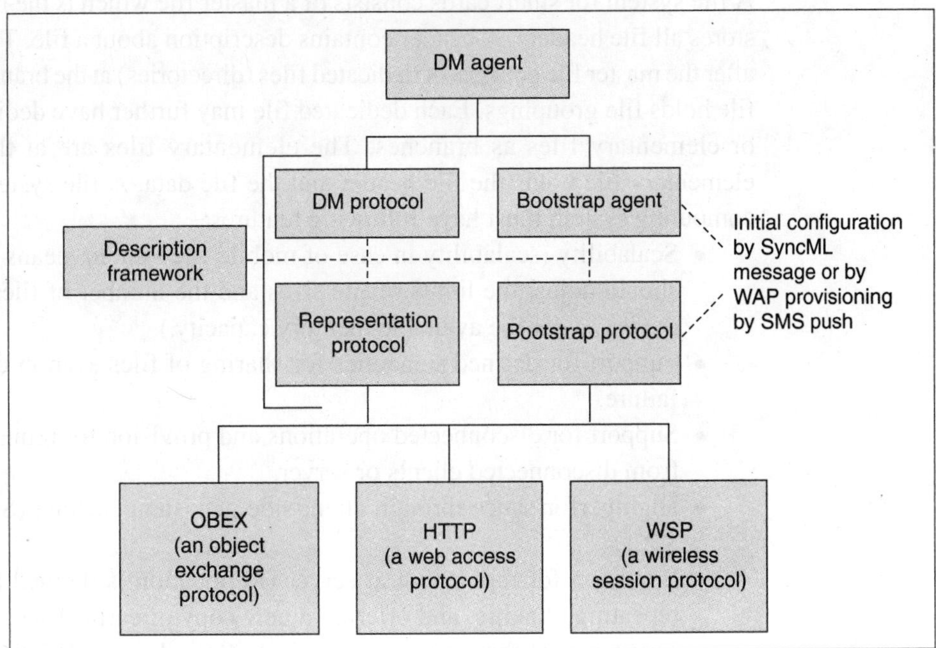

Fig. 10.12 SyncML DM stack

OMA DM is a one-way synchronization. The server initiates the setup phases, requests the client device for data, and updates data at the server as well as downloads data to the device.

10.7 Mobile File Systems

When a file system is designed for a mobile device, then it is called mobile file system. Mobile device may be a phone or smartcard. A file system is a basic middleware which glues applications to an OS. It can be defined as a method of organizing and storing files on a storage device at computer or a mobile device. The files may not be present on a single storage device. They may be distributed as in case of a distributed file system or present at different nodes in a network system. Some features of a file system are:

- Hierarchical organization, storage, modification, navigation, access, and retrieval of files
- Easy search and access of files
- Maintenance of physical location of files on a storage device (e.g., memory stick, or hard disk)
- The system functions use the data from the file server software in response to client requests
- Technology similar to that of databases

A file system for smart cards consists of a master file which is the root directory. It stores all file headers. A header contains description about a file. The second layer after the master file consists of dedicated files (directories) at the branches. Dedicated file holds file groupings. Each dedicated file may further have dedicated files and/or elementary files as branches. The elementary files are at third layer. The elementary file holds the file header and the file data. A file system for a mobile computing system must have following features:

- Scalability (scalability in case of mobile file system means that the system should adjust the limits of file sizes and the number of files in the storage device as per the available memory capacity.)
- Support for defined semantics for sharing of files even in case of network failure.
- Support for disconnected operations and provision for reintegration of data from disconnected clients or server.
- High performance through client-side persistent caching (Sections 7.2 and 8.2.1).
- Provision for replication at server (Replication is defined as a process of repeating, making, and offering a new copy of earlier ones. Replication by server means that server repeats and offers (broadcasts) the set of records using broadcast disk model).
- Security, access control, authentication, and encryption.

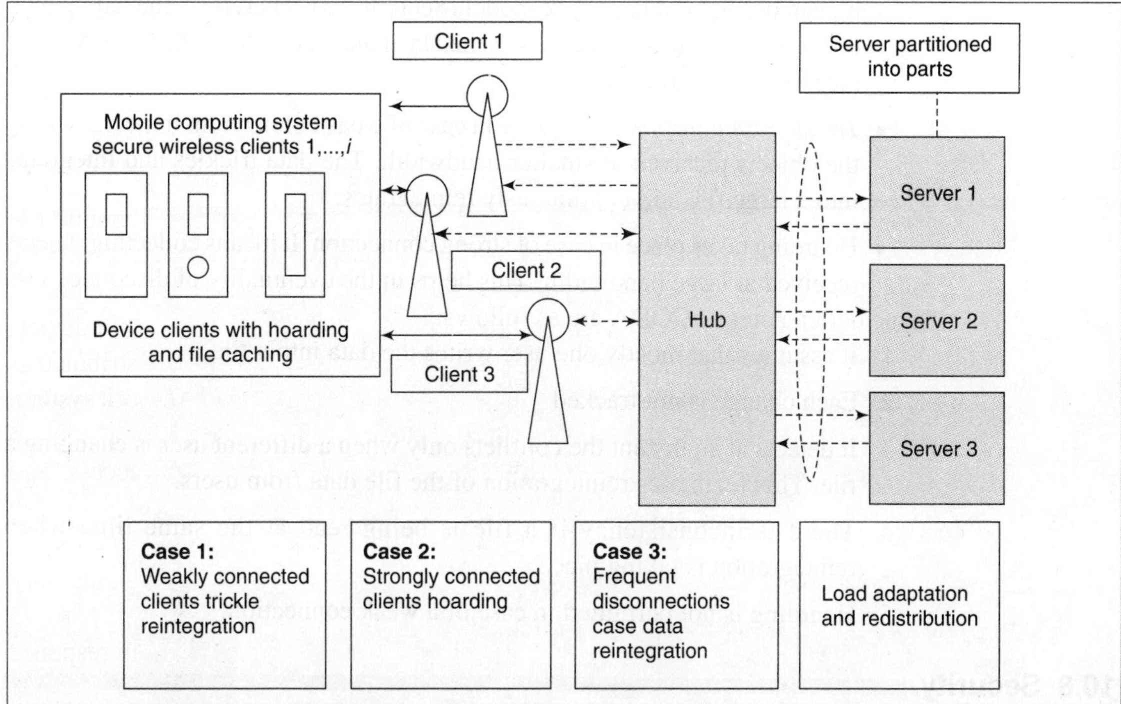

Fig. 10.13 CODA file system architecture

- Continuous operation even in case of partial failure of network connectivity (partial failure means disconnection between server and a few clients).
- Network which adapts to the bandwidth available at a given instant. An application performance improves if the file system adapts to the bandwidth variations.

CODA is a distributed file system developed at CMU (Carnegie Mellon University). It possesses all the features of a file system. Figure 10.13 shows CODA file system architecture. Files on a CODA server are organized by server partitioning. These partitions will contain files, which are grouped into volumes. The volume is a logical unit of files. A single server would have some hundreds of volumes. Files in each volume have a directory structure, that is, a tree-like structure. There is a root directory for the volume. The root has the branches. Similarly, root directory has subdirectories at the branches. Each branch has several further branches and the last branch is called leaf. The volume is much smaller than a partition but bigger than a single directory. Its size is chosen as per the manageable file data.

The three states of connection to the clients to which files are distributed by a server are as follows—(i) disconnection, (ii) weak (low bandwidth) connection, and (iii) strong (large bandwidth) connection.

- *Integration* takes place in case of disconnection. It means merging objects received from a connection at different instants. The server sends the data repeatedly in broadcast disk model (Section 8.3). Therefore, the data which could not be cached earlier gets cached later and integrates with the previously received file records.

- *Trickle reintegration* takes place in case of weak connection. It means adding the objects received at smaller bandwidth. The data trickles and integrates into a large file and continuously reintegrates.
- Hoarding takes place in case of strong connection. It means collecting objects received at large bandwidth. This helps in the eventuality of disconnection.

Some deficiencies of CODA are as follows:

1. It assumes that mostly one user writes the data into a file.
2. Each change is not tracked.
3. It detects at an instant the conflicts only when a different user is changing a file. This facilitates reintegration of the file data from users.
4. There is inconsistency if a file is being read at the same time when reintegration is taking place.
5. Hoarding is not permitted in case of a weak connection.

10.8 Security

Section 1.8 described security. It is important for maintaining privacy and for mobile e-business transactions. Table 10.1 describes some of the security problems in mobile and wireless computing systems and Table 10.2 gives the solutions for such problems.

Table 10.1 Security problems in mobile and wireless computing systems

Security Problems	*Description*
Confidentiality	Only destined user must be able to read data. (Encryption of the data before transmission and deciphering it at the user end is a method employed for ensuring confidentiality.)
Integrity	Data integrity needs to be maintained or else the user receives a manipulated message. System integrity needs to be maintained or else system can issue the message to wrong node.
Pre-keying	In order to decipher the encrypted messages, a key for deciphering is first exchanged between transmitter and receiver. If a private key is used, key exchanges over wireless systems increase the risk of key trapping.
Availability	There may be a *denial of service* attack. A source can block the availability of data at the user end. For example, the packets sent can be prevented from reaching the destination by some intermediate router misdirecting them due to the attack.

(*Contd*)

(*Contd*)

Non-repudiation	Non-repudiation means that a sender is unable to deny having sent a message or information. For example, if a user books an air ticket using a credit card and a mobile device, he is unable to deny this fact.
Resource constraint	Resource constraints of a mobile system include—(a) CPU runs at a slow speed compared to a conventional PC, (b) smaller memory availability, and (c) limited battery life. An attack form is exhausting of device-power by forcing it to transmit or receive data continuously or exhaustion of device-memory due to caching and hoarding irrelevant data from the attacker. Such an attack if occurs in between routers in the network, it seriously affects the whole network.
Power of detection	A mobile device may not detect the signals and therefore get data or message due to attack by jamming signals (A solution is Frequency hopping of the modulation signal which has high background noise.)
Interception	The interception of the signals (CDMA FHSS can be a solution to this problem.)
Replay	After studying the authentication requests and the client responses, attacker can replay the same sequence repeatedly.
Stealing of the subscribed service	Hijacking of user name and password by an attacker results in getting a service subscribed by another client.
Mobility risks	Changed location results in signals routing through paths, which cannot be relied upon.
Spoofing (impersonating address)	A node can impersonate an address in a mobile ad hoc network (Sections 1.5.3, 11.1, and 11.2). A common node to several paths can lead to choking of all routes.
Reconfiguration	An attack can be network configuration (e.g., manipulation of routing table). Network reconfiguration at different periods prevents such attacks.
Eavesdropping	Unsolicited messages from another source during a talk between two sources is called eavesdropping.
Traffic analysis	A security attack can occur by extracting information form the analysis of network traffic.

Table 10.2 Solutions for the security problems in mobile and wireless computing systems

Solution	*Description*
Direct signalling with restricted signal strengths	Using the directed signals with the signal strengths set such that these are just sufficient to reach and detected successfully to reduce security risks from other directions as well as sources at farther distances in the same direction.
Hardware technique	FHSS is an example in which hardware is used for reducing security risks.
Hash	The hash of a message is a set of bits obtained after applying the hash algorithm (or function). This set of bits is altered in case the data is modified during transmission. It checks data integrity.
MAC	MAC (Message authentication code) is a combination of hash and secret key.
Encryption	Public key and private key encryptions—DES, AES, and RSA cryptographic algorithms (Table 1.5).

(*Contd*)

(Contd)

SSL	Secure socket layer (SSL) is a protocol that runs between HTTP and TCP (application and transport layer protocols) for secure transactions between client and Web server. The protocol HTTP + SSL is called HTTPS and the website address now starts with www.https://. The sub layers of SSL are handshake and record protocols. Handshake protocol authenticates both client and server and allows both of them to negotiate the data security protocols to be used later. The record protocol transfers the payload. SSL supports (i) hash function MD5 (message-digest algorithm 5) and SHA (secure hash algorithm), (ii) digital signatures, (iii) RSA and DES , (iv) MAC (medium access control) on SHA and MD5, (v) data encryption algorithms—DES and triple DES, and other encryptions.
Checksum or Parity	Checksum and parity are the primitive methods to check message integrity.
IPSec	IPSec (Internet protocol for security) protocols (i) Authentication Header (AH) is a method for message integrity check. (ii) ESP (Encapsulating Security Payload) is a protocol for confidentiality. (iii) Internet Key Exchange (IKE) protocol to exchange the authentication keys with protection.
CHAP	CHAP (challenge handshake authentication protocol) is a method for authentication of point-to-point communication.
RADIUS	RADIUS (remote authentication dial in user service) is a service for sending the message that the client stands authenticated.
AAA	AAA (authentication, authorization, and auditing) is a strategy for security.

Keywords

Advertisement: An advertisement means sending messages to all. A user node then need not search for a message, service, or address. A new device, establishing connection with network, waits for listening to an advertisement.

Application server: An application server is a software which is executed on a server and serves the application-level logic of the business functions. It gets requests from the collaborating or independent mobile devices of an enterprise or from a distributed mobile computing system. It processes these requests and generates responses. The application server also handles presentation services to the devices or computing systems.

CODA: A mobile file system for facilitating reintegration of the data of a file from users in case of three states of connection—strong, weak, and disconnection. It is the system which has been found to be very effective when mostly one user writes the data into a file, changes are not required to be tracked, and conflict detection is only at the instant of a different user changing a file.

Connection gateway: Software to connect two application layers using two different protocols. It connects client application and application server.

Device management: It means configuring at initialisation (bootstrapping), monitoring current configuration, processing maintenance requests, and taking care of location and handover of each device. In pervasive computing environment, it means managing the infrastructure of a large number of networks at the same time. It also refers to updating, uploading, and downloading software for running applications.

Device support infrastructure: A solution, called Tivoli from IBM for providing gateway, device management agent and management support for ATMs, handheld devices, set-top boxes, and cable modems. *Description*: It specifies a service. A device sends description to the control point when registering its services. Each description has a URL which the control point provides to other devices requiring that service. The descriptions do not include classes and the point provides just the device-description.

File system: A method of organizing and storing files on a computer hard disk or CD or a mobile device memory.

Gateway: Software to enable adaptation or conversion of the codes and data of a source end on transmission to a destination end which is deploying different communication protocol, data formats, or application software.

Integrity: Maintaining integrity of data means provisioning of using or transmitting of messages such that the user does not get a manipulated message as well as only right users get the messages.

Jini: An open architecture network technology with core protocols for service discovery, join, and lookup. One special feature of this technology is that it can not only discover the description of device service(s) but can also get the classes to run these services.

Lookup service: A service to find and register a new device. Jini consists of service to lookup as well as provide the Java classes (program codes) to a client device. It just provides URLs and descriptions of other devices.

Mobile agent: An autonomous software which runs on a host with some data and dynamically moves to another host (which has other required data) as and when required.

Mobile file system: A file system with adaptability to available bandwidth, scalability, and support for defined semantics for sharing of files even in case of network failure. Its other features include high performance through client-side persistent caching support for disconnected operations, and provision for replication at server, and for reintegration of data from disconnected clients or server.

Multicast: A multicast means a client or server transmitting the messages using TCP/IP or any other protocol on a network of devices or nodes such that a group of destined nodes or devices can listen and receives the messages. In contrast, in a broadcast any node or device on the network can listen and receive the messages.

Portal: A Web portal is a website that provides application-specific programs and personalized capabilities like business transactions to client device(s) or computing systems or databases. It also provides a path to other content related to the portal personalization criteria and enables running of distributed or collaboration applications.

Proxy: Software which runs on behalf of a client application and has conversion, computational, and analyzing capabilities.

Residential gateway: Residential gateway is a gateway between a network of home devices and the outside world.

Security problem: A problem of breach of confidentiality or integrity, spoofing, and other attacks to a communication.

Self-administering network: A network with software for starting the operating system, allocating network and system access addresses, initiating accesses, establishing and terminating the connections, and making secure connections by the provisions by a system on its own without using a network administration software when connecting to a network.

Self-configuration network: A network with software for establishing and modifying the route information for the connections on its own.

Self-healing network: A network with the characteristics that the network can establish, make alternative route when a connection or enroute node breaks.

Service discovery: An adaptable middleware in a device (or a mobile computing system) that dynamically discovers services.

SLP (Service Location Protocol): A protocol to dynamically discover a service. Its stack consists of SLP header and SLP service URL which enables other devices to use the service of the client device. It advertises its service needs in multicast mode and does not use directory service.

Spoofing: A method by which a node can impersonate an address in a mobile ad hoc network.

STT (speech-to-text conversion): Speech is converted into a text message. This enables a mobile device to get commands directly from user and dial the desired number.

Tier: A stage where requests are made or where the responses are generated, or where the processing of requests or responses takes place.

Transcoding: Transcoding is an abbreviation for transformation-cum-coding. Transcoding means adaptation, conversions, and changes using software which renders the web responses, messages or records in formats and representations required and acceptable at a device. Similarly, the device requests are adapted, converted, and changed into the required formats acceptable at the server by transcoding software.

Transcoding gateway: Software to provide transcoding between data transfers between two nodes using different protocols, applications, and software.

TTS (text-to-speech conversion): A text message is converted into speech. This facilitates an automobile driver to get messages and alerts directly.

Unicast: A unicast means a node or server transmitting the messages using TCP/IP or any other protocol on a network of devices or nodes such that a single destined node or device can listen and receives the messages. In contrast, in a broadcast any node or device on network can listen and receive the messages.

UPnP (Universal Plug and Play): A Microsoft solution for service discovery, description, control, and eventing.

WebSphere MQe: An application server from IBM for Web applications on mobile devices with the feature of MQe (Messaging and Queuing Everyplace). Everyplace means in any mobile device including sensors, phones, and PDAs. The server includes IBM WebSphere Voice Server, text-to-speech (TTS) software, and IBM WebSphere RFID Premises Server.

Review Questions

1. Describe the functions of a mobile agent. Why does an agent move from tier to tier during an application? Distinguish between an agent and a server.
2. Show an application-server-based *N*-tier architecture ($N \geq 3$) with requests processed at the application server using backend database(s) and system. Describe the key role of application server in mobile computing system.
3. Explain IBM WebSphere Application Server for Web applications on mobile devices with the feature of MQe (Messaging and Queuing Everyplace).
4. Describe Oracle 9i Application Server.
5. Describe Sun Java System Web Server 6.
6. Explain the role of a gateway in connecting networks using different protocols. Describe a transcoding gateway and its applications in mobile computing systems.
7. Describe a residential gateway.
8. What are portal and portlets? How is a business-to-business (B2B) portal used?
9. Describe service discovery mechanisms used in Jini and UPnP.
10. Compare the use of unicasting, multicasting, and advertising in service discovery.
11. Describe OMA DM.
12. Explain the requirements of a mobile file system over the conventional one.
13. Describe CODA file system.
14. Describe security problems in mobile computing systems and networks. Also, suggest their solutions.

Objective Type Questions

Pick the correct or most appropriate statement among the choices given:

1. Mobile agent software—(i) migrates to a mobile device from the server and functions as an agent of the server to provide device connectivity and device information to the server (ii) displays mobility of its code and data from one computing system (host) to another (iii) does not have the ability to learn in order to adapt its code and data to a host computing system, to create its clone or extend, or dispose itself of at the host after its role is over (iv) is compatible to the host (v) continuously and autonomously processes requests and sends responses and alerts.

Which of these are true?
(a) (i), (ii), and (iv) are true. (b) Only (ii) and (iv) are true.
(c) Only (i) is true. (d) All are true except (i) and (iii).

2. Application server is a software (a) for an application which runs on a mobile device (b) which is executed on a server and serves the application-level-logic of the business functions, processes the requests and generates responses and also handles the presentation services to the devices or computing system (c) which runs a set of applications on a server for the mobile device (d) which is executed on a server and serves the application-level logic of the business functions only.
3. (a) (i) IBM WebSphere Application Server with specialized mobile Web-computing application server supports J2EE web applications and XML databases (ii) IBM Domino Application Server supports workgroup, e-mail applications, and handheld and Windows CE devices. (b) None of the servers, IBM WebSphere or IBM Domino, supports Windows CE devices. (c) None of the servers, IBM WebSphere or IBM Domino, supports J2EE. (d) None of the servers, IBM WebSphere or IBM Domino supports J2EE and e-mail applications.
4. Sun Web Server 6 (i) supports large business applications (ii) functions on different operating systems (iii) provides a platform for deployment of CGIs and PHPs (iv) provides a platform for deployment of ColdFusion (v) has a content engine for publishing Web-application response.
(a) (i), (ii) and (v) are true. (b) (i) to (iv) are true.
(c) All are true. (d) All are true except (iv).
5. Sun Web Server 6 engine (i) has a search engine (ii) has a content management engine (iii) has an HTTP engine (iv) creates content for Web, (v) does addition and storage using HTML (vi) renders JSP and ASP pages (vii) maintains integrity of client contents with server. Which of these are true?
(a) (ii) to (vi) are true. (b) (i) to (v) are true.
(c) All are true. (d) All are true except (iii) and (vii).
6. Orcale9i Application Server (a) has J2EE EJB container for applications and integrates business logic, management, security, e-business, and portal functions (b) does not have J2EE EJB container and portal functions (c) does not have e-business and portal functions (d) does not have J2EE EJB container.
7. Transcoding gateway software (i) connects two application layers using two different protocols (ii) converts formats of multimedia data from server into mobile TV, Internet TV, WAP phone, and Smartphone as per the client device (iii) filters, compresses, or decompresses (iv) can be configured to provide authoring and deployment (v) facilitates TTS (text-to-speech) conversion (vi) facilitates STT (speech-to-text) conversion (vii) converts XML style sheets to WML or HTML, video to image, image to hyperlink, and one image format to another and also does scaling, mapping, and colour adaptation. Which of these are true?
(a) (i), (v) and (vi) are true. (b) (ii) to (vii) are true.
(c) (v) to (vii) are true. (d) All are true except (ii) and (iii).
8. (a) A client device using SLP can dynamically discover a service by stacking. An SLP protocol stack consists of SLP header and SLP service URL (b) SLP deploys directory service (c) SLP is similar to lookup service database in Jini or a service catalogue (d) SLP provides Java classes (program codes) to a client device and need not provide description of the other devices.
9. (a) Tivoli DSI architecture has a Device Gateway which has a device management agent to connect devices at one end and the gateway at other end (b) Tivoli DSI Gateway need not include Tivoli Management Gateway (c) Tivoli DSI architecture proxy connects to device management server of the service provider (d) Tivoli DSI architecture is not used but OMA DM is used as a device management agent.
10. Lightweight Directory Access Protocol is (a) a networking protocol for access and querying of WAP directory services, (b) an open source networking protocol for accessing, modifying, and querying TCP/IP directory services, (c) an open source networking protocol for mobile devices, (d) for accessing, modifying, and querying XML databases.

11. Device Manager can be assigned functions of (i) configuring at initialisation (bootstrapping) (ii) monitoring current configuration (iii) processing maintenance requests (iv) location management (v) handover management (vi) updating, uploading, and downloading software for running applications (vii) managing the infrastructure of a large number of networks for each device at the same time in pervasive computing environment. Which of these are true?
(a) (i), (ii) and (vi) are true. (b) (i) and (iv) to (vi) are true.
(c) (i) to (vii) are true. (d) All are true except (vi).
12. An OMA DM (Open Mobile Alliance Device Management) object is a parent node of data synchronization account management object and has DM specification in SyncML. It has the nodes—Addr (address) (Addr is accessible by Add and Get functions), AddrType (address type), PortNbr (Port number), Name, DB (database), clientUserName, clientPW, ToNapID (default physical reference by relative URI for connectivity information stored elsewhere in the management tree), and ServerID (data synchronization server ID or DSS ID).
(a) Addr and all of these have one occurrence and must exist and be configured for all of these, (b) Addr has one occurrence and must exist and be configured, and all remaining ones have zero or one occurrence, (c) Addr has zero, one, or two occurrences as per the path and must exist and be configured for all of these and all others have one occurrence, (d) Addr, AddrType, and PortNum have one occurrence each.
13. (a) Advertisement means broadcasting device description (b) Advertisement means broadcasting device services and data (c) Advertisement in a service discovery protocol means sending messages to all so that a user node need not search by itself a message, service, or address (d) A new device establishing a connection to a network need not wait for listening to an advertisement in service discovery protocol.
14. (a) XML database file system facilitates reintegration of the data of a file from users in case of weak connection (b) CODA is not used for reintegration of files in case there are disconnections (c) CODA is a mobile file system which facilitates reintegration of the data of a file from users in case of three states of connection—strong, weak and disconnection (d) CODA system is not very effective when mostly one user writes data in a file, when there is no need to track each change, and when there is detection of conflict only at an instant when a different user is changing the file.
15. (a) FHSS is not an example in which hardware is used for reducing security risks (b) Eavesdropping is a result of password hacking (c) Spoofing is a method in which, in order to decipher the encrypted messages, a key for deciphering is first exchanged between transmitter and receiver and that key is hacked. (d) RADIUS (remote authentication dial in user service) is used as service for sending the information that the client stands authenticated by the service and secure socket layer (SSL) is used between HTTP and TCP (application and transport layer protocols) for secure transactions between client and Web server.
16. UPnP (Universal Plug and Play) (i) is for control and eventing by an external computing system (ii) is for service and description discovery by an external computing system (iii) uses XML to describe services (iv) uses XML to describe the capabilities of a device (v) uses TCP/IP as communication protocol between the device and server (vi) uses remote method invocations for communication between the device and server (vii) is a Microsoft solution (viii) is a Sun Java based solution for service discovery, description, control and eventing. Which of these are true?
(a) All are true except (i), (v), and (vii). (b) All are true except (vi) and (viii).
(c) All are true except (iii), (iv), and (vii). (d) Only (i), (ii), and (vi) are true.

Chapter 11

Mobile Ad-hoc and Sensor Networks

This chapter describes mobile ad-hoc networks (MANETs) and wireless sensor networks. It covers the following topics:

- Basic concept of ad-hoc networks
- MANETs properties
- Spectrum requirements of MANETs
- Learning how the routing of messages takes place using reactive and proactive protocols
- Security problems special to ad-hoc networks
- Applications of ad-hoc networks
- Ad-hoc network of wireless sensors and their applications

11.1 Introduction to Mobile Ad-hoc Network

A fixed infrastructure network has access-points, base stations, or gateways networked together using switches, hubs, or routers. The locations of these switches, hubs, or routers are fixed. A mobile device or wireless sensor has to be moved in the vicinity (connectivity range) of an access-point in case it has to connect to and access the network. A cellular network is an example of fixed infrastructure network.

An ad-hoc network is a network in which the locations of the switches, hubs, or routers can be mobile, the number of routers available at an instant can increase or decrease, and the available routing paths can change. The mobile devices or wireless sensors as well as the access-points have switches or routers. The routes available to the mobile devices or wireless sensors can thus change at any time and depend on presence and locations of other wireless devices in their vicinity (connectivity range). Figure 11.1 represents the fixed infrastructure and the mobile ad-hoc network (MANET) architectures, respectively.

In Fig. 11.1(a), each mobile device or sensor connects to an access-point, base station, or gateway with a switch, hub, or router *A*. A switch is used to provide connectivity between the two, a hub functions as a central switching exchange, the

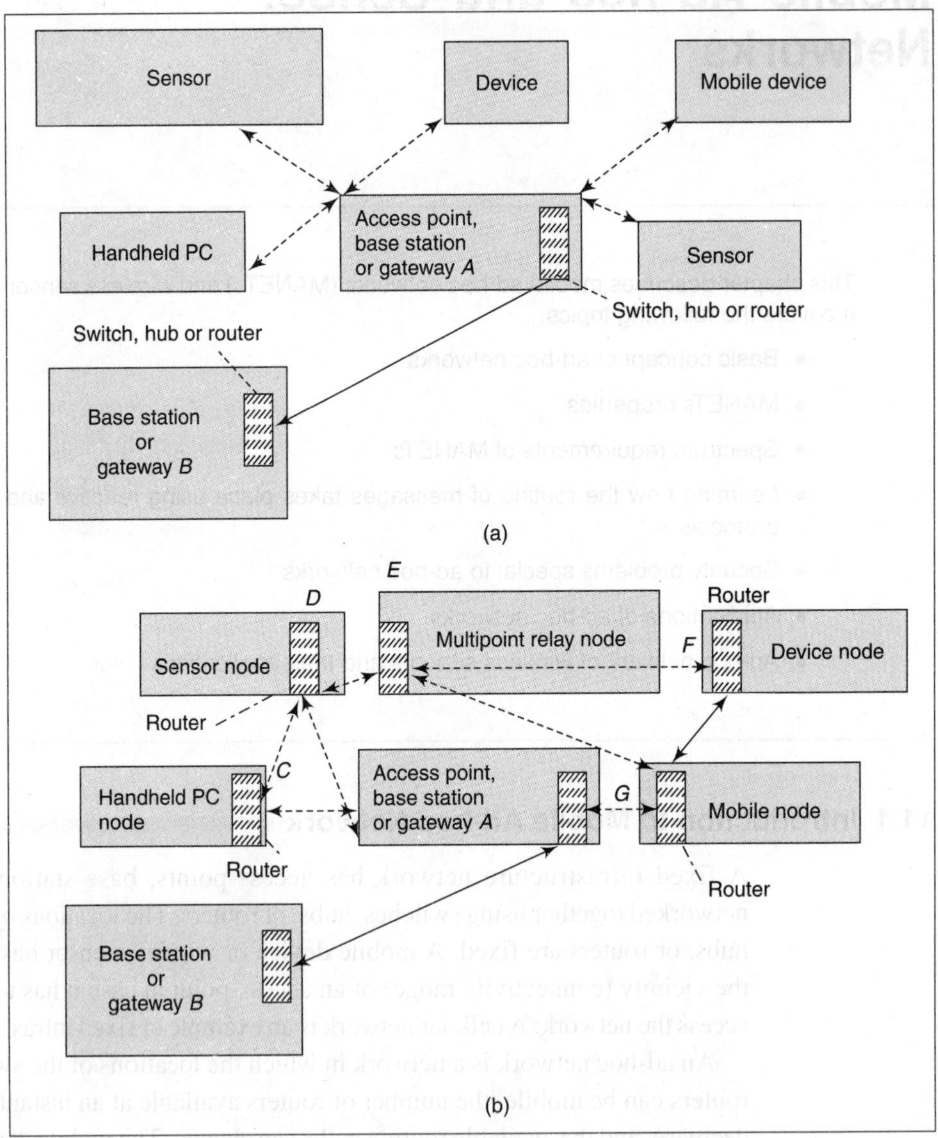

Fig. 11.1 (a) Fixed infrastructure network architecture (b) Mobile ad-hoc network architecture

routers provide two or more paths to route a message or packet so that the available path can be used at an instant. They function as the nodes of the network. A mobile device or sensor has to acquire an access-point or node of the fixed infrastructure network before being able to connect to another.

Consider the case of cellular phones in a fixed infrastructure network. The connectivity between two cellular phones is ensured through GSM public land mobile PSTN, ISDN, or PSPDN networks as shown in Fig. 3.1. Figure 3.2 demonstrated that a three-subsystems architecture consisting of radio subsystem (RSS), network subsystem (NSS), and operational subsystem (OSS) is used in a GSM system. Each cellular phone has to first connect to a RSS of GSM public land mobile network.

The problem with fixed infrastructure network is that if a wireless sensor or mobile device moves out of the range of access-point, base station, or gateway having the switch or router, it gets disconnected from the network and thus unable to communicate through the network even though there may be another wireless device in its vicinity connected to network. Moreover, the fixed infrastructure network is not usable in operations like disaster relief.

MANET is a self-configuration wireless ad-hoc network of mobile nodes. Each node has a router or a switch connected by the wireless connection. The union of connections is in an arbitrary topology. Network can function independently or connect to Internet IPv4 or IPv6. The MANET organization depends upon the location of the nodes, their connectivity, their service discovery capability, and their ability to search and route messages using nearest node or nearby nodes.

Figure 11.1(b) demonstrates the ad-hoc network formed by the nodes *A*, *C*, *D*, *E*, *F*, and *G*. It shows that each mobile device or sensor functions as a node with a switch or router. An important characteristic of ad-hoc network architecture is that its organization can change due to movement of a device or sensor. In other words, the ad-hoc networks are self-organizing. The following points illustrate how MANETs are established and how they reorganize themselves.

- Consider Fig. 11.1(b) once again. The network organization will change if *D* and *E* move away from each other such that they reach out of the range of wireless coverage. Two new ad-hoc networks will then be formed by (i) *A*, *C*, and *D* and (ii) *A*, *G*, *F*, and *E*. The devices on two networks can still connect to each other through the common node *A*.
- Consider a Bluetooth-enabled mobile device, a Bluetooth-enabled computer, and Internet with WiFi connection at home. Assume that there is a Bluetooth-enabled computer connected to TCP/IP Internet and also to Jini client printer at office. When the user carrying the device moves from office to home and handheld PDA mobile device reaches near the home computer, an ad-hoc network is established between the mobile device at home and printer at

office through intermediate nodes—WiFi, Internet, and office computer. Using service discovery, the services of Jini, TCP/IP Bluetooth, and WiFi are discovered and a MANET is established. When the user of same mobile device goes to an airport with WiFi connectivity, a MANET is again established with the office printer.

- Consider a library with smart labels on each book. The library also has a set of wireless access-points and wireless Internet connectivity to a central server computer. In order to understand how a MANET is established, recapitulate Section 2.5.2. The smart labels are networked together to a central reading server through wireless access-points, Internet, and a centralized server. Cluster of labels form a MANET similar to a LAN network and the MANET connects to central server through wireless access-point and Internet. When a labeled book is removed from the shelf, the smart label on the neighbouring book discovers service through another neighbouring label and MANET is reorganized and establishes connection to central server is established. Smart labels are sensors of radiation from wireless access-points. Therefore, the MANET of smart labels is also called wireless sensors ad-hoc network.

11.2 MANET

This section deals with distinct properties of a MANET, which provide a seamless interacting and ubiquitous mobile computing environment. Each MANET node has much smaller frequency spectrum requirements than that for a node in a fixed infrastructure network.

The following subsections describe various properties, spectrum requirements, and applications of MANETs. Special routing strategies and algorithms needed for MANETs are also discussed. Security problems of MANETs which are more severe than those of fixed infrastructure networks are also explained.

Similar to MANETs, are the vehicular ad-hoc networks (VANETs) which are used for communication among vehicles and between vehicles and roadside equipment.

11.2.1 Properties

One of the important characteristics of a MANET node is the neighbour discovery for data reception and transmission. It also possesses data routing abilities, that is, data can be routed from a source node to a neighbouring node. It has flexible network architecture and variable routing paths to provide communication in case of the limited wireless connectivity range and resource constraints. Table 11.1 lists the properties of MANETs.

Table 11.1 Properties of MANETs

Property	*Description*
Flexibility	MANET enables fast establishment of networks. When a new network is to be established, the only requirement is to provide a new set of nodes with limited wireless communication range. A node has limited capability, that is, it can connect only to the nodes which are nearby. Hence it consumes limited power.
Direct communication through nearby node and neighbour discovery	A MANET node has the ability to discover a neighbouring node and service. Using a service discovery protocol, a node discovers the service of a nearby node and communicates to a remote node in the MANET.
Peer-to-peer connectivity	MANET nodes have peer-to-peer connectivity among themselves.
Computations decentralization	MANET nodes have independent computational, switching (or routing), and communication capabilities.
Limited wireless connectivity range	MANETs require that a node should move in the vicinity of at least one nearby node within the wireless communication range, else the node should be provided with the access-point of wired communication. In other words, the wireless connectivity range in MANETs includes only nearest node connectivity.
Weak connectivity and remote server latency	Unreliable links to base station or gateway. The failure of an intermediate node results in greater latency in communicating with the remote server.
Resource constraints	Limited bandwidth available between two intermediate nodes (Section 11.2.2) becomes a constraint for the MANET. The node may have limited power and thus computations need to be energy-efficient.
No access-point requirement	There is no access-point requirement in MANET. Only selected access points (e.g., *A* and *B* in Fig. 11.1(b)) are provided for connection to other networks or other MANETs.
Requirement to solve exposed or hidden terminal problem	There is a requirement of a mechanism to solve exposed or hidden terminal problem (Section 4.1.3).
Diversity	MANET nodes can be the iPods, Palm handheld computers, Smartphones, PCs, smart labels, smart sensors, and automobile-embedded systems.
Protocol diversity	MANET nodes can use different protocols, for example, IrDA, Bluetooth, ZigBee, 802.11, GSM, and TCP/IP.
Data caching, saving, and aggregation	MANET node performs data caching, saving, and aggregation.
Seamless interaction and ubiquitous mobile computing	MANET mobile device nodes interact seamlessly when they move with the nearby wireless nodes, sensor nodes, and embedded devices in automobiles so that the seamless connectivity is maintained between the devices.

11.2.2 Spectrum

An access-point-based fixed infrastructure network (Fig. 11.1(a)) connects as a centralized server to a large number of devices. The bandwidth requirement in that case becomes too high. Suppose there are N devices which can communicate simultaneously using FDMA in duplex transmission, then the required bandwidth will be $2 \times N \times f_{\text{bw0}}$, where f_{bw0} is the bandwidth allotted to one device for sending a packet to its access-point. Using TDMA and SDMA, the bandwidth (spectrum) requirement is reduced (recapitulate Section 3.2).

In a MANET, a node itself is a router for all the packets coming from or going to the other nodes. For example, node *D* at a given instant in Fig. 11.1(b) can get incoming packets from *E*, *F*, *G*, and *A* and can send packets to *C* and *A* or vice versa. Nodes are themselves mobile. Therefore, bandwidth available to any node at any instant is variable.

However, MANET enables spectrum reuse. Each wireless link provides a limited bandwidth. MANET communication is multi-hop. When node *D* transmits to *G*, it is through the three hops—(i) *D–E*, (ii) *E–F*, and (iii) *F–G*. Assume that each node has low and adaptable transmission power which is optimized to have signal strength just sufficient to carry the signal up to single hop. Hops can therefore occur simultaneously using the same frequency band. In other words, there is spatial reuse of bandwidth. Also, the bandwidth depends on surrounding environment. Following example clarifies this fact.

Example 11.1

(a) List the reusable spectrum paths in MANET shown in Fig. 11.1(b).

(b) Which node will need higher bandwidth than the other and why? How can the bandwidth requirement be reduced?

Solution:

(a) *D–E*, *F–G*, and *C–A* reuse the bandwidth.

(b) Assume FDMA mode access by a node. The bandwidth of each node depends on the number of next hop neighbours. The node with higher number of next hop neighbours will require higher bandwidth in FDMA. The nodes *D*, *E*, and *G* shall need higher bandwidth compared to *C* and *F*. Bandwidth required will be $2\times 3\times f_{\text{bw0}}$ when node *D*, *E*, or *G* can transmit in full duplex mode to all the nodes and will be $2\times 2\times f_{\text{bw0}}$ for *C* or *F*. When each node path and each direction hop is scheduled to operate at different instants, as in TDMA, then the bandwidth required will be just f_{bw0}.

11.2.3 Applications

MANET has many applications. It can be used for synchronizing the contents at one device with those of another. Ability to discover neighbouring service in a mobile environment is used for distributing the contents over a network. It can be used for multicasting and broadcasting using multicast tree network topology. It can be configured in mesh topology. It is used for images acquisition, processing, and distribution over the network. It is used for provisioning of seamless interaction and ubiquitous mobile computing environment to a mobile device. Some examples of these applications are discussed below.

11.2.3.1 Content Distribution and Synchronization

Assume that in an enterprise, there are a number of Bluetooth-enabled mobile handheld devices, PCs, laptops, and WiFi access-points. MANET is used for content-distribution, PIM, other information dissemination, information fusion, and file sharing in the enterprise.

11.2.3.2 Multicast Network

MANET nodes in multicast tree topology disseminate data packets and form a multicasting network. Clusters of the nodes are used to give a multicast tree topology in a MANET. Section 11.2.4.4 will describe the CGSR protocol for clustering.

11.2.3.3 Mesh Networking and Mobile Service Provider Network

Mesh-based mobile networks offer highly dynamic autonomous topology segments for the robust IP-compliant data services within the mobile wireless communication networks and inexpensive alternatives or improvement to infrastructure-based cellular CDMA or GSM mobile service provider. Figure 11.2 shows a mesh network.

Protocol for unified multicasting through announcements (PUMA) is a protocol that builds a mesh that connects MANET nodes with each other. A multicast tree network differs from mesh as it provides only a single path between a sender and a receiver. On the other hand mesh provides multiple paths between sender and receiver nodes. As mentioned before, MANETs have nodes which may be mobile and the wireless links are error-prone. As a result certain packets may not be delivered to receivers through a multicast tree topology. Mesh-based protocols like PUMA which send packets from senders to receivers through multiple paths may thus have a greater packet delivery ratio. However, in benign conditions (low mobility and traffic load which leads to lesser collisions), sending packets is not needed and is a waste. In such situations PUMA is able to reduce the redundancy. In other words, PUMA adapts the amount of redundancy in the network depending on need.

11.2.3.4 Image Acquisition, Processing, and Distribution using MANET

Consider that there are a number of imaging devices forming a MANET—low cost digital still camera with a wireless network interface, Wireless WebCam, mobile device connected to a digital still camera, mobile phones, and pocket PCs equipped with an image acquisition sensor. Six applications to which such MANETs have been applied are as follows:

- Remote viewfinder by security personnel in an office
- Remote processing on a computer for a video stream from wireless WebCam and other devices

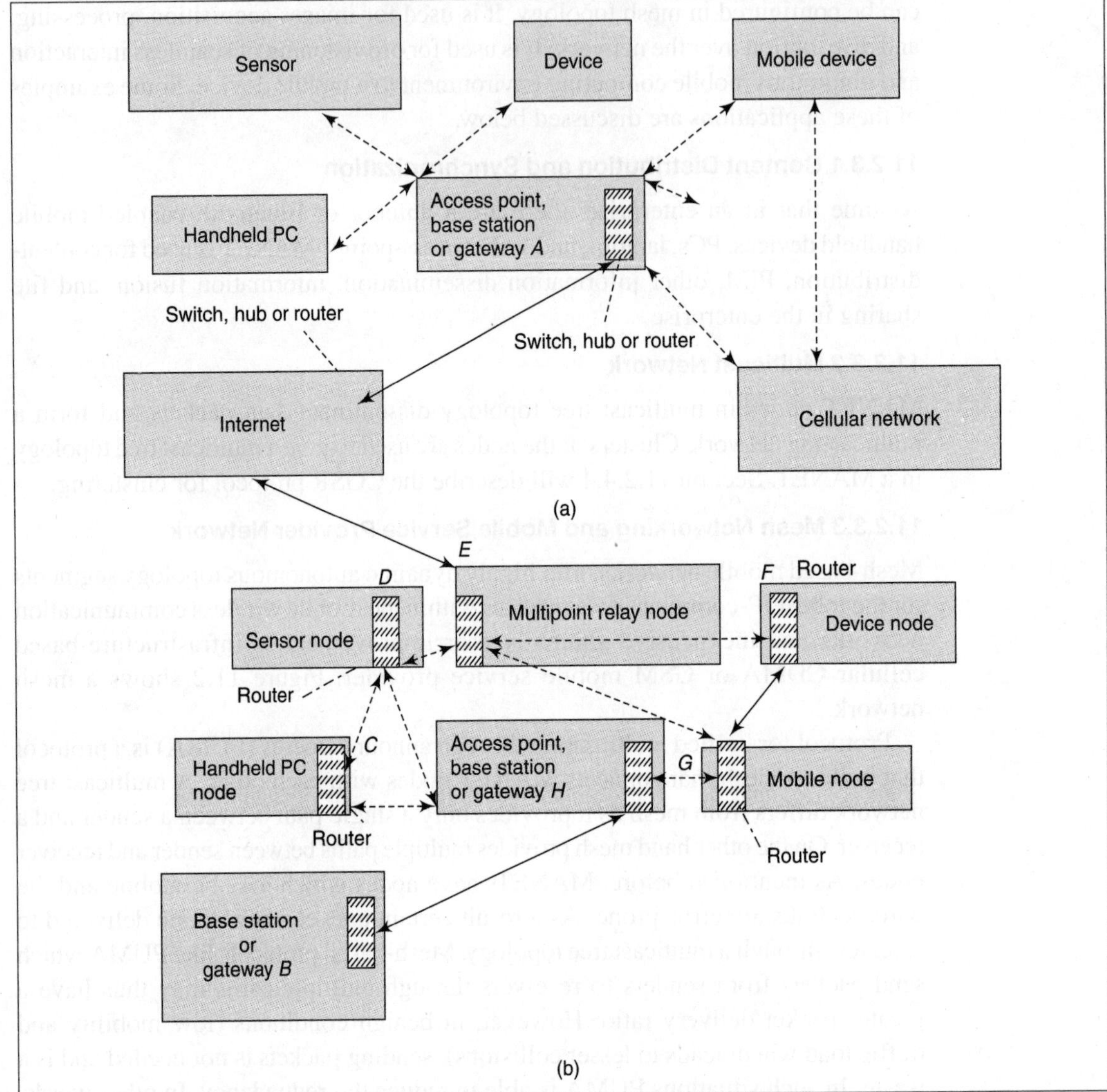

Fig. 11.2 Mesh network formed by interconnected MANETs and fixed networks

- Image file transfer
- Messaging and data transmission to remote devices using 802.11b (Section 1.1.5)
- Remote controlling

11.2.3.5 IPv6 Integration and Wireless Sensor Networks

IPv6 is a new generation Internet and is used for Internet radio and real time video over the Internet. IPv6 can be integrated with MANET and wireless sensor networks.

CDMA handsets and Apple iPhone are the examples of integration of the mobile phones with IPv4 and IPv6. Classical Internet IPv4 addresses are made up of 32 bits ($2^{32} = 4.3 \times 10^9$ addresses). New generation Internet IPv6 addresses are of 128 bits ($2^{128} = 3.4 \times 10^{38}$ addresses). MANET consists of mobile devices as well as wireless sensor nodes. Therefore, integration of IPv6 with MANET enables assignment of IPv6 addresses to each sensor or device node. IPv6 deploys packet encryption and source authentication. It enables real-time traffic, peer-to-peer applications, and dissemination by push (Section 8.2.1). Flow label defined in IPv6 provides the granular QoS support for multimedia real-time applications.

Pervasive computing devices, MANET, and wireless sensors need large number of addresses and the above features. IPv6 integrated devices offer new applications of mobile Internet devices. Integration of IPv6 with IPv4 and WLAN devices offers innovative applications of the MANETs and sensor networks.

11.2.4 Routing and Routing Algorithms

The ad-hoc networks are deployed for routing, target detection, and service discovery. Following are the routing protocols which can be used for routing by MANET nodes.

11.2.4.1 Dynamic Source Routing Protocol

The dynamic source routing (DSR) protocol deploys *source routing* which means that each data packet includes the routing-node addresses also. It is a reactive protocol, that is, it reacts to the changes and dynamically maintains only the routing addresses from source to destination, which are the active paths to a destination at a given instant. It performs unicast routing which means routing packets to a single destined address.

In addition, each node *caches* the specified route to destination during source routing of a packet through that node. This enables a node to provide route-specification when a packet source routes from that node. Each node *deletes* the specified route to destination during routing of error packet in reverse path to the source in case of observing a disconnection during forward path to destination. The deletion of link shown in a table or cache is called link reversal.

Let us first understand two phases of the protocol in order to understand source routing header, caching, and reversal processes.

Phase 1 in DSR Protocol Source node initiates a route discovery process. It broadcasts the packets, each with a header. It then expects return of acknowledgement from each destination. The packets are called route request (RREQ) packets. DSR uses flooding (sends multiple RREQs).

A header for each route request packet has the unique request number and source and destination addresses. This enables identification of request at each intermediate node in the request and acknowledged packet(s).

When the process starts, initially only the source address is given in the header. When the packet reaches a neighbour, that is, any intermediate node, the node adds its own address in the header if it is able to successfully send the packet to its next neighbour. When the packet reaches the destined address, its header therefore has all addresses of the nodes in the path.

Example 11.2 Consider the MANET shown in Fig. 11.1(b). Assume that node *D* is a source and *G* is a destination and the path *D–E–F–G* is not known. In such a case the path cannot be placed in the header. Explain the route discovery process using an RREQ from a node *D* in the network shown in Fig. 11.1(b).

Solution:

1. Source header from *D* puts the sequence number *q* and source address *D* in the packet header and sends the packet to its next neighbour. When the packet reaches *E*, its header is (*q*, *D*). Assume that no route error packet bounced back from neighbour *E*.
2. The packet is then transmitted to *F*. When the packet reaches *F*, its header is (*q*, *D*, *E*). Assume that no route error packet bounced back from neighbour *F*.
3. The packet is transmitted from *F* to *G*. Assume that no route error packet bounced back from neighbour *G*. When the packet reaches *G*, its header is (*q*, *D*, *E*, *F*).

When packet reaches the destination, a route reply (RREP) for the sequence used in RREQ is generated. On return path, the route cache builds up at each intermediate node for deployment at a later instant of phase 2.

Example 11.3 Assume that the packet reaches the destination *G* in Example 11.2 for the MANET shown in Fig. 11.1(b). Show how the route cache builds up using route reply packet for sequence *q* to enable source routing to *G* without employing a new discovery process later and by just using route caches at *D*, *E*, *F*, and *G*.

Solution:

1. Node *G* caches the routes for *F*, *E*, and *D* as *G–F*, *G–F–E*, and *G–F–E–D*, respectively, in route cache table. RREP header from *G* has the packet header = (*q*, *G*, *F*, *E*, *D*) when it reaches *F*.
2. Node *F* caches from RREP the routes for *E* and *D* as *F–E* and *F–E–D*, respectively, in route cache table. RREP header from *F* has the packet header (*q*, *F*, *E*, *D*) when it reaches *E*.
3. Node *E* caches from RREP the routes for *D* as *E–D* in route cache table. RREP header from *E* has the packet header (*q*, *E*, *D*) when it reaches *D*.
4. Node *D* caches from RREP the routes for *E*, *F*, and *G* as *D–E*, *D–E–F* and for *D–E–F–G*, respectively.

Phase 2 in DSR Protocol When any source node desires to send a message, it first looks at its route cache. If the required route is available in cache, the source node puts all the addresses of the nodes for the path to destination in the header.

Example 11.4 Show source routing addresses in DSR assuming that there is a message from a MANET node *D* in the network shown in Fig. 11.1(b).

Solution: Table 11.2 gives a DSR routing table for source routing addresses in cache at *D*.

Table 11.2 Source routing addresses in cache with node *D* as source

Node	*Destination*	*Cached path*	*Destination*	*Cached path*
D	*A*	*D—C—A*	*B*	*D—C—A—B*
D	*F*	*D—E—F*	*G*	*D—E—F—G*

Note: The path specified at a source is dynamically modified when an alternative path to a destination has to be followed. The alternative path is cached when there is route error packet generated in the reply.

When a disconnection occurs during transmission from a node to its neighbour, a route error packet is generated and sent along the reverse path to the source. All route addresses containing the broken link addresses (as shown by route error packet) are deleted from the caches. When an erroneous packet reaches back to source, new address discovery process is initiated for the given destination.

Example 11.5 Assume that a packet for destination *G* did not reach *F* when the source *D* sends a data packet using route cache at *G* in Example 11.2 for the MANET shown in Fig. 11.1(b). Show how the route error is generated from *E* on noticing the disconnection during source routing *D–E–F–G* in phase 2.

Solution:

1. Node *E* deletes the paths *E–F* and *E–F–G* from route cache. Links *E–F* and *E–F–G* are therefore reversed.
2. Node *E* sends a route error packet (RERR) to *D* through the reverse path; RERR packet header has sequence [(*e1*, *E*, *D*), *F*]. The node *D* route cache table deletes entries *D–E–F* and *D–E–F–G*, respectively. Links *D–E–F* and *D–E–F–G* are therefore reversed in sequence *e1*.

11.2.4.2 Ad-hoc On-demand Distance Vector Routing Protocol

The ad-hoc on-demand distance vector routing protocol (AODV) is a reactive protocol. It means that it reacts to the changes and maintains only the active routes in the caches or tables for a pre-specified expiration time. These routes are found and are expected to be available at a given instant. It also performs unicast routing. Distance vector means a set of distant nodes, which defines the path to destination. For example, *D–E–F–G* is a distance vector for source–destination pair *D* and *G*. In AODV, a distance vector is provided on demand during forwarding of a packet to destination by a node in the path and not by the route cache providing path through the header in the source data packet (recapitulate phase 2 in DSR).

Every node keeps a next-hop routing table, which contains the destinations to which it currently has a route. A routing table entry expires if it has not been used or reactivated for a pre-specified expiration time. AODV adopts the destination sequence number technique.

AODV does not deploy flooding (multiple RREQs). It stores the next hop routing information of the active routes in the routing caches (tables) at each node. Therefore, it has small header size and thus reduces the network traffic overhead.

Phase 1 in AODV Protocol The next hop routing table is generated as follows: A node uses hello messages to notify its existence to its neighbours. Therefore, the link status to the next hop in an active route is continuously monitored. When any node discovers a link disconnection, it broadcasts a route error (RERR) packet to its neighbours, who in turn propagate the RERR packet towards those nodes whose routes may be affected by the disconnected link. Then, the affected source can be informed.

Example 11.6 Consider the MANET shown in Fig. 11.1(b). Assume that it deploys AODV routing protocol for discovering the distance vector *D–E–F–G*. How do *D* and other nodes issue the hello messages on demand in the network to find the path to *G*? Let the period for sending repeat hello messages be *T*, where *T* is the time to live (TTL) for a link address at a node.

Solution:

1. Node *D* periodically sends three hello messages to three nodes *E*, *C*, and *A* at the next hops.
2. Node *E* periodically sends two hello messages to two nodes *F* and *G* at the next hops.
3. Node *F* periodically sends one hello message to *G* at the next hop.

Phase 2 in AODV Protocol A source node initiates a route discovery process if no route is available in the routing table. It broadcasts the demand through the RREQ packets. Each RREQ has an ID and the addresses of the source and destination in its header. It expects return acknowledgement from destination. A node identifies the last observed sequence number of the destination from the ID. Each RREQ starts with a small TTL value. If the destination is not found during the TTL, the TTL is increased in subsequent RREQ packets. The node also identifies sequence number of source node. Sequence numbers ensure loop-free and up-to-date routes. Loop-free means free from bouncing of a packet to a node after intermediate hops. Each node rejects the RREQ which it had observed before. This reduces flooding which means it reduces too many RREQs present in the network at a given instant.

Route tables keep entries for a specified period and each node maintains a cache. The cache saves the received RREQs. Only the RREQ of highest sequence numbers are accepted and previous ones are discarded. The cache also saves the return path for each RREQ source. When a node having a route to the destination or the destined node receives the RREQ, it checks the destination sequence number it currently knows and the one specified in the RREQ. A RREP packet is created and forwarded back to the source only if the destination sequence number is equal to or greater than the one specified in RREQ. It guarantees the updation of routing cache information. AODV uses only symmetric links and a RREP follows the reverse path of the respective RREQ. Each intermediate link on receiving the RREP packet updates the table after discarding the previous entry and forwards back RREP to neighbour in the path to the source.

11.2.4.3 Temporally Ordered Routing Algorithm

The temporally ordered routing algorithm (TORA) is a reactive protocol. It means that it reacts to the changes and link reversals. It is employed for highly dynamic MANETs and provides an improved partial link reversal process. This protocol discovers the network portions showing the link reversal(s). It has a feature that it stops the non-productive link reversals in a given portion of the network.

TORA assumes addresses of the routers in the path and of source and destination for one set of input route. Each node provides only one set of subsequent route addresses. In other words, network shows a directed acrylic graph topology. (Acrylic graph topology means one input path and one outgoing path.) TORA possesses network capacity such that many nodes can send packets to a given destination. It guarantees loop-free routes and supports multicasting (from one source to multiple destinations).

Unlike AODV, but similar to DSR, TORA supports unidirectional links and also provides multiple routing paths. It does not exchange hello messages periodically to listen to disconnected links as done by AODV. Phases 1, 2, and 3 in TORA are route creation, route maintenance, and productive (useful) link reversal(s) (vector discarding).

An *i*th each node has a metric h_i called height which is a function of three parameters defining a reference level—(i) t^i_f is the logical time of a link failure (longer is the time, greater is the height), (ii) r^i_L bit is the reflection of a packet (availability of path used earlier from *i*th node) and (iii) ID_i.

A link direction is from h_i to h_j where h_j defines a new reference level. Each node also has an ordering parameter for propagation of packets. Destination node height is taken as 0.

11.2.4.4 Cluster-head Gateway Switch Routing

The cluster-head gateway switch routing (CGSR) is a hierarchical routing protocol. It is a proactive protocol. When a source routes the packets to destination, the routing tables are already available at the nodes. A cluster higher in hierarchy sends the packets to the cluster lower in hierarchy. Each cluster can have several daughters and forms a tree-like structure in CGSR.

CGSR forms a cluster structure. The nodes aggregate into clusters using an appropriate algorithm. The different clusters can be assigned to different band of frequencies in FDMA (Section 4.1.5) or different spreading CDMA codes (Section 4.3). The algorithm defines a cluster-head, the node used for connection to other clusters. It also defines a gateway node which provides switching (communication) between two or more cluster-heads. There will thus be three types of nodes—(i) internal nodes in a cluster which transmit and receive the messages and packets through a cluster-head, (ii) cluster-head in each cluster such that there is a cluster-head which dynamically schedules the route paths. It controls a group of ad-hoc hosts, monitors broadcasting within the cluster, and forwards the messages to another cluster-head, and (iii) gateway node to carry out transmission and reception of messages and packets between cluster-heads of two clusters.

The cluster structure leads to a higher performance of the routing protocol as compared to other protocols because it provides gateway switch-type traffic redirections and clusters provide an effective membership of nodes for connectivity. Phases 1, 2, and 3 of CGSR are routing path discovery and caching, maintaining update, and distribution, respectively. The basic processes of CGSR are cluster definitions and selection of clusters for routing. Algorithms are used for both the processes.

Phase 1: Routing Path Discovery and Caching Each node maintains a routing table that determines the next hop to reach other clusters. An algorithm that can be used in this phase is the destination-sequenced distance vector (DSDV) algorithm. It means that it finds the next hop to the distant destinations. With the help of the algorithm, cluster member table is created using the sequence numbers in RREQ packets which are broadcast periodically.

Phase 2: Maintaining Routing Information A clustering algorithm used for maintenance of routing information is least cluster change (LCC). A cluster-head changes only when one of the nodes moves out of the range of all the cluster-heads in the cluster or two cluster-heads come within one cluster. A token-based algorithm can be used to control the transmission within a cluster. A cluster-head should get more chance to transmit as it is in charge of broadcasting within the cluster and of forwarding packets between nodes. The cluster-heads are given priority such that route path utilization is maximized and packets are transmitted with minimum delay. Another algorithm is the priority token scheduling (PTS). It gives higher priority to the neighbouring nodes from which a packet was received recently.

Each node dynamically maintains a cluster member table. An algorithm is such that a node will periodically change the entries in its cluster member table when a new sequence number entry is received from the neighbour after successful RREP. The table rows map each node to its own cluster-head. A node transmits its cluster member table periodically. After receiving broadcasts from other nodes, a node uses the DSDV algorithm to update its own cluster member table.

Example 11.7 Consider a MANET shown in Fig. 11.1(b). Show how hierarchical clustering can take place.

Solution:

1. Nodes *D*, *C*, and *A* form a cluster *A*. Nodes *E*, *F* and *G* form another cluster *B*.
2. When CGSR is deployed and nodes *D* and *G* are source and destination, respectively, then cluster *A* is higher and *B* is lower in hierarchy.
3. Packets route along *D–G* by switching (establishing connection) between the two gateways containing cluster *A* and cluster *B*. In present MANET (Fig. 11.1(b)), the gateway connects only one cluster at each end.

11.2.4.5 Flat Routing Table Driven Protocol

Routing cache table used earlier was a routing table which builds by caching the RERP and RERR packets. Flat routing table driven protocol is a proactive protocol. This means that routing table will be available in advance at a node. In the proactive protocol, the routing table is available at each node shows available routes from itself to target destination node, is dynamically modified to show available routes, and has rows for all destined targets irrespective of whether they will eventually be needed or not. The packet does not specify route in the header and the routes need not be discovered after the demand is raised.

Example 11.8 Show a flat routing table for a MANET node *D* in the network shown in Fig. 11.1(b).

Solution: Table 11.3 gives a flat routing table:

Table 11.3 Routing table for node *D*

Node	*Destination*	*Path to next hop*[1]	*Status*[2]	*Destination*	*Path to next hop*	*Status*
D	*A, B*	*D–C*	*Y*	*A, B*	*D–A*	*Y*
D	*C*	*D–C*	*Y*	*E*	*D–E*	*Y*
D	*F*	*D–E*	*Y*	*G*	*D–E*	*Y*

[1]*The path to the neighbour for next hop for onward transmission to destination.*

[2]*At some instant, a status given as Y in the table can be N. A status represents whether the node utilizes the given path to next hop to transmit or receive at present instant. (Y means yes and N means No).*

The table dynamically modifies to show available or unavailable routes. It has the rows for all destined targets, irrespective of whether they will eventually be needed or not.

The method has minimum route acquisition delay and offers an alternative path when the existing path is not available due to disconnection or is busy in service to another destination.

The disadvantages of the method are—(i) overhead (additional efforts and network messages) in notifying the modifications in the row(s) of the routing table when a path that was previously unavailable becomes available or vice versa, (ii) computations even for routes that are not required, and (iii) there may be redundant paths shown in the table.

11.2.4.6 Optimized Link State Routing Protocol

Optimized link state routing protocol (OLSR) has characteristics similar to those of link state flat routing table driven protocol, but in this case, only required updates are sent to the routing database. This reduces the overhead control packet size and numbers. Further, there are multi-point nodes for relay of data. A node selects independently a multi-point node which also relays the route tables. The node provides bi-directional links such that the routes provided by the multi-point relay routing neighbouring node is also taken into account.

Example 11.9 Show a multi-point relay node in the MANET shown in Fig. 11.1(b).

Solution: Figure 11.1(b) shows that node *E* can be selected by a node as a multi-point relay node as it has the paths *E–F*, *E–G*, and *E–D*.

11.2.5 Security

Section 10.8 described the security aspects of mobile devices. Ad-hoc network have some additional security requirements compared to those for fixed infrastructure networks. These are listed in Table 10.4.

11.3 Wireless Sensor Networks

Mobile devices as MANET nodes have sophisticated hardware, software, and

Table 10.4 Additional security problems in ad-hoc mobile and wireless computing systems

Security problems	*Description*
Increased threat of eaves-dropping	The probability that a MANET or sensor node transmits unsolicited messages while moving in the wireless region of two nodes is increased in ad-hoc networks. Each node attempts to identify itself with a new node moving in its vicinity and during that process eaves-dropping occurs.
Unknown node caching the information	An unknown node can move into the network and thus rigorous authentication is required before the node is accepted as a part of MANET.
Denial of service attacks	A number of transmission requests can be flooded into the system by the attacking nodes. Since for each request an authentication process is initiated, which requires exchange of messages, the flooding of the message-exchanges chokes the MANET and denies the required services to genuine nodes.
Authenticated node becoming hostile	A previously authenticated device can be used for security attacks.

features. Section 2.5.4 described the sensors. Each node has an analog sensor with signal conditioner circuit. Sensing can be of the light level, temperature, location shift, time stamps of GPS satellites (Section 2.9), vibration, pressure, weather data, noise levels, traffic density, and nearby passing vehicles. The smart sensors have computational, communication, and networking capabilities but are constrained by their small size, limited energy availability, and limited memory. Since greater computational speed needs greater energy, these sensors operate at limited computational speed. Moreover, these have limited bandwidth. Wireless sensor network is a MANET or a network of smart sensors having computational, communication, and networking capabilities. Figure 11.3 shows a wireless sensor network.

A smart wireless sensor architecture consists of an RF transceiver for communication, a microcontroller [CPU, memory, and ADC (analog-to-digital converter)], and an energy source or a power supply. A charge pump in the sensor can trap the charge from the radiations (e.g., from WiFi transceiver or an access-point) and supply power for computations and communication. Alternatively, an energy-harvesting module can be used to trap solar radiation and store the energy. The RF transceiver enables a node to receive data packets from nearby nodes and route these to next hop of the packet. A wireless sensor node disseminates

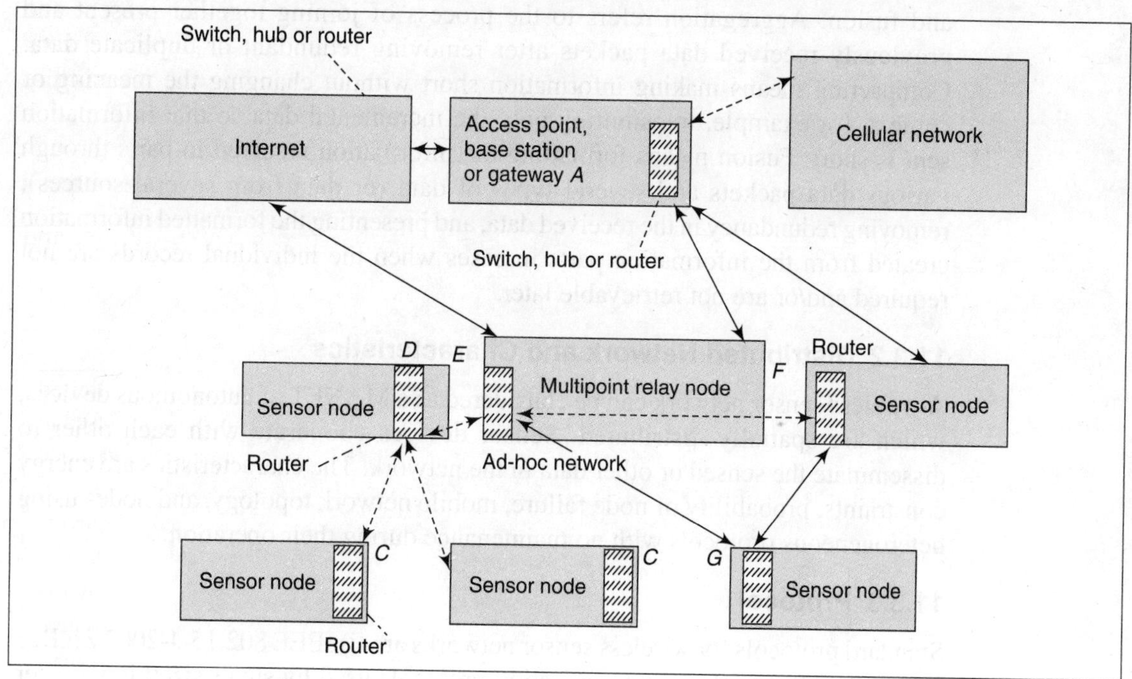

Fig. 11.3 Wireless sensor network

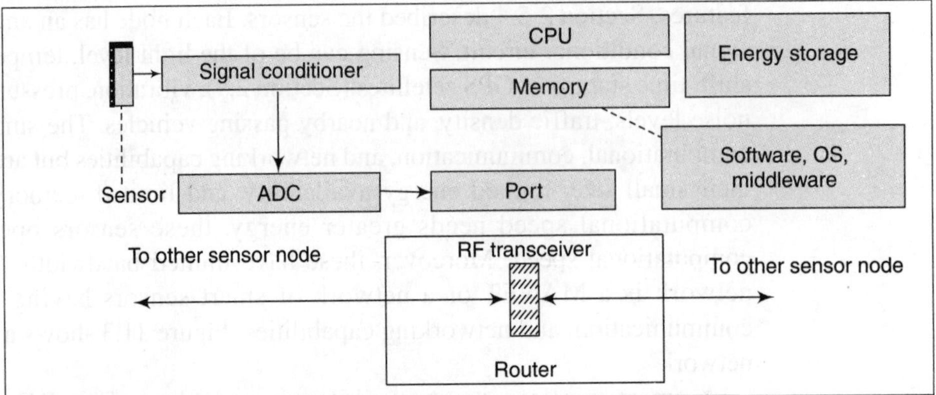

Fig. 11.4 Smart wireless sensor architecture

information to the network, central computer, or controller. Figure 11.4 shows a smart wireless sensor architecture. Some related aspects of wireless sensor networks including the protocols and software employed are discussed below.

11.3.1 Data Dissemination after Aggregation, Compaction, and Fusion

Data dissemination by sensor nodes is carried out after aggregation, compaction, and fusion. Aggregation refers to the process of joining together present and previously received data packets after removing redundant or duplicate data. Compacting means making information short without changing the meaning or context, for example, transmitting only the incremental data so that information sent is short. Fusion means formatting the information received in parts through various data packets and several types of data (or data from several sources), removing redundancy in the received data, and presenting the formatted information created from the information parts in cases when the individual records are not required and/or are not retrievable later.

11.3.2 Distributed Network and Characteristics

A wireless sensor network can be considered as a MANET of autonomous devices, which are spatially distributed. Sensor devices cooperate with each other to disseminate the sensed or other data in the network. The characteristics are energy constraints, probability of node failure, mobile network topology, and nodes using heterogeneous protocols with no maintenance during their operation.

11.3.3 Protocol

Standard protocols for wireless sensor networks are (i) IEEE 802.15.4-2003 ZigBee (Section 12.9) (a set of communication protocols used by small-sized low power digital radio embedded in the sensors and home devices) and (ii) 6lowpan (IPv6 over low power wireless personal area network which permits communication of IPv6 packets in sensor network)

11.3.3.1 Software

Software once embedded into the sensor node is for lifetime. It should be robust, fault tolerant, and should provide maximum features and middleware The software should have features of security, self-healing, and self-configuration. Examples of a few specific operating systems are tinyOS and CORMOS (communication-oriented runtime system for sensor networks). Examples of a few specific middleware (agent or database adapter) are tinyDB, SensorWare, and GSN (global sensor networks) application-oriented middleware.

11.3.3.2 Router

Sensor networks deploy special routing protocols such as CGSR, DSR, or AODV (Section 11.2.4).

11.4 Applications

Some applications of wireless sensor networks and MANETs are described below:

- Industrial plant wireless sensor networks—Industrial plants use large number of sensors in instruments and controllers. Wireless sensor network proves to be powerful for an industrial plant.
- Pervasive computing networks—Mobile pervasive computing means a set of computing devices, sensors, or systems or a network having the characteristics of transparency, application-aware adaptation, and environment sensing. Wireless sensor networks and MANETs are therefore an important element of pervasive computing.
- Traffic monitoring using traffic density wireless sensor networks—Traffic can be monitored at different points in a city and traffic density information can be aggregated at a central server. The server relays this information to motorists on wireless Internet. A traffic control server sends the traffic reports on Internet. The automobile owner can subscribe to a traffic control service provider which provides SMS messages about traffic slowdowns and blockades at various city points. It enables a driver to select the roads with the least hurdles. TTS (text to speech) converters can also give voice messages to the drivers.
- Medical applications of wireless sensor networks—Patients can be monitored by the sensors attached to them. When a patient moves, the sensors form a MANET.
- Military applications of wireless sensor networks—The voice of a person can be sensed by a wireless sensor network deployed in remote border areas. This monitors the enemy troop and machines movements.
- Smart labels and RFID-based wireless sensor network—It is used worldwide for monitoring movement of goods, movement of books in library, and supply chain management systems.

- Environmental monitoring wireless sensor network—Environmental parameters like temperature, pressure, light, rainfall, and seismic activity are sensed and communicated over a wireless network.
- Home automation—Home automation including security is possible using a wireless sensor network.

Keywords

Access-point: A centralized communication node of a service provider to which number of mobile devices can connect before establishing communication. Transmission and reception of data packets occurs through that access-point only in fixed infrastructure network. Access-point also provides connectivity between two MANETs.

Ad-hoc on demand: When route information is required, then a demand is raised for it and the path information is created on ad-hoc basis at each node. The node discovers the available next hop as follows—It first sends the hello messages to all neighbours and waits for acknowledge-ments. The acknowledging node is the next hop which is available and can be used for routing.

Aggregation: Process of joining together present and previously received data packets after removing redundant or duplicate data.

Cluster: Including several routing nodes forms a cluster structure. The nodes aggregate into clusters using appropriate algorithm.

Cluster algorithm: An algorithm which defines a cluster-head, the node used for connection to another cluster, and a gateway node which provides switching (communication) between two or more cluster-heads.

Compaction: Process of making information short without changing the meaning or context, for example, transmitting only the incremental data so that information sent is short.

Fusion: Formatting the parts of information received through various data packets and several types of data (or data from several sources), removing redundancy in the received data, and presenting the formatted information created from the information parts in cases when the individual records are not required and/or are not retrievable later.

Hop: When a data packet is transmitted from a node as per the routing table or information in cache, it reaches another node where again it uses a route information. This process of transfer of a data packet from one node to another is called hop. Data packet reaches the destination by one or more hops.

MANET: A peer-to-peer wireless network which transmits data packets from one computer to another without the use of a central base station (access-point). Routing from one node to another in such a network typically uses an "on-demand routing protocol" such as DSR or AODV, which performs route discovery only when a station initiates transmission of a data packet.

Ordered routing: Routing path from a node to a destination pre-calculated by a routing algorithm such that the path is as per the routing path followed by a packet before reaching the node.

Peer-to-peer connectivity: Connectivity between two nodes such that both nodes are capable of sending requests for authentication and data and obtain the responses from each other.

Proactive protocol: A protocol in which a routing node maintains the routing addresses for all possible routes such that when a source routes the packets to destination, the routing tables are already available at the node.

Reactive protocol: A protocol in which a routing node needs to maintain the routing addresses about the active paths only.

Route discovery: A process in which a source seeking route information broadcasts the packets, each with a header. The source expects return acknowledgement from destination. The route information pertaining to a destination is cached from the packet header in reply to the source. Caching is done at the source as well as intermediate nodes.

Route Error (RERR) Packet: The packet sent back by a node in case of a disconnection during route discovery for making the next hop for a destination.

Route Maintenance: Storing and dynamically updating route cache or routing table for a destination or cluster member table.

Route Reply (RREP) Packet: The packet sent by destination during route discovery. The nodes cache the route information for destination using these packets.

Route Request (RREQ) Packet: The packets sent for route discovery.

Router: A router is a device at a node that networks the nodes consisting of mobile devices, sensors and access-points, base stations, or gateways. It provides at the node two or more paths to route a message or packet so that the available path can be selected and used at an instant to route a message or packet from a source to destined node.

Seamless connectivity: Connectivity between the mobile nodes in a MANET such that the connectivity is always ensured or is made endless by deploying interconnections between the mobile device nodes when they move, the nearby wireless nodes, sensor nodes, and embedded devices in automobiles.

Self-configuring: Software or routing algorithm at a node selecting the part of the codes and data by itself as per present requirement.

Self-healing: Software or routing algorithm at a node correcting part of its codes and data as per present requirement with no need of maintenance by system administrator.

Source routing: Each data packet from source has a header and includes the routing-node addresses also.

Spectrum: Range of frequencies deployed for communication.

Switching exchange: A system through which two nodes establish a connection till the connection terminates.

Ubiquitous mobile computing: Computing by maintaining connectivity through the mobile nodes in a MANET irrespective of the protocol deployed for interaction between mobile device nodes when they move with the nearby wireless nodes, sensor nodes, and embedded devices in automobiles so that endless computing and communication is performed by a device.

Weak connectivity: Unreliable links to base station or gateway.

Wireless sensor network: A MANET or network of smart sensors having computational, communication, and networking capabilities. Smart sensors are constrained by their small size, limited energy availability, and limited memory.

Review Questions

1. Describe MANET. How does a MANET differ from a fixed infrastructure network?
2. Explain how MANETs are deployed in various applications.
3. Describe the properties of MANETs.
4. Compare the reactive and proactive routing protocols. Describe DSR and AODV protocols.
5. Describe TORA. Compare the features of TORA with the DSR and AODV protocols.
6. Describe an application of defining the cluster of nodes and features of CGSR protocol.
7. What are the security threats to a MANET? Why a MANET faces greater security threats than a fixed infrastructure network?
8. Describe wireless sensor networks. Explain the similarities in MANET and wireless sensor networks.
9. What are the advantages of MANETs and wireless sensor networks integrated with IPv6?
10. Write about the application of wireless sensor networks in home personal area networking as in industrial plants.

Objective Type Questions

Pick the correct or most appropriate statement among the choices given:

1. GSM mobile phones
   (a) connect to MANET infrastructure network provided by the service provider and form a WPAN when similar phones with Bluetooth, IrDA, or ZigBee come in their vicinity.
   (b) connect to fixed infrastructure network provided by the service provider and form a fixed infrastructure WPAN when similar phones with Bluetooth, IrDA, or ZigBee come in their vicinity.

(c) connect to fixed infrastructure network provided by the service provider and form a MANET when similar phones with Bluetooth, IrDA, or ZigBee are in vicinity in a WPAN.
(d) MANET is not formed when GSM phones with Bluetooth are used.

2. (a) MANET is established using service discovery and then the services of Jini, TCP/IP, Bluetooth, and WiFi are discovered.
(b) MANET discovers services by establishing Bluetooth and WiFi for service discovery.
(c) MANET is established using service location protocol.
(d) MANET is established using ad-hoc services discovery protocol.

3. Consider mobile devices, iPods, Palm handheld computers, Smartphones, PCs, smart labels, smart sensors, tokens, set-top boxes, and automobile-embedded systems. MANET nodes
(a) cannot be formed in the iPods and set-top boxes.
(b) cannot be formed in the smart labels, smart sensors, smart labels, tokens, and set-top boxes.
(c) cannot be formed in the iPods, smart labels, smart sensors, tokens, and set-top boxes.
(d) can be formed in all except in the set-top boxes.

4. WLAN 802.11 and satellite-based phone networks are called
(a) mobile ad-hoc networks.
(b) fixed infrastructure network with additional ad-hoc network support, and fixed infrastructure network, respectively.
(c) fixed WLAN infrastructure, mobile satellite, and fixed infrastructure land networks.
(d) wireless sensor and MANET networks.

5. (a) MANET is a configurable wireless ad-hoc network of mobile nodes.
(b) Network functions independently and does not provide connection to the Internet IPv4 or IPv6.
(c) Sensors or devices which are mobile can form an ad-hoc network in which each node has router or switch connected by wireless connection and the union of connections has an arbitrary topology.
(d) WLAN-based networks form a MANET.

6. (i) A MANET node discovers the node that can connect it to network and discovers the service availability through it. (ii) Using a service discovery protocol, a node discovers the service of nearby node and communicates to remote targeted node in the MANET. (iii) MANET nodes have client–server connectivity from a node to neighboring ones. (iv) MANET node has unreliable links to base station or gateway. (v) An intermediate node failure results in greater latency to the remote server. (vi) MANET nodes have independent computational, switching (or routing), and communication capabilities.
(a) All are true except (i) and (vi). (b) All are true except (vi).
(c) (i), (ii), and (vi) are true. (d) All are true except (iii).

7. A mobile device can at best find six mobile devices reaching in its vicinity. Let there be FDMA mode access by a node. Assuming that f_{bw0} is the bandwidth requirement between two neighbours, what bandwidth will be needed when all next hop neighbours communicate in full duplex mode and in same time slots?
(a) f_{bw0} (b) $2 \times f_{bw0}$ (c) $6 \times f_{bw0}$ (d) $2 \times 6 \times f_{bw0}$

8. (i) Content distribution and synchronization (ii) Mesh networking (iii) Mobile service provider network (iv) Multicast network (v) Image acquisition and processing (vi) Remote viewfinder by security personnel in an office (vii) Remote processing on a computer for a video stream from wireless WebCam and other devices (viii) Image file transfer (ix) Messaging and data transmission to remote using 802.11b (x) Remote controlling (xi) IPv6 integration with wireless sensor networks. MANET applications can be for
(a) all except (i) and (xi) (b) (i) to (xi)
(c) all except (iv), (v), and (vi) to (viii) (d) all except (ix) to (xi).

9. (a) MANET node performs data caching, saving, and aggregation and does not use Bluetooth or 802.11 as means of inter-node communication.
(b) MANET nodes can use IrDA, Bluetooth, ZigBee, 802.11, GSM, TCP/IP, and any other protocol for communication and the node performs data caching, saving, and aggregation.
(c) MANET nodes can use IrDA, ZigBee, or 802.11 protocol and the node performs data caching and saving.

(d) MANET nodes can use SLP and SDP and perform data aggregation.

10. (a) TinyOS or CORMOS can be used as OS and tinyDB or SensorWare or GSN can be used as agent or database adapter.
    (b) TinyOS can be used as OS and tinyDB or WebSphere Everyplace Application Server can be used as agent or database adapter.
    (c) WebSphere Everyplace Application Server or CORMOS can be used as OS and tinyDB or GSN can be used as agent or database adapter.
    (d) CGSR can be used as agent or database adapter in a MANET.

11. (i) A smart wireless sensor architecture has an RF transceiver which enables a node to receive data packets from nearby node and route these to next hop of the packet. (ii) A wireless sensor node disseminates information to the network. (iii) It disseminates information to the central computer. (iv) It disseminates information to the controller. (v) At least one sensor or node connects to access-point, base station, or gateway to provide Internet connectivity.
    (a) (i) to (iii) and (v) are true.
    (b) (i) to (v) are true.
    (c) (i), (ii), and (vi) are true.
    (d) (i), (ii), and (iv) are true.

12. CGSR is
    (a) hierarchical and dynamic routing protocol in which the routing tables at the nodes are already available and a cluster higher in hierarchy need not send the packets to the cluster lower in hierarchy.
    (b) dynamic distance routing protocol in which the routing tables at the nodes are already available and a cluster lower in hierarchy sends the packets to the cluster higher in hierarchy.
    (c) hierarchical and proactive routing protocol in which the routing tables at the nodes are already available and a cluster higher in hierarchy sends the packets to the cluster lower in hierarchy.
    (d) non-hierarchical and non-proactive routing protocol in which the routing tables at the nodes are to be formed and obtained from a cluster.

13. (a) AODV is a reactive protocol. It means that it reacts to the dynamic changes and maintains only the active routes in the caches or tables for a pre-specified expiration time. These routes are found and are expected to be available at a given instant.
    (b) AODV is a proactive protocol and uses the tables for the active routes which are found and are expected to be available at a given instant.
    (c) TORA does not react to the changes and link reversals and functions in less dynamic MANETs.
    (d) DSR deploys *source routing* which means that each data packet includes all the routing tables up to the destination which reacts to the changes and dynamically maintains only the routing addresses for the source to destination.

14. Flat routing table driven protocol is (i) proactive (ii) and provides the routing table which is available in advance at a node (iii) a packet specifies route in the header and (iv) the routes have to be discovered after the demand is raised. Which of these are correct?
    (a) All are correct.
    (b) (i) and (iii) are correct.
    (c) (i) and (iv) are correct.
    (d) (i) and (ii) are correct.

15. Software or routing algorithm at a node is self-configuring if the node selects the part of the codes and data by itself as per its present requirement and it is self-healing if it corrects part of the codes and data as per present requirement.
    (a) Self-configuring and self-healing networks need maintenance by system administrator.
    (b) Wireless sensor ad-hoc networks should be self-healing.
    (c) Wireless sensor ad-hoc networks should be self-configuring.
    (d) Wireless sensor ad-hoc networks should be self-configuring and self-healing.

16. Consider a process of making information short without changing the meaning or context, for example, transmitting only the incremental data so that information sent is short in a wireless ad-hoc network. Process is
    (a) called compaction (b) called compression (c) called fusion (d) an action before aggregation

Chapter 12

Wireless LAN, Mobile Internet Connectivity, and Personal Area Network

This chapter describes wireless communication over local area networks and wireless application protocol (WAP). Just as HTTP is deployed over wired Internet, WAP is a protocol deployed over wireless mobile communication on Internet. The chapter also describes data synchronization, exchange, and network over short range wireless personal area network (WPAN). It covers the following:

- Understanding of wireless LAN (WiFi) architecture
- Study of IEEE 802.11 WLAN protocol
- Learning of WAP 2.0 architecture, wAP layers, and XHTML-MP
- Study of the protocol for Bluetooth-enabled WPAN mobile devices and systems
- Understanding salient features of IrDA in a WPAN
- Learning ZigBee for low power short range wireless personal area network, routing of messages using ZigBee to enable applications in big scale automation and remote controls, formation of a mesh network, and reactive and proactive protocols

12.1 Wireless LAN (WiFi) Architecture and Protocol Layers

Local area network (LAN) means a set of computers, computational systems, units, and devices, for example, mobile phones, printers, laptops, smart sensors, and smart labels, networked using a standard suite of protocols.

Local refers to some defined area or a set of nearby or distant stations. The networking is such that all units in the set can address and communicate with each other. Section 1.1.5 introduced wireless LAN (WLAN), also called WiFi (wireless fidelity). Figure 1.11 demonstrated a simplified model of WLAN. A set of popular standards (IEEE 802.11a, 802.11b, and 802.11g) was listed in Table 1.4. These standards are recommended for WLAN in mobile communication and for establishing communication between mobile devices and Internet or other networks.

A description of WLAN architecture and the protocol layers in 802.11 is given next.

12.1.1 WLAN Architecture

Figure 12.1 shows two service sets—basic service set (BSS) and independent basic service set (IBSS) in the WLAN architecture. Set *A* consists of one station STA_A. Set *B* has several stations STA_B, STA_C, ... Set *A* also has nodes which connect to an access-point. The stations in set *B* do not connect to any access-point and their nodes connect among themselves. Each station can have the nodes forming a set of computers, computational systems, units, and devices, for example, mobile phones, printers, laptops, smart sensors, and smart labels. Each node of a station uses the same frequency band if it is at a distance from another station or a distinct frequency band if it is not distant enough from another station. Node at a station can communicate directly to an access-point (in BSS) and to another node at another station through the access-point. Nodes of a station can communicate among themselves after forming an ad-hoc or any other type of network (e.g., Bluetooth) using same frequency band for each node.

BSS devices in each set (Fig. 12.1(a)) interconnect to the access-point using 802.11 and form a single station STA_A of WLAN using same frequencies for radio and the station interconnects to other stations through access-points.

IBSS (Fig. 12.1(b)) has devices which network with each other using 802.11 protocol. These devices either communicate directly with one another or communicate among themselves after forming an ad-hoc network. They form a set of stations (STA_B, STA_C, ...) in a WLAN. Each station uses same frequency band for radio coverage. Therefore, each station within an IBSS does not connect to another station even of same IBSS. IBSS does not connect to an access-point also. A station can have several devices and can form an ad-hoc network among the nodes which interact through peer-to-peer communication. The 802.11 protocols suite does not specify the protocols for the nodes for data routing, exchanging, or supporting exchange of network topology information. Thus, there can be Bluetooth exchanges (Section 12.4) between the nodes or the nodes can be the ZigBee devices (Section 12.9).

Figure 12.2 shows 802.11 LAN station access-points *A* to *G* networked together forming extended service set (ESS), which functions as a distribution system possessing an ID, called ESSID. For example, Internet can be deployed by WLAN distribution system. The access points, A, B, C, D, E, F, and G form the ESS. An

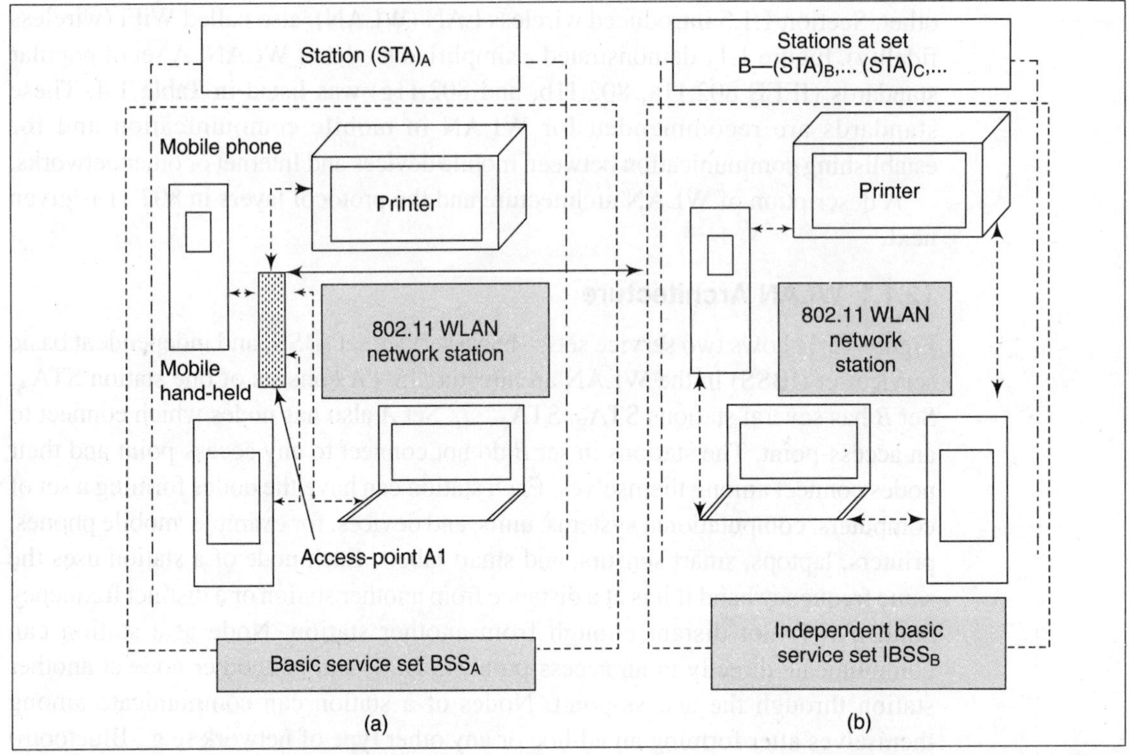

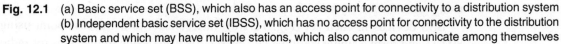

Fig. 12.1 (a) Basic service set (BSS), which also has an access point for connectivity to a distribution system (b) Independent basic service set (IBSS), which has no access point for connectivity to the distribution system and which may have multiple stations, which also cannot communicate among themselves

access-point can also exist at a base station or gateway, for example, *A* in Fig. 12.2. An access-point can also be present at a multi-point relay node, for example, *E* in Fig. 12.2. The 802.11 provides the definition for ESSID, but the distribution system network protocols are not defined within 802.11. The protocols will depend on how the BSSs interoperate in a service provider servicing set up. These protocols may or may not be TCP/IP or IPv6. Also a node can be mobile and can move from one BSS to another such that its service access-point becomes different on moving (roaming).

The following examples highlight a standard basic feature of 802.11 that it supports both access-point-based fixed infrastructure WLAN network using BSSs and ad-hoc peer-to-peer data routing network using IBSS stations.

- Stations in a given IBSS—(i) Consider a mobile phone, TV with a set-up box, security system, and computer at home. These form a WLAN station and can use the same frequency band for radio. Since it does not have an access-point to a distribution system or ESS, the station is a part of an IBSS. These devices can also have Bluetooth OBEX exchange between mobile phone and computer. (ii) Consider the mobile phones, computers, and printers

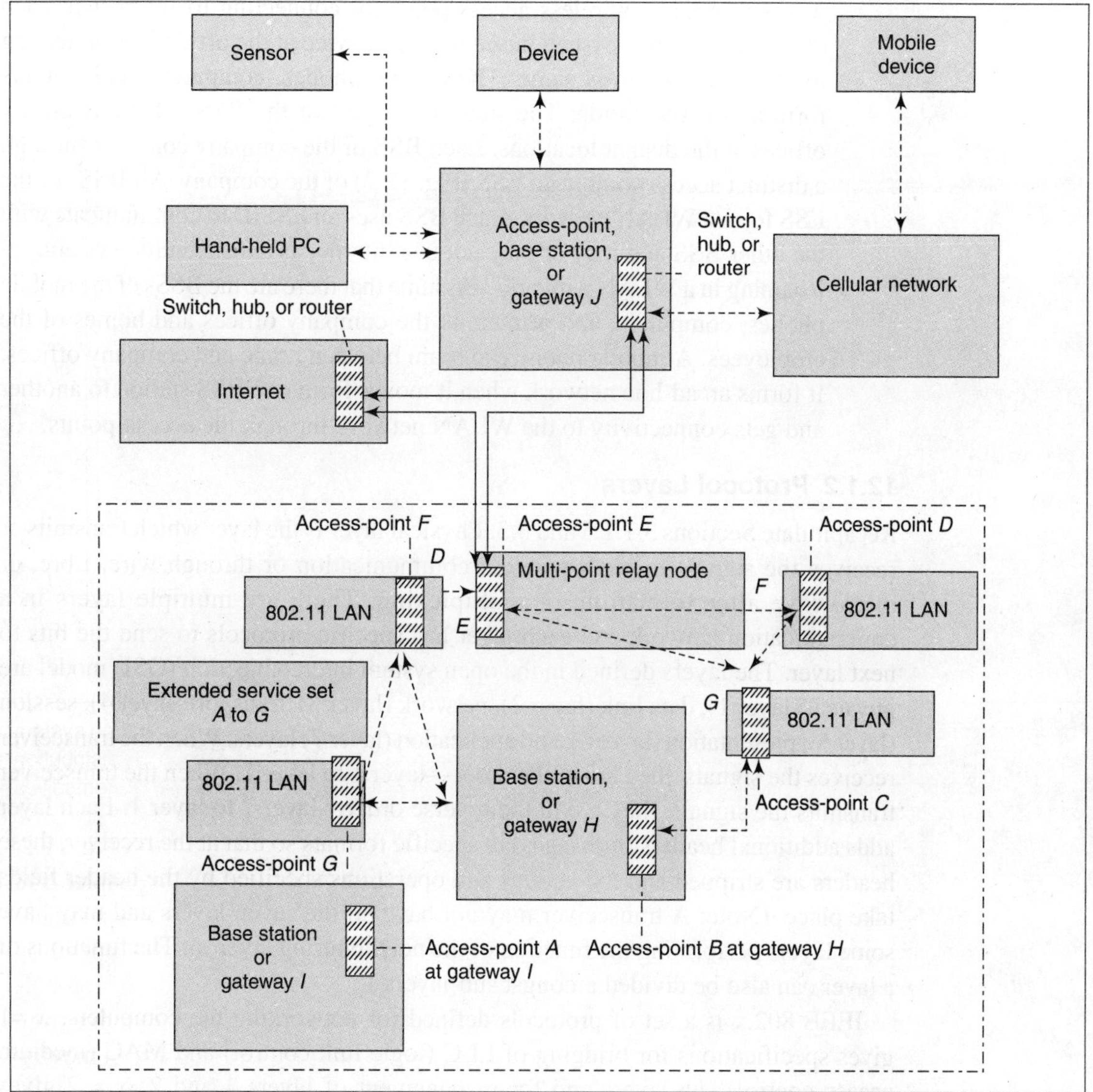

Fig. 12.2 802.11 LAN access points networked together using extended service set (ESS), which functions as a distribution system

at a company office having independent workspace for each set of a mobile phone, a computer, and a printer. Each set forms a WLAN station. Each station uses same frequency band for radio if the frequencies do not interfere and distinct frequency band if the frequencies are too close to each other. All stations together form an IBSS, which is distinct from the IBSS at home.

- WLAN network having stations at the BSSs in an ESS—Consider the mobile phones, computers, and printers at the company office having independent workspace for each set of a mobile phone, a computer, and a printer. Each workspace has a wireless access-point for connecting to Internet in each office. The frequency-band used by each device at the office for connecting to the access point is same. The mobile phones, computers, and printers form a WLAN station. The station is a part of the BSSs of the company offices at the distant locations. Each BSS of the company connects through a distinct access point in an ESS (Fig. 12.2) of the company. All BSSs of the ESS form a WLAN network. Each BSS uses an ESSID to communicate with the other BSS and may or may not use Internet as a distribution system.

- Roaming in a WLAN network—Assume that there are the BSSs of the mobile phones, computers, and printers at the company offices and homes of the employees. A mobile phone can roam between home and company offices. It forms an ad-hoc network when it moves from one BSS station to another and gets connectivity to the WLAN network through the access-points.

12.1.2 Protocol Layers

Recapitulate Sections 3.1.1.5 and 3.3. Physical layer is the layer which transmits or receives the signals through wireless communication or through wire, fibre, or microwave after formatting or multiplexing. There are multiple layers in a communication network and each layer has specific protocols to send the bits to next layer. The layers defined in the open system interconnection (OSI) model are physical (layer 1), data link (layer 2), network (layer 3), transport (layer 4), session (layer 5), presentation (layer 6), and application (layer 7) layers. When the transceiver receives the signals, they are in the order—layer 1 to layer 7. When the transceiver transmits the signals, they are in the reverse order—layer 7 to layer 1. Each layer adds additional headers (messages) in specific formats so that at the receiver, these headers are stripped and the actions and operations specified by the header fields take place. (Note: A transceiver may not have all the seven layers and may have some layers performing the function of the neighbouring layer(s). The functions of a layer can also be divided amongst sub-layers.)

IEEE 802.x is a set of protocols defined for networking the computers. $x = 1$ gives specifications for bridging of LLC (logic link control) and MAC (medium access control) sub-layers and for management of layers 1 and 2. $x = 2$ gives specifications for LLC sub-layer at layer 2. $x = 1$ and 2 specifications are common to all standards in 802.x for $x = 3$ and above. $x = 3$ gives the specifications for MAC sub-layer of layer 2 and physical layer for wired LAN, called Ethernet. Upper layers are common in protocols 820.x. $x = 10$ gives the security specifications for layers 2 and above and is common in protocols 802.1y. [$y = 1$ means 802.11, 5 means 802.15 and 6 means 802.16.]

Figure 12.3(a) shows the functions of physical layer and MAC sub-layer of layer 2 (data link layer). Figure 12.3(b) shows basic protocols layers of the transmitter and receiver in IEEE 802.11. The 802.11 standard is a suite of WLAN protocols for

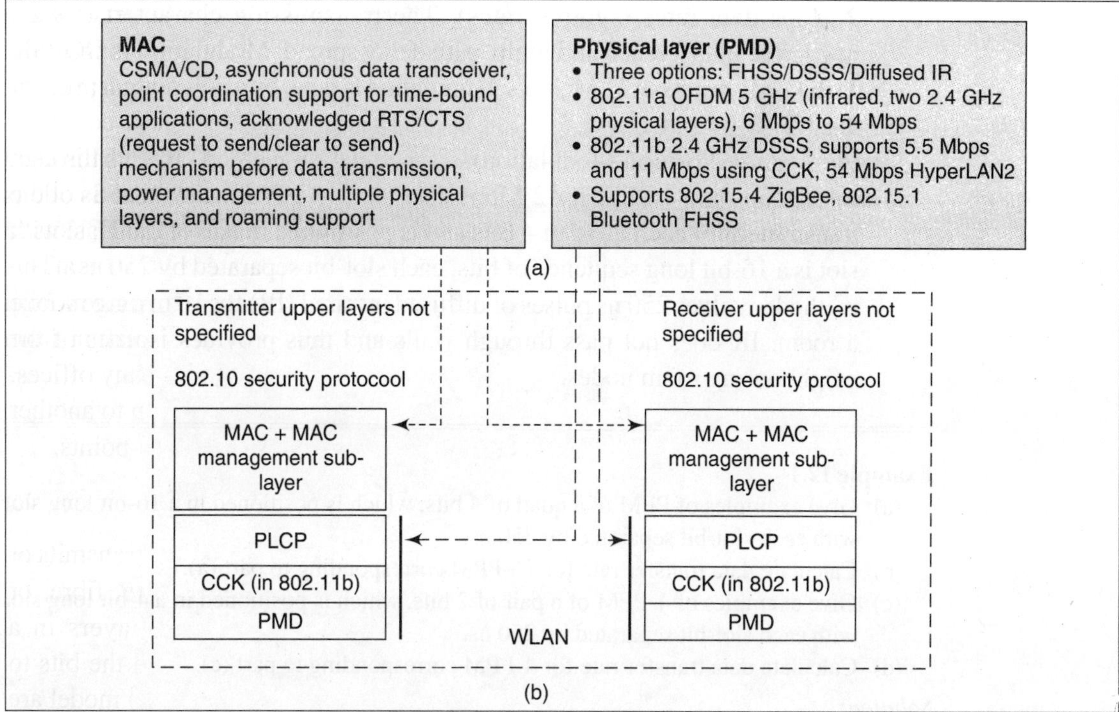

Fig. 12.3 (a) Functions of MAC and physical layer (b) Basic protocols layers in 802.11 in receiver and transmitter

the MAC sub-layer of layer 2 and physical layer (layer 1), which includes security 802.10 specifications.

The 802.11 standard includes previous specifications in 802.x (x = 1–10). The 802.11 WLAN Internet access can therefore be either direct or through existing wired LAN because the higher layers are identical to all previous 802.x standards.

Physical layer has two sub-layers in 802.11a—PMD (physical medium dependent) sub-layer and PLCP (physical layer convergence protocol) sub-layer. There is an additional sub-layer in 802.11b—CCK (complementary code keying) for data rates of 5.5 Mbps by QPSK to map 4 bits and 11 Mbps 8-QPSK to map 8 bits simultaneously (Section 4.1.1). PMD, PLCP, MAC, and MAC management sub-layers are explained in the following subsections.

12.1.2.1 PMD Sub-layer

PMD protocol specifies the modulation and coding methods. (There is a service access-point with 1 Mbps or 2 Mbps data rate to MAC layer). Three different methods are briefly described below:

(a) FHSS (Section 4.3.2)—radiated at 10 mW, 100 mW, and 1 W as per country-specific restrictions. Modulation is 1 Mbps Gaussian BPSK or 2 Mbps Gaussian QPSK.

(b) DSSS—using 11-bit Barker code (Sections 4.3.1 and 4.4.1.1) radiated at 10 mW, 100 mW, and 1 W as per country-specific restrictions and 1 Mbps or

2 Mbps data rates (symbol rates). The transmission characteristics are negligible interference and multi-path delay spread. Modulation is DQPSK. 11-bit code results in 11 Mchip/s. Scrambling is done by a polynomial (Section 4.4.1.2) $G_Q = z^7 + z^4 + 1$.

(c) PPM (Pulse Position Modulation)—a modulation method. 16-PPM is used for 1 Mbps and 4-PPM for 2 Mbps data rate. 16-PPM means that a code is transmitted for each quad of 4 bits and is positioned in one of the 16 slots (a slot is a 16-bit long sequence of bits, each slot-bit separated by 250 ns). This method involves 250 ns pulses of diffused infrared (IR) for 10 m range within a room. IR does not pass through walls and thus provides isolation from neighbouring room nodes.

Example 12.1

(a) Give examples of PPM of a quad of 4 bits, which is positioned in a 16-bit long slot with each slot-bit separated by 250 ns.
(b) Calculate data transfer rate for 16-PPM corresponding to part (a).
(c) Give examples of 4-PPM of a pair of 2 bits, which is positioned in a 4-bit long slot with each slot-bit separated by 250 ns.
(d) Calculate data transfer rate for 4-PPM corresponding to part (c).

Solution:

(a) (i) Consider a quad of 4-bits as $0000_b = 0_d$. It means that at 0th position (counting positions from 0, 1, 2, ...), there will be 1. Hence the 16-bit sequence will have 0th slot-bit or lsb as 1. The transmitted bits after PPM will therefore be 0000 0000 0000 0001. (ii) Consider a quad of 4 bits as $0100_b = 4_d$. It means that at 4th position (counting positions from 0, 1, 2, ...), there will be 1. Hence, the 16-bit sequence will have the 4th slot-bit as 1. The transmitted bits after PPM will therefore be 0000 0000 0001 0000. (iii) Consider a quad of 4 bits as $1111_b = 15_d$. The transmitted bits after PPM in this case will be 1000 0000 0000 0000.

(b) 16 quad bits are transmitted in 16×250 ns = 4000 ns. Each quad actually carries 4 bits of data. Hence data transfer rate = $(4000 \text{ ns} \div 4)^{-1}$ = 1 Mbps.

(c) Consider a pair of 2 bits as $00_b = 0_d$. It means that at 0th position (counting positions as 0, 1, 2, and 3), there will be 1. Hence the 4-bit sequence will have 0th slot-bit or lsb as 1. The transmitted bits after PPM will therefore be 0001. (ii) Consider a pair of 2 bits as $11_b = 3_d$. It means that at 3rd position (counting positions as 0, 1, 2, and 3), there will be 1. Hence the 4-bit sequence will have 3rd slot-bit = 1. The transmitted bits after PPM will therefore be 1000.

(d) 4 slot bits are transmitted in 4×250 ns = 1000 ns. Each actually carries 2 bits of data. Hence data transfer rate = $(1000 \text{ ns} \div 2)^{-1}$ = 2 Mbps.

12.1.2.2 PLCP Sub-layer

PLCP specifies sensing of the carrier at the receiver and packet formation at the transmitter. The different transmission and reception protocols (FHSS, DSSS, and diffused IR) are specified for the PMD. Therefore, a convergence protocol sub-layer is required in between the PMD and MAC sub-layers. PLCP sub-layer protocol prescribes the standard procedure for convergence of PMD to MAC at receiver and from MAC to PMD at transmitter.

The header and payload in each frame are defined for different cases as follows:

(a) PLCP for FHSS:

1. First 80 bits (010101…) are for synchronization.
2. Next 16 bits (0000 1100 1011 1101) are for SFD (start frame delimiter) for synchronization between the frames.
3. Next 12 bits specify the data length in bytes. This includes the error check (CRC) bits in the payload (PLCP-PDU-LW (protocol data unit with length in words) of 1 byte each).
4. Next 4 bits are for PSiF (PLCP signalling field) (0000 for 500 kbps, 0010 for 2Mbps, and …. the maximum is 8.5Mbps).
5. Next 16 bits are for checksum, called HEC (header error check).
6. Remaining bits in the frame are for the payload.

(b) PLCP for DSSS:

1. First 128 bits are for synchronization, gain set, energy detection, and frequency offsets.
2. Next 16 bits (1111 0011 0100 0010) are for SFD (start frame delimiter) for synchronization between the frames.
3. Next 8 bits are for PSiF (0000 1010 for 1 Mbps DBPSK and 0001 0100 for 2 Mbps DQPSK, and so on for higher data rates).
4. Next 8 bits (0000 0000) are for PSF (PLCP service field) for 802.11 standard frame.
5. Next 16 bits are for PLF (PLCP length field) to specify the data length of MPDU (maximum protocol data units) in μs.
6. Next 16 bits are for checksum of PSiF, PSF, and PLF. The sum is the HEC.
7. Remaining bits in the frame are for the payload.

(c) PLCP for diffused IR:

1. 57–73 bits (010101…) are for synchronization and frequency offsets.
2. Next 4 bits (1001) are for SFD for synchronization between the frames.
3. Next 3 bits for PSiF (000 for 1 Mbps 16-PPM and 001 for 2 Mbps 4-PPM).
4. Next 32 bits are for PSF for DC level adjustment in the frame.
5. Next 16 bits are for PLF to specify data length of MPDU in μs.
6. Next 16 bits are for checksum of PLF, PSF, and PSiF. The sum is the FCS (frame checksum).
7. Remaining bits in the frame are for the payload (MPDU less than 2500 bytes).

12.1.2.3 MAC and MAC Management Sub-layers

A MAC sub-layer specifies CSMA/CD (CSMA/CollissionDetect) (Section 4.1.2), RTS/CTS, and point coordination function (PCF) mechanisms. Another sub-layer specifies MAC management. CSMA/CD is a protocol in which the presence or absence of a carrier on the network is sensed by a transmitter before transmitting the signals and it is checked whether a collision is detected at the transceiver while attempting to transmit. This protocol is analogous to the one used by the nodes in Ethernet LAN. This technique permits connectionless data-oriented transceivers. Connectionless means no connection establishment and termination related header field exchanges between the source and addressed destination. Medium access control provides point coordination support for the time-bound applications and there is an acknowledgement mechanism before transmission of data (packet). A transceiver makes a request to send and waits for clear to send. This takes care of the hidden terminal problem (Section 4.1.3).

MAC layer for medium access control has the following features:

- CSMA/CD
- Point coordination support for time-bound applications
- Acknowledged RTS/CTS (request to send/clear to send) mechanism before the data transmission
- Power management
- Multiple physical layers for same MAC support
- Mobile node roaming within an ESS by registration and node association, dissociation, and re-association on moving to another BSS.

MAC frame format of 802.11 in each frame is as follows:

1. 16 bits for frame control with specifications as follows:
   (a) 2-bit field for protocol version.
   (b) 2-bit field for frame type—(i) 00 means that the frame is for management (registration, handover, power management) (802.11i security using AES and DES), (ii) 01 means that the frame is for control, and (iii) 10 means that the frame is for data.
   (c) 4-bit field for subtype (for the management frame, it is 1100_b for CTS, 0000 for request for association, and 1011_b for RTS) and subtype = 0000 when a user data frame is transmitted.
   (d) 2-bit field to specify one of the four possible addresses when transmission is between mobile stations and access-point (within BSS nodes and devices) or between the access-points over a data source (DS) (using the ESS).
   (e) 1-bit field which is 1 when more data fragment(s) is to follow for a present service data units frame of the MAC protocol (before transmission, a service data units frame may be fragmented and therefore another fragment may be required to follow the present one. This field

is set to 1 when another fragment is to follow the present one and to 0 when none is to follow.)

(f) Four 1-bit fields for retry, power management, more data expected from sender, and WEP (wired equivalent privacy in 802.11b).

(g) 1-bit field which is 1 for *order* (means receiver must strictly process as per the order in which bits are received).

2. Next 16 bits for duration/ID field (msb 0 means duration in μs for frame and 1 means ID) for SFD for synchronization between the frames (Also used as NAV (net allocation vector)).
3. Total 26 × 8 bits for three 48-bit address fields to specify addresses, 16-bit sequence control field, and fourth 48-bit address field. (Sequence number in control field is necessary due to separate duplicate frame or acknowledgements.)
4. Data bits (<2312 bytes) in a frame are for the payload (MPDU less than 2500 bytes).
5. 32 bits CRC as in Ethernet or other 802.x data frames.

Functions of MAC management sub-layer are as follows:

- Roaming management which is done using scanning of the nodes (devices) moving into a new area. The scanning is done by the access-points and detects the new devices or the loss of devices in the area. The access-point registers or deregisters the devices after the scanning. New device registration provisions for device association at new access-point when it roams into the new area from another access-point area.
- Internal receiver clocks are synchronized, which is necessary.
- Generation of beacon signals is also part of management functions. (Beacon signals are used for helping or warning others by indicating the presence of a device or an access-point. A BSS periodically sends beacon signals, which contain—(i) time stamp for synchronizing node clock and (ii) power management and roaming data.)
- Transmitter switches to power-save mode after each successful data transmission for power management periodically activating the sleep mode. Buffering by a receiver and start of processing after enough data is received in buffer also saves power.

12.2 WAP 1.1 and WAP 2.0 Architectures

Recapitulate Section 4.4. I-mode provides integrated services for voice, data, Internet, picture, music attachment to mail, gaming applications, ring tone downloads, remote monitoring, and control services. WAP offers open development platform for similar applications.

Figure 12.4 shows the architectures for communication between a WAP client and HTTP (hypertext transfer protocol) server. WAP 2.0 is wireless protocol for

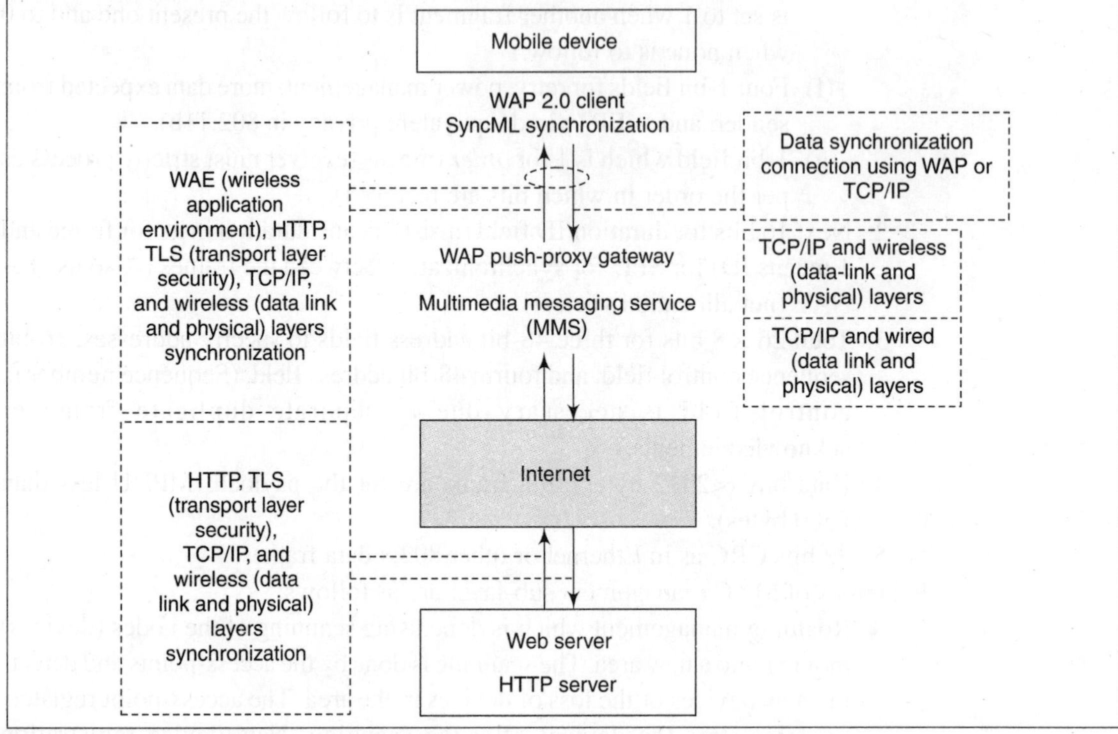

Fig. 12.4 WAP 2.0 client, gateway, and web HTTP server architecture

synchronization of WAP client computers and WAP or HTTP server. This is derived from the architecture of the synchronization between HTTP client (browser) and the HTTP server (web server) in wired Internet. A WAP gateway connects WAP client to HTTP servers. HTTP server serves the websites on the Internet. HTTP layer in TCP/IP protocol suite acts as application layer protocol when connecting to Internet on a wired network. WAP 2.0 have three important features over WAP 1.1—(a) SyncML synchronization, (b) WAP push service, and (c) MMS service.

WAP 1.1 architecture does not provide HTTP-TLS-TCP-IP layers at the WAP client for interacting with the WAP gateway. [TLS (transport layer security) is a security protocol.] WAP gateway, therefore, uses a decoder for sending request to web HTTP server and an encoder for sending response to WAP client. But WAP 2.0 has backward compatibility and supports WAP 1.1 devices.

Section 12.2.1 gives all layers of WAP in brief. Section 12.2.2 describes the physical and networking layers. Section 12.2.3 describes wireless transport layer security in WAP. Sections 12.2.4 and 12.2.5 describe wireless transaction and session protocols (WTP and WSP), respectively. WSP layer (Section 12.2.5) in WAP 2.0 acts as application layer protocol when connecting to Internet on a wireless network. SyncML codes bind with WSP for connectivity to the Internet. Synchronization is through SyncML codes binding over the WAP application layer client or server.

Section 12.2.6 describes wireless application environment (WAE). Hypertext means a text which can embed the links to any other text, image, audio clip, or video clip through the URLs and hence enable navigation through these URLs. HTTP transfers a hypertext, text, data, or voice tagged using HTML. Similarly, WAP transfers a text written in WML (wireless markup language).

HTTP client uses CGI scripts which are sent to the HTTP server in order to request for the contents required by an application running at the client. WMLScript is a WAP 1.1 script language in WML for retrieving the application data after running the script codes at the server. It enables the server to get the response which is sent back to the client through a gateway. Mobile devices are now used for a large number of innovative applications and WAP provides a powerful framework for these applications.

12.2.1 Layers of WAP

Figure 12.5(a) lists basic functions of WAE at a client device or mobile computing system. Figure 12.5(b) shows protocols layers WAP 1.1/WAP 2.0 transmitter and gateway/proxy, respectively. Figure 12.5(a) shows that WAE in WAP 1.1 consists of the following components:

- WML (wireless markup language)
- WMLScript
- WBXML (WAP binary XML)
- WTA (wireless telephony application)
- Data formats [vCard 2.1, vCalendar 1.0, address book, pictures (jpg, gif, ..)…].

These are explained in Section 12.2.6.

Figure 12.5(b) shows that WAE in WAP 2.0 consists of deploying XHTMLMP and browser. Moreover, it can directly deploy HTTP. The lower layers at the client devices are TLS and UDP (or ICMP) or TCP instead of WSP, WTP, WTLS (wireless TLS), and WDP (wireless datagram protocol) or WCMP (wireless Internet control message protocol) in WAP 1.1. WAP 1.1 gateway encoders and decoders are also not required in WAP 2.0.

WAP gateway provides *access to the wireless and wired networks* and also builds *caches* due to frequent disconnections in the wireless environment. It ensures security in wireless and wired networks. An HTTP server environment provides only pull mode service (Section 8.2.2) but WAP gateway has provisions for both push and pull mode services from the servers. Pull mode means client sends a request and pulls the data from server as a response.

12.2.1.1 WAP 1.1 Gateway

A WAP 1.1 gateway is required for *protocol conversions* between two ends—mobile client device and HTTP server. Gateway converts WAE 1.1, WSP, WTP, WTLS,

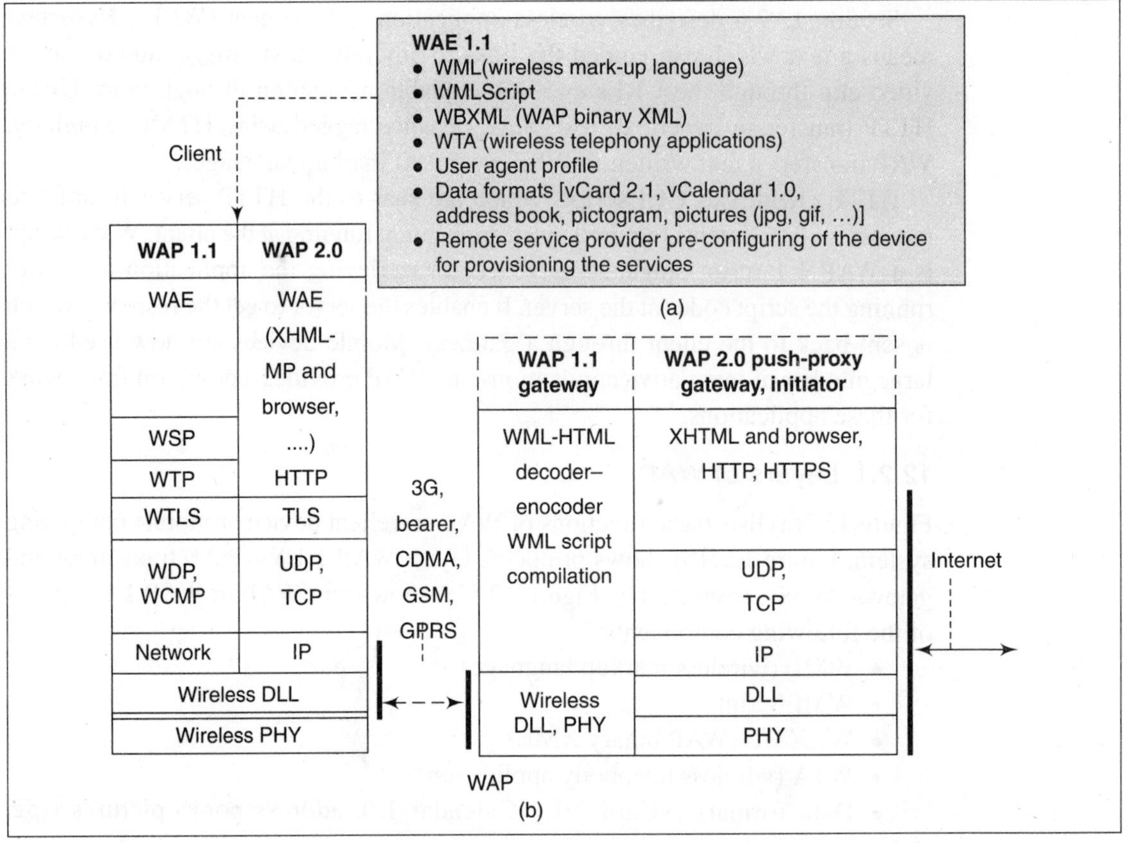

Fig. 12.5 (a) Functions of WAE (b) Protocols layers WAP 1.1/WAP 2.0 transmitter and gateway/proxy, respectively

and UDP layers encoded data packets into the HTTP, TLS (in HTTPS), and UDP layers encoded data packets when the device transmits data to server. It does decoding when vice versa is done, that is when the server sends data to client through the gateway.

WAP 1.1 gateway also has a *WML encoder–decoder* so that the application written in WML gets converted to HTML when WAE application is sending request to HTTP server. The response of the server, which is in HTML gets decoded into WML when the server responds to the request.

Another function that the gateway performs i*WMLScript compilation* into CGI script which runs at the HTTP server to get HTML response which is sent to the client application.

12.2.1.2 WAP Push-Proxy Gateway

WAP 2.0 uses XHTMLMP (Extensible Hypertext Markup Language Mobile Profile) in place of WML. Therefore encoding and compilation is not required and only a WAP 2.0 proxy suffices.

As mentioned above, HTTP web servers function in pull mode. That is, an HTTP client application sends a request to the server and the HTTP server sends the response. Mobile device applications require data dissemination by server in push mode, pull mode, and push-pull hybrid mode. A WAP 2.0 proxy is required for pull mode. A push-proxy gateway is used to exchange data packets between a mobile device through wired Internet and Web servers. The role of WAP 2.0 gateway is restricted to provisioning for push and pull mode services from the servers.

WAP transmission physical layer can be HSCSD (Sections 3.1.2, 3.8 to 3.10), SMS, and GPRS in GSM, CDPD, and 3G bearer (SMS, MMS, ...) services. Following subsections explain various protocol layers.

12.2.2 Physical and Networking Layers

Recapitulate Section 5.1.1.9. A datagram gives independent information and is stateless. The data of a datagram is sent by a connectionless protocol. WDP (wireless datagram protocol) is one such protocol to send the connectionless information as a datagram. A WDP in WAP suite is for datagram service, similar to UDP in TCP/IP suite. It can be used for multicasting a datagram on the network. WCMP (wireless control message protocol) is similar to ICMP (Section 5.1.1.9). It is another connectionless protocol, which is a part of the WAP protocol suite. WCMP employs a datagram with a WCMP header when sending the messages for querying to find information, reporting errors, making route address advertisement (Section 5.3.1.1), and for a router seeking (soliciting) messages.

Transmitted WDP datagram has a header and then user data which is received from upper layers at the device. The header consists of a source port, a destination port (optional), source address (an identifier IP address or telephone number), destination address (optional), length of data, and checksum bytes for the header (to check erroneous receipt of header). An error-code as per the error is also reported to the upper layer, for example, in case the datagram could not reach its destination. Figure 12.6 shows the transfer of a WDP datagram.

12.2.3 Wireless Transport Layer Security

When data transaction occurs between client device and gateway, wireless transport layer security (WTLS) assures integrity and privacy in transactions and device authentication. WTLS layer maps to SSL (secure socket layer) in HTTPS. [SSL is also called TLS (transport layer security).] WTLS supports TCP (transport layer protocols), WDP, and WCMP.

A secure session must be established before the data from upper layers (WAE, WSP, and WTP) that are above the WTLS layer is transmitted through gateway or proxy to other end peer and received through gateway or proxy to the upper layers (WTP, WSP, and WAE). WTLS specifies the following sequence of peer-to-peer message exchanges for establishment of the secure session:

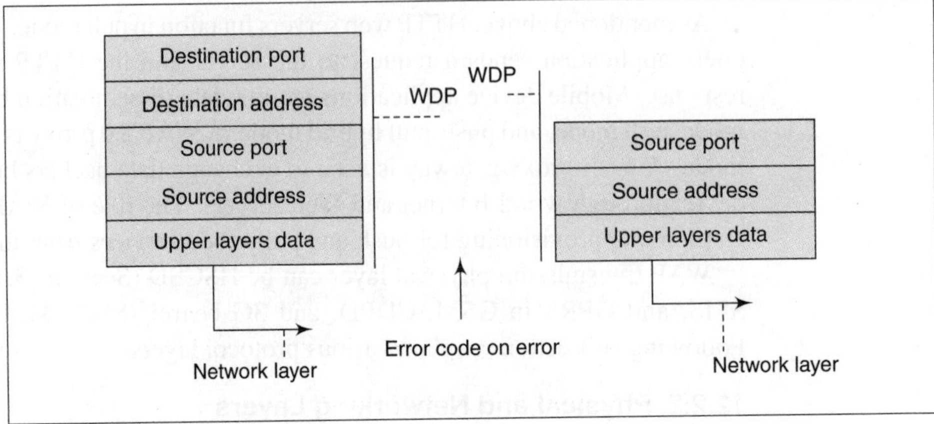

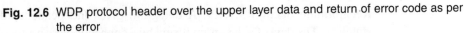

Fig. 12.6 WDP protocol header over the upper layer data and return of error code as per the error

1. Source device messages to *create* a secure channel as follows—(i) source address and port, (ii) destination address and port, (iii) RSA or ECC (a proposed suite of algorithms for key exchange), (iv) IDEA or DES (a proposed suite of algorithms for ciphering the data), and (v) compression method for data compression.
2. Other end messages for secure channel exchange for confirmation of *create* process as follows—(i) sequence number mode, (ii) how many times key is refreshed and exchanged again, (iii) identification of session after establishment of the session, (iv) RSA or ECC (a chosen suite of algorithms for key exchange), (v) IDEA or DES (a chosen suite of algorithms for ciphering the data), and (vi) chosen compression method for data compression.
3. On request from the other end, source device messages for secure channel public key authentication by a client certificate.
4. Source device messages to commit request (Section 7.4 explained meaning of committed transaction).
5. Other end peer messages for commit confirmation request.

After the above exchanges, the user data exchanges start as shown in Fig. 12.7.

12.2.4 Wireless Transaction Protocol

Wireless transaction protocol (WTP) transmits data to WTLS in case of secure transactions and directly to WDP or WCMP. WTP supports joining (fusion) of the messages and enables asynchronous transactions. A transaction is considered as aborted if it is initiated and started but it stops before completion and brings the state back to initial one. WTP supports abortion of the transactions and provides the information about the success or failure of a transaction to the sender. WTP is

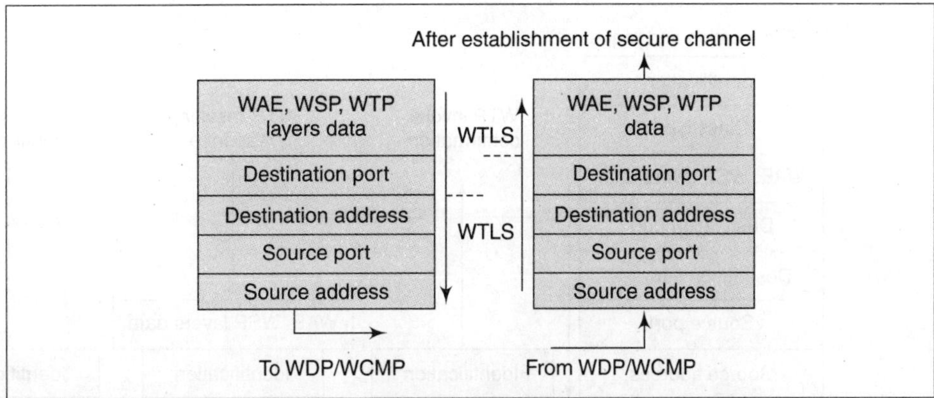

Fig. 12.7 WTLS protocol header over the upper layer data in requests and WTLS protocol header from the lower layers in responses

an interface to ensure reliability of transactions. There are three WTP service classes—0, 1, and 2.

1. Class 0—In this class, a source sends the messages with no response from the other end.
2. Class 1—This class is for reliable data transfer which takes place in the following manner—Source first invokes a transaction along with the request. Device then obtains the confirmation of invocation. This is followed by the transaction for the resulting response. The device sends the acknowledgement. The transaction removes duplicate data, provides retransmission as well as a transaction identifier. It provides push services in which there is no acknowledgement of data by user, except that there is confirmation of invocation.
3. Class 2—This class is also for reliable data transfer occurring as follows—Source first invokes a transaction along with the request. Device then obtains the acknowledgement of data (through gateway or proxy) from user. This is followed by a transaction for the resulting response. The device sends the acknowledgement. The transaction removes duplicate data, provides retransmission as well as a transaction identifier. It provides acknowledgement of two types—user acknowledgement and automatic acknowledgement.

Figure 12.8 shows WTP headers when sending WTP invocation and request for results, confirmation of WTP invocation, WTP resulting response, and WTP result confirmation.

12.2.5 Wireless Session Protocol

Wireless session protocol (WSP) transmits data to WTP in case of thin client transactions or directly to WDP or WCMP. WTLS serves as a layer above WDP

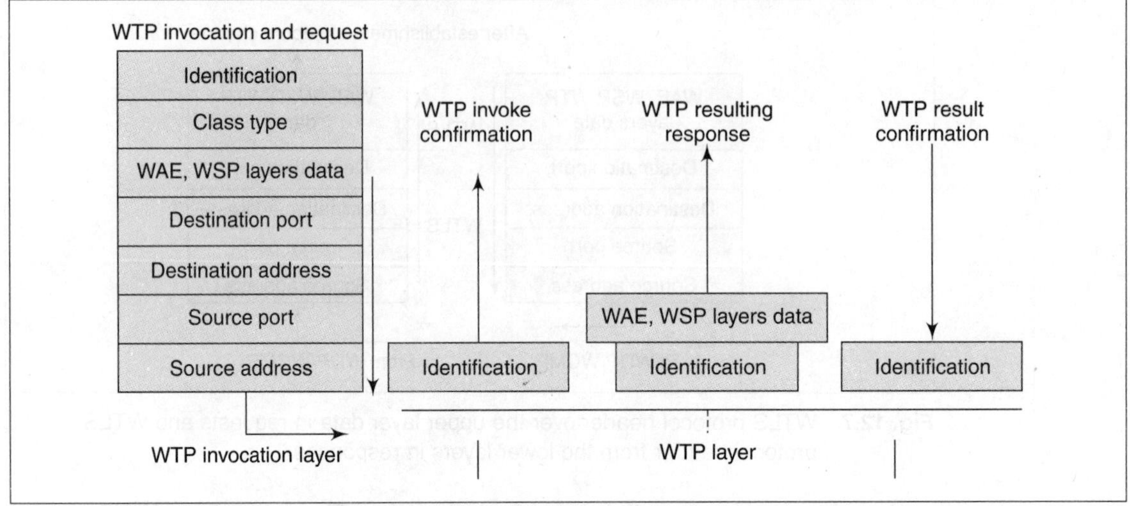

Fig. 12.8 WTP headers when sending WTP invocation and request for results, confirmation of WTP invocation, WTP resulting response and WTP result confirmation

when a secure transfer is required for a datagram. Just as HTTP, WSP supports stateless data transfers. This enables a browser to get the packets from the server in any sequence.

WSP also supports asynchronous exchanges, multiple requests, push and pull mechanisms of data dissemination (Section 8.2), capability negotiation, content encoding, content type definitions, and WBXML (WAP binary XML). It possesses HTTP functionality. WSP manages sessions as follows:

1. A session is first established. It can use the functions of agreed common protocol. Figure 12.9 shows WSP protocol session connection establishment and resulting response headers.

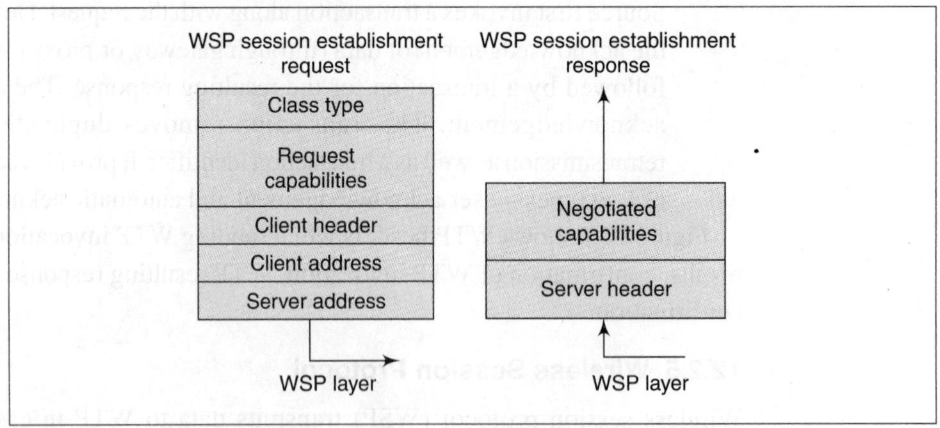

Fig. 12.9 WSP protocol session connection establishment and resulting response headers

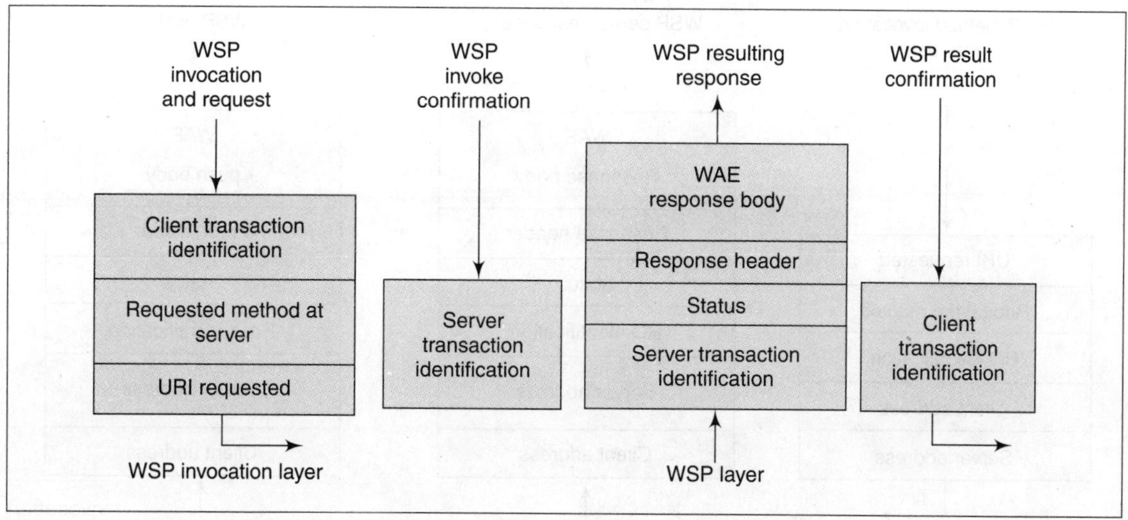

Fig. 12.10 WSP headers when sending WSP invocation and request for results, confirmation of WSP invocation, WSP resulting response and WSP result confirmation

2. An established session can be suspended and then resumed from the point at which it was suspended.
3. A session can be terminated (released).

There are three WSP service classes—0, 1, and 2. Figure 12.10 shows WSP headers when sending WSP invocation and request for results, confirmation of WSP invocation, WTP resulting response, and WSP result confirmation.

1. Class 0—This class is for a source sending the unconfirmed push. It supports session suspension, resumption, and management. The messages sent from the source do not get any response from the other end.
2. Class 1—for a source sending the confirmed push.
3. Class 2—for a source supporting session invocation, suspension, and resumption.

Figure 12.11 shows WSP headers when connectionless WSP exchanges take place for WSP in the following sequence:

1. A method is invoked and server is requested for results.
2. The method runs at the server and generates a response.
3. The server pushes the response.

12.2.6 Wireless Application Environment

WAP 1.1 has two sets of software—WAE (wireless application environment) user agent and WAE services. Section 12.2.1 mentioned that WAE in WAP 1.1 consists of WML, WBXML, WTA, WMLScript, and data formats. Examples of data formats are vCard 2.1, vCalendar 1.0, address book, and pictures (jpg, gif, …)

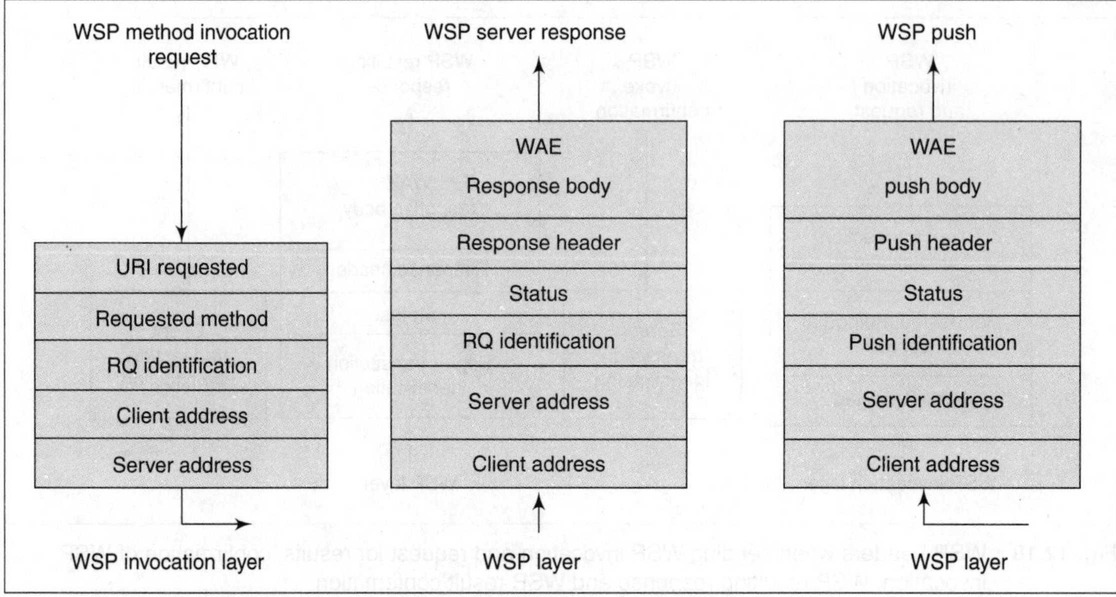

Fig. 12.11 WSP headers when connectionless WSP exchanges take place for WSP invocation and request for results, method resulting response and server pushes, respectively

12.2.6.1 WML

Recapitulate Sample Codes 9.1 to 9.5 for applying the XML and XML-based language (SyncML or SMIL). Sample Codes 7.1 and 9.2 demonstrated the simple applications for creation of an XML database or document. These codes showed how the markup by the tags is done in between a text to give the characteristic information or a specific meaning to the text or to define the specific function or action of the text. The codes also showed how the definition of attribute along with a tag provides the data or additional characteristic information for the text within a pair of start and end tags.

A mobile device has following characteristics—(a) narrow bandwidth network connection with intermittent loss of connectivity, (b) string parameterization and state management, (c) constraints of (i) limited user input and output facilities [for example, T9 keypad input, text presentation in a layout with small display screen, or image or pictogram presentation on a screen with small resolution], and (ii) computational resources and limited memory. WML is an XML-based language which takes into account these constraints while programming an application required for running on the device.

State management is an important feature in mobile devices. Concept of the states and changes in the state can be understood from the following example for a change in displayed state on the device screen.

Example 12.2 Show how the states of the strings on display change on the screen of a mobile device.

Solution:

1. State 1—Display state is a string 'Message'. Display state becomes 'Text message' in top line, 'Voice message' in second line, 'Minibrowser messages' in third line, and so on. Top line is highlighted and is the present position of the cursor on the screen.
2. Switch to state 2—If a down cursor-key is pressed, then 'Voice message' will become the top line. The state will become 2. If option-select key is pressed, then menu for 'Voice Message' will appear on the screen and state will become 2A.
3. Switch to state 1A from state 1—If option-select key is pressed, then display menu changes to that for text messages. Top line will be 'Create message', second line 'Inbox', and so on.

One is well-aware of an *HTML page* which is downloaded daily while surfing the Web. A page contains text, image presentation, forms, and hypertext links. These links enable a client to navigate from one page to another. The link is defined using a URI (universal resource identifier) or URL (universal resource locator).

WML is used to create the cards for mobile application(s). Two versions of WML are WML 2.x and WML 1.x. WML 2.x includes XHTML-MP which includes XHTML. Section 12.3 will describe XHTML. WML 1.x does not include XHTML. A scaled down set of procedural elements are used to control navigation between cards. A card represents an interaction with the user and the deck contains the cards. A WML parser parses the tags, attributes, and underlying text within the tags present within the deck or card. [The parser is a part of a browser or server.] A WML card has following features:

- Provides the *content* (e.g., a program, command, data, string, or image).
- Supports variety of formatting commands as well as layout commands. The commands are defined by tags and attributes.
- Provides user interface for mobile devices with constraints as mentioned in the preceding text.
- Organizes similar to deck and cards. All information in WML is a collection of decks and cards. A WML deck is saved in a file with extension wml. Each file contains one deck. For example, a *welcome deck* can be saved in a WML file *welcome.wml.* Figure 12.12 shows the format of a WML deck and card. Each deck can have number of cards. There is a navigational link from one card to another. WML provides for management of the navigation between cards and decks.

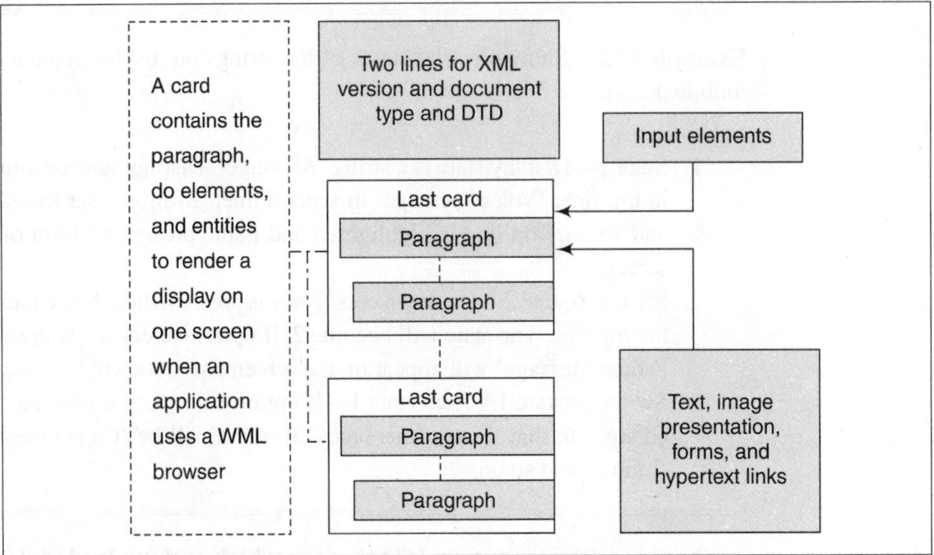

Fig. 12.12 Format of WML deck and WML card, and paragraph contents

Sample Code 12.1 This codes is an example of a WML deck which is saved in a file called *example12_3.wml*.

First line specifies the XML 1.0 version on which the WML card is based. Second line specifies the document type and DTD for the document. Third line has the tag `<wml>` to indicate the start of a WML document which will end at the tag `</wml>`. Fourth line specifies the start of a *card*. One of the attributes of the card is an *id* which is defined here by *welcome*. Another attribute is *title* which is defined here by *First Card*. The card specifications are upto the line `</card>`. The paragraph tag is *p*. Another attribute called *mode* is defined here as wrap so that text wraps when displayed by browser. Text in the present example is WELCOME TO ABC MOBILE. The code for the deck is shown below:

```
<?xml version="1.0"?>
<!DOCTYPE wml PUBLIC "-//PHONE.COM//DTD WML 1.1//EN"
"http://www.phone.com/dtd/wml13.dtd" >
<wml>
<card id="welcome" title="First Card">
<p mode="wrap"> WELCOME TO ABC MOBILE </p>
</card>
</wml>
```

A WML card is first validated against its declared document type using WML 1.3 DTD (document type definition) before parsing. Parsed data, information, and contents are used to give input to a Java program for the application or server which runs *method*(s) at the browser or server. [A program object consists of the *methods* (called functions in C).] Browser program runs at the client. WinWAP has an Emulator which is an alternative program used for emulating the actual run at the mobile client and runs on a PC. WinWAP is for a computing system running on PocketPC, WindowsMobile 2003, or Windows operating system.

The following explains how an application runs at a mobile device in WAE. A WML card containing a client-request is transmitted and is decoded at the gateway. After decoding, the gateway communicates the request to the server in HTML format. The server generates an HTML response by employing the method(s) which run at the server. The response is encoded in WML form at the gateway. The gateway transmits the WML response to mobile-device client which runs an application after parsing the WML. The application uses the methods which run at the client.

The element *do* is used to process the text within the do tags. The element *label* is an attribute which defines a text, the purpose of which is simply to specify the incoming text or action. Label text is not the input to any program or processing element. For example, label = "Show Welcome Message". This means that next action in the sequence of actions for navigation to the card is showing the welcome message. The navigation is by `<go href = "#FirstCard">`.

Sample Code 12.2 This code is an example to show how the displayed text switches to welcome message while processing a WML document at client. Use the card defined in Example 12.3.

A possible solution is as under:

```
<do>
<p mode="wrap" label = "Show Welcome Message">
<go href = "#FirstCard">
</p>
</do>
```

WML has tags `<b>`, `<u>`, `<i>`, `<big>`, `<small>`, `<strong>`, and `<em>` for bold, underline, italic, big, small, strong, and emphasis rendering, respectively, for a text-display. HTML anchor tag `<a with attribute href = ....>` is used for linking and navigation in WML also. `<timer>` is another tag. The actions on events are by the tags `<ontimer>`, `<oneventbackward>`, `<oneventforward>`, and `<oneventpick>`. The renderings after the actions on the events occur as per the interior paragraph entities.

12.2.6.2 WMLScript

A WMLScript is a script language in which each line is loaded in computer and is executed at run time only. There is no pre-compilation. WMLScript in WAP is similar to JavaScript and is used for client-side scripting. It obviates the need to communicate with the server by sending a request and waiting for the response generated by an application running at the server. WMLScript can embed the markups in WML. WAP browser displays the page having WML and WMLScript. WMLScript is used to open dialog box so that the user can input data or text. It is also used for generating error messages. The execution of WMLScript is fast because it is compiled at the gateway and the byte code is sent to client. An example of WMLScript is given below followed by a description of standard library functions.

Sample Code 12.3 This code shows how can the division $z = x \div y$ can be carried out using WMLScript.

```
extern function divide (varCompute x, y)
{
    var z = x/y;
    WMLBrowser.setvar (varCompute, z)
    WMLBrowser.refresh ( );
}
```

Standard Library Functions Standard library functions in any programming language help a programmer to directly use the library functions. WMLScript standard libraries have WMLScript functions which cannot be easily implemented in case a person decides to code these himself. A brief description of the different libraries and the corresponding functions is as follows:

- *WMLBrowser* library has the functions to control the WML browser or to get information from the browser.
- *WMLDialogs* library has the functions which display the input boxes to users. It also provides for alert and confirmation messages.
- WMLLang library has the core WML functions. For example, converting a data type integer to string character.
- *WMLString* library has the functions that help in concatenation, truncation, picking of select portions, and manipulation or finding the length of the strings. An example is the *find()* function. It helps in knowing whether a sub-string is a part of a string. If yes, then the function returns the index of the first character of the match in the string, otherwise it returns –1. String.find ("09229122230", "30") returns 9 which is the index of first character of the match in the string. String.find ("09229122230", "39") returns-1 since there is no match between sub-string characters and the string. Another example is var *strlen* = String.length ("WELCOME TO ABC MOBILE). It returns 21 because number of string characters are 21 (Space is also a character.).
- WMLURL has functions for using relative URLs or absolute URLs for finding the port number or for testing whether a URL is valid or not. [For example, http://www.microsoft.com/msoffice/winword/ is a relative URL.

Full form http://www.microsoft.com/msoffice/winword/newfile.doc in which the file name newfile.doc is also mentioned in the end after the winword/ is called absolute URL.]

- *WMLFloat* library has the functions that help in performing floating-point arithmetic operations in case a specific WAP device supports floating-point operations, conversions, and calculations.

12.2.6.3 WBXML

WAP 1.1 provides for communication of client with gateway or proxy using WBXML. XML and WML page document are not compact. WBXML is a specification in binary representation so that XML or XML-based language can be transmitted in compact format. A binary number can represent a tag in place of characters. Another binary number can represent an attribute in place of characters. For example, attribute ID needs two characters. It is represented by a single byte. Attribute title needs five characters. It is also represented by a single byte.

12.2.6.4 WTA

The specific telephonic features are call set up, call accept, call forwarding, caller line ID, connected line ID, closed user group, multiparty groupings, call hold, call waiting, call barring, operator restrictions, call charge advice, contacts entry in phone book, call hold, conferencing, ring tones, speed dial, telephone/fax, SMS up to 160 characters, emergency number, MMS—gif, jpg, wbmp, teletext, and videotext access. These features are defined by WATI (WTA interface). It provides the interfaces for the features using WML browser. A WTA URI can be wtai://wap.mcard: followed by a telephone number. This is identical to port number specifications provided in the URL.

A WTA server can push (Section 8.2.1) the WMLScript or deck contents. A WTA event handler can handle events. An example of an event is change in data-field content. A persistent storage interface helps in storing the data on device when the content is modified. WTA also provides security interface. Only authorised gateway or proxy can access the data at the server.

12.2.6.5 User Agent Profile

User agent is software used by the user to give input using VUI (voice user interface) and GUI (graphic user interface) and to interact with mini browser (browser with limited screen size). It executes the WMLScript at the client and displays the results. User agent displays the WML decks received as response from the server. User Agent Profile provides small screen device characteristics, font, and display capabilities. The profile also enhances the input capabilities, for example, the use of T9 keypad, stylus, and touch screen is enabled.

12.2.6.6 Data Formats

The data displayed on a mobile device is in special data formats. vCard 2.1 is the format for visiting card. vCalendar 1.0 is the format for calendar. Also a mobile device provides pictogram which is a small picture of very low resolution that cannot be split and can be placed along with the text. A pictogram is used for displaying logo.

12.2.6.7 Remote Service Provider Pre-configuring of the Device for Provisioning of Services

WAP 2.0 provides for SyncML (Section 9.3) which is used for data synchronization between server and mobile devices. WAP 1.1 also provides the functions and WML methods at the server for pre-configuration of the device from the server.

12.2.6.8 WAP Push

The constraints of mobile devices are low computing capability and narrow bandwidth network connection with intermittent loss of connectivity. Sections 8.1 to 8.4 described push models and methods for push and push-pull for data dissemination. WAP provides simple protocol stacks also. WSP in WAE can be used for WAP user agent push service. WTA events can be used to push the data to the device.

WAP provides Push OTA (over the air) which is a simple protocol sub-layer in WSP. It provides authentication of the push initiator (server) and also helps in selection of the pushed contents. The protocol handles push-session-request, connect, suspend, resume, and disconnect functions. It also handles push, server-confirmed push, abort, and unit-push functions.

12.3 XHTML-MP (Extensible Hypertext Markup Language Mobile Profile)

HTML is not rigid in syntax. For example, if the end tag corresponding to a start tag or an inverted comma is missing or some other syntax error is present, then also the HTML browser will interpret it somehow. It will make the required assumptions. HTML has limited number of tags. It does not support very small or multiple screen sizes or reduced pixel resolution required to fit an image or pictogram on the screen. HTML does not support multiple ways of user inputs. That is, it does not let the user give inputs by using different keys, keypads, stylus with touch screen, and voice while interacting with the device. Moreover, HTML does not allow state transitions based on user inputs and displayed information.

XML is rigid in syntax and needs the document to be *well-formed* (syntactically correct). It does not make any assumptions. It uses the corresponding DTD for validation of the document. XML can thus be encoded into other formats and then retrieved back into original format. For example, it can be encoded in WBXML for transmission and WBXML can be decoded back into XML on reception.

XML is in document form and is textual and structured. It is thus extremely valuable in developing GUIs, VUIs, or any other human interface with computer.

Due to the feature of extensibility, XML supports the use of very small or multiple screen sizes or reduced pixel resolution. It also supports multiple ways of user inputs, that is, inputs from different keys, keypads, stylus with touch screen, and voice inputs when a user is interacting with the device. Due to its extensibility feature, XML has unlimited number of tags and allows state transitions based on user inputs and displayed information. XLINK feature of XML facilitates the state transitions and display of nested menus.

For all the above reasons, XML-based languages SyncML, Funambol, SMIL, and WML are used (Sections 9.3, 9.4, 9.5, and 12.2.6, respectively). These enable the development of the data synchronization applications, Java applications, multimedia interchange applications, and WAP device applications, respectively.

XHTML has two forms—XHTML basic and XHTML-MP. XHTML basic is HTML with strict syntax. XHTML-MP is for mobile devices and PDAs. An XHTML document can be parsed using a standard XML library and processed like any other standard XML document but using the same set of tags as in HTML. XHTML 1.1 has been a W3C recommendation since middle of 2001.

XHTML can be partitioned into modules and provides for navigation from one module to another. This feature is characteristic of XLINK. For example, WML deck is partitioned into cards and navigation from one card to another is permitted through hyperlinks (Fig. 12.12). XHTML basic does not support scripting but supports the basic *forms* with POST/GET features as in case of HTML. It supports tables but not the client-side style sheets since small devices are assumed to possess small computation resources. Frames are also not supported in basic XHTML.

New features are added to the HTML family of markup languages by XHTML 2.0. The elements can function as a hyperlink or with src attribute. [The src is attribute which specifies a source file, for example, a jpg image file.] This feature is similar to the cards in WML. XHTML 2.0 is only a draft form and not yet a recommended standard. A new version of a language or software is considered as full backward compatible if that version supports and includes all features in the codes written earlier. When recommended, XHTML 2.0 will be the new version with break from full backward compatibility with XHTML 1.1. XHTML 2.0 user agent parser is compatible with XHTML 1.1. HTML form will be replaced by XML-based user input specification and can be displayed as per user specfications and on the appropriate display devices. These are called XFORMs. XFrames have new features compared to HTML frames. The document object model (DOM) events are now XMLEvents and will use XML DOM.

XHTML 2.0 has a number of modules. Examples are edit, presentation, meta information (abstract or title of information, which specify how and what the full information are) of a document, text, list, table, forms, structures, targets, bi-directional text, style-sheets, server-side image map, client-side image map, scripting, and link.

12.4 Bluetooth-enabled Devices Network

The name Bluetooth is derived from the name of a Danish king. The king Harald Blatand which means Bluetooth in English reigned before 1000 A.D. Bluetooth devices can form a network known as *piconet* with the devices within a distance of about 10 m. Various piconets form an ad-hoc network called *scatternet* within 100 m through a Bluetooth-enabled bridging device.

Bluetooth 2.0 (from 2004) has backward compatibility with Bluetooth 1.x. It provides for improved BER (bit error rate) performance, three times enhanced maximum data rate of 3.0 Mbps over 100 m, and simplification of piconet multi-link transactions due to higher bandwidth (3 MHz) availability.

Section 12.4.1 describes synchronization in a WPAN and remote computing systems. Section 12.4.2 describes piconet and scatternet in a Bluetooth WPAN. Table 12.1 gives the basic features of Bluetooth.

Table 12.1 Basic features of Bluetooth

Property	*Description*
Frequency band	2.4 GHz with Bluetooth radio characteristics given in Table 12.2
Bluetooth protocol layers	(a) Layer 1—baseband and radio (b) Data link layer—layer 2—RF-communication, L2CAP (logical link control and adaptation protocol) and LM (link manager) or host-controller interface, layer 2—L2CAP, or layer 2—audio (c) Layer 6—session—object exchange (OBEX) (d) Layer 7—security and application software layers are as specified by Bluetooth Sync, vCal (for Calendar), vCard [for contact (visiting card)] or Object Push (PIM) or Binary File Transfer or audio applications.
Sessions, object exchange, and other protocols	SyncML client, SyncML engine, and OBEX (Fig. 12.13(a))
Network characteristics	(a) Connection-oriented communication (b) Master–slave communication within same piconet (Section 12.4.2) (c) Negligible interference between piconets as each uses distinct channel-frequency hopping sequences (d) Ad-hoc network peer-to-peer communication between two devices on two different piconets in a scatternet
Bluetooth features	• Used for low power short range transmission • Employed for wireless short range exchanges in mobile environment within 10 m network in master–slave mode and within 100 m in scatternet. Bluetooth radiations between piconets are omni-directional. • Network connection latency—3 s • Bit rate—less than 1 Mbps • Protocol stack—larger than IrDA or Bluetooth • Code size—2% to 50% more compared to that for a Zigbee device • Bluetooth radio—FHSS

(*Contd*)

(*Contd*)

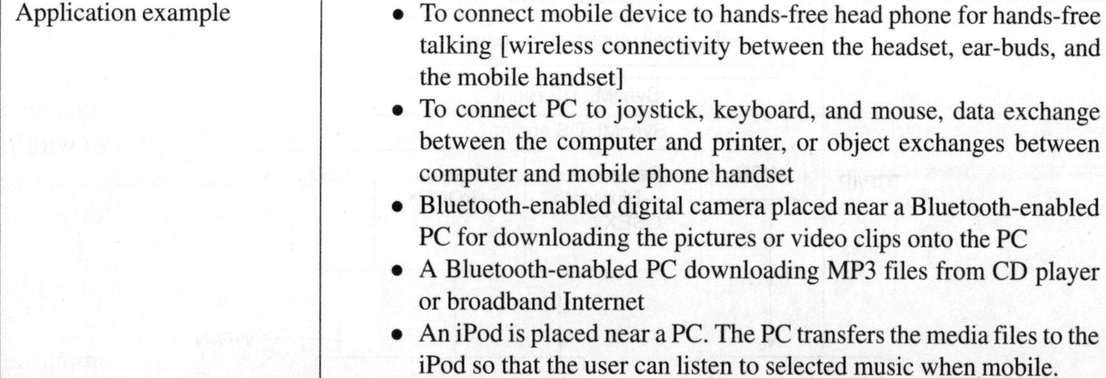

Application example	• To connect mobile device to hands-free head phone for hands-free talking [wireless connectivity between the headset, ear-buds, and the mobile handset] • To connect PC to joystick, keyboard, and mouse, data exchange between the computer and printer, or object exchanges between computer and mobile phone handset • Bluetooth-enabled digital camera placed near a Bluetooth-enabled PC for downloading the pictures or video clips onto the PC • A Bluetooth-enabled PC downloading MP3 files from CD player or broadband Internet • An iPod is placed near a PC. The PC transfers the media files to the iPod so that the user can listen to selected music when mobile.

12.4.1 WPAN Synchronization

Figure 12.13(a) shows how the data synchronization (Section 9.1) takes place between the mobile devices and computing systems in a WPAN and how remote computing systems connect to each other through Internet. Figure 12.13(a) shows how synchronization occurs between the personal and remote area devices and systems using TCP/IP, WSP, Bluetooth, IrDA, and USB.

Figure 12.13(b) demonstrates synchronization between Bluetooth devices within short distance (1 to 100 m range as per the radio) and remote computing systems. The data rates between the wireless electronic devices are up to 1 Mbps. Bluetooth protocol is used for wireless personal area synchronization among mobile devices and Bluetooth-enabled PCs. Bluetooth protocol is connection-oriented protocol using Bluetooth object exchange OBEX (a protocol for transport layer in Bluetooth). An object can be a file, address book, or presentation with a specification for a method which runs a specific task.

Synchronization can be through SyncML codes. SyncML uses the message CONNECT, followed by PUT or GET, and then ABORT, either automatically at periodic intervals, or initiated by device or PC (Section 9.4). Alternatively, synchronization is carried out by configuring a Bluetooth ActiveSync partnership or first configuring virtual COM port at Bluetooth and then synchronizing through ActiveSync. Bluetooth synchronizes PIM data (calendar, e-mail, business card, text messages, and phone book) (Section 9.3.1).

12.4.2 Bluetooth Networks

Figure 12.14(a) shows an example of a piconet of Bluetooth devices with master–slave architecture. Any device can function as master or slave. The device which first establishes a piconet becomes master and others which discover the master become slaves in the piconet. Slave means that the clock of the master functions as reference for synchronization (Section 9.1).

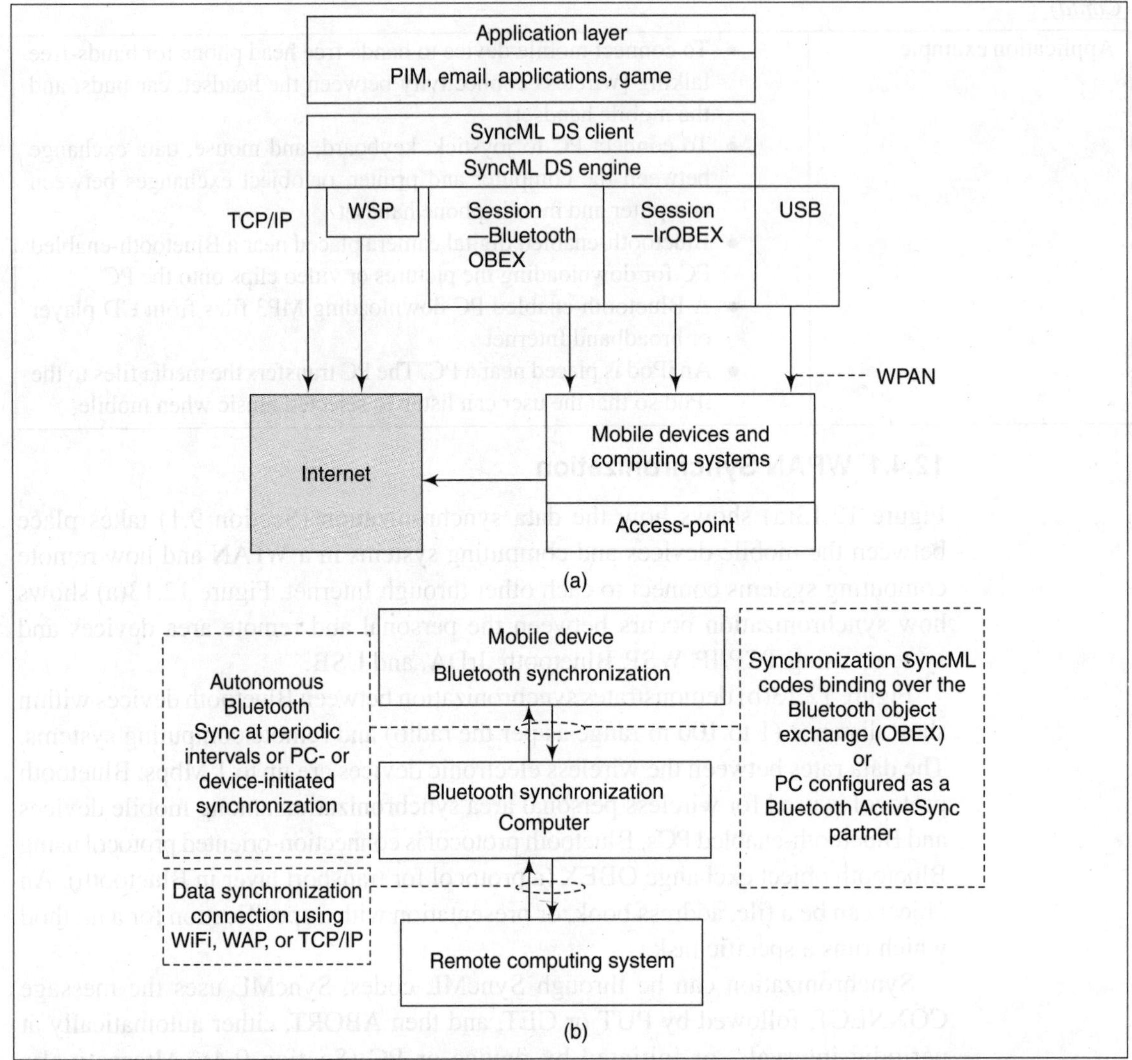

Fig. 12.13 (a) Data synchronization (Section 9.1) between the mobile devices and computing systems in a WPAN takes place and how remote computing systems connect to each other through Internet (b) Synchronization between the Bluetooth devices within short distance (1 m to 100 m range as per the radio) and remote computing systems

A master synchronizes all active devices and there are identical hopping sequences of their frequencies for each device *radio* (transmitter) (Section 4.3.2). A hopping sequence defines one channel. There can be a maximum of eight devices with a master in piconet and a maximum of 79 channels in Bluetooth networks.

The devices in a piconet can be present in the following states:

(i) Standby state—When a device is in standby state, it is actually waiting to discover the master and thus the piconet and no RF signal communication is taking place). The device is yet to be assigned an address in the piconet.

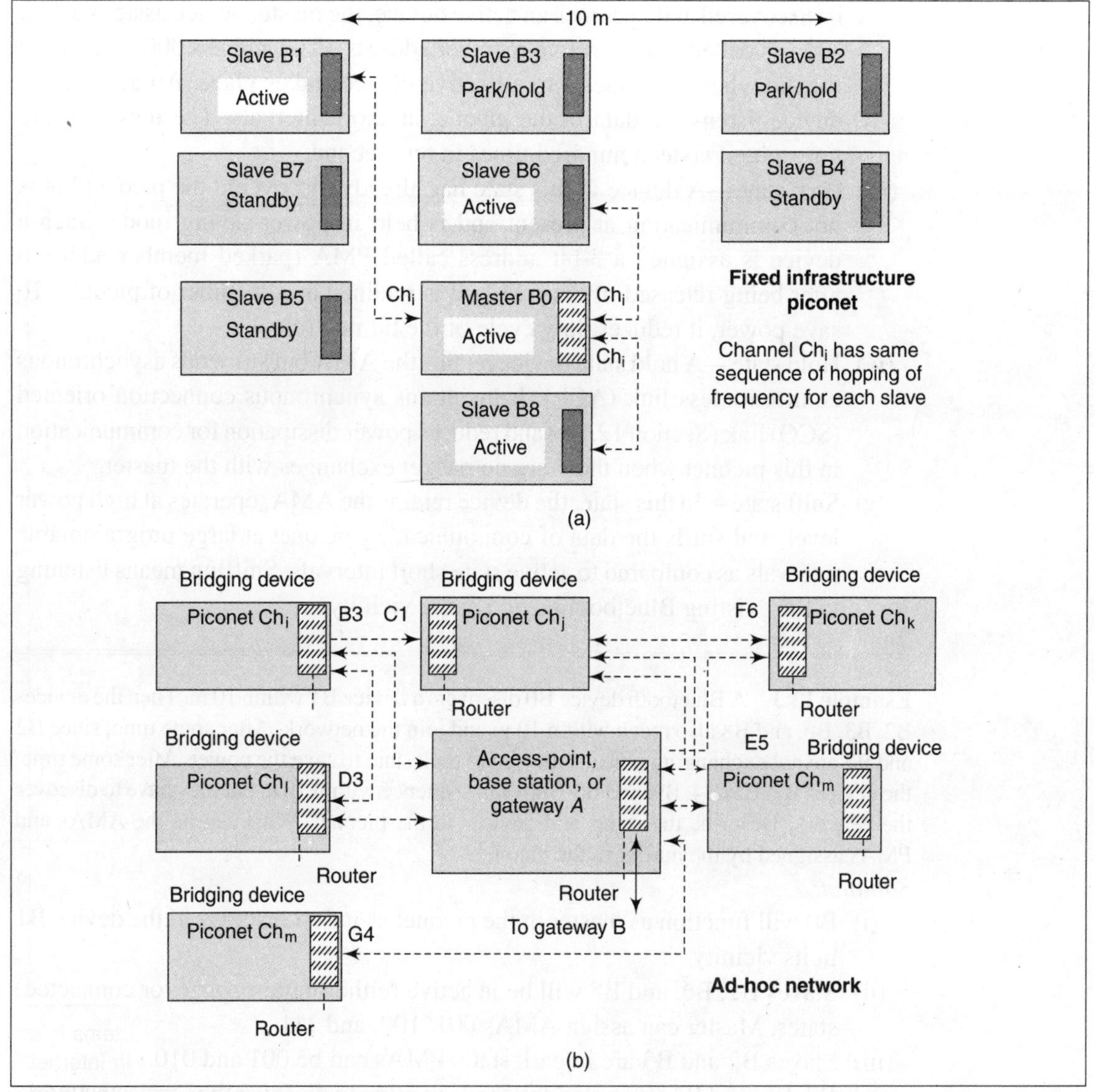

Fig. 12.14 (a) Piconet of three active Bluetooth devices and a master, each synchronized and using same hopping sequence of their frequencies forming a fixed infrastructure network architecture. Two are in park/hold state and three are in standby mode
(b) Bluetooth enabled devices ad-hoc network architecture forming a scatternet through a bridging devices of the piconets

(ii) Active state—An active device can be in one of the three modes—(a) *inquiring* (carrying out discovery broadcasting in all neighbourhood and listening to the response for finding out a radio channel to connect) (b) *paging* (sending a page specifying the relationship with master after discovering the channel), and (c) connected and performing data transactions. When a device is discovered and becomes an active device, the master device assigns a 3 bit address called AMA (active member address). The address 000 is reserved for use when a broadcast to all active devices takes place. An active state device listens the data in the piconet at short intervals. The messages are transmitted sixteen hundred times in one second.

(iii) Park state—A device in this state has already discovered the piconet but is not communicating at present and is held in power-saving mode. Such a device is assigned a 3-bit address called PMA (parked member address) after being released of the AMA. It is retained as a member of piconet. To save power, it reduces duty cycle of the bit rate (clock).

(iv) Hold state—A hold state device retains the AMA but suspends asynchronous connectionless link (ACL). It maintains synchronous connection oriented (SCO) link (Section 12.5.1) and reduces power dissipation for communication in this piconet when there are no packet exchanges with the master.

(v) Sniff state—In this state, the device retains the AMA, operates at high power level, and sniffs the data of communicating piconet at large programmable intervals as compared to active state short intervals. Sniffing means listening to the existing Bluetooth device in the vicinity.

Example 12.3 A Bluetooth device B0 discovers a device B1 within 10 m. Then the devices B2, B3, B6, and B8 also reach within 10 m and join the network. After some time, since B2 and B3 are not exchanging objects, they go to park state to save the power. After some time, the devices B7, B5, B4, B9, and B10 move in sequence within 10 m but they have to discover the network. Describe the states and devices in the piconet. What can be the AMAs and PMAs assigned by the master in the piconet?

Solution:

(i) B0 will function as master in the piconet as it first discovered the device B1 in its vicinity.

(ii) Slaves B1, B6, and B8 will be in active (either inquire, page, or connected) states. Master can assign AMAs 001, 100, and 111.

(iii) Slaves B2 and B3 are in park state. PMAs can be 001 and 010.

(iv) B7, B5, and B4 are in standby mode waiting to discover the service network. These devices are yet to be assigned member address by the master.

(v) B9 and B10 cannot be a part of this piconet and will form another piconet operating at another channel with another sequence of frequency hops.

Master and slave devices can move out to another piconet. Master becomes slave in a new piconet. When it moves to another established piconet, the communication in the previous piconet freezes. The device which rediscovers another device then becomes master in the previous piconet. When a slave moves to another area, it communicates its unavailability to its master. It then synchronizes with a new piconet.

Scatternet is an ad-hoc network formed by the bridging devices (Fig. 12.14(b)). A bridging device connects two piconets. Any Bluetooth device can function as a bridge in order to form the ad-hoc network. In such a network, two devices in two piconets communicate as in a peer-to-peer communication. The two devices use two different channels (two different hopping sequences from among the 79 provided). Each piconet uses FH-CDMA so that there is a distinct hopping sequence with respect to other piconets in the scatternet. Therefore, there are no collisions between signals in two piconets.

Figure 12.14(b) shows an example of a scatternet of Bluetooth piconets. It demonstrates an ad-hoc network architecture of Bluetooth-enabled devices formed by the bridging devices. There are six piconets consisting of the devices B0, …, B7, C0, …, C7, D0, …, D7, E0, …, E7, F0, …, F7, and G0, …, G7. Figure 12.14(b) shows that scatternet is formed with B3, C1, D3, E5, F6, and G4 as bridging devices.

12.5 Layers in Bluetooth Protocol

Bluetooth uses protocol IEEE 802.15.1 specifications. A Bluetooth device has number of protocols that can be used at the application, presentation, session, transport, network, data link, and physical layers (layers 7–1). For example, vCard, vCal, telephonic network protocol, device management protocol, SyncML, SyncML Client, SyncML Engine are the protocols at the application layer. A host-controller interface manages the link between the upper layer and baseband and radio sub-layers of the physical layer. Figure 12.15(a) lists the functions of the protocol layers and Fig. 12.15(b) shows the protocol layers in Bluetooth.

12.5.1 Physical Layer

A physical layer (layer 1) in a network is responsible for transmitting header fields encapsulating the payload and payload. (Payload means data to be transmitted after passing through the layers 7 to 2 and retrieved at other end after passing through layers 2 to 7. A sub-layer of physical layer is responsible for interfacing with the upper layers. Another sub-layer generates the baseband signal used for transmission and yet another generates the radio signals as per the data bits from or for the MAC (layer 2).

Bluetooth device physical layer has three sub-layers—radio, baseband, and link manager or host controller interface. These sub-layers are described below.

12.5.1.1 Radio

Radio sub-layer transmits packets received from baseband sub-layer. At the other end of the link, it receives and sends the received signals for formatting packets at

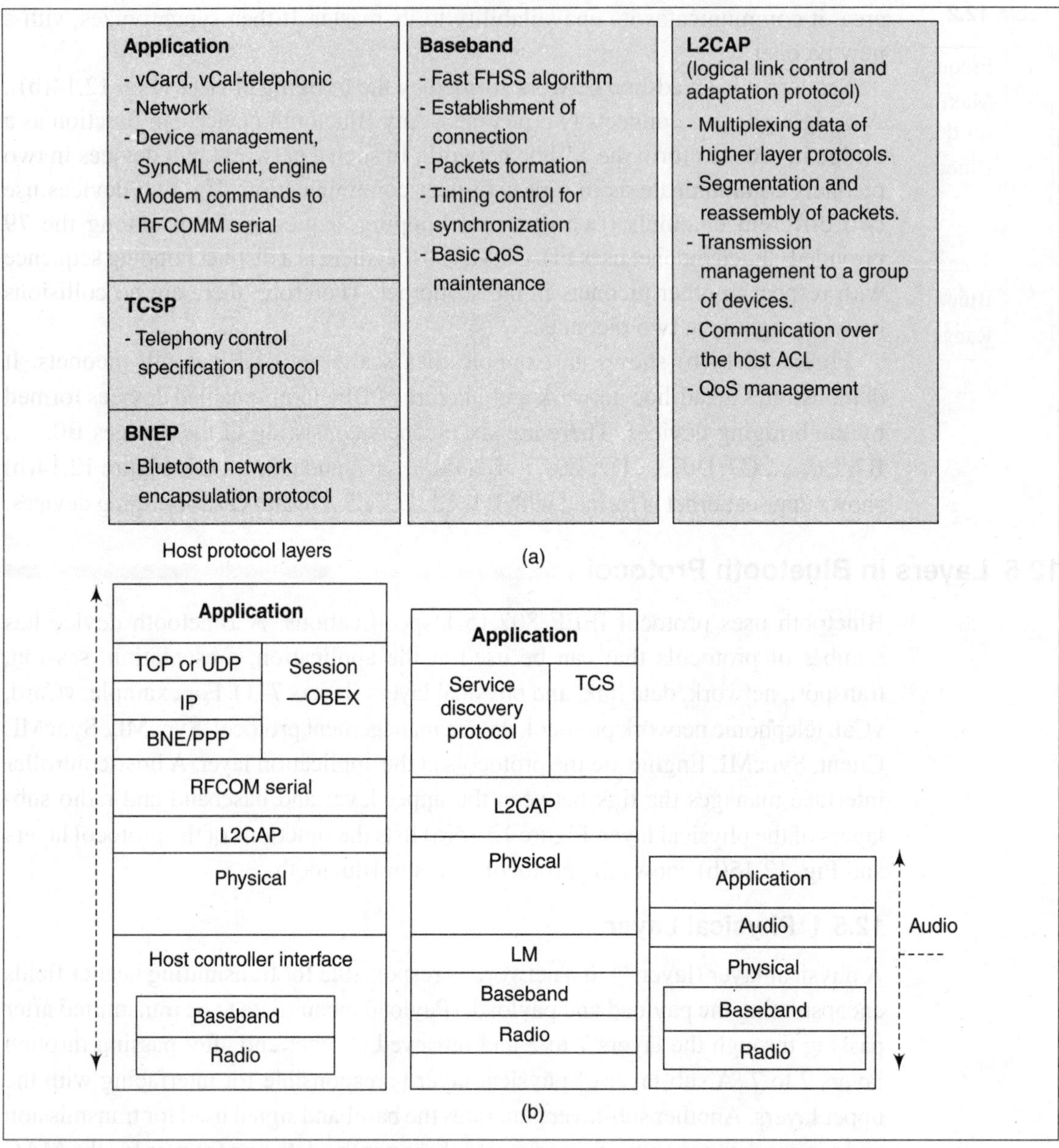

Fig. 12.15 (a) Functions of protocol layers (b) Host protocol layers and other protocol layers between the object client, device and server

baseband sub-layer. Bluetooth devices operate at the least required power levels so that the transactions may not be detected by a distant receiver of signals, thus maintaining confidentiality. Moreover, frequency hopping CDMA ensures negligible interference and least risk of jamming by distant sources. Table 12.2 gives the characteristics of the radio sub-layer.

Table 12.2 Radio sub-layer characteristics

1.	Frequency band	2.4 GHz
2	Maximum power levels for devices	1, 2, or 3 corresponding to 100 mW, 2.5 mW, or 1 mW
3.	Bluetooth radio	FHSS (Section 4.3.2) with each carrier separated by 1 MHz, 1600 frequency hops per second among 79 carriers, and frequency changing after every 1/1600 s or 625 μs
4	Transceiver modulation	GFSK (Section 4.1.1)
5	Bluetooth data transfer rates	1 Mbps or less
6	Range	10 m in a piconet and 100 m in a scatternet

12.5.1.2 Baseband

Baseband uses a fast frequency hopping algorithm (Section 4.3.2) and after 625 μs, the frequency of a channel changes according to a code which defines the sequence in which the carrier frequency has to change. Two types of links can be present between two Bluetooth devices—ACL (asynchronous connectionless link) and SCO (synchronous connection oriented) link. ACL link provides best effort traffic. SCO links provide real-time voice traffic using reserved bandwidth.

A packet is of about 350 bytes and has a payload of 2744 bits or 343 bytes, b0-b2743. It has 68 bits access code plus 4 bit trailer to access code (or alternatively 72-bit channel access code) at the beginning. It is followed by a 54-bit packet header. Packet deploys FEC (Section 3.1.1.5) for correcting the transmission errors. A 72-bit packet access code has following fields:

1. A 4-bit word to specify preamble.
2. A 64-bit synchronization field having 24 bits from lower significant bits of link access address [called lower address part (LAP)]. LAP is from the 48 bit global address of the master.
3. A 4-bit word for trailer bits.
4. Instead of 64 plus 4 bits referred to above, a 72-bit channel-access code can be used.

A 54-bit packet header transmits the 18 bits with 36-bit FEC bits. The 18 bits consist of the following fields:

1. 3-bit AMA or PMA field for slave address (Section 12.4)—When a piconet device sends the data to master, then slave address suffices. When master sends data to the device, then also it will use AMA of the slave. Address 000 means master is broadcasting to all the slaves whether they are in active, park, or hold state.
2. 4-bit packet-type field— A packet can be transmitted in fragments. A 32-bit word specifies an ID for the packet, flags, and fragment offset for the fragments of same ID.
3. 1-bit control field for asynchronous flow of data. When it is 1 or 0, the transmission and reception starts or stops, respectively.

4. 1-bit sequence number field (SEQN) which alternates between 0 and 1
5. 1-bit acknowledgement number field [for automatic repeat request sequence number ARQN] which alternates between 0 and 1.
6. 8-bit HEC—header check field for checksum of the header. It is an error control field.

12.5.1.3 Link Manager (LM)

Baseband and radio provide a link between master and slave. The functions of the link manager are—(i) supervision, (ii) monitoring of power, synchronization, state, and mode of transmission, (iii) exchange of QoS parameters (e.g., packet flow latency, peak data rate, average data rate, and maximum burst size) for L2CAP and higher layers and capability information exchange, (iv) handling device pairing, and (v) handling data encryption and device authentication. Table 12.3 lists the functions of the link manager.

Table 12.3 Functions of link manager

Function	*Description*
Supervision	Administration and control of all the link actions. It provide for setting of a new SCO link. It monitors the failure of the link.
Monitoring power	Finding the signal strength S_r of the received packets and decreasing the transmitted power S_t if S_r is high. Transmitted power is adjusted so that it is just sufficient to maintain the link. [Further, if the link is no longer active, the transmitter goes into power-save mode and AMA becomes PMA.]
Monitoring synchronization	Enabling the clock of the slave device clock to continuously remain in phase with the clock of the master device and continuously adjusted the phase differences (time-offsets). This ensures synchronous connection link. Link manager provides for exchanging synchronization packet for timing information to synchronize two adjacent piconets. It also analyses received special packets for synchronization to maintain synchronization.
Monitoring state changes	The state of a device can change from slave to master (when master moves out of piconet region) and an established connection can end in connection-oriented connection which means that there are header exchanges for establishing a connection before the start and for terminating the connection after completion of transfer of data as done in the TCP connection. The manager monitors these changes.
Monitoring transmission modes	An ACL can change to SCO or vice versa. The manager monitors these changes.
QoS parameters exchange	If 2 bits out of 3 are FEC bits (Section 3.1.1.5), data protection and integrity is maintained. Manager provides (2/3) option when data protection is required. Parameters which define the QoS are polled, exchanged and monitored by manager.

(Contd)

(Contd)

Capability information exchanges	Facilitates the exchange of information related to device capabilities. This includes the information of voice encoding, encryption, park/hold state mode, or multi-slot packets mode.
Device pairing	Devices are paired by assigning a link key which can be modified, accepted, or not accepted by the manager.
Handling data encryption	Handling the key size, random speed, and set modes for encryption by higher layers— no encryption mode, broadcast mode, and unicast mode.
Handling device authentication	Monitoring the pseudo-random number exchanges and digitally signed responses

12.5.1.4 Host Controller Interface

It is hardware abstraction layer in place of link manager. Host controller interface (HCI) interfaces RF communication serial [3 wire UART (universal asynchronous receiver and transmitter) which is an RS232 emulation in RF communication] line and mode through L2CAP software layer.

12.5.2 MAC Layer

Data link control is carried out using L2CAP (Fig. 12.15). The function of L2CAP is to provide logical link control and provide an adaptation mechanism using Bluetooth. It passes the segmented or reassembled packets directly to the link manager or HCI in case of host-controller-based system.

A logical-link adaptaion-request is received from upper layers at client device. It is then multiplexed and segmented as per available maximum transferable units (MTUs) at the baseband. After this, the logical link adaptaion confirmation is sent to higher protocol layers and then a link program request is sent to lower layer HCI or LM. The link program protocol confirmation is received back at L2CAP layer.

A link program protocol indication is received at the server device from lower layers. The baseband packets are then demultiplexed and reassembled. The logical link adaptaion indication is then sent to higher protocol layers which on its successful receipt send the logical link control response. The link protocol response is then sent from L2CAP layer.

L2CAP facilitates segmentation of packets while transmitting and reassembling of packets on reception. It multiplexes the data between different higher layer protocols (Fig. 12.15) and manages forward transmission from a Bluetooth device to other devices. It does QoS management for higher layer data. [A program file needs error free transfer. A picture being transferred can contain errors but still look good. It means ensuring the expected level of service.] It provides connection-establishment-based communication after a host ACL is established.

L2CAP supports three logical channels. (A logical channel is between two softwares, one sending input and other receiving the input.) These are—(i) signalling messages between L2CAP at transmiiter and receiver devices, (ii) bi-directional connection-oriented with support for QoS parameters from higher layers, and (iii) unidirectional connectionless broadcast from master to slave.

12.6 Security in Bluetooth Protocol

Bluetooth devices have the ability to detect the signals from the select transmitting device only. These devices can operate in secured as well as unsecured modes and can enforce link-level as well as service-level security. Bluetooth protocol provisions for highly secure personal area wireless communication using various keys, encryptions, and pseudo-random number generation. Bluetooth security features are listed in Table 12.4.

Table 12.4 Security features in Bluetooth-enabled devices and PC

Security features	*Description*
Ability to detect by select transmitting device only	Bluetooth devices operate at the least required power levels so that their transactions may not be detected by the distant receiver of signals, thus maintaining confidentiality and frequency hopping. CDMA ensures negligible interference and least risk of jamming by distant sources.
Security modes	1. Link-level enforced security—authentication of the linked device and authorization by the linking device. 2. Service-level enforced security—a servicing device can deny the service to a device which has discovered its services. 3. Non-secure—for common devices, for example, printer, headphone, and ear-buds.
Link layer security	Uses a private key of 8 to 128 bits for encryption and a 128 bit pseudo-random key for authentication. Device generates 128 bit pseudo-random number and *has a 48 bit* public address. Both together ensure *the link-level* enforced security.
Keys	Initialization key before device installation, unit key for the device after installation, a unique combination key for two Bluetooth devices communicating with each other, and *master key* instead of current link layer key when a master communicates with multiple slaves.
Three encryption modes	Master key encryption for all data exchanges, master key encryption for data exchanges except broadcasts (e.g., for service discovery), and no encryption.

12.7 IrDA

Infrared (IR) rays are invisible radiations of wavelength higher than that of red. An LED or a solid-state laser emits IR rays when given 10–20 mA current from a low power battery or power source. When the 1s and 0s are transmitted, the IR source current is modulated as per the 1s and 0s. Direct line-of-sight IR from an LED is detected at the receiver (photodetector) to get the data. The detector has 30° (±15°) window to detect the incoming radiation. The rays emitted by laser suffer diffuse reflections from obstacles which are present at the receiver. IR rays are used for remote control of TV or IrDA (infrared data association) which is a protocol for personal communication area network deploying infrared rays. Section 1.1.4 introduced Infrared-based communication for short distances when there are no obstacles, such as walls, between the devices. For example, IrDA. Table 12.5 gives the basic features of IrDA.

Table 12.5 Basic features of IrDA

Property	*Description*
Wavelength	900 nm
IrDA device levels of communication	Five levels of communication—minimum, access, index, sync, and SyncML (Levels 1–5). [A level specifies a method of communication from simple to SyncML-based.]
IrDA data transfer rates	IrDA 1.0 protocol for data rates up to 115 kbps. IrDA 1.1 supports data rates of 1.152 Mbps to 4 Mbps (16 Mbps draft recommended).
Sessions, object exchange, and other IrDA protocols	IrLAN (for Infrared LAN access) , IrBus (for access to serial bus by joysticks, keyboard, mice, and game ports), IrMC (IrDA mobile communication and telephony protocol), IrTran (IrDA transport protocol for image or file transfers), IrComm [IrDA communication protocol by emulating serial (for example RS232C COM) or parallel port] , and IrOBEX (for object exchange)
IrDA protocol layers	(a) Layer 1—physical (b) Data link layer— 2a—IrLAP (link access protocol), 2b—IrLMP (link management protocol) (c) Layer 3-4—transport layer—tiny TP (transport protocol) or IrLMIAS (link management information access service protocol) (d) Layer 5—session—IrLAN, IrBus, IrMC, IrTran, IrComm, and IrOBEX (object exchange) (e) Layer 6-7—security and application software layers as specified by the IrDA Alliance Sync (PIM), object push (PIM), or binary file transfer
Network characteristics	Point to shoot communication from peer to peer
Dissimilarity with Bluetooth	• Bluetooth is for wireless short range exchanges in mobile environment within 10 m network and IrDA is for exchanges within a range of one meter in the vicinity of line-of-sight. Bluetooth has small form factor in radiation pattern and has 30° (±15°) window • Network connection Latency—3 s for Bluetooth and a few ms for IrDA • Bit rate—1 Mbps for Bluetooth and 1.152 Mbps to 4 Mbps for IrDA • Protocol stack—250K bytes • Code Size—50% down to 2% as compared to a Bluetooth device • FHSS for Bluetooth and serial synchronous, asynchronous, or PPM for IrDA

(*Contd*)

(*Contd*)

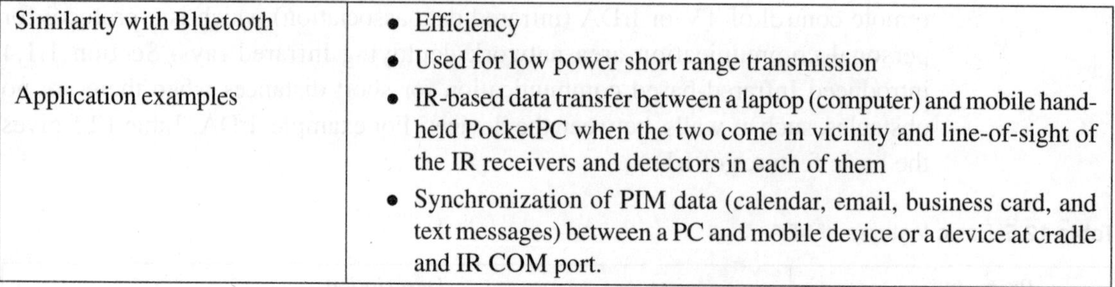

Similarity with Bluetooth	• Efficiency • Used for low power short range transmission
Application examples	• IR-based data transfer between a laptop (computer) and mobile hand-held PocketPC when the two come in vicinity and line-of-sight of the IR receivers and detectors in each of them • Synchronization of PIM data (calendar, email, business card, and text messages) between a PC and mobile device or a device at cradle and IR COM port.

Figure 12.16 shows functions of protocol layers along with all protocol layers between the peer-to-peer transmitter and receiver devices. OBEX supports security by encryption and decryption at transmitter and receiver, respectively. It

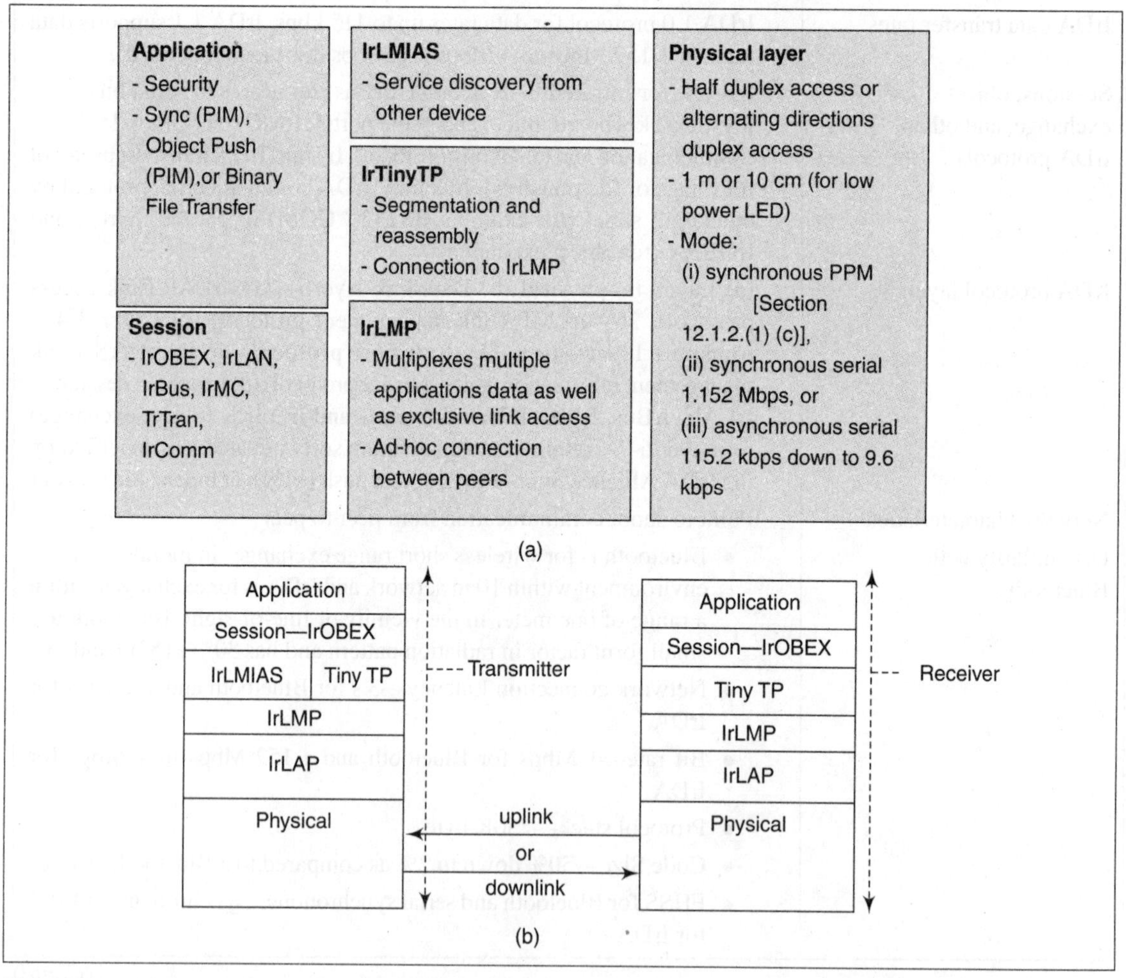

Application
- Security
- Sync (PIM), Object Push (PIM),or Binary File Transfer

Session
- IrOBEX, IrLAN, IrBus, IrMC, TrTran, IrComm

IrLMIAS
- Service discovery from other device

IrTinyTP
- Segmentation and reassembly
- Connection to IrLMP

IrLMP
- Multiplexes multiple applications data as well as exclusive link access
- Ad-hoc connection between peers

Physical layer
- Half duplex access or alternating directions duplex access
- 1 m or 10 cm (for low power LED)
- Mode: (i) synchronous PPM [Section 12.1.2.(1) (c)], (ii) synchronous serial 1.152 Mbps, or (iii) asynchronous serial 115.2 kbps down to 9.6 kbps

(a)

Transmitter: Application | Session—IrOBEX | IrLMIAS, Tiny TP | IrLMP | IrLAP | Physical

Receiver: Application | Session—IrOBEX | Tiny TP | IrLMP | IrLAP | Physical

uplink or downlink

(b)

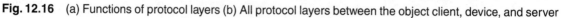

Fig. 12.16 (a) Functions of protocol layers (b) All protocol layers between the object client, device, and server

communicates and exchanges binary data by establishing a client–server network between two IR devices. The OBEX client makes a request and server returns the response. The OBEX client and server either establish connection-oriented connection before data transmission or send requests and responses through connectionless connection.

12.9 ZigBee

Section 1.1.4 introduced ZigBee. Figure 12.17 shows a network of ZigBee sensors, end-devices, and ZigBee router-devices. ZigBee is a suite of high-level communication protocols. ZigBee 1.0 specification was released in December 2004. ZigBee devices conform to the IEEE 802.15.4-2003 Wireless Personal Area Network (WPAN) standards for operations at low data rates and low power dissipation. ZigBee devices form a personal area (home) network of embedded sensors, industrial controllers, or medical data systems. It consists of three type of ZigBee devices which are as follows:

(a) ZigBee coordinator—root node at each ZigBee network tree. It can connect to other networks and has full network information along with a store of the security keys for the ZigBee network nodes (ZA in Fig. 12.17).

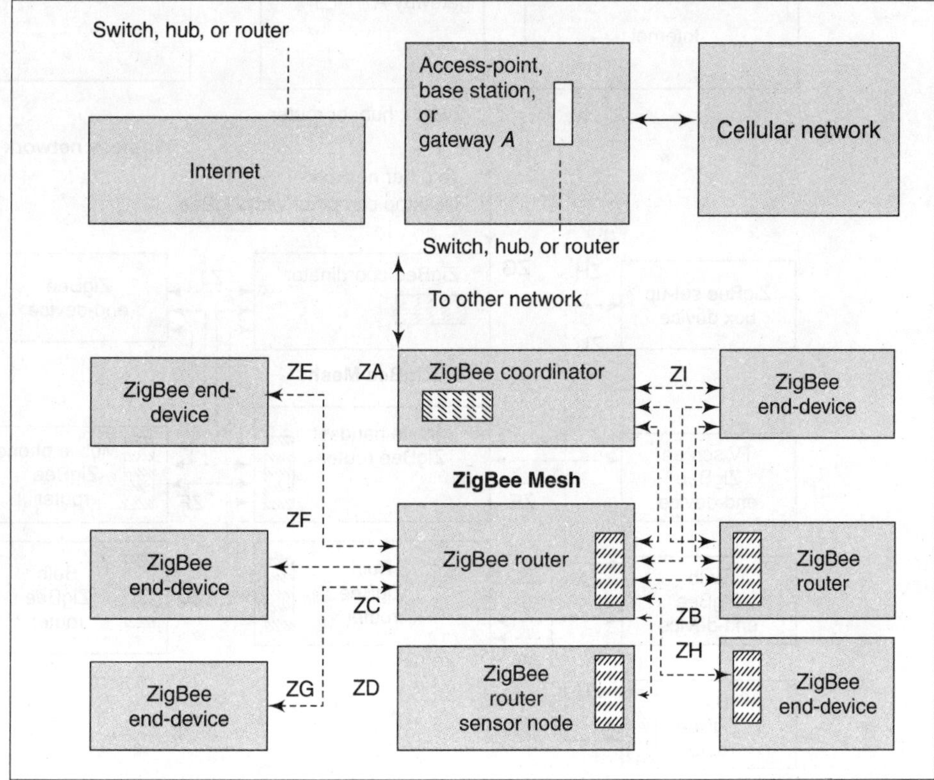

Fig. 12.17 ZigBee sensors, end-devices, and ZigBee router-devices networks

(b) ZigBee router node—responsible for transfer of packets from the neighboring source to nearby node in the path to destination (ZB, ZC, and ZD in Fig. 12.17).

(c) ZigBee end-device—receives packet from a nearby node in the path from a source (ZE, ZF, ZG, ZH, and ZI in Fig. 12.17).

A ZigBee network can be of two types:

- *Peer-to-peer*—For example, ZC— ZD—ZH network in Fig. 12.17 in which each node has a single path to neighbouring node only
- *Mesh*—For example, ZA— ZB— ZC network in Fig. 12.17 in which each node has a path to every other node.

Example 12.4 Give examples of a wireless personal area network assuming that each sensor, device, or node uses ZigBee.

Solution: Figure 12.18 shows an exemplary network in which each sensor, device, or node uses ZigBee.

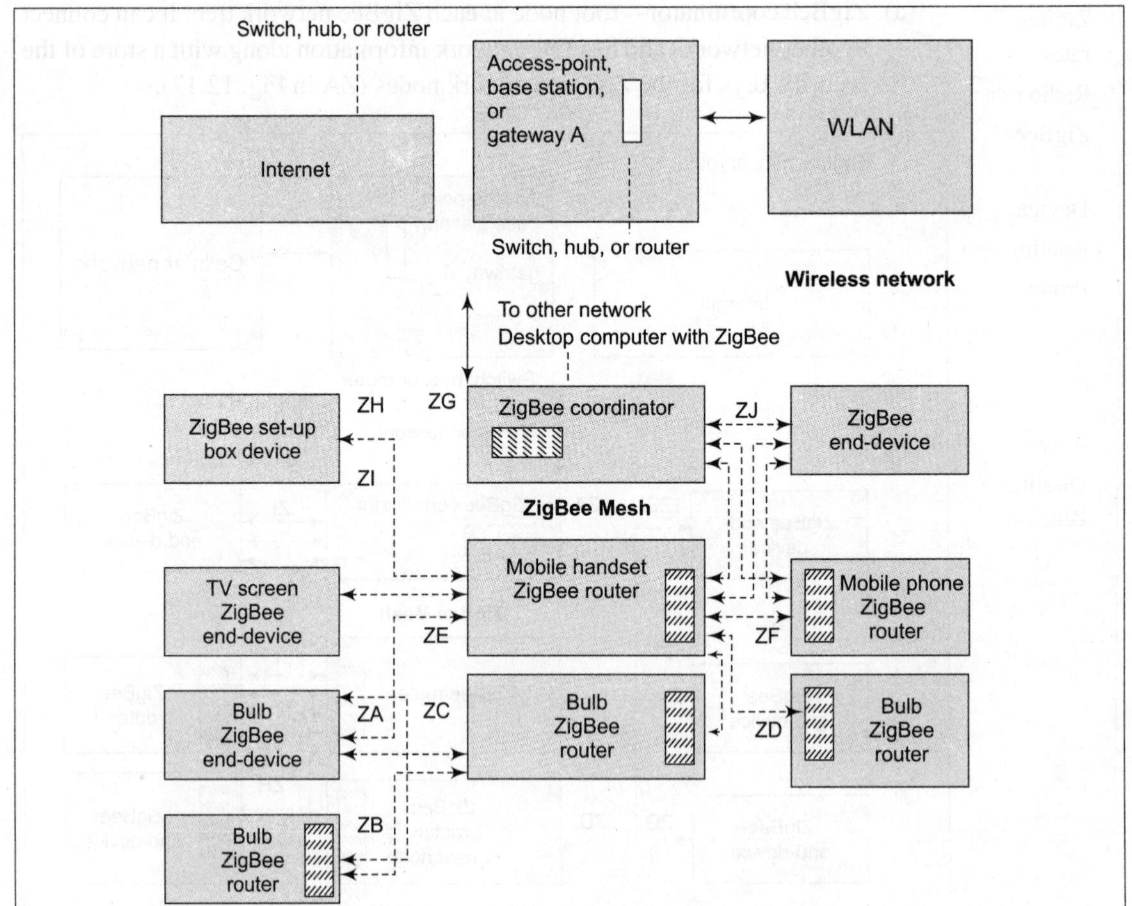

Fig. 12.18 ZigBee sensors, end-devices, and ZigBee router-devices networks

In Fig.12.18 the ZigBee end-device is ZA. A set of electric bulbs associates with the ZigBee routers ZB, ZC, and ZD. Each router is in parallel. The set forms a peer-to-peer connection network (ZE—ZD—ZC—ZB—ZA) with last one being ZigBee end-device (ZA). A ZigBee coordinator (ZG) will connect this network with other ZigBee networks (e.g., of mobile hand-held devices ZE and ZF). The coordinator ZG also connects the access-point for WLAN and provides Internet connectivity to router ZJ for security system, to cellular phone network and set-up box device ZH, and screen ZI. ZigBee network router nodes ZE, ZF, and ZG use mesh network connections. Table 12.6 lists the basic features of ZigBee.

Table 12.6 Basic features of ZigBee

Property	*Description*
Radio frequency bands and modulation methods	ISM bands—2.4 GHz orthogonal QPSK , 915 MHz (USA) BPSK, and 868 MHz (USA) BPSK
ZigBee device channels	For 2.4 GHz, there are 16 ZigBee channels. Each channel has freuency band $(2400 + 5\times n) \pm 1.5$ MHz, where $n = 1, 2, ..., 15$, or 16
ZigBee data transfer rates	2.4 GHz at 250 kbits/s per channel, 915 MHz bands at 40 kbit/s per channel, and 868 MHz bands at 40 kbit/s per channel.
Radio interface	DSSS (Section 4.3.1)
ZigBee protocol layers	Physical and a DLL (data link layer) part, called MAC (media access control)
Device types	Coordinator, router, and end-device types
Routing protocol	AODV (Section 11.2.4.2)
Protocol layers	• Physical layer as provided in IEEE 802.15 • MAC layer as provided in IEEE 802.15 • Security and application software layers as specified by the ZigBee Alliance
Network characteristics	Self-organization, peer-to-peer, and mesh networks
Dissimilarity with Bluetooth	• Bluetooth used for wireless short range exchanges in mobile environment and ZigBee for big scale mesh-network-based automation and remote control • Network connection latency—3 s for Bluetooth and 20 ms for ZigBee • Bit rate—1 Mbps for Bluetooth and 250 kbps for ZigBee • Protocol stack—250 kB for Bluetooth and 28 kB for ZigBee • Code size—50% down to 2% as compared to a Bluetooth device • FHSS used for Bluetooth and DSSS for ZigBee

(Contd)

(*Contd*)

Similarity with Bluetooth	• Both conform to IEEE 802.15 set of standards • Use of spread spectrum modulation results in spectrum efficiency in both • Use of 2.4 GHz (in USA) in both • Used for low power short range transmission • Both have small form factors radiation pattern
Application examples	• A ZigBee-enabled electric meter communicates electricity consumption data to the mobile meter reader • A ZigBee-enabled home security system alerts the mobile user of any security breach at home

Keywords

Basic service set (BSS): A set of computing systems in which each node at a station can communicate directly with access-point (in BSS) and with another node at another station through the access-point using 802.11x and the nodes of a station can communicate among themselves by forming an ad-hoc or any other type of network (e.g., Bluetooth) using same frequency band for each node.

Bluetooth: A protocol which derives its name from the name of a Danish king (Harald Blatand, which means Bluetooth in English) who reigned before 1000 A.D. Bluetooth is popularly used in mobile devices for data synchronization and communication between them by service discovery in self-organizing networks. Bluetooth devices can form a network known as *piconet* with the devices within a distance of about 10 m. Various piconets form an ad-hoc network called *scatternet* within 100 m through a Bluetooth-enabled bridging device.

Complementary code keying (CCK): An additional sub-layer in 802.11b for data rates of 5.5 Mbps by QPSK to map 4 bits and 11 Mbps 8-QPSK to map 8 bits simultaneously.

Extended service set (ESS): A set which is formed by the 802.11 LAN station access-points networked together and functioning as a distribution system.

Independent basic service set (IBSS): A set of devices which network with each other using 802.11 protocol. These devices either communicate directly with one another or among themselves after forming an ad-hoc network and form a set of stations in a WLAN. Each station uses same frequency band for radio coverage. Therefore, each station within an IBSS does not connect to another station even of same IBSS. IBSS does not connect to an access-point also.

Infrared data association (IrDA): A protocol for personal communication area network deploying infrared rays for short distance communication when there are no obstacles, such as walls, between the devices.

Local area network (LAN): A set of computers, computational systems, computational units, and computational devices networked using a standard suite of protocols such that all units in the set can address and communicate with each other using a common set of lines (wires).

Medium access control (MAC): A sub-layer which supports in 802.11x, the CSMA/CD and asynchronous data transceiver, and has point coordination support for time-bound applications. Some of its features are acknowledged RTS/CTS (request to send/clear to send) mechanism before the data transmission, power management, multiple physical layers and roaming support.

Physical layer convergence protocol (PLCP): A protocol which specifies sensing of the carrier at receiver and packet formation at the transmitter. PLCP sub-layer protocol prescribes the standard procedure for convergence of PMD layer data units with the upper layer MAC data units at the receiver end and of MAC layer with the lower PMD layer at the transmitter end.

Physical medium dependent (PMD) protocol: A protocol for a sub-layer of physical layer, which specifies the modulation and coding methods.

Wireless application protocol (WAP): A protocol which in its 2.0 version—(a) provides HTTP-TLS-TCP-IP layers at the WAP client for interacting with the Internet, (b) supports SyncML synchronization, (c) provisions for the WAP push service, and (d) provisions for MMS service.

Wireless Binary XML (WBXML): A specification in binary representation so that XML or XML-based language can be transmitted in compact format.

Wireless LAN (WLAN): A set of computers (computational systems, computational units, and computational devices) networked using a wireless protocol such that communication is possible from one point to any other point using a common access-point in an Independent Basic Service Set (IBSS) and also with other sets of computers in distributed system in a Basic Service Set (BSS) through 802.11x protocol.

Wireless markup language (WML): A language used by a WAP 1.1 client or gateway just as HTML is used by a browser, and HTML client or server. It is used to write the data for presentation and application for a mobile device and uses the concept of cards and deck.

Wireless session protocol (WSP): A layer in WAP 1.1 which supports stateless data transfers similar to HTTP, supports asynchronous exchanges, multiple requests, and push and pull mechanisms of data dissemination.

Wireless transaction protocol (WTP): A layer in WAP 1.1 which transmits data to WTLS in case of secure transactions and directly to WDP or WCMP. WTP supports the joining (fusion) of the messages and enables asynchronous transactions. It supports abortion of the transaction. It provisions for information about the success or failure of a transaction to the sender. WTP functions as an interface to ensure reliability of transactions.

WMLScript: A WAP 1.1 script language in WML which is used for retrieving the application data after running the script at the server and which enables the server to put the response which is sent back to the client through a gateway.

XHTML: A language in two forms—XHTML basic and XHTML-MP. XHTML basic is HTML with strict syntax. XHTML-MP is XHTML with mobile profile and is for mobile devices and PDAs. An XHTML document can be parsed using a standard XML library and processed like any other standard XML document but using the same set of tags as in HTML.

ZigBee: A suite of high-level communication protocols which conform to IEEE 802.15.4-2003 WPAN standards for operations at the low data rates and low power dissipation and is used by devices which are in a personal area (home) network of embedded sensors, industrial controllers, or medical data systems. The network consists of three types of ZigBee devices—ZigBee coordinators, routers, and end-devices.

Review Questions

1. Describe WLAN-based architecture. How does a WLAN function?
2. Describe protocol layers of WLAN. Explain PMD, PLCP, and CCK layers.
3. Compare WAP 1.1 and 2.0 architectures. Describe WSP.
4. What are the features added to WML 1.1 as compared to HTML? What are the new features in WML 2.0?
5. Give the uses of WMLScript.
6. What are the functions of WAP 1.1 gateway? How does these differ from those of WAP 2.0 proxy/gateway? How does WAP provide connectivity to a web server?
7. Explain the physical and networking layers of WAP.
8. Explain XHTML-MP. How does it differ from HTML? Describe the features recommended in new XHTML versions.
9. Show the architecture for synchronization of devices in a WPAN and its connection to Internet.
10. Show master–slave architecture in a piconet of Bluetooth devices. What are the states in which a Bluetooth device can be found?

11. Describe how a piconet forms when Bluetooth devices come in the vicinity of each other. How does a Bluetooth scatternet function as an ad-hoc network?
12. Explain the functions of radio, baseband, and link manager in Bluetooth.
13. Describe L2CAP. What are the security features in Bluetooth devices?
14. Describe IrDA. What are the five levels of IR communication in IrDA? Compare the applications of IrDA with those of Bluetooth.
15. What are the features of ZigBee as compared to Bluetooth and IrDA in a WPAN?

Objective Type Questions

Pick the correct or most appropriate statement among the choices given:

1. 802.11 standard basic features are that it supports (i) access-point-based fixed infrastructure WLAN network using base service sets, (ii) ad-hoc peer-to-peer data routing network using independent base service stations, (iii) ad-hoc client-to-server data routing network using independent base service stations, (iv) access-point-based fixed infrastructure WLAN network using base service sets. Which of these are true?
   (a) (i) to (iv) are true.
   (b) (i) and (ii) are true.
   (c) (iii) and (iv) are true.
   (d) (i) and (iii) are true.
2. Physical media dependent layer in 802.11 has
   (a) the options FHSS/DSSS/Diffused IR and supports 802.15.4 in ZigBee, 802.15.1 FHSS in Bluetooth.
   (b) the options TDMA/FDMA/Diffused IR and supports 802.15.1 FHSS in Bluetooth.
   (c) the options SDMA/USB/IrDA and does not support 802.15.1.
   (d) FDMA/OFDM and supports 802.15.4 in ZigBee and 802.15.1 FHSS in Bluetooth.
3. 802.11a OFDM operates at
   (a) 2.4 GHz (DSSS) and supports 54 Mbps HyperLAN2.
   (b) 5 GHz (infrared, three 1.28 GHz physical layers) and has 2 Mbps to 10 Mbps data transfer rates.
   (c) 5 GHz (infrared, two 2.4 GHz physical layers) and has 6 Mbps to 54 Mbps data transfer rates.
   (d) 4 GHz (two 2.4 GHz physical layers) and has 1 Mbps to 8 Mbps data transfer rates.
4. 802.11b operates at
   (a) 5 GHz (infrared, three 1.28 GHz physical layers) and has 2 Mbps to 10 Mbps data transfer rates.
   (b) 2.4 GHz (DSSS) and supports 5.5 Mbps and 11 Mbps using CCK and 54 Mbps HyperLAN2.
   (c) 5 GHz (infrared, two 2.4 GHz physical layers) and has 6 Mbps to 54 Mbps data transfer rates.
   (d) 4 GHz (two 2.4 GHz physical layers) and has 1 Mbps to 8 Mbps data transfer rates.
5. (i) WAP 2.0 is a wireless protocol for synchronization of WAP client computers and WAP or HTTP server. (ii) WAP 1.1 has architecture derived from HTTP client (browser) synchronization with the HTTP server (web server) in wired Internet. (iii) WAP 1.1 supports XHTML-MP language, HTML browser, and HTTP. (iv) WAP 2.0 devices and services have three important features over those of WAP 1.1—SyncML synchronization, WAP push service, and MMS service. (v) WAP 1.1 architecture does not provide HTTP-TLS-TCP-IP layers at the WAP client for interacting with the WAP gateway. Which of these are true?
   (a) All are true.
   (b) All are true except (ii) and (iii).
   (c) All are true except (v).
   (d) All are true except (iv) and (v).
6. When data transaction occurs between client device and gateway, wireless transport layer security (WTLS)
   (a) assures transaction-privacy and device authentication and maps to SSL in HTTPS and WAP 1.1 does not support TCP.

(b) assures integrity of transactions and device authentication, does not map to SSL in HTTPS and does not support TCP, WDP, and WCMP.
(c) assures SSL security in HTTPS and supports WDP and WCMP.
(d) assures integrity of transactions, transaction-privacy, and device authentication, maps to SSL in HTTPS, and supports TCP, WDP, and WCMP.

7. WAE in WAP 2.0 consists of—(i) deploying XHTML-MP and browser, (ii) can directly deploy HTTP, and (iii) lower layers at the client devices, which are TLS, UDP (or ICMP), or TCP in place of lower layers of WAP 1.1, which are WSP, WTP, WTLS, and WDP (or WCMP), (iv) WAE in WAP 1.1 deploys WAP 1.1 gateway encoders and decoders. (v) Gateway encoders and decoders are not required in WAP 2.0. Which of these are correct?
   (a) (iii) is incorrect.
   (b) All are correct.
   (c) (i) is incorrect.
   (d) Only (ii) and (iii) are correct.
8. Bluetooth provides
   (a) connectionless-oriented communication.
   (b) peer-to-peer slave communication within same piconet with negligible interference between piconets as each uses distinct channel-frequency hopping sequences.
   (c) ad-hoc network peer-to-peer communication when two devices are on two different piconets specifying a scatternet.
   (d) wireless LAN connectivity.
9. Bluetooth
   (a) Layer 1 provides for Baseband and radio and layer 7 the Object Exchange OBEX.
   (b) Layer 3 sub-layers provide RF communication, L2CAP, and LM or host-controller interface.
   (c) Layer 7 provides for security and application software layers are as specified by thė Bluetooth Sync, vCal (for Calendar), vCard [for contact (visiting card)] or Object Push (PIM) or Binary File Transfer or audio applications.
   (d) Layer 7 has L2CAP audio sub-layer.
10. (a) Bluetooth provides serial synchronous, asynchronous, or PPM and IrDA communication is using FHSS.
   (b) Bluetooth devices have radiation pattern with large form factors and 130° (±65°) window, wireless short range exchanges in mobile environment within 100 m network, and IrDA for line-of-sight within a range of one meter in the vicinity.
   (c) Bluetooth and IrDA network connection latencies are 3 s and a few ms, bit rates1 Mbps and 1.152 Mbps to 4 Mbps, respectively.
   (d) IrDA code size is about 70% down to 62% as compared to a Bluetooth device.
11. (a) Bluetooth devices operate at the least required power levels so that the transactions may not be detected by a distant receiver of signals.
   (b) Confidentiality can not be maintained in Bluetooth due to synchronization between the devices.
   (c) DSSS CDMA ensures negligible interference and thus least risk of jamming by distant sources.
   (d) Application layer in Bluetooth deploys SyncML client, SyncML engine, and OBEX for the session between the devices.
12. Bluetooth devices use the (i) *unit key* before device installation, (ii) *Initialisation key* for the device after installation, (iii) master key for two communicating Bluetooth devices, (iv) a unique *combination key* instead of current link layer key when a master communicates with multiple slaves
   (a) Only (i) and (ii) are incorrect.
   (b) (iii) is incorrect.
   (c) (iv) is incorrect.
   (d) All are incorrect.

13. Bluetooth devices security levels are—(i) non-secure printer, headphone, and ear-buds, (ii) service-level enforced security, (iii) link-level enforced security by authentication of the linked device and authorization by the linking device
    (a) (ii) and (iii) are true.
    (b) All are true.
    (c) Only (iii) is true.
    (d) Only (ii) is true.
14. IrDA
    (a) data link layer deploys IrLAP or IrLMP.
    (b) transport layer deploys TCP, UDP, or IrLMIAS.
    (c) network layer deploys IrLAN, IrBus, IrMC, TrTran, IrComm, and IrOBEX.
    (d) session security and application software layers specified by the IrDA Alliance are the Sync (PIM), object push (PIM), or binary file transfer.
15. IrDA wavelength and IrDA levels of communication—(i) 700 nm (ii) minimum (iii) access (iv) index (v) sync (vi) SyncML (vii) IrDA 1.0 supports data rates up to 500 kbps (viii) IrDA 1.1 supports data rates of 1.152 Mbps to 4 Mbps (ix) latest recommended—12 Mbps
    (a) All are correct except (i), (vii), and (ix).
    (b) All are correct except (iv), (v), and (vi).
    (c) Only (vi) and (viii) are correct.
    (d) All are correct except (i).
16. (a) Bluetooth devices deploy DSSS and ZigBee devices FHSS.
    (b) Bluetooth is for wireless short range exchanges in mobile environment and ZigBee for big scale mesh-network-based automation and remote controlled network connection.
    (c) Latency of Bluetooth devices is 2 s and of ZigBee devices 1 s.
    (d) Bit rate for Bluetooth is 1 Mbps and for ZigBee 800 kbps and the protocol stack size is 500 kB for Bluetooth and 280 kB for ZigBee.

Chapter 13

Mobile Application Languages—XML, Java, J2ME, and JavaCard

Mobile cellular phones, in addition to teleservices and supplementary services (Section 3.1.1), are used for many applications. This chapter deals with platform-independent languages used for application-development. It covers the following topics:

- Learning of XML which is a standard language for platform-independent creation of contents and information for presentation
- Understanding the use of an XML document for developing applications for a given mobile device or server in the mobile computing system
- Understanding Java for processing the information and contents in the mobile applications
- Learning two versions of Java—J2ME and JavaCard which are used in mobile devices

13.1 Introduction

Software processes the information and contents present in a database or file. In other words, it processes the data and information. Software is generated by a set of (a) statements, functions, service routines, threads, objects, and classes written in a high-level language (e.g., C, C++, Java, Visual Basic, or Visual C++), (b) codes using a framework (e.g., Symbian or .NET), or (c) codes using a development tool (e.g., PalmOS software development kit or Metrowerks CodeWarrior). Most popular high-level languages for mobile application information, contents, and data interchange are XML and XML-based languages (e.g., SyncML, VoiceXML, SMIL, and XHTMLMP).

In addition to phone which includes SMS, MMS, other additions to teleservices, and various supplementary services (Section 3.1.1), there are many mobile applications. Most popular ones are as follows:

- Music [e.g., Nokia N91 8 GB handset for 6000 tracks music]
- 3D gaming
- Video uploads
- Content sharing, file sharing, search, and on-air song recognition
- Email
- WiFi-enabled Internet
- Enterprise solutions
- FM transmitter streaming music
- 3″ TFT screen digital TV for home viewing
- Video recording on memory card

A mobile application can be developed through any one of the following three approaches:

- Using a language like Java that compiles and runs on diversified operating systems and hardware platforms. In case of Java, compiled byte codes run using a virtual machine called Java virtual machine (JVM). Sections 13.3 to 13.5 describe Java, J2ME (Java 2 Micro Edition), and JavaCard.
- Using a language supported by the operating system which provides a development platform such that compiled program can run on the hardware supported by that OS. An example of such an operating system is Symbian OS (Section 1.4). In this case, application can be written using C/C++ software development kit (SDK). Another example is Symbian OS with PersonalJava, J2ME, Visual basic, Python, OPL, or Simkin.
- Using a common framework and OS platform which supports different languages. For example .NET framework on Windows as common OS platform.

13.2 XML

XML (extensible markup language) is a derivative of SGML (standard generalized markup language). Extensible means that the special instances of the tag-based languages can be defined such that each instance observes the fundamental rules of representing and structuring the XML document. An XML-based language uses the extensible property of XML to define the standardized sets of instances of the tags, attributes and their representation and behaviour, and other characteristics. SyncML, Funambol, and SMIL are the examples (Sections 9.3 to 9.5).

A markup language is for presentation of the marked (tagged) textual content. An encapsulated text is processed, displayed, or printed as per the tag. A browser does the presentation. However, an XML or XML-based language not only encapsulates the data and metadata but can represent a behaviour or set of actions.

XML document can have a text with the tags. A tag in the document specifies the meaning of the text encapsulated within the start and corresponding end tag. Example 13.1 shows this aspect of XML.

Example 13.1 Give an example of XML document which can be used as Contacts in a mobile Smartphone.

Solution: Figure 13.1 shows an example with two Contacts.

XML document can represent a database with tags and a pair of start and end tags in the document specifies the start and end of a record in the database. A

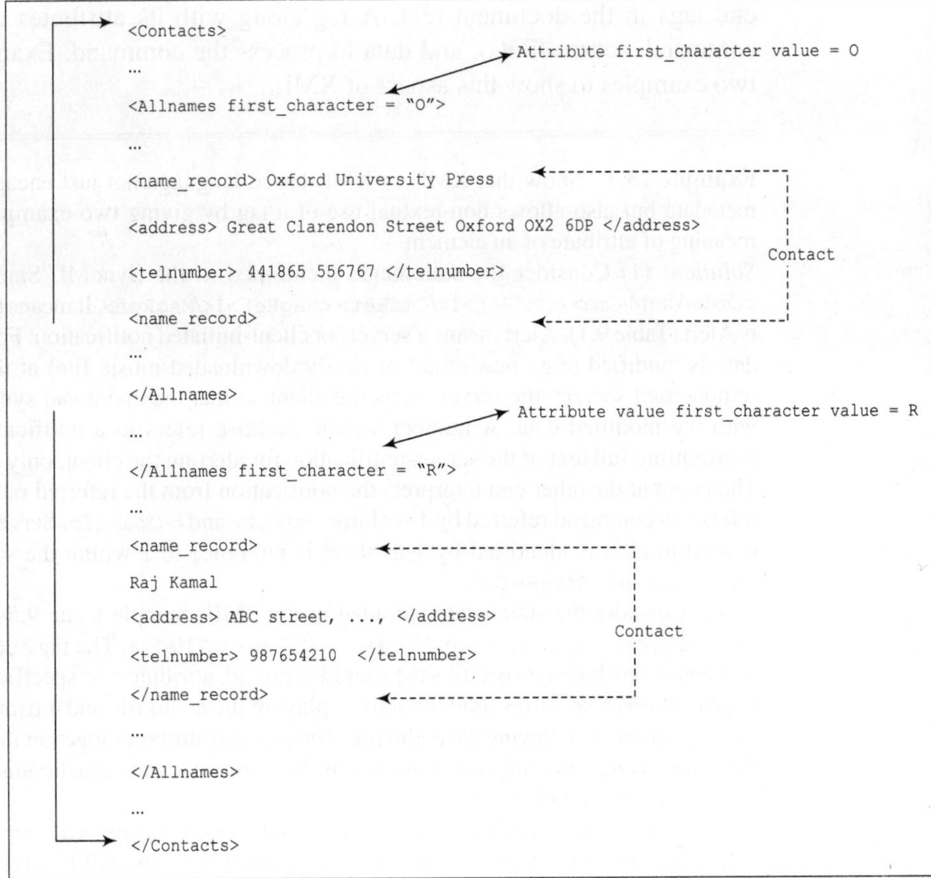

Fig. 13.1 Example of XML document which can be used as Contacts in a mobile Smartphone

database can be used to retrieve the specific record or set of records by querying the database or by business logic transaction (Section 7.1.1). Example 13.2 shows this aspect of XML.

Example 13.2 Show that XML is a language for textual document to represent a database.
Solution: Refer to Example 13.1. It demonstrated that XML uses the elements which describe a textual content. Consider `<name_record>` Raj Kamal `<address>` ABC Street, ... `</address>` `<telnumber>` 9876543210 `</telnumber>` `</name_record>` in an XML document. The elements are `name_record`, `address`, and `telnumber`. Above set of elements describe a single contact. It also exemplifies a hierarchical structure. The single contact element `name_record` has a textual content (Raj Kamal) and two elements `address` (ABC Street, ...) and `telnumber` (9876543210) within it. A document formed by XML document creates a database named `Contacts` in a mobile application.

An XML document can make non-textual use of text. A tag can represent a command to process the data using the command name within the pair of start and end tags in the document text. A tag along with its attributes can specify the command, source file(s), and data to process the command. Example 13.3 gives two examples to show this aspect of XML.

Example 13.3 Show that XML or XML-based language not just encapsulates data and metadata but also allows non-textual use of a tag by giving two examples. Illustrate the meaning of attribute of an element.
Solution: (1) Consider the statements mentioned in the SyncML Sample Code 9.3—`<Cmd>`Alert`</Cmd>` `<CmdRef>`1`</CmdRef>` `<MsgRef>`1`</MsgRef>`. It means that the command is Alert (Table 9.1). Alert means a server- or client-initiated notification. For example, when data is modified (e.g., new email or newly downloaded music file) at the server (PC or remote mail server), the server alerts the client so that the client can synchronize its data with the modified data. A number within `CmdRef` refers to a notification. In place of transmitting full text of the server-notification for alerting the client, only a number is sent. The client at the other end interprets the notification from the referred number. `Command` refers to a command referred by 1 within `<CmdRef>` and `</CmdRef>`. Server message which is used to alert is identified by 1 as there is reference to 1 within the start and end tags `<MsgRef>` and `</MsgRef>`.

(2) Consider the statements included in the SMIL Sample Code 9.5—`<audio id =` "`my_id`" `src` = "`myaudio.wav`" `begin` = "0s" `dur` = "10s"/>. The tag `audio` specifies the command, attribute `id` specifies the user id is my_id, attribute `src` specifies the source file, attribute `begin` specifies time for start of playing the audio file and attribute `dur` specifies time for duration of playing the audio file. The tags and attributes together therefore represent the system behaviour, that is, it plays a .wav file "myaudio.wav" and begins it from 0 second mark for a duration of 10 seconds.

(3) In the above example, `audio` is the element with attributes `id`, `begin`, and `dur`. The values of attributes `id`, `begin`, and `dur` are myaudio.wav, 0s, and 10s, respectively.

13.2.1 Document Type Definitions and Parsing of XML Documents

An XML document is given Document Type Definitions (DTDs) either internally or in a separate DTD file. The DTD file or DTDs within the document includes the specification of the role of text elements in a model document. It specifies the attributes associated with an element as well as their valid values. The functions of DTDs are as follows—(i) to enable validation of a document, (ii) to specify which document structures can be used for authoring of the document, and (iii) to specify which structures a parser must handle.

Parser is a software written in Java or any other language. Its functions are to validate and parse the tags, their attributes, and the text within each pair of start and end tags and to enable extraction of specified information either from the document file or using document and DTD files.

For example, a parser can parse the document `contacts.xml` in Sample Code 7.1 and Example 13.1, the parser can create a hash table with `name_record` as the key and corresponding texts for the name, address, and telnum as values of the key. [Hash table is a table with key in first column and values in subsequent columns in each row.] A parser precedes the execution of an *application*.

A *parser* first validates an XML document and then handles the specified document structures. It just needs the DTD for validation of the document. Validation means investigating about (i) whether the document contains a root element with the same name as of DTD, (ii) whether it contains the header information for the version, encoding, and reference to other files, and (iii) whether it contains the DTD with declaration of the markups in the document or in a linked external DTD document. The XML document has an extension `.xml`. The external DTD document has file `extension.dtd`.

Java and J2ME-based mobile devices use KVM (kilo virtual machine) in which JVM is of a few kilobytes (a JVM micro version) and kXML parser. (JVM requires 1 MB ROM and RAM). KVM is a J2ME virtual machine. kXML parser is a parser which parses XML document using J2ME and KVM in mobile devices. Microsoft Windows Mobile- and Windows CE-based devices use .NET XML parsers. WAP 1.x supports XML parser. This enables processing of WML (wireless markup language).

13.2.2 Models of an XML Document and Parsing

There are two models of an XML document and correspondingly two types of parsers. One model is called SAX (simple API for XML) model in which each set of elements within the tags is independent and need not be considered as a tree-like structure. (A tree structure is one in which each element derives from a root element.) SAX model parser does serial access of the XML data in the API and is modelled as a stream parser, with an event-driven API. Each tag in the set of tags generates an event for parsing and processing. SAX provides the API a serial access and event-driven mechanism for reading and parsing data from an XML document.

The other model is called document object model (DOM) in which each set of elements is dependent and derives from a root element and whole document forms a tree-like structure.

13.2.2.1 SAX Model of an XML Document

Figure 13.2 shows SAX model of parsing of an XML document corresponding to Contacts. It demonstrates a SAX parser of XML document and an alternative arrangement of elements for each contact in the Contacts which helps in fast processing by the parser. SAX model document is such that its parser can generate the series of events which proceeds from beginning to end. Assume that XML contacts file has many contacts. An application needs to dial a contact with name record Raj Kamal and dial the corresponding telnumber.

SAX parser generates an event on `name_record` as it parses through the record and on event accepts the value 'Raj Kamal'. Then it generates another event on telnumber as it parses through the document and on event accepts the value

```
<Contacts>
...
<Allnames first_character = "O">
...
<name_record> Oxford University Press </name_record>
<address> Great Clarendon Street Oxford OX2 6DP </address>
<telnumber> 441865 556767 </telnumber>
...
</Allnames>
...
<Allnames first_character = "R">
...
<name_record>
Raj Kamal </name_record>
<address> ABC street, ..., </address>
<telnumber> 987654210  </telnumber>
</name_record>
...
</Allnames>
...
</Contacts>
```

SAX Parser

From begin serially **Parse** name `name_record`,
Generate event `Ename_record`
Parse `name telnumber`,
Generate `event telnumber to end`

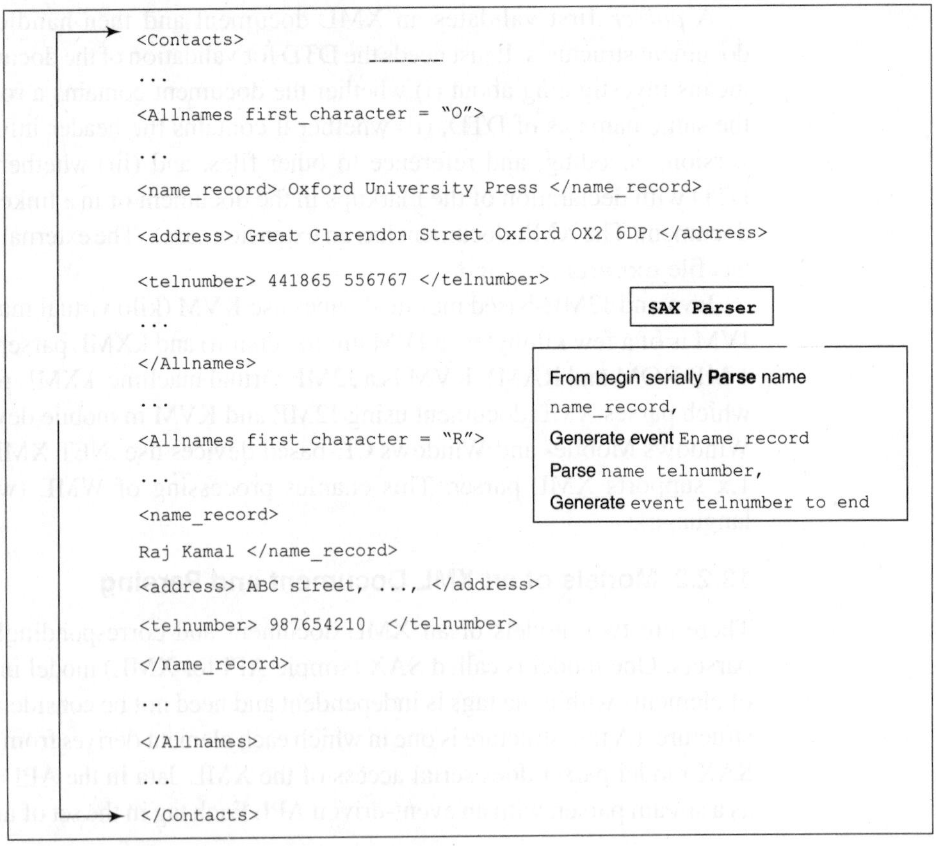

Fig. 13.2 SAX parser of XML document

'9876543210'. An advantage of XML SAX parser is that the entire document need not be first parsed thoroughly and all the needed data need not be extracted, as is the case with DOM model of parsing explained next. SAX parsing and processing is fast.

13.2.2.2 DOM Model of an XML Document

Figure 13.3 shows DOM model and parsing of an XML document corresponding to Contacts. DOM model document is always hierarchically arranged. The whole document is parsed initially to create a hash table of keys and corresponding values for each key. An advantage of DOM parsing is that the structure is well-defined and the same parser can be used for parsing all XML documents and later the interpreter or processing program is able to extract the desired information by simply using the keys.

Extracting and parsing a program for a DOM model document and parser is not complex. However, in an intermittently connected wireless environment or in case

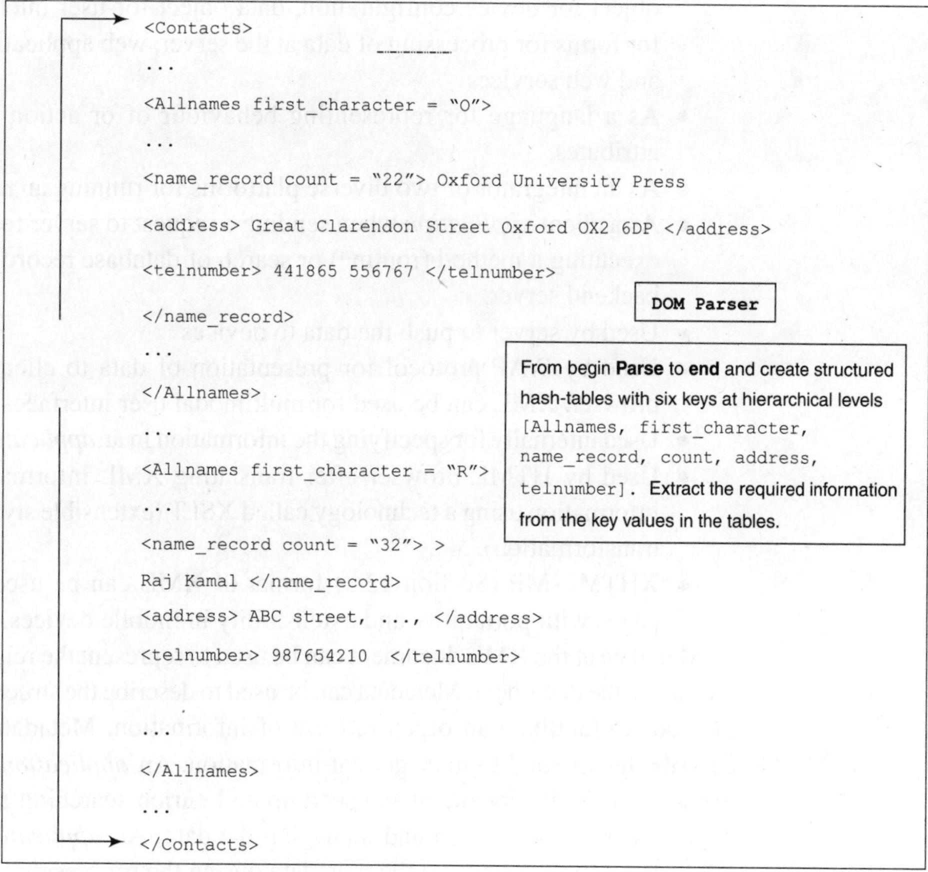

Fig. 13.3 DOM parser of XML document

of a long document with many levels of hierarchy, it could take a long time before whole document is received at the parsing end.

13.2.3 Applications of Parsed Information and Data

An *application* uses the extracted output information and data from the parser after a further processing using a programming language. (Refer Sample Code A2.3 in Appendix 2.) The parsed data is interpreted at the *application*. An *application* can create ascending or descending order tabular data as per selection by the user. Another *application* can count the number of contacts. A third *application* can display the desired contact using appropriate GUI by deploying up-down menu keys in the keypad.

XML has many applications in mobile computing systems. The following are the examples of XML usage:

- As a language for preparing a textual document for platform-independent application data, for example, database, data object for synchronization, data object for device configuration, data object for user interface, data object for forms for processing of data at the server, web applications at the server, and web services.
- As a language for representing behaviour of or action by a tag and its attributes.
- As an integrator of two diverse platforms for running an *application*.
- As a client *application* when sending a request to server for data or result of executing a method (routine) or search of database record at the associated backend server.
- Used by server to push the data to devices.
- Used by WAP protocol for presentation of data to client using an XML browser. XML can be used for multimodal user interfaces.
- Used internally for specifying the information in an *application* or *framework*.
- Used by HTML browser after translating XML information into HTML information using a technology called XSLT (extensible style-sheet language transformation).
- XHTML-MP (Section 12.3) format of XML can be used as HTML web pages with portability and extensibility in mobile devices.

Metadata at the XML document can be used to represent the relationships among the data in the document. Metadata can be used to describe the structure and workable methods to facilitate an organized use of information. Metadata can be used to describe the method to manage that information. An *application* can also use the metadata in XML document to speed up and enrich searching for the resources, organising the information and managing the data. An *application* in addition to XML data can thus also use the metadata during the processing.

13.2.4 XML-based Standards and Formats for Applications

XML is an extensible language. The tags and attributes, and markup language format can be standardized for *applications*. An *application* may generate a form for answering the queries at server end, which needs to be filled and submitted at the client end. XForms provides a UI (user interface). XForms is a form in XML format which specifies a data processing model for XML data and UIs for the XML data. For example, web forms which can also be used in a stand-alone manner or with presentation languages other than the ones for a set of common data manipulation tasks or user interface.

A form has the fields (keys) which are initially specified. They either have no field values or default field values. A form may have several UIs like text area fields, buttons, check-boxes, and radios. It is presented to a user using a browser or presentation software and user interactions take place. The user supplies the values for these fields by entering text into the fields, checking the check-boxes, and selecting the radio. The user then submits the form which is transmitted from client to server for carrying out the needed form processing. The server program can be at remote web server or at the device computing system.

XForms controls are `<textarea>`, `<input>`, and `<secret>` and a string data type is attached with these fields. Other XForms controls are `<selectOne>` (for a radio), `<choices>` (for check-boxes), `<item>`, and `<itemset>`. The attributes of XForms input control are `ref` and `xml:lang`. `xml:lang = "en"` enables use of XML. The other attributes are `accessKey`, `navIndex`, and `class`. The elements of XForms are `<alert>`, `<caption>`, `<hint>`, and `<extension>`. There are client-side as well as server-side validations before processing of the XForms.

XML user interface (XUI) is a framework in XML and Java for mobile and mobile handheld device applications. WBXML (WAP binary markup language) is a format in which WAP presents XML document as binary numbers. A binary number represents a string of characters for each element and attribute. This reduces the size of the document and hence the transmission time from client to gateway or gateway to client.

VoiceXML (VXML) 2.0 is a version of an XML-based language in use from 2004 for interactive talking between a human being and a computer. There is a voice browser analogous to an HTML display browser (e.g. NetScape or IE 5.0). The voice browser interfaces with mobile or landline telephone network. VXML provides a voice user interface (VUI) to a voice browser. There are a large number of applications of VXML. A few examples are bank account enquiry, report of loss of credit card, and directory and flight enquiry. The tags in VXML are for playing a sound file from audio book, automatic speech recognition, dialog management, and speech synthesis by voice browser. For example, `<prompt>` Hello Lucy `</prompt>` will output the message Hello Lucy using speech synthesizer with the voice browser. VXML is used with HTTP protocol on a network analogous to that for HTML. IBM WebSphere supports VXML.

Many applications need conversion of text to speech (TTS). For example, the displayed text message from a website for control of traffic and congestions at nearby places. In a moving automobile, it is required that the site data is downloaded and read and spoken to a driver. SSML (speech synthesis markup language) is used in TTS (text to speech converter). It is for synthesizing the speech by interpretation of a text by a synthesizer. It creates audio book. It has tag `s` with attribute `xml:lang`. For example, to specify attribute `English`, `<s xml:lang="en">`. It specifies the synthesis source for voice as `<voice name="Mary" gender="female" age="22">`. The text is then synthesized by spoken English in voice of Ms. Mary of age 22.

Speech recognition grammar specification (SRGS) is used in SSML. It lets speech recognizer to define the pattern of sentences. SSML also specifies *emphasis* of the voice. For example, it specifies when the voice is normal and when it is loud. SSML can be embedded in VXML scripts. The scripts provide an interactive telephone.

Call control extensible markup language (CCXML) is a standard for an XML-based language, which can be used independently or with VXML for telephony support. CCXML commands the browser to handle the calls of the voice channel. Examples of commands are initial call set-up or call disconnect between a caller and voice browser. CCXML 2.0 handles the events and state transitions (e.g., transition from call set-up to call connected state or call connected to disconnected state).

13.3 JAVA

C and C++ with in-line-assembly, Visual Basic, and Visual C++ are the programming languages used with Windows platform. Java is a language which compiles into byte codes assuming a hypothetical CPU which possesses stack-organized architecture. The codes run on a virtual machine. Therefore, these can run on any operating system and hardware platform on which JVM is installed. The JVM provides the execution engine and native interfaces to run the codes on the given OS and hardware. Figure 13.4 shows Java application, classes, and application threads compiled as byte codes and JVMs for platform-independent execution. Java is, therefore, the most popular language used for mobile computing. Java 2 Standard Edition (J2SE) is standard edition.

J2SE is used to program the application, threads, applets, servlets, and aglets. An application is a program which runs on the user system. Application threads are the programs of the application, each one of which is assigned a priority and runs concurrently. Servlets are the programs running on a web server. The lifecycle of a servlet consists of the sequential steps—initiate, service, and destroy. [The `init` is

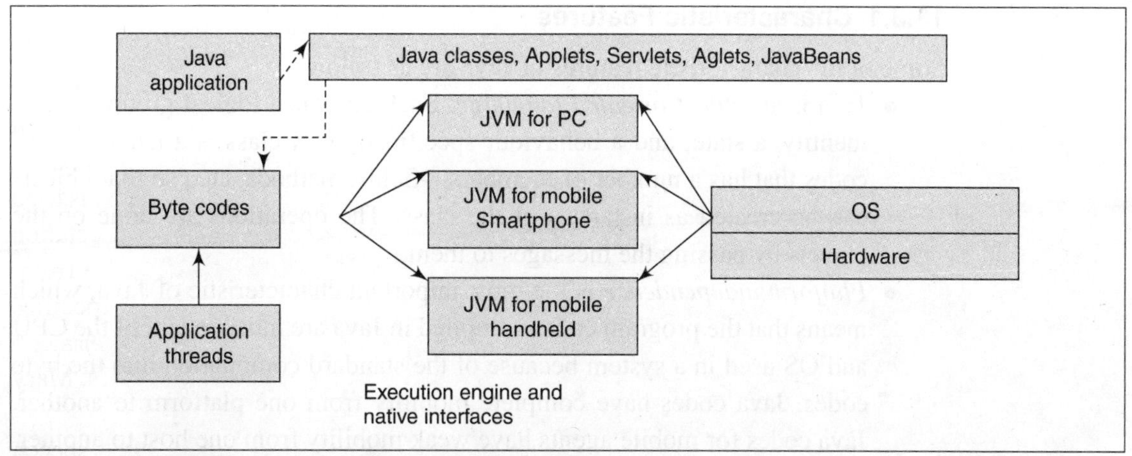

Fig. 13.4 Java application, classes, and application threads compiled as byte bytes and JVMs for platform-independent execution

analogous to `main` in a Java program. Just as `main( )` is executed first in Java program, `init( )` is executed first in servlets.] Applets are the programs running on a client. The lifecycle of an applet consists of the sequential steps—initiate, start, stop, and destroy. [The `init` in applet is analogous to `main` in a Java program. Just as `main( )` is executed first in Java program, `init( )` is executed first in Applet.] aglets are the mobile agents (Section 10.1) which run on a host. The lifecycle of an Aglet consists of the sequential steps—creation, dispatch, activation, deactivation, and disposition. Aglet can be retracted or cloned and can be used for messaging. It supports weak mobility. This means that an aglet can retain its state when it migrates from one host to another.

An application is developed by a set of Components. Database or server can be connected to the UI components using the APIs. Java applications are developed using open software Java 2 SDK from Sun, JBuilder, Microsoft Visual Studio J++, and IBM Visual Age are the popular Java application development tools. IBM Visual Age provides a visual programming method to a programmer. The Visual Age tool provides source code and compiled byte codes.

Java and Java-based languages possess characteristic features which make them the most used languages for mobile applications. Section 13.3.1 describes these features.

A Class is a construct in Java or C++, which is used to define a common set of variables, fields, and methods and whose instances give the objects. A JavaBean is a set of Java classes, that is, many objects group into a single object called JavaBean. JavaBean is reusable software component which can be manipulated by a Software builder tool. Section 13.3.2 explains the Java Classes and JavaBeans.

Java 2 has an edition for enterprises where there are different servers and tiers for processing client requests. This edition, called J2EE (Java 2 enterprise edition), is described in Section 13.3.3.

13.3.1 Characteristic Features

Some of the characteristic features of Java are as follows:

- Java is an *object-oriented language.* Each class is a logical group with an identity, a state, and a behaviour specification. A class is a named set of codes that has a number of members—fields, methods, etc., so that objects can be created as instances of the class. The operations are done on the objects by passing the messages to them.
- *Platform independence* is the most important characteristic of Java, which means that the program codes compiled in Java are independent of the CPU and OS used in a system because of the standard compilation into the byte codes. Java codes have complete mobility from one platform to another. Java codes for mobile agents have weak mobility from one host to another.
- *Robustness* is yet another characteristic of Java. It is due to the fact that Java does not let the user assign pointers and write code for pointer manipulations. All references to memory, freeing of the memory (garbage collection), de-allocation (de-referencing), and validation of object types and array indices are done internally at the compile and run time.
- *Standard APIs* (application programming interfaces) enforcement is another characteristic of Java. APIs help a Java program to connect an application to a program, database, distributed object, or server developed on other platforms.

13.3.2 Classes and Beans

Java has packages consisting of the classes and interfaces. The packages help in fast development of the codes. Some of the widely used packages are:

1. java.lang is the package for fundamental classes of Java.
2. java.io is the package for input, output, and file access.
3. java.math is the package for classes for the mathematical methods.
4. java.awt is the package for Java foundation classes for creating GUIs.
5. java.swing is the package for Java swing classes for creating lightweight GUIs. [Lightweight means not depending on the platform [system (OS) resources (for example, buttons)] unlike foundation classes which depend on the system resources.]
6. java.net is the package for network related classes, for example, TCP/IP, UDP, HTTP, and FTP.
7. java.security is the package for classes in Java security framework. The examples are classes for RSA, DSA, and certificates.
8. java.sound is a useful package for audio processing.

9. java.io.Serializable has serializable interfaces. When an object in Java is serialized, its state, as defined by the heap in memory, is written in memory as a serial byte stream. Therefore, when the object migrates, it is received as serial byte stream and after de-serializing, the original object is restored into memory.

JavaBean has a property assigned to the fields in a class. JavaBeans are integrated into a container. Any change in property is listened to by a listener and the action method is executed. For example, property can be address. Any change in address has to be listened by all related Beans in the container. JavaBean is valuable for developing reusable software components. The application, applets, or servlets integrate the JavaBeans. java.beans package contains the classes for developing the JavaBeans.

13.3.3 Java 2 Enterprise Edition (J2EE)

Figure 13.5 shows an *N*-tier architecture in which a client device connects to the enterprise databases. The Java enterprise edition (J2EE) is used for web- and enterprise-server-based programming of the applications. A J2EE application is used for distribution across the multiple computing tiers.

Enterprise Java bean (EJB) is a Component at a server in an enterprise application. It is managed such that it leads to a modular design. Each EJB encapsulates business

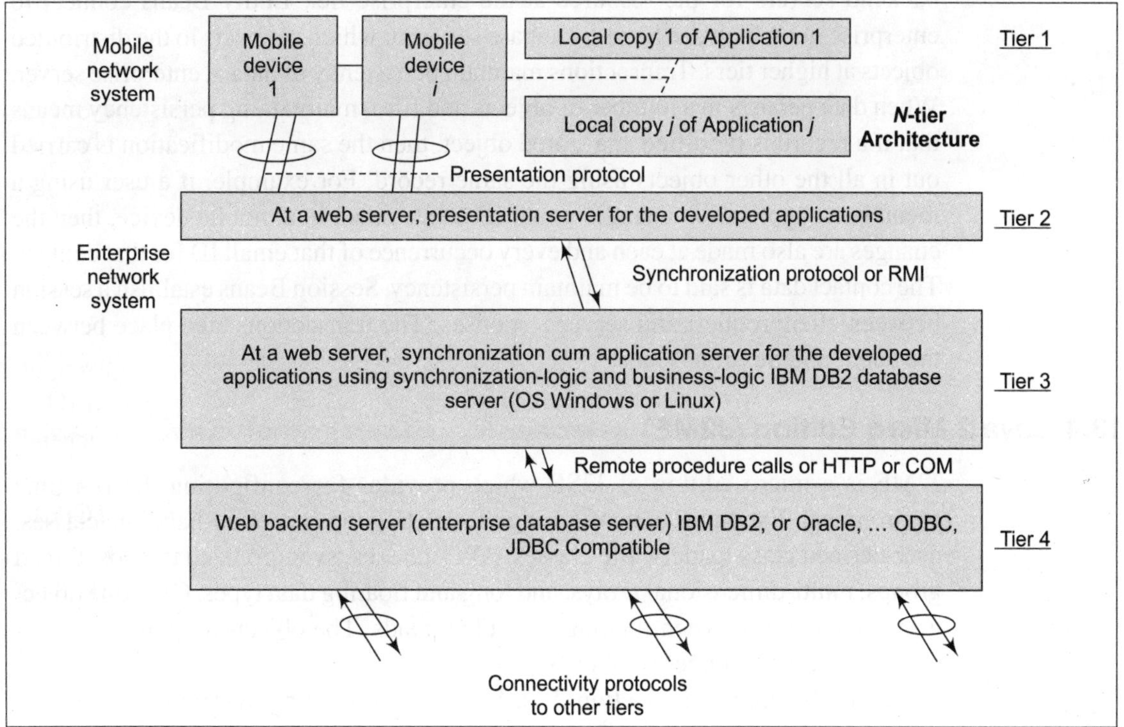

Fig. 13.5 *N*-tier architecture in which a client-device connects to the enterprise databases

logic of the application. EJBs are configurable components. EJBs enable persistency of the application components. An application server manages the security, scalability, transactions, and concurrent execution of the applications deploying the EJBs.

J2EE supports the API specifications for the following:

1. XML
2. Web services
3. Email
4. RMI (remote method invocation) by using the package java.rmi
5. JMS (Java messaging service) [A middleware API for interchanging messages between two or more clients]
6. JDBC (Java database connectivity)
7. JTA (Java transaction API)
8. Servlets (to dynamically render the response to the client request), Java server pages (JSPs) (to create dynamic contents), and Java Portlet
9. JNDI (Java Naming and Directory Interface) to provide lookup services to enterprise directory.

A Container has EJBs as Session Beans and Entity Beans. It provides transactions, security, scalability, pooling of the resources, and concurrent execution of the applications. Resources are pooled by transactions between Entity Beans and backend servers for the resource at the enterprise tier. Entity Beans connect to enterprise tier which can have a database server or which connects to the distributed objects at higher tiers. Transactions maintain persistency of data at enterprise server. When data persists in a number of objects and files, maintaining persistency means that if a record is modified at a stored object, then the same modification is carried out in all the other objects using the same record. For example, if a user using a menu in an application changes email ID of a contact in a mobile device, then the changes are also made at each and every occurrence of that email ID for that contact. The contact data is said to be maintain persistency. Session Beans establish a session between client request and server responses. The transactions take place between the client and Session Beans.

13.4 Java 2 Micro Edition (J2ME)

J2ME is a micro edition of J2SE which provides for configuring the run time environment. Examples of configuring are deleting the exception-handling classes, user-defined class loaders, file classes, AWT classes, synchronized threads, thread groups, multi-dimensional arrays, and long and floating data types. Only one object is created at a time when running multiple threads. The objects are reused instead of using a larger number of objects.

Mobile Smartphone (Section 2.1), handheld computing system (Section 2.3), and set-top box (Section 2.5.8) have constraints of memory. The mobile devices may also have network and other resource constraints. J2SE needs 512 kB ROM and 512 kB RAM in a device. J2ME is a set of Java APIs which require small memory while developing Java applications. It is also a platform for development of mobile phone games. Windows mobile devices do not support J2ME or Java-based virtual machine. J2ME platform binary implementations and virtual machine implementation are done by another source not from Windows.

Initially, PersonalJava provided a Java application development platform. It has mostly Java 1.1.8 APIs. Now, J2ME, CLDC (connected limited device configuration), and CDC (connected device configuration) provide the development platform for small memory devices and systems. Sections 13.4.1 and 13.4.2 describe the profiles and configurations for the development of applications for the devices using JME.

13.4.1 Profiles

A profile means a standardized agreed-upon subset and interpretation of a specification. A profile may also mean a specification for a set of configuration settings and other data which are used in the APIs for a device, user, or group of devices. A profile may also mean a standarized specification for the APIs for a device, user, or group of devices. J2ME-framework-based devices use Java APIs specified in a profile.

Foundation profile contains APIs of J2SE without GUIs. PersonalProfile is profile for embedded devices, for example, PDAs and set-top boxes and contains the APIs of foundation profile, complete AWT as well as lightweight GUIs, and applet classes.

Mobile information device profile (MIDP) is the profile for mobile devices with small screen option for GUIs, wireless connectivity, and greater than 128 kB flash memory. MIDP is a specification of APIs for mobile information, Smartphone, and gaming devices. A MIDlet is a J2ME application (similar to an Applet) for embedded devices which runs with MIDP. MIDlet main class is a subclass of javax.microedition.midlet. MIDlets are programmed to run games and phone applications. Also they are compiled once and are platform-independent. A MIDlet has to fulfill certain requirements in order to run on a mobile phone. Table 13.1 gives the MIDP source packages and the sets of the Java class libraries.

MIDP 3.0 is the latest profile version. MIDP 3.0 (www.opensource.motorola.com) is a profile for special-featured phones and handheld devices. It provides improved UIs, UI extensibility, and interoperability between the devices. Moreover, it supports multiple network interfaces in a device, IPv6 (Internet protocol version 6 for broadband Internet), large display devices, and high performance games.

MIDP 3.0 has provisioning for MIDlets using SyncML DM/DS (device management and device synchronization) protocol, Bluetooth, removable media (e.g., memory stick or card), and MMS. It enables—(i) auto-starting of MIDlets on device booting, (ii) running several MIDlets concurrently and sharing the class

Table 13.1 MIDP source packages and sets of Java class libraries

	Source package	*Sets of Java class libraries*
1.	java.lang	Standard java types and classes for String, Integer, Math, Thread, Security, and Exception.
2.	java.io	Standard java types and classes for input and output streams.
3.	javax.microedition.lcdui	LCDUI for mobile devices with no Internet connectivity. The API provides a limited set of UIs in mobile devices. These are TextBox, Form, List, and Canvas (low-level graphics as well as full screen games graphic mode). Graphics are needed for games. MIDP controls the GUIs.
4.	java.util	A set of classes such as Timers, Calendars, Dates, Hashtables, Vectors, and others.
5.	javax.microedition.rms	A record management system (RMS) API to retrieve and save data and limited querying capability.
6.	javax.microedition.pim	Personal information management API (optional), access the device's address book.
7.	javax.microedition.pki	Secure connections authenticate APIs.
8.	javax.microedition.messaging	Wireless messaging APIs used when sending SMS and MMS messages.

libraries for MIDlets, (iii) running background MIDlets (MIDlets without a UI), (iv) specifications of the runtime behaviour, (v) proper fire walling and lifecycle managing functions for MIDlets, (vi) enabling MIDlets to draw to secondary display(s), and (vii) specifications of the behaviour of MIDlets in the CLDC and CDC. [CLDC and CDC are the configurations of MIDP, described in Section 13.4.2.]

MIDP enables inter-MIDlet communication (similar to inter-process communication between the processes controlled by an OS) which in turn allows querying of device capabilities required to be done when a server service is to discover the device services.

Development tools are used to develop MIDP applications. For example, NetBeans is a development tool. It provides a UI using which the APIs are taken into the application. A mobile 3D graphics (M3G) API creates 3D graphics. The API provides for Java-accelerator-hardware-based as well as simple-software-based Java implementation.

Information module profile (IMP) is for embedded devices for example, security systems or vending machines that have no display UIs and game APIs. An IMlet is an application created from IMP APIs. IMlet is inherited from MIDlet. Auto profile (AutoP) is a profile for automobile application development. TV profile (TVP) is a profile for TV set-top box application development.

13.4.2 Configurations

A configuration is a subset of profile. CLDC (connected limited device configuration) is a configuration for limited connected devices. The CLDC defines a base set of APIs and VM for the resource-constrained mobile phones or handheld computers. The configuration is a subset of MIDP. It is used for developing Java applications. Usually CLDC can just connect to mobile application service provider and have less than 64 kbps data transfer rate.

CDC (connected device configuration) is a configuration for connected devices. CDC is a Java framework for developing an *application* that can be shared in networked devices, for example, set-top boxes. Table 13.2 gives the features available in subsets of Java class libraries in CLDC and CDC. CLDC has minimal needed subset of the Java class libraries running on virtual machine for a CLDC .

CLDC provides a configuration with an API called LCDUI. A mobile phone with games is an example of CLDC. It provides limited networking capabilities. There is a separate javax.mircoedition.io package in CLDC configuration and provides for IO to flash and persistent data and IO streams. It uses LCDUI and has limited security framework for MIDlet and CLDC application.

CDC provides a J2ME framework for applications which run on wirelessly connected devices and APIs for HTTP. The devices will need 2.5 MB flash or ROM and 2 MB RAM. Set-top box is an example of connected device.

Table 13.2 gives source profiles, packages for the configurations of CLDC, and CDC, and the required virtual machine.

KVM does not support weak reference which means full object reference to the object must be used for reaching to the object in the application program. KVM is a virtual machine which has no floating point mathematical operations support (supports only integers), limited exceptional handling, no automatic garbage collection (memory freeing), no native support using JNI (Java native interface) to use C/C++ application interface with the Java APIs, and no ThreadGroup.

Table 13.2 Source profiles, packages for the configurations of CLDC and CDC, and the required virtual machine

	Feature	*CLDC*	*CDC*
1	Profiles and source packages	A configuration for the MIDP, which does not provide for the applets, awt, beans, math, net, rmi, security, sql, and text packages in the java.lang	A configuration for the Foundation and Personal Profiles, TV Profile, or Auto Profile, which includes classes inherited from a limited number of classes at *net*, *security*, *io*, *reflect*, *security.cert*, *text*, *text.resources*, *util*, *jar*, and *zip* packages
2	Virtual machine	KVM	CVM (coherent virtual machine) for multi-protocol and multi-threading support

13.5 JavaCard

A card has CPU with limited processing command and low clock frequency operations. An IC card has 4 kB ROM and a microprocessor card has 18 kB flash read only memory. JavaCard (Java for card) is a micro-edition of Java for the cards. JavaCard is a limited-memory sized edition for cards, labels, tokens, and similar devices which have limited memory as well as processing capacity.

JavaCard 2.2.2 provides interoperability for cards and APIs for highly memory-efficient applications. It has multiple communication interfaces for inter card-host contact/contactless (wireless) communication APIs. Also, it has ISO7816-based extended length APDU (Section 2.5.1) support. The maximum number of logical communication channels supported is 20. JavaCard 2.2.2 also provides standard communication with the host using latest SIM cards. The javacard.security and javacard.crypto supports digital-signature-based message recovery and advanced cryptography class libraries for HMAC-MD5, HMAC-SHA1, and SHA-256.

JavaCard framework (javacard.framework) provides the library functions, Card interfaces, PIN (personal identification number), and ISO 7816 APIs for Card applet. JCardSystem is a class in JavaCard. It has a method makeTransientArray ( ). The method creates a transient array and it persists till power is down.

Card profile is a set of limited class libraries. JavaCard profile is smartcard profile which has a separate virtual machine called card VM. The card VM is different for different card OSs. JavaCard Executive provides communication I/O streams between the JavaCard APIs and supports interoperability among the cards from different hardware and different card OSs.

Card virtual machine (Card VM) has an instruction set for a subset of Java language. It installs applets and libraries into JavaCard-based devices. Card VM does not support weak reference as already mentioned for KVM. Card VM is a virtual machine which has no char, double precision, or single precision and no floating point mathematical operations support [supports only Boolean, 8-bit Byte, BCD, 16-bit integer (short) (32-bit integer support optional)]. It has limited exceptional handling, no object clones, no String class libraries, no automatic garbage collection (memory freeing), no native support using JNI to use C/C++ application interface with the Java APIs, no SecurityManager class libraries, no multi-threading, and no ThreadGroup. The application executes Card applets and Java class libraries of the application. JavaCard is a technology which supports a secure environment for smart cards and small-devices applets. Multiple applets can be deployed on a single card and new ones can be added any time at the user end.

The card applet creates card-specific byte code on compilation. Java applet has lifecycle starting from initiate, followed by start, stop, and destroy. Card applet differs from Java applet as it can reside permanently on card. Objects created by card applet are persistent. When applet is deselected the transient array disposes.

An applet for communication uses select method for selecting the process APDU (Section 2.5.1) whether it is a command APDU or response APDU. Command APDU identifies a class for instruction, finds specific instruction from the class, uses two bytes for two parameters of the instruction, and specifies the length of optional data with the APDU. Response APDU has data (optional) and two byes for status words.

JCRE (JavaCard runtime environment) interprets the card byte codes and implements them using JavaCard virtual machine. It does not support inter-communication between different card applets and provides runtime support to the services like selection and deselection of applets.

Keywords

Aglet: An application code running on a mobile-agent host, which can migrate from one host to another and which has a life cycle—creation, dispatch, activation, deactivation, and disposition, and which can be retracted or cloned and used for messaging.

Applet: An application which runs at a client and which has a life cycle starting from initiate followed by start, stop, and destroy.

Byte codes: The compiled code generated from a Java program which can run anywhere at any place

Call control extensible markup language (CCXML): A standard for an XML-based language, which can be used independently or with VoiceXML (VXML) for telephony support in which VXML provides a VUI to the voice browser. CCXML commands the browser for handling the calls of the voice channel.

Card applet: A card-specific byte code generated on compilation, which can reside permanently on card, the objects created by which are persistent, and in which, when the applet deselects, the transient array disposes. It is an applet for communication, which uses select method for selecting the process APDU.

Card profile: A set of limited class libraries, which has a separate virtual machine called card VM which is different for different card OSs.

CDC: A J2ME framework and applications which run on wireless connected devices, APIs for HTTP, and devices with minimum 2.5 MB flash or ROM and 2 MB RAM.

Check-box: A set of labelled text with a circle nearby. The text represents a choice. A click at the circle selects the choice and next click deselects.

CLDC: A configuration with an API called LCDUI and for programming the applications on a mobile phone with game device, which provides for IO to flash and persistent data and IO streams and which also provides a limited security framework for MIDlet application.

Command APDU: A data unit which identifies a class for instruction, finds specific instruction from the class, uses two bytes for two parameters of the instruction, and specifies the length of optional data.

Document Type Defintions (DTDs): It specifies the attributes associated with an element, their valid values, and the role of text elements in a model document.

DOM: A model for XML document in which the information is arranged in hierarchical order and first whole document is parsed to create a hash table of keys and corresponding values for each key and then the required information is found.

Framework: A utility which saves much of the coding while building an application. A framework supports validation of data, event handling, plug-ins, interactive creation of user interface (GUI or VUI), and binding of data.

Input: It is a button along with a form, for example, a button to send reset or submit.

Java card executive: APIs to provide communication I/O streams between the JavaCard APIs and to support interoperability among the cards from different hardware and different card OSs.

JCRE (Java Card Runtime Environment): A runtime program which interprets the card byte codes and implements them using JavaCard virtual machine. It does not support the inter communication between different card applets and provides runtime support to the services like selection and deselection of applets.

JVM: A machine which takes Java byte codes in the input and with the help of the execution engine and native interfaces runs the codes on the given OS and hardware.

Metadata: Metadata are data that describe other data and a set of metadata describes a resource (single set of data). An XML tag also encapsulates metadata.

MIDlet: A J2ME application (similar to an applet) for embedded devices, which runs with MIDP, the main class of which is a subclass of javax.microedition.midlet, and which is programmed to run games and phone-applications

Mobile information device profile (MIDP): A profile for mobile devices with small screen option for GUIs, wireless connectivity, greater than 128 kB flash memory, and a specification of APIs for mobile information, Smartphone, and gaming devices.

Parser: A program to extract the specified information from the document either from the document file or using document and DTD files.

Radio: A set of choices out of which only one can be selected. When one of these is selected, the others are deselected automatically.

SAX model: A model for XML document or XML parsing in which the parser can generate the series of events which are created as processing proceeds from beginning to end.

Servlet: An application in Java for running on a web-server and which has a lifecycle—initiate, service, and destroy.

Speech recognition grammar specification (SRGS): Along with SSML, it lets speech recognizer to define pattern of sentences.

Speech synthesis markup language (SSML): A language for use in TTS (text to speech) converter and which is also used to command the synthesis of speech by interpretation of a text by a synthesizer and to create an audio book.

State transitions: Transition from one state to another, for example, transition from call set-up to call connected state or call connected to disconnect state.

VoiceXML (VXML): An XML-based language for interactive talking between a human being and a computer for which there is a voice browser analogous to an HTML display browser (e.g., NetScape or IE 5.0). The voice browser interfaces with mobile or landline telephone network.

XForms: A form in XML format which specifies a data processing model for XML data and UIs for the XML data. It has fields (keys) which are initially specified. They either have no field values or default field values. A form may have several UIs like text area fields, buttons, check-boxes, and radios.

XML (Extensible markup language): A derivative of SGML (standard generalized markup language), which is extensible. Extensible means that the special instances of the tag-based languages can be defined such that each instance observes the fundamental rules of representing and structuring the XML document.

Review Questions

1. Why are XML-based languages used in mobile application? Give examples of SyncML and SMIL tags to explain the tags and attributes.
2. Give examples of XML-based UIs.
3. Compare DOM and SAX models of the documents. How are these parsed? Explain with examples.
4. What are the main characteristics of Java which make it a language for anywhere and anyplace?
5. What are the packages in J2SE? Explain the APIs and class libraries in each.
6. Explain J2EE architecture.
7. Describe J2ME features.
8. What is a profile? How is MIDP defined from J2ME? Describe features of MIDP 3.0.
9. Explain CLDC and CDC configurations.
10. What are the requirements for JavaCard virtual machine? Explain JavaCard technology. How does a card applet differ from an applet?

Objective Type Questions

Pick the correct or most appropriate statement among the choices given:

1. (i) SyncML, Funambol, SMIL, and VoiceXML are the XML-based languages. (ii) XML extends HTML tags and is a derivative of HTML which is a derivative of SGML. (iii) An XML document is given

document type definitions (DTDs). (iv) The role of text elements in a model document is always specified in a DTD file separately in a document. (v) DTDs specify the attributes associated with an element as well as their valid values. (vi) Only functions of DTDs are to enable validation of a document and to specify which structures a parser must handle. Which of these are correct?
(a) (ii), (iv), and (vi) are incorrect. (b) All are correct. (c) All are incorrect except (iii) and (iv).
(d) SyncML, Funambol, SMIL, and not VoiceXML are XML-based languages and (iii) and (iv) are correct.

2. (i) A parser first validates an XML document. (ii) A well-formed document need not be a validated during parsing. (iii) A parser handles the specified document structures before validation but after testing whether the document is well-formed. (iv) A parse just needs DTD for validation of the document. (v) The parser output is analogous to a hash table output and the output has the keys or key pairs and corresponding key values. (vi) A tag and its attribute pair define a pair of keys for which there is one key value. (vii) XML parser parses a document which may contain the DTD with declaration of the markups in the document or in a linked external DTD document. Which of these are correct?
(a) All are correct except (iii) and (vii). (b) All are correct except (v) and (vi).
(c) All are correct except (iii). (d) All are correct.
3. A mobile application is developed
(a) using a language such that it compiles and runs on diversified operating systems and hardware platforms.
(b) using a language supported in the operating system which provides a development platform such that compiled program can be run on the hardware supported by that OS.
(c) using a common framework and OS platform which support different languages.
(d) using (a), (b), or (c).
4. (a) Symbian OS is an example where compiled byte codes run on diversified OS and hardware platforms using a virtual machine (JVM).
(b) Java, J2ME, and JavaCard compiled byte codes run on the virtual machines which are installed and are as per the given OS and hardware platform.
(c) .NET is an example where OS provides the development platform for an application which can be written in C/C++ software development kit (SDK), .NET with PersonalJava, J2ME, Visual basic, Python, OPL, or Simkin.
(d) Visual C++ framework on Windows as common OS platform is an example of using a common framework and OS platform which support different languages.
5. (a) SSML (speech synthesis markup language) lets speech recognizer to define the pattern of sentences and SRGS (speech recognition grammar specification) specifies *emphasis* of the voice.
(b) SSML is used in TTS (text to speech converter). It is used for synthesizing the speech by interpretation of a text by a synthesizer. It creates audio book and SRGS is used in SSML.
(c) SSML cannot be embedded in VoiceXML scripts for provisioning an interactive telephone.
(d) VoiceXML can be used independently or with CCXML for telephony support, provides a VUI to the voice browser, and commands the browser for handling the calls of the voice channel.
6. (i) Applets are the programs running on a client with a lifecycle consisting of the steps—initiate, start, stop and destroy and do not retain their state on web-page transition. (ii) Aglets are the mobile agents which run on a host. (iii) The lifecycle of an aglet consists of the steps—creation, dispatch, activation, deactivation, and disposition. (iv) Aglet can retain its state when it migrates from one host to another. (v) An Aglet can be retracted or cloned. Which of these is correct?
(a) None is correct except (i). (b) All are correct except (ii) and (iv).
(c) All are correct. (d) All are correct except (iii) and (v).
7. (a) EJBs (Enterprise Java Bean) are non-configurable installed components.
(b) EJBs do not enable persistency of the application components.
(c) EJB is a component at a client in an enterprise application.
(d) Each EJB encapsulates business logic of the application.

8. (a) In a J2EE application JNDI which provides lookup services to enterprise directory, is not the required API.
   (b) Transactions maintain persistency of data of the enterprise at enterprise server.
   (c) The container provides transactions, security, and scalability but no pooling of the resources and nested execution of EJBs.
   (d) Entity Beans connect to web tier which can have a database server or which connects to the distributed objects at lower tiers.
9. (a) J2ME is a set of Java APIs which require small memory when developing Java applications and is also a platform for development of mobile phone games.
   (b) Windows Mobile devices support J2ME or java-based virtual machine.
   (c) J2ME platform binary implementations and virtual machine implementation are available from Sun.
   (d) java.swing is the package for Java Swing Classes for creating heavyweight GUIs which depend on the system (OS) resources.
10. (a) PersonalJava provides a Java application development platform and deploys mostly J2SE APIs.
   (b) Foundation profile contains APIs of J2SE without GUIs.
   (c) java.lang provides standard java types and classes for String, Integer, Math, and Thread but not Security and Exception classes.
   (d) LCDUI for mobile devices with no Internet connectivity is an API in java.io package.
11. MIDP 3.0 (i) is the latest profile version, (ii) is a profile for special-featured phones, (iii) is a profile for handheld devices, (iv) does not provide UI extensibility, (v) provides interoperability between the devices, (vi) does not support multiple network interfaces in a device, (vii) supports IPv6 (Internet Protocol version 6 for broadband Internet), (viii) supports large display devices, and (ix) does not support high performance games.
   (a) All are correct except (vii) and (viii).
   (b) All are correct except (vi) and (vii).
   (c) except (iv), (vi), and (viii).
   (d) All are correct.
12. MIDP 3.0 MIDlets use
   (a) SyncML DM/DS protocol
   (b) Bluetooth
   (c) removable media
   (d) all three (a), (b), and (c).
13. (a) CDC has minimum needed subset of the Java class libraries running on Java virtual machine.
   (b) CDC provides a configuration with an API called LCDUI.
   (c) CLDC has TCP/IP networking but no IPv6 capabilities.
   (d) A configuration is a subset of profile, and CLDC and CDC are subsets of MIDP profile.
14. (a) JavaCard framework (javacard.framework) provides the library functions, PIN (personal identification number), and ISO 7816 APIs for Card applet.
   (b) Card profile is a set of J2ME class libraries.
   (c) JavaCard profile has an identical virtual machine called card VM for different OSs.
   (d) JavaCard class method is makeTransientArray ( ) and the method creates a transient array and it persists permanently.
15. Card VM is a virtual machine which has (i) char, (ii) double precision, or (iii) single precision, (iv) floating point mathematical operations support, (v) Boolean, (vi) 8-bit byte, BCD, (vii) 16-bit integer (short), (viii) optional 32-bit integer support, (ix) object clones, (x) String class libraries, (xi) automatic garbage collection (memory freeing), (xii) native support using JNI, (xiii) SecurityManager class libraries, (xiv) multi-threading, and ThreadGroup.
   (a) (i) and (v) to (xii) are correct.
   (b) All are correct except (ii) and (iii).
   (c) (v) to (viii) are correct.
   (d) All are correct except (ii).
16. (i) The card applet creates card specific byte code on compilation. (ii) Card applet resides permanently on card. (iii) Objects created by card applet are persistent. (iv) When applet is deselected, the transient array disposes. (v) An applet for communication uses select method for selecting the process APDU whether it is a command APDU or response APDU. Which of these are correct?
   (a) (iv) and (v) are incorrect.
   (b) All are correct except (v).
   (c) All are correct.
   (d) (ii) and (iii) are incorrect.

Chapter 14

Mobile Operating Systems

In the preceding chapters we have discussed that mobile computing has *N*-tier architecture in which only limited database is cached at the mobile device and intensive computations are carried out at the $N \geq 2$ tier(s). The server(s) tier ($N \geq 2$) may be present at personal area network, WPAN, remote server(s) at mobile service provider, enterprise, or Internet. A mobile device tunes to pushed data or records or pulls the data on demand. The device caches and hoards the database(s) which can then be used in the applications which run at the device.

The device does limited computations at its end in *N*-tier architecture. It functions just as a thin or thick client in wirelessly connected client–server architecture. *Thin client* means that the computational resources at the device are very limited and in this case, the architecture requires strong connectivity. *Thick client* means computational resources are limited, the applications run at the client device itself, and only the applications that need intensive computation run at the server tier(s).

The *client framework* at the client device consists of a user interface, a client engine, and an adapter. The object exchange, synchronization, and wireless protocols are required when a thin or thick client provisions the mobile distributed computing system for running of applications. A markup language and a browser are used to develop user interfaces. In thick-client configuration, applications can be running independently at the client device.

We also learnt that for seamless interaction and platform-independence, the application databases can be in XML. The synchronization, device management, multi-modal data streams, and user interfaces can be in XML-based languages and applications can be in Java. The application program uses the parsed information from the program and data document files. We learnt that platform-independent languages could be used for application-development. Java is an example for application-development platform-independent language. J2ME in CLDC and CDC configurations is used for application development in devices having limited memory. The MID profile has classes for configuration and application development in a mobile device. The smartcard, AutoPro, TV, and other profiles are used for configuration and application development using J2ME for smartcards, automobiles, and TV set-top boxes.

A virtual machine enables execution of the application at different platforms (operating system and hardware). Java, J2ME, and JavaCard use JVM, VM, and CVM, respectively, as the virtual machines.

This chapter deals with application development for a mobile device for a mobile-device-specific operating system (OS) which provides a platform for the application development. Mobile OS controls a mobile device system and manages all software and hardware resources. Three popular mobile OSs are Palm, Windows CE, and Symbian.

This chapter deals with the following:

- The functions of an OS and its provisioning and applications.
- Study of the three popular mobile OSs—Palm, Windows CE, and Symbian.

14.1 Operating System

An operating system (OS) is the master control program in a system that manages all software and hardware resources. It controls, allocates, frees, and modifies the memory by increasing or decreasing it. In addition to all this, it also manages files, disks, and security, provides device drivers and GUIs for desktop or mobile computer, other functions, and applications. Moreover, the OS enables the assignment of priorities for requests to the system and controls IO devices and network.

An OS has the utility programs, for example, file manager and configuration of OS (memory and resource allocation and enabling and disabling the use of specific resources and functions). An OS can be accompanied by a specific suite of applications, for example, Internet Explorer and MS Office. Section 14.1.1 introduces certain specific terms to be used in the subsequent sections and also discusses some of the related points.

14.1.1 Process, Task, Thread, ISR, and IST

A *process* is a program unit which runs when scheduled to do so by OS and each state of which is controlled by OS. A process can be in any of the states—created, active, running, suspended, pending for a specified time interval, and pending for want of a specific communication (e.g., signal, semaphore (Section 14.1.2), mailbox-message, queue-message, or socket) from another process through the OS. It can call a function (method) but cannot call another process directly.

A *task* is an application process which runs according to its schedule set by the OS and each state of which is controlled by OS. It can be a real-time task which has time constraints or maximum defined latency within which it must run or finish.

A *thread* is an application process or a process unit (when a process or task has multiple threads) which runs as scheduled by the OS, each state of which is controlled by OS, and which runs as a light-weight process. Light-weight means it does not depend on certain system resources, for example, memory management unit (MMU), GUI functions provided by the OS, or the functions which need running of other processes or threads for their implementation.

An *interrupt service routine* (ISR) is a program unit (function, method, or subroutine) which runs when a hardware or software event occurs and running of which can be masked and can be prioritized by assigning a priority.

Some examples of hardware events are time-out of a timer (clock tick), division by zero, overflow or underflow detection by hardware during computation, finishing of DMA (direct memory access by a peripheral) transfer, data abort, external FIQ (fast interrupt request through a pin input), external IRQ (interrupt request through a pin input), a memory buffer becoming full, port, transmitter, receiver, or device buffer becoming half filled, buffer with at least one memory address filled, and buffer becoming empty. Buffer may be associated with the memory addresses for the LCD, printer, serial or USB port, keypad, or modem.

A software event can be *exception* (detection of a certain condition during computations or error while logging in) or illegal operation code provided to CPU.

An interrupt service thread (IST) is a special type of ISR or ISR unit (function, method, or subroutine) which initiates and runs on an event and which can be prioritized by assigning a priority. The type of IST depends on the specific OS. For example, Windows CE IST is one which is placed in a priority queue so that the ISTs execute turn by turn—FIFO (first-in first-out). An IST is initiated and put in the FIFO according to ISR priority after the execution of an ISR starts on an event. The ISR, therefore, has an event-initiated short piece of code. It runs only the critical part of the code and rest of the code runs at the IST initiated by it. Section 14.3.4 will describe the ISTs in case of the mobile OS, Windows CE.

Page is a unit of memory which can load from a program stored in a hard drive or from any other storage device to the program memory, RAM, before the execution of a program. It is a contiguous memory address block of 4 kB (in x86 processors), 2 kB, or 1 kB.

A *page table* is used for address mapping. The pages of memory are spread over the memory-address space leading to fragmentation of codes and data in physical memory space. A page table provides the mapping of fragmented physical memory pages with the pages of the virtual addresses which are the memory addresses assumed by the programmer and system MMU for the pages between the memory addresses 000000000_h and $FFFFFFFF_h$ with no consideration of actual physical addresses provided in the system and assuming that no other program or data exists. The MMU creates and maintains the page table and hence performs address mapping and translation. A program during execution first translates the accessed address (virtual address) into a physical address using the page table at the MMU and then accesses the physical address and fetches the code or data.

A process or thread which is to provide a waiting object to a higher priority process or thread gets the priority of that process or thread. This mechanism is called *priority inheritance*. Some examples of waiting object are signal, semaphore, queue, mailbox message, or bytes from a pipe. A pipe is a virtual device which sends the bytes from a thread to another thread.

Priority inversion is said to take place when a process or thread which is to provide a waiting object to a higher priority process or thread gets preempted by a middle priority process or thread and, thus, the middle priority process or thread starts running on obtaining the object for which it was waiting and higher priority process or thread keeps waiting for wait object. Enabling the priority inheritance provision protects the system from priority inversion.

14.1.2 Mobile Operating System

Mobile systems have specialized hardware, GUIs, VUIs, and the provision for mobile communication. These systems have many constraints, for example, the constraints of CPU speed, memory, battery life, display, and size of input devices. Mobile OS is an OS which enables running of application tasks taking into account such constraints of hardware. Mobile OS enables a programmer to develop application without considering the specifications, drivers, and functionalities of the hardware of the system. (Application is an abbreviation for application software.) A driver is software component which enables the use of a device, port, or network by configuring (for *open*, *close*, *connect*, or specifying a buffer size, mode, or control word) and sends output or receives input.

Mobile OS enables an application to run by simply abstracting the mobile system hardware. In other words, it enables the programmer to abstract the devices such that the application need not know full details of the font and font size of the mobile device display and how the message will be displayed by the LCD hardware.

Example 14.1 Give an example of hardware abstraction by the OS and example of program line.

Solution: Assume that keypad, LCD display, serial input, and serial output devices are abstracted by an application as the input and output devices with device numbers 1 2, 3, and 4, respectively. Then a program line in application can be *write* (1, 'Welcome to ABC Telecom') when a message Welcome to ABC Telecom is sent in the output for display. The line can be *write* ('Welcome to ABC Telecom') when display device is taken as default output device.

Mobile OS facilitates execution of software components on diversified mobile device hardware. Application need not be aware of the details of the LCD driver and memory at which the CPU will send the message for display. The mobile OS also provides interfaces for communication between processes, threads, and ISRs at the application and middleware layers and middleware and system hardware. The OS provides management functions (such as creation, activation, deletion, suspension, and delay) for tasks and memory, enables running of processes and also helps the processes in obtaining access to system resources.

An application consists of application tasks. The OS provides the functions used for scheduling the multiple tasks in a system. This is carried out through synchronization of the tasks by using semaphores (tokens). A task may have multiple threads. The OS provides for synchronization of the threads and their priority allocation. Also, it accomplishes real-time execution of the application tasks and threads.

An application uses the system resources, for example, CPU, memory keypad, display unit, modem interface, USB or serial port, and battery. These resources are shared concurrently by the applications running on the system. User application's GUIs (graphic user interfaces), VUI (voice user interface) components, and phone API are a must in many user-operated devices. Mobile OS provides configurable libraries for the GUI in the device. It provides for multi-channel and multi-modal user interfaces.

14.2 PalmOS

PalmOS is one of the most popular OSs for handheld devices. It is designed for highly efficient running of small productivity programs for devices with a few application tasks. The OS offers high performance due to a special feature that it supports single process which controls all computations by the event handlers and thus there is no multiprocessing or multitasking which greatly simplifies the kernel of the OS. There is an infinite waiting loop in the only process that kernel runs. The process does the polling for events at specific intervals and each polled event, if occurs, sends interrupt signal which is handled by an event handler. This is equivalent to non-maskable non-prioritized ISRs. Polling for the events is for a request to run an application or sub-application, for example, a search program requesting the processing of a query, notifications (like time-out alarm), and GUI actions (such as touching or tapping the screen with stylus).

PalmOS has following features:

- It compiles for a specific set of hardware and its performance is very finely tuned.
- It is optimized to support a very specific range of hardware, CPU, controller chips, and smaller screens of PalmOS-based devices.
- Memory space is partitioned into program memory dynamic heap for process stacks and global variables and multiple storage heaps for data and applications.

- A file is in format of a database, having multiple records and information fields about the file—name, attributes, and version of the database for the application.
- IP-based network connectivity and WiFi (in later version only).
- Integration to cellular GSM/CDMA phone.

PalmOS supports 16 MB memory, 256 MB internal flash (non-volatile ROM), and 256 MB card consisting of flash memory which the user inserts into the device. It provides simple APIs for developing the buttons, menus, scroll bar, dialogs, forms, and tables GUIs. PalmOS supports the following:

- Simple APIs compared to Windows CE.
- Desktop for Windows and Mac both and other essential software.
- PIM, address book, data book for task-to-do and organization, memo pad, SMTP (simple mail transfer protocol) email download, offline creation and sending of POP3 (post office protocol 3) email, Internet browsing functions using Blazer (a browser for handhelds), Windows organizer, and PDA (personal digital assistant).
- Query development support platform—Palm query applications (PQA) written using HTML and ported at Palm device
- Client-side application, GUI development support on C/C++ platform using Palm SDK, for Java application using J2ME and advanced tools, for example, Metrowerk CodeWarrior.
- Multimedia applications such as playing music (Palm Tungsten).
- Wireless communication protocols.

Table 14.1 lists the features of PalmOS.

Table 14.1 Highlights of PalmOS

Property	*Description*
Strength	Efficient running
OS basic functions	• Single process (no multi-processing and multi-threading) • Compiled for a specific set of hardware, performance very finely tuned • Memory space partitioned into program memory and multiple storage heaps for data and applications • A file is in format of a database • IP-based network connectivity and WiFi (in later version only) • Integration to cellular GSM/CDMA phone
Memory	• OS memory requirement is ~16 MB • 256 MB internal flash memory • Memory address space—256 MB in a card

(*Contd*)

(*Contd*)

GUIs	APIs for the buttons, menus, scroll bar, dialogs, forms, and tables using HTML markup language
User interface (GUI) display resolution	Generally wide screen—160 × 160 pixels with optimized layout of desktop programs displayed on screen and 256 colour touch screen (higher resolution support depends on new versions)
Software	• Simple APIs compared to Windows CE desktop for Windows and Mac both and other essential software • Simple APIs compared to Windows CE • PIM and e-mail and Internet , address book, data book for task-to-do and organizationing, memo pad, SMTP (simple mail transfer protocol) e-mail download, offline creation and sending of POP3 (post office protocol 3) e-mail, Internet browsing functions using Blazer (a browser, for handhelds), Windows organizer, and PDA (personal digital assistant) • GUI development support on C/C++, Java platforms using Palm SDK and for Java application using J2ME • Query development support—Palm query applications (PQA) written using HTML and are ported at Palm device • Support to client-side application and GUI development support on C/C++ platform using Palm SDK and for Java application using J2ME • Support to multimedia applications such as playing music (Palm Tungsten) wireless communications
Desktop program examples	SMS, Address, Card-Info, HotSync, To-Do-List, SMS, Security, Date Book/Calendar, Calc, Welcome, and Clock
Ports support	Serial and infrared ports for communication with mobile phones and external modems and for synchronizing a PC personal area computer using HotSync after resolving the conflicts in different versions of files during data exchange
HotSync synchronization	HotSync synchronizes a personal area computer PC through a serial port or infrared port (HotSync)
Personal area computerPC to handheld connection and file transfers	IrDA or serial device mounted on a cradle connects to computerPCs through IR or serial port. (A cradle is an attachment on which the handheld device can rest near a PC and connects to the PC via a USB or infrared.) A device is assumed as a new flash drive of a PC—the computer and HotSync facilitates drag and drop of files from device to PC and vice versa
Cards support	MMC (multimedia card), SD (secure digital) memory card, and SDIO (secure digital input/output) memory card

(*Contd*)

(*Contd*)

Third party common applications support	Examples—games, travel and flight planner, calculators, graphic drawings, and preparing slide shows
No support to multitasking	• Instead of multi-tasking, PalmOS provides for running a sub-application from within an application • Not an ideal platform for running multimedia applications because due to PalmOS is not for designing real-time systems • Does not offer much expandability
No adaptability	• Inability to adapt to different sorts of hardware may also be considered a limitation for this operating system

Figure 14.1 gives the architectural layers of PalmOS. The lowest level layer in OS is a kernel. This layer directly interfaces the assembler, firmware (software installed in the hardware devices in the system), and hardware. PalmOS has a micro-kernel.

PalmOS 4.x adds improved security, GUIs, VUIs, telephony libraries, and standard interfaces for access to the external SD cards for the files. PalmOS 5.x supports—(i) a standardized API for high resolution screen, (ii) dynamic input areas, (iii) instead of persistent battery-backed RAM, a non-volatile file system using

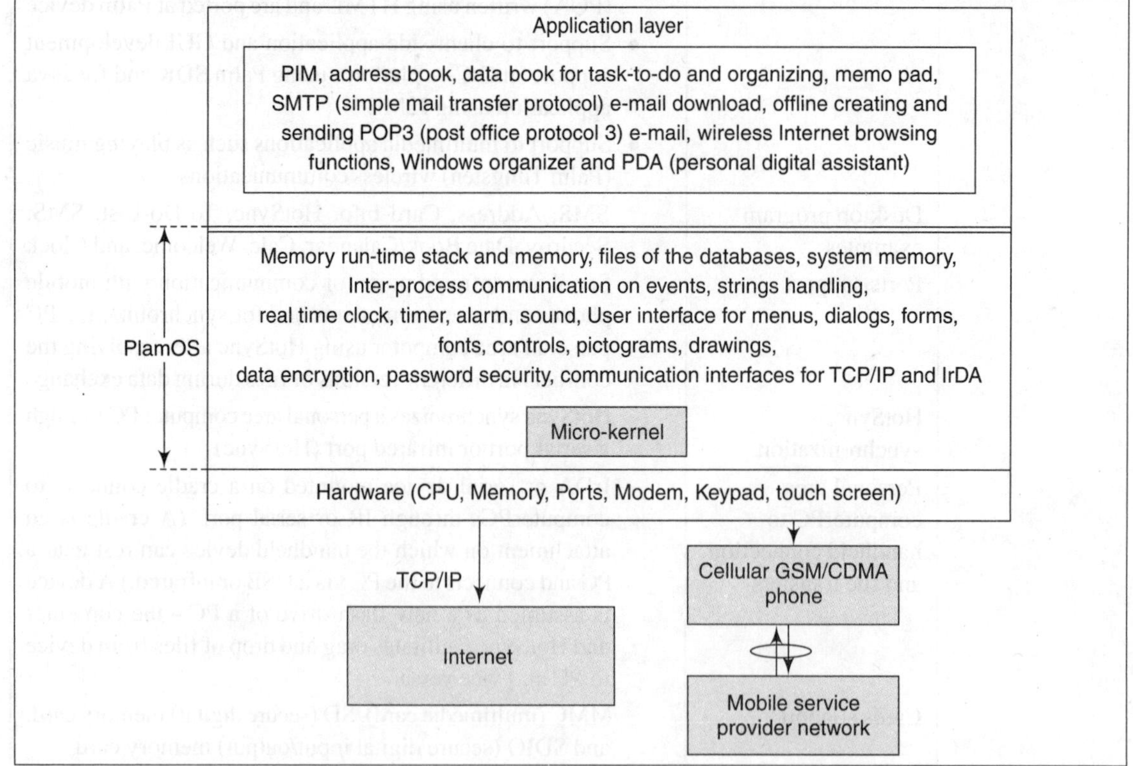

Fig. 14.1 Architectural layers of PalmOS with OS layer in between the applications and hardware

flash memory that saves the files and data in case the battery charge is draining out, and (iv) ARM, the processor which provides efficient code and an energy-efficient architecture.

A Palm handheld is now advanced. It has merged PDA and Smartphones. It has the feature to double as a hard drive using USB cable to PC. This enables the drag and drop of files between the Palm and PC in a manner similar to the drag and drop functions in a PC between C: and D: drives. Lately, a few Windows mobile handheld devices in use are Palm look-alikes. But these do not deploy PalmOS platform.

14.2.1 Memory Management

PalmOS assumes that there is a 256 MB memory card(s). The card can have RAM, ROM, and flash memories. A memory card has logical hexadecimal addresses from 00000_h to $3FFFF_h$. RAM is used for stacks of processes and for the global variables in the running processes of the *application*(s). ROM or flash is used for permanent resident application programs and OS. Flash is used for storage of non-volatile data.

The functions of the memory manager of PalmOS are shown in Fig. 14.2(a). Figure 14.2(b) shows how the manager partitions memory. There is a dynamic heap (within 96 kB for PalmOS 3.x) of application process stacks (3 kB), TCP/IP stack (32 kB), OS functions stack, applications and OS functions dynamic-memory spaces, system global variables (2.5 kB), and application global variables. Direct memory address spaces in card are used instead of allocated dynamic memory buffers for the I/O port devices (e.g., LCD display, keypad input, and modem I/Os). PalmOS 3.x and 4.x (not 5.x) provides execute-in-place system due to static allocation of the memory addresses for the application-installed programs and data storage.

14.2.2 File Management

PalmOS file manager manages each file as a database which has multiple records and information fields. Each record can have following attributes—protected record, deleted record (similar to deleted file), locked record (in use by application process or OS), and updated record. Deleted record attribute helps in data recovery by a recovery program.

The info fields of each record have record ID and record attributes. Info fields about the file have—(i) name, (ii) file attributes, (iii) version of application database, (iv) modification number (number of times modified) and access counter for number of times accessed), and (v) file local ID. [Usually file ID has several characters, thus requiring several bytes. The file local ID is a number used to identify the file locally when an application is running. A local file sorting table uses the file local ID to sort the file in the required order. For example, the ordering may be based on the time and date of the last modification made in the file.]

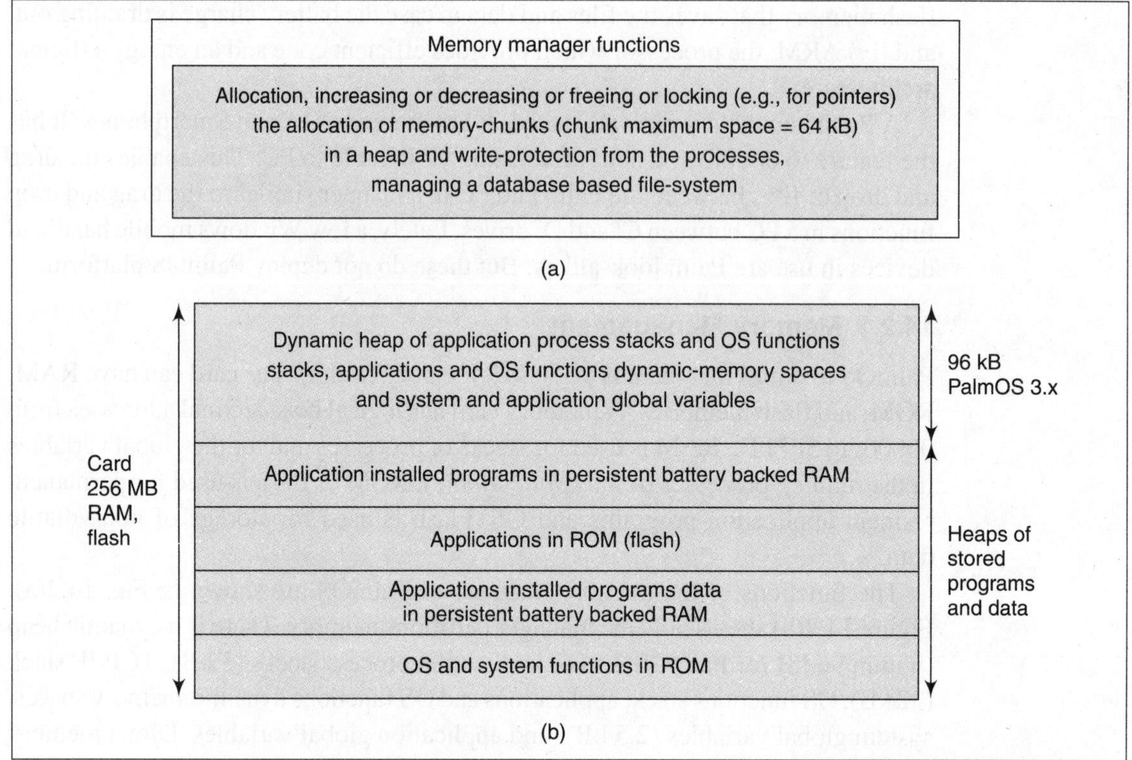

Fig. 14.2 (a) PalmOS memory manager functions, (b) PalmOS memory partitioning

14.2.3 Communication APIs

PalmOS provides communication and network APIs for serial, IrDA, and TCP/IP communication. These three types of communication are discussed below.

Serial communication uses a cradle. A serial manager (SM) provides interface to the device on cradle with the RS232C COM port of the PC. Connection management (CM) protocol, modem manager (MM), or serial link protocol (SLP) interacts with SM to transmit the data to PC. A device receives the data from the other end through serial manager and SLP, MM, or CM. SLP has on top of it, a packet assembly and disassembly protocol (PADP). A desktop link protocol (DLP) transmits data to PADP when serial device is sending data to PC and receives data from PADP when serial device is receiving data from the PC. (MM is for deploying a dial up-modem. CM carries out exchanges for establishing connection, baud rate selection, and finding version number. SLP is for packet communication on serial line.)

IrDA asynchronous serial (115 kbps) or synchronous serial communication (1.152 or 4 Mbps) uses exchange manager as session layer and IrDA library functions (for IrDA protocol layers) at lower level (Table 12.5). Exchange manager enables data interchange directly without HotSync. Exchange manager and application use a set of launch codes to generate appropriate events.

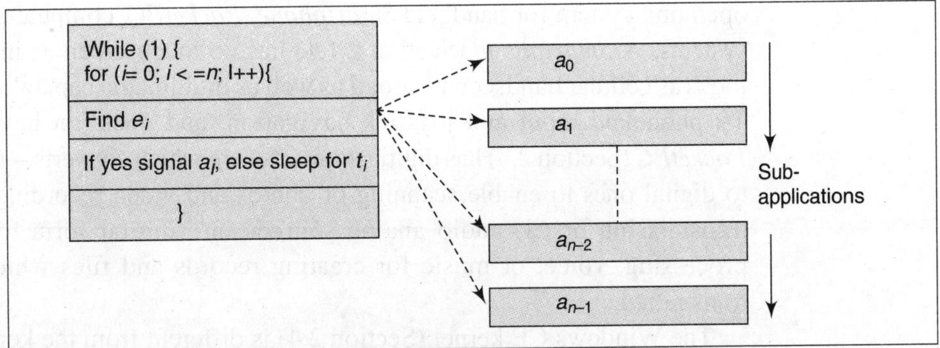

Fig. 14.3 Application program running on PalmOS platform as event-driven number of sub-applications

14.2.4 Network Library Functions

PalmOS uses TCP/IP network library functions (for UDP and TCP) to send stack to a net protocol stack (NPS) and provides a socket API. Berkeley Socket APIs are also supported by a Palm device. PalmOS uses HTTP/HTTPS net library for Internet connectivity.

14.2.5 Application Development Method

Corresponding to each event, there is an event handler. Application development means defining additional events and coding for the corresponding handlers. An application can be assumed to be divisible into sub-applications a_m to a_{n-1} along with the existing event handlers a_0 to a_{m-1}. Assume that a_i runs on the events e_i and an event e_i is polled at a sleep interval of t_{i-1} in an infinite while loop. Figure 14.3 shows how an application program runs on PalmOS platform as number of event-driven sub-applications.

14.2.6 Application Development Packages

PalmOS supports development packages—Palm SDK (software development kit) and CDK (conduit development kit). A conduit is a path. A CDK conduit of Palm provides a two-way path for data exchanges and synchronization between a desktop PC application (for Windows or Solaris) and the applications running on a device (SMS, Address, Card-Info, HotSync, To-Do-List, SMS, Security, Date Book/ Calendar, Calc, Welcome, and Clock).

14.3 Windows CE

Section 2.4.1 described Windows CE-based devices. Windows CE is an operating system from Microsoft. It is a component-based, embedded, real-time operating system with deterministic interrupt latency. It can be configured as a real-time operating system for handheld *Smartphone*, *PocketPC*, computers, and embedded systems. A *Smartphone* (Section 2.1.1) has no touchscreen as in a PDA, closely looks as cellular handset with email as well as multimedia capabilities, uses keypad T9 phonePad input and joystick navigation, and does not have a digitizer. A *PocketPC* (Section 2.3) has digitization software which converts—(i) analog signals to digital ones to enable scanning of photos and video recordings for storage or transmission or (ii) audio analog sources into digital form to enable speech processing, voice, or music for creating records and files which are stored or transmitted.

The Windows CE kernel (Section 2.4) is different from the kernel of Windows 98, 2000, and other desktop versions of Windows. It is meant for computing devices with low storage and can run with about 1 MB of memory. However, a Windows CE-based device needs larger memory as compared that needed by PalmOS-based device (Section 14.2).

Windows CE is nowadays one of the most popular OSs for the handheld devices. Microsoft design precepts for Windows CE are *compact*, *connectable*, *compatible*, *companion*, and *efficient*.

The latest version of Windows CE is Windows Embedded CE 6.0. Its features are as follows:

- Open, scalable, 32-bit operating system (OS) with small-footprint and advanced Windows technologies.
- Provides hard real-time capabilities, with a redesigned kernel and embedded-specific development tools.
- Can provide devices for home as well as work places.
- Devices have provisions for media and shared presentations.
- Devices can connect to cellular networks.

Windows Mobile 6 is a platform for mobile devices such as PocketPC. The platform manages Visual C# and Visual Basic .NET codes. The platform is based on Windows CE and hardware such as personal digital assistants (PDAs) and Smartphones. Microsoft Visual Studio 2005 and Windows Mobile SDK are used for creating software for the platform. The code is developed in Visual C++.

The following subsections provide a detailed description of the highlights and architecture of Windows CE, the management of memory and its files and communication in Windows CE, and the application development methods using Windows CE

14.3.1 Highlights and Architecture

The highlights of Windows CE kernel are as follows:

- The user gets a personal-computer-like feel and Windows-like GUIs when using a Windows CE device.
- Large number of Windows-based applications are available at the device.

- The OS code is customized for each specific hardware and processor in order to fine-tune the performance.
- Windows CE supports multi-processing or multi-tasking and multi-threading like Symbian and unlike Palm.
- Kernel is divided into two sublayers. One sublayer consists of large part of the OS and is offered in the form of source code first. This sublayer is the Microsoft source code for Windows CE kernel. Then the OS is adjusted according to the device hardware by adding the remaining part of the OS. Second sublayer is called hardware abstraction layer.
- Shared source is licensed with controlled access to full or limited parts of the source code for a product. Windows CE 5.x and 6.0 developers have the freedom to modify down to the kernel level without the need to share their changes with Microsoft or competitors.

Table 14.2 gives the basic highlights of Windows CE.

Table 14.2 Highlights of Windows CE

Property	*Description*
OS	• A 32-bit OS • Customized for each specific hardware and processor in order to fine-tune the performance • Compatible with a variety of processor architectures • Compiled for a specific set of hardware, its performance is very finely tuned • Kernel consisting of source code and hardware abstraction sublayer • Shared source and source code access • Modular/componentized to provide the foundation of several classes of devices and supports addition of features of other components for Windows, DCOM, and COM • Windows CE is real time OS, and thus supports multi-tasking. Windows CE 6.0 provisions for each simultaneously running process to have 2 GB of virtual memory space • Memory requirement is large but scales to the requirement of the device peripherals • Memory space partitioned • Data is in formats of a database or object file • File automatically compresses when stored and decompresses when loaded • Visual C/C++ platform integrates use of web • .NET XML parsing (trimmed version)

(*Contd*)

(*Contd*)

Windows Mobile 5.0	• Windows CE 5.0 with a set of specific applications and GUIs/VUIs and for a specific set of processors which are deployed in (i) Smartphone, (ii) handheld PocketPC which features the *digitizer* in the human computer interface (HCI), and (iii) portable media player • PDA with Microsoft Smartphone phone device, touch screen, touchpad, or directional pad
Memory	Minimum footprint of Windows CE is 350 kB (Section 14.3.2) Windows Mobile 5.x—all user data in persistent (flash) memory and RAM to be used only for running applications
GUIs and VUIs	GUI development support using markup language as well as C/C++ language and embedded complex APIs provided in Windows CE. It gives the user a PC-like feel and Windows-like GUIs (window resizing not provided.), VUIs (in PocketPC and automotive PC), buttons, shortcut icons, menus, scroll bar, dialogs, forms, and tables
User interface (GUI) display resolution	High resolution colour/display, touch screen, and stylus keypad with Windows layout of desktop programs displayed on coloured touch screen, built-in microphone for voice recording
Strength	Efficient running of most programs and support for multi-tasking real-time and multimedia applications
Software	Desktop for Windows and other essential software, PIM, Contacts, Task-to-do, Smartphone, and multimedia applications such as playing music. (Refer Fig. 14.4 for application layer software)
Desktop program, tool bar, start menu	• Today calendar, owner, number of messages not read, tasks, and present hour subject. Button and tool bar for task start menu [today calendar, contacts, Internet explorer, messages, phone, pocket MSN, album, MSN messenger, camera, programs, settings, and help], phone mode indicator (on/off), signal strength status, speaker status (on/off), and time • Two context-sensitive soft buttons at the bottom of the screen, which can be mapped to hardware buttons on any specific device
Ports support	USB and infrared port support for communication of a device with mobile phones and for synchronizing a PC using ActiveSync after resolving the conflicts due to different versions of object files during data exchange Bluetooth and/or TCP/IP. WiFi or Ethernet LAN interface
ActiveSync synchronization	Synchronization of mobile device data with PC using a USB, Bluetooth, and PC infrared port

(*Contd*)

(*Contd*)

PC to handheld connection and file transfers	USB 2.0 in Windows CE 5.0 PocketPC conform as the USB mass storage class, the storage on device can be accessed, and drag and drop menu can be used from any USB port of PC, which considers the handheld device just another flash drive. A cradle connects to PC
Memory support	External memory stick (strip) (e.g., 2 GB)
Multi-tasking real-time OS feature	Windows CE is real-time OS allowing multi-tasking on handheld devices with defined real-time constraints for each task and a deterministic latency (period within which it will definitely run) Kernel supports 32 simultaneous or concurrent running of—(i) the processes and (ii) threads of a process
Threading feature	Thread is basic unit of computation. A process can have any number of threads. Which can also run simultaneously. A device based on CE 6.0 can run a larger number of complex applications and can run as many as 32,000 simultaneous processes
Inter-process communication	Threads are synchronized by synchronization objects, for example, a *mutex* (semaphore) wait and critical section or wait object for a *signal*
Third party common applications support	Very large support (about 20000) for games, applications, mobile e-commerce, and stock-trading
No support to simultaneous multi-modality	Cooperative running of multi-threading does not support simultaneous multi-modal user interfaces (data by multiple modes, for example, text as well as speech)
Poor adaptability	Adapts to different sorts of hardware limits mainly because of two reasons—(i) compiled for a specific set of hardware for very fine-tuned Windows CE performance, (ii) large parts of OS offered in the form of source code first and then adjusted to the hardware by the manufacturer

Figure 14.4 gives the architectural layers of Windows CE with OS layer in between the applications and hardware in the form of kernel and hardware abstraction layers.

Application layer supports the following APIs—Bluetooth, WiFi, Word mobile with embedded graphics, Excel mobile with charts, tables, ordered lists, PowerPoint mobile, set of hardware application (e.g., camera) buttons, email download, offline creation and sending of POP3 (post office protocol 3) email, Internet Explorer functions on handheld mobile system, data call API, telephony API and wireless Smartphone support, digital camera card, Direct3D (three-dimesional graphics), games, and Microsoft Windows Media and other players.

Next to application layer is Windows CE layer which consists of memory and virtual memory managers, process manager, thread and interrupt handlers, and user

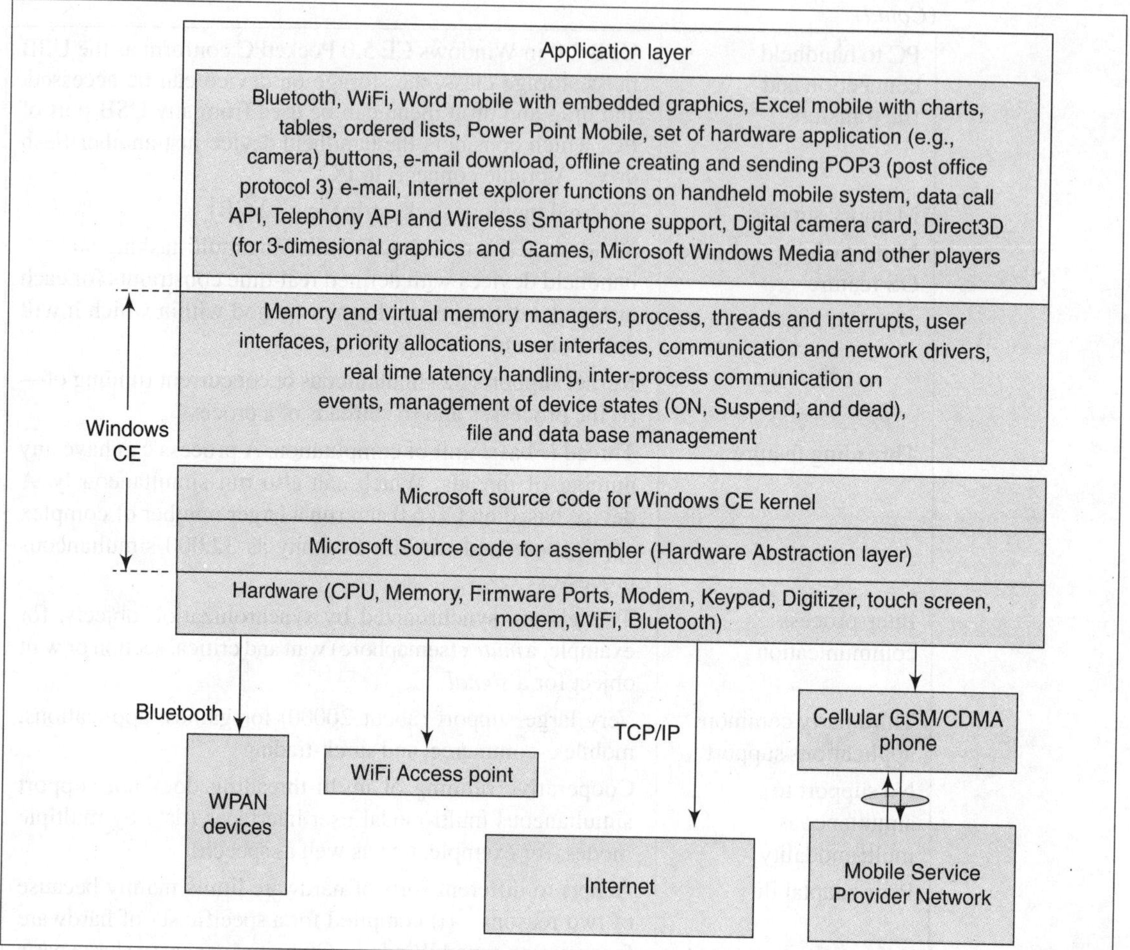

Fig. 14.4 Architectural layers of Windows CE with OS layer in between the applications and hardware through the kernel hardware abstraction layers

interfaces. Windows CE supports priority allocation to the processes and interrupt service threads. It consists of communication and network drivers. It handles real-time constraints and real-time latency. OS has inter-process communication functions to pass the events. OS manages events, device states (ON, suspend, and dead), files, and databases.

The Windows CE layer next to the above is kernel source code. Hardware abstration layer is layer added for customizing Windows CE for the device hardware. Hardware layer consists of CPU, memory, firmware ports, modem, keypad, digitizer, touch screen, modem, WiFi, and Bluetooth.

Just as PalmOS 4.x, Windows CE 4.x adds improved security, GUIs, VUIs, telephony libraries, and standard interfaces for access to the external SD cards for the files. Windows CE 5.x supports a non-volatile file system using flash memory which are nowadays used instead of persistent battery-backed RAM. Windows CE supports a new file system that supports larger file sizes, removable media encryption, and larger storage media. The flash file system saves the files and data in case the battery charge is draining out.

Windows CE is customized for each specific hardware, for example, for a processor deployed from amongst a large array of processor architectures and compiled for a specific set of hardware. It means that its performance is very finely tuned. Moreover, a Window CE-based device connected to a personal area computer using a USB cable doubles as a hard drive thus enabling the drag and drop of files between the computer and the device. The drag and drop action is similar to that in a PC between C: and D: drives.

Window CE device can be in three states—(i) ON with clock frequency lowered in idle state, (ii) suspend with power to unused system units and port peripherals disconnected, memory data persistent, CPU idle till next interrupt, and clock running, and (iii) dead with power disconnected.

Windows CE with a specific suite of applications and GUI/VUIs is called Windows Mobile OS. There are many different versions of Windows Mobile. A brief description of this OS follows.

14.3.1.1 Windows Mobile

Handheld PC using Windows Mobile OS is also known as PocketPC. Windows mobile is a suite of basic applications for handheld devices along with a compact operating system. Mobile 5.0 enforces device password for security. An application can be written in JavaScript and ActiveX. It provides remote wipe. It enables PIM, Internet Explorer, Windows Media Player, various audio and video formats, Voice over IP, pocket (small screen display version) Excel, Pocket Word, Microsoft Outlook Mobile, and familiar Microsoft Office Mobile software.

The Windows Mobile Microsoft .NET platform is an open platform. About 20,000 Windows Mobile applications are available from third-party developers. Mobile operators, networking, and synchronization technologies also deploy Windows Mobile Microsoft .NET platform.

14.3.2 Memory and File Management

Windows CE assumes that there is a 4 GB virtual memory. A system can have RAM, ROM, and flash memories. A memory has logical hexadecimal addresses from 00000000_h to $FFFFFFFF_h$. The functions of the memory manager of Windows CE are shown in Fig. 14.5(a). Figure 14.5(b) demonstrates how the manager partitions memory. A ROM image is the footprint of OS, data, and the applications at the permanently installed ROM (or flash memory). When an OS is configured and customized for an embedded application(s), then OS footprint is the ROM image. The customization reduces the OS footprint. Windows CE needs a minimum footprint of 350 kB. Device applications are optimized so that the devices need

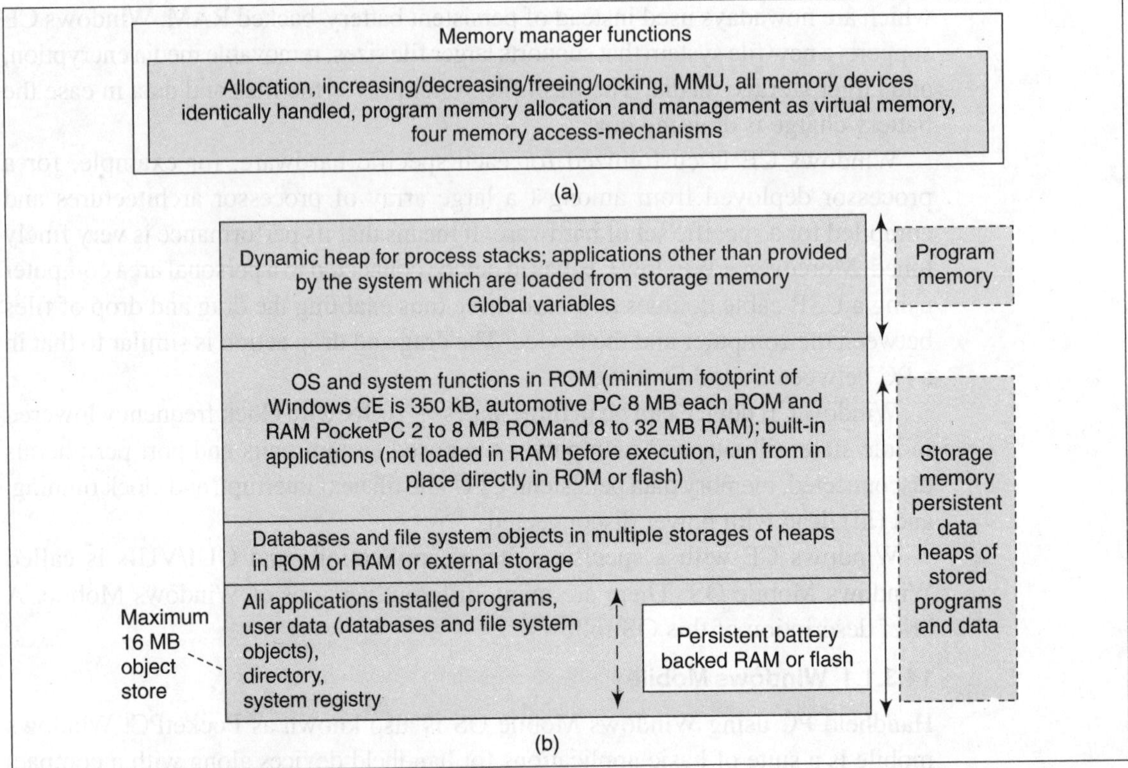

Fig. 14.5 (a) Windows CE memory manager functions, (b) Windows CE memory partitioning

minimal storage below 1 MB with no disk storage. Windows CE footprint is burned in ROM and is configured as it does not allow end-user extension. (Windows CE 5.x developers have the freedom to modify down to the kernel level, without the need to share their changes with Microsoft or competitors.)

There is a dynamic heap of application process stacks, TCP/IP stack, OS functions stack, applications and OS functions dynamic-memory spaces, system and application global variables, Bluetooth stack, and WiFi stack in the *program memory* (Fig. 14.5).

OS and system functions are saved in ROM. As pointed out above, the minimum footprint of Windows CE is 350 kB. An automotive PC with Windows CE has 8 MB ROM and RAM each. The PocketPC has 2–8 MB ROM and 8–32 MB RAM in *storage memory*. Built-in applications are also in storage memory and not loaded in RAM before execution so that a process runs these directly from storage memory addresses (called execution in place (XIP) directly from process codes and data in ROM or flash). Windows CE 3.x and 4.x (not 5.x) provides XIP due to static allocation of memory addresses for the application-installed programs and data storage.

Windows CE memory manager has four different types of access mechanisms. These are as follows:

- Simple—There is no use of paging or cache.
- Cache—Cache is used but paging is not used. Processor cache is used along with the RAM/ROM/flash memory.
- Page—Paging is used but not cache.
- Sophisticated—both cache and paging are employed.

Windows CE assigns contagious pages in virtual address space to one of the 32 memory slots. Each slot is of 32 MB. Memory manager allocates a distinct slot to a distinct process among the maximum 32 concurrently running processes. Each process is one among the multiple processes (tasks) of an application. Allocation of memory slots reduces fragmentation of pages to a significant extent as pages of processes are at the contagious addresses in the slot. Fragmentation occurs only when the memory needed by the process code and data is more than 32 MB.

Windows CE *file manager* manages data as a database or object file. A file system has a root directory with which the file folders associate in a tree-like structure. The file system divides all file folders into volumes. A volume is a unit which can be loaded from the device to the computer or can be stored on the device from the computer. Each volume has a root directory which has directories and file folders. Each directory can have further divisions into subdirectories and file folders. Each subdirectory can have further divisions into files and subdirectories till the leaf node which has a single file folder.

14.3.3 Communication, Network, Device, and Peripheral Drivers

Windows CE provides communication and network APIs for serial, IrDA, TCP/IP, Bluetooth stack, and WiFi stack. Windows CE provides IP-based connectivity to the network and WiFi-based connectivity in later versions. Window CE integrates Microsoft Smartphone software. It enables the application of Windows CE device as cellular GSM/CDMA phone.

Serial communication uses a cradle. A serial manager (SM) provides interface to the device on cradle with the RS232C COM port of the PC. Connection management is by using the serial link interface protocol (SLIP) and point-to-point (PPP) protocol. IrDA asynchronous serial (115 kbps) or synchronous serial communication (1.152 or 4 Mbps) uses ActiveSync.

Network connectivity is by radio transceiver and LAN adapter. A NDIS (network driver interface specification) is used for the drivers other than the driver loaded at the hardware. Windows CE uses TCP/IP network library functions (for UDP and TCP) to transmit or receive the stacks and application [HTTP, HTTPS, FTP, or RAS (remote access service)] requests and response on or from the network.

Windows CE provides device driver and peripheral driver functions for low-level drivers at the kernel. (Kernel resides in ROM image.) USB connectivity is provided for the peripherals. Windows CE deploys the CSP (cryptographic service providers) server for actual implementation of the encryption and security. The CSP deploys CAPI (cryptographic application programming interface).

14.3.4 Application Development

Windows CE considers thread of a process to be a fundamental unit of execution and providing access of CPU. Applications are developed by coding for the threads. It supports 256 priority levels for assignment to the threads. Windows CE provides protection from priority inversion as it provides for priority inheritence mechanism.

Corresponding to each event, there is an event handler. (Events are asynchronous.) An event sends interrupt signal which is checked for source, that is, whether the source is a hardware event, software exception, user-action-based event, or kernel event (e.g., a kernel command, WM_HIBERNATE). Event handler is called interrupt service thread (IST) in Windows CE device.

An IST can have one of the eight priority levels. Highest priority level means *time critical* priority (to run in hard real-time mode when latency is least and thread execution critical). Lowest priority level means *idle* priority (to run when system is idle). If priorities are same, each thread is allotted a time slot. Initiated ISTs form a priority-ordered queue of routines and are executed as per priorities.

An external interrupt is notified to the OS and for each request, there is an ISR assigned by the OS on notice. ISR is a short kernel code which identifies the IST for service and thus activates the IST corresponding to the request (event). Using the concept of ISR, thread priority, and ISTs, Windows CE provides the OS for a hard real-time system with multiple processes.

An application development needs to define processes and each process has threads. The application development also needs to define user events and coding for the corresponding interrupt service threads or functions for handling the events. An application can be assumed to be divisible into processes consisting of threads th_m to th_{n-1} of varying priorities p_0 to p_{m-1} and ISTs as event handlers th_0 to th_{m-1}. A th_i runs on event e_i if $p_i < p_j$ where p_j is priority of the thread presently running.

Figure 14.6 shows how an application program runs on Windows CE platform as event-driven number of ISTs ($i, j, \ldots$) and application threads ($p, q, \ldots$).

Development tools for the application development are Visual basic and Visual C++. Using PlatformBuilder, a programmer can develop a Windows CE application by carrying out OS image creation and integration. PlatformBuilder provides an integrated environment for customized operating system designs based on Windows CE. A platform builder is used for CE 6.0, which plugs into the development environment for Microsoft Visual Studio 2005. Windows CE provides a two-way path for data exchanges and synchronization between a desktop PC application and the applications (Table 14.2) running on a device.

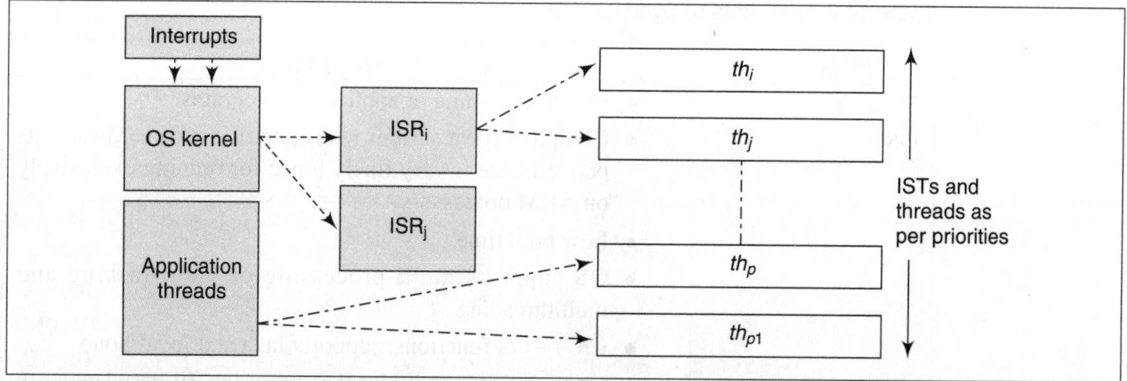

Fig. 14.6 Application program running on Windows CE platform as event driven number of ISTs (*i*, *j*, ..) and application threads (*p*, *q*, ...)

14.4 Symbian OS

Symbian is one of the popular OSs for handheld Smartphones and mobile handhelds with phone and multi-modal communication (Section 9.5) features. *Multi-modal* means usage of different modes—text, image, video, or audio. Multi-modal communication integrates and synchronizes multimedia. Symbian OS supports application development in C/C++ as well as Java and many communication protocols. Symbian OS performance is fine-tuned to ARM processors which are used in about more than 70% of mobile devices, for example, Nokia mobile devices. Symbian OS supports multi-processing or multi-tasking and multi-threading like Windows CE and unlike Palm.

Symbian supports application architecture consisting of GUI, application view, and application engine. It also supports applications in C/C++, Java, and many other development platforms. Personal Java and Symbian Everywhere are application development tools employed while using Java and Symbian. Symbian C++ Software development kit (SDK) is used for developing applications in C++. Symbian emulator provides application development using Windows Metrowerk CodeWarrior. Symbian provides powerful development platforms and GUIs. Symbian-based devices use SyncML synchronization and can also deploy C/C++-based synchronization software.

Table 14.3 gives the basic highlights of Symbian OS.

Figure 14.7 gives the architectural layers of Symbian. The lowest level layer in OS is a kernel which directly interfaces the assembler, firmware, and hardware.

Symbian has improved security, MD5, RSA, and many other additional features. It has GUIs, VUIs, and telephony standard interfaces. Symbian supports—(i) a standardized API for high resolution screen, (ii) non-volatile file system using flash memory to save the files and data instead of saving them in persistent battery-backed RAM, and (iii) ARM, the processor which provides efficient code and an energy-efficient processor architecture.

Table 14.3 Highlights of Symbian OS

Property	*Description*
Strength	• Efficient running of application programs
OS	• Compiled for a system with specific set of hardware, its performance is very finely tuned for running exclusively on ARM processors • Low boot time • OS supports multi-processing or multi-tasking and multithreading • C/C++ OS functions, supports Java and JavaPhone • Internet connectivity for Web browsing, IP-based network connectivity, and WiFi (in later version only) • JavaPhone and integration to cellular GSM/CDMA phone.
Memory	Large storage memory which includes 80 MB of built-in memory in a multimedia card (MMC)
GUIs/VUIs	APIs for the buttons, menus, advanced voice features such as a hands-free speakerphone, and conference calling capability
User interface (GUI) display resolution	Generally small screen 160 × 160 pixels with optimized layout of desktop programs displayed on screen and 256 colour touch screen (higher resolution support depends on new versions)
Software	Desktop both for Lotus and Windows programs, other essential software, push-to-talk, high-end security enhancement features, graphics support including support for 3D rendering, simple APIs as compared to Windows CE, PIM, JavaPhone, JVM, MIDP (mobile information device profile), contacts, SyncML, Office, address book, spreadsheet, calendar, agenda, word processor, text-to-speech converter, browsing, messaging (SMS, MMS, email, and IMAP4), WAP push Microsoft Office formats (MS Office 97 onwards) and slide shows, email download, offline creation and sending of POP3 (post office protocol 3) email, and Internet browsing functions, GUI development support on C/C++ and Java platform, Java application using J2ME, multimedia applications such as playing music (Palm Tungsten), wireless communications Support for WLAN, Adobe Reader for accessing PDF files, Symantec Client Security 3.0 and the Fujitsu mProcess Business Process Mobilizer, corporate solutions such as IBM WebSphere Everyplace Access, BlackBerry Connect, Oracle Collaboration Suite, and secure mobile connections via VPN Client
Ports support	Serial, USB, infrared ports, telephony, and Bluetooth for communication with mobile phones and external modems

(*Contd*)

(Contd)

SyncML and other synchronization software	SyncML PC synchronization feature, synchronizes and chains to a PC in the vicinity. Also supports synchronization software developed on C/C++ platforms
Multi-threading support	Supports multi-threading, MMC (multimedia card), SD (secure digital) memory card, and SDIO (secure digital input/output) memory card
Third-party common applications support	Large number of examples—games, travel and flight planner, enterprise solutions, calculators, graphic drawings, and preparing slide shows.
Support to multi-tasking	• Multi-tasking Symbian allows running of multiple processes of an application and pre-emption of low priority process by a higher priority one • Platform for running multimedia applications due to Symbian is not for designing real-time systems 3

14.4.1.1 Communication and network APIs

Symbian provides communication using WAP, WiFi, and network APIs for serial, Bluetooth, IrDA, and TCP/IP, communication APIs for HTTP, TCP/IP, DNS, SSL, WAP, PPP, GUI/VUI framework, and communication interfaces for TCP/IP and

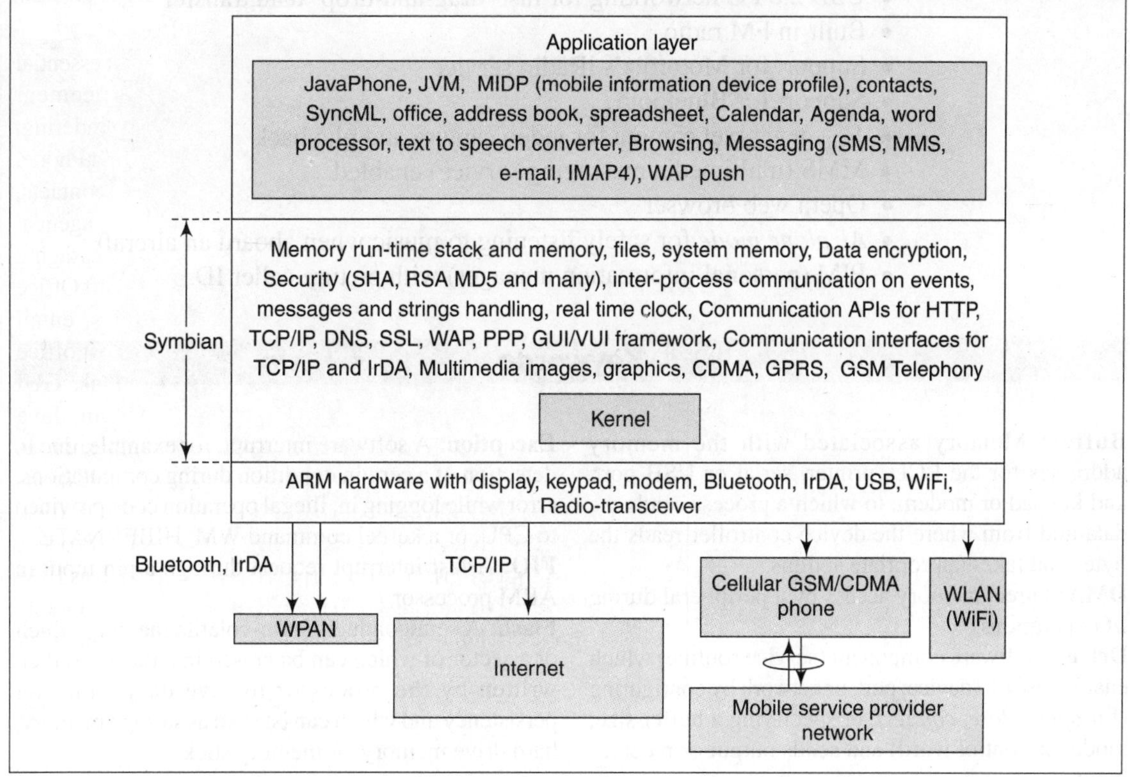

Fig. 14.7 Architectural layers of Symbian with OS layer in between the applications and hardware

IrDA, multimedia images, graphics, CDMA, GPRS, and GSM Telephony. Symbian uses TCP/IP network library functions (for UDP and TCP) to send stack.

14.5 Linux for Mobile Devices

Most operating systems used in mobile devices are designed for use on specific hardware and offer a platform for only select software applications. Linux, however, can be modified easily to suit different sorts of hardware and software applications. Being an open source OS, it enables the user to customize their device to suit their specific needs. The Embedded Linux Consortium (ELC) is an association which promotes Linux and develops standards for Linux in embedded systems. They also develop standards for designing user interfaces, managing power consumption in devices, and real-time operation of embedded Linux. The ELC platform specification (ELCPS) is also a result of this consortium. Linux is also considered to be more secure than most other operating systems. Also Linux support is easily available from many forums and associations that promote this OS. Many international mobile phone manufacturers are turning to Linux for their OS requirements. The Motorola Rokr E2 music phone is an example of a Linux-based mobile phone. Its main features are listed below:

- A 240 × 320 TFT display with 262,000 colours
- USB 2.0 PC networking for fast 'drag-and-drop' data transfer
- Built-in FM radio
- Support for Motorola's iRadio service
- Support for Bluetooth.
- 1.3-megapixel camera for video capture and playback
- MMS (multimedia messaging service) enabled
- Opera web browser
- *Airplane mode* for safely listening to music when aboard an aircraft
- PIM (personal information manager) with picture caller ID

Keywords

Buffer: Memory associated with the memory addresses for the LCD, printer, serial or USB port, and keypad or modem, to which a process sends the data and from where the device controller reads the bytes and takes appropriate actions.

DMA: Direct memory access by a peripheral during I/O operations.

Driver: Software component (service routine) which enables use of a device, port, or network by configuring (for *open*, *close*, *connect*, or specifying a buffer size, mode, or control word) and sends output or receives input.

Exception: A software interrupt, for example, due to detection of a certain condition during computations, error while logging in, illegal operation code provided to CPU, or a kernel command WM_HIBERNATE.

FIQ: A fast interrupt request through a pin input in ARM processor.

Flash: A semiconductor non-volatile memory which or a sector of which can be erased in a flash and then written by the processor to save data requiring persistency and which can be used as storage memory, hard drive memory, or memory stick.

Hardware interrupt: An interrupt from hardware units or devices, in case of hardware events, for example, time-out of a timer (clock tick), division by zero, overflow or underflow detection by hardware during computation, finishing of DMA (direct memory access by a peripheral) transfer, data abort, external FIQ, external IRQ, a memory buffer becoming full, port, transmitter, receiver, or device buffer becoming half filled, buffer with at least one memory address filled, and buffer becoming empty.

Interrupt service routine (ISR): A program unit (function, method, or subroutine) which runs when a hardware or software event occurs and running of which can be masked and can be prioritized by assigning a priority.

Interrupt service thread (IST): A special type of ISR or ISR unit (function, method, or subroutine) which initiates and runs on an event and which can be prioritized by assigning a priority.

IRQ: An interrupt request through a pin input.

Kernel: The lowest level layer in OS which directly interfaces the assembler, firmware (software installed in the hardware devices in the system), and hardware.

Light-weight process: Thread or process which does not depend on certain system resources, for example, memory management unit (MMU), GUI functions provided by the OS, or the functions which need running of other processes or threads for their implementation.

MMU: A unit for managing the memory called memory management unit which creates and maintains the page table and hence performs address mapping and translation. A program during execution first translates the accessed address (virtual address) into a physical address using the page table at the MMU and then accesses the physical address and fetches the code or data.

Page: Page is a unit of memory which can load from a program stored in a hard drive or from any other storage device to the program memory, RAM, before the execution of a program. It is a contiguous memory address block of 4 kB (in x86 processors), 2 kB, or 1 kB.

Page table: A table used for address mapping.

Priority inheritence: A mechanism in which a process or thread which is to provide a wait object to a higher priority process or thread gets the priority of that process or thread.

Priority inversion: Priority inversion is said to take place when a process or thread which is to provide a wait object to a higher priority process or thread gets preempted by a middle priority process or thread and thus the middle priority process or thread starts running on obtaining the object for which it was waiting and higher priority process or thread keeps waiting for wait object.

Process: A program unit which runs when scheduled to run by a master control program in a system, called OS and each state of which is controlled by OS.

Process states: Created, active, running, suspended, and pending.

Real-time: System time once set and then used for synchronization of all processes and threads in order to synchronize the tasks and time constraints and control latencies.

Real-time task: An application process which has time constraints or a maximum defined latency within which it must run or finish.

Stack: A collection of bytes for transmission on a network, a collection of received bytes from a network, a last-in-first-out data structure in memory, the memory used by local variables of a thread or function, or memory used to save the task information in PCB (process control block) so that on switching of task, the system retrieves the saved information for the task.

Software interrupt: A software code, instruction, or routine related interrupt (also called exception), for example, due to detection of a certain condition during computations, error while logging in, illegal operation code provided to CPU, or a kernel command WM_HIBERNATE.

Task: An application process which runs when scheduled to run by the OS and each state of which is controlled by OS.

Thread: An application process or a fundamental execution unit of a process (when a process or task have multiple threads) which runs when scheduled to run by the OS, each state of which is controlled by OS, and which runs as a light-weight process.

Virtual address: The memory addresses assumed by the programmer and system MMU for the pages between the memory addresses 000000000_h and $FFFFFFFF_h$ with no consideration of actual physical addresses provided in the system and assuming that no other program or data exists.

Wait object: An object (a signal, semaphore, queue or mailbox message or piped device byte) for which a proces waits.

Review Questions

1. What are the features desired in a mobile OS as compared to a desktop PC?
2. What are the features of PalmOS? Give the highlights of a PalmOS device. Write the functions of dynamic heap and multiple storage heaps in PalmOS.
3. Is PalmOS a multi-tasking system? If yes, explain how. If no, explain how is it possible to run multiple event handlers in a Palm device.
4. Describe memory manager in PalmOS. How does data store in a database file in PalmOS?
5. Explain how cradle and USB or serial are used to synchronize PalmOS files between a mobile device and PC.
6. How do you develop applications for Palm device?
7. How will you describe a multi-tasking OS? What do you mean by real-time OS?
8. How are the event handlers used to run the applications in a Palm device?
9. Is Palm OS a multi-tasking system? How is it possible to run multiple event handlers in a Palm device? Is Windows CE a multi-tasking system? If yes, explain how the events are handled and priorities assigned to threads.
10. Explain how cradle and USB or serial are used to synchronize Windows CE PIM data and files between a mobile device and PC.
11. Explain the similarity between the dynamic heap in PalmOS and program memory in PalmOS.
12. How does data store in Windows CE databases and files?
13. Explain memory manager features in Windows CE. Write the functions of program and storage memories.
14. How are the applications developed for a Windows CE device?
15. Describe the highlights of Symbian OS. Explain Symbian OS architecture.
16. How are the applications developed for a Symbian device?

Objective Type Questions

Pick the correct or most appropriate statement among the choices given:

1. Windows CE interrupt service thread is
   (a) a thread placed in a priority FIFO (queue) of threads which is initiated and put in the FIFO when the ISR execution starts on an event.
   (b) a thread created on an interrupt.
   (c) a priority FIFO (queue) of ISRs.
   (d) sequentially executing ISRs.
2. Mobile OS enables a programmer
   (a) to develop *application* considering the specifications of mobile system hardware during coding and to use the optionally provided management functions at OS (such as creation, activation, deletion, suspension, and delay) for tasks and memory.
   (b) to develop *application* without considering the specifications, drivers, and functionalities of mobile system hardware, to use the configurable libraries for the GUI (graphic user interface) in the device and multi-channel and multi-modal UIs thus enabling an *application* to run by simply abstracting the mobile system hardware.
   (c) to define the constraints and functionalities of mobile system hardware and does not provide the interfaces for communication between processes, threads, and ISRs at the application and middleware layers, and middleware and system hardware.
   (d) to define the device driver functions of mobile system and to use the APIs for the GUI (graphic user interface) in the device and multi-channel and multi-modal UIs.

3. PalmOS
   (a) supports only one process and no multi-processing or multi-tasking.
   (b) is optimized and supports only a very specific range of hardware, CPU, controller chips, and similar screens of PalmOS-based devices.
   (c) compiles codes for only a specific set of hardware and therefore its performance is very finely tuned.
   (d) has all three of these features.
4. PalmOS
   (a) memory space optionally partitions into the storage heaps and program memory dynamic heap for process stacks and global variables.
   (b) does not provide for multiple storage heaps for data and applications in order to save memory.
   (c) file-system file is in the format of a database having multiple records and information fields about the file—name, attributes, and version of the database for the application.
   (d) has OS memory requirement ~2 MB and supports only 64 MB internal flash memory.
5. PalmOS
   (a) enables the running of a sub-application from within an application, instead of multi-tasking, and sub-applications run on polling for the events, which are the requests to run sub-application, for example, a search program requesting a query processing, notifications (like time-out alarm), and GUI actions (touching or tapping the screen with stylus).
   (b) provides an ideal platform for running multimedia applications.
   (c) is designed for real-time systems.
   (d) has the ability to adapt to different sorts of hardware.
6. (i) There are a dynamic heap (for PalmOS 3.x within 32 kB), (ii) application process stacks (3 kB), (iii) TCP/IP stack (32 kB), (iv) and system global variables (2.5 MB), (v) PalmOS provides for direct memory address spaces in card, which are used instead of the allocated dynamic memory buffers for the I/O port devices, (vi) PalmOS 3.x and 4.x (not 5.x) provides XIP system due to static allocation of the memory addresses for the application-installed programs and data storage. Which of these are correct?
   (a) (i) and (iv) are incorrect.
   (b) All are correct but PalmOS 5.x also provides XIP with no dynamic memory buffers.
   (c) (iii) and (iv) are incorrect.
   (d) Only (v) is correct.
7. Windows CE (i) is modular/componentized to provide the foundation of several classes of devices, (ii) does not support additional features of other components—windowing, DCOM, and COM, (iii) is a real-time OS, supports multi-tasking, (iv) has large memory requirement but scales to the requirement of the device peripherals (v) has memory space partitioned, (vi) has data in the format of a database or object file, (vii) has file automatically compressed when stored and decompressed when loaded. Which of these are correct?
   (a) (ii) is incorrect.
   (b) (i) to (v) are correct.
   (c) (iii) to (vi) are correct.
   (d) All are true.
8. Windows CE has following features—(i) It supports Visual C/C++ platform. (ii) It integrates use of Web. (iii) It supports .NET XML parsing in a truncated version. (iv) The OS is customized for a specific set of hardware deployed among a sets of processors chosen. (v) Large parts of OS are offered in source code form first and then they are adjusted according to the hardware at the user end. (vi) The OS has shared source (licensed with controlled access to full or limited parts of the source code for a product). (vii) Windows CE developers cannot modify down to the kernel level and have to share their changes with Microsoft. Which of these are correct?
   (a) (iii) is incorrect.
   (b) (i) to (v) are correct.
   (c) (iii) to (vi) are correct.
   (d) All are correct except (iv) and (vii).

9. Windows Mobile 5.0 supports (i) PDA with Microsoft Smartphone phone device, touch screen, touchpad, or directional pad, (ii) GUIs/VUIs for Portable Media player, (iii) GUIs/VUIs for a specific set of processors, (iv) GUIs/VUIs for Smart phone, and (v) GUIs/VUIs for PocketPC handheld which features the *digitiser* in the human computer interface (HCI). (vi) The OS consists of Windows CE 5.0 with a set of specific applications.
   (a) (iii) and (ii) are incorrect. (b) All correct except (i) and (ii).
   (c) All are true. (d) All are correct except (v) and (ii).
10. Windows CE
    (a) applications are developed by coding for the interrupt service threads.
    (b) supports 32 priority level assignment to the threads.
    (c) provides protection from priority inversion as it provides for prioirty inheritence mechanism.
    (d) assumes event handlers as fundamental units of execution and providing access of CPU.
11. Symbian OS (i) is for a system with specific set of hardware, its performance is very finely tuned for running exclusively on ARM processors, (ii) has low boot time, (iii) supports single process with multiple sub-applications and multiple threads, (iv) C/C++ and .NET functions, (v) Java and JavaPhone, (vi) Internet connectivity for Web browsing IP-based network connectivity, (vii) WiFi, and (viii) integration to cellular GSM/CDMA phone. Which of these are correct?
    (a) (i) is incorrect. (b) All are correct except (iv).
    (c) All are correct except (iii) and (iv). (d) All are correct except (v) and (vi).
12. Consider Symbian OS devices corporate solutions. It
    (a) does not support IBM WebSphere Everyplace Access.
    (b) does not support BlackBerry Connect.
    (c) supports Oracle Collaboration Suite.
    (d) does not support Secure Mobile Connections via VPN Client.
13. Which of the following supports—(i) Desktop both for Windows and Mac and other essential software, (ii) simple APIs, (iii) PIM, address book, data book for task-to-do and organization, (iv) memo pad, (v) SMTP email download, (vi) offline creation and sending of POP3 email, (vii) Internet browsing functions, (viii) handhelds, Windows organizer, (ix) PDA, and (x) GUI development support on C/C++ platform and Java applications using J2ME?
    (a) Windows CE memory manager. (b) Symbian OS memory management system.
    (c) PalmOS. (d) Windows XP embedded.
14. Which of the following supports four different type of access mechanism—(i) no paging and no use of caches, (ii) cache but no use of paging, (iii) cache not used but paging used, and (iv) both cache and pages used?
    (a) Windows CE memory manager (b) Symbian OS memory management system
    (c) Palm OS (d) Windows XP embedded
15. Latest version of Symbian is Symbian v9. Its new features include
    (a) codes written in C++, Visual C++, or in Java
    (b) large screens and mobile TV
    (c) Windows Mobile
    (d) Digital rights management (DRM) systems which allow content owners to specify and control the usage policy for their content on mobile phones.
16. Symbian OS v7.0 supports (i) multi-tasking, (ii) a UI framework, (iii) advanced graphics support, (iv) application engines in 3G mobile phones, (v) integrated telephony, (vi) communications protocols, and (vii) data management. Which of these are true?
    (a) All are true except (ii). (b) All are correct except (vii).
    (c) All are true except (iv). (d) All are true.

Appendix 1

Mobile Satellite Communication Networks

A1.1 Satellite Networks

A satellite orbits around the earth. A satellite system consists of a number of orbiting satellites which receive signals from earth stations (base stations and gateways) and routes them back over a very wide area covering almost whole of the globe. The system provides a network for communication. The satellite system network functions along with a terrestrial network to provide global communication. Along with the terrestrial network, the satellite systems are nowadays being increasingly used for mobile communication networks.

Section A1.1.1 describes the basic parameters of a satellite. Section A1.1.2 provides a description of GEO, MEO, LEO, and HEO configurations of the satellites, and Section A1.1.3, the capacity allocation in the satellites.

A1.1.1 Basic Parameters

(a) *Rotational frequency f*—Figure A1.1(a) demonstrates a geometry of satellite rotation. Rotational frequency, f, is reciprocal of period T_{orbit} of one rotation of a satellite orbiting round the earth. Assume that the distance of satellite from earth's center is r and orbital altitude, d_{al}, is the distance of surface of the satellite from the surface of earth. R and d_{al} are related to f by the following equation:

$$r = R + d_{al} = [gR^2/(4\pi^2 f^2)]^{1/3} = [gR^2(T_{orbit})^2/(4\pi^2)]^{1/3} \quad \text{(A1.1)}$$

where g is the acceleration due to gravity and R is the earth's radius. Both parameters g and R are constants of the earth and are approximately equal to 0.00981 km/s^2 and 6,370 km, respectively. r increases with increase in T_{orbit}. When T_{orbit} is 24 hr, d approaches 36,000 km.

Example A1.1 A satellite with geosynchronous earth orbit (GEO) rotates in an orbit at 42,156 km from earth's center and thus has T_{orbit} as 24 hr. How much will be the duration of one orbit for a medium earth orbit (MEO) satellite at 18,000 km?

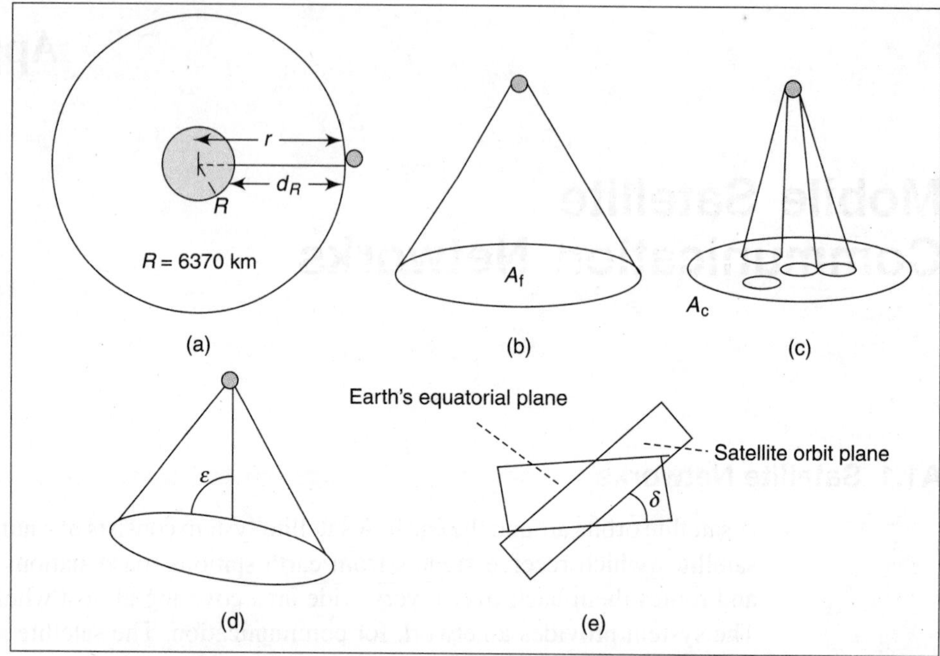

Fig. A1.1 (a) Geometry of satellite rotation (b) Footprint area A_f falling on earth's surface (c) Cell areas A_{ci} on earth's surface (d) Elevation angle ε (e) Inclination angle δ

Solution: From Eqn (A1.1), $r \propto (T_{orbit})^{2/3}$.

Therefore, $r_{GEO}/r_{MEO} = (T_{orbit})^{2/3}_{GEO}/(T_{orbit})^{2/3}_{MEO}$.

A GEO orbits once in 24 hours. Therefore, $(T_{orbit})^{2/3}_{MEO} = (24)^{2/3}\ [r_{MEO}/r_{GEO}] = (24)^{2/3}$ (18,000/42,156).

$T_{orbit\ MEO} = (24)(18{,}000/42{,}156)^{3/2}$ hr = 6 hr 42 min. The MEO at an altitude of (18,000 – 6,370) km = 11,630 km is expected to complete one orbit in 6 hr 42 min.

(b) Footprint area A_f—The beams emitted by a satellite antenna fall on the surface of the earth. This area on the surface of earth, called footprint, is represented by Fig. A1.1(b). Each point on the boundary of the footprint is isoflux, that is, has same intensity. The intensity (flux) is 0 dB outside the circle and the area inside is also isoflux with a finite non-zero intensity. As the satellite rotates with respect to the earth's rotation, the footprint also moves. A GEO will have stationary footprint. This is because the relative rotational velocity of GEO is zero.

(c) Cell area A_{ci}—A satellite may use several directed antennae inclined to each other such that each emits several spot beams within the satellite footprint. Figure A1.1(c) shows that there will be cells corresponding to each antenna. Each cell has 0 dB intensity (flux) outside the circle and the region inside the circle is isoflux, that is, has a finite non-zero intensity. The cell area A_{ci} is the area of ith cell on earth's surface.

(d) Elevation angle ε—Figure A1.1(d) shows the angle ε geometrically. The elevation angle ε is the angle between the plane tangent to the earth's surface and the line along the axis of the conical beam emitted by the satellite and forming a footprint or cell.

(e) Inclination angle δ—Inclination angle δ is the angle between two planes (Fig. A1.1(e)), equatorial plane and the plane of the orbit in which a satellite is rotating.

(f) Satellite line-of-sight distance, d_ε, from a signal-transmitting or receiving point Q is given by the following equation:

$$d_\varepsilon = [r^2 - (R\cos\phi)^2]^{1/2} - R\sin\phi, \tag{A1.2}$$

where ϕ is the angle between the tangential line on the surface of earth at Q and the line joining Q and the satellite.

(g) Signal transmission delay t_{tr}—One-way signal transmission delay t_{tr} for traversed beam to earth's surface is given by following equation:

$$t_{tr} = (d_\varepsilon/c) \tag{A1.3}$$

where d_ε is the path traversed in km at satellite beam elevation angle ε and c is the electromagnetic radiation velocity = 3×10^5 km/s. The delay is least when $\varepsilon = 90°$ and thus $d_\varepsilon = d$.

(h) Power loss P_L—P_L is due spread of the beam over a greater surface area when the beam propagates up to earth's surface at distance d. P_L is given by the following equation:

$$P_L = (4\pi d/\lambda)^{-2} \tag{A1.4}$$

where λ is the wavelength of the signal and is given by $\lambda = c/f$.

(i) Atmospheric power loss P_{La}—P_{La} is the power loss due to atmospheric conditions. It becomes very high in fog and rainy conditions and is a function of elevation angle ε. As ε decreases, P_{La} increases and the increase is proportional to $(d_a/\sin\varepsilon)^2$, where d_a is the distance traversed within atmosphere and increasing with ε.

(j) Atmospheric attenuation percentage α—α is given by $(100\, P_{La}/P_L)$.

Example A1.2 Assume that a Teledesic satellite (LEO satellite) rotates at d_{al} = 700 km altitude. (a) What is the minimum two-way delay (latency) between two stations communicating through the Teledesic? Velocity of electromagnetic waves is $c = 3 \times 10^5$ km/s. (b) Assume that Teledesic channels transmit at f = 19 GHz. What is the power loss P_L in a Teledesic channel between the satellite and earth if a non-directed antenna is used? What should be the gain of the directed antenna to receive P_{rec}= 3.22 nW of power when transmitted power is P_{trans} = 1 kW?

Solution:

(a) Two-way minimum delay occurs when the beam falls perpendicular to earth's surface. In such a case, delay = $2 \times 700/(3 \times 10^5)$ s = 4.7 ms.

(b) The formula for P_L is as follows:

$$P_L = (4\pi d/\lambda)^{-2}$$
$$= [4\pi d \times (f/c)]^{-2}$$

$$= [4 \times 3.141 \times 700 \times 19 \times 10^9/(3 \times 10^5)]^{-2}$$
$$= 3.22 \times 10^{-18}$$
$$\text{Gain} = P_{rec}/(P_{trans} \times P_L)$$
$$= 3.22 \times 10^{-9}/(10^3 \times 3.22 \times 10^{-18})$$
$$= 10^6$$

A1.1.2 Configurations

Satellites can be placed in orbits having distinct d_{al} and hence distinct T_{orbit} (Eqn A1.1). These orbits can be classified into four configurations and each configuration exhibits distinct characteristics. Each configuration has its own features, advantages, and disadvantages. The four configurations of satellite orbits are GEO, MEO, LEO, and HEO. Following subsections describe these configurations.

A1.1.2.1 GEO

Figure A1.2(a) shows a geosynchronous (geostationary) earth orbit (GEO) configuration of a satellite. GEOs orbit at a distance r such that T_{orbit} from Eqn (A1.1) is identical to the rotational period of earth, that is, 24 hours. Hence, a GEO satellite appears stationary from any point on earth's surface. The distance d_{al} is close to 42,000 km and d is close to 36,000 km. A large distance from earth results in a large footprint on earth's surface. The Inmarsat 4-series of satellites is

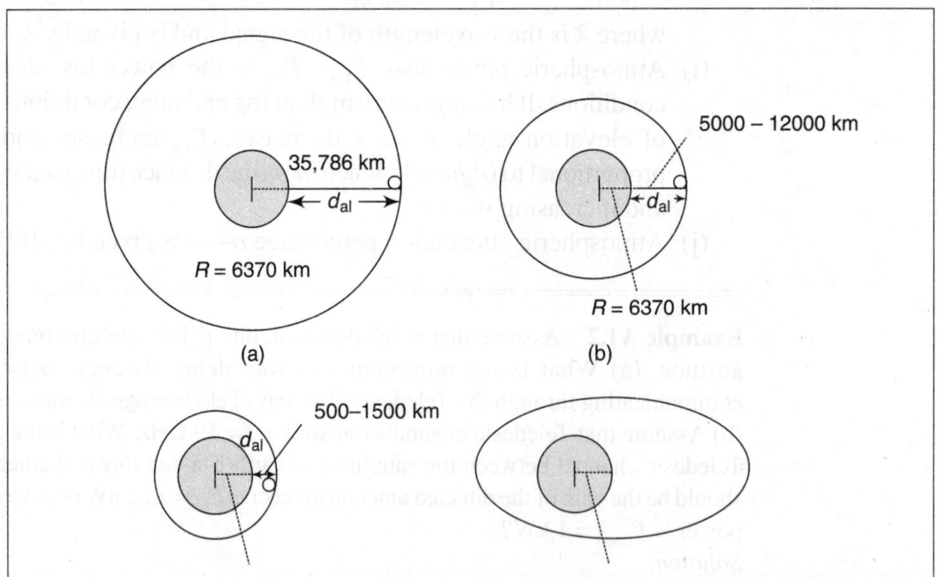

Fig. A1.2 Satellite configurations—(a) Geosynchronous (geostationary) earth orbit, (b) Medium earth orbit (MEO) (c) Low earth orbit (LEO) (d) High elliptical orbit (HEO)

an example of GEO satellites. Based on Inmarsat, new mobile telecommunication as well as voice and data solutions and services for terrestrial, maritime, and aeronautical applications are being continuously developed.

One of the advantages of using GEOs is that because of large footprints only three GEO satellites, in principle, can cover almost all points on the surface of the earth. The Doppler shift of frequencies is negligible due to geostationary feature. The beam transmitters and receivers on earth use fixed direction antennae because of geostationary position of satellite. Fixed directed antenna can be made with large arrays and thus can be used in case of very low signal powers. The GEOs are suitable for radio and TV signal transmission over the globe.

On the other hand, the GEOs have disadvantages due to large r and thus large distance from earth's surface. There is a need of large antennas due to high power loss P_L (Eqn (A1.4)). Further, there is high atmospheric loss P_{La} at large ε, for example, near the poles. Hence, GEO satellites cannot be used by the mobile phone and systems for directly receiving signals from satellites. Further, the one-way t_{tr} delay between transmitter and receiver is large and is about 0.25 s (Eqn (A1.3)). The two-way latency of 0.5 s is perceptible to human ears and communication through satellite phones have to take into account the latency. The large footprint of a GEO has another disadvantage that same band of frequencies may not be re-deployable over a footprint for other communication needs.

A1.1.2.2 MEO

Figure A1.2(b) shows a medium earth orbit (MEO) configuration of a satellite. MEO satellites orbit at a distance r between 18,000 km and 11,000 km such that T_{orbit} from Eqn (A1.1) is between 6 hr 40 min and 3 hr 10 min, respectively. Hence, an MEO satellite does not seem to be stationary from any point on the surface of the earth. The distance d_{al} is between 12000 km and 5000 km. The medium distance from earth results in medium footprint on earth's surface. As a result, same band of frequencies may not be re-deployable, over a smaller footprint as compared to that for GEO, for other communication needs. Another advantage of MEO satellites is that about 12–24 such satellites, in principle, can cover almost all points on the surface of the earth. This is only because of smaller footprints as compared to those of GEOs. The Doppler shift of frequencies is not negligible and also not large due to medium earth orbit. The beam transmitters and receivers on earth either use slowly rotating directional antennae or communication system can deploy in between satellite handovers with directional antennae. An MEO can deploy small footprints and larger transmitting power compared to a GEO. By increasing the inclination angle δ, MEO footprint can be increased to cover a larger area. Further, the one-way t_{tr} delay between transmitter and receiver is medium and is about 60 ms down to 37 ms (Eqn (A1.3)). The two-way latency of 120–148 ms is less perceptible to human ears and communication through satellite phones need not take into account the latency.

ICO (intermediate circular orbit) from Global communications is an example of the applications of the MEO. ICO has plans to use the constellation of MEOs along with a global ground telecommunications network. It plans to provide the global mobile communications services at 2 GHz and 2.2 GHz (uplink and downlink) and 5.2 GHz and 7 GHz (uplink and downlink) at 4.8 kbps using 4,500 channels.

A1.1.2.3 LEO

Figure A1.2(c) illustrates a low earth orbit (LEO) configuration of a satellite. LEO satellites orbit at a distance r between 8,000 km and 7,000 km such that T_{orbit} from Eqn (A1.1) is between 2 hr and 1 hr 40 min. The distance d_{al} is between 1,600 km and 600 km. Hence, an LEO satellite does not appear to be stationary from any point on the surface of the earth. If a directional antenna is used for transmission and reception to increase the power levels over the cell, for mobile communication, then the effective visibility duration will be as low as 10 min.

Very small footprint of an LEO offers the advantage, in comparison to the case of an MEO and GEO, that same band of frequencies is re-deployable in different footprint areas or cells for other communication needs. The small distance from earth results in small footprint on earth's surface and gives the advantage of higher power of signals from the satellite due to very low P_L as compared to that for GEO. Hence, LEO satellites can be used for mobile communication with light-weight devices. Another advantage of an LEO is that it can deploy very small footprints and can have very large transmitting power (Example A1.2(b)). Further, the one-way t_{tr} delay between transmitter and receiver is very small and is about 5 ms down to 2 ms (Eqn (A1.3) and Example A1.2(a)). The two-way latency of 4–10 ms is very small for being perceptible to human ears and communication through satellite phones need not take into account the latency. LEOs can cover large elevation without high atmospheric power loss P_{La} and are therefore suitable for polar regions also.

A disadvantage of LEOs is that because of very small footprint as compared to that of GEO, many LEO (48–200) satellites are required to cover almost all points on the surface of the earth. The exact number depends on the effective visibility period. [For example, when the rotation speed is large and the period is 2 hr, the satellite will be moving away from visibility in the footprint 12 times in a day. However, if directional antennae are used for transmission and reception, then effective visibility can be just 10 m. Another disadvantage is that Doppler shift of frequencies is high. The beam transmitters and receivers on earth either have rotating antennae or communication system can deploy in between satellite handovers with slowly rotating antennae. Another disadvantage is that fast movement of the satellites requires handover between them. At short distances, UV radiations damage the solar panels and other circuits of the satellite. The atmospheric drag makes the life span of LEO small, almost close to 8 years. The latency in case of multiple handovers, needed to communicate from one part of globe to another, becomes high and is proportional to the number of handovers in case of inter-satellite handovers and to the number of gateways in case of terrestrial gateways for satellite-to-satellite communication.

Motorola Iridium and Teledesic are FDMA/TDMA-based LEOs. Qualcom Globalstar is CDMA-based LEOs. Iridium and Teledesic provide global mobile, voice, data, and fax communications services at 2.4 kbps over 4,000 channels and at 64 Mbps over 4000 channels. Globalstar provides 9.6 kbps over 2,700 channels.

Example A1.3 A GEO satellite rotates in an orbit at 42,156 km from earth's center. (a) How much will be the duration of one orbit for Iridium (an LEO) at an altitude of 780 km? (b) Iridium satellite system has 72 satellites with six in reserve. How much should be the duration of visibility of each satellite to a fixed antenna with a directed transmitter if a 24 hr communication channel is maintained using at least one satellite at an instant?

Solution:

(a) From Eqn (A1.1), $r \propto (T_{orbit})^{2/3}$.

Therefore, $r_{GEO}/r_{Iridium} = (T_{orbit})^{2/3}_{GEO}/(T_{orbit})^{2/3}_{Iridium}$

A GEO satellite orbits once in 24 hr. Therefore, $(T_{orbit})^{2/3}_{Iridium} = (24)^{2/3}\ [r_{Iridium}/r_{GEO}] = (24)^{2/3}\ (r_{Iridium}/42{,}156)$. Now, earth's radius is 6,370 km. Therefore, $r_{Iridium}$ = 6,370 km + 780 km = 7,150 km.

Therefore, $T_{orbit\ Iridium} = (24)(7{,}150/42{,}156)^{3/2}$ = 1 hr 42 min. The Iridium at an altitude of 780 km is expected to complete one orbit in 1 hr 42 min.

(b) The number of Iridium satellites used in communication are 72 – 6 = 66. Duration of visibility of each satellite to a fixed antenna with a directed transmitter if a 24 hr communication channel is maintained using at least one satellite at an instant is given by 24 hr/66 = 22 min.

Example A1.4 A GEO rotates in an orbit at 42,156 km from earth's center. (a) How much will be the duration of one orbit for Teledesic (an LEO) at an altitude of 700 km? (b) Teledesic satellite system has 288 satellites. How much should be the duration of visibility of each satellite to a fixed antenna with a directed transmitter if a 24 hr communication channel is maintained using at least one satellite at an instant?

Solution:

(a) From Eqn (A1.1), $r \propto (T_{orbit})^{2/3}$.

Therefore, $r_{GEO}/r_{Teledesic} = (T_{orbit})^{2/3}_{GEO}/(T_{orbit})^{2/3}_{Teledesic}$.

A GEO orbits once in 24 hr. Therefore, $(T_{orbit})^{2/3}_{Teledesic} = (24)^{2/3}\ [r_{Teledesic}/r_{GEO}] = (24)^{2/3}\ (r_{Teledesic}/42{,}156)$. Now, earth's radius is 6,370 km. Therefore, $r_{Teledesic}$ = (6,370 + 700) km = 7,070 km.

$T_{orbit\ Teledesic} = (24)(7{,}070/42{,}156)^{3/2}$ = 1 hr 39 m. The Teledesic at an altitude of 700 km is expected to complete one orbit in 1 hr 39 min.

(b) Number of Teledesic satellites used in communication are 288. Duration of visibility of each satellite to a fixed antenna with a directed transmitter if a 24 hour communication channel is maintained using at least one satellite at an instant is given by 24 hr/288 = 5 min.

A1.1.3 Capacity Allocation

A number of channels are transmitted by an antenna to a satellite and vice versa. The capacity, C_s, of a satellite is defined as the number of channels that can be transmitted and received through it. Out of C_s, C_a are allocated to a satellite. C_a depends on the bandwidth. In L/S band, the bandwidth is 15 MHz. Different channels transmitted in a given bandwidth are separated by guard bands. This can be understood as follows—L band is 1.3 GHz and S band is 2.32/2.45 GHz. An Iridium satellite functioning in L/S band (around L and S band) has uplink frequency between 1.610 and 1.625 GHz and downlink frequency between 2.4835 and 2.5000 GHz. It has a capacity of 4,000 channels and can transmit each channel separated by 15 MHz/4000 = 3.75 kHz. If 0.6725 kHz is the guard band on each side of a channel, the usable band for data transmission per channel is (3.75 kHz – 2×0.6725 kHz) = 2.4 kHz. The data transmission rate per channel = 2.4 kbps.

The range for the functioning of a Teledesic satellite is within Ka band. It has uplink frequencies between 27.5 and 31 GHz and downlink frequencies in the two ranges—between 18.3–18.8GHz and between 19.7–20.2GHz. The operation frequency for a Teledesic for uplink is 28.8 GHz and for downlink is approximately 19 GHz. It has a capacity of 2,500 FDMA/TDMA channels, each with a data rate of 2×64 Mbps for uplink channels and 64 Mbps for downlink channels.

A1.2 Integration of GEO, LEO, and MEO Satellites and Terrestrial Mobile Systems

Figure A1.3 shows the LEOs communication network with the gateway links and inter-satellite links. A satellite earth station and broadcast service 1 (SESBS-1) uplinks and sends signals to a satellite S_1 from mobile base stations, data services, TV broadcast, and radio broadcast services. S_1 returns the signals to the cells within its footprint F_1. The mobile sets in the cells within F_1 can communicate with each other through satellite S_1. Another terrestrial SESBS-2 within F_1 of the satellite S_1

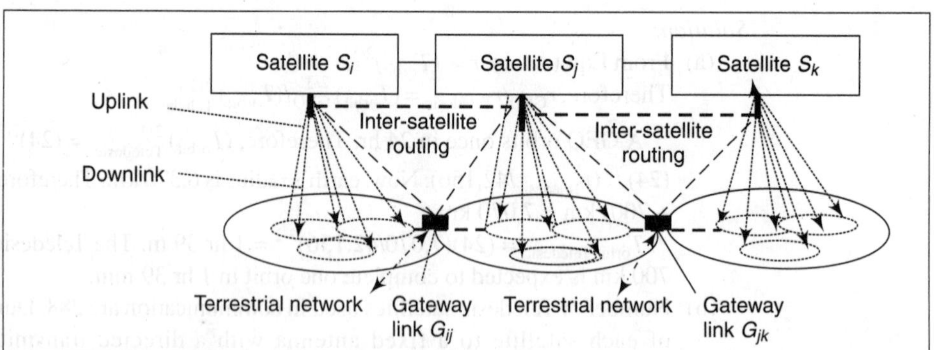

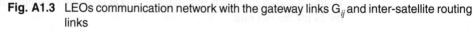

Fig. A1.3 LEOs communication network with the gateway links G_{ij} and inter-satellite routing links

receives these signals and retransmits to another satellite S_2. The SESBSs may have the linking circuits to connect LOS signals of two satellites. The links are called gateway and are said to provide a gateway link. S_2 returns the signals to the cells within its footprint F_2. The mobile systems in the cells within F_2 can communicate with each other through the satellite S_2. Another terrestrial SESBS-3, within the footprint F_2 of the satellite S_2, receives these signals and retransmits to another satellite S_3.

The figure shows two gateway links, G_{ij} and G_{jk}. G_{ij} provides an inter-satellite link between the satellites S_i and S_j. G_{ij} is present in SESBS-2. The S_i cell signals are first communicated to satellite S_i which then communicates to G_{ij}. The G_{ij} then links these signals to satellite S_j. The S_j then communicates to S_j cells. Thus the signals from S_i cells reach S_j cells through S_i, G_{ij}, and S_j. G_{jk} is in the SESBS-3. It links the signals from S_j cells with S_k cells through S_j, G_{jk}, and S_k.

For example, Iridium LEOs have 66 satellites in space. When 1 LEO satellite leaves the signal LOS (visibility) of SESBS and another LEO enters this range, handover of the signals to the latter takes place. Therefore, the 66 SESBSs maintain the services.

In case of cellular services, the handover (Fig. 1.14) occurs between two base stations and the phones move from one area to another. In case of LEOs, the SESBSs are stationary but the satellite moves. The handover of SESBS signals from a satellite S_i to another satellite S_j takes place when S_i moves away from signal LOS and S_j enters the LOS. The SESBSs can be connected with one another either through terrestrial network, for example, a fibre network or through GEOs.

Four types of handovers are present. *Inter-satellite handover* occurs when the user of a mobile set leaves a cell in footprint F_1 and moves to a cell in F_2 or when the satellite leaves a cell area in footprint F_1 and another satellite moves into the cell in F_1 to cover the user. *Intra-satellite handover* occurs when the user of mobile set leaves a cell in footprint F_1 and moves to another cell within F_1 itself. This happens when the satellite has several directed antennae which are creating several cells. *Gateway handover* occurs when the user of mobile set does not leave a cell in footprint F_1 but the satellite moves so that it sends the signals to user from a new gateway and not from the earlier one. *Inter-system handover* occurs when the user of mobile set getting service through one satellite system switches over to another system (e.g., landline system or cellular system) in case of some path obstruction due to a flying object or increase in P_{La} due to bad atmospheric conditions. Reverse handover can occur when the earlier system becomes available.

Mobile sets, which directly communicate with the LEO satellites, are heavy compared to mobile phone sets which are in use nowadays. Advanced mobile satellite-services are becoming available through integrated mobile satellite service and ancillary terrestrial component (MSS/ATC) networks. For example, France telecom mobile satellite communications provides GEOs, Inmarsat- and Thuraya-

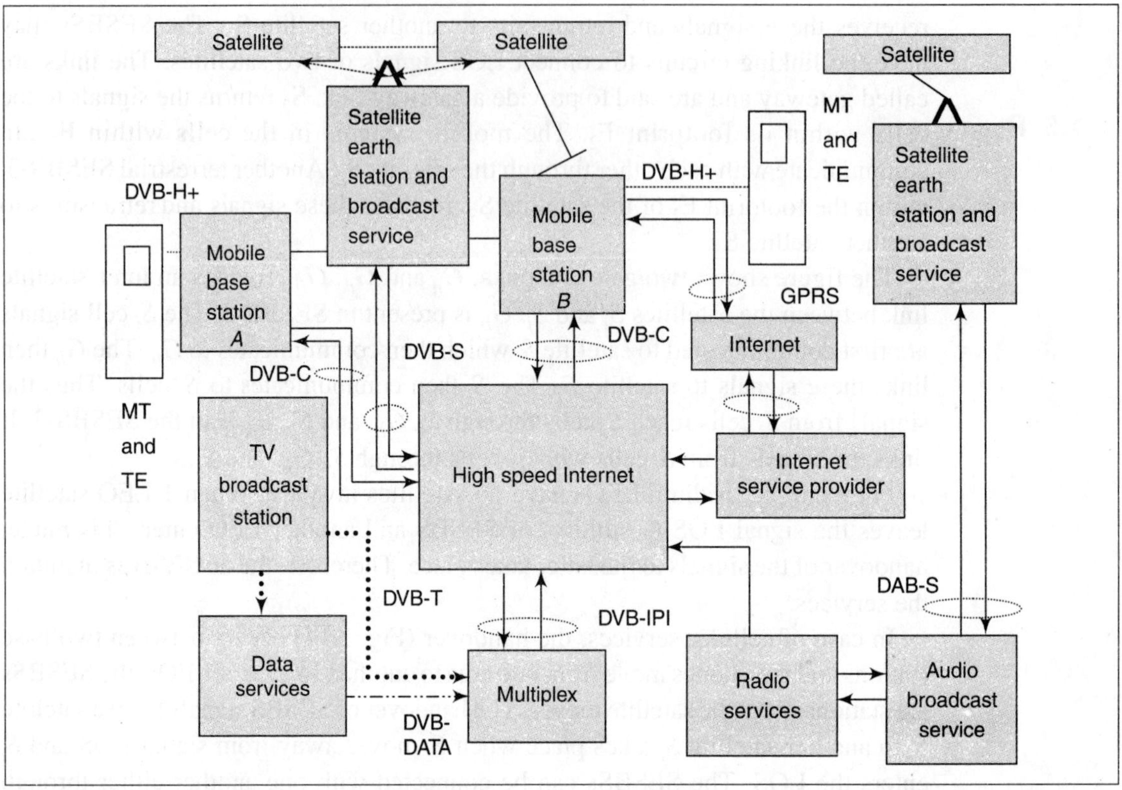

Fig. A1.4 Integrated DVB-H+ and convergence of mobile, Internet and broadcasting network architecture

and LEOs, Iridium- and Globalstar-based mobile satellite services. Mobile satellite services integrate the satellites LEO, MEO or GEO, terrestrial all-IP mobile communication networks, and mobile service providers existing networks. A description of the use of mobile communication networks and the concept of advanced mobile communication systems using satellites follows.

Figure A1.4 shows an integrated satellite, Internet, TV, radio, and mobile terrestrial system. The figure shows that a satellite earth station or broadcast service (SESBS) connects to a mobile base station *A*. *A* connects to high speed Internet through DVB-C protocol. The mobile service base station *B* can directly connect to a satellite. *B* also connects to high speed Internet through DVB-C protocol. Using DVB-H+ protocol, the service provider base stations *A* and *B* connect to the mobile handheld devices. The SESBS connects to high speed Internet employing DVB-S protocol.

The handheld devices, therefore, get integrated service through Internet, satellite, and mobile service provider, GSM or CDMA. The high speed Internet networks to TV broadcast service and data services using DVB-T and DVB-data, respectively, through a MUX. Radio service is provided by DAB-S connectivity of radio broadcast service to an SESBS. Radio service networks with the high speed Internet. The handheld devices, therefore, get further integrated with TV, data, and radio services.

A1.3 Personal Satellite Communication Programs

Two L bands are L1 at 1.269 GHz and L2 at 1.268 GHz. The S band ranges from 2 to 4 GHz. Personal satellite communication is in L/S band (uplink 1.610–1.625 GHz and downlink 2.4835–2.5000 GHz). Commercial video or TV broadcast, data casting, and audio broadcast services deploy L band [1.3 GHz in frequency band between 1.3 GHz ± 950 kHz and 1.3 GHz ± 1,450 kHz], S band (2.32 GHz—2.45 GHz), and Ku band (uplink 11.7–12.2 GHz and downlink 14.0–14.5 GHz) satellite communication and use DAB-S and DVB-S protocols (Sections 8.5 and 8.6). Thuraya and Iridium are GEO-based systems for mobile telecommunication. Thuraya permits handheld-size mobile satellite communication using rotating steering spot beams (cells). Globalstar deploys LEO. It uses CDMA. Iridium uses FDMA/TDMA. Further, the call must be made within same footprint area as for gateway station. Inter-gateway handover is not provided in Globstar. Iridium provides inter-gateway handover.

Personal satellite communication programs also deploy GPS services for automotive (Section 2.7) and other applications. The personal communication services from Inmarsat, Iridium, Thuraya, and Globalstar are described below.

A1.3.1 Inmarsat, Iridium, MSAT, VSAT, DBS, and Orbcumm Satellite Services

Inmarsat is a network of GEO satellites for voice and data services, mobile broadband communications network, and Internet covering the globe for terrestrial, aeronautical, and marine applications. It offers four active and five emergency satellites. Mobile service providers of terrestrial networks offer the mobile services for telephone, email, fax, and data file transfers using Inmarsat satellite network.

Thuraya is a network of GEO satellites for voice and data services for terrestrial, aeronautical, and marine applications. Mobile service providers of terrestrial networks offer mobile services for telephone, email, fax, and data file transfers using Iridium satellite network.

Iridium is a network of 66 LEO satellites and 6 emergency LEO satellites for voice and data services for terrestrial, aeronautical, and marine applications. Mobile service providers of terrestrial networks offer mobile services for telephone, email, fax, and data file transfers using Iridium satellite network.

Globalstar is the LEOs-based network for telephone and low speed data communications. It has CDMA gateways. It provides connectivity to PSTN lines also. It functions as repeaters and has no inter-satellite handovers.

Appendix 2

Java Programs for Mobile Computing

A2.1 Directory-Based Caching of Records

Section 8.3.2 described directory-based data dissemination. A directory can hold a group of data records which are of interest to a number of client devices. Assume that the records of each directory are broadcast as character stream using a broadcast disk model. In sample code let us use Telnet between two windows though in actual practice a server broadcasts a stream of bytes and the mobile devices listen to the broadcast stream. Sample Code A2.1 StreamReadDirectory.java runs on local host. StreamReadDirectory.java is a mobile device program that reads stream of characters from server, as soon as it finds *start flag* in the stream. *start flag* consists of two consecutive characters # #. Flag alerts the mobile device that a broadcast cycle has started. Next to these start flag characters (# #), the program finds an *offset* from the stream. The offset is between 000 and 999. It, thus, consists of 3 characters. The device waits till the requisite numbers of bytes specified by offset are finished. After this the device starts reading the records (or messages) or buckets from the stream. *End flag* indicates end of the records. The character $ is used as end flag.

To run StreamReadDirectory.java, we first compile it and create StreamReadDirectory.class file. Then we open two windows. In the first window, we interpret the class file using the command java StreamReadDirectory <portno.>, e.g., java StreamReadDirectory 9500.

The program requires port number (at which device listens to the stream) through command line argument. The second window runs Telnet. To run Telnet, use the command Telnet <IP Address> <Port No.>, e.g., Telnet 127.0.0.1 9500 **/

Sample Code A2.1

```
/** Program: StreamReadDirectory.java.
```

```
import java.io.*;
import java.net.*;
/** ThreadHandler class to define a Socket object named newsock and
run the thread **/
class ThreadHandler extends Thread
{
    Socket newsock;
    int n;
   /**Using ThreadHandler constructor for new socket s with number v
**/
    ThreadHandler (Socket s, int v)
    {
        newsock = s;
        n = v;
    }

    public void run()
    {     /** more_data = true means more characters to be read from
the stream **/
        boolean more_data = true;
      try{
            System.out.println("SERVER STARTED");
        // Infinite loop for listening to stream at client
            while(more_data)
            { /* Construct object inp from DataInputStream. inp gets
the characters from InputStream  at Socket newsock */
                DataInputStream inp = new
DataInputStream(newsock.getInputStream());
        byte buffer[] = new byte[500];
        char frameBuffer[] = new char[500];
        char stx[] = {'#','#'};      // start-flag ##
        int i=0;
             int nBytesRead =0 ;
             int offset=0;  // Initially frame offset = 0
             boolean flag = false , dataFlag = false;
        do    {
                 if(!flag && !dataFlag)
                 {
                   nBytesRead = inp.read(buffer);
                   System.out.write(buffer, 0, nBytesRead);
```

```
                                    if((char)buffer[0] == stx[i])
                                    {
                                        i++;

                                        do  {

                                            nBytesRead = inp.read(buffer);

                                            System.out.write(buffer, 0, nBytesRead);
                                            if((char)buffer[0] == stx[i])
                                            {
                                                 System.out.print("String Matched.
        ");

                                                 flag = true;

                                            }

                                              break;

                                        } while(nBytesRead >0);

                                    }
                                    i=0;
                                  }
                                  else if (flag)
                                  {/* read three characters and find offset in
decimal value */
                                    byte n1 = inp.readByte();
                                    byte n2 = inp.readByte();
                                    byte n3 = inp.readByte();
                                    offset = (((byte)(n1-48)*100)+((byte)(n2-
48)*10)+ ((byte)n3-48));
                                   flag = false;
                                   dataFlag = true;
                                   break;
                                  }
                              } while (nBytesRead < 5); /* Wait till start flag and
offset reading finishes */
                              int cnt=0;
                              System.out.println(offset);
                           /* To read characters after skipping the number of
characters = offset */
```

```
do {
    if(dataFlag)
    {
        nBytesRead = inp.read(buffer);
frameBuffer[cnt] = (char)buffer[0];
    }
    cnt++;
} while ((frameBuffer[cnt-1] != '$') && (cnt<
frameBuffer.length) );

cnt = offset;
do{
    System.out.print(frameBuffer[cnt]);


    cnt++;
} while (cnt < frameBuffer.length);
//try{Thread.sleep(10000);}
catch(InterruptedException e) { }
        } // end of infinite while loop
    newsock.close();
    }
    catch(IOException e){System.out.println("In Exception");}
    }
}
public class StreamReadDirectory
{
    public static void main(String args[])
    {
        int nreq=1;
        System.out.println("Welcome . . . . . ");
        try
        {
        if(args.length != 1 )
        {
            System.err.println("Requires port no. as command line
argument.");
            System.exit(1);
        } /* Construct a Server socket object sock */
        ServerSocket sock = new
```

```
ServerSocket(Integer.parseInt(args[0]));
                for(;;)
                {
                    Socket newsock = sock.accept();
                    System.out.println("Creating Thread" + nreq);
                    Thread t = new ThreadHandler(newsock,nreq);
                    t.start();
                    nreq++;
                }
            }
            catch(Exception e)
            {
                System.out.println("Error " + e);
            }
        }
    }
```

A2.2 Selective Tuning of Indices-Based Caching of Records

Section 8.4 described selective tuning and indexing techniques. Selective tuning is a process by which client device selects only the required pushed buckets or records, tunes to them, and caches them. Tuning means getting ready for caching at those instants and intervals when a selected record of interest is broadcast. Assume that the tuned records or buckets are broadcast as a character stream using a broadcast disk model. Sample Code A2.2 is a device client program `StreamReadIndex.java`. The client at device reads a stream of bytes which are broadcast from a source. As soon as the client finds a start flag, it looks for three frames of three characters each. The start flag is indicated by two consecutive # characters. It is assumed that the source sends three sets of buckets at three different indices. Each bucket is assumed to have 10 characters. The client thus reads offset of three frames of 3 characters each. [The offset is assumed to be between 000 and 999.] Then it starts reading data from the buckets sent with the offsets. Each bucket ends with an end flag. The end flag is assumed to be the character $. The program requires port number (at which device listens to source stream) through command line argument.

A device program reads stream of bytes from source. As soon as it gets start flag, it reads offset of three sets of 3 characters each. From the characters, the offset is converted to a decimal numbers between 000 and 999. Then it starts reading data for each set from the offsets till the end flag is encountered. The program requires port number (at which server listens to client request) through command line argument. Start flag is ##. End of line is $

To run the program, the file `StreamReadIndex.java` is first compiled and `StreamReadIndex.class` file is created. We open two windows. In the first window, interpret the class file using the command `java StreamReadIndex <portno.>`, e.g., `java StreamReadIndex` 9500. In the second window, run Telnet using the command `telnet <IP address> <port no.>`, e.g., telnet 127.0.0.1 9500 (to run on local host)

Sample Code A2.2

```
/** Program: StreamReadIndex.java.
    import java.io.*;
    import java.net.*;
    // size of each set in the stream is 10 characters.
    //Start flag is ## , End of line is $
    class ThreadHandler extends Thread
    {
        Socket newsock;
        int n;

        ThreadHandler (Socket s, int v)
        {
            newsock = s;
            n = v;
        }

        public void run()
        {
            boolean more_data = true;
      try{
              System.out.println("DEVICE CACHING STARTED");

          // Infinite loop for listening to source
              while(more_data)
              {
                  DataInputStream inp = new
DataInputStream(newsock.getInputStream());

            byte buffer[] = new byte[500];
            char frameBuffer[] = new char[500];
```

```
        char stx[] = {'#','#'};  // Start Flag
        int i=0;
            int nBytesRead =0 ;
            int offset[]= {0,0,0}; // Offset of three frames
            boolean flag = false , dataFlag = false;
    do  {
            if(!flag && !dataFlag)
            {
                nBytesRead = inp.read(buffer);

              System.out.write(buffer, 0, nBytesRead);
              if((char)buffer[0] == stx[i])
              {
                  i++;

                  do   {

                   nBytesRead = inp.read(buffer);

                   System.out.write(buffer, 0, nBytesRead);
                   if((char)buffer[0] == stx[i])
                   {

                      System.out.print("String Matched...");
                      flag = true;
                   }
                      break;

                  } while(nBytesRead >0);

              }
              i=0;
            }
            else if (flag)
            {
              for (int k=0 ;k<3; k++)
                {
                  byte n1 = inp.readByte();//first digit of index
                  byte n2 = inp.readByte();//second digit of index
```

```
                                        byte n3 = inp.readByte();//third digit of index
                                        offset[k] = (((byte)(n1-
48)*100)+((byte)(n2-48)*10)+ ((byte)n3-48));//construct a number of 3 digits
                                    }

                                    flag = false;
                                    dataFlag = true;
                                    break;

                                }
                            } while (nBytesRead < 5);

                            int cnt=0;

                            // To read incoming data
                            do {
                               if(dataFlag )
                               {
                                  nBytesRead = inp.read(buffer);
                            frameBuffer[cnt] = (char)buffer[0];
                               }
                               cnt++;
                            } while ((frameBuffer[cnt-1] != '$') && (cnt<
frameBuffer.length) );

                            // To separate three frames—offset value of each frame
is used.
                            for (int k=0;k<3 ;k++)
                            {
                               int noOfChar = 0;
                               System.out.println();
                               cnt = offset[k];
                               do{
                                  System.out.print(frameBuffer[cnt] );

                                  cnt++;
                                  noOfChar++;
                               } while ((cnt < frameBuffer.length) && (noOfChar
< 10));
                            }
```

```
                              //try{Thread.sleep(10000);}
        catch(InterruptedException e) { }
                              } // end of infinite while loop
                    newsock.close();
                 }
                     catch(IOException e){System.out.println("In Exception");}
                 }
            }
            class StreamReadIndex
            {
                public static void main(String args[])
                {
                     int nreq=1;
                     System.out.println("Welcome . . . . . ");
                     try
                     {
                   if(args.length != 1 )
                        {
                             System.err.println("Requires port no. as command line
        argument.");
                             System.exit(1);
                        }
                        ServerSocket sock = new
        ServerSocket(Integer.parseInt(args[0]));
                        for(;;)
                        {
                             Socket  newsock = sock.accept();
                             System.out.println("Creating Thread" + nreq);

                             Thread t = new ThreadHandler(newsock,nreq);
                             t.start();
                             nreq++;
                        }
                     }
                     catch(Exception e)
                     {
                          System.out.println("Error " + e);
                     }
                }
            }
```

A2.3 XML Parser

Section 7.1.1 described XML databases. An XML database is parsed using a Java language parser. Figures 13.1, 13.2, and 13.3 demonstrated XML database for contacts and parser models. Sample Code A2.3 gives the codes for a DOM parser, which reads XML database file `Contacts` and generates a hash table. The parser converts the tags to primary keys in the hash table. Sample XML database file is in Sample Code A2.4.

To run the program, we compile the file `xmlParser.java`. It creates `xmlParser.class` file.

Check that `Contacts.xml` is present in current working directory.

Interpret the class file using the command `java xmlParser**/`

Sample Code A2.3 (To parse an XML file, get its contents in hash table, and print them.)

```
/** Program: xmlParser.java.

import java.io.*;
import javax.xml.parsers.*;
import org.w3c.dom.*;
import java.util.*;
class Address{
        public String name1,email1,num1;

        Address(String str1, String str2, String str3)
        {
                name1  = str1;
                email1 = str2;
                num1   = str3;
        }
}
public class xmlParser{
        public static void main(String args[]){
        Hashtable ht = new Hashtable(); /* Create hash table object ht */
        try{
            DocumentBuilderFactory fact1 =
    DocumentBuilderFactory.newInstance();
            DocumentBuilder build1=fact1.newDocumentBuilder();
```

```
                String cont="contacts.xml";
                Document conDocument = build1.parse(new File(cont));
                Element conElement = conDocument.getDocumentElement();
                NodeList cnodes = conElement.getElementsByTagName("contact");
              //Retrieving elements from XML
                for(int i=0;i<cnodes.getLength();i++)
                {
                        Element addr =(Element) cnodes.item(i);
                        NodeList nameList=addr.getElementsByTagName("name");
                        Text nameNode = (Text) nameList.item(0).getFirstChild();
                        NodeList emailList= addr.getElementsByTagName("emailId");
                        Text emailNode = (Text) emailList.item(0).getFirstChild();
                        NodeList phoneList= addr.getElementsByTagName("number");
                        Text phoneNode = (Text) phoneList.item(0).getFirstChild();
                        //Adding elements to HashTable
                        ht.put(nameNode.getData(),(new
    Address(nameNode.getData(),emailNode.getData(),phoneNode.getData())));
          }
        }

        catch(Exception e){
                System.out.println("parsing error "+ e.getMessage());
        }


    Enumeration en = ht.elements();
    while(en.hasMoreElements()){
        System.out.println(ad.name1 + "\n");
         System.out.println(ad.email1 + "\n");
         System.out.println(ad.num1 + "\n");

     }

       }
    }
```

Sample Code A2.4 (For XML database file)

/** To be parsed XML database file Contacts format is as follows: **/

```
<contacts>
   <contact>
      <name>R__t</name>
      <emailId>r__t@yahoo.com</emailId>
      <number>93xxxx0111</number>
   </contact>
   <contact>
      <name>M__y</name>
      <emailId>m__y@gmail.com</emailId>
      <number>930210xyxx</number>
   </contact>
 = <contact>
      <name>H__i</name>
      <emailId>h__i@hotmail.com</emailId>
      <number>xxxxy71001</number>
   </contact>
 = <contact>
      <name>J____h</name>
      <emailId>j____h@rediffmail.com</emailId>
      <number>9826xxxx07</number>
   </contact>
 = <contact>
      <name>J..n</name>
      <emailId>j..n@yahoo.com</emailId>
      <number>9xxyyy0999</number>
   </contact>
</contacts>
```

A2.4 XML Database File Transfer from a Source to Client

Sections 7.1.1 and A2.3 provide examples of XML database. Section 9.2 described synchronization software and protocols for mobile devices. As a first step, a synchronizer gets the database to be synchronized from the server. Sample Code A2.5.

`XMLTransferClient` is the client program which can be a part of synchronizer and transfers an XML file. The name of the XML file, IP address, and port number of source (server) are given through command line argument. **/

Sample Code A2.5

```
/** Program: XMLTransferClient.java.- A client program to send an
XML file to client
import java.net.*;
import java.io.*;
public class XMLTransferClient
{
    public static void main(String args[]) throws IOException
    {
 XMLTransferClient client = new XMLTransferClient();
 try{
    if(args.length != 3 )
            {
                System.err.println("Require XML Filename, IP Address &
    Port No.");
                System.exit(1);
            }
            String fileName = args[0];
    String host = args[1];                    // IP Address of server
    int PORT = Integer.parseInt(args[2]);     // Port number

    client.sendFile(fileName,host, PORT);
    }
          catch(Exception e){
   e.printStackTrace() ;
}
    }

    public void sendFile(String fileName, String cHost, int cPort)throws
   Exception
    {
 Socket socket = new Socket(cHost,cPort);//define socket object
 OutputStream sos = socket.getOutputStream();//define socketstream object
        File file = new File(fileName);//define file object
 int length = (int)file.length();

 FileInputStream fis = new FileInputStream(file);
 BufferedInputStream bis = new BufferedInputStream(fis);
```

```
byte[] buffer = new byte[1024];
int bytesRead;
while((bytesRead = bis.read(buffer)) != -1)
        {
            sos.write(buffer);
        }
    }

}
```

A2.5 XML Database File Transfer from a Source to Server

Section 9.2 described synchronization software and protocols for mobile devices. As a second step a synchronizer gets the database to be synchronized from the client. Sample Code A2.5 described the code for file transfer to client. Sample Code A2.6 describes the code for file transfer to server.

Steps for running the program are as follows:

(i) Compile the file `XMLTransferServer.java` to create `XMLTransferServer.class` file

(ii) Compile the file `XMLTransferClient.java` to create `XMLTransferClient.class` file

(iii) Run server on one window, client on the other using the commands:

(a) java XMLTransferServer <portnum.>, e.g., java XMLTransferServer 9500

(b) java XMLTransferClient <xml_file_name> <IP address> <port no.>, e.g., java XMLTransferClient Contacts.xml 127.0.0.1 9500. The program copies `Contacts.xml` from source to server as `Temp1.xml`

Sample Code A2.6

```
/** Program: XMLTransferServer.java.
** A server program to receive an XML file from client.
** The program listens to client request at port no. provided through command
line argument.
** It saves the received xml file with the name "Temp(counter_number).xml"
** Program : XMLTransferClient.java
** A client program to transfer an XML file.
** Provide XML File, IPaddress, and port number of server through command line
argument.
**/
/** Program: XMLTransferClient.java code is same as in Sample code A2.5.
```

`Program: XMLTransferServer.java` codes are as follows:

```
 **/
import java.io.*;
import java.net.*;
import java.util.*;

public class XMLTransferServer
{
   public static void main(String[] args)
   {
      if(args.length != 1 )
      {
         System.err.println("Requires port no. ");
         System.exit(1);
      }
      try
      {
         int i=1;
         ServerSocket s=new ServerSocket(Integer.parseInt(args[0]));
         //define server socket
         for(;;)
         {
   System.out.println("Welcome......");
 ((
   Socket incoming =s.accept();//define incoming socket object
   System.out.println("Server Started .. . . .. ");
   System.out.println("Client No. " + i);
   Thread t=new Threader(incoming,i);
   t.start();
 ((
   i++;
         }
 }
 catch (Exception e)
 {
            e.printStackTrace() ;
 }
   }
}
```

```
class Threader extends Thread
{
    private Socket incoming ;
    private int counter;

    public Threader (Socket i,int c)
    {
        incoming =i;
counter=c;
    }

    public void run()
    {
        int i=0;
byte[] buffer = new byte[1024];
FileOutputStream fos;//define file output stream object

System.out.println("Connected. . . . . ");
try{
  DataInputStream is = new DataInputStream(incoming.getInputStream());
  while (true)
                {
                    i++;
                    fos = new FileOutputStream(new File("Temp" + counter +
  ".xml"));
                    while(is.read(buffer)!=-1)//read data input buffer
                    {
                        fos.write(buffer);//write file buffer
                    }
                    incoming.close();//close incoming socket

    }
}
catch(Exception ex)
        {
            ex.printStackTrace() ;
}
    }

}
```

A2.6 Synchronization

We learnt the sample codes in Java for transfer of XML databases from server to client and from client to server in Sections A2.4 and A2.5. IBM alphaWorks offers an XML Diff and Merge tool. The requirements for the tool are JDK 1.2.x and XML4J 2.0.15 (this is included in the lib\xml4j.jar, therefore no need to reinstall JDK.). Installation of Diff and Merge tool is by the command `edit bin\xmldiff.bat` on Windows, and `xmldiff.sh` on Unix. Also the JDK environment variable needs to be set to point to jdk1.2.2 install and the XMLDIFF environment variable to the installation directory for this driver. To launch the tool with a GUI, invoke `bin\xmldiff.bat`. To launch the tool without a GUI, invoke `bin\xmldiff2.bat` inputFile1 inputFile2. This compares the two input files and gives the result on the standard output. This program finds the difference between two XML databases, inserts the new records in `Temp(counter_number).xml`, and replaces the previous records with modified records in `Temp(counter_number).xml`.

A2.7 Search Tool

Section 13.3 described Java. Recently mobile computing devices have been popularly used for searching locations of nearby ATM, Pizza shop, bank, or shopping mall. Sample Code A2.7 can be used as a tool to search for nearby ATM(s), pizza shop(s), or shopping mall(s). When a mobile user selects search option, a menu asks the user to select ATM, Pizza, or Mall. When the user selects the option ATM, a menu asks the user to select the present location. The device computes and displays names of ATMs in present location and if none is there in the present location, then those of nearest ones. Similar actions take place when the user selects Pizza or Mall option.

Sample Code A2.7

```
/** Program: SearchTool.java.
 ** A GUI-based program that defines nine locations A0, A1, A2, ..., A8.
 ** It takes current location of the user as input and gives the nearest ATM,
Pizza shop, or mall, respectively. If these search options are not available
in that area, it gives the respective options from neighbouring location. **/

import java.awt.*;
import java.awt.event.*;
import java.util.Hashtable;
import java.util.Vector;
import java.util.Enumeration;

class Screen1 extends Frame implements ItemListener
{
 List list;
        static String selection;
```

```
public Screen1(String s)
{
  setLayout(new FlowLayout(FlowLayout.LEFT));//screen layout setting
  setTitle(s);
  Label l1 = new Label("Choose Service ");//label to a menu list
  add(l1);
        list = new List();
  list.add("Search ATM");//add a menu item list
  list.add("Search PizzaShop");//add a menu item list
  list.add("Search Mall");// add a menu item list
  list.add(" ");
  add(list);

            list.addItemListener(this);
}

public void itemStateChanged(ItemEvent ie)
{
  String str = list.getSelectedItem();//create object to get selected choice
  System.out.println("Menu selected : " + str);
            selection = str;
            setVisible(false);//Before new screen, present
                              screen invisible
  Frame sc2 = new Screen2("Current Location");//Frame object
  sc2.setSize(750,500);//define screen frame size
  sc2.setVisible(true);//screen frame visible

}
}

class Screen2 extends Frame implements ItemListener
{
Choice ch;
      String resultLocation ="" , q = "";
public Screen2(String s)
{
  setLayout(new FlowLayout(FlowLayout.LEFT));//screen layout
  setTitle(s);
  Label l2 = new Label("Enter Current Location");
  add(l2);
```

```
        ch = new Choice();

        ch.add("A0");
        ch.add("A1");
        ch.add("A2");
        ch.add("A3");
        ch.add("A4");
        ch.add("A5");
        ch.add("A6");
        ch.add("A7");
        ch.add("A8");
        add(ch);

                ch.addItemListener(this);

        }

            public void itemStateChanged(ItemEvent ie)
        {

        String str = ch.getSelectedItem();
        System.out.println("You are currently in  " + str + "  location.");

        Vector v = new Vector(SearchTool.htLocation.keySet());

                int i = v.size();
                int j=0;



                for (Enumeration e = SearchTool.htLocation.keys();
        e.hasMoreElements();)
                {
                  String str1 = (String)e.nextElement() ;
                   if((str.equals(str1)) &&
        (Screen1.selection.equals("Search ATM")))//ATM Search
                      {
                        Location atmLoc =
        (Location)SearchTool.htLcpAtm.get(v.elementAt(j));
                        Location neighbourLoc =
        (Location)SearchTool.htLocation.get(v.elementAt(j));
```

```
                    System.out.println("Nearest ATMs are :-");
                    if (atmLoc.length() == 0)
                    {
                       resultLocation = "";
                        for (int k=0;k<neighbourLoc.length();k++ )
                        {
                            Location neighbourLocAtm =
(Location)SearchTool.htLcpAtm.get(neighbourLoc.a[k]);//neighbour location search
                            resultLocation +=
neighbourLocAtm.getLocation();
                        }
                        break;
                    }
                    resultLocation = atmLoc.getLocation();
                }
                if((str.equals(str1)) &&
(Screen1.selection.equals("Search PizzaShop")))  //Pizza shop search
                {
                    Location psLoc =
(Location)SearchTool.htLcpPs.get(v.elementAt(j));
                    Location neighbourLoc =
(Location)SearchTool.htLocation.get(v.elementAt(j));

                    System.out.println("Nearest Pizza Shop's are :-");
                    if (psLoc.length() == 0)
                    {
                       resultLocation = "";
                        for (int k=0;k<neighbourLoc.length();k++ )
                        {
                            Location neighbourLocPs =
(Location)SearchTool.htLcpPs.get(neighbourLoc.a[k]);
                            resultLocation +=
neighbourLocPs.getLocation();
                        }
                        break;
                    }
                    resultLocation = psLoc.getLocation();
                }
                if((str.equals(str1)) &&
(Screen1.selection.equals("Search Mall")))//mall search
                {
                    Location mallLoc =
```

```
(Location)SearchTool.htLcpMall.get(v.elementAt(j));
                         Location neighbourLoc =
(Location)SearchTool.htLocation.get(v.elementAt(j));

                         System.out.println("Nearest Pizza Shop's are :-");
                         if (mallLoc.length() == 0)
                         {
                            resultLocation = "";
                              for (int k=0;k<neighbourLoc.length();k++ )
                              {
                                     Location neighbourLocMall =
  (Location)SearchTool.htLcpMall.get(neighbourLoc.a[k]);
                                     resultLocation +=
  neighbourLocMall.getLocation();
                                }
                                break;
                              }
                              resultLocation = mallLoc.getLocation();
                         }

                         j++;
                       }
                     System.out.println(resultLocation);
                     repaint();

                }

                public void paint(Graphics g)
                {
                    g.setFont(new Font("TimesNewRoman",Font.BOLD, 16));
                    g.drawString("Please Visit :",50,150);
                    g.setFont(new Font("TimesNewRoman",Font.PLAIN, 16));
                    g.drawString(resultLocation,50,200);
                }
}

public class SearchTool
{
 public static Hashtable htLcpAtm, htLcpPs, htLcpMall;
 public static Hashtable htLocation ;
```

```
public static void main(String a[])
{
   String A0[] = {"A1","A3","A4"};//define all neighbours of AD
   String A1[] = {"A0","A2","A3","A4","A5"};
   String A2[] = {"A1","A4","A5"};
   String A3[] = {"A0","A1","A4","A6","A7"};
   String A4[] = {"A0","A1","A2","A3","A5","A6","A7","A8"};
   String A5[] = {"A1","A2","A4","A7","A8"};
   String A6[] = {"A3","A4","A7"};
   String A7[] = {"A3","A4","A5","A6","A8"};
   String A8[] = {"A4","A5","A7"};
   /* ATM0 and ATM 1 are the ATMs nearest to location A0 */
   String ATM0[] = {"ATM0","ATM1"};
   String ATM1[] = { }; // No ATM at location A1
   String ATM2[] = {"ATM2","ATM6","ATM8"};
   String ATM3[] = {"ATM9","ATM10"};
   String ATM4[] = {"ATM11","ATM12","ATM13","ATM14"};
   String ATM5[] = {"ATM21","ATM23"};
   String ATM6[] = { }; // No ATM at location A6
   String ATM7[] = {"ATM45","ATM46"};
   String ATM8[] = {"ATM57","ATM58"};
   /* PIZZA0 is pizza shop nearest to location A0 */
   String PIZZA0[] = {"PIZZA0"};
   String PIZZA1[] = {"PIZZA1"};
   String PIZZA2[] = { };// No Pizza shop near to location A2
   String PIZZA3[] = {"PIZZA4"};
   String PIZZA4[] = {"PIZZA5"};
   String PIZZA5[] = {"PIZZA6"};
   String PIZZA6[] = { };
   String PIZZA7[] = { };
   String PIZZA8[] = {"PIZZA7","PIZZA8"};
   /* MALL0 is pizza shop nearest to location A0 */
   String MALL0[] = {"MALL0"};s
   String MALL1[] = { };
   String MALL2[] = {"MALL1"};
   String MALL3[] = { };
   String MALL4[] = {"MALL2","MALL3","MALL4","MALL5"};
   String MALL5[] = {"MALL6","MALL7"};
   String MALL6[] = { };
   String MALL7[] = {"MALL10","MALL11"};
   String MALL8[] = {"MALL13"};
```

```
htLocation = new Hashtable();
htLocation.put("A0", new Location(A0));
htLocation.put("A1", new Location(A1));
htLocation.put("A2", new Location(A2));
htLocation.put("A3", new Location(A3));
htLocation.put("A4", new Location(A4));
htLocation.put("A5", new Location(A5));
htLocation.put("A6", new Location(A6));
htLocation.put("A7", new Location(A7));
htLocation.put("A8", new Location(A8));

htLcpAtm = new Hashtable();
htLcpAtm.put("A0", new Location(ATM0));
htLcpAtm.put("A1", new Location(ATM1));
htLcpAtm.put("A2", new Location(ATM2));
htLcpAtm.put("A3", new Location(ATM3));
htLcpAtm.put("A4", new Location(ATM4));
htLcpAtm.put("A5", new Location(ATM5));
htLcpAtm.put("A6", new Location(ATM6));
htLcpAtm.put("A7", new Location(ATM7));
htLcpAtm.put("A8", new Location(ATM8));


htLcpPs = new Hashtable();
htLcpPs.put("A0", new Location(PIZZA0));
htLcpPs.put("A1", new Location(PIZZA1));
htLcpPs.put("A2", new Location(PIZZA2));
htLcpPs.put("A3", new Location(PIZZA3));
htLcpPs.put("A4", new Location(PIZZA4));
htLcpPs.put("A5", new Location(PIZZA5));
htLcpPs.put("A6", new Location(PIZZA6));
htLcpPs.put("A7", new Location(PIZZA7));
htLcpPs.put("A8", new Location(PIZZA8));

htLcpMall = new Hashtable();
htLcpMall.put("A0", new Location(MALL0));
htLcpMall.put("A1", new Location(MALL1));
htLcpMall.put("A2", new Location(MALL2));
htLcpMall.put("A3", new Location(MALL3));
htLcpMall.put("A4", new Location(MALL4));
htLcpMall.put("A5", new Location(MALL5));
htLcpMall.put("A6", new Location(MALL6));
htLcpMall.put("A7", new Location(MALL7));
htLcpMall.put("A8", new Location(MALL8));
```

```
      Frame sc1= new Screen1("Location-Based Service");
      sc1.setSize(750,500);
      sc1.setVisible(true);
   }
  }

class Location
{
 String a[];

 public Location(String[] a1) //,String a2,String a3,String a4,String a5)
 {
         this.a = a1 ;
 }

      String getLocation()
 {
         String str = "";
         //System.out.println(a.length);
         for(int i=0 ;i< a.length; i++)
         {
   str += a[i];
            str += "   ";
         }
         return str;
 }

      int length()
      {
         return a.length;
      int length()
      {
         return a.length;
      }
 }
```

Certain new mobile computing devices have GPS within them. In this case, the device need not ask the present location from the user. It can itself discover its present location without using GUI for selecting location in Sample Code A2.7. Moreover, mobile computing devices may have WiFi or Bluetooth within them. In such cases also, the device need not ask the present location from the user and also need not save data records for the ATM, pizza shops, and shopping mall locations. It can itself discover its location by connecting to Internet using WiFi. If the device is Bluetooth-enabled, it can discover a nearby computer with Bluetooth. The computer connects to Internet, gets the server databases, and transfers search results to the device by connecting to Bluetooth interface between the two.

Appendix 3

Solutions to Objective Type Questions

Ans	Ch 1	Ch 2	Ch 3	Ch 4	Ch 5	Ch 6	Ch 7	Ch 8	Ch 9	Ch 10	Ch 11	Ch 12	Ch 13	Ch 14
1	b	b	a	c	b	a	d	c	a	d	c	b	a	a
2	a	c	d	a	b	a	c	d	a	b	a	a	c	b
3	b	b	b	d	a	c	b	b	b	a	d	c	d	d
4	a	d	a	b	d	a	c	b	c	c	b	b	b	c
5	c	a	b	d	c	d	a	b	d	c	c	b	b	a
6	d	a	a	c	c	b	a	d	d	a	d	d	c	a
7	a	c	a	a	d	c	d	c	b	b	d	b	d	a
8	a	a	d	a	b	b	c	d	c	a	b	c	a	d
9	a	c	c	c	a	d	b	a	a	a	b	c	a	c
10	a	d	b	b	d	d	c	b	b	b	a	c	b	c
11	c	b	c	d	a	a	b	c	a	c	b	a	c	c
12	a	a	b	d	a	c	b	d	d	b	c	d	d	c
13	c	—	d	b	c	a	d	b	c	c	a	b	d	c
14	a	—	a	a	b	b	c	c	b	c	d	a	a	a
15	—	—	c	c	c	d	a	b	c	d	d	a	c	d
16	—	—	c	a	d	c	b	a	d	b	a	b	c	d
17	—	—	a	b	—	—	—	—	—	—	—	—	—	—
18	—	—	b	d	—	—	—	—	—	—	—	—	—	—
19	—	—	c	b	—	—	—	—	—	—	—	—	—	—
20	—	—	a	c	—	—	—	—	—	—	—	—	—	—

Select Bibliography

Books

Adelstein, F., S.K.S. Gupta, G.G. Richard III, and L. Schwiebert, *Fundamentals of Mobile and Pervasive Computing*, McGraw-Hill, 2005, Reprint, Tata McGraw-Hill, 2005.

Aggelou, G., *Mobile Ad Hoc Networking: Design and Integration*, McGraw-Hill, 2004.

Agrawal, D.P., *Qing-An Zeng*, Thomson Brooks/Cole, 2003.

Alesso, P. and C. Smith, *Intelligent Wireless Web*, Addison Wesley, 2001.

Andriessen, J.H. (Ed.), *Mobile Virtual Work: A New Paradigm*, Springer-Verlag, 2005.

Barbará , D. (Ed.), *Databases and Mobile Computing*, Springer-Verlag, 1996.

Basagni, S. (Ed.), *Ad hoc Networking*, Wiley, 2004.

Behravanfar, R., *Mobile Computing Principles—Designing and Developing Mobile Applications with UML and XML*, Cambridge University Press, 2005.

Bellavista, P. and A. Corradi, *The Handbook of Mobile Middleware*, Auerbach, 2006.

Blahut, R.E., *Algebraic Codes for Data transmission*, Cambridge, 2003.

Blanchet, M., *Migrating to IPv6: A Practical Guide for Implementing IPv6 in Mobile and Fixed Networks*, Wiley, 2006.

Boukerche, A. (Ed.), *Handbook of Algorithms for Wireless Networking and Mobile Computing*, Chapman and Hall/CRC, 2006.

Bulusu, N. and S. Jha (Eds.), *Wireless Sensor Network Systems*, Artech, 2005.

Burkhardt, J., H. Henn, S. Hepper, K. Rindtorff, and T. Schack (Eds.), *Pervasive Computing: Technology and Architecture of Mobile Internet Applications*, Addison-Wesley, 2001.

Chittaro, L. (Ed.), *Human–Computer Interaction with Mobile Devices and Services*, 5th International Conference on Mobile HCI, Italy, Springer, 2003.

Cordeiro, Carlos De Morais, *Ad Hoc and Sensor Networks*, World Scientific, 2006.

Correia, L.M., *Mobile Broadband Multimedia Networks Techniques, Models and Tools for 4G*, Academic Press Elsevier, 2006.

Coyle, F., *Wireless Web—A Manager Guide*, Addison Wesley/Longman Inc., 2001.

Crooks, Clayton E. II, *Mobile Device Game Development*, Charles River Media, 2005.

Eady, F., *Implementing 802.11 with Microcontrollers: Wireless Networking for Embedded Systems*, Elsevier/Newnes, 2005.

Fox D., *Building Solutions with the Microsoft.Net Compact Framework: Architecture and Best Practices for Mobile Development*, Pearson Education, 2003.

Gast, M., *802.11 Wireless Networks: The Definitive Guide*, O'Reilly, 2005.

Geier, J., *Wireless LANs: Implementing IEEE 802.11 Networks*, Sams, 2001.

Gibson, J.D. (Chief Ed.), *Mobile Communication Handbook*, CRC/IEEE Press, 1999.

Giguère, E., *Java 2 Micro Edition*, Wiley, 2000.

Goyal, V., *Pro JAVA ME MMAPI: Mobile Media API for Java Micro Edition*, Academic Press, Springer, 2006.

Guthery, S.C., *Mobile Application Development with SMS and the Sim Toolkit*, McGraw-Hill Companies, November 2001.

H˜fele, Claus *Mobile 3d Graphics: Learning 3d Graphics + the Java Micro Edition*, Course Technology, 2007.

Hac, A., *Mobile Telecommunications Protocols for Data Networks*, Wiley, 2003.

Hansmann, U., L. Merk, M.S. Nicklons, and T. Stober, *Principles of Mobile Computing*, Springer, 2003.

Heijden, M.V.D. and M. Taylor, *Understanding WAP*, Artech, 2000.

Held, G., *Data over Wireless Networks—Bluetooth, WAP and Wireless LANs*, McGraw-Hill, 2000.
Hillborg, M., *Wireless XML Developer's Guide*, McGraw-Hill/Osborne, 2002.
Hoboken, N.J., in: Hüseyin Arslan, Zhi Ning Che (Eds.), *Ultra Wideband Wireless Communication*, Wiley-Interscience, 2006.
Ilyas, M. and I. Mahgoub (Eds.), *Mobile Computing Handbook*, CRC Press, 2004.
Imielinski, T. and H.F. Korth, in: T. Imielinski (Ed.), *Mobile Computing*, Kluwer, 1996.
Jaokar, A., *Mobile Web 2.0: The Innovator's Guide to Developing and Marketing Next Generation Wireless/Mobile Applications*, Futuretext, 2006.
Keogh, J., *J2ME: The Complete Reference*, Tata McGraw-Hill, 2003.
Kupper, A., *Location-based Services: Fundamentals and Applications*, Wiley, 2005.
Laberge, R. and S. Vujosevic, *Building PDA Databases for Wireless and Mobile Development*, Wiley, 2003.
Lin, Yi-Bing and Ai-Chun Pang, *Wireless and Mobile all-IP Networks*, Wiley, 2005.
Longoria, R. (Ed.), *Designing Software for the Mobile Context*, Springer, 2004.
Mahmoud, Q.H., *Mobile Computing Middleware*, Wiley, 2004.
May, P., *Mobile Commerce—Opportunities, Applications and Technologies*, Cambridge University Press, 2001.
Mikkonen, T., *Programming Mobile Devices: An Introduction for Practitioners*, Wiley, 2007.
Othman, M., *Principles of Mobile Computing and Communications*, Auerbach, 2007, in press.
Pahlavan, K. and P. Krishnamurthy, *Principles of Wireless Network—A Unified Approach*, Pearson, 2002.
Pareek, D., *WiMAX: Taking Wireless to the MAX*, Auerbach, 2006.
Perkins, C. (Ed.), *Ad hoc Networks*, Addison Wesley, 2000.
Perkins, C., *Mobile IP*, Addison Wesley, 1999.
Rubin, E., *Microsoft.NET Compact Framework Kick Start*, Sams, 2003.
Salmre, I., *Writing Mobile Code: Essential Software Engineering for Building Mobile Applications*, Pearson Education, 2005.
Schiller, J., *Mobile Communications*, 2nd Ed., Addison Wesley, 2003, Indian Reprint Pearson Education, 2003.
Schwartz, M., *Mobile Wireless Communications*, Cambridge University Press, 2005.
Stallings, W., *Wireless Communication and Networks*, Prentice Hall, 2001.
Steele, R., C. Lee, and P. Gould, *GSM, cdmaOne and 3G Systems*, Wiley, 2001.
Stojmenovic, I., *Handbook of Wireless Networks and Mobile Computing*, Wiley, 2002.
Szymans, B.K. (Ed.), *Advances in Pervasive Computing and Networking*, Springer, 2005.
Taluker, A.K. and R.R. Yavagal, *Mobile Computing: Technology, Applications, and Service Creation*, McGraw-Hill, 2006.
Tisal, J., *GSM Network: GPRS Evolution: One Step Towards UMTS*, Wiley, 2001.
Tse, D. and P. Viswanath, *Fundamentals of Wireless Communication*, Cambridge University Press, 2005.
Vacca, J., *Wireless Broadband Networks Handbook*, McGraw-Hill, 2001.
Wei-Meng, .NET Compact Framework Pocket Guide, O'Reilly Media, 2004.
Wheeler, W., *Integrating Wireless Technology in the Enterprise: PDAs, Blackberries, and Mobile Devices*, Elsevier, 2003.
Wigley, A., *Microsoft Mobile Development Handbook*, Microsoft Press, 2007.
Williams, V., *Wireless Computing Primer*, M&T Books, 1996.
Zheng, P. and L. Ni, *Smartphone and Next-Generation Mobile Computing*, Morgan Kaufmann, 2006.

Websites

http://www.computer.org (Pervasive Computing (Mobile and Ubiquitous Computing) Magazine from IEEE).
http://www.hpcmag.com (Handheld PC Magazine).
http://www.mobilecomputing.com (Mobile Computing Magazine).
http://www.networkcomputing.com (Network Computing Magazine).
http://www.synchrologic.com (Mobile Computing Newsletter).
http://www.wirelessinternet.com (Wireless Internet Magazine).

Index

Introduction to Data Structures and Algorithms with C++